# Hydrogen Materials Science and Chemistry of Carbon Nanomaterials

edited by

## T. Nejat Veziroglu
International Association for Hydrogen Energy, University of Miami, FL, U.S.A.

## Svetlana Yu. Zaginaichenko
Institute of Hydrogen and Solar Energy, Kiev, Ukraine

## Dmitry V. Schur
Institute for Problems of Materials Science of NAS, Kiev, Ukraine

## B. Baranowski
Institute of Physical Chemistry of PAS, Warsaw, Poland

## Anatoliy P. Shpak
Institute for Metal Physics of NAS, Kiev, Ukraine

and

## Valeriy V. Skorokhod
Institute for Problems of Materials Science of NAS, Kiev, Ukraine

**Kluwer Academic Publishers**

Dordrecht / Boston / London

Published in cooperation with NATO Scientific Affairs Division

Proceedings of the NATO Advanced Research Workshop on
Hydrogen Materials Science and Chemistry of Carbon Nanomaterials
Sudak, Crimea, Ukraine
14–20 September 2003

A C.I.P. Catalogue record for this book is available from the Library of Congress.



ISBN 1-4020-2667-6 (HB)
ISBN 1-4020-2669-2 (e-book)


Published by Kluwer Academic Publishers,
P.O. Box 17, 3300 AA Dordrecht, The Netherlands.

Sold and distributed in North, Central and South America
by Kluwer Academic Publishers,
101 Philip Drive, Norwell, MA 02061, U.S.A.

In all other countries, sold and distributed
by Kluwer Academic Publishers,
P.O. Box 322, 3300 AH Dordrecht, The Netherlands.




*Printed on acid-free paper*

# Contents

# Preface

The 2003 International Conference "Hydrogen Materials Science and Chemistry of Carbon Nanomaterials" (ICHMS'2003) was held in September 14-20, 2003 in the picturesque town Sudak (Crimea, Ukraine) known for its sea beaches. In the tradition of the earlier ICHMS conferences, this 8th ICHMS'2003 meeting served as an interdisciplinary forum for the presentation and discussion of the most recent research on transition to hydrogen-based energy systems, technologies for hydrogen production, storage, utilization, materials, energy and environmental problems. The aim of ICHMS'2003 was to provide an overview of the latest scientific results on research and development in the different topics cited above. The representatives from industry, public laboratories, universities and governmental agencies could meet, discuss and present the most recent advances in hydrogen concepts, processes and systems, to evaluate current progress in these areas of investigations and to identify promising research directions for the future.

The ICHMS'2003 was the conference, where a related new important topic of considerable current interest on fullerene-related materials as hydrogen storage was included into the conference program. This meeting covered synthesis, structure, properties and applications of diverse carbon materials ranging from nanotubes and fullerenes to carbon fiber composites and sorbents. Thus, the ICHMS'2003 conference was unique in bringing together hydrogen and carbon materials researchers and engineers from developed countries of Europe and America, new independent states of FSU and other countries for discussions in advanced materials development and applications.

The ICHMS'2003 format consisted of invited lectures, oral and poster contributions and also the conference representatives took part in the exhibition of new materials and equipment.
This book with ICHMS'2003 Proceedings brings together the research

papers that were presented. We hope that they will serve as both a useful reference and resource material for all the participants and for those whose interest in the subject matter may develop after the event.

Finally, this conference was generously supported by the Scientific and Environmental Affairs Division of NATO as an Advanced Research Conference within the Physical and Engineering Science and Technology Area of the NATO Science Programme. Their contribution is gratefully acknowledged and the Organizing and all ARW participants want to overflow with effusive thanks to NATO Committee for the financial support of our 8th ICHMS'2003 Conference and to **Mr. Jean Fournet**, Assistant Secretary General, Chairman of NATO Science Committee, and **Mr. Fausto Pedrazzini**, Programme Director, NATO Scientific Affairs Division, for the displayed mutual understanding and the comprehension of significance of problems under discussions at the ICHMS'2003 conference.

**T. Nejat Veziroglu**
**Svetlana Yu. Zaginaichenko**
**Dmitry V. Schur**
**Anatoliy P. Shpak**
**Valeriy V. Skorokhod**

# INTERNATIONAL ADVISORY AND ORGANIZING COMMITTEE OF ICHMS'2003

**Chairperson:**

| | | |
|---|---|---|
| Prof. B. Baranowski | Institute of Physical Chemistry of PAS, Warsaw | Poland |

**Honour Chairperson:**

| | | |
|---|---|---|
| Prof. T.N.Veziroglu | President of International Association for Hydrogen Energy, Coral Gables | USA |

**Co-Chairpersons:**

| | | |
|---|---|---|
| Prof. A.P.Shpak | Institute for Metal Physics of NAS, Kiev | Ukraine |
| Prof. V.V.Skorokhod | Institute for Problems of Materials Science of NAS, Kiev | Ukraine |
| Prof. Yu.A.Ossipyan | Institute of Solid State Physics of RAS, Chernogolovka | Russia |
| Prof. V.V.Lunin | Moscow State University, Moscow | Russia |

**Members:**

| | | |
|---|---|---|
| Prof. S.A. Firstov | Institute for Problems of Materials Science of NAS, Kiev | Ukraine |
| Prof. I.M. Astrelin | National Technical University "Kiev Polytechnical Institute" | Ukraine |
| Prof. R.O.Loutfy | Materials & Electrochemical Research (MER) Corporation, Tucson | USA |
| Prof. V.I.Shapovalov | Materials & Electrochemical Research (MER) Corporation, Tucson | USA |
| Prof. Z.A.Matysina | Dnepropetrovsk State University | Ukraine |
| Prof. P. Vajda | Laboratoire des Solides Irradies, Palaiseau cedex | France |
| Prof. Y. Carmel | Science and Technology Center in Ukraine | Canada |
| Prof. E. Manninen | Science and Technology Center in Ukraine | EU |
| Dr. B.A. Atamanenko | Science and Technology Center in Ukraine | Ukraine |
| Prof. R.T. Turner | Columbian Chemicals Company, Marietta | USA |
| Dr. A.L.Shilov | Institute of General and Inorganic Chemistry, Moscow | Russia |
| Prof. U.M.Mirsaidov | Academy of Sciences of Tajikistan | Tajikistan |
| Prof. M. Groll | Stuttgart University | Germany |
| Prof. I.R. Harris | University of Birmingham | UK |
| Prof. Dr. J.Schoonman | Delft University of Technology, Delft | Netherlands |
| Dr. H.J. Bauer | Ludwig-Maximilians-Universität, Munich | Germany |
| Dr. L.Grigorian | Fundamental Research Lab, Honda R&D Americas, Inc. | USA |
| Prof. Dr. A.Mekhrabov | Middle East Technical University, Ankara | Turkey |
| Prof. B.Ibrahimogly | Middle East Technical University, Ankara | Turkey |

| Prof. D. Hui | University of New Orleans | USA |
|---|---|---|
| Prof. L.A.Avaca | Inst. de Quimica de Sao Carlos, Universidade de Sao Paulo | Brasil |
| Prof. M.Genovese | European Commission INCO COPERNICUS, Brussels | Belgium |
| Prof. B.Rao | Virginia Commonwealth University, Richmond | USA |
| Prof. D.Tománek | Michigan State University, East Lansing | USA |
| Prof. D.K.Slattery | Florida Solar Energy Center | USA |
| Prof. O.A.Ivashkevich | Belarusian State University, Minsk | Belarus |
| Prof. E.M.Shpilevsky | Belarusian State University, Minsk | Belarus |
| Prof. P.Catania | International Energy Foundation | Canada |
| Prof. S.K.Gordeev | Central Research Institute of Materials, St. Petersburg | Russia |
| Prof. I.E.Gabis | St. Petersburg State University, Institute of Physics | Russia |
| Dr. B.P.Tarasov | Institute of Problems of Chemical Physics of RAS | Russia |
| Dr. O.N. Efimov | Institute of Problems of Chemical Physics of RAS | Russia |
| Dr. Yu.M. Shul'ga | Institute of Problems of Chemical Physics of RAS | Russia |
| Dr. M.V.Lototsky | Institute for Energy Technology, Kjeller | Norway |
| Dr. A.P.Mukhachev | Plant "Zirconium", Dneprodzerjinsk | Ukraine |
| Dr. A.P. Pomytkin | National Technical University "Kiev Polytechnical Institute" | Ukraine |
| Dr. N.S. Astratov | National Technical University "Kiev Polytechnical Institute" | Ukraine |
| Dr. V.E. Antonov | Institute of Solid State Physics of RAS, Chernogolovka | Russia |
| Prof. O.Savadogo | Ecole Polytechnique de Montreal | Canada |
| Dr. V.V.Kartuzov | Institute for Problems of Materials Science of NAS, Kiev | Ukraine |
| Dr. Yu.F.Shmal'ko | Institute of Mechanical Engineering Problems of NAS, Kharkov | Ukraine |
| Dr. A.A.Moskalenko | Institute of Thermal Physics of NAS, Kiev | Ukraine |
| Dr. T.A.Iljinykh | Cabinet of Ministry of Ukraine | Ukraine |
| Prof. V.N.Verbetsky | Moscow State University | Russia |
| Prof. C.Sholl | University of New England, Armidale | Australia |
| Prof. A.Switendick | ACS Associates, Albuquerque | USA |
| Prof. B.Timoshevskiy | Ukrainian State Maritime University, Nikolaev | Ukraine |

# PROGRAM COMMITTEE

Dr. D.V.Schur           – chairperson      (Ukraine)

Dr. S.Yu.Zaginaichenko     – vice-chairperson   (Ukraine)

Dr. B.P.Tarasov         – vice-chairperson  (Russia)

Dr. Yu.M. Shul'ga       – vice-chairperson  (Russia)

Mr. V.K. Pishuk
Mrs. K.A. Lysenko
Mrs. T.N. Golovchenko
Mr. I.A. Kravchuk
Mrs. N.S. Anikina
Mr. A.G. Dubovoy
Mr. A.D. Zolotarenko
Mr. A.Yu. Vlasenko
Mrs. L.I. Kopylova
Mrs. T.I. Shaposhnikova
Mr. A.F. Savenko
Mr. A.A. Rogozinskiy
Dr. K.A. Meleshevich
Mr. V.M. Adejev
Mr. M.I. Maystrenko
Mr. V.I. Tkachuk
Mrs. A.A. Rogozinskaya
Dr. M.S. Yakovleva

# FIRST PRESIDENT OF UKRAINE

### Dear conference participants and guests !

Allow me to greet all of you at this remarkable place of the Earth.

With great satisfaction I want to establish the fact that by the end of XX century Ukraine become at last the reality for the whole world. At the attainment of its position on political arena our country remains nevertheless the unknown practically from the viewpoint of contribution to the world science in spite of its vast scientific potential.

Virtually nothing strange is in that because Ukrainian scientists had no way to declare themselves over a long period of time for the well-known for all of us reasons.

The time has come to show for world community our scientific achievements and by every new fact and experiment, by every result to demonstrate ourselves and all world that Ukrainian science exists really.

The science in civilized world is cosmopolitical by its nature. And your ICHMS'2003 Conference confirms this generally known fact. The co-operation of scientists from different world countries has met here and chemists, physicists, production engineers and others are deeply involved in studying of such global problem as alternative power sources. All of you gather together in order to inform one another about new results, to discuss problems, to find the new approaches, to see something in smb's eyes and above all to create new knowledge in archimportant field of science as power engineering.

The power engineering is the most significant element of civilization life without which it is impossible to imagine both the present and the future of mankind. Everything that we use now in order to live, drink, eat, work, move in time and space is bound up in any case with energy. All's that exist owing to the realization of scientific knowledge gained tens and hundreds years ago is not enough to look ahead with confidence. The gas, oil and especially bituminous coal are inexhaustible power sources. The power engineering of the future is first of all the new ideas and ways of solar energy transformation and then the materials allowing the realization of ideas. All over the world thousands of peoples work in this area. Therefore the co-operation of scientists is important for solution of global problems for all mankind. Having such opportunity I want to wish all of you that these several days of ICHMS'2003 Conference in such beautiful site of Crimea will become useful for you and the new acquaintances and contacts will make the beginning for new programs and projects.

**First President of Ukraine**
**L.M. Kravchuk**

international association for hydrogen energy

POST OFFICE BOX 248266 • CORAL GABLES • FLORIDA 33124 • USA

# WELCOME TO THE PARTICIPANTS OF ICHMS'2003

At The Hydrogen Economy Miami Energy (THEME) Conference in March 1974, a handful of scientists from around the world proposed the Hydrogen Economy. Since then, over the first quarter of a century, the foundations for the Hydrogen Energy System have been established through the hard and ingenious work of researchers from many countries.

Conversion to the Hydrogen Energy System began early this century. The Toyota and Honda companies have started leasing hydrogen fuel cell cars in Los Angeles, Tokyo and Yokohama. General Motors is planning to test a hydrogen-fueled delivery van in the streets of Tokyo. Daimler-Chrysler has begun manufacturing hydrogen buses. Although, at the moment, they are more expensive than diesel buses, many large cities are buying these hydrogen-fueled buses in order to fight pollution in the city centers. The Siemens-Westinghouse Company is marketing hydrogen fuel cell power plants for electric utilities. Hydrogen hydride electric batteries have already been commercialized. The Airbus Company is working on hydrogen-fueled air transport.

Japan has earmarked four billion dollars in order to acquire all the hydrogen energy technologies by the year 2020. Europe has initiated a vigorous hydrogen energy program. They are going to spend five billion Euros during the next five years on Hydrogen Energy R & D. President Bush of the United States, in his State of the Union address, February 2003, referred to hydrogen as the 'freedom fuel,' which will free the world from dependence on petroleum. The U.S. Government has earmarked 1.7 billion dollars for commercializing hydrogen fuel cell vehicles, and 1.2 billion dollars for $CO_2$ free hydrogen production from coal.

International conferences, such as the ICHMS'2003, will help speed up this transformation. At the conference, recent research findings on hydrogen materials science and metal hydrides chemistry will be presented and discussed. The chemistry of metal hydrides and hydrogen materials science will play an important role in hastening the conversion to the Hydrogen Energy System. The research endeavors of the scientists and engineers participating in this conference will make significant contributions to facilitate this milestone conversion.

I take this opportunity to congratulate the organizers of this important series of International Conferences on Hydrogen Materials Science and Chemistry of Carbon Nanomaterials, and wish all of the participants a very productive conference and pleasant days in the beautiful Crimea.

**T. Nejat Veziroglu**
**Honorary Chairman, ICHMS'2003**
**President, International Association for Hydrogen Energy**

xix

# NATIONAL ACADEMY

# OF SCIENCE OF UKRAINE

54 Volodymyrs'ka str,. Kyiv, 01601 Ukraine,
Tel: 380-44-226-2347, Fax: 380-44-228-5522

## Dear colleagues!
## Delegates of ICHMS'2003!

Allow me to greet the ICHMS'2003 Conference on behalf of National Academy of Sciences of Ukraine.

The conference subjects testify about the scale and importance of problems to be considered at this scientific forum. The carbon nanotubes as well as compositions based on hydrideforming alloys and carbon nanostructures are perspective materials with high hydrogen capacity. These materials and their properties inspire hydrogen scientists with certain optimism because application of carbon nanomaterials in energetics and automotive transport helps in handling the important problems and first of all energy and ecological problems. The materials science subject will receive the attention it deserves, the decision of these questions requires the wide application of energetic installations based on fuel cells both for large-scale generation of energy and for autonomous power supply of separate objects and transport.

I do not want to minimize the importance of a valuable contribution of all world association, but I should like to remind that Ukrainian Academy of Sciences does not stand aside of investigations in this field.

At present time much attention is being given within the framework of Academy to the study and development both of hydrideforming materials and of various carbon nanostructures. Leading Institutes of National Academy of Sciences of Ukraine make the considerable contribution for solving the problems of hydrogen materials science and chemistry of carbon nanomaterials.

Thus, it is no coincidence that Ukraine become the country where scientists from many countries come for exchange of experience and knowledge in this prospective science field for the second tens years.

Take this opportunity I want to wish You the successful scientific work and further creative initiative. Let such conferences, as ICHMS'2003, unite scientists eliminating the geographical and language barriers.

I wish You the fruitful scientific work, every success and the fine rest in the bright Crimean sun.

**B.E. Paton**
**President**
**of National Academy of Sciences of Ukraine**

# Russian Academy of Science

The twenty-first century will be known as the century during which the Hydrogen Energy System replaces the present fossil fuels system. Hydrogen is going to be the permanent answer to the twin global problems: (1) the rapid depletion of fossil fuels, and (2) the environmental problems caused by their utilization, such as the greenhouse effect, climate change, acid rains, ozone layer depletion, pollution and oil spills.

Hydrogen is already making inroads into the fossil fuel realm in every direction. Especially over the past two years, there has been an increase in activities. Siemens-Westinghouse announced that they will have a 1 MW $H_2$ power plant available for sale soon, having a 70% efficiency. There are hydrogen fueled bus demonstration projects in several cities of the world. A $H_2$ fueled Mercedes bus will be on the market in two years' time. All of the major car companies have announced that they will offer $H_2$ fueled cars to the public by the year 2004. Hydrogen hydride electric batteries are already available for lap-top computers and electric cars. The Airbus Company is developing a $H_2$ fueled air transport. The United States and Japan are working on $H_2$ fueled hypersonic passenger planes. The Shell Oil company has established a Hydrogen Division. No doubt the other petroleum companies will follow suit.

In the tradition of the earlier conferences, the 8th ICHMS'2003 Conference is providing an international forum for the presentation and discussion of the latest R&D results in field of hydrogen materials science and carbon nanomaterials, covering hydrogen production, storage, distribution, i.e. engines, fuel cells, catalytic combustion, hydride applications, aerospace applications, hydrogen fuelled appliances, environmental impact and economies. I am sure that proceedings of this conference will bring together research papers and they will serve as a useful reference and resource material for all the participants.

On behalf of Russian Academy of Sciences I would like to extend my deepest appreciation to all delegates and participants who come from many different countries to make this conference a success.

**Academician**
**Yu.A. Ossipyan**

HOLDER OF NATIONS FRIENDSHIP'S ORDER

# ACADEMY OF SCENCES,
## REPUBLIC OF TAJIKISTAN

33, Rudaki Ave., Dushanbe, Tajikistan, 734025
Tel: (992372) 21 50 83
Fax: (992372) 21 49 11

## Dear colleagues!

I am honoured to welcome all of you on the behalf of the Presidium of the Academy of Sciences of the Republic of Tajikistan.

Every year hydrogen power engineering takes more and more important role in the life of society and in the life of each state. This problem is especially actual for Tajikistan, which has not gas and oil resources.

Investigations in the filed of power-intensive substances and hydride chemistry are being successfully carried out in Tajikistan since 1968. Needs of promptly developing new techniques and technology for substances with such properties considerably promoted the development of a number of new fields of chemistry, including hydrogen materials science and chemistry of carbon nanomaterials.

Creation of the ecologically clean transport by the way of application of different non-traditional fuels, including hydrogen, is the general direction of air basin protection.

Hydrogen is one of the perspective fuels for transport. Many properties of hydrogen give him the first place in future expectations. The last stage of ecologically safe hydrogen cycle is water. Consequently, the source f hydrogen on the Earth is practically inexhaustible.

Today's hydrogen power engineering is very broad notion.

It is pleasant that we discuss many aspects of hydrogen and carbon chemistry.

Our conference is traditional one owing to the great efforts of Organizing Committee and I would like to thank them heartily for their great job on conference organization.

I hope that the ICHMS'2003 Conference will allow us to work fruitfully, to learn a lot of useful information and will give the opportunity to establish joint projects.

I heartily greet the participants of the Conference and wish them enjoyable and fruitful time in Crimea.

**U.M. Mirsaidov**

**President**
of Academy of Sciences
of the Republic of Tajikistan
**Chairman of the Commission**
**of Majlisi Milli (the Parliament)**
**of the Republic of Tajikistan**

# SCIENCE AND TECHNOLOGY CENTER IN UKRAINE
# НАУКОВО-ТЕХНОЛОГІЧНИЙ ЦЕНТР В УКРАЇНІ

21 Kamenyariv str., Kyiv, 03138 Ukraine,
Tel./Fax: +380 (044) 490 7150

## Dear participants of ICHMS'2003 !

The Science and Technology Center in Ukraine (STCU) welcome in my person the participants of the VIII[th] International Conference "Hydrogen Materials Science and Chemistry of Carbon Nanomaterials".

I am glad that a lot of world scientists and investigators want to take part in this conference in Ukraine and present the papers in many aspects of hydrogen in metals, alloys, carbon nanomaterials from fundamental to applications. Hydrogen sorbing properties of newly discovered carbon nanostructural materials inspire hydrogen scientists with certain optimism. In the development of all new energy options, hydrogen necessarily will play an important role because of its ability to supplement any energy stream and to be applied to any load. Given the significance of energy in the environmental problems of our world, it is urgently necessary that the leaders in civic and industrial societies have a more thorough understanding and appreciation of the existing states of energy systems and their related technologies.

The search of alternative power sources is the most actual theme of today. I hope that your work will accelerate the substitution of existing power systems which use fossilized fuels for inexhaustible and ecologically clear Hydrogen Systems. The creation and development of such systems will give an opportunity to harmonize the pragmatic human's treatment of nature.

I very hope that Ukrainian science, as well as in the developed countries, in spite of all difficulties and problems, will have its future and will develop including such an important and considerable fields as hydrogen materials science and carbon nanomaterials.

STCU provide financial support to carry out of projects and to hold a number of conferences devoted to consideration of environmental and energy problems, among which is ICHMS conference.

On behalf of STCU let me wish you the fruitful work and every success in discussing vital important problems for all humanity.

**B.A. Atamanenko**
**Senior Deputy of**
**Executive Director of STCU**

# VIII INTERNATIONAL CONFERENCE
## *"Hydrogen Materials Science & Chemistry of Carbon Nanomaterials"*

### Sudak, Crimea, UKRAINE, September 14 – 20, 2003

## Dear Colleagues, guests, ladies and gentlemen!

The ICHMS'2003 Organizing Committee is glad to welcome you in Sudak and we consider it an honour that a lot of outstanding scientists and investigators from every corner of the world want to take part in this conference in Crimea. Our best wishes to all participants and visitors of ICHMS'2003.

We are especially obliged to our sponsors, as NATO Science Committee, Science and Technology Center in Ukraine, Columbian Chemicals Company and others that provide us the means to carry out this representative forum.

Two years have passed since our last meeting in Alushta. Certainly, this is a short space of time but we are filled with expectation of new discoveries and excellent results. We are sure that new substantial scientific results will be presented here and they will permit us to extend our's knowledge mainly in the strategically important field for the future, as hydrogen energy and in directly connected with it fields of hydrogen materials science and nanostructural carbon.

Our conference demonstrates the present-day state of affairs in 4 conference topics, which are perspective and quickly developed directions of modern materials science with a view to stimulate the new ideas, to support and ensure their realization.

The fact that scientists of various schools, directions and tendencies get together in Crimea two tens of years will favour the active discussions, fruitful contacts and new knowledge gaining. We hope that both the beauty of Crimean nature and fine weather will inspire scientists on the active work and deserved rest after ICHMS'2003.

We very hope to hear the qualitatively new results that will permit us to move from the pure scientific investigations to their wide practical implementation. This is not only our wish but also our scientific duty, result of work, which many of us devote all their efforts.

We would like to wish all scientists success and good luck.

**Organizing Committee of ICHMS'2003**

| | |
|---|---|
| **A.P. Shpak,** | **V.V. Skorokhod,** |
| **B.P. Tarasov,** | **S.Yu. Zaginaichenko,** |
| **D.V. Schur,** | **Yu.M. Shulga** |

# PHASE TRANSFORMATIONS IN CARBON MATERIALS

Z.A. MATYSINA, D.V. SCHUR, S.Yu. ZAGINAICHENKO,
V.B. MOLODKIN, A.P. SHPAK
*Dnepropetrovsk National University, 72, Gagarin str.,*
*Dnepropetrovsk 49000 Ukraine*
*Institute of Hydrogen and Solar Energy, P.O. Box 195, Kiev, 03150 Ukraine*
*Institute for Problems of Materials Science of Ukrainian Academy of*
*Sciences, 3, Krzhizhanovsky str., Kiev, 03142 Ukraine*
*E-mail: shurzag@materials.kiev.ua*
*Institute for Metal Physics of Ukrainian Academy of Sciences,*
*36 Academician Vernadsky Blvd., 03680 Kiev-142, Ukraine*

The statistical theory of ordering and phase transformations in allotropic modifications of carbon has been developed. The ordering temperatures and pressures for phase transitions have been calculated, their mutual influence has been clarified. The phase diagram has been plotted. The temperature and pressure ranges, in which each of carbon phases (diamond, graphite, carbyne, fullerite) is realized, have been evaluated.

## 1. Introduction

At present, four allotropic modifications of carbon (diamond, graphite, carbyne, fullerite) are known.

Variety in properties of carbon materials is caused by the electron structure of a carbon atom. Redistribution of electron density, electron clouds of different modifications forming around the atoms, hybridization of orbits (formation of the mixed ones) ($sp^3$-, $sp^2$-, sp- hybridization in diamond, graphite, carbyne respectively) are responsible for existence of different crystalline allotropic phases. For diamond, orbits hybridization provides the strong chemical bond in the nearest tetrahedral surrounding of every atom, in graphite - the strong bond in the basal planes, in carbyne - the strong bond in three 6-atom chains-cells, and in fullerite - the strong bond of carbon atoms in the spherical fullerene molecule. The bond between the basal planes in graphite, between the cells in carbyne and between the fullerene molecules in fullerite is due to weak Van der Waals forces. In conformity with above, the diamond structure is three-dimensional, spatial, the graphite one is quasi-two-dimensional, schistose, planar, the carbyne one is quasi-unidimensional, threadlike, linear.

The intermediate degree in hybridization of $sp^m$ carbon atoms is also possible when $m \neq 1$; 2 or 3 [5], what can cause the formation of new structural carbon modifications such as peapods, graphanes, gliters, khenocombs, clasrits, cubans etc. [6]. As noted in [7], there exists *fcc* carbon forms when the hybridized electron orbits are absent.

1

*T.N. Veziroglu et al. (eds.),*
*Hydrogen Materials Science and Chemistry of Carbon Nanomaterials, 1-24.*
© *2004 Kluwer Academic Publishers. Printed in the Netherlands.*

The mechanism of reconstruction in the crystal lattices is determined by forming different carbon modifications and consists in changing configurations of external valence electron clouds. This mechanism is controlled by the charge state of carbon atoms and their different valence [5-9]. Carbon atoms can be tetra-, tri- and bivalent. The type of the bond between carbon atoms can be different (single-, two- and tri-paired).

Difference in the electron density of carbon atoms can cause carbon atoms of different rating $C_1$, $C_2$ present in the crystal what, in turn, can determine the ordering process of these atoms.

Fig.1 illustrates the experimental phase diagram for carbon with the temperature and pressure ranges in which the phases of diamond D, graphite G, melt M and the probable metal phase MP are realized [10-18]. The diagram does not show other known crystalline carbon forms: carbyne C and fullerite F.

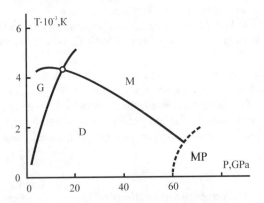

Fig.1. Experimental state diagram for carbon phases. D - diamond, G - graphite, M - melt, MP - probable metal phase. The triple site is marked with the circle.

Thermodynamic calculation [19-23] of the dependence for the equilibrium temperature for diamond and graphite phases is in a good agreement with the experimental data given in Fig.1.

In searching new active elements for microelectronics and optoelectronics the diamond-like films were synthesized. These films simultaneously contain diamond, graphite, Carbyne phases in which there exist area-clusters with the short-range order and $sp^3$ -, $sp^2$ -, sp – hybridization of carbon bonds, respectively [24]. With increasing temperature graphitization occurs in the films.

The study of known allotropic carbon modifications, calculation of free energy and thermodynamic potentials of phases, their comparison for different carbon structures, evaluation of energy parameters to determine conditions for possible phase transitions from one modification to the other, plotting the phase diagram for carbon are of interest.

In the present paper the thermodynamical functions of diamond, graphite, carbyne, fullerite have been calculated with consideration for ordering process in phases and effect of pressure on the ordering process. The temperature dependences of thermodynamical potentials has been plotted for different values of the external all-round pressure. The energetic parameters, at which the plots above intersect, have been evaluated and the phase diagram has been plotted using these intersection sites.

The dependence of equilibrium order parameters for the phases on temperature and pressure has been calculated. The ordering temperatures and ordering pressures, their mutual dependence have been evaluated.

The calculations have been carried out using molecular-kinetic theory [25, 26] ignoring the substitution of the lattice sites by atoms of different sorts (in fullerite - by different $C_{60}$, $C_{70}$ molecules) and with the assumption on geometrical perfection of phase crystal lattices and single-domain crystal structure. Interatomic (intermolecular) interaction is considered in two coordination spheres.

## 2. Diamond. Calculation of free energy

The diamond phase is metastable under normal conditions [1,3,27] but it can exist for the indefinitely long time. Synthetic diamonds are produced from graphite under high pressures. The chemical synthesis of diamond has been developed at moderate pressures [28, 29]. On heating to 1300-2100 K, diamond graphitization begins.

The diamond crystal lattice is face-centered cubic (Fig.2) with the parameter $a$=0,357 nm. The lattice can be depicted as two *fcc* lattices combined in such a way that one of them is displaced by a quarter of the spatial diagonal. The elementary cell is a cub with 8 atoms, each of them is the center of the tetrahedron in the nearest surrounding.

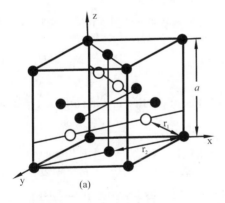

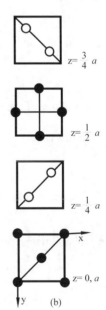

Fig.2. The elementary cell of the diamond crystal lattice in the space representation (a) and in the projection on the plane (x, y) (b) for different values of the $z$ coordinate
● – sites of the first type corresponding to $C_1$ atoms,
○ – sites of the second type corresponding to $C_2$ atoms.

In the ordered state we assume the layered distribution of $C_1$ and $C_2$ carbon atoms, as shown in Fig.2. In the elementary cell we have four sites of the first and the second types corresponding atoms of $C_1$ and $C_2$ sorts, respectively.

Free energy is calculated with regard to interaction between $C_1$ and $C_2$ atoms in two coordination spheres that are at the following distances from each atom, respectively:

$$r_1 = a\sqrt{3}/4, \qquad\qquad r_2 = a/\sqrt{2}, \qquad\qquad (1)$$

which for diamond equal to $r_1$=0,155 nm and $r_2$=0,252 nm.

For the first coordination sphere the site of each type has four nearest sites of other type, and for the second coordination sphere each site has 12 sites of the same type at the distance of $r_2$.

The free energy of the crystal is calculated by the formula

$$F_D = E_D - kT \ln W_D, \qquad\qquad (2)$$

where $E_D$ is configuration energy, $W_D$ is thermodynamic probability, $k$ is Boltzmann's constant, $T$ is absolute temperature, also we introduce the following designation:

$N$ is the number of all sites (atoms) of the crystal,

$N_1 = N_2 = \dfrac{1}{2}N$ is the number of sites of the first and the second types, respectively,

$N_{c_1}$, $N_{c_2}$ are the numbers of atoms of $C_1$, $C_2$ sorts

$$c_1 = N_{c_1}/N, \qquad\qquad c_2 = N_{c_2}/N \qquad\qquad (3)$$

are their atomic concentrations,

$$c_1 + c_2 = 1. \qquad\qquad (4)$$

$N_{c_1}^{(1)}$, $N_{c_1}^{(2)}$, $N_{c_2}^{(1)}$, $N_{c_2}^{(2)}$ are the numbers of atoms of $C_1$, $C_2$ sorts in the sites of the first and the second types,

$$P_{c_1}^{(1)} = N_{c_1}^{(1)}/N_1, \qquad\qquad P_{c_1}^{(2)} = N_{c_1}^{(2)}/N_2,$$
$$P_{c_2}^{(1)} = N_{c_2}^{(1)}/N_1 \qquad\qquad P_{c_2}^{(2)} = N_{c_2}^{(2)}/N_2 \qquad\qquad (5)$$

are probabilities of the substitution of sites of the first and the second types by $C_1$ and $C_2$ atoms.

In this case

$$P_{c_1}^{(1)} + P_{c_2}^{(1)} = 1, \qquad\qquad P_{c_1}^{(2)} + P_{c_2}^{(2)} = 1,$$
$$P_{c_1}^{(1)} + P_{c_1}^{(2)} = 2c_1, \qquad\qquad P_{c_2}^{(1)} + P_{c_2}^{(2)} = 2c_2. \qquad\qquad (6)$$

The degree of crystal ordering is defined by the order parameter

$$\eta = 2\left(P_{c_1}^{(1)} - c_1\right) \qquad\qquad (7)$$

Then probabilities $P_{c_i}^{(i)}$ ($c_i$=$c_1$, $c_2$, $i$=1; 2) are defined by the relations

$$P_{c_1}^{(1)} = c_1 + \frac{1}{2}\eta, \qquad\qquad P_{c_1}^{(2)} = c_1 - \frac{1}{2}\eta,$$
$$P_{c_2}^{(1)} = c_2 - \frac{1}{2}\eta, \qquad\qquad P_{c_2}^{(2)} = c_2 + \frac{1}{2}\eta. \qquad\qquad (8)$$

The configuration internal energy $E_D$ is defined by the sum of energies of paired interaction between $C_1$ and $C_2$ atoms at the distances $r_1$, $r_2$

$$E_A = -N_{c_1c_1}(r_1)v'_{11} - N_{c_2c_2}(r_1)v'_{22} - N_{c_1c_2}(r_1)v'_{12} -$$
$$- N_{c_1c_1}(r_2)v''_{11} - N_{c_2c_2}(r_2)v''_{22} - N_{c_1c_2}(r_2)v''_{12}, \tag{9}$$

where $v'_{ij}$, $v''_{ij}$ $(i,j=1;2)$ are the energies of interaction between $C_1$ and $C_2$ atoms with the opposite sign at the distances $r_1$, $r_2$, respectively, and $N_{c_ic_j}(r_1)$, $N_{c_ic_j}(r_2)$, are the numbers of pairs of $C_1$ and $C_2$ atoms at the same distances.

The calculation of the numbers of pairs gives the following formulae:

$$N_{c_1c_1}(r_1) = 2NP_{c_1}^{(1)}P_{c_1}^{(2)},$$
$$N_{c_2c_2}(r_1) = 2NP_{c_2}^{(1)}P_{c_2}^{(2)},$$
$$N_{c_1c_2}(r_1) = 2N\left(P_{c_1}^{(1)}P_{c_2}^{(2)} + P_{c_1}^{(2)}P_{c_2}^{(1)}\right),$$
$$N_{c_1c_1}(r_2) = 3N(P_{c_1}^{(1)^2} + P_{c_1}^{(2)^2}), \tag{10}$$
$$N_{c_2c_2}(r_2) = 3N(P_{c_2}^{(1)^2} + P_{c_2}^{(2)^2}),$$
$$N_{c_1c_2}(r_2) = 6N(P_{c_1}^{(1)}P_{c_2}^{(1)} + P_{c_1}^{(2)}P_{c_2}^{(2)}).$$

According to the rules of combinatorics thermodynamic probability $W_D$ can be written as

$$W_D = \frac{N_1!}{N_{c_1}^{(1)}!N_{c_2}^{(1)}!} \cdot \frac{N_2!}{N_{c_1}^{(2)}!N_{c_2}^{(2)}!} \tag{11}$$

considering the above and using equation (5) and Stirling formula $lnX!=X(lnX-1)$ correct for the large numbers, we find

$$W_D = -\frac{1}{2}N\left(P_{c_1}^{(1)}lnP_{c_1}^{(1)} + P_{c_1}^{(2)}lnP_{c_1}^{(2)} + P_{c_2}^{(1)}lnP_{c_2}^{(1)} + P_{c_2}^{(2)}lnP_{c_2}^{(2)}\right). \tag{12}$$

Substituting the equations (9), (12) in (2) with regard to formulae (8), (10), we find the free energy for the diamond crystal as follows

$$F_D = -2N[c_1^2(v'_{11} + 3v''_{11}) + c_2^2(v'_{22} + 3v''_{22}) + 2c_1c_2(v'_{12} + 3v''_{12}) +$$
$$+ \frac{1}{4}\eta_D^2(\omega'_D - 3\omega''_D)] + \tag{13}$$
$$+ \frac{1}{2}kTN\left[(c_1 + \frac{1}{2}\eta_D)ln(c_1 + \frac{1}{2}\eta_D) + (c_1 - \frac{1}{2}\eta_D)ln(c_1 - \frac{1}{2}\eta_D) +\right.$$
$$\left. + (c_2 - \frac{1}{2}\eta_D)ln(c_2 - \frac{1}{2}\eta_D) + (c_2 + \frac{1}{2}\eta_D)ln(c_2 + \frac{1}{2}\eta_D)\right],$$

where $\eta_D$ is the order parameter in the diamond crystal, and values

$$\omega'_D = 2v'_{12} - v'_{11} - v'_{22}, \qquad \omega''_D = 2v''_{12} - v''_{11} - v''_{22} \tag{14}$$

are the ordering energies of $C_1$ and $C_2$ atoms in the first and the second coordination spheres, respectively. The sign (-) in front of the value $\omega''_D$ in (13) shows that the considered interatomic interaction in the second coordination sphere decreases the resulting energy for the crystal ordering which is defined by the difference $\omega'_D - 3\omega''_D$.

The derived formula (13) defines the dependence of the free energy for diamond on the crystal composition, temperature, order parameter and energetic constants. Further we shall investigate the diamond phase of carbon using this formula.

## 3. Graphite. Free energy

Under normal conditions graphite is the thermodynamically stable phase [1,3,27]. The graphite structure is packed hexagonal, layered [30]. The crystal lattice can be depicted as two simple hexagonal sublattices positioned in such a way that in the neighboring layers (in the basis planes of different sublattices at the distance $c/2$, $c=0.671$ nm) the sites in one layer are opposite the centers of hexagons in the other (Fig.3). The elementary cell is a prism with a rhomb in its base. The side of the rhomb equals to $a=0.246$ nm. The distance between the nearest atoms in the basis plane and in the adjacent planes are defined by the following relations

$$r_1 = a/\sqrt{3} \ , \qquad r_2 = \sqrt{4a^2 + 3c^2}/2\sqrt{3} \ . \tag{15}$$

For the graphite crystal these distances equal to $r_1=0.142$ nm and $r_2=0.364$ nm, respectively. The significant difference in these distances, which define the interatomic distances in the first and the second coordination spheres, is due to the strong interaction between atoms at the distance $r_1$ (covalent forces) and the weak interaction at the distance $r_2$ (Van der Waals forces) what explains the easy slipping of the crystal layers relative to each other [31]. Therefore, the graphite structure is assumed to be quasi-two dimensional.

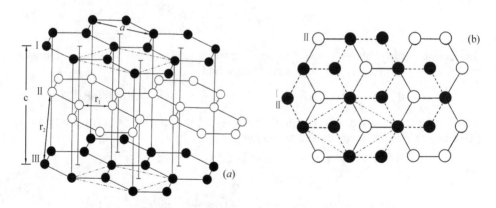

Fig.3. The crystal lattice of graphite
● - sites of the first type corresponding to $C_1$ atoms,
○ -sites of the second type corresponding to $C_2$ atoms.
        The dot-and-dash line marks the bases of the prism - the elementary cell of the crystal.
(a)   The space image of the crystal lattice.
(b) The projection of the lattice on the basis plane. The sites in basis plane II are connected with the solid straight lines. The sites in planes I and III, above and below the plane II, are connected with the dotted lines. The projections of some white sites of plane II coincide with the black ones of plane I.

For the ordered state we assume that the sites of the sublattices are the sites of different types corresponding to $C_1$ and $C_2$ atoms, respectively. Thus, the elementary cell consists of four sites: two of the first type and two of the second one. Each site of the crystal in the basis plane is surrounded by three sites of the same type at the distance $r_1$, and in the planes below and above the considered basis plane each sites is surrounded by 6 sites of the other type at the distance $r_2$.

The free energy is calculated with regard to the interatomic interaction in two coordination spheres.

The number of atomic pairs $C_iC_j$ $(i,j=1;2)$ at the distances $r_1$, $r_2$ equals to, respectively,

$$
\begin{aligned}
& N_{c_1c_1}(r_1) = \frac{3}{4}N(P_{c_1}^{(1)^2} + P_{c_1}^{(2)^2}), \quad N_{c_2c_2}(r_1) = \frac{3}{4}N(P_{c_2}^{(1)^2} + P_{c_2}^{(2)^2}), \\
& N_{c_1c_2}(r_1) = \frac{3}{2}N(P_{c_1}^{(1)}P_{c_2}^{(1)} + P_{c_1}^{(2)}P_{c_2}^{(2)}), \\
& N_{c_1c_1}(r_2) = 3N\,P_{c_1}^{(1)}P_{c_1}^{(2)}, \quad N_{c_2c_2}(r_2) = 3N\,P_{c_2}^{(1)}P_{c_2}^{(2)}, \\
& N_{c_1c_2}(r_2) = 3N(P_{c_1}^{(1)}P_{c_2}^{(2)} + P_{c_1}^{(2)}P_{c_2}^{(1)}).
\end{aligned}
\tag{16}
$$

The configuration energy $E_G$ of graphite and thermodynamic probability $W_G$ are determined by the same formulae (9) and (11) as for diamond. Considering these formulae and the number of pairs (16), we find the free energy of the graphite crystal is expressed in the possibilities $P_{c_i}^{(i)}$, as follows:

$$
\begin{aligned}
F_G = & -\frac{3}{4}N[(P_{c_1}^{(1)^2} + P_{c_1}^{(2)^2})\upsilon_{11}' + (P_{c_2}^{(1)^2} + P_{c_2}^{(2)^2})\upsilon_{22}' + 2(P_{c_1}^{(1)}P_{c_2}^{(1)} + P_{c_1}^{(2)}P_{c_2}^{(2)})\upsilon_{12}'] - \\
& -3N[P_{c_1}^{(1)}P_{c_1}^{(2)}\upsilon_{11}'' + P_{c_2}^{(1)}P_{c_2}^{(2)}\upsilon_{22}'' + (P_{c_1}^{(1)}P_{c_2}^{(2)} + P_{c_1}^{(2)}P_{c_2}^{(1)})\upsilon_{12}''] + \\
& +\frac{1}{2}kTN(P_{c_1}^{(1)}lnP_{c_1}^{(1)} + P_{c_1}^{(2)}lnP_{c_1}^{(2)} + P_{c_2}^{(1)}lnP_{c_2}^{(1)} + P_{c_2}^{(2)}lnP_{c_2}^{(2)}),
\end{aligned}
\tag{17}
$$

where $\upsilon_{ij}'$, $\upsilon_{ij}''$ are energies of interaction between $C_1$, $C_2$ atoms, as in the case of diamond, at the distances $r_1, r_2$.

Substituting probabilities (8) into (17), we find

$$
\begin{aligned}
F_G = & -\frac{3}{2}N[c_1^2(\upsilon_{11}' + 2\upsilon_{11}'') + c_2^2(\upsilon_{22}' + 2\upsilon_{22}'') + 2c_1c_2(\upsilon_{12}' + 2\upsilon_{12}'') + \frac{1}{4}\eta_G^2(-\omega_G' + 2\omega_G'')] + \\
& +\frac{1}{2}kTN[(c_1 + \frac{1}{2}\eta_G)ln(c_1 + \frac{1}{2}\eta_G) + (c_1 - \frac{1}{2}\eta_G)ln(c_1 - \frac{1}{2}\eta_G) + \\
& + (c_2 - \frac{1}{2}\eta_G)ln(c_2 - \frac{1}{2}\eta_G) + (c_2 + \frac{1}{2}\eta_G)ln(c_2 + \frac{1}{2}\eta_G)],
\end{aligned}
\tag{18}
$$

where $\omega_G'$, $\omega_G''$ are ordering energies of $C_1$, $C_2$ atoms, respectively, they are determined by the same formulae (14), as for diamond; in graphite for the first and the second coordination spheres, respectively, and $\eta_G$ is the degree of the long-range order in the graphite crystal. As seen, in (18) $\omega_G'$, $\omega_G''$ have different signs. The sign (-) in front of $\omega_G'$ indicates that ordering in graphite occurs due to interaction between $C_1$, $C_2$ atoms in the second coordination sphere.

Below we shall study formula (18) defining dependence of free energy for graphite on its composition, temperature, order parameter and energetic constants.

## 4. Carbyne. Calculation of free energy

Under normal pressure the carbyne form of carbon appears at higher temperature than that for diamond, i.e. higher than $(1-3) \cdot 10^3$ K. It has been found that the natural conditions, under which carbyne and diamond are formed, are close. The X-ray diffraction pattern of diamond showed the structural lines of carbyne. Diamond contained the carbyne phase embedded.

Crystals are also produced from graphite [33-38]. As found out in [39], the graphite-carbyne phase transition occurred at 2600 K. At 3800 K and prolonged heating carbyne forms and melts so that its triple site is determined by the temperature of 3800 K and pressure of $2 \cdot 10^4$ Pa.

In natural conditions carbyne was found in the meteorite crater. The linear form of carbon crystals - rod-shaped molecules was synthesized in the laboratory conditions. It this case, carbyne is formed under milder conditions than those at which diamond is formed. Thermodynamically carbyne is more stable than graphite. When heated to ~ 2300 K, carbyne transforms into the graphite phase.

As found, the carbyne phase is stable at high temperature (above 1000 K) and low pressure ($\sim 1,5 \cdot 10^{-2}$ Pa).

Carbyne is formed by crystals of small linear size about $10^2$ nm. The elementary cell consists of three 6-atom carbon chains arranged along the straight parallel lines [32]. The filar quasi-onedimensional crystals of carbyne are formed due to the $sp$-hybrid electron structure of carbon atoms. The dissemination of atoms with other ($sp^2$, $sp^3$) hybrid bonds into the carbyne crystal can change the periodicity in cross-linking chains-cells what, however, is not considered in calculations.

Fig.4 shows the elementary cell of the carbyne crystal lattice. Its parameters equal to (in nm):

$$a = 0,508, \quad b = 0,295, \quad c = 0,780 \tag{19}$$

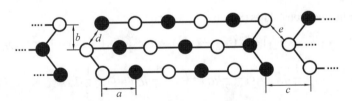

Fig.4. The elementary cell of the carbyne crystal lattice

● – sites of the first type corresponding to $C_1$ atoms,
○ – sites of the second type corresponding to $C_2$ atoms.

We assume that elementary cells are at the distance $c$ from each other not only in the direction of the $c$ axis, but also in the directions perpendicular to the $b$ and $c$ axis.

For the ordered state, we divide the sites of the elementary cell into two types corresponding to atoms of $C_1$, $C_2$ sorts, respectively. The elementary cells contains 18 sites, nine of them are of the first type and the rest of them is of the second type. We assume that both the cell shown in Fig.4 and the cell, in which sites of the first and the second types change their positions, can be formed in the ordered state.

Calculating free energy, we consider interaction between carbon atoms at the distances

$$r_1 = d = \sqrt{a^2 + 4b^2} / 2 = 0{,}389 \ nm, \qquad r_2 = a = 0{,}508 \ nm. \tag{20}$$

The numbers of atomic pairs $C_i C_j$ (i, j=1;2) at these distances equal to

$$N_{c_1 c_1}(r_1) = \frac{11}{18} N(P_{c_1}^{(1)} + P_{c_1}^{(2)})^2, \quad N_{c_2 c_2}(r_1) = \frac{11}{18} N(P_{c_2}^{(1)} + P_{c_2}^{(2)})^2,$$

$$N_{c_1 c_2}(r_1) = \frac{11}{9} N(P_{c_1}^{(1)} + P_{c_1}^{(2)})(P_{c_2}^{(1)} + P_{c_2}^{(2)}),$$

$$N_{c_1 c_1}(r_2) = \frac{5}{9} N P_{c_1}^{(1)} P_{c_1}^{(2)}, \qquad N_{c_2 c_2}(r_2) = \frac{5}{9} N P_{c_2}^{(1)} P_{c_2}^{(2)}, \tag{21}$$

$$N_{c_1 c_2}(r_2) = \frac{5}{9} N (P_{c_1}^{(1)} P_{c_2}^{(2)} + P_{c_1}^{(2)} P_{c_2}^{(1)}).$$

The configuration internal energy $E_C$ of carbyne, evaluated by the formula analogous to (9), is expressed through probabilities $P_{c_i}^{(j)}$ by the relation

$$E_C = -\frac{11}{18} N[(P_{c_1}^{(1)} + P_{c_1}^{(2)})^2 \upsilon_{11}(d) + (P_{c_2}^{(1)} + P_{c_2}^{(2)})^2 \upsilon_{22}(d) +$$

$$+ 2(P_{c_1}^{(1)} + P_{c_1}^{(2)})(P_{c_2}^{(1)} + P_{c_2}^{(2)})\upsilon_{12}(d)] - \tag{22}$$

$$-\frac{5}{9} N[P_{c_1}^{(1)} P_{c_1}^{(2)} \upsilon_{11}(a) + P_{c_2}^{(1)} P_{c_2}^{(2)} \upsilon_{22}(a) + (P_{c_1}^{(1)} P_{c_2}^{(2)} + P_{c_1}^{(2)} P_{c_2}^{(1)})\upsilon_{12}(a).$$

The thermodynamic probability $W_C$ is evaluated by formula (11) in which the order parameter should be written for carbyne, $\eta_C$.

Considering formulae (8), we find the free energy for carbyne as follows

$$F_C = -\frac{2}{9} N[c_1^2 (11\upsilon_{11}' + \frac{5}{2}\upsilon_{11}'') + c_2^2 (11\upsilon_{22}' + \frac{5}{2}\upsilon_{22}'') + 2c_1 c_2 (11\upsilon_{12}' + \frac{5}{2}\upsilon_{12}'')] -$$

$$-\frac{5}{36} N(2\upsilon_{12}'' - \upsilon_{11}'' - \upsilon_{22}'')\eta_C^2 +$$

$$+\frac{1}{2} kTN[(c_1 + \frac{1}{2}\eta_C)\ln(c_1 + \frac{1}{2}\eta_C) + (c_1 - \frac{1}{2}\eta_C)\ln(c_1 - \frac{1}{2}\eta_C) +$$

$$+ (c_2 - \frac{1}{2}\eta_c)\ln(c_2 - \frac{1}{2}\eta_C) + (c_2 + \frac{1}{2}\eta_C)\ln(c_2 + \frac{1}{2}\eta_C)], \tag{23}$$

where

$$\upsilon_{11}' = \upsilon_{11}(d), \ \upsilon_{22}' = \upsilon_{22}(d), \ \upsilon_{12}' = \upsilon_{12}(d),$$
$$\upsilon_{11}'' = \upsilon_{11}(a), \ \upsilon_{22}'' = \upsilon_{22}(a), \ \upsilon_{12}'' = \upsilon_{12}(a). \tag{24}$$

As formulae (23) show, the ordering process in carbyne is caused by interatomic interaction at the distance $r_2 = a$. This results from the fact that at the distance $r_1 = d$ the site of each type has two nearest sites of the first and the second types. At the distance $r_2 = a$ every site is surrounded by two sites of the other type.

Below we shall study the free energy $F_C$ of carbyne according to formula (23) which defines the dependence of free energy on the composition, temperature, degree of the long-range order and energetic constants.

### 5. Fullerite. Free energy of fullerite

Fullerite is a molecular crystal formed from $C_{60}$, $C_{70}$ molecules-clusters and the other ones called fullerenes.

Since the discovery of fullerenes [40-46], and especially after the development of methods for fullerenes production in preparative amounts [47-52], fullerite attracts more and more scientists' attention.

A number of works [53-64] including monographs [65-70] are dedicated to current methods for preparing fullerenes, studying their properties and applications.

Fullerenes easily add and donate hydrogen isotopes [71-80]. Therefore hydrofullerenes can be used as hydrogen accumulators. Wide investigations are being performed to solve practical problems on the search of mobile systems to storage hydrogen, ecologically clean energy source. Both fullerenes and their metal derivatives can be hydrogen sorbents [81-88].

Depending on the conditions for developing the random composition by components $F_1=C_{60}$, $F_2=C_{70}$, fullerite has been produced [54, 89-93]. Evaluation showed that solid solutions of fullerite were more stable than the mixture of $F_1$, $F_2$ fullerenes [89]. It should be noted that solid solutions of $F_1$-$F_2$ fullerenes are of interest as model objects to develop the theory of molecular crystals.

Under normal conditions fullerite has the *fcc* structure [51, 79, 94] given in Fig.5.

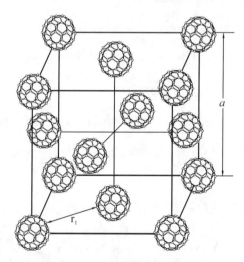

Fig.5. The elementary cell of the fullerite molecular crystal

 - lattice sites in which molecules of $F_1=C_{60}$, $F_2=C_{70}$ fullerenes are distributed

The orientation ordering takes place in fullerite that was experimentally studied in papers [95-97] and at ~ 260 K the *fcc* lattice was formed from the simple cubic one due to this orientation ordering. In this case the *fcc* lattice in its ordered state consists of four simple cubic sublattices [98]. Fullerenes are bonded in fullerite due to Van der Waals forces. Both the mechanism of the interfullerene bond and the ordering process are controlled by the charge state of fullerenes [94].

Free energy of fullerite is calculated in the model of spherically symmetrical stiff balls [99] with regard to interaction between fullerene molecules in two coordination spheres in spite of the fact that, as evaluation showed, the interaction in the second, the third and other coordination spheres contributes into the free energy no more than 5% [100].

We assume that the fullerite crystal lattice is of $Ll_0$ structure (CuAu type). Thus, in the ordered state $F_1$, $F_2$ fullerenes are mainly distributed in the alternating planes, and for the stoichiometric composition the $c_1$, $c_2$ concentrations of $F_1$, $F_2$ fullerenes are equal. In this case the order parameter and substitution probabilities of sites of the first and the second types are determined by formulae (7) and (8), respectively. Thermodynamical probability $W_F$ is also determined by derived formula (12).

For determination of the configuration internal energy $E_F$ the numbers of fullerene pairs have been calculated at the following distances:

$$r_1 = a / \sqrt{2} \quad \text{and} \quad r_2 = a \ , \tag{25}$$

where $a = 1,417$ nm is the fullerite lattice constant, in this case $r_1 = 1,002$ nm, $r_2 = 1,417$ nm.

In the structure $Ll_0$ every site of the first type in the first coordination sphere has four nearest sites of the first type and eight sites of the second type at the distance $r_1$. The site of the second type is surrounded by four nearest sites of the second type and by eight sites of the first type. In the second coordination sphere every site is surrounded by six sites of the same type at the distance $a$.

The calculation of the numbers of $F_1F_1$, $F_2F_2$, $F_1F_2$ fullerene pairs at the distances $r_1$, $r_2$ gives the following result:

$$N_{F_1F1}(r_1) = N[(P_{F_1}^{(1)} + P_{F_1}^{(2)})^2 + 2P_{F_1}^{(1)}P_{F_1}^{(2)}] \ ,$$

$$N_{F_2F2}(r_1) = N[(P_{F_2}^{(1)} + P_{F_2}^{(2)})^2 + 2P_{F_2}^{(1)}P_{F_2}^{(2)}] \ ,$$

$$N_{F_1F_2}(r_1) = 2N[(P_{F_1}^{(1)} + P_{F_1}^{(2)}) \ (P_{F_2}^{(1)} + P_{F_2}^{(2)}) + P_{F_1}^{(1)} \ P_{F_2}^{(2)} + P_{F_1}^{(2)}P_{F_2}^{(1)})] \ ,$$

$$N_{F_1F_1}(r_2) = \frac{3}{2}N(P_{F_1}^{(1)^2} + P_{F_1}^{(2)^2}) \ ,$$

$$N_{F_2F_2}(r_2) = \frac{3}{2}N(P_{F_2}^{(1)^2} + P_{F_2}^{(2)^2}) \ , \tag{26}$$

$$N_{F_1F_2}(r_2) = 3N(P_{F_1}^{(1)}P_{F_2}^{(1)} + P_{F_1}^{(2)}P_{F_2}^{(2)}) \ ,$$

where $N$ is the number of sites (fullerenes) in the crystal.

The configuration energy $E_F$ determined by the sum of interaction energies between fullerenes equal to

$$E_F = -N\{[(P_{F_1}^{(1)} + P_{F_1}^{(2)})^2 + 2P_{F_1}^{(1)} P_{F_1}^{(2)}]\upsilon'_{11} + [(P_{F_2}^{(1)} + P_{F_2}^{(2)})^2 + 2P_{F_2}^{(1)} P_{F_2}^{(2)}]\upsilon'_{22} +$$

$$+ 2[(P_{F_1}^{(1)} + P_{F_1}^{(2)})(P_{F_2}^{(1)} + P_{F_2}^{(2)}) + P_{F_1}^{(1)} P_{F_2}^{(2)} + P_{F_1}^{(2)} P_{F_2}^{(1)}]\upsilon'_{12}\} - \tag{27}$$

$$- \frac{3}{2} N[(P_{F_1}^{(1)^2} + P_{F_1}^{(2)^2})\upsilon''_{11} + (P_{F_2}^{(1)^2} + P_{F_2}^{(2)^2})\upsilon''_{22} + 2(P_{F_1}^{(1)} P_{F_2}^{(1)} + P_{F_1}^{(2)} P_{F_2}^{(2)})\upsilon''_{12}],$$

where $\upsilon'_{ij}$, $\upsilon''_{ij}$ are energies of interaction between $F_i F_j$ fullerene pairs at the distances $r_1$, $r_2$ ($i$, $j$=1;2).

Thermodynamical probability $W_F$ has the same form as for diamond, graphite, carbyne, only with the order parameter $\eta_F$ for fullerite. In this case the calculation of free energy for fullerite in dependence on the composition, degree of ordering and temperature gives the following formula

$$F_F = -3N[c_1^2(2\upsilon'_{11} + \upsilon''_{11}) + c_2^2(2\upsilon'_{22} + \upsilon''_{22}) + 2c_1 c_2(2\upsilon'_{12} + \upsilon''_{12})] - \frac{1}{4}N\eta_F^2(2\omega'_F - 3\omega''_F) +$$

$$+ \frac{1}{2}kTN[(c_1 + \frac{1}{2}\eta_F)\ln(c_1 + \frac{1}{2}\eta_F) + (c_1 - \frac{1}{2}\eta_F)\ln(c_1 - \frac{1}{2}\eta_F) + \tag{28}$$

$$+ (c_2 - \frac{1}{2}\eta_F)\ln(c_2 - \frac{1}{2}\eta_F) + (c_2 + \frac{1}{2}\eta_F)\ln(c_2 + \frac{1}{2}\eta_F)],$$

where

$$\omega'_F = 2\upsilon'_{12} - \upsilon'_{11} - \upsilon'_{22}, \quad \omega''_F = 2\upsilon''_{12} - \upsilon''_{11} - \upsilon''_{22}$$

are the ordering energies of fullerenes for the first and the second coordination sphere, respectively.

As the derived formula (28) shows, fullerenes are ordered due to their interaction in the first coordination sphere. When interaction between fullerenes in the second coordination sphere is taken into account, the resulting ordering energy will be decreased (the value of $\omega''_F$ has the sign (-)). However, as noted above, the estimation showed that this decrease was insignificant.

## 6. Discussion of theoretical results, comparison with experiment. Temperature and pressure of ordering. The constitution diagram

Comparing equations (13), (18), (23), (28) for the free energies of carbon phases, we can write the overall formula for free energy of diamond, graphite, carbyne, fullerite for one lattice site. This formula has the following form:

$$f_i = e_i - \omega_i \eta_i^2 + \frac{1}{2}kT\Delta_i, \tag{29}$$

where index $i$ denotes one of the D, G, C or F phases and

$$\Delta_i = (c_1 + \frac{1}{2}\eta_i)ln(c_1 + \frac{1}{2}\eta_i) + (c_1 - \frac{1}{2}\eta_i)ln(c_1 - \frac{1}{2}\eta_i) +$$

$$+ (c_2 + \frac{1}{2}\eta_i)ln(c_2 + \frac{1}{2}\eta_i) + (c_2 - \frac{1}{2}\eta_i)ln(c_2 - \frac{1}{2}\eta_i). \tag{30}$$

The $e_i$ and $\omega_i$ values in (29) for D, G, C, F phases equal to, respectively:

$$e_D = -2[c_1(v'_{11} + 3v''_{11}) + c_2(v'_{22} + 3v''_{22}) + c_1 c_2(\omega'_D + 3\omega''_D)],$$

$$\omega_D = \frac{1}{2}(\omega'_D + 3\omega''_D) \qquad \text{for D phase,} \qquad (31)$$

$$e_G = -\frac{3}{2}[c_1(v'_{11} + 2v''_{11}) + c_2(v'_{22} + 2v''_{22}) + c_1 c_2(\omega'_G + 2\omega''_G)],$$

$$\omega_G = \frac{3}{8}(-\omega'_G + 2\omega''_G) \qquad \text{for G phase,} \qquad (32)$$

$$e_C = -\frac{2}{9}[c_1(11v'_{11} + \frac{5}{2}v''_{11}) + c_2(11v'_{22} + \frac{5}{2}v''_{22}) + c_1 c_2(11\omega'_C + \frac{5}{2}3\omega''_C)],$$

$$\omega_C = \frac{5}{36}\omega''_C \qquad \text{for C phase,} \qquad (33)$$

$$e_F = -3[c_1(2v'_{11} + v''_{11}) + c_2(2v'_{22} + v''_{22}) + c_1 c_2(2\omega'_F + \omega''_F)],$$

$$\omega_F = \frac{1}{4}(2\omega'_F - 3\omega''_F) \qquad \text{for F phase.} \qquad (34)$$

External all-round pressure P decreases the interatomic distance

$$r = r_0 - \alpha P, \qquad \alpha \sim 10^{-2} \ \text{GPa}, \qquad (35)$$

and, therefore, changes the energetic parameters and ordering energy of the crystal.

Dependence of energetic parameters and ordering energy on the external pressure can be represented as follows [25]:

$$e_i = e_{0i}[1 + \frac{P}{P'} + n(\frac{P}{P'})^2], \qquad \omega_i = \omega_{0i}[1 + \frac{P}{P'} + n(\frac{P}{P'})^2], \qquad (36)$$

where values of P' and $n$ are constants which can be both positive and negative.

Depending on the numerical value of external pressure P, only linear terms, only quadratic ones or both of them can be considered in formulae (36). The terms in (36), proportional to $P^2$, will play significant part only under sufficiently high pressures. The value of external pressure equal to $P^* = 10\,GPa$ was estimated in [25]. Here, at $P \ll P^*$ only linear terms are considered in (36) and at $P \gg P^*$ only quadratic terms are considered.

Experimentally measured temperature and pressure ranges, in which different carbon phases exist (Fig.1), give the values of pressure in the order of 10 GPa. This means that in the studied case both linear terms and quadratic ones should be considered in (36).

The thermodynamical potential for one lattice site in terms of (36) can be written as follows:

$$\varphi_i = (e_{0i} - \omega_{0i}\eta_i^2)[1 + \frac{P}{P'} + n(\frac{P}{P'})^2] + \frac{1}{2}kT\Delta_i + \omega_{0i}(1 - \text{æ}P)P, \qquad (37)$$

where $\omega_{0i}$ is the atomic volume of the crystal at $P=0$, $\text{æ} = -\frac{1}{V}\frac{\partial V}{\partial P} \approx 10^{-2} \ GPa^{-1}$ is the compressibility of the crystal. Formula (37) determines the dependence of thermodynamical potential on the temperature, pressure, order parameter and energetic constants.

14

We have restricted ourselves to the case when both the volume $V$ and the compressibility æ slightly depend on the order parameter and we disregard this dependence [25].

For different values of temperature and pressure the equilibrium properties of the crystal are defined by the minimization of thermodynamical potential (37). In this approximation the equilibrium condition $\partial \varphi_i / \partial \eta_i = 0$ can be changed by the condition

$$\partial f_i / \partial \eta_i = 0. \tag{38}$$

After substitution of free energy (29) into (38) for the stoichiometric composition of the crystal, when $c_1 = c_2 = \frac{1}{2}$, we get the equilibrium equation as follows:

$$kTln\frac{1+\eta_i}{1-\eta_i} = 4\omega_{0i}\eta_i\left[1+\frac{P}{P'}+n(\frac{P}{P'})^2\right], \tag{39}$$

which defines the equilibrium value of the degree of long-range order depending on temperature and pressure.

Assuming in (39) $\eta_i \to 0$, we find the ordering temperature $T_0$ and the ordering pressure $P_0$:

$$kT_0 = 2\omega_0\left[1+\frac{P}{P'}+n(\frac{P}{P'})^2\right], \tag{40}$$

$$\frac{P_0}{P'} = \frac{1}{2n}(\sqrt{1-4n+\frac{2nkT}{\omega_0}}-1). \tag{41}$$

These equations (40), (41) also define the interrelation between the ordering temperature and the ordering pressure. This dependence is given in Fig.6. Fig.6 shows that at $P' > 0$, n = 1 the ordering temperature is increased with increasing pressure and at $P' < 0$, n = -1 the growth of external pressure decreases the ordering temperature.

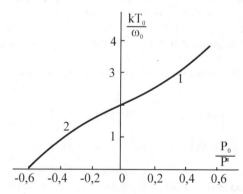

Fig.6. The plot of interrelationship between the ordering temperature and ordering pressure for values of $P' > 0$, n = 1 (section 1) and $P' < 0$, n = -1 (section 2).

Fig. 7 and 8 show the curve plot for the equilibrium values of the order parameter in dependence on temperature and pressure. The curves were plotted according to formula (39).

The dependence $\eta(T)$ (Fig.7) shows that with increasing temperature, at the certain value of pressure, the order parameter is decreased. In this case external pressure increases the order parameter at $P' > 0$ and decreases it at $P' < 0$ compared to its value at $P = 0$.

The dependence $\eta(P)$ also shows (Fig.8) that external pressure increases the order parameter at all values of temperature if $P' > 0$, and decreases it at $P' < 0$.

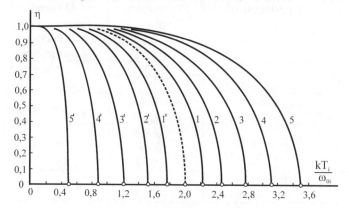

Fig.7. The curve plots of the temperature dependence of equilibrium order parameters constructed by formula (39) for different values of P/P'=0; ±0,1; ±0,2; ±0,3; ±0,4; ±0,5 (curves 1-5 for P'>0, $n$=1 and curves 1' - 5' for P'< 0, $n$=-1). The dotted line corresponds to P/P'= 0. The ordering temperatures for different pressures are marked with circles on the abscissas axis.

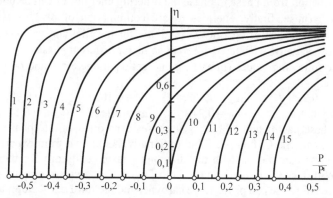

Fig. 8. The curve plots of the dependence of equilibrium order parameters on pressure constructed by formula (39) for P'>0, $n$=1 and P'< 0, $n$= -1, and for different temperatures equal to kT/$\omega_0$ = 0.2; 0.4; 0.6; 0.8; 1; 1.2; 1.4; 1.6; 1.8; 2; 2.2; 2.4; 2.6; 2.8; 3 (curves 1-15). The ordering pressures at different temperatures are marked with circles on the abscissas axis.

To evaluate temperature and pressure for possible phase transitions by formula (37), the curves for temperature dependences of thermodynamic potential were plotted for different values of external pressure. The intersection points correspond to the points of phase transformations.

The energetic parameters $e_i$ and $\omega_i$ were chosen according to the relationships between atomic volumes and average values of atomic radii of carbon phases. The structural parameters of crystal lattices for all carbon modifications are given in Table.

TABLE. Structural parameters of crystals for different allotropic carbon phases

| Crystal | Crystal lattice | Parameters, nm | Atomic (molecular for fullerite) volume, nm3 | Atomic radius, nm |
|---------|-----------------|----------------|----------------------------------------------|-------------------|
| Diamond | fcc, three-dimensional. | $a = 0,357$ | $a^3/8 = 0,006$ | 0,179 |
| Graphite | hcp, quasi-two-dimensional | $a = 0,246,$ $c = 0,671$ | $a^2 c \sqrt{3}/8 = 0,009$ | 0,206 |
| Carbyne | hcp, quasi-one-dimensional | $a = 0,508,$ $b = 0,295,$ $c = 0,78$ | $2abc/3 = 0,078$ | 0,427 |
| Fullerite | fcc, three-dimensional | $a = 1,417$ | $a^3/4 = 0,711$ | 0,893 |

As Table shows, the atomic volume and atomic radius are increased at transformation from diamond to graphite and further to carbyne and fullerite. In this case the interatomic interaction (intermolecular for fullerite) and, consequently, the energetic parameters $e_i$ must be decreased. In transformation from phase to phase the ordering energies $\omega_i$ can be both increased and decreased because, firstly, energies $\omega_i$ are combination of $\omega = 2\upsilon_{12} - \upsilon_{11} - \upsilon_{22}$ type and, secondly, they consist of energies $\omega'$ and $\omega''$ for the first and the second coordination spheres, respectively, wich have different signs for different phases.

The following values (in eV) for energetic parameters $e_i$ and ordering energies $\omega_i$ were chosen:

$$e_1 = 0,2, \qquad e_2 = 0,168, \qquad e_3 = 0,16, \qquad e_4 = 0,09,$$
$$\omega_1 = 0,2, \qquad \omega_2 = 0,21, \qquad \omega_3 = 0,16, \qquad \omega_4 = 0,17. \qquad (42)$$

These parameters of 1, 2, 3, 4 phases can correspond to phases of diamond D, graphite G, carbyne C, fullerite F but optionally. As known from experimental data, the carbyne phase can coexist with the diamond phase; conditions for their formation are similar. With increasing temperature both D and C phases are graphitized. This means that the second phase can prove to be carbyne, and the third one - graphite.

According to the chosen values for energetic parameters (42) and revealed regularities in the change of order parameters depending on temperature and pressure, the numerical values of order parameters have been evaluated for every of four phases.

The case $P' > 0$ and $n = 1$ has been studied.

17

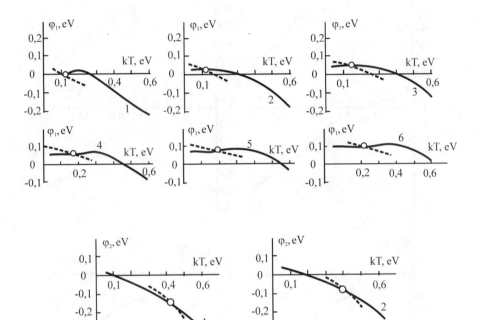

Fig. 9.

18

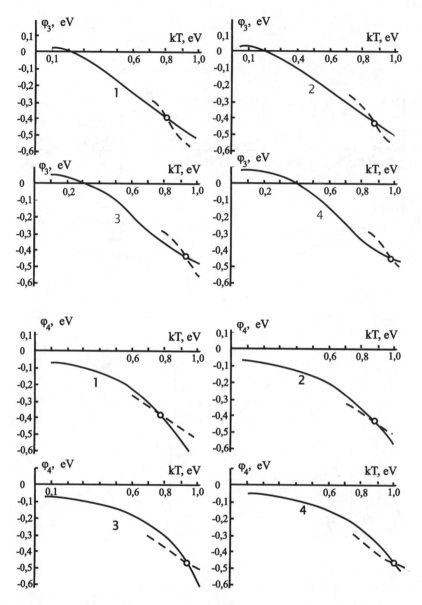

Fig. 9. The curve plots of the temperature dependence of thermodynamic potentials for carbon phases constructed by formula (37) for energetic constants (42) and different pressures $P/P' = 0$; 0.1; 0.2; 0.3; 0.4; 0.5 (curves 1-6). The intersection points of functions $\varphi_1(T)$ and $\varphi_2(T)$, $\varphi_2(T)$ and $\varphi_3(T)$, $\varphi_3(T)$ and $\varphi_4(T)$, $\varphi_4(T)$ and $\varphi_3(T)$ are marked with circles. The dotted lines around the intersection sites show the sections in the plots for thermodynamic potentials with which the intersection occurs.

19

The curves plots for temperature dependence of thermodynamical potentials $\varphi_i$ (i=1, 2, 3, 4), plotted by formula (37) for different pressures are shown in Fig.9. As indicated by Fig. 9, for all the phases i = 1, 2, 3, 4 the thermodynamical potential is decreased with increasing temperature. On the other hand, the growth of pressure increases the thermodynamical potential: curves $\varphi_i$(T,P) fall with increasing temperature and rise with increasing pressure. This means that the elevation of temperature stabilizes each phase: the temperature range, in which the phase exists, can be increased. The rise of pressure decreases the phase stability: the pressure range, in which the phase exists, can be decreased. Moreover, as seen in Fig. 9, at the temperatures above the points of $\varphi_i$ curves intersection, the values of thermodynamical potentials $\varphi_i$, in transforming from the first phase to the second one and further to the third one and the fourth one, are decreased what indicates the increase of stability of the fourth phase in comparison with to the third one, the third phase in comparison with the second one, the second phase in comparison with the first one.

The phase diagram given in Fig. 10 has been plotted by the intersection points of $\varphi_i$(T) curves for different pressures. Fig. 10 shows that beginning from the values $P / P' = 0{,}15$ temperatures of phase transitions rise with increasing pressure what corresponds to experimental data for the diamond-graphite transition. At low pressures, the temperature of phase transition can slightly drop with increasing pressure as it was found out for the equilibrium curve between phases 2 and 3. The temperature range, in which every of the phases exists, is increased with growing pressure (because from the first phase to the fourth one the slope of equilibrium curves is increased). The exception is the second phase at low pressures, for wich with increasing $P / P'$ to 0.1 the temperature range of its realization is decreased, but further at $P / P' > 0{,}1$ this range is increased.

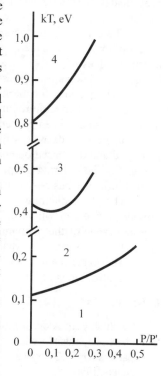

Fig. 10. The phase diagram calculated for carbon.

We can suppose that phase 1 corresponds to the diamond phase. Like in the experimental diagram (Fig. 1), the growth of pressure increases the temperature at which the diamond phase is realized. In this case the pressure range of this phase existence is decreased with increasing temperature, and the temperature range is increased with pressure rise. On the contrary, for the graphite phase the temperature range, in which this phase is realized, is decreased with growing pressure and the pressure range is increased with increasing temperature.

The experimental data for carbyne and fullerite phases are not known. Experimental verification of the obtained results for these phases is of interest.

20

## 7. Conclusions

The molecular-kinetic theory for phase transitions in allotropic carbon modifications has been developed. The possible ordering of atoms and fullerenes in phases has been considered. The statistical calculation of free energies and thermodynamic potentials for carbon phases: diamond, graphite, carbyne, fullerite has been made, their dependences on the phase composition, temperature, pressure, order parameter and energetic constants have been established.

The curves for temperature dependence of thermodynamic potentials at different values of external pressure have been plotted using the chosen energetic parameters. The possible values of temperature and pressure for phase transitions have been determined by the intersection sites of these curves.

The equations for thermodynamic equilibrium, defining dependence of the order parameter on temperature and pressure, have been derived. The curves of these dependences have been plotted. As illustrated, the increase of temperature at the certain constant pressure decreases the order parameter. The degree of ordering can be both increased and decreased due to pressure depending on the values of energy constants of the phase. The ordering temperatures and ordering pressures have been calculated, their mutual influence has been clarified.

The phase diagram has been constructed with estimation of the temperature and pressure ranges, in which different carbon phases can be realized. As found out, the pressure range of the diamond phase existence narrows with temperature increase. For other phases this range is expanded.

The experimental verification of obtained regularities of carbyne and fullerite phases realization is of interest to physicist-theorists.

## 8. References

1. Fizicheskiy entsiklopedicheskiy slovar'. V. 1. M.: Sovetskaya entsiklopediya. 1960. 664 p.
2. Fizicheskiy entsiklopedicheskiy slovar'. V. 5. M.: Sovetskaya entsiklopediya. 1966. 576 p.
3. Bar'yakhtar V.G. Fizika tverdogo tela. Entsiklopedicheskiy slovar'. V. 1. K.: Naukova dumka. 1996. 652 p.
4. Bar'yakhtar V.G. Fizika tverdogo tela. Entsiklopedicheskiy slovar'. V. 2. K.: Naukova dumka. 1996. 648 p.
5. Heimann R.B., Evsyukov S.E., Koga Y. Carbon allotropes: a suggested classification scheme based on valence orbital hybridization //Carbon. 1997. Vol. 35. P. 1654-1657.
6. Belenkov E.A. Klassifikatsiya ugleroda struktur //Vodorodnoe materialovedenie i khimiya uglevodnikh nanomaterialov. K.: IHSE. 2003. P. 730-733.
7. Guseva M.B., Babaev V.G., Konyashin I.Yu., Savchenko N.F., Khvostov V.V., Korobov Yu.A., Guden' V.S., Dement'ev A.P. Novaya faza ugleroda c GTsK strukturoy //Poverhnost. Rentgenovskie, sinhrotronnye i neitronnye issledovaniya. 2004.Vol. 3 P. 28-35.
8. Koulson Ch. Valentnost'. M.: Mir. 1965. 426 p.
9. Levin A.A. Vvedenie v kvantovuu khimiyu tverdogo tela. M.: Khimiya. 1974. 240 p.
10. Bundy F.P., Bovenkerk H.P., Strong H.M., Wentorg P.H. Diamond-graphite equilibrium line from growth and graphitization of diamond. // Journ. Chem. Phys. 1961. Vol. 35. N 2. P. 383-391.

11. Bundy F.P. Melting of graphite at very high pressures //Journ. Chem. Phys. 1963. Vol. 38. N 3. P. 618-630.
12. Bundy F.P. Direct conversion of graphite to diamond in static pressure apparatus. Ibid. P. 631-643.
13. Strong H.M., Hanneman R.E. Crystallization of diamond and graphite // Journ. Chem. Phys. 1967. Vol. 49. N 9. P. 3668-3676.
14. Bundy F.P. Direct phase transformation in carbon //Reactivity of solids: Proceed. 6th Int. Symp. New York: Wiley-Intersi. 1969. P. 817-825.
15. Strong H.M., Chrenko R.M. Further studies of diamond growth rates and physical properties of laboratory-made diamond // Journ. Phys. Chem. 1971. Vol. 75. N 12. P. 1838-1843.
16. Kennedy C.S., Kennedy G.C. The equilibrium boundary between graphite and diamond // Journ. Geophys. Res. 1976. Vol. 81. N 14. P. 2467-2470.
17. Novikov N.V. Sintez sverkhtverdikh materialov. K.: Naukova dumka. 1986. 280 p.
18. Novikov N.V. Fizicheskie svoystva almaza. Spravochnik. 1987. 192 p.
19. Leypunskiy O.I. Ob iskusstvennikh almazakh // Uspekhi khimii. 1939. № 8. P. 1519-1534.
20. Berman R., Simon F. On the graphite-diamond equilibrium // Z. Electrochem. 1955. Vol. 55. N 5. P. 333-338.
21. Vereschagin L.F., Yakovlev E.N., Buchnev L.M., Dymov B.K. Usloviya termodinamicheskogo ravnovesiya almaza s razlichnymi uglerodnymi materialami // Teplofizika vysokikh temperatur. 1977. № 2. P. 316-321.
22. Kurdumov A.V., Pilyankevich A.N. Fazovie prevrascheniya v uglerode i nitride bora. K.: Naukova dumka. 1979. 188 p.
23. Andreev V.D., Malik V.R., Efimovich L.P. Termodinamicheskiy raschet krivoy ravnovesiya grafit-almaz // Sverkhtverdie materially. 1984. № 2. P. 16-20.
24. Bakiy A.S., Strel'nitskiy V.E. Strukturnie i fazovie svoystva uglerodnikh kondensatov, poluchennikh osagdeniem potokov bistrikh chastits. M.: 1984.
25. Smirnov A.A. Molekulyarno-kineticheskaya teoriya metallov. M.: Nauka. 1966. 488 p.
26. Matysina Z.A. Atomniy poryadok i svoystva splavov. Dnepropetrovsk: DGU. 1981. 112 p.
27. Spitsin B.V., Alekseenko A.E. Vozniknovenie, sovremennie vozmozhnosti i nekotorie perspektivy razvitiya sinteza almaza iz gazovoy sredy // Almaznie plenki i plenki rodstvennikh materialov. Kharkov. 2002. P. 122-147.
28. Spitsin B.V., Alekseenko A.E. Khimicheskiy sintez almaza // Uglerod: fundamental'nie problemy nauki, materialovedenie, tekhnologiya. M.: MGU. 2003. P. 35.
29. Khimicheskaya entsiklopediya. V. 1. M.: Sovetskaya entsiklopediya. 1988. P. 1189.
30. Shaskol'skaya M.P. Kristallographiya. M.: Vysshaya shkola. 1984. 375 p.
31. Radchenko I.V. Molekulyarnaya fizika. M.: Nauka. 1965. 479 p.
32. Korshak V.V., Kudryavtsev Yu.P., Sladkov A.M. Karbinovaya allotropnaya forma ugleroda // Vesn. AS USSR. 1978. № 1. P. 70-78.
33. Babaev V.G., Guseva M.B., Zhuk A.Z., Lash A.A., Fortov V.V. Udarno-volnovoy sintes kristallicheskogo karbina // DAN. 1995. V. 343. № 2. P. 176-178.
34. Borodina T.I., Fortov V.E., Lash A.A., Zhuk A.Z., Guseva M.B., Babaev V.G. Shok-induced transformation of carbyne//Jour. Appl. Phys. 1996. vol. 80. №7. p. 3757-3759.
35. Zhuk A.Z., Borodina T.I., Fortov V.E., Lash A.A., Valiano G.E. High Pressure Res. 1997. vol.15. p.245.

22

36. Babina V.M., Busti M., Guseva M.B., Zhuk A.Z., Migo A., Milyavskiy V.V. Dinamicheskiy sintes ristallicheskogo karbina iz grafita i amorfnogo ugleroda // Teplofizika vysokih temperatur 1999. V. 37. № 4. P. 573-581.
37. Zhuk A.Z., Borodina T.I., Milyavskiy V.V., Fortov V.V. Udarno-volnovoy sintes karbina is grafita//DAN. 2000. V. 370. № 3. P. 328-331.
38. Zharkov A.S., Borodina T.I., Zhuk A.Z., Milyavskiy V.V., Mozdykov V.A., Babaev V.G., Guseva M.B. Udarno-volnovoy sintez kristallicheskogo karbina. // Uglerod: fundamental'nie problemy nauki, materialovedenie, tekhnologiya. M.: MGU. 2003. P. 101.
39. Klimovskiy I.I., Markovets V.V. Fizicheskie yavleniya, soprovozhdayuschie tverdofazoviy perekhod grafit-karbin pri nagreve grafita.// M.: MGU. 2003. P. 119.
40. Kroto H.W. et al. Nature. 1985. Vol. 318. P. 162.
41. Curl R.F., Smalley R.E. Probing $C_{60}$ // Science. 1988. Vol. 242. P. 1017-1021.
42. Kerl R.F., Smolli R.E. Fullereny // V mire nauki. 1991. № 12. P. 14-24.
43. Curl R.F. Pre- 1990 evidence for the fullerene proposal //Carbon. 1992. Vol.30. N 8. P. 1149-1155.
44. Smolli R.E. Otkryvaya fullereny // Uspekhi fiz. nauk. 1998. V. 168. №3 P. 323-330.
45. Kerl R.F. Istoki otkrytiya fullerenov, experiment i gipoteza // Tam zhe. P. 331-342.
46. Kroto G. Simmetriya, kosmos, zvezdy i $C_{60}$ // Tam zhe. P. 343-344.
47. Kratschmer W. Solid $C_{60}$: a new form of carbon // Nature. 1990. Vol. 347. P.354-388.
48. Kratschmer W., Lamp L.D., Fotiropoulos K., Huffman D.R. Nature. 1991. Vol. 347. P. 354.
49. Kratschmer W., Huffman D.R. Fullerites: new form of crystalline carbon // Carbon. 1992. Vol. 30. N 8. P. 1143-1147.
50. Parker D.H., Chatterjee K., Wurz P., Lykke K.P., Pellin M.J., Stock L.M., Hemminger J.C. Fullerenes and giant fullerenes: synthesis, separation and mass-spectrometric characterization // Carbon. 1992. Vol. 30. N 8. P. 1167-1182.
51. Eletskiy A.V., Smirnov B.M. Fullereny // Uspekhi fiz. Khimii. 1993. V. 163. № 2. P. 33-60.
52. Bubnov V.P., Kramnitskiy I.S., Laukhina E.E., Yakubskiy E.B. Poluchenie sazhy s vysokim soderganiem fullerenov $C_{60}$ i $C_{70}$ metodom elektricheskoy dugi // Izv. RAN. Ser. Khim. 1994. № 5. P. 805-809.
53. Fink J., Solmen E., Merkel M., Masaki A., Romberg H., Alexander M., Knupfer M., Golden M.S., Adelmann P., Renker B. // Proceed. of the "First Italian Workshop on Fullerenes: Status and Perspectives". Bologna. 1992.
54. Sokolov V.I., Stankevich I.V. Fullereny – novie allotropnie formy ugleroda:struktura, elektronnoe stroenie I khimicheskie svoystva // Uspekhi khimii. 1993. V. 62. № 5. P. 455-473.
55. Vol'pin M.E. Fullereny – novaya forma ugleroda //Vestnik RAN. 1993. № 1. P. 25-30.
56. Osip'yan Yu.A., Kveder V.V. Fullereny – novie veschestva dlya sovremennoy tekhniki //Materialovedenie. 1997. № 1. P. 2-6.
57. Sokolov V.I. Khimiya fullerenov – novikh allotropnikh modifikatsiy ugleroda // Izv. RAN. Ser. Khim. 1999. № 9. P. 1211-1218.
58. Zarubenskiy G.M., Bitskiy A.E., Andrianova L.S., Stavinskiy E.N., Zvyagina A.B. Osnovnie napravleniya issledovaniy v oblasti polucheniya, izucheniya svoystv i prakticheskogo ispol'zovaniya fullerenov // Zhurn. Prokladnoy khimii. 1999. V. 72. № 5. P. 870-875.
59. Shteyman E.A. Issledovanie prirody i svoystv strukturnikh defektov v fullerite $C_{60}$ // Nauka proizvodstvu. 2001. № 2. P. 61-70.

60. Bagenov A.V., Bashkin I.O., Kveder V.V. Fullereny // Nauka proizvodstvu. 2001. № 2. P. 4-11
61. Kveder V.V., Shteyman E.A. Issledovanie prirody I svoystv strukturnikh defektov v fullerite $C_{60}$ // Nauka proizvodstvu. 2001. № 2.
62. Fullereny i fullerenopodobnie struktury. Sb. Nauchn. Trudov. Minsk: BGU. 2000. 210 p.
63. Tarasov B.P. Accumulation of hydrogen in carbon nanostructures // Int. Sc. J. Alternative Energy and Ecology. – 2000. – Vol. 1. – P. 168-173.
64. Tarasov B.P., Efimov O.N. Vodorod v uglerodnyh nanostrukturah // Nauka proizvodstvu.- 2000.- N 10.- P. 47-50.
65. Hirsch A. The chemistry of the fullerenes. Georg Thieme Verlag. Stuttgart. – 1994. – 203 p.
66. Dresselhaus M.S., Dresselhaus G., Eklund P.C. Science of fullerenes and carbon nanotubes. San Diego: Academic Press. – 1995.
67. Shiner T. (Ed.) Optical and electronic properties of fullerenes and fullerene-based materials. New York: J. Dekker. – 1999.
68. Hirsch A. (Ed.) Fullerenes and related structures. Berlin – Heidelberg: Springer Verlag. – 1999.
69. Eklund P.C. (Ed.) Fullerene polymers and fullerene polymer composites. Berlin: Springer. Springer series in materials science. – 2000.
70. Trefilov V.I., Schur D.V., Tarasov B.P., Shul'ga Yu.M., Chernogorenko A.B., Pishuk V.K., Zaginaychenko S.Yu. Fullereny – osnova materialov buduschego.- K.: ADEF.- 2001.- 146 p.
71. Jin C., Hettich R.L., Compton R.N., Jouce D., Blencoe J., Burch T Direct solid – phase hydrogenation of fullerenes // Journ. Phys. Chem. – 1994. – Vol. 98. – P. 4215-4217.
72. Avent A.G., Birkett P.R., Darwish A.D., Kroto H.W., Taylor R., Walton D.R.M. Fullerene and atomic clusters. Proceed. Int. Workshop IWFAC'95. St.Peterburg. – 1995. – P. 7.
73. Lobach A.S., Tarasov B.P., Shul'ga Yu.M., Perov A.A., Stepanov A.N. Reaktsiya $D_2$ s fulleridom polladiya $C_{60}Pd_{4,9}$ // Izv. RAN. Ser. Khim.- 1996.- P. 483-484.
74. Tarasov B.P., Fokin V.N., Moravsky A.P., Shul'ga Yu.M., Yartys V.A., Schur D.V. Proceed. of 12th World Hydrogen Energy Conf. Buenos Aires. Argentina. – 1998. – Vol. 2. – P. 1221-1230.
75. Tarasov B.P., Fokin V.N., Moravsky A.P., Shul'ga Yu.M. Sintez i svoistva kristallicheskih hydridov fullerenov // Izv. RAN. Ser. Khim.- 1998.- P. 2093-2096.
76. Shul'ga Yu.M., Tarasov B.P., Fokin V.N., Vasilets V.N. Kristallicheskiy deiterid fullerena $C_{60}D_{24}$: issledovnie spektralnymi metodami // Fiz. Tv. Tela.- 1999.- V. 41.- P. 1520-1526.
77. Gol'dshleger N.F., Tarasov B.P., Shul'ga Yu.M., Perov A.A., Roschupkina O.S., Moravskiy A.P. Vzaimodeeistvie fullerida platiny $C_{60}Pt$ s deiteriem // Izv. RAN. Ser. Khim.- 1999.- P. 999-1002.
78. Shul'ga Yu.M., tarasov B.P. Kristallicheskie gidrofullereny: poluchenie i svoystva // Fullereny i fullerenopodobnie struktury. Minsk: BGU. 2000. P. 14-19.
79. Tarasov B.P. Akkumulirovanie vodoroda v uglerodnikh nanostrukturakh // Tam zhe. P. 113-120.
80. Tarasov B.P., Gol'dshleger N.F., Moravskiy A.P. Vodorodosodergaschie soedineniya uglerodnikh nanostruktur: sintez i svoystva // Uspekhi khimii. 2001. V. 71. Jin C., Hettich R.L., Compton R.N., Jouce D., Blencoe J., Burch T. Journ. Phys. Chem. – 1994. – Vol. 98. – P. 4215. 2. P. 149-166.
81. Hall L.E., McKenzie D.A., Attalla M.I., Vassallo A.M., Davis R.L., Dunlop J.B., Cockayne D.J.H. The structure of $C_{60}H_{36}$ // Journ. Phys. Chem. 1993. Vol. 97. P.5741-5744.

24

82. Bubnov V.P., Koltover V.K., Laukhina E.E., Esterin Ya.P., Yagubskiy E.B. Sintez i identifikatsiya endometallofullerenov La@C82 // Izv. RAN. Ser. Khim. 1997. P. 254-259.

83. Tarasov B.P., Fokin B.N., Moravskiy A.R., Shul'ga Yu.M. Hydrirovanie fullerita v prisutstvii intermetallidov i metallov // Izv. RAN. Ser. Khim. 1997. N 4. P. 679-683.

84. Tarasov B.P., Fokin V.N., Moravsky A.P., Shul'ga Yu.M., Yartys V.A. Hydrogenation of fullerenes $C_{60}$ and $C_{70}$ in presence of hydride forming metals and intermetallic compounds // J. Alloys and Compounds. 1997. Vol. 25. P. 253-254.

85. Tarasov B.P., Fokin V.N., Moravsky A.P., Shul'ga Yu.M. Prevrascheniya v sistemah fullerit-intermetallid-vodorod // Zhurn. Neorganicheskoi. Khimii. 1997. V. 42. P. 920-922.

86. Tarasov B.P. Mehanizm hydrirovaniya fullerit-metallicheskih kompozitsiy // Zhurn. Obschey khimii. 1998. V. 68. P. 1245-1248.

87. Tarasov B.P., Fokin V.N., Fokina E.E., Rumynskaya Z.A., Volkova L.S., Moravskiy A.P., Shul'ga Yu.M. Syntes hydridov fullerenov vzaimodeistviem fullerita s vodorodom, vydelyauschimsya is hydridov intermetallicheskih soedineniy // Zhurn. obschey khimii. 1998.. V. 68. P. 1585-1589.

88. Matysina Z.A., Schur D.V. Vodorod i tverdofaznie prevrascheniya v metallakh, splavakh, fulleritakh. Dnepropetrovsk: Nauka I obrazovanie. 2002. 420 p.

89. Boba M.S., T.S.L., Balasubramaian R. et al. Studies on the thermodynamics of the $C_{60}$-$C_{70}$ binary system // Journ. Phys. Chem. 1994. Vol. 98. P. 1333-1340.

90. Kriaz K., Fisher J.E., Girifalco L.A. et al. Fullerene alloys // Sol. State Comm. 1995. Vol. 96. P. 739-743.

91. Havlik D., Schranz W., Haluska M., Kuzmany H., Rogl P. Thermal expansion measurements of $C_{60}$-$C_{70}$ mized crystals // Sol. State Comm. 1997. Vol. 104. P. 775-779.

92. Bezmel'nitsin V.N., Eletskiy A.V., Okun' M.V. Fullereny v rastvorakh // Ukr. Fiz. Zhurn. 1998. V. 168. P. 1195-1221.

93. Vovk O.M., Isakina A.P., Garbuz A.S., Kravchenko Yu.G. Tverdie rastvory fullerenov // Fullereny i fullerenopodobnie struktury. Minsk.: BGU. 2000. P.70-76.

94. Makarova T.L. Elektronnie i opticheskie svoystva monomernikh i polimernikh fullerenov (obzor) // Fizika i tekhnika poluprovodnikov. 2001. V. 35. № 3. P. 257-293.

95. David W.J., Ibberson R.M., Dennis T.J.S., Harr J.P., Prassides K. Europhys. Lett. 1992. Vol. 18. P. 219.

96. Blinc R., Selinger J., Dolinsek J., Arcon D. Two-dimensional C NMR study of orientational ordering in solid $C_{60}$ // Phys. Rev. B. 1994. Vol. 49. P. 4993-5002.

97. Brazhkin V.V., Lyapin A.G. Prevraschenie fullerita $C_{60}$ pri vysokih davleniyah i temperaturah // Uspekhi phiz. Nauk. 1996. V. 166. P. 893-897.

98. Tareeva E.E., Schelkacheva T.I. Orientatsionnoe uporyadochenie v tverdom $C_{60}$: metod vetvleniy // Teoriya i matematich. Fizika. 1999. V. 121. № 3. P. 479-491.

99. Girifalco L.A. Molecular properties of $C_{60}$ in the gas and solid phases // Journ. Phys. Chem. 1992. Vol. 96. P. 858-861.

100. Martinovich V.A., Shpilevskiy E.M. Kristallicheskaya struktura allotropnikh form ugleroda – grafita, almaza, fullerita // Fullereny i fullerenopodobnie struktury. Minsk: BGU. 2000. P. 2001-209.

# HYDROGEN SOLUBILITY IN FCC FULLERITE

D.V. SCHUR, Z.A. MATYSINA, S.Yu. ZAGINAICHENKO
*Institute of Hydrogen and Solar Energy, P.O. Box 195, Kiev 03150 Ukraine*
*Dnepropetrovsk National University, 72 Gagarin str., Dnepropetrovsk*
*49000 Ukraine*
*Institute for Problems of Materials Science of Ukrainian Academy of*
*Sciences, 3 Krzhizhanovsky str., Kiev 03142, Ukraine*
*E-mail: shurzag@materials.kiev.ua*

## Abstract

The hydrogen solubility in ordering fcc fullerite has been calculated by the configuration method with regard to hydrogen atoms arrangement in octahedral, tetrahedral, trigonal and bigonal interstices. The character of hydrogen atoms distribution in positions of different types has been clarified for different relations between energy constants. The special cases of hydrogen atoms arrangement in interstitial sites of two and three types have been studied. The dependence of hydrogen solubility on fullerite composition, temperature and degree of long-range order has been established.

## 1. Introduction

The solid-phase hydrogenation of fullerenes can be the reliable method for hydrogen storage. Hydrogen is ecologically clean energy source. Therefore, hydro-(deutero-) fullerenes attract the attention of scientists.

Hydrogen atoms in the solid-phase fullerite can exist in two states occupying the interstitial sites of crystal lattice (lattice hydrogen) or are bound with carbon atoms of fullerene molecules (fullerited hydrogen), respectively. It has been found that formation of lattice hydrogen is more energetically favorable in comparison to fullerited hydrogen [1].

When lattice hydrogen atoms are distributed in the interstitial sites of different types, there exists possibility to produce $FH_x$ ($F=C_{60}$ or $C_{70}$) hydrofullerenes with the hydrogen concentration within the range $0 \le x \le 18$ [2-4].

As found out experimentally, at temperatures above 260 K the crystal structure of solid-phase fullerite can be manifested, depending on production methods, both of body-centered centered cubic (bcc) [5-7] and of face-centered cubic (fcc) [8-10]. The orientation ordering in fullerite was studied in [11-15]. In this case, the fcc lattice forming at high temperatures is divided into four simple cubic sublattices, in which instead of free rotation, fullerene molecules rotate along the certain axes with different orientation in sublattices. As this takes place, external pressure as well as decreasing temperature slow down and brake the rotation of fullerene molecules what induces transition of fullerite into the simple cubic phase [16].

25

*T.N. Veziroglu et al. (eds.),*
*Hydrogen Materials Science and Chemistry of Carbon Nanomaterials,* 25-44.

The hydrogen solubility in bcc fullerite was calculated in papers [10,17-19]. As presented in these works, the hydrogenation of low-temperature fullerite stimulates its phase transition from the simple cubic phase into the bcc one.

In this paper the hydrogen solubility in the fcc phase is studied on the assumption that hydrogen atoms are placed in octahedral O, tetrahedral $\Theta$, trigonal Q and bigonal D interstices. As it was found out experimentally [2,8,9], hydrogen sorption did not change the fcc structure of fullerite but only increased the lattice constant $a$ from 1.417 nm at x=0 to 1,448 nm at x=18 (Fig.1). As a result, the average radius of the fullerene volume in the fcc crystal is increased from 0.711 nm to 0.759 nm. The nearest distances between fullerenes and hydrogen atoms are changed in the following ranges: (0.709-0.724) nm, (0.614-0.627) nm, (0.578-0.594) nm, (0.502-0.513) nm for O, $\Theta$, Q, D interstitial sites respectively. Note that the large volume of octa-interstices and the small size of hydrogen atoms due to their weak bond can favour the arrangement of hydrogen atoms only in $\Theta$, Q, D interstitial sites or several hydrogen atoms in O interstitial sites. In the second case the hydrogen concentration $x$ is increased and can be higher than 18 (all the O, $\Theta$, Q, D interstices in the fullerite crystal per one fullerene molecule). However, the main factor defining distribution of interstitial atoms (hydrogen atoms) proves to be not geometrical but energy one [20-22]. Calculations will take into consideration the possibility to arrange one hydrogen atom in each interstitial site. Two-component fullerite from $F_1=C_{60}$ and $F_2=C_{70}$ will be studied. The solid-phase fullerite $F_1$-$F_2$ of any composition was produced experimentally in [24-28]. As it was substantiated in [24], due to the entropy term the $F_1$-$F_2$ fullerite crystal is more stable thermodynamically compared to the pure fullerene components or their mixture.

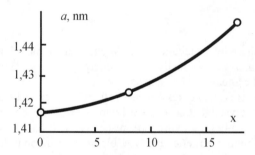

Fig.1. Lattice parameter of hydro- (deutero-) $FH_x$ fullerite vs. hydrogen concentration $x$. Circles mark experimental points.

The free energy of fullerite is calculated for solving the raised problem. Calculations are carried out by the configuration method, i.e. taking into account the possibility of formation of different configurations of $F_1$-$F_2$ fullerenes around the hydrogen atoms. This approach is more successive compared to the method of average energies and makes possible to have more correct information on the functional dependences of hydrogen solubility in fullerite.

We assume that the crystal lattice of fullerite in its ordered state is the one of the $Ll_0$ type (Fig.2). In this lattice the sites of the first and the second types valid for $F_1$ and $F_2$ fullerenes, respectively, alternate in layers. In this case O, $\Theta$, Q, D interstices, depending on their surrounding by the lattice sites of first and second types, are divided into $O_1$, $O_2$ $\Theta$, $Q_1$, $Q_2$, $D_1$, $D_2$, $D_3$ interstitial sites, respectively. In this case, the number of lattice sites of the first and the second types in the nearest surrounding of all the interstices is the following:

$O_1$ has four lattice sites of the first type and two of the second type,

$O_2$ has two lattice sites of the first type and four of the second type,

$\Theta$ has two lattice sites of the first type and two of the second type,

$Q_1$ has two lattice sites of the first type and one of the second type,       (1)

$Q_2$ has one lattice site of the first type and two of the second type,

$D_1$ has one lattice site of the first type and one of the second type,

$D_2$ has two lattice sites of the first type,

$D_3$ has two lattice sites of the second type.

The distances between hydrogen atoms and the nearest fullerenes for O, $\Theta$, Q, D interstitial sites are equal to:

$$r_O = 0,5a, \qquad\qquad r_Q = a/\sqrt{6} \approx 0,408a, \qquad (2)$$

$$r_\Theta = a\sqrt{3}/4 \approx 0,433a, \qquad\qquad r_D = a/2\sqrt{2} \approx 0,354a.$$

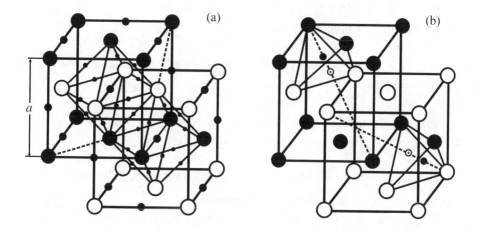

Fig.2. $Ll_o$ crystal lattice of hydrofullerite

● ○ -lattice sites of the first and the second types occupied mainly by $F_1$, $F_2$ fullerenes, respectively. Interstitial sites: ●- octahedral O (a), ● -tetrahedral $\Theta$ (b), ◉ - trigonal Q (b), ● - bigonal D (a)

## 2. Free energy of hydrofullerite. Equilibrium equations

The free energy of the crystal is calculated by the formula:

$$F=E-kT\ln W, \qquad (3)$$

where E is the configuration internal energy, W is the thermodynamical probability, k is Boltzmanns constant, T is absolute temperature. Also we introduce the following symbols:

N is the number of lattice sites (fullerenes) in the crystal,

N,2N,8N,7N are the numbers of O, $\Theta$, Q, D interstitial sites, respectively,

18N is the number of all interstitial sites,

$N_O^{(l)}, N_\Theta^{(l)}, N_Q^{(l)}, N_D^{(l)}$ are the numbers of hydrogen atoms in O, $\Theta$, Q, D interstitial sites with $l^{th}$ configuration of $F_1$, $F_2$ fullerenes. In this case some of interstitial sites can not be occupied by hydrogen atoms;

$G_O^{(l)}, G_\Theta^{(l)}, G_Q^{(l)}, G_D^{(l)}$ are the numbers of O, $\Theta$, Q, D interstitial sites, respectively, with $l^{th}$ configuration of fullerenes,

$$N_{H\to O} = \sum_l N_O^{(l)}, \quad N_{H\to\Theta} = \sum_l N_\Theta^{(l)}, \quad N_{H\to Q} = \sum_l N_Q^{(l)}, \quad N_{H\to D} = \sum_l N_D^{(l)} \qquad (4)$$

are the numbers of hydrogen atoms in O, $\Theta$, Q, D positions with any configuration,

$$N_H = N_{H\to O} + N_{H\to\Theta} + N_{H\to Q} + N_{H\to D} \qquad (5)$$

is the total number of hydrogen atoms,

$$c_O = N_{H\to O} / N, \quad c_\Theta = N_{H\to\Theta} / N, \quad c_Q = N_{H\to Q} / N, \quad c_D = N_{H\to D} / N \qquad (6)$$

are hydrogen atoms concentrations in O, $\Theta$, Q, D interstices in relation to the number of lattice sites (fullerenes) of the crystal

$$c = c_O + c_\Theta + c_Q + c_D \qquad (7)$$

is the total hydrogen concentration determining its solubility,

$c_1$, $c_2$ are $F_1$, $F_2$ fullerenes concentrations,

$$c_1 + c_2 = 1. \qquad (8)$$

$P_{F_1}^{(1)}, P_{F_1}^{(2)}, P_{F_2}^{(1)}, P_{F_2}^{(2)}$ are the a priori probabilities of the substitution the lattice sites of the first and second types with $F_1$, $F_2$ fullerenes. In this case

$$P_{F_1}^{(1)} + P_{F_2}^{(1)} = 1, \quad P_{F_1}^{(2)} + P_{F_2}^{(2)} = 1, \quad P_{F_1}^{(1)} + P_{F_1}^{(2)} = 2c_1, \quad P_{F_2}^{(1)} + P_{F_2}^{(2)} = 2c_2. \qquad (9)$$

$$\eta = 2(P_F^{(1)} - c) \qquad (10)$$

is the order parameter in fullerene distribution in the lattice sites of the crystal.

Equations (9) and (10) allow one to find the possibilities as follows:

$$P_{F_1}^{(1)} = c_1 + \frac{1}{2}\eta, \quad P_{F_1}^{(2)} = c_1 - \frac{1}{2}\eta, \quad P_{F_2}^{(1)} = c_2 - \frac{1}{2}\eta, \quad P_{F_2}^{(2)} = c_2 + \frac{1}{2}\eta. \qquad (11)$$

$\upsilon_O^{(l)}, \upsilon_\Theta^{(l)}, \upsilon_Q^{(l)}, \upsilon_D^{(l)}$ are energies with the opposite sign for interaction of hydrogen atoms in O, $\Theta$, Q, D interstices with the nearest fullerenes in the case of fullerenes with $l^{th}$ configuration around these interstitial sites.

Each O, $\Theta$, Q, D interstitial site in its nearest surrounding has 6, 4, 3 and 2 fullerenes, respectively. The $l^{th}$ configuration of fullerenes around the interstitial site points to the number of $F_1$ fullerenes around the hydrogen atom. The rest of $F_2$ fullerenes around the hydrogen atom will be: 6-$l$ for O, 4-$l$ for $\Theta$, 3-$l$ for Q and 2-$l$ for D interstitial sites.

Then the hydrogen atoms energies in each interstitial site, determined by the sum of interaction energies with the nearest fullerenes, will be equal, respectively:

$$
\begin{aligned}
\upsilon_O^{(l)} &= l\alpha_1 + (6-l)\alpha_2, \\
\upsilon_\Theta^{(l)} &= l\beta_1 + (4-l)\beta_2, \\
\upsilon_Q^{(l)} &= l\gamma_1 + (3-l)\gamma_2, \\
\upsilon_D^{(l)} &= l\delta_1 + (2-l)\delta_2,
\end{aligned}
\tag{12}
$$

where $\alpha_i$, $\beta_i$, $\gamma_i$, $\delta_i$ (i=1; 2) are the energies of interaction between hydrogen atoms and $F_1$ and $F_2$ fullerenes, respectively:

$$
\begin{aligned}
\alpha_1 &= \upsilon_{iF_1}(r_O), & \alpha_2 &= \upsilon_{iF_2}(r_O), & \beta_1 &= \upsilon_{iF_1}(r_\Theta), & \beta_2 &= \upsilon_{iF_2}(r_\Theta), \\
\gamma_1 &= \upsilon_{iF_1}(r_Q), & \gamma_2 &= \upsilon_{iF_2}(r_Q), & \delta_1 &= \upsilon_{iF_1}(r_D), & \delta_2 &= \upsilon_{iF_2}(r_D).
\end{aligned}
\tag{13}
$$

Using the taken symbols, we get the formulae for configuration energy E and thermodynamical probability W:

$$
E = -\sum_{l=0}^{6} N_O^{(l)}\upsilon_O^{(l)} - \sum_{l=0}^{4} N_\Theta^{(l)}\upsilon_\Theta^{(l)} - \sum_{l=0}^{3} N_Q^{(l)}\upsilon_Q^{(l)} - \sum_{l=0}^{2} N_D^{(l)}\upsilon_D^{(l)},
\tag{14}
$$

where

$$
N_O^{(l)} = N_{O_1}^{(l)} + N_{O_2}^{(l)}, \quad N_Q^{(l)} = N_{Q_1}^{(l)} + N_{Q_2}^{(l)}, \quad N_D^{(l)} = N_{D_1}^{(l)} + N_{D_2}^{(l)} + N_{D_3}^{(l)},
\tag{15}
$$

$$
\begin{aligned}
W &= \prod_{l=0}^{6} \frac{G_{O_1}^{(l)}!}{N_{O_1}^{(l)}!(G_{O_1}^{(l)} - N_{O_1}^{(l)})!} \cdot \prod_{l=0}^{6} \frac{G_{O_2}^{(l)}!}{N_{O_2}^{(l)}!(G_{O_2}^{(l)} - N_{O_2}^{(l)})!} \times \\
&\times \prod_{l=0}^{4} \frac{G_\Theta^{(l)}!}{N_\Theta^{(l)}!(G_\Theta^{(l)} - N_\Theta^{(l)})!} \cdot \prod_{l=0}^{3} \frac{G_{Q_1}^{(l)}!}{N_{Q_1}^{(l)}!(G_{Q_1}^{(l)} - N_{Q_1}^{(l)})!} \cdot \prod_{l=0}^{3} \frac{G_{Q_2}^{(l)}!}{N_{Q_2}^{(l)}!(G_{Q_2}^{(l)} - N_{Q_2}^{(l)})!} \times \\
&\times \prod_{l=0}^{2} \frac{G_{D_1}^{(l)}!}{N_{D_1}^{(l)}!(G_{D_1}^{(l)} - N_{D_1}^{(l)})!} \cdot \prod_{l=0}^{2} \frac{G_{D_2}^{(l)}!}{N_{D_2}^{(l)}!(G_{D_2}^{(l)} - N_{D_2}^{(l)})!} \cdot \prod_{l=0}^{2} \frac{G_{D_3}^{(l)}!}{N_{D_3}^{(l)}!(G_{D_3}^{(l)} - N_{D_3}^{(l)})!},
\end{aligned}
\tag{16}
$$

where P is the product of factors by $l$.

Substituting values of E (14) and W (16) in equation (3) in terms of (15) and using the Stirling formula $\ln X! = X(\ln X - 1)$ for large X numbers, we find energy as follows:

$$F = -\sum_{l=0}^{6}(N_{O_1}^{(l)} + N_{O_2}^{(l)})\upsilon_O^{(l)} - \sum_{l=0}^{6}N_{\Theta}^{(l)}\upsilon_{\Theta}^{(l)} -$$

$$- \sum_{l=0}^{3}(N_{Q_1}^{(l)} + N_{Q_2}^{(l)})\upsilon_Q^{(l)} - \sum_{l=0}^{2}(N_{D_1}^{(l)} + N_{D_2}^{(l)} + N_{D_3}^{(l)})\upsilon_D^{(l)} +$$

$$+ kT\{\sum_{l=0}^{6}(G_{O_1}^{(l)}\ln G_{O_1}^{(l)} + G_{O_2}^{(l)}\ln G_{O_2}^{(l)}) + \sum_{l=0}^{4}G_{\Theta}^{(l)}\ln G_{\Theta}^{(l)} +$$

$$+ \sum_{l=0}^{3}(G_{Q_1}^{(l)}\ln G_{Q_1}^{(l)} + G_{Q_2}^{(l)}\ln G_{Q_2}^{(l)}) +$$

$$+ \sum_{l=0}^{2}(G_{D_1}^{(l)}\ln G_{D_1}^{(l)} + G_{D_2}^{(l)}\ln G_{D_2}^{(l)} + G_{D_3}^{(l)}\ln G_{D_3}^{(l)}) -$$

$$- \sum_{l=0}^{6}(N_{O_1}^{(l)}\ln N_{O_1}^{(l)} + N_{O_2}^{(l)}\ln N_{O_2}^{(l)}) - \sum_{l=0}^{4}N_{\Theta}^{(l)}\ln N_{\Theta}^{(l)} - \qquad (17)$$

$$- \sum_{l=0}^{3}(N_{Q_1}^{(l)}\ln N_{Q_1}^{(l)} + N_{Q_2}^{(l)}\ln N_{Q_2}^{(l)}) -$$

$$- \sum_{l=0}^{2}(N_{D_1}^{(l)}\ln N_{D_1}^{(l)} + N_{D_2}^{(l)}\ln N_{D_2}^{(l)} + N_{D_3}^{(l)}\ln N_{D_3}^{(l)}) -$$

$$- \sum_{l=0}^{6}[(G_{O_1}^{(l)} - N_{O_1}^{(l)})\ln(G_{O_1}^{(l)} - N_{O_1}^{(l)}) + (G_{O_2}^{(l)} - N_{O_2}^{(l)})\ln(G_{O_2}^{(l)} - N_{O_2}^{(l)})] -$$

$$- \sum_{l=0}^{4}(G_{\Theta}^{(l)} - N_{\Theta}^{(l)})\ln(G_{\Theta}^{(l)} - N_{\Theta}^{(l)}) -$$

$$- \sum_{l=0}^{3}[(G_{Q_1}^{(l)} - N_{Q_1}^{(l)})\ln(G_{Q_1}^{(l)} - N_{Q_1}^{(l)}) + (G_{Q_2}^{(l)} - N_{Q_2}^{(l)})\ln(G_{Q_2}^{(l)} - N_{Q_2}^{(l)})] -$$

$$- \sum_{l=0}^{3}[(G_{D_1}^{(l)} - N_{D_1}^{(l)})\ln(G_{D_1}^{(l)} - N_{D_1}^{(l)}) + (G_{D_2}^{(l)} - N_{D_2}^{(l)})\ln(G_{D_2}^{(l)} - N_{D_2}^{(l)}) +$$

$$+ (G_{D_3}^{(l)} - N_{D_3}^{(l)})\ln(G_{D_3}^{(l)} - N_{D_3}^{(l)})]\}.$$

The equilibrium state of fullerite is determined from the conditions of free energy, that are easily found by the method of the indeterminate Lagrange factor $\lambda$. Factor $\lambda$ is correlated with the equation of the relation between numbers $N_s^{(l)}(s = O_1, O_2, \Theta, Q_1, Q_2, D_1, D_2, D_3)$

$$\varphi \equiv N_H - \sum_{l=0}^{6}(N_{O_1}^{(l)} + N_{O_2}^{(l)}) -$$

$$- \sum_{l=0}^{4}N_{\Theta}^{(l)} - \sum_{l=0}^{3}(N_{Q_1}^{(l)} + N_{Q_2}^{(l)}) - \sum_{l=0}^{2}(N_{D_1}^{(l)} + N_{D_2}^{(l)} + N_{D_3}^{(l)}). \qquad (18)$$

In this case the conditions for minimum of free energy are written as follows:

$$\frac{\partial F}{\partial N_{O_1}^{(l)}} + \lambda\frac{\partial \varphi}{\partial N_{O_1}^{(l)}} = 0, \quad \frac{\partial F}{\partial N_{O_2}^{(l)}} + \lambda\frac{\partial \varphi}{\partial N_{O_2}^{(l)}} = 0, \quad \frac{\partial F}{\partial N_{\Theta}^{(l)}} + \lambda\frac{\partial \varphi}{\partial N_{\Theta}^{(l)}} = 0,$$

$$\frac{\partial F}{\partial N_{Q_1}^{(l)}} + \lambda\frac{\partial \varphi}{\partial N_{Q_1}^{(l)}} = 0, \quad \frac{\partial F}{\partial N_{Q_2}^{(l)}} + \lambda\frac{\partial \varphi}{\partial N_{Q_2}^{(l)}} = 0, \qquad (19)$$

$$\frac{\partial F}{\partial N_{D_1}^{(l)}} + \lambda\frac{\partial \varphi}{\partial N_{D_1}^{(l)}} = 0, \quad \frac{\partial F}{\partial N_{D_2}^{(l)}} + \lambda\frac{\partial \varphi}{\partial N_{D_2}^{(l)}} = 0, \quad \frac{\partial F}{\partial N_{D_3}^{(l)}} + \lambda\frac{\partial \varphi}{\partial N_{D_3}^{(l)}} = 0.$$

After substitution of free energy F into (19), we find

$$-\upsilon_0^{(l)} + kT[lnN_{0_1}^{(l)} - ln(G_{0_1}^{(l)} - N_{0_1}^{(l)})] - \lambda = 0, \qquad l = 0,\ldots,6,$$

$$-\upsilon_0^{(l)} + kT[lnN_{0_2}^{(l)} - ln(G_{0_2}^{(l)} - N_{0_2}^{(l)})] - \lambda = 0, \qquad l = 0,\ldots,6,$$

$$-\upsilon_\Theta^{(l)} + kT[lnN_\Theta^{(l)} - ln(G_\Theta^{(l)} - N_\Theta^{(l)})] - \lambda = 0, \qquad l = 0,\ldots,4,$$

$$-\upsilon_Q^{(l)} + kT[lnN_{Q_1}^{(l)} - ln(G_{Q_1}^{(l)} - N_{Q_1}^{(l)})] - \lambda = 0, \qquad l = 0,\ldots,3,$$

$$-\upsilon_Q^{(l)} + kT[lnN_{Q_2}^{(l)} - ln(G_{Q_2}^{(l)} - N_{Q_2}^{(l)})] - \lambda = 0, \qquad l = 0,\ldots,3, \tag{20}$$

$$-\upsilon_D^{(l)} + kT[lnN_{D_1}^{(l)} - ln(G_{D_1}^{(l)} - N_{D_1}^{(l)})] - \lambda = 0, \qquad l = 0,1,2,$$

$$-\upsilon_D^{(l)} + kT[lnN_{D_2}^{(l)} - ln(G_{D_2}^{(l)} - N_{D_2}^{(l)})] - \lambda = 0, \qquad l = 0,1,2,$$

$$-\upsilon_D^{(l)} + kT[lnN_{D_3}^{(l)} - ln(G_{D_3}^{(l)} - N_{D_3}^{(l)})] - \lambda = 0, \qquad l = 0,1,2.$$

Using these equations, we find the numbers $N_s^{(l)}$ of hydrogen atoms in the interstitial sites s= $O_1, O_2, \Theta, Q_1, Q_2, D_1, D_2, D_3$ with $l^{\text{th}}$ configuration of $F_1, F_2$ fullerenes

$$N_{O_1}^{(l)} = A\,G_{O_1}^{(l)} \exp\frac{\upsilon_O^{(l)}}{kT}/(1 + A\exp\frac{\upsilon_O^{(l)}}{kT}), \quad N_{O_2}^{(l)} = A\,G_{O_2}^{(l)} \exp\frac{\upsilon_O^{(l)}}{kT}/(1 + A\exp\frac{\upsilon_O^{(l)}}{kT}),$$

$$N_\Theta^{(l)} = A\,G_\Theta^{(l)} \exp\frac{\upsilon_\Theta^{(l)}}{kT}/(1 + A\exp\frac{\upsilon_\Theta^{(l)}}{kT}),$$

$$N_{Q_1}^{(l)} = A\,G_{Q_1}^{(l)} \exp\frac{\upsilon_Q^{(l)}}{kT}/(1 + A\exp\frac{\upsilon_Q^{(l)}}{kT}), \quad N_{Q_2}^{(l)} = A\,G_{Q_2}^{(l)} \exp\frac{\upsilon_Q^{(l)}}{kT}/(1 + A\exp\frac{\upsilon_Q^{(l)}}{kT}), \tag{21}$$

$$N_{D_1}^{(l)} = A\,G_{D_1}^{(l)} \exp\frac{\upsilon_D^{(l)}}{kT}/(1 + A\exp\frac{\upsilon_D^{(l)}}{kT}), \quad N_{D_2}^{(l)} = A\,G_{D_2}^{(l)} \exp\frac{\upsilon_D^{(l)}}{kT}/(1 + A\exp\frac{\upsilon_D^{(l)}}{kT}),$$

$$N_{D_3}^{(l)} = A\,G_{D_3}^{(l)} \exp\frac{\upsilon_D^{(l)}}{kT}/(1 + A\exp\frac{\upsilon_D^{(l)}}{kT}),$$

where $A = \exp\dfrac{\lambda}{kT}$ is chemical potential defined the hydrogen atoms activity, i.e. the increase in the state function of the crystal due to each additional hydrogen atom appeared from surroundings and dissolved in this crystal.

Summing up the attained numbers (21) for all configurations $l$, we can find the total number $N_H$ of hydrogen atoms in the crystal by equation (18) or the hydrogen solubility in fullerite by Eqs (6), (7).

We shall examined special cases of hydrogen solubility in fullerite.

# 3. Hydrogen solubility in single-component fullerite

Let F fullerene of one sort ($C_{60}$ or $C_{70}$) be contained in the crystal. Then we shall have only four types of O, $\Theta$, Q, D interstitial sites. Each O, $\Theta$, Q, D interstice has the only configuration: surrounding by the nearest F fullerenes.

Instead of eight equations (21), we shall have four. In this case

$$\alpha = \alpha_1 = \alpha_2, \quad \beta = \beta_1 = \beta_2, \quad \gamma = \gamma_1 = \gamma_2, \quad \delta = \delta_1 = \delta_2, \tag{22}$$

and $\upsilon_s^{(l)}$ energies of hydrogen atoms in interstices s = O, $\Theta$, Q, D will be equal to

$$\upsilon_O^{(l)} = 6\alpha, \quad \upsilon_\Theta^{(l)} = 4\beta, \quad \upsilon_Q^{(l)} = 3\gamma, \quad \upsilon_D^{(l)} = 2\delta. \tag{23}$$

Values of $G_O^{(l)}, G_\Theta^{(l)}, G_Q^{(l)}, G_D^{(l)}$ determine the numbers of interstices $G_O$, $G_\Theta$, $G_Q$, $G_D$ and equal, respectively

$$G_O=N, \quad G_\Theta=2N, \quad G_Q=8N, \quad G_D=7N. \tag{24}$$

The numbers $N_s^{(l)}$ of hydrogen atoms in the interstices s = O, Θ, Q, D from (21) we attain in the following form

$$N_0 = 2N(1 + \frac{1}{A}\exp\frac{-6\alpha}{kT})^{-1},$$

$$N_\theta = 2N(1 + \frac{1}{A}\exp\frac{-4\beta}{kT})^{-1}, \tag{25}$$

$$N_Q = 8N(1 + \frac{1}{A}\exp\frac{-3\gamma}{kT})^{-1},$$

$$N_D = 7N(1 + \frac{1}{A}\exp\frac{-2\delta}{kT})^{-1},$$

and the hydrogen concentration in the fullerite interstices will be equal to

$$c_0 = 1/(1 + \frac{1}{A}\exp\frac{-6\alpha}{kT}), \quad c_\theta = 2/(1 + \frac{1}{A}\exp\frac{-4\beta}{kT}),$$

$$c_Q = 8/(1 + \frac{1}{A}\exp\frac{-3\gamma}{kT}), \quad c_D = 7/(1 + \frac{1}{A}\exp\frac{-2\delta}{kT}). \tag{26}$$

Each individual formula from (26) determines the hydrogen solubility in fullerite for hydrogen atoms distribution only in O, Θ, Q or D interstitial sites, respectively.

When hydrogen atoms are arranged in all interstitial sites, hydrogen solubility will be determined by the sum of $c_O$, $c_\Theta$, $c_Q$, $c_D$ concentrations according to equation (7).

Eqs.(25), (26) show that at the infinitely low temperature T→0 there exists possibility of maximum hydrogen solubility in fullerite, determined by the sum of interstitial sites (24) when all interstices are occupied with hydrogen atoms. The maximum hydrogen solubility is equal to

$$c_m=c_O+c_\Theta+c_Q+c_D=(G_O+G_\Theta+G_Q+G_D)/N=18. \tag{27}$$

This case corresponds to the formation of $FH_{18}$ fullerite.

However, by virtue of the large differences in the distances (2) between hydrogen atoms and fullerenes, as it was noted above, hydrogen atoms can be arranged in the interstitial sites of two or three types. We shall consider these cases below.

## 4. Hydrogen atoms distribution in octahedral and tetrahedral interstitial sites of single-component fullerite

In this case the maximum possible the concentration of hydrogen atoms in fullerite at T→0 is defined by the relative number of interstitial sites and equals

$$c_m=(G_O+G_\Theta)/N=1+2=3.$$

At any temperatures the hydrogen atoms concentration in O, Θ interstitial sites and the total concentration are determined by the following equations

$$c_0 = (1+\frac{1}{A}\exp\frac{-6\alpha}{kT})^{-1}, \quad c_\theta = (1+\frac{1}{A}\exp\frac{-4\beta}{kT})^{-1}, \quad c = c_0 + c_\theta. \quad (28)$$

Using these equations, we easily derive the following relations

$$\frac{c_0}{c_\theta} = \frac{1-c_0}{2-c_\theta}\varepsilon_1, \quad c_\theta = c - c_0, \quad \varepsilon_1 = \exp\frac{6\alpha-4\beta}{kT}, \quad (29)$$

whence find

$$c_0 = \{c(1-\varepsilon_1)-2-\varepsilon_1+\sqrt{[c(1-\varepsilon_1)-2-\varepsilon_1]^2+4c\varepsilon_1(1-\varepsilon_1)}\ \}/2(1-\varepsilon_1), \quad c_\theta = c-c_0. (30)$$

Formulae (30) allows us to evaluate the hydrogen atoms distribution over O and Θ interstitial sites that is determined by temperature, the relation between $\alpha$ and $\beta$ energy parameters and the total concentration of hydrogen atoms.

Based on the relations (30), at $T\to0$ we find for different hydrogen concentrations in the range $0\le c\le3$

$$c_0 = \begin{cases} c \text{ at } 0\le\tilde{n}\le1 \\ 1 \text{ at } 1\le\tilde{n}\le3 \end{cases}, \quad c_0 = \begin{cases} 0 \text{ at } 0\le\tilde{n}\le1 \\ \tilde{n}-1 \text{ at } 1\le\tilde{n}\le3 \end{cases}, \text{ if } 6\alpha\text{-}4\beta>0, \quad (31)$$

and

$$c_0 = \begin{cases} 0 \text{ at } 0\le\tilde{n}\le2 \\ c-2 \text{ at } 2\le\tilde{n}\le3 \end{cases}, \quad c_0 = \begin{cases} c \text{ at } 0\le\tilde{n}\le2 \\ 2 \text{ at } 2\le\tilde{n}\le3 \end{cases}, \text{ if } 6\alpha\text{-}4\beta<0, \quad (32)$$

The dependences (31), (32) for hydrogen atoms distribution in O and Θ interstitial sites at $T\to0$ are shown by solid broken lines in the Fig. 3. The dotted lines show the hydrogen atoms distribution in O and Θ interstitial sites when $6\alpha\text{-}4\beta\lessgtr0$ and $T\ne0$.

The maximum hydrogen concentration $c_m=3$ corresponds to the formation of $FH_3$ hydrofullerite, i.e. in this case the atomic hydrogen concentration in reference to the number of fullerenes cannot exceed $x=3$.

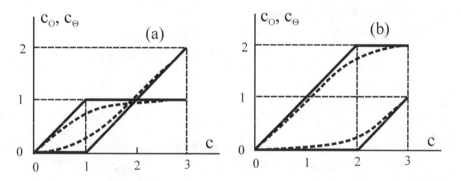

Fig.3. The hydrogen atoms distribution in octahedral O and tetrahedral Θ interstitial sites at $6\alpha - 4\beta > 0$ (a) and $6\alpha - 4\beta < 0$ (b). The solid broken lines correspond to $T\to0$, the dotted lines correspond to the temperature $T\ne0$.

34

## 5. Hydrogen atoms distribution in trigonal and bigonal interstitial sites of single-component fullerite

Note that in virtue of small radii of hydrogen atoms and large volumes of interstitial sites in fullerite, hydrogen atoms distribution in trigonal Q and bigonal D interstitial sites can appear more probable. Let us consider this case.

The hydrogen atoms concentration in Q and D interstitial sites and the total concentration are determined by the following equations

$$c_Q = 8(1+\frac{1}{A}\exp\frac{-3\gamma}{kT})^{-1}, \quad c_D = 7(1+\frac{1}{A}\exp\frac{-2\delta}{kT})^{-1}, \quad c = c_Q + c_D, \quad (33)$$

and the maximum hydrogen concentration in fullerite can be equal to the number of Q and D interstitial sites, i.e.

$$c_m = (G_Q + G_D)/N = 7 + 8 = 15. \quad (34)$$

Using (33), we easily find the following relations

$$\frac{c_Q(7-c_D)}{c_D(8-c_Q)} = \exp\frac{3\gamma-2\delta}{kT}, \qquad c_D = c - c_Q, \quad (35)$$

from which we have

$$c_Q = \{c(1-\varepsilon_2)-7-8\varepsilon_2 + \sqrt{[c(1-\varepsilon_2)-7-8\varepsilon_2]^2 + 32c\varepsilon_2(1-\varepsilon_2)}\}/2(1-\varepsilon_2),$$

$$c_D = c - c_Q, \quad \varepsilon_2 = \exp\frac{3\gamma-2\delta}{kT}. \quad (36)$$

Formulae (36) define the character of hydrogen atoms distribution in Q and D interstitial sites depending on the temperature, the total hydrogen concentration $c$ and the relation between $\gamma$ and $\delta$ energetic constants.

For the infinitely low temperature T$\to$ 0 we find

$$c_Q = \begin{cases} c & at \ 0\le c\le8 \\ 8 & at \ 8\le c\le15 \end{cases}, \quad c_D = \begin{cases} 0 & at \ 0\le c\le8 \\ c-8 & at \ 8\le c\le15 \end{cases}, \quad if \ 3\gamma-2\delta>0, \quad (37)$$

and

$$c_Q = \begin{cases} 0 & at \ 0\le c\le7 \\ 7 & at \ 7\le c\le15 \end{cases}, \quad c_D = \begin{cases} c & at \ 0\le c\le7 \\ c-7 & at \ 7\le c\le15 \end{cases}, \quad if \ 3\gamma-2\delta<0. \quad (38)$$

These hydrogen atoms distributions over Q and D interstitial sites are shown in Fig.4. The broken lines correspond to T$\to$ 0, the dotted lines correspond to T$\neq$0.

When hydrogen atoms are arranged in trigonal and bigonal interstitial sites, there exists possibility to form FH$_{15}$ hydrofullerite in which the number of hydrogen atoms is x=15 times as great as the number of fullerenes.

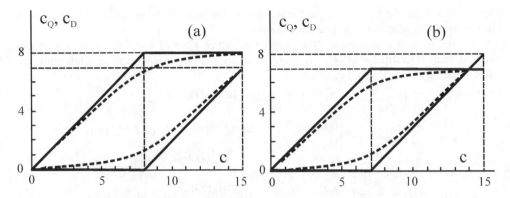

Fig.4. The hydrogen atoms distribution in trigonal Q and bigonal D interstitial sites at $3\gamma - 2\delta > 0$ (a) and $3\gamma - 2\delta < 0$ (b). The solid broken lines correspond to the infinitely low temperature $T \rightarrow 0$, the dotted lines correspond to $T \neq 0$

## 6. Small concentration of hydrogen atoms in $F_1$-$F_2$ combined fullerite

Let us elucidate the functional dependences of hydrogen solubility in $F_1$-$F_2$ combined fullerite at the low concentration of hydrogen atoms when $N_H \leq N$. In this case formulae (21) are simplified and have the form

$$
\left.
\begin{aligned}
N_{O_1}^{(l)} &= AG_{O_1}^{(l)} \exp\frac{\upsilon_O^{(l)}}{kT}, \quad l = 0,\ldots,6,\\
N_{O_2}^{(l)} &= AG_{O_2}^{(l)} \exp\frac{\upsilon_O^{(l)}}{kT}, \quad l = 0,\ldots,6,\\
N_\Theta^{(l)} &= AG_\Theta^{(l)} \exp\frac{\upsilon_\Theta^{(l)}}{kT}, \quad l = 0,\ldots,4,\\
N_{Q_1}^{(l)} &= AG_{Q_1}^{(l)} \exp\frac{\upsilon_Q^{(l)}}{kT}, \quad l = 0,\ldots,3,\\
N_{Q_2}^{(l)} &= AG_{Q_2}^{(l)} \exp\frac{\upsilon_Q^{(l)}}{kT}, \quad l = 0,\ldots,3,\\
N_{D_1}^{(l)} &= AG_{D_1}^{(l)} \exp\frac{\upsilon_D^{(l)}}{kT}, \quad l = 0,1,2,\\
N_{D_2}^{(l)} &= AG_{D_2}^{(l)} \exp\frac{\upsilon_D^{(l)}}{kT}, \quad l = 0,1,2,\\
N_{D_3}^{(l)} &= AG_{D_3}^{(l)} \exp\frac{\upsilon_D^{(l)}}{kT}, \quad l = 0,1,2.
\end{aligned}
\right\}
\tag{39}
$$

In the ordered fullerite the $l^{\text{th}}$ configuration of $F_1$, $F_2$ fullerenes is defined by the configuration of fullerenes in the lattice sites of the first and second type, respectively, i.e. the $l$ value is determined by the sum $l = i + j$, where i is the number of $F_1$ fullerenes in the lattice sites of the first type, and j is the number of the same $F_1$ fullerenes in the lattice sites of the second type around each interstitial sites.

Taking into consideration the number of lattice sites of the first and second types around each interstitial site (1), the number of interstitial sites with $l^{th}$ configuration are determined by the following equations

$$
\begin{aligned}
G_{O_1}^{(l)} &= \frac{1}{2} N \frac{4!}{i!(4-i)!} P_{F_1}^{(1)i} P_{F_2}^{(1)(4-i)} \frac{2!}{j!(2-j)!} P_{F_1}^{(2)j} P_{F_2}^{(2)(2-j)}, \ i=0,...4, \ j=0,1,2, \\
G_{O_2}^{(l)} &= \frac{1}{2} N \frac{2!}{i!(2-i)!} P_{F_1}^{(1)i} P_{F_2}^{(1)(2-i)} \frac{4!}{j!(4-j)!} P_{F_1}^{(2)j} P_{F_2}^{(2)(4-j)}, \ i=0,1,2, \ j=0,...,4, \\
G_{\Theta}^{(l)} &= 2N \frac{2!}{i!(2-i)!} P_{F_1}^{(1)i} P_{F_2}^{(1)(2-i)} \frac{2!}{j!(2-j)!} P_{F_1}^{(2)j} P_{F_2}^{(2)(2-j)}, \ i=0,1,2, \ j=0,1,2, \\
G_{Q_1}^{(l)} &= 4N \frac{2!}{i!(2-i)!} P_{F_1}^{(1)i} P_{F_2}^{(1)(2-i)} \frac{1!}{j!(1-j)!} P_{F_1}^{(2)j} P_{F_2}^{(2)(1-j)}, \ i=0,1,2, \ j=0;1, \\
G_{Q_2}^{(l)} &= 4N \frac{1!}{i!(1-i)!} P_{F_1}^{(1)i} P_{F_2}^{(1)(1-i)} \frac{2!}{j!(2-j)!} P_{F_1}^{(2)j} P_{F_2}^{(2)(2-j)}, \ i=0;1, \ j=0,1,2, \\
G_{D_1}^{(l)} &= 4N \frac{1!}{i!(1-i)!} P_{F_1}^{(1)i} P_{F_2}^{(1)(1-i)} \frac{1!}{j!(1-j)!} P_{F_1}^{(2)j} P_{F_2}^{(2)(1-j)}, \ i=0;1, \ j=0;1, \\
G_{D_2}^{(l)} &= \frac{3}{2} N \frac{2!}{i!(2-i)!} P_{F_1}^{(1)i} P_{F_2}^{(1)(2-i)}, \ i=0,1,2, \\
G_{D_3}^{(l)} &= \frac{3}{2} N \frac{2!}{j!(2-j)!} P_{F_1}^{(2)j} P_{F_2}^{(1)(2-j)}, \ j=0,1,2,
\end{aligned}
\tag{40}
$$

where the $\frac{1}{2}N$, $\frac{1}{2}N$, $2N$, $4N$, $4N$, $4N$, $\frac{3}{2}N$, $\frac{3}{2}N$ numbers determine the number of interstitial sites in the fullerite crystal of $O_1$, $O_2$, $\Theta$, $Q_1$, $Q_2$, $D_1$, $D_2$, $D_3$ types, respectively.

In formulae (39) the energies of hydrogen atoms in the interstitial sites with regard to the numerical values of i and j, given in (40), are equal to, respectively

$$
\begin{aligned}
\upsilon_{O_1}^{(l)} &= \upsilon_{O_1}^{(i+j)} = i\alpha_1 + (4-i)\alpha_2 + j\alpha_1 + (2-j)\alpha_2, \\
\upsilon_{O_2}^{(l)} &= \upsilon_{O_2}^{(i+j)} = i\alpha_1 + (2-i)\alpha_2 + j\alpha_1 + (4-j)\alpha_2, \\
\upsilon_{\Theta}^{(l)} &= \upsilon_{\Theta}^{(i+j)} = i\beta_1 + (2-i)\beta_2 + j\beta_1 + (2-j)\beta_2, \\
\upsilon_{Q_1}^{(l)} &= \upsilon_{Q_1}^{(i+j)} = i\gamma_1 + (2-i)\gamma_2 + j\gamma_1 + (1-j)\gamma_2, \\
\upsilon_{Q_2}^{(l)} &= \upsilon_{Q_2}^{(i+j)} = i\gamma_1 + (1-i)\gamma_2 + j\gamma_1 + (2-j)\gamma_2, \\
\upsilon_{D_1}^{(l)} &= \upsilon_{D_1}^{(i+j)} = i\delta_1 + (1-i)\delta_2 + j\delta_1 + (1-j)\delta_2, \\
\upsilon_{D_2}^{(l)} &= \upsilon_{D_2}^{(i+j)} = i\delta_1 + (2-i)\delta_2, \\
\upsilon_{D_3}^{(l)} &= \upsilon_{D_3}^{(i+j)} = j\delta_1 + (2-j)\delta_2.
\end{aligned}
\tag{41}
$$

Substituting the $G_s^{(l)}$ numbers (40) and $\upsilon_s^{(l)}$ energies into (39) and summing up the $N_s^{(l)}$ numbers for all the configurations $l = i+j$, we find the number of hydrogen atoms in the interstitial sites s = $O_1$, $O_2$, $\Theta$, $Q_1$, $Q_2$, $D_1$, $D_2$, $D_3$.

In summation we use Newton binomial

$$
(a+b)^n = C_n^0 a^n + C_n^1 a^{n-1} b + C_n^2 a^{n-2} b^2 + ... + C_n^m a^{n-m} b^m + ... + C_n^{n-1} ab^{n-1} + C_n^n b^n, \tag{42}
$$

where $C_n^m$ is the number of combinations from $n$ elements by $m$

$$C_n^m = \frac{n!}{m!(n-m)!}, \tag{43}$$

at this $C_n^o = C_n^n = 1$.

In our case

$$\left. \begin{aligned} a &= P_{F_1}^{(1)} \exp\frac{\alpha_1}{kT}, \quad P_{F_1}^{(1)} \exp\frac{\beta_1}{kT}, \quad P_{F_1}^{(1)} \exp\frac{\gamma_1}{kT}, \quad P_{F_1}^{(1)} \exp\frac{\delta_1}{kT}, \\ b &= P_{F_2}^{(1)} \exp\frac{\alpha_2}{kT}, \quad P_{F_2}^{(1)} \exp\frac{\beta_2}{kT}, \quad P_{F_2}^{(1)} \exp\frac{\gamma_2}{kT}, \quad P_{F_2}^{(1)} \exp\frac{\delta_2}{kT}, \end{aligned} \right\} \tag{44}$$

or

$$\left. \begin{aligned} a &= P_{F_1}^{(2)} \exp\frac{\alpha_1}{kT}, \quad P_{F_1}^{(2)} \exp\frac{\beta_1}{kT}, \quad P_{F_1}^{(2)} \exp\frac{\gamma_1}{kT}, \quad P_{F_1}^{(2)} \exp\frac{\delta_1}{kT}, \\ b &= P_{F_2}^{(2)} \exp\frac{\alpha_2}{kT}, \quad P_{F_2}^{(2)} \exp\frac{\beta_2}{kT}, \quad P_{F_2}^{(2)} \exp\frac{\gamma_2}{kT}, \quad P_{F_2}^{(2)} \exp\frac{\delta_2}{kT}, \end{aligned} \right\} \tag{45}$$

As a result of summation of the numbers (39) for all the configurations we get the numbers of hydrogen atoms in the interstitial sites, respectively

$$\begin{aligned} N_{H\to O_1} &= \sum_{i=0}^{4}\sum_{j=0}^{2} N_{O_1}^{(l)} = \frac{1}{2}NAK_1^{(O)^4}K_2^{(O)^2}, \\ N_{H\to O_2} &= \sum_{i=0}^{2}\sum_{j=0}^{4} N_{O_2}^{(l)} = \frac{1}{2}NAK_1^{(O)^2}K_2^{(O)^4}, \\ N_{H\to\Theta} &= \sum_{i=0}^{2}\sum_{j=0}^{2} N_{\Theta}^{(l)} = 2NAK_1^{(\Theta)^2}K_2^{(\Theta)^2}, \\ N_{H\to Q_1} &= \sum_{i=0}^{2}\sum_{j=0}^{1} N_{Q_1}^{(l)} = 4NAK_1^{(Q)^2}K_2^{(Q)}, \\ N_{H\to Q_2} &= \sum_{i=0}^{1}\sum_{j=0}^{2} N_{Q_2}^{(l)} = 4NAK_1^{(Q)}K_2^{(Q)^2}, \\ N_{H\to D_1} &= \sum_{i=0}^{1}\sum_{j=0}^{1} N_{D_1}^{(l)} = 4NAK_1^{(D)}K_2^{(D)}, \\ N_{H\to D_2} &= \sum_{i=0}^{2} N_{D_2}^{(l)} = \frac{3}{2}NAK_1^{(D)^2}, \\ N_{H\to D_3} &= \sum_{j=0}^{2} N_{D_3}^{(l)} = \frac{3}{2}NAK_2^{(D)^2}, \end{aligned} \tag{46}$$

where

$$K_1^{(s)} = P_{F_1}^{(1)} \exp\frac{\omega_1^{(s)}}{kT} + P_{F_2}^{(1)} \exp\frac{\omega_2^{(s)}}{kT}, \quad K_2^{(s)} = P_{F_1}^{(2)} \exp\frac{\omega_1^{(s)}}{kT} + P_{F_2}^{(2)} \exp\frac{\omega_2^{(s)}}{kT}, \tag{47}$$

$$\omega_1^{(s)} = \alpha_1, \beta_1, \gamma_1, \delta_1, \quad \omega_2^{(s)} = \alpha_2, \beta_2, \gamma_2, \delta_2 \tag{48}$$

for $s = O, \Theta, Q, D$, respectively.

At present we find the hydrogen concentration in O, $\Theta$, Q, D interstices, respectively

$$c_O = (N_{H\to O_1} + N_{H\to O_2})/N = \frac{1}{2}AK_1^{(O)^2}K_2^{(O)^2}(K_1^{(O)^2} + K_2^{(O)^2}),$$

$$c_\Theta = N_{H\to\Theta}/N = 2AK_1^{(\Theta)^2}K_2^{(\Theta)^2},$$

$$c_Q = (N_{H\to Q_1} + N_{H\to Q_2})/N = 4AK_1^{(Q)}K_2^{(Q)}(K_1^{(Q)} + K_2^{(Q)}),$$

$$c_D = (N_{H\to D_1} + N_{H\to D_2} + N_{H\to D_3})/N = A(4K_1^{(D)}K_2^{(D)} + \frac{3}{2}K_1^{(D)^2} + \frac{3}{2}K_2^{(D)^2}).$$

$$(49)$$

The derived formulae (49) determine the hydrogen solubility, if hydrogen atoms are distributed in one type of O, $\Theta$, Q, D interstitial sites.

When hydrogen atoms are distributed in all the interstitial sites, the hydrogen solubility is defined by the sum of concentrations (49) and will be equal to

$$c = c_O + c_\Theta + c_Q + c_D = A[\frac{1}{2}K_1^{(O)^2}K_2^{(O)^2}(K_1^{(O)^2} + K_2^{(O)^2}) + 2K_1^{(\Theta)^2}K_2^{(\Theta)^2} +$$

$$+ 4K_1^{(Q)}K_2^{(Q)}(K_1^{(Q)} + K_2^{(Q)}) + 4K_1^{(D)}K_2^{(D)} + \frac{3}{2}K_1^{(D)^2} + \frac{3}{2}K_2^{(D)^2}].$$

$$(50)$$

Formulae (49), (50) with regard to the relations (47) and (11) determine the corresponding hydrogen solubilities in dependce on the fullerite composition ($c_1$, $c_2$ concentrations), the temperature, the degree of the long-range order in the fullerenes distribution over the lattice sites and the energetic constants.

In the absence of atomic order in the crystal sites we have

$$K^{(s)} = K_1^{(s)} = K_2^{(s)} = c_1 \exp\frac{\omega_1^{(s)}}{kT} + c_2 \exp\frac{\omega_2^{(s)}}{kT},$$

$$(51)$$

and the hydrogen solubilities, according to Eqs.(49), (50), will be equal to

$$\left.\begin{array}{ll} c_O^o = (c_O)_{\eta=0} = AK^{(O)^6}, & c_\Theta^o = (c_\Theta)_{\eta=0} = 2AK^{(\Theta)^4}, \\[2mm] c_Q^o = (c_Q)_{\eta=0} = 8AK^{(Q)^3}, & c_D^o = (c_D)_{\eta=0} = 7AK^{(D)^2}, \end{array}\right\}$$

$$(52)$$

$$c^o = c_{\eta=0} = A[(c_1\exp\frac{\alpha_1}{kT} + c_2\exp\frac{\alpha_2}{kT})^6 + 2(c_1\exp\frac{\beta_1}{kT} + c_2\exp\frac{\beta_2}{kT})^4 +$$

$$+ 8(c_1\exp\frac{\gamma_1}{kT} + c_2\exp\frac{\gamma_2}{kT})^3 + 7(c_1\exp\frac{\delta_1}{kT} + c_2\exp\frac{\delta_2}{kT})^2].$$

$$(53)$$

Formulae (52), (53) define the temperature and concentration dependence of hydrogen solubility, the character of which depends to a large extent on the relations between $\omega_1^{(s)}$ and $\omega_2^{(s)}$ energetic parameters (48).

The dependence of hydrogen solubility on the degree of fullerite ordering is easily found out for the relative values

$$f_O = c_O/c_O^o, \quad f_\Theta = c_\Theta/c_\Theta^o, \quad f_Q = c_Q/c_Q^o, \quad f_D = c_D/c_D^o.$$

$$(54)$$

Let us introduce and consider the following parameters

$$\text{æ}_s = \frac{1}{2}\eta\,\frac{\exp\dfrac{\omega_1^{(s)}}{kT} - \exp\dfrac{\omega_2^{(s)}}{kT}}{c_1\exp\dfrac{\omega_1^{(s)}}{kT} + c_2\exp\dfrac{\omega_2^{(s)}}{kT}}\,, \qquad s = O,\,\theta,\,Q,\,D. \tag{55}$$

The range for possible values of $\text{æ}_S$ parameters is easily evaluated by Eq.(55). In fact, the order parameter has the maximum possible value $|\eta| = 1$ ($\eta=-1$ corresponds to changing roles between lattice points of the first and the second types). For crystals with the stoichiometric composition, when $c_1=c_2=0.5$, and for slightly differed energetic constants, when $\omega^{(S)} \approx \omega_1^{(S)} \approx \omega_2^{(S)}$, the values $\text{æ}_S$ are defined by the function of hyperbolic tangent

$\text{æ}_S \approx \text{th}\,\dfrac{\omega_{(S)}}{kT}$. As it is known, this function has values in the range $[-1, +1]$. Consequently, the $\text{æ}_S$ parameters can change in the range from $-1$ to $+1$, i.e. $-1 \leq \text{æ}_S \leq 1$. It is easy to see that the ratios between values of $K_1^{(S)}$, $K_2^{(S)}$ (47) and $K^{(S)}$ (51) are equal

$$K_1^{(S)}/K^{(S)} = 1+ \text{æ}_S, \qquad K_2^{(S)}/K^{(S)} = 1 - \text{æ}_S, \tag{56}$$

and according to (49), (52), the relative solubilities $f_O$, $f_\Theta$, $f_Q$, $f_D$ will be determined by the following equations

$$f_O=(1 - \text{æ}_O^4)\,(1 - \text{æ}_O^2), \quad f_\Theta=(1 - \text{æ}_O^2)^2, \quad f_Q= 1 - \text{æ}_Q^2, \quad f_D = 1-\frac{1}{7}\,\text{æ}_D^2. \tag{57}$$

The dependence of $f_O$, $f_\Theta$, $f_Q$, $f_D$ values on the corresponding $\text{æ}_S$ parameters, which are proportional to the degree of the long-range order according to (55), defines the character of the atomic order effect during $F_1$, $F_2$ fullerenes distribution over the fullerite lattice sites on the hydrogen solubility when hydrogen atoms are arranged in the interstitial site of any one type $s = O, \Theta, Q, D$. The plots of these dependences are shown in Fig.5.

As Fig.5 shows, the atomic order decreases the hydrogen solubility in fullerite. In this case for the low value of $\text{æ}_S$, which is proportional to the order parameter $\eta$, the hydrogen solubility slightly depends on the degree of the long-range order. However, with increasing $\text{æ}_S$ parameters, effect of the long-range order on the hydrogen solubility during hydrogen atoms distribution over $O$, $\Theta$, $Q$ interstitial sites can be significant. At hydrogen atoms arrangement in $D$ interstitial sites, the order has a weak effect on hydrogen solubility in the whole range of $\text{æ}_D$ values.

If hydrogen atoms are arranged in the interstitial sites of two, three or four types, the dependence of hydrogen solubility on order parameterwill be defined by some average function from $f_O$, $f_\Theta$, $f_Q$, $f_D$ functions, respectively, represented by the plots in Fig. 5. In particular, when hydrogen atoms are arranged in the trigonal and bigonal interstices, the relative hydrogen solubility will be defined by some curve on the $f(\text{æ}_Q\,\text{æ}_D)$ surface shown in Fig. 6. The hydrogen solubility can either slightly or significantly vary with the change of order parameter depending on the run of the curve on this surface what will be defined by relations between $\gamma_1$, $\gamma_2$, $\delta_1$, $\delta_2$ energetic constants.

Below for some specific cases we will find out the behaviour of hydrogen solubility depending on the fullerite composition and its temperature.

40

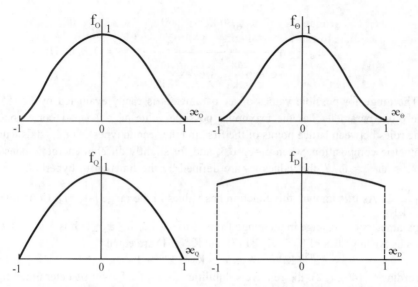

Fig.5. The curves for relative hydrogen solubility in fullerite plotted by formulae (57) in dependce on the parameter $æ_S$ (s=O, $\Theta$, Q, D), which is proportional to the degree of the long-range order, for octahedral O, tetrahedral $\Theta$, trigonal Q and bigonal D interstitial sites, respectively

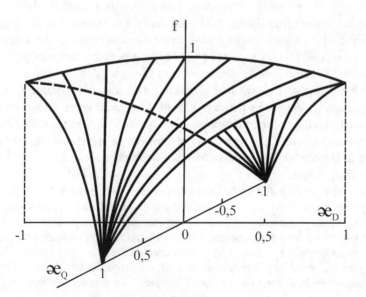

Fig.6. The $f(æ_Q, æ_D)$ surface of the curve for relative hydrogen solubility in fullerite, defined the solubility dependence on order parameters at hydrogen atoms distribution in trigonal and bigonal interstitial sites.

## 7. Dependence of hydrogen solubility on fullerite composition and temperature at hydrogen atoms distribution in tetrahedral, trigonal and bigonal interstitial sites

As it follows from (2), the nearest distance between hydrogen atoms and fullerenes for octa-interstices significantly exceed such distances for tetrahedral, trigonal and bigonal interstitial sites while these distances for the latter differ from one another slightly. Therefore, one can suppose that hydrogen atoms in fullerite are arranged in $\Theta$, $Q$, $D$ positions. In this case in the absence of ordering ($\eta=0$) the solubility is defined by the function

$$c^0 = A[2(c_1 \exp\frac{\beta_1}{kT} + c_2 \exp\frac{\beta_2}{kT})^4 + 8(c_1 \exp\frac{\gamma_1}{kT} + c_2 \exp\frac{\gamma_2}{kT})^3 + 7(c_1 \exp\frac{\delta_1}{kT} + c_2 \exp\frac{\delta_2}{kT})^2] \quad (58)$$

The study of this function $c^0 = c^0(c_1)$ on the extremum by $c_1$ concentration gives the following condition

$$4(c_1 e_1 + \exp\frac{\beta_2}{kT})e_1 + 12(c_1 e_2 + \exp\frac{\gamma_2}{kT})e_2 + 7(c_1 e_3 + \exp\frac{\delta_2}{kT})e_3 = 0, \quad (59)$$

where

$$e_1 = \exp\frac{\beta_1}{kT} - \exp\frac{\beta_2}{kT}, \quad e_2 = \exp\frac{\gamma_1}{kT} - \exp\frac{\gamma_2}{kT}, \quad e_3 = \exp\frac{\delta_1}{kT} - \exp\frac{\delta_2}{kT}. \quad (60)$$

We have derived the cubic equation (59) in relation to $c_1$, that can be solved, for example, using Cardan formulae. The attained result is inconvenient and not given here.

However, one can assert that the $c^0(c_1)$ function can be extreme (has the minimum) if dependences of interaction energies of H-F pairs on distances are so that the plots of these dependences intersect with changing distance from $r_D$ to $r_Q$ and further to $r_\Theta$ so that at $r=r_D$ we will have, for example, $\upsilon_{iF_2}(r_D) > \upsilon_{iF_1}(r_D)$ but

$$\upsilon_{iF_2}(r_Q) < \upsilon_{iF_1}(r_Q) \quad \text{and} \quad \upsilon_{iF_2}(r_\Theta) < \upsilon_{iF_1}(r_\Theta).$$

Each term in Eq.(58) gives the monotonous dependence on $c_1$ concentration. With changing energy $\upsilon_{HF1}$, $\upsilon_{HF2}$ depending on the distance in the H-F pair, the first two terms in (58) will give the increasing curves with $c_1$ concentration, and the third term will give the descending curve with increasing $c_1$. The resulting curve $c^0 = c^0(c_1)$ can be extreme. This case, as an example, is given in Fig.7 when

$$\exp\frac{4\beta_1}{kT} = 0,875 \quad , \qquad \exp\frac{4\beta_2}{kT} = 0,25 \quad ,$$

$$\exp\frac{3\gamma_1}{kT} = 0,5 \quad , \qquad \exp\frac{3\gamma_2}{kT} = 0,795 \quad , \qquad (61)$$

$$\exp\frac{2\delta_1}{kT} = 0,143 \quad , \qquad \exp\frac{2\delta_2}{kT} = 0,857 \quad .$$

The study of $c^0 = c^0(T)$ functions (58) has shown that the extremum can appear on the case of change of interaction character between H-F pairs. In this case, with changing distance r from $r_D$ to $r_\Theta$, interaction in this pair will change from attraction to repulsion or inversely what is highly improbable. Therefore, one can suppose that the temperature dependence of hydrogen solubility in the disordered fullerite must be monotonous.

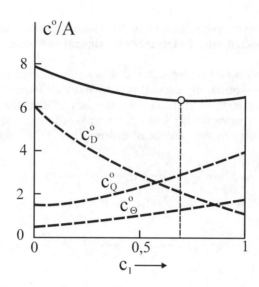

Fig.7. The concentrational dependence of hydrogen solubility in the disordered fullerite (the solid curve), plotted by Eq.(58), and energetic parameters (61). The circle marks the minimum hydrogen solubility at $c_1=0.7$. The dotted curves correspond to hydrogen solubility when hydrogen atoms are arranged only in tetrahedral $\Theta$, only in trigonal Q or bigonal D interstitial sites.

The interaction character between H-F pairs is repulsive what corresponds to the negative values of $\beta_1$, $\beta_2$, $\gamma_1$, $\gamma_2$, $\delta_1$, $\delta_2$ energies. For this case, as an example, Fig.8 shows the plots for $c^\circ=c^\circ(T)$ dependence for fullerite crystals of the stoichiometric composition ($c_1=c_2=0.5$) and for the energies defined at $kT=0.86 \cdot 10^{-1}$ eV by relations

$$\exp\frac{\beta_1}{kT} = 0{,}967 \ , \qquad \exp\frac{\beta_2}{kT} = 0{,}707 \ ,$$

$$\exp\frac{\gamma_1}{kT} = 0{,}795 \ , \qquad \exp\frac{\gamma_2}{kT} = 0{,}572 \ , \qquad (62)$$

$$\exp\frac{\delta_1}{kT} = 0{,}378 \ , \qquad \exp\frac{\delta_2}{kT} = 0{,}926 \ .$$

It is seen from Fig.8 that hydrogen solubility $c^\circ=c^\circ(T)$ in fullerite (as well as terms $c^\circ_\Theta(T)$, $c^\circ_Q(T)$, $c^\circ_D(T)$) is increased with temperature rise. In this case, at the beginning, at low temperatures the solubility is sharply increased with increasing temperature, and further the growth of solubility is determined by the flattened plot.

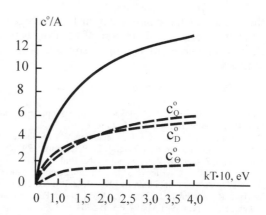

Fig.8. The temperature dependence of hydrogen solubility in the disordered fullerite (the solid curve), plotted by formulae (58), and energetic parameters (62). The dotted curves correspond to hydrogen solubility when hydrogen atoms are arranged only in tetrahedral $\Theta$, only in trigonal Q or bigonal D interstitial sites.

## 8. Conclusions

Hence the statistical and thermodynamical theory of the lattice hydrogen solubility in fullerite with consideration for the hydrogen atoms distribution over the interstitial sites of different types has allowed us to explain and justify the formation of $FH_x$ hydrofullerites with high hydrogen concentration when $0 \le x \le 18$. It has been found that hydrogen solubility depends on the fullerite composition, its temperature, the order parameter in $F_1=C_{60}$, $F_2=C_{70}$ fullerenes distribution over the lattice sites, the energetic constants characterizing the interaction between H-F pairs at the different distances.

The specific cases have been studied. The character of hydrogen atoms distribution in the interstitial sites of different types has been studied at the infinitely low temperature $T \to 0$ and at $T \ne 0$.

The possibility of appearing extremum in the concentrational dependence of hydrogen solubility has been established. It has been shown that the temperature dependence of hydrogen solubility is the function monotonically increasing with temperature rise. It has been elucidated that the order in $F_1$, $F_2$ fullerenes distribution in the lattice sites decreases hydrogen solubility for each type of interstitial sites. In this case the low degree of ordering is scarcely affected by solubility, however, the large values of the order parameter, depending on the type of interstititial sites in which hydrogen atoms are arranged, can decrease the hydrogen solubility down to zero. At hydrogen atoms distribution in the interstitial sites of different types depending on the degree of fullerite ordering, the hydrogen solubility is defined by some average curve on the surface in the five-dimensional space $f(æ_O, æ_\Theta, æ_Q, æ_D)$. Behaviour of this curve must depend on the energetic parameters of H-F pairs interaction at the different distances corresponding to the hydrogen atoms arrangement in O, $\Theta$, Q, D interstitial sites.

In summary it should be noted that the pursued investigations can be refined due to taking into account: the volume effects induced by hydrogen atoms, local distortions of the crystal lattice around the hydrogen atoms (deformation effects), the interaction between H-F pairs in the next coordination spheres, the fullerenes ordering and formation of four

44

sublattices [15], the change of the phase states in the hydrogen atoms system (possible isotopic ordering), the effect of external pressure of gas from which the hydrogen atoms go in fullerite, the electron properties of hydrogen solvent, possible presence of fullerited hydrogen in hydrofullerite.

## 9. References

1. Tkachenko L.I., Lobach A.S., Strelets V.V. Redoks-indutsiruemyi perenos vodoroda s hydrofullerena $C_{60}H_{36}$ na fulleren $C_{60}$ // Isv. RAN. Ser. him. 1988. №6. P. 1136-1139.
2. Bezmelnitsyn V.N., Glazkov V.B., Zhukov V.P., Somenkov V.A., Shilshtein S.Sh. Abstr. 4th Biennial Int. Workshop in Russia "Fullerene and Atomic Clusters". St. Petersburg. 1999. P. 70.
3. Glazkov V.P., Zhukov V.P., Somenkov V.A., Ivanova T.N. Ibid. P. 71.
4. UgovenukT.S., Udalov Yu.P., Gavrilenko I.B. Vzaimodeistvie fullerenovoi sazhi s vodorodnoi neravnovesnoi plazmoi// Jurnal prikladnoi khimii. 1999. V.72. P. 1728-1733.
5. Hall L.E., McKenzie D.R., Attalla M.J., Vassallo A.M., Davis R.L., Dunlop J.B., Cockayne D.J.H The structure of $C_{60}H_{36}$ // Journ. Phys. Chem. 1993. Vol. 97. P.5741-5744.
6. Kolesnikov A.I., Antonov V.E., Bashkin I.O., Grosse G., Moravsky A.P., Muzychka A.Yu., Ponyatovsky E.G., Wagner F.E. Neutron spectroscopy of $C_{60}$ fullerene hydrogenated // Journ. Phys. Condens. Matter. 1957. Vol. 9. P.2831.
7. Goldshleger N.F., Moravskiy A.p., Hydridy fullerenov: poluchenie, svoystva, structura // Uspekhi khimii. 1997. V. 66. №4. P. 353-375.
8. Shulga Yu.M, Tarasov B.P. kristallicheskie hydrofulleri: polucheni i svoystva. Fullereni i fullerenopodobnie stuctury. Minsk: BGU. 2000. P. 14-19.
9. Trefilov V.I., Schur D.V., Tarasov B.P., Shulga Yu.M., Chernogorenko A.V., Pishuk V.K., Zaginaichenko S.Yu. Fullereny – osnova materialov buduschego K.: ADEF. 2001. 148 p.
10. Matysina Z.A., Schur D.V. Vodorod I tverdofaznie prevrascheniya v metallah, splavah i fulleritah. Dnepropetrovsk: Nauka i obrazovanie. 2002. 420 c.
11. David W.I., Ibberson R.M., Dennis T.J.S., Harr J.P., Prassides K. Europhys. Lett. 1992. Vol. 18. P. 219.
12. Blinc R., Selinger J., Dolinsek J., Arcon D. Two – dimensional C NMR study orientational ordering in solid $C_{60}$. //Phys. Rev. 1994. Vol. 49. P. 4993-5002.
13. Brazhkin V.V., Lyapin A.G. Prevraschenie fullerita $C_{60}$ pri vysokih davleniyah i temperaturah// Uspehi fizicheskih nauk. 1996. V. 166. P. 893-897.
14. Tareeva E.E., Schelkacheva T.I. Orientatsionnoe uporyadochenie v tverdom $C_{60}$: metod vetvleniy //Tereticheskaya i matematicheskaya fizika. 1999. V. 121. № 3. P. 479-491.
15. Makarova T.L. Elektricheskie i opticheskie svoystva monomernyh i polimernyh fullerenov (obzor) // Fizika i tehnika poluprovodnikov. 2001. V. 35. № 3. P. 257-293.
16. Sundqvint B. Adv. Phys. 1999. Vol. 48. P. 1.
17. Matysina Z.A., Schur D.V. Structurnie prevrascheniya pri hydrirovanii fullerita. Rastvorimost vodoroda// Dep. № 3188-B99. M.: VINITI. 1999. 17 p. Izvestiya vuzov. Phisika. 2000. № 2. P. 100.
18. Zaginaichenko S.Yu., Matysina Z.A., Schur D.V., Chumak V.A. The calculation of constitutional diagrams and hydrogen solubility in simplest hydrofullerenes // Proceed. 13th World Hydrogen Energy Conf. Beijing. 2000. P. 1216-1221.
19. Zaginaichenko S.Yu., Matysina Z.A., Schur D.V., Chumak V.A., Pishuk V.K. Teoreticheskoe issledovanie perehoda mezhdu fazami PK-OCK tipa pri nasyschenii fullerenov vodorodom // Mezhdunarodnyi nauchnyi jurnal «Alternativnaya energetika i ekologiya». 2002. № 1. P.45-55.
20. Somenkov V.A., Shilshtein C.Sh., Fazovie prevrascgeniya vodoroda v metallah. M.: IAE. 1978. 80 p.
21. Smirnov A.A. Teoria splavov vnedreniya. M.: Nauka. 1979. 388 p.
22. Smirnov A.A. Teoria diffuzii v splavah vnedrenia. K.: Naukova dumka. 1982. 167 p.
23. Smirnov A.A. Teoria fazovyh prevrascheniy i razmescheniya atomov v splavah vnedreniya K.: Naukova dumka. 1992. 280 c.
24. Boba M.S., T.S.L., Bolasubramajan R., Studies on the thermodynamics of $C_{60}$-$C_{70}$ binary system // Journ. Phys. Chem. 1994. Vol. 98. P. 1333-1340.
25. Kniaz K., Fisher J.E., Girifalco L.A., Fullerene alloys // Sol. State Comm. 1995. Vol. 96. P. 739-743.
26. Havlik D., Schranz W., Haluska M., Kuzmany H., Rogl P. Thermal expansion measurements of $C_{60}$-$C_{70}$ mized crystals // Sol. State Comm. 1997. Vol. 104. P.775-779.
27. Bezmelnitsin V.N., Eletskiy A.V., Okun M.V. Fullereny v rastvorah // Uspehi fizicheskih nauk. 1995. V. 168. P. 1195-1221.
28. Vovk O.M., Isakina A.P. Garbus A.S., Kravchenko Yu.S. Tverdye rastvory fullerenov $C_{60}$-$C_{70}$, poluchennye iz geksan-toluolnogo rastvora. Fullereny i fullrenopodobnnie structury. Minsk: BGU. 2000. P. 70-76.

# CONTROLLING ROLE OF ELECTRON CONCENTRATION IN PLASMA-CHEMICAL SYNTHESIS

G.N. CHURILOV
*L.V. Kirensky Institute of Physics SB RAS, Akademgorodok, Krasnoyarsk, 660036 Russia*
*E-mail: churilov@iph.krasn.ru*

*Keywords:* A. Fullerene; B. Arc discharge; C. Computational chemistry, modeling

## 1. Introduction

In 1990 W. Kraetschmer et al. had published the article in Nature journal about fullerene synthesis and extraction [1]. In this work the methods of synthesis and extraction of the new carbon allotrope were shown.

The theorists at once took an interest in the fullerene formation mechanism and many hypotheses and models of their formation were built. But, in my opinion, the work of Bochvar and Gal'perin [2], which was published long before efficient fullerene synthesis discovery, remains the most significant up to now.

The experimenters attempted to improve Kraetschmer's method and detected that the new carbon allotrope practically does not form at pressures greater than 200 torr.

Before the fullerenes discovery we repeatedly observed the self-focusing and self-blowing carbon plasma jet arising at special form of electrodes and at a current feed of kHz frequency (Fig.1). The jet in open space achieved the length of 0.75m. In 1988 the work about its generation method and plasma parameters was published [3].

There was nothing but to use water-cooling chamber and to investigate carbon condensate forming on chamber walls. Consequently it was established that fullerene synthesis is possible at atmospheric pressure [4].

## 2. Results and discussion

In Fig.2 one can see the chromatogram of fullerene extract. The synthesis leads to higher percentage of $C_{70}$ (up to 30%) and higher fullerenes (up to 15%). Because the fullerene synthesis at atmospheric pressure was substantially unsuccessful, there is vital necessity to explain successful fullerene synthesis in our setup.

It is well known that in plasma at pressures less then 200 torr differ in ionization instability. Ionization instability often appears as electron density oscillations, which can be observed as discharge stratification. In such case glow discharge in a tube is separated on alternate areas of different brightness. The view of alternate dark and bright areas can be moving or static. This phenomenon named discharge stratification and it is the visual display of ionization instability.

45

*T.N. Veziroglu et al. (eds.),*
*Hydrogen Materials Science and Chemistry of Carbon Nanomaterials,* 45-51.
© 2004 *Kluwer Academic Publishers. Printed in the Netherlands.*

46

Figure 1. Self-focusing plasma jet at open space

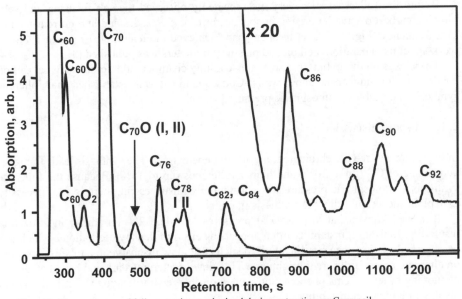

Figure 2. Chromatogram of fullerene mixture obtained during extraction on Cosmosil
Buckyprep column.

We had carried out the discharge between graphite electrode and hole water-cooling copper rod in argon flow. The discharge current (44 kHz) was co-phased with the rotation angle of high-speed photorecorder mirror. It was achieved due to the choice of frequency and phase of the mirror rotation as a reference signal for the discharge current [5]. In this experiment forced ionization waves first were registered at atmospheric pressure.

The experiments described above had shown that plasma both in Kraetschmer's method and in discharge of kHz frequency range is unstable toward arising of electron density oscillations. Thus, a model of fullerene formation has to take into account electron concentration of ionized carbon vapor.

We used quantum-chemical methods to calculate bonding energies, ionization potentials and electron affinities for different carbon clusters. At first for every carbon cluster geometry optimization was carried out by semiempirical method PM3. Obtained geometry was used in calculations of bonding energy of a cluster with different charges in VASP package [6]. In this DFT calculations forevery atom Vanderbilt pseudopotential was used [7]. The results are shown in the Table 1.

TABLE 1. Ionization potential (IP), electron affinity to first (EA1) and second (EA2) electron

| k | IP, eV | EA1, eV | EA2, eV |
|---|--------|---------|---------|
| 2 | 11.666 | 4.752 | 1.562 |
| 4 | 9.963 | 5.073 | 1.676 |
| 10 | 8.194 | 1.924 | 1.046 |
| 14 | 7.583 | 3.365 | 1.426 |
| 18 | 6.825 | 4.559 | 2.575 |
| 20 | 6.86 | 4.11 | 2.01 |
| 31 | 6.27 | 4.624 | 3.078 |
| 40 | 5.85 | 3.97 | 2.41 |
| 60 | 5.47 | 3.79 | 2.62 |
| 70 | 5.354 | 3.964 | 2.622 |
| 80 | 5.074 | 3.92 | 2.804 |

The classic collision theory gives the formula for collision cross section of charged clusters [8]:

$$\sigma_{12} = \sigma_0 \left( 1 - \frac{q_1 q_2 e^2}{r_{12} \varepsilon_{kin}} \right),$$

where $\sigma_0 = \pi r_{12}^2$ – classical collision cross section of neutral particles, $q_1 q_2 e^2 / r_{12} = E_{12}$ – energy of Coulomb interaction of particles and $\varepsilon_{kin} = 3/2 \cdot k_B T$ – average kinetic energy of relative moving of particles. From this formula one can see the greater difference between charges of colliding particles the greater cross section. In the case of local thermodynamical equilibrium concentrations of charged components of a cluster defined by Saha equation:

$$\frac{n_k^{q+1} \cdot n_e}{n_k^q} = \frac{Z_k^{q+1}}{Z_k^q} a \cdot \exp\left(- IP_k^q / k_B T\right), \qquad q = -2, -1, 0, +1 \quad ,$$

where $n_k^q$ is the concentration of cluster $C_k$ with charge q, $n_e$ is the electron concentration, $a = 2\left(m_e kT / 2\pi\hbar^2\right)^{3/2}$, $Z_k^q(T) = \sum_{n=1}^{n_{max}} g_n \exp[-(\varepsilon_n - \varepsilon_1)/k_B T]$ is the electronic statistical sum of cluster $C_k$ with charge $q$ at temperature T, $IP_k^q$ is the ionization potential of cluster $C_k^q$ with charge $q$. Average charge of a cluster at given electron concentration $n_e$ and temperature $T$ was equal to $\langle q_k \rangle = \left(n_k^+ - n_k^- - 2n_k^{-2}\right)/\left(n_k^+ + n_k^0 + n_k^- + n_k^{-2}\right)$.

On Fig.3 the dependencies of carbon cluster charges on electron concentration at constant temperature (2500K) are shown.

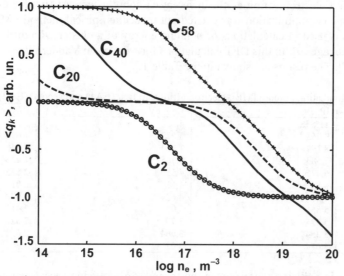

Fig.3. Dependence of average charges of different clusters on electron concentration at temperature 2500K.

On Fig.4 one can see that formation rate of different size clusters has maximum at different electron concentrations at the same temperature (2500K).

We considered formation of $C_{60}$ and had found that an area of values of temperature and electron concentration exists where formation rate is greater more then order then average rate for all electron concentrations and temperatures in considered range. The oscillations of electron concentration increase formation rate of $C_{60}$ up to few times at temperatures lower than 3000K and at electron concentrations exceed electron concentration of equilibrium plasma without any impurities [9].

The ideas represented above were developed to estimate formation rates of endohedral fullerenes depending on temperature for self consisted electron concentration, which are shown in Fig.5 (also see [10]).

These calculations qualitatively correspond to experimental results. We carried out the synthesis in the setup described above at atmospheric pressure with different metals and obtained fullerene mixtures with different contents of metals: Fe – 0. 03%, Sc – 0.0015% and Pt – 0.00%.

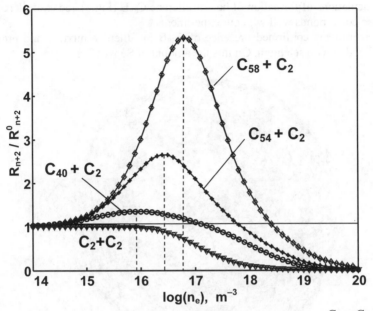

Fig.4. Dependence of relative rate of association of a cluster $C_n$ and small cluster $C_2$ ( $C_n + C_2 \rightarrow C_{n+2}$ ) on electron concentration at temperature 2500K.

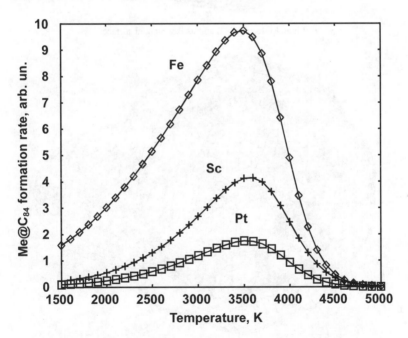

Fig.5. Rate of metallofullerenes $Me@C_{84}$ formation depending on temperature at self-consistent electron concentration.

50

The calculations of formation of heterofullerene $C_{59}B$ (Fig.6) and exofullerene $C_{60}Se$ (Fig.7) also corresponds well with our experiments.

Our experiments confirmed presence of $C_{59}B$ in fullerene mixture and presence of fullerene bonded with selenium. On this the amount of Se was 3%.

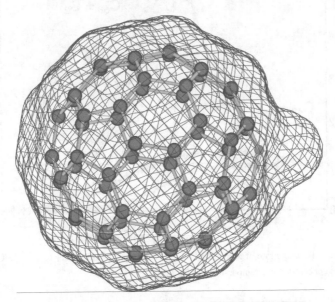

Fig.6. Molecule $C_{59}B$ and its isopotential surface

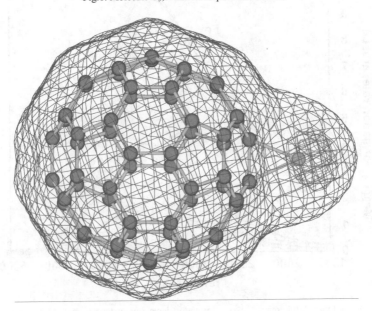

Fig.7. Molecule $C_{60}Se$ and its isopotential surface

## 3. Conclusions

In prospect we plan to improve the fullerene formation model considering electron concentration especially in calculations of endo- and heterofullerenes, to investigate experimentally and theoretically conditions of their maximum yield, to extract and to investigate isolated fullerene derivatives.

## 4. Acknowledgments

The work was supported by INTAS (01-2399) and RFBR (03-03-32326), and with support of President RAS program (dir. 9, proj. 1) and Russian state science and technical program.

## 5. References

1. Kratschmer, W., Fostiropoulos, K. and Huffman, D.R. (1990) The success in synthesis of macroscopic quantities of $C_{60}$, *Chem. Phys. Lett.* **170**, 167.
2. Bochvar, D.A. and Gal'perin, E.G. (1973) About hypothetical systems: carbo-dodecahedron, s-icosahedron, carbo-5-icosahedron, *Dokl. Akad. Nauk SSSR, ser. chem.* **209**, 610.
3. Ignat'ev, G.F., Churilov, G.N. and Pak, V.G. (1988) Investigation of plasma jet generating by plasmatron of kilohertz frequency range, *Izvestia SO AN SSSR, ser tekh nauk* **4**(15), 93-95.
4. Churilov, G.N., Solovyov, L.A., Churilova, Y.N., Chupina, O.V. and Malcieva S.S. (1999) Fullerenes and other structures of carbon synthesized in a carbon plasma jet under helium flow, *Carbon* **37**, 427-431.
5. Churilov, G.N., Novikov, P.V., Taraban'ko, V.E., Lopatin, V.A., Vnukova, N.G. and Bulina, NV. (2002) On the mechanism of fullerene formation in a carbon plasma, *Carbon* **40**, 891-896.
6. Kresse, G, Furthmuller, J. (1996) Efficient iterative schemes for ab-initio total energy calculations using a plane-wave basis set, *Phys Rev B* **54**, 11169.
7. Vanderbilt, D. (1990) Soft self-consistent pseudopotentials in a generalized eigenvalue formalism, *Phys Rev B* **41**, 7892.
8. Landau, L.D., Lifshitz, E.M. Mekhanika (Mechanics). Moscow: Nauka, 1973. Translated into English Oxford: Pergamon Press, 1980.
9. Churilov, G.N., Fedorov, A.S., Novikov, P.V. (2003) Influence of electron concentration and temperature on fullerene formation in a carbon plasma, *Carbon* **41**, 173-8.
10. Fedorov, A.S., Novikov, P.V. and Churilov G.N. (2003) Influence of electron concentration and temperature on endohedral metallofullerene $Me@C_{84}$ formation in a carbon plasma, *Chem Phys* **293**(2), 253-261.

# OXYGEN SOURCE FOR ISOLATED FUEL CELLS

R. LOUTFY, V. SHAPOVALOV, E. VEKSLER
*MER Corporation,*
*7960 Kolb Road, Tucson, Arizona, 85706, USA*
*Fax: (520) 574-1983*
*E-Mail: mercorp@mercor.com*

## Abstract

A compressed gas such as oxygen or hydrogen is stored at elevated pressure by introducing it into a plurality of storage cells having a common gas distribution manifold thereby maintaining the same pressure in all cells. The flow of gas between the manifold in any given cell is blocked in the event of a sudden pressure drop in such cell such as may occur in the event of a leak in the cell. Preferably, the cells are constructed in modular form to include a plurality of the cells and the modular material is preferably made of porous gasar material. The low pressure, i.e., leaky cell, is preferably blocked by a closure formed of pliable rubber like material that is forced against the passageway leading from the manifold to the cell by higher gas pressure in the manifold, which occurs when the pressure becomes lower in the given cell due to a leak.

*Keywords:* oxygen; tubular cell; gasar material cell; compressed gas; safety gas storage

## 1. Introduction

The overall technical objective of this works is to establish the feasibility of using the proposed *gasar* materials [1] as well *tube cell* modules [2] for the development of high performance oxygen storage system. The feasibility study was to include synthesis and full characterization of both systems in terms of improved gas storage capacity, mechanical integrity, as well as safety and reliability. The basic engineering object of the project is creation of a new oxygen high pressure storage, which will combine safety with high capacity and yet be inexpensive and convenient, for use in aircraft and spacecraft. They can be utilized also for ground vehicles as a hydrogen source for fuel cell electric generation.

## 2. Technical, economical and environmental considerations

We has refined the fabrication process [2] and performed initial testing of the tubular oxygen storage system comprised of separate small diameter thin-wall metal tubes reinforced with carbon fiber wrapping (Figure 1). The measured oxygen storage capacity for wrapped stainless steel tubes was found to be close to that calculated theoretically (Figure 2), which allowed further extension of the modeling onto other light-weight materials, for example, aluminum. In the course of design optimization, several

*T.N. Veziroglu et al. (eds.),*
*Hydrogen Materials Science and Chemistry of Carbon Nanomaterials,* 53-58.

modifications of selected system components were generated. Based on these suggested improvements aimed at weight reduction, the overall oxygen storage capacity of the proposed system utilizing aluminum tubes is expected to reach ~ 58 % by weight, which doubles the capacity of the system as compared with currently utilized options (for example, oxygen candle).

Figure 1. The outward of tube module oxygen storage

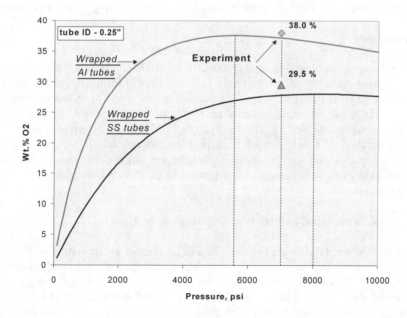

Figure 2. Results of theoretical simulation and experimental testing of oxygen storage capacity based on tubular storage cells

In addition, variety of safety tests were designed and conducted to demonstrate an outstanding reliability of the proposed system. During the course of this research, a great deal of attention was paid to optimizing specific components of the system (manifold, tube end connectors, sealing mechanism, etc.) as well as the process of wrapping the tubes to ensure their uniform and reliable reinforcement in both radial and axial directions which should enable withstanding of repeated gas charge and discharge cycles under reasonable pressure.

It is important to emphasize, that due to uniqueness of oxygen behavior under high pressure, continuing increase in pressure does not necessarily result in continuing increase in oxygen storage capacity **Fig. 2.** This can be explained by variation of oxygen compressibility factor, which increases with pressure causing less gas storage as compared to ideal gas.

## 3. Results and discussion

In accordance with the main idea, by providing a storage system comprising of plurality of gas tight cells, preferably in modules, rather than a single gas cylinder of equivalent capacity, the danger of sudden total gas release will be reduced and the safety margin can therefore be reduced accordingly. The result is a much higher weight fraction of storage. The idea of using a storage system with a plurality of cells formed of micro tube cells [2] or gasar-material having cellular structure [1] as the basic gas storage block (cellular module) is the scientific and technical basis of a preferred aspect of this work.

High-strength porous materials have only recently become available at reasonable costs due to the innovative, gas eutectic transformation process. The high strength of these materials, known as *gasar-material*, results from the smoothly rounded pore shape, the mirror smooth pore surface finish and the absence of porosity in the interpore spaces. An additional advantage is the cleanliness of the base material. The base material is free of the nonmetallic inclusions, which are usually present in porous metal materials created by sintering together oxide or carbide coated fine powders.

Ellipsoidal, tapered and cylindrical steel gasar ingot's shapes were studied to create best variants of gas storage unites. It is shown that steel-based gasar ingots was not result in porosity over 50% and qualified ingot's surface in all cases under studied parameters. To improve this results modification of equipment must be curried out.

It was suggested to use light magnesium alloy instead steel gasar-based oxygen storage. The magnesium alloy with manganese and cerium featuring in high corrosion resistance and strength was selected for investigation.

Ellipsoidal and cylindrical magnesium alloy – based gasar ingots was studied. It was created that the better structure, higher porosity and acceptable ingot's surface in magnesium-based gasar form already under pressure 0,12 - 0,2 MPa. In proper conditions magnesium–based gas storage units may be applied.

Structures with radial continuous pores are preferable for use in gasar cellular storage (GCS), since the charging/discharging channels are all clustered together where it is relatively easy to block or unblock them by the same mechanism, as shown in Figure 3,4.

On the bases of conducted research magnesium alloy is recommended for oxygen storage. The samples of ellipsoidal and cylindrical magnesium alloy–based gasar ingots was prepared for tests.

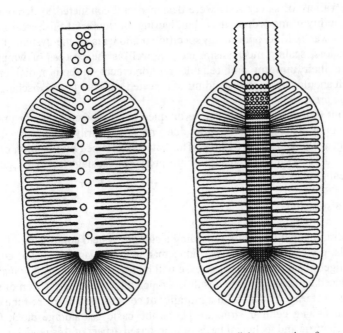

Figure 3. Schematic inner structure changing of the radial gasar casting after machining:
left – initial structure, right – after machining.

Figure 4. An ellipsoidal steel-based gasar ingot:
left- general view; right- longitudinal section

Increased safety is achieved by a special design element, which we have named "smart" rubber. It is a thin (about 1-mm thick) rubber film, which covers the internal pore apertures. In a case where outer wall destruction causes oxygen to escape from one or several pores. This "smart" rubber will nestle tightly into these open-ended apertures and block release of oxygen from the undamaged pores.

For initial manufacture of the GCS in laboratory conditions we have chosen an austenitic stainless steel. It has the greatest margin of safety at the same oxygen storage amount and due to its corrosion resistance is unlikely to weaken with time due to corrosion.

## 4. Conclusions

In conclusion, the performance of the proposed system was compared with other commercially available options, for example a $NaClO_3$ oxygen candle, which generates oxygen as a result of a high temperature chemical reaction between sodium chlorate and iron powder. Examples of various o The main disadvantages of such systems are high temperature operation, which poses a fire hazard, and impossibility of recharging the system.

Comparative data between existing oxygen candles and the proposed system based on calculation results conducted for aluminum tubes for better weight savings are given in Table 1. Oxygen candle devices as well as schematic diagram are shown in **Table** below.

TABLE 1. Comparative characteristics of MER's prototype and commercial oxygen candle

| Characteristic | MER Prototype | Oxygen candle |
|---|---|---|
| Amount of generated $O_2$ | 2900 L (4.14 kg) | 2900 L (4.14 kg) |
| $O_2$ source | compressed gas | Chemical reaction |
| $O_2$ purity | up to 100% (< 5 ppm CO) | Up to 99% (< 25 ppm CO) |
| System weight | 7.14 kg (Aluminum) | 15 kg |
| Wt. percent $O_2$ | 58 % | 27.6 % |
| Start-up | Instantaneous | Delayed |
| Operating conditions | room temperature | 600 degrees |
| Discharge rate | as required | 64.4 liters/min maximum |
| Duration | Controllable | 30 – 45 min |
| Volume | Conformable to available space | Fixed |
| Suitability for other gases | Yes ($H_2$, air, etc.) | No |
| Rechargeability | Yes | No |
| Fire hazard | No | Yes |

Based on comparative characteristics shown above, the proposed system shows definite advantages over existing systems and will be further investigated in Phase II program.

58

## 5. References

1. Shapovalov, V. Method for manufacturing porous articles. US patent #5,181,549, January 26, 1993.
2. Shapovalov, V. and Loutfy, R. Method and apparatus for storing compressed gas US Patent # 6520219, February 18, 2003.

# MICROSCOPIC CHARACTERISTICS OF H DIFFUSION AND DIFFUSE SCATTERING OF RADIATIONS IN H.C.P.-*Ln*–H (FROM THE DATA ON ELECTRICAL-RESISTIVITY RELAXATION)

V.A. TATARENKO, T.M. RADCHENKO, V.B. MOLODKIN
*G.V. Kurdyumov Institute for Metal Physics, N.A.S.U.,*
*36 Academician Vernadsky Blvd., UA-03680 Kyyiv-142, Ukraine*

## Abstract

The further theoretical study of the short-range order kinetics of H-atoms at tetrahedral interstices in h.c.p. lanthanoids (*Ln*), for instance, lutetium (Lu), is developed. It is studied by the use of available data of measurements of residual electrical resistivity for interstitial solid solutions h.c.p.-Lu–H during the isothermal annealing. Within the framework of two models (with one and two relaxation times), the relaxation times of this process in some α-Lu–H solid solutions at temperatures 160–180 K are evaluated on a base of the available experimental results. Within the framework of the hypothesis about an identity of maximum characteristic relaxation times of diffuse radiation scattering and relaxation times of electrical resistivity, the time evolution of normalized change of radiation diffuse-scattering intensity is predicted and corresponds to the wave-vector star, which dominates in mapping a structure of the short-range order (at different quenching and annealing temperatures) for polycrystalline $LuH_{0.180}$ and $LuH_{0.254}$ solid solutions. The pre-exponential factor in the Arrhenius-type law for the relaxation time and energy of migration activation for H atoms in these solid solutions are obtained.

## 1. Introduction

Short-range ordering (as well as long-range ordering) exerts a strong influence on many physical properties of the rare earth–hydrogen systems (see Ref. [1, 2]).

The present paper is dedicated to the further theoretical study of the short-range order kinetics of H-isotopes at tetrahedral interstices (see Fig. 1) in h.c.p. lanthanoids (*Ln*), in particular, lutetium (Lu), studied by electrical-resistance measurements.

To investigate the short-range order (SRO) kinetics, a direct approach consists of determining the SRO structure by diffuse-scattering experiments at a given temperature, *i.e.* reciprocal-space distribution of diffuse-scattering intensity as a function of time. This supposes a large number of measurements, which are not easily performed by such a method.

Another solution is to follow a physical property affected by the SRO evolution. The most used is electrical resistivity, which allows obtaining the results with more simplicity and accuracy. Nevertheless, such measurements could only be considered as macroscopic

*T.N. Veziroglu et al. (eds.),*
*Hydrogen Materials Science and Chemistry of Carbon Nanomaterials,* 59-66.
© 2004 *Kluwer Academic Publishers. Printed in the Netherlands.*

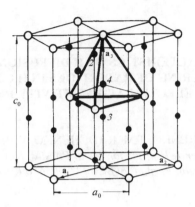

*Figure 1.* Location of tetrahedral interstices (●) forming four interpenetrating sublattices (1, 2, 3, 4) within the h.c.p. lattice with parameters $a_0$ and $c_0$ (○—metal atoms at sites forming two interpenetrating sublattices); $\mathbf{a}_1$, $\mathbf{a}_2$, $\mathbf{a}_3$—basic translation vectors.

information compared with the determination of each Warren–Cowley parameter. Indeed the different effects contributing to the variations of the electrical resistivity are difficult to separate. This led to the development of several kinetic models for predicting the evolution of SRO [3, 4].

Because of its macroscopic character, it is clear that method of 'electrical-resistance change' cannot bring an information as detailed as that deduced from diffuse-scattering intensity measurements [5–13]. However, the high accuracy of the resistivity measurement and its great sensitivity to any change in the atomic distribution, combined to its relative experimental simplicity, makes this technique a useful tool in studying both long-range order and short-range order kinetics [14–16].

Kinetics of the short-range order relaxation is controlled by diffusion of atoms on intersite distances, and it is realizing during comparatively short times (*i.e.* high temperatures are not necessary for its realization). Thus, investigation of the relaxation of short-range order allows determining microscopic characteristics of bulk diffusion— probabilities of atomic jumps in elementary diffusion acts, *etc.*

## 2. Model

Because of the macroscopic character of the resistivity relaxation, our model is phenomenological and macroscopic. It is based on the hypothesis that, for any interstitial mechanism of hydrogen diffusion in h.c.p.-$LnH_c$ system ($0 \leq c \leq 0.5$), the reciprocal of relaxation time for the *i*-type channel of relaxation process, $\tau_i^{-1}$, is directly proportional to the hydrogen jump frequency, $\nu_{Hi}$, with an efficiency factor $\chi_i$:

$$\tau_i^{-1} = \chi_i (1-c)\nu_{Hi}. \tag{1}$$

As assumed, the relaxation time corresponds to the time required for the *i*-type channel migration of hydrogen to change the SRO-parameter value in $e$ times. The temperature dependence of $\nu_{Hi}$ has its usual definition given by a Boltzmann distribution,

$$\nu_{Hi} = \nu_{0i} exp\{- E_{mi}/(k_B T)\}.$$  (2)

In Eq. (2), $\nu_{0i}$ is a pre-exponential factor, $k_B$—Boltzmann constant, $T$—annealing temperature, $E_{mi}$—energy of the $i$-type channel migration activation of H atoms. In a case of spatial redistribution of H atoms between (tetrahedral) interstices, $E_{mi}$ corresponds to H-atom total activation energy, $E_{ai}$: $E_{ai} \approx E_{mi}$. Therefore, the temperature dependence of $\tau_i$ follows the so-called Arrhenius law [1]:

$$\tau_i = \tau_{0i} exp\{E_{ai}/(k_B T)\} \text{ with } \tau_{0i} = \{\chi_i(1-c)\nu_{0i}\}^{-1}.$$  (3)

In this case, experimental data have to be described with only two parameters $\tau_{0i}$ and $E_{ai}$ for each ($i$th) contribution to relaxation mechanism.

In residual-resistivity studies, the rate of isothermal variations of the degree of SRO is generally considered proportional to a 'thermodynamical force' as a function $F$ of the distance to equilibrium.

Therefore, the kinetic equation could be written in the form [3, 16]

$$\frac{\partial\rho(t,T)}{\partial t} = -F(\{\tau_i^{-1}\}; \rho(t,T)-\rho_\infty(T))$$  (4)

where $\rho(t,T)$ is the instantaneous residual resistivity and $\rho_\infty(T)$—the equilibrium residual resistivity at annealing temperature $T$.

Under stationary conditions, $i.e.$ constant concentration of H atoms as interstitial defects, several different kinetic treatments are commonly used.

The simplest one is the first-order kinetics model with

$$F(\tau^{-1}; \rho(t,T)-\rho_\infty(T)) = \tau^{-1}(\rho(t,T)-\rho_\infty(T)),$$  (5)

where the SRO relaxation is described by a single-exponential behaviour of the normalized resistivity change [5, 17],

$$\frac{\rho(t,T)-\rho_\infty(T)}{\rho_0(T)-\rho_\infty(T)} = exp\left[-\frac{t}{\tau(T)}\right]$$  (6)

($\rho_0(T)$—the initial residual resistivity at annealing temperature $T$) and characterised by a unique relaxation time $\tau(T)$, which is defined in terms of the hydrogen jump frequency (mobility), $\nu_H$, with an efficiency factor $\chi$, and the hydrogen concentration, $c$, in $Ln$ (see Eq. (1)).

Higher-order ($n$) kinetics model resulting in a sum of several exponential relaxations is postulated as follows [14, 18–20]:

$$\frac{\rho(t,T)-\rho_\infty(T)}{\rho_0(T)-\rho_\infty(T)} = \sum_{i=1}^{n} A_i exp\left[-\frac{t}{\tau_i(T)}\right],$$  (7)

with

$$\sum_{i=1}^{n} A_i = 1,$$  (8)

where $A_i$—the weight of the $i$th channel of relaxation process. Such more complex relaxation behaviour, in the manner of a Khachaturyan approach [21] to the description of SRO-relaxation kinetics, might be also explained in the model of disperse order [18, 19]. As the number of independent fit parameters increases as $3n$ 1, in order to keep the analysis trustworthy, the number of exponentials ($n$) necessary to fit the data, as a rule, should not exceed 2 or 3 [15].

## 3. Results

The kinetics of short-range order of H in h.c.p. Lu was studied in Ref. [22] by electrical resistivity measurements during isothermal ($T$=const) anneals performed with two specimens. Their hydrogen concentrations, given as the ratio $c$ of H to Lu atoms, were determined from pressure drop in the reaction volume during loading (charging) and during degassing after the measurements as follows: $c=0.180$ and $c=0.254$; another estimates of corresponding H-to-Lu ratios for these two specimens were derived from measurement of the quenched-in resistivities $\rho$ at 4.2 K and 296 K and are $c_\rho=0.259$ and $c_\rho=0.280$, respectively (*i.e.* above the $T$-independent solubility limit, $c^{max} \approx 0.20$, in the metastable low-$T$ short-range-ordered $\alpha^*$-phase [1]). Considerable changes in resistivity $\rho(t,T)$ of LuH$_c$ ($c=0.180$ and $c=0.254$) during isothermal annealing within reasonable times, $t$, were only observed at temperatures between about 150 K and 200 K [22]. As evaluated, an activation enthalpy of 0.45 eV corresponds to the migration of H atoms at temperatures from about 160 K to about 190 K [22].

In Ref. [4], the relaxation times, $\{\tau_i\}$, for alloys LuH$_{0.180}$ and LuH$_{0.254}$ were estimated by using experimental results of Ref. [22] within the framework of the first-order kinetics model and the (more realistic) second-order kinetics model. Besides, the normalized resistivity changes in these solid solutions during isothermal annealing at the given temperatures were also obtained in Refs. [3, 4].

TABLE I. Parameters obtained by the least-squares method within the framework of the models with one (first-order kinetics model) and two (second-order kinetics model) relaxation times for two $\alpha$-Lu–H solid solutions at issue.

| Alloy | First-order kinetics model | | Second-order kinetics model | | | |
|---|---|---|---|---|---|---|
| | Pre-exponential parameter $\tau_0$, s | Activation energy of migration $E_a$, eV | First pre-exponential parameter $\tau_{01}$, s | First activation energy of migration $E_{a1}$, eV | Second preexponential parameter $\tau_{02}$, s | Second activation energy of migration $E_{a2}$, eV |
| LuH$_{0.180}$ | $1.464 \cdot 10^{-6}$ | 0.326 | $2.930 \cdot 10^{-5}$ | 0.304 | $7.622 \cdot 10^{-6}$ | 0.291 |
| LuH$_{0.254}$ | $1.969 \cdot 10^{-8}$ | 0.383 | $1.142 \cdot 10^{-7}$ | 0.373 | $8.193 \cdot 10^{-8}$ | 0.348 |

The Arrhenius-type dependences (with parameters $\{\tau_{0i}\}$ and $\{E_{ai}\}$; see Table I) of the relaxation times $\tau$ (within the framework of the first-order kinetics model) and $\tau_1, \tau_2$ (within the framework of the more realistic second-order kinetics model) on a temperature, $T$, for

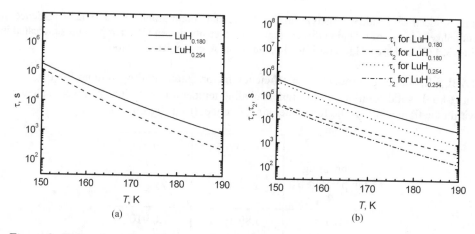

*Figure 2.* Relaxation times, $\tau$, $\tau_1$ and $\tau_2$, on annealing temperature, $T$, within the framework of the first-order (a) and second-order (b) kinetics models for h.c.p.-Lu–H solid solutions.

alloys $LuH_{0.180}$ and $LuH_{0.254}$ are presented in Fig. 2. In any case, the obtained values of activation energies of migration $\{E_{ai}\}$ (0.291–0.383 eV at $T \in$ (160 K, 180 K); Table I) differ from comparatively high-temperature ($T \in$ (380 K, 540 K)) data [23] about diffusion of H atoms in the enough weak solution $LuH_{0.05}$ with $E_a = 0.574 \pm 0.015$ eV (see Refs. in [1]). This result maybe caused by the fact that the H atoms in Lu migrate more easily at lower temperatures, *e.g.*, from about 135 K to 170 K (as in Ref. [24]), due to tunnelling-mechanism diffusion contribution.

From second-moment NMR experiments carried out on $LuH_{0.17}$ between 170 and 420 K, the authors of the study (Ref. [25]) evaluated the activation energy of $\approx 0.28$ eV for H diffusion. A similar values for the activation energies were obtained from resistance-relaxation measurements after quenching at low-temperature electron irradiation [26, 27]. For temperatures around 160 K, the measurements yielded activation energies between 0.22 eV and 0.29 eV for H. These values are not consistent with outcomes ($0.428 \pm 0.003$ eV and $0.433 \pm 0.003$ eV) of Ref. [24], but are more close to our simulated results.

Product of an efficiency factor, $\chi$, and hydrogen jump frequency, $\nu_H$, for $\alpha$-Lu–H solid solutions at different annealing temperatures are presented in Table II. They were calculated within the framework of the kinetic model with one relaxation time (see Eq. (6)) taking into consideration Eq. (1).

Within the framework of the hypothesis about an identity of maximum characteristic relaxation times of diffuse radiation scattering and relaxation times of electrical resistivity, the time evolution of normalized change of radiation diffuse-scattering intensity was predicted and corresponds to the wave-vector star $\{k^*\}$, which dominates in mapping a structure of the short-range order (at different quenching and annealing temperatures) for polycrystalline $LuH_{0.180}$ and $LuH_{0.254}$ solid solutions. Time dependence of normalized change of radiation diffuse-scattering intensity is presented in Fig. 3 (for one and two relaxation times).

It is necessary to note that, using temperature dependences of the SRO-induced relaxation time of electrical residual resistivity in Fig. 2, one can obtain graphs of the time dependence of normalized change of radiation diffuse-scattering intensity not only for temperatures indicated in Figs. 3, 4 (161.8 K, 167.3 K, and 180.2 K), but for another temperatures too.

TABLE II. Values of product of an efficiency factor, $\chi$, and hydrogen jump frequency, $\nu_H$, for $\alpha$-Lu–H solid solutions at different annealing temperatures. Values $\chi\nu_H$ were calculated within the framework of the first-order kinetic model (6) using Eq. (1).

| Annealing temperature $T$, K | LuH$_{0.180}$ $\chi\nu_H$, $\sigma^{-1}$ | LuH$_{0.254}$ $\chi\nu_H$, $\sigma^{-1}$ |
|---|---|---|
| 161.8 | $5.89\cdot10^{-5}$ | $7.83\cdot10^{-5}$ |
| 167.3 | $1.21\cdot10^{-4}$ | $1.86\cdot10^{-4}$ |
| 180.2 | $1.02\cdot10^{-3}$ | $2.68\cdot10^{-3}$ |

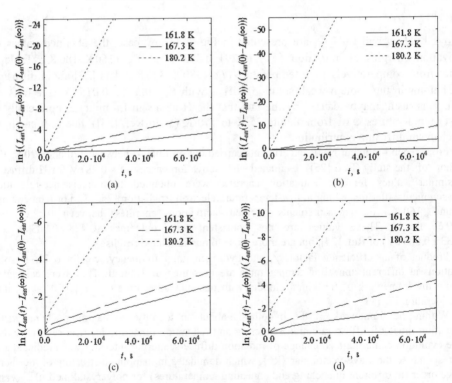

Figure 3. Logarithm of the normalized radiation diffuse-scattering intensity changes versus isothermal-annealing time within the framework of the first-order (a, b) and second-order (c, d) kinetics models for the interstitial solid solutions h.c.p.-LuH$_{0.180}$ (a, c) and h.c.p.-LuH$_{0.254}$ (b, d) at the given annealing temperatures.

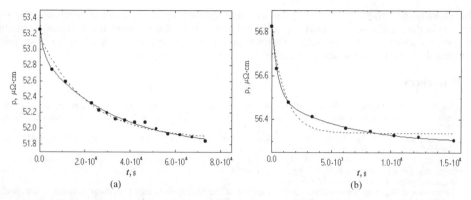

*Figure 4.* Dependence of residual electrical resistivity, $\rho$, on the time of isothermal annealing, $t$, within the framework of the first-order (dashed-line curve) and second-order (solid curve) kinetics models for polycrystalline h.c.p.-LuH$_{0.180}$ solid solution at annealing temperatures 161.8 K (a) and 180.2 K (b) ($\bullet$—experimental data from Ref. [22]).

## 4. Summary

By help of above-mentioned hypothesis on the example of the interstitial solid solution h.c.p.-Lu–H, we attempted to unify direct (diffuse radiation scattering) and indirect (residual electrical resistivity) methods of the SRO relaxation study.

Experimental results obtained in Ref. [22] for $\rho = \rho(t,T)$ in interstitial solid solutions h.c.p.-Lu–H have been described by its authors within the framework of the *power*-law kinetics model (see [3, 4, 22]), which is most probably justified at long times and/or high temperatures. In a given paper, we propose to describe these experimental data (and maybe data of Ref. [24]) within the framework of the alternative first-order and second-order kinetics models, which are justified starting with small times (early stages) when SRO-induced relaxation of $\rho(t,T)$ has the single-*exponential* or two-*exponential* behaviour. Standard deviation of theoretical curve from the experimental data is a criterion of the reliability of above-mentioned kinetic exponential models (see Fig. 4 and Refs. 3, 4).

The parameters ($\{\tau_{0i}\}$ and $\{E_{ai}\}$) of the Arrhenius-type temperature dependence of relaxation times, $\{\tau_i\}$, corresponding to the times required for the migration of hydrogen to change the characteristic SRO value in $e$ times, are evaluated for LuH$_{0.180}$, LuH$_{0.254}$ solid solutions. The values of $\{E_{ai}\}$ revealed here for h.c.p.-Lu–H solutions containing much more amount of H ($c = 0.180$, 0.254 [22]) as compared with such a solid solution ($c = 0.05$) studied in Ref. [23] are clearly smaller than a value from Ref. [23]. This decrease of $\{E_{ai}\}$ values maybe caused by the switching on of the tunnelling-diffusion mechanism, which takes place predominantly at lower temperature [24].

Increase in an efficiency factor, $\chi$, in more concentrated LuH$_{0.254}$ solid solution (in comparison with less concentrated LuH$_{0.180}$) promotes the decrease in relaxation time for this solution (see Table II, Eq. (1), and Table I in Ref. [4]). This result is consistent with results in Ref. [24], where SRO-relaxation time for LuH$_{0.12}$ is less then for LuH$_{0.06}$.

Characterization of residual electrical resistivity (and short-range order) in a given polycrystalline solutions by two relaxation times maybe proved by the different probabilities of jumps of interstitial H atoms along the preferential directions of axes '$a_0$' and '$c_0$' of h.c.p. lattice of any Lu crystallite (even in a case of the absence of macroscopic appearance of anisotropy of electrical resistivity of polycrystalline h.c.p.-LuH$_c$ as a whole).

The time evolution of normalized change of radiation diffuse-scattering intensity is predicted here and corresponds to the wave-vector star $\{\mathbf{k}^*\}$, which dominates in mapping a structure of the short-range order [1–3, 10–13] (at different quenching and annealing temperatures) for polycrystalline $LuH_{0.180}$ and $LuH_{0.254}$ solid solutions. It is done within the framework of the hypothesis about an identity of maximum characteristic relaxation times of diffuse radiation scattering and relaxation times of electrical resistivity.

## 5. References

1. Vajda, P. (1995) Hydrogen in rare earth metals, including $RH_{2x}$-phases, in K.A. Gschneidner, Jr. and L. Eyring (eds.), *Handbook on the Physics and Chemistry of Rare Earths* **20**, Elsevier Science, Amsterdam, ch. 137, pp. 207–292.
2. Fukai, Yuh. (1993) *The Metal–Hydrogen System: Basic Bulk Properties*, in *Springer Series in Materials Science* **21**, Springer-Verlag, Berlin–Heidelberg–New York.
3. Tatarenko, V.A. and Radchenko, T.M. (2002) Direct and indirect methods of the analysis of interatomic interaction and kinetics of a short-range order relaxation in substitutional (interstitial) solid solutions, *Uspehi Fiziki Metallov* **3**, 111–236 (in Ukrainian).
4. Tatarenko, V.A., Radchenko, T.M. (2002) Kinetics of the hydrogen-isotopes short-range order in interstitial solid solutions h.c.p.-*Ln*–H(D,T), in N. Veziroglu et al. (eds.), *Hydrogen Materials Science and Chemistry of Metal Hydrides: NATO Science Series, Series II: Mathematics, Physics and Chemistry* **82**, Kluwer Academic Publishers, Dordrecht, pp. 123–132.
5. Cook, H.E. (1969) The kinetics of clustering and short-range order in stable solid solutions, *J. Phys. Chem. Solids* **30**, 2427–2437.
6. Chen, H. and Cohen, J.B. (1977) *J. Physique* **C7**, 314.
7. Bley, F., Amilius, Z., and Lefebvre, S. (1988) Wave vector dependent kinetics of short-range ordering in $^{62}Ni_{0.765}Fe_{0.235}$, studied by neutron diffuse scattering, *Acta. metall.* **36**, 1643–1652.
8. Bonnet, J.E., Ross, D.K., Faux, D.A., and Anderson, I.S. (1987) Long-range and short-range ordering of hydrogen or deuterium atoms in solid solution in yttrium, *J. Less-Common Metals* **129**, 287–295.
9. Metzger, T.H., Vajda, P., and Daou, J.N. (1985) Debye temperature and static displacements in $LuH_x$ single crystals from energy dispersive x-ray diffraction, *Z. Phys. Chem. Neue Folge* **143**, 129–138.
10. Blaschko, O., Krexner, G., Pleschiutschnig, J., Ernst, G., Daou, J.N., and Vajda, P. (1989) Hydrogen ordering in α-$LuD_x$ investigated by diffuse neutron scattering, *Phys. Rev. B* **39**, 5605–5610.
11. Blaschko, O., Krexner, G., Daou, J.N., and Vajda, P. (1985) Experimental evidence of ordering of deuterium in □-$LuD_x$, *Phys. Rev. Letters* **55**, 2876–2878.
12. Pleschiutschnig, J., Schwarz, W., Blaschko, O., Vajda, P., and Daou, J.N. (1991) Hydrogen ordering and local mode in □-$LuD_x$ at low concentrations, *Phys. Rev. B* **43**, 5139–5142.
13. Blaschko, O. (1991) Hydrogen ordering in metal–hydrogen systems, *J. Less-Common Metals* **172–174**, 237—245.
14. Dahmani, C.E., Cadeville, M.C., and Pierron-Bohnes, V. (1985) Temperature dependences of atomic order relaxations in Ni–Pt and Co–Pt alloys, *Acta metall.* **33**, 369–377.
15. Afyouni, M., Pierron-Bohnes, V., and Cadeville, M.C. (1989) SRO relaxation in dilute <u>Ni</u>Al solid solution through 4 K resistometry, *Acta metall.* **37**, 2339–2347.
16. Sitaud, B. and Dimitrov, O. (1990) Equilibrium and kinetics of local ordering in Ni(Al) solid solutions, *J. Phys: Condens. Matter* **2**, 7061–7075.
17. Kidin, I.I. and Shtremel', M.A. (1961) Kinetics of changes of the short-range order in binary alloys, *Fiz. Met. Metalloved.* **11**, 641–649 (in Russian).
18. Shtremel', M.A. and Satdarova, F.F. (1969) The short-range order in alloys with a body centred cubic lattice (equilibrium and kinetics), *Fiz. Met. Metalloved.* **27**, 396–407 (in Russian).
19. Püschl, W. and Aubauer, H.P. (1980) The kinetics of stable heterogeneous binary systems, *Phys. Status Solidi b* **102**, 447–457.
20. Bessenay, G. (1986) *Doctoral dissertation*, Univ. of Paris VI.
21. Khachaturyan, A.G. (1983) *Theory of structural transformations in solids*, Wiley, New York.
22. Jung, P. and Lässer, R. (1992) Short-range ordering of hydrogen isotopes in lutetium, *J. Alloys & Compounds* **190**, 25–29.
23. Völkl, J., Wipf, H., Beaudry, B.J., and Gschneidner, K.A., Jr. (1987) Diffusion of H and D in lutetium, *Phys. Status Solidi b* **144**, 315–327.
24. Yamakawa, K. and Maeta, H. (1997) Diffusion of hydrogen in Lu at low temperatures, in *Defect and Diffusion Forum* **143–147**, pp. 933–938.
25. Barrère, H. and Tran, K.M. (1971) *C.R. Acad. Sci. Paris* **273B**, 823.
26. Daou, J.N., Vajda, P., Lucasson, A., Lucasson, P., and Burger, J.P. (1986) Low-temperature defect study of □-$LuH(D)_x$ solid solutions, *Phil. Mag. A* **53**, 611–625.
27. Vajda, P., Daou, J.N., Burger, J.P., Kai, K., Gschneidner, K.A., and Beaudry, B.J. (1986) Investigation of the anisotropic deuterium ordering in α-LuD single crystals by resistivity and heat-capacity measurements, *Phys. Rev. B* **34**, 5154–5159.

# INVESTIGATION OF CONTENT OF ENDOMETALLOFULLERENES EXTRACTS

I.E. KAREEV, Yu.M. SHULGA, V.P. BUBNOV, V.I. KOZLOVSKI,
A.F. DODONOV, M.V. MARTYNENKO, K.B. ZHOGOVA,
E.B. YAGUBSKII
*Institute of Problems of Chemical Physics RAS, Chernogolovka,
Moscow region, 142432 Russian Federation
Fax: (096) 514 3244, E-mail: shulga@icp.ac.ru
M.V.Lomonosov Moscow State University, Moscow,
119899 Russian Federation
Branch of Institute for Energy Problems of Chemical Physics RAS,
Chernogolovka, Moscow region, 142432 Russian Federation
Russian Federal Nuclear Center – All-Russia Scientific Research Institute of
Experimental Physics, Sarov, Nizhny Novgorod region,
607190 Russian Federation*

## Abstract

Gases evolved from endometallofullerene (EMF) extracts (M = La or Y) upon heating has been investigated. It has been shown that solvent molecules used to release EMF are observed in extracts up to high temperatures. It has been shown that solvent molecules used in the extraction of EMF remain in extracts up to high temperatures. It has been found that in the dimethylformamide (DMFA) extract the EMF molecules are present mainly as the [EMF]⁻ anions.

*Keywords:* endometallofullerenes, electric arc synthesis, extraction, mass spectrometry, thermogravimetry.

## 1. Introduction

Endohedral metallofullerenes (EMF) are compounds comprising one or more metal atoms inside the fullerene cage. The formation of such compounds is most characteristic of fullerene $C_{82}$ with metals of III group (Sc, Y, La) and lanthanides[1-3]. The isolation of EMF from soot by extraction with various organic solvents is now the most conventional procedure. A set of solvents used to extract EMF is wide enough: toluene, o-xylene[4,5], carbon disulfide[6], N,N-dimethylformamide[7,8], pyridine[9] and others[2,3]. The efficiency of the extraction of EMF by polar solvents possessing high dipole moments is essentially higher than that by other ones. Solid precipitates or extracts formed as a result of solvent evaporation from EMF solutions turned out to contain noticeable amounts of impurities. Possibly major impurities in the extracts are solvent molecules, which form stable solvates with EMF and, therefore, modify their properties[10]. Vacuum heat treatment reduces the solvent share in the extract. However, its full removal is a rather complicated

*T.N. Veziroglu et al. (eds.),*
*Hydrogen Materials Science and Chemistry of Carbon Nanomaterials,* 67-74.
© 2004 *Kluwer Academic Publishers. Printed in the Netherlands.*

task[11,12] and the study of gas evolution from extracts and the determination of their content are urgent now.

## 2. Experimental

### 2.1. PREPARATION OF ENDOMETALLOFULLERENES

Soot containing La@$C_{82}$ or Y@$C_{82}$ was prepared by electric arc evaporation of corresponding metal- graphite electrodes to which La or Y were added[7,13]. The electrodes were evaporated under the following conditions: helium pressure of 120 Torr, current of 90 A, voltage of 28÷30 V, the distance between the electrodes of 5 mm, that between the arc and a cooled reactor wall of 50 mm, and an anode evaporation rate of 1 mm/min.

### 2.2. PREPARATION OF ENDOMETALLOFULLERENE EXTRACTS

EMF was isolated from soot by an extraction method in argon atmosphere in boiling solvents[14].

The product free of empty fullerenes was prepared using a three-step extraction scheme. The first step consisted in that soot was treated with o-xylene, which efficiently removed empty fullerenes ($C_{60}$, $C_{70}$ and higher fullerenes: $C_{76}$, $C_{78}$, $C_{82}$, $C_{84}$) and very weakly dissolved EMF. O-dichlorobenzene (ODCB) used in the second step allows one to remove residual empty fullerenes from soot since it much better dissolves empty fullerenes in contrast to o-xylene. O-dichlorobenzene also partly dissloves EMF. At the third step the soot pretreated by o-xylene and then by o-dichlorobenzene was extracted with DMFA. As a result, we prepared the EMF solution free of empty fullerenes. Further studies were performed using extracts prepared by evaporating solutions in vacuum at ~100° C.

### 2.3. ELEMENTAL ANALYSIS OF EMF EXTRATCS

Analysis of C, H, and M content has been carried out by oxygen burning approximately 10 mg sample under study at 1400 K. Carbon and hydrogen content nas been calculated using data on $CO_2$ and $H_2O$ weights measurement. Metal content has been determined from unburned residual which was presented as metal oxide ($La_2O_3$ or $Y_2O_3$). Nitrogen content was measured be means of so–called gasometrical method. Results are presented in Table 1.

### 2.4. TG ANALYSIS OF EMF EXTRACTS

Thermogravimetric analysis was performed on a "Mettler M3 TG" instrument in argon atmosphere at temperatures ranging from 25 °C up to 950 °C, a heating rate was 10 °C/min, a sample weight was 10 mg.

### 2.5. MASS SPECTROMETRY

The spectra of gases evolved upon heating the samples were measured from m/z=1 up to m/z=205 using a MI 1201B mass-spectrometer. Gas was ionized by an electron impact (electron energy was 70 eV). Positively charged ions were registered.

The gas phase was prepared as follows: the EMF extract (20 mg) was put at room temperature in a quartz ampoule preliminarily annealed in air at 700°C for 5 hours. The ampoule was then put in a pyrolysis device and hermetically connected to inlet syatem of the mass-spectrometer. The accuracy of temperature maintenance was equal to 10 degrees. Prior to mass-spectrometric measurements the ampoule was evacuated down to $\sim 1 \times 10^{-7}$ Torr with diffusion mercury pumps, which prevented from peaks attributed to hydrocarbons in residual mass-spectra. The evacuated ampoule was heated up to required temperature at which the collection of gas was accomplished and mass-spectrometric analysis was done.

Mass-spectrometric analysis of the solutions of EMF was performed also using an electrospray time-of-flight high resolution mass-spectrometer with an orthogonal injection of ions[15], which allows the study of solutions containing ionic compounds in "mild" conditions without additional ionization and ion fragmentation. The solution of the DMFA-extract of EMF in o-dichlorobenzene or DMFA diluted by acetonitrile (5:1) was loaded to an ionic source through a metallic capillary of an internal diameter of 0.1 mm at a flow rate of 2 mcL/min. The voltage of 2 kV applied to the capillary provided a fine-dispersed sputtering of the analyzed solution and the final step was a field isolation of ions contained in the liquid. A portion of the ions was loaded to the mass-analyzer through a vacuum interface with differential pumping-out.

## 3. Results and Discussion

Fig.1 shows the electrospray mass spectra of the solutions of extracts under study. It is seen that the basic components of the DMFA extracts are the $La@C_{82}^-$ and $Y@C_{82}^-$ anions. The spectra also manifest peaks attributed to the $M@C_{80}^-$ and $M_2@C_{80}^-$ anions. The formation of EMF anions is probably associated with the reduction of neutral EMF by dimethylamine, which are formedas a result of the decomposition of DMFA in the course of repeated cyclic extraction in boiling DMFA. This assertion is supported by that the addition of aqueos solution of dimethylamine to the EMF-containing soot plased in THF at room temperature results in its partial dissolution due to the formation of the EMF anions.

Fig.2 shows the curve of weight losses upon heating the La-extract in the TGA regime. The curves were similar for all the samples and, only one of them is presented. It is seen that the sample weight monotonically decreases. However, there are four particular portions in the curve. Heating up to 100°C can be considered as a preliminary one when unintentional weakly coupled impurities, which could be sorbed on the sample during its short contacting with ambient atmosphere, are desorbed. At 100°C - 200 °C one observes a large enough weight loss that is clearly pronounced in the derived curve (Fig.1). The line at 230-400° C also has a local minimum in the derived curve. It could be assumed that below 400°C the solvent molecules, which are most bound to the EMF molecules are removed. At temperatures higher than 400° C the curve of weight loss is more flat. However, the weight loss is still noticeable.

All the solvents used in the extraction procedure described above have boiling points lower than 200°C and, therefore, provided the absence of strong bonds with the EMF molecules, the solvent ones had to remove from the sample at temperatures less than 200°C.

70

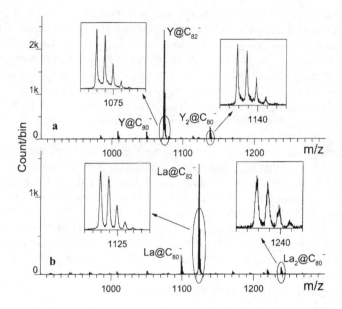

Figure 1. Electrospray mass-spectra of EMF extract solutions

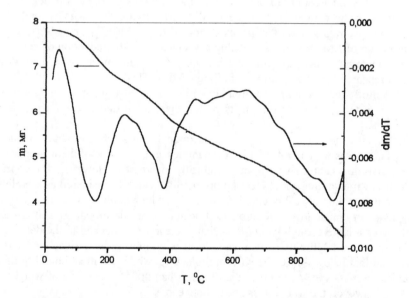

Figure 2. Characteristic thermogravimetric curve of endometallofullerene extract

Fig.3 shows the mass-spectra of gases evolved from the EMF extracts in the above mentioned temperature ranges. It should be noted that the metal located inside the fullerene cage does not affect the character of the spectra.

It was found that the spectra 1-3 contain well-pronounced peaks attributable to the presence (in gas phase) of all three solvents used in extraction! One could assume that $o$-xylene used, as it was mentioned above, to remove the major part of empty fullerenes does not form strong solvates with EMF.

Nevertheless, up to 500°C one could observe peaks in the range close to that of molecular ions of $o$-xylene (the most intense peak is that with m/z = 106).

Fig.4 shows measured intensities of the peaks, which are stipulated mainly by molecular ions of the solvents used in the experiments. It is seen that the maximal content of the solvents in gas phase over the sample falls in the 100 - 230° C range.

The peak intensity with m/z = 15 monotonically grows with temperature (Fig.5). The lack of synchronous changes in peak intensities with m/z = 15 and 73 seems to indicate that the peak with m/z = 15 is formed only as a result of fragmentation of the $N,N$-dimethylformamide molecules. A similar behavior of the peaks with m/z = 15 and 16 (Fig.4) and low intensity of that with m/z = 32 ($O_2$) and other peaks, which could be attributed to oxygen containing molecules, indicates that the peaks with m/z = 15 and 16 correspond to methane molecules. At 400 - 550° C the sample evolves $O_2$ as well the peak with m/z = 2, Fig.3). The lack of the peaks with m/z = 15 and 2 in background spectrum indicates that the $CH_4$ and $H_2$ molecules are formed as a result of the decomposition of carbon containing substances involved in the samples. Since hydrogen containing compounds were not used in the preparation of soot, hydrogen and methane are possible formed as a result of pyrolysis of solvent molecules, which are very strongly bound to EMF. The weight losses at temperatures higher than 400° C are probably associated with the decomposition of the EMF anionic complexes. The origin of ions in these complexes has not been clarified unambiguously. complexes. The origin of a counterion in these complexes has not been clarified yet. It is most possibly that a cation of dimethyamine, $N^+H_2(CH_3)_2$ performs as a counterion and the increase in hydrogen and methane content in the pyrolysis products at 400-500°C is due to the decomposition of these cations.

## 4. Conclusions

It was found that the solid extract prepared using the above described three-step extraction scheme involves EMF donor-acceptor complexes with solvent molecules used in the extraction procedure. The molecules of the solvents used in the isolation and purification of EMF remain in the extract up to high temperatures ($\geq$400 °C).

## 5. Acknowledgements

The work was supported by Russian Foundation for Basic Research (grants ##02-03-33352 and 03-03-06164), ISTC (project # 2511) and INTAS Young Scientist Fellowship # YSF 2002-327/F3.

72

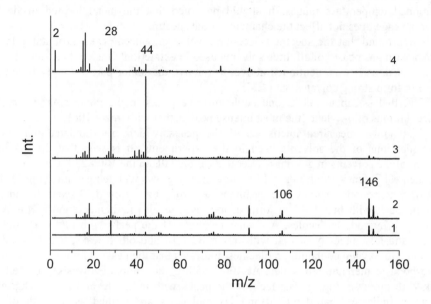

Figure 3. Mass-spectra of gas phase over EMF extract. The gas phase was collected at 20-100°C (1); 100-230 °C (2); 230-400 °C (3); 400-550 °C (4)

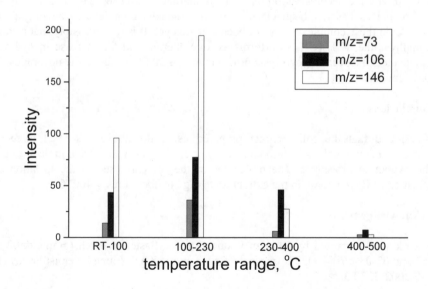

Figure 4. Dependency of peak intensities (m/z = 73, 106 and 146) on extract heating temperature

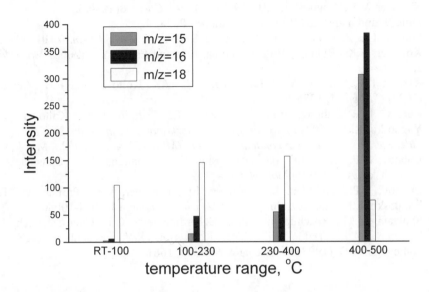

Figure 5. Dependency of peak intensities (m/z = 15, 16 and 18) on extract heating temperature

## 6. Tables

Table 1. Results of elemental analysis of EMF extractes*

| sample | Metal, wt.% | Carbon, wt.% | Hydrogen, wt.% | Nitrogen, wt.% |
|---|---|---|---|---|
| La-contained extract | 9.4 ÷9.7 | 83 ÷84 | 1.0 ÷1.5 | 2.2 ÷2.3 |
| Y-contained extract | 6.5 ÷6.8 | 85 ÷86 | 0.9 ÷1.4 | 2.2 ÷2.4 |

* only DMFA was used for extraction of EMF from soot

## 7. References

1.  Bethune, D.S., Johnson, R.D., Salem, J.R., de Veles, M.S. and Yannoni, C.S. (1993) *Nature* **336**, 123.
2.  Eletskii, A.V. (2000) *Uspekhi Fizicheskikh Nauk* **170**, 113.
3.  Shinohara, H. (2000) *Rep. Prog. Phys.* **63**, 843.
4.  Bartl, A., Dunsch, L., Kirbach, U. and Schandert, B. (1995) *Synthetic Metals* **70**, 1365.
5.  Bubnov, V.P., Koltover, V.K., Laukhina, E.E., Estrin, Y.I., Yagubskii, E.B. (1997) *Russ. Chem.Bull.* **46**, 240.

6. Lian, Y., Shi, Z., Zhou, X., He, X. and Gu, Z. (2000) *Carbon* **38**, 2117.
7. Laukhina, E.E., Bubnov, V.P., Estrin, Y.I., Golub, Y.A., Khodorkovskii, M.A., Koltover, V.K., Yagubskii, E.B. (1998) *J. Mater. Chem.* **8**(4), 893.
8. Ding, J. and Yang, S. (1996) *Chem. Mater.* **8**, 2824.
9. Sun, D., Liu, Z., Guo, X., Xu, W. and Liu, S. (1997) *J. Phys. Chem. B* **101**, 3927.
10. Koltover, V.K., Bubnov, V.P., Laukhina, E.E. and Estrin, Y.I. (2000) *Mol. Mater.* **13**, 239.
11. Kareev, I.E., Bubnov, V.P., Laukhina, E.E., V.K. Koltover and E.B. Yagubskii (2003) *Carbon* **41**, 1375.
12. Kareev, I.E., Bubnov, V.P., Laukhina, E.E., Fedutin, D.N., Koltover, V.K., Yagubskii, E.B. (2002) *Proceedings of II International symposium «Fullerenes and fullerene-like structures in condensed media» (Minsk, June 4-8, 2002.)*, Minsk, p.97.
13. Bubnov, V.P., Krainskii, I.S., Yagubskii, E.B., Lauhkina, E.E., Spitsina, N.G., Dubovitskii, A.V. (1994) *Mol. Mater.* **4**, 169.
14. Bubnov, V.P., Laukhina, E.E., Kareev, I.E., Koltover, V.K., Prokhorova, T.G., Yagubskii, E.B. and. Kozmin, Y.P. (2002) *Chem. Mater.* **14**, 1004-1008.
15. Shulgam, Y.M., Roschupkina, O.S., Dzhardimalieva, G.I., Chernushevich, I.V., Dodonov, A.F., Baldokhin, Y.V., Kolotyrlin, P.Y., Rosenberg, A.S. and Pomogailo, A.D. (1993) *Russ. Chem. Bull.* **42**, 1661.

# AN OVERVIEW OF HYDROGEN STORAGE METHODS

V.A.YARTYS[1] and M.V.LOTOTSKY[1,2]
[1]Institute for Energy Technology, POB 40, N–2027, Kjeller, Norway
[2]Institute for Problems of Material Science of National Academy of Sciences
of Ukraine, Lab.#67, 3 Krzhyzhanovsky Str, 03142, Kiev, Ukraine
e-mail Volodymyr.Yartys@ife.no

## Abstract

Hydrogen is an attractive, pollution-free energy carrier, which is characterised in addition by a flexible and efficient energy conversion. Technology of hydrogen production is well developed and has advantage of practically unlimited basis of the raw materials. However, low density of hydrogen gas, low temperature of its liquefaction, as well as high explosive risk in combination with its negative influence on the properties of design materials bring to the forefront the problems of the development of effective and safe hydrogen storage systems. Namely these problems suppress at present the development of hydrogen power engineering and technology.

This paper reviews the existent and prospective hydrogen storage methods, observes the dynamics of their development in the last three decades. The special attention is paid to metal hydrides, both as to their direct usage in the hydrogen storage systems and concerning the maintenance of the alternative methods, to improve their efficiency.

On the basis of the analysis of available data, it is concluded that the best competitive position in the future will have combined systems realising several methods of hydrogen storage and processing. The methods could include small- and medium-scale metal hydride hydrogen storage units, as well as thermal sorption compressors providing gas-cylinder and cryogenic hydrogen storage systems. Buffer units for hydrogen storage, purification and controlled supply to a consumer could be in demand as well.

## 1. Introduction

The deficit of natural resources of hydrocarbon fuels, in combination with global environment problems, stipulated a great interest to use hydrogen as a universal synthetic energy carrier for stationary and mobile applications. Such an approach is caused by unlimited basis of the raw materials for hydrogen production, the high energy output of hydrogen, technological flexibility and environment safety of the energy conversion processes using hydrogen.

The concept of hydrogen energy system, proposing the transfer of power engineering, industry and transport to hydrogen, was born in mid 1970s, against a background of the worldwide oil crisis [1–4]. The changing in the state of world energy supply market in the second half of the 1980s resulted in some decrease of growth rates in hydrogen R&D activities. First of all, it was caused by economic factors, mainly by high price of commercial hydrogen as compared to conventional energy carriers. As a conclusion, the main attention was paid at that time to the improvement of technical and economic indices of hydrogen production, as well as to the increase of efficiency of its consumption [5]. At the same time, the problems of hydrogen storage and processing

T.N. Veziroglu et al. (eds.),
Hydrogen Materials Science and Chemistry of Carbon Nanomaterials, 75-104.

remained in the shadow. However, since the moment when the factor of environment hazard began to be introduced into economic calculations, the ecological cleanness of hydrogen made its usage in a number of production processes potentially profitable [6]. Simultaneously, hydrogen storage problems and transportation became of special topicality.

The present status of hydrogen energy and technology in the world is characterised by the stable growth rates which are mainly conditioned by environment factor. The corresponding activities are mainly financed from state and international public budgets. The Hydrogen Program of the USA Department of Energy (DOE) [7,8], the Program of International Energy Agency (IEA) [9,10] and a number of others are the typical examples. Apart from state and non-government public organisations, a number of commercial companies, mainly from the motor-car industry [11,12], invests the hydrogen energy R&D activities. The main priorities in these investments are hydrogen fuel cells, as well as improvement of existent and creation of new hydrogen storage methods.

Utilization of hydrogen as a fuel or energy carrier (Table 1) is based on the highly exothermic reaction of hydrogen with oxygen. When hydrogen is produced by water electrolysis, and the efficiency of an electrolyser is of 60–75%, the real power consumption for the production of 1 $m^3$ (STP) of hydrogen gas is 4 to 5 kW·h. When 1 $m^3$ (STP) of hydrogen gas is burned in a power installation with the efficiency of 15–20%, the yield is 0.45 to 0.6 kW·h. Similarly, the yield of a fuel cell with the efficiency of 40–60% is 1.2 to 1.8 kW·h/$m^3$ $H_2$. Therefore, the total efficiency of a hydrogen energy cycle is 10 to 15% when a heat engine is used as a final consumer, and 24 to 45% if hydrogen is utilised electrochemically. The advantages of hydrogen as a fuel include its high heating value, completeness of combustion practically in the whole range of "fuel / oxidant" ratios, high flame temperature, high heat efficiency (for hydrogen fuelled internal combustion engines it is 50% higher than for petrol fuelled ones), and the absence of air pollutants emissions.

The specific (per weight unit) energy value of hydrogen is much more than the one for all natural or synthetic fuels, at the lowest environment pollution (Fig. 1). At the same time, hydrogen is the lightest known substance, so it is problematic to store it in small containers. The main disadvantage of hydrogen is too low volumetric energy density, because at usual conditions it exists in the gaseous form having very low boiling point (–252.8°C) and critical temperature (–239.96 °C) [19]. 1 kilogram of hydrogen gas at room temperature and atmospheric pressure has volume of 11 $m^3$, so as to provide 100 km run for the electric car with fuel cells it is necessary to have on board about 33 $m^3$ of gaseous hydrogen [20]. Thus the development of the effective methods of compact storage of hydrogen is a key problem of its utilization.

Nowadays the annual hydrogen production is about 50 billions $m^3$ (STP) or 45 millions tons. Despite of wide hydrogen usage in the industry (main consumers are ammonia and methanol production, as well as petrochemical industry), the fraction of marketable hydrogen is only 5% of its total production [21,22]. Such low value is caused by hydrogen storage and transportation problems, so as the large consumers prefer to have their own hydrogen production facilities, rather than to buy hydrogen from specialised producers [5,13].

The problem of hydrogen storage is closely connected to the necessity of development of reliable and effective methods of hydrogen compression. Hydrogen compressors having wide spectrum of technical characteristics are also in great demand for chemical industry, metallurgy, cryogenic engineering, air-space industry and other fields.

As it was mentioned earlier, hydrogen production by water electrolysis is rather power-consuming process. The less power-consuming methods of hydrogen production by coal or hydrocarbons conversion [5,13] are accompanied by $CO_2$ emissions. Moreover,

TABLE 1. Characteristics of hydrogen as a fuel or energy carrier [13–18]

| Theoretical energy equivalent for the reaction: $2 H_2 + O_2 \rightarrow 2 H_2O$ | | MJ / kg $H_2$ | 120.6 |
|---|---|---|---|
| | | kW·h / kg $H_2$ | 33.5 |
| | | kW·h / $m^3$ $H_2$[(1)] | 3.00 |
| Electrochemical decomposition potential of pure water to $H_2 + O_2$, V[(2)] | | | 1.24 |
| Heating value, MJ / kg $H_2$ | | Lower | 114.5–119.9 |
| | | Higher | 135.4–141.8 |
| Hydrogen content in the stoichiometric combustible mixture, vol.% | | Hydrogen – air | 29.53 |
| | | Hydrogen – oxygen | 66.67 |
| Temperature of self-ignition of the combustible mixture, °C | | Hydrogen – air | 510–580 |
| | | Hydrogen – oxygen | 580–590 |
| Flame temperature, °C | | $H_2$ – air, stoichiometric mixture | 2235 |
| | | $H_2$ (73.0 vol.%) – $O_2$ | 2525 |
| Concentration limits of hydrogen burning in the mixture, vol. %: | | Hydrogen – air | 4.1–72.5 |
| | | Hydrogen – oxygen | 3.5–94.0 |
| Concentration limits of detonation in the mixture, vol. %: | | Hydrogen – air | 13–59 |
| | | Hydrogen – oxygen | 15.5–93 |
| Maximum flame propagation speed, m/s | | Hydrogen – air | 2.6–3.4 |
| | | Hydrogen – oxygen | 9–13.5 |
| Extinguishing distance for hydrogen – air stoichiometric mixture, mm[(2)] | | | 0.6–0.8 |
| Minimum energy for the ignition of stoichiometric mixture $H_2 + O_2$, mJ | | | 0.02 |

Notes:  [(1)] – at normal conditions: P=1.01308 bar, T=0 °C
[(2)] – P=1.010085 bar, T=25 °C

hydrogen produced by these methods is much less pure (main impurities are $CO_2$ and CO) than one produced by electrolysis. At the same time, there exist the significant additional resources to produce hydrogen by its extraction from waste and technological gases of chemical, petrochemical and by-product-coking industries, and metallurgy. These resources can be big enough: according to the data of the Institute of Mechanical Engineering Problems of National Academy of Science of Ukraine for 2000 [23], in the Ukraine, only on the chlorine and fertilisers plants, the annual losses of hydrogen were of 11.8 thousands tons. On the Ukrainian by-product-coking plants 350 thousands tons of commercial hydrogen which could be produced from coke-oven gas were lost. The other chemical enterprises had significant resources as well. The possibility to use the similar resources in hydrogen energy and technology directly depends on the solution of the problem of selective hydrogen extraction from gas flows and related problem of hydrogen purification.

The above-mentioned problems of hydrogen storage, compression and extraction / purification are far from being the complete list of the problems of hydrogen energy-technological cycle. At the same time, namely these problems determine the possibility of its successful implementation on the stage which connects hydrogen production and end-user consumption. The most difficult here is to solve the hydrogen storage problem. Below we present the more detailed review of the actual state of the art, including both commercial hydrogen storage methods and the prospective ones, being on the stage of technological developments.

78

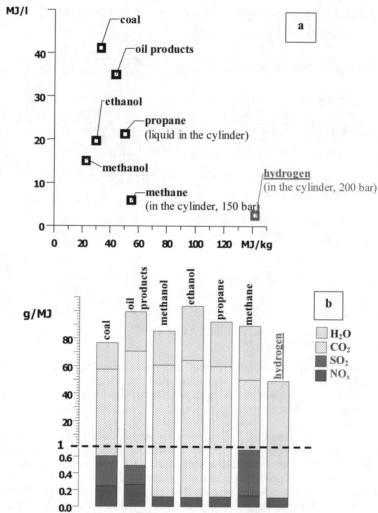

*Figure 1. Specific heating value (a) and emissions of combustion products (b) for natural and synthetic fuels [5,13–17]*

## 2. Methods of hydrogen storage

The requirements to hydrogen storage systems first of all are determined by the character of corresponding applications. So, the low total weight is important for mobile application, but it is not so critical for stationary systems, e.g. filling stations. Similarly, the low volume is essential for small transport, but it is less important for stationary applications. A fuel cell car with the peak hydrogen consumption of 1–3 g/s to provide necessary acceleration, has to have about 5 kg $H_2$ on board (is equivalent to 600 MJ) to have the same running distance as a petrol-consuming car [11]. DOE [7,8] established the goals for on-board hydrogen storage systems which require the weight H storage capacity not less than 6.6%, and the volumetric one not less than 63 g/l. According to DOE requirements, the mobile hydrogen storage systems must also be effective as to their cost / performances relation: the reversible hydrogen storage capacity must be more than 75% of the total one, and the storage cost must be less than 50% of the cost of stored hydrogen. The self-discharge time, or dormancy, should be less than one month [11]. The motor-car H storage system has to operate at the temperature not more than 100 °C having the good dynamics of hydrogen input / output, especially in the course of its recharging. The mechanical strength (resistance to a damage when accidental strike occurs) is obligatory. It is also desirable that the system could operate at moderate pressures, would have low heat losses, and would be serviceable if charged with hydrogen contaminated with impurities of oxygen, water vapours, traces of methane, carbon dioxide and carbon monoxide.

According to DOE classification [7], hydrogen storage methods can be divided into two groups:

The first group includes **physical** methods which use physical processes (compression or liquefaction) to compact hydrogen gas. Hydrogen being stored by physical methods contains $H_2$ molecules which do not interact with a storage medium. Now the following physical methods of hydrogen storage are available:

> Compressed hydrogen gas:
  • Gas cylinders;
  • Stationary storage systems including underground reservoirs;
  • Hydrogen storage in pipelines;
  • Glass microspheres.
> Liquid hydrogen: stationary and mobile cryogenic reservoirs.

**Chemical** (or physical-chemical) methods provide hydrogen storage using physical-chemical processes of its interaction with some materials. The methods are characterised by an essential interaction of molecular or atomic hydrogen with the storage environment. The chemical methods of hydrogen storage include:

> Adsorption:
  • Zeolites and metal-organic adsorbents;
  • Active carbon;
  • Carbon nanomaterials.
> Bulk absorption in solids (metal hydrides).
> Chemical interaction:
  • Liquid (organic) hydrides;
  • Fullerenes;
  • Ammonia and methanol;
  • Water-reacting metals (sponge iron, Al and Si-based alloys).

At present these methods are on the stage of R&D or pilot industrial implementation.

This Section considers the hydrogen storage methods which are alternative as to metal-hydride one.

## 2.1. STORAGE OF THE COMPRESSED HYDROGEN GAS (CGH$_2$)

### 2.1.1. High-pressure cylinders

Hydrogen storage in high-pressure cylinders is the most convenient and industrially-approved method. Usually, steel gas cylinders of the low (up to 12 l) or medium (20 to 50 l) capacity are in use for the storage and transportation of the moderate quantities of compressed hydrogen at the environment temperature from −50 to +60 °C. The industrial cylinders of larger capacity, up to several cubic meters, are produced as well. Gas pressure in the commercial cylinders is 150 atmospheres for the countries of the former USSR and

200 bar for West Europe and USA. Even for cylinders of large volume, the weight hydrogen storage capacity does not exceed 2–3% [13].

The main way to improve the characteristics of hydrogen storage in cylinders is to reduce cylinder weight by the usage of lighter metals and / or composite materials, and to increase the working pressure. The progress in material science since mid 1970s called for the development of new generation of composite gas cylinders (Fig. 2) consisting of thin aluminium or plastic shell which is covered outside by composite plastic being reinforced by fibre glass wrapped around the shell [7,24–27]. Such cylinders are commercially available now, mainly for motor-car applications. The cylinders provide hydrogen storage under the pressure up to 350 bar. The further improvement of the material and layout of the high-pressure cylinders (mainly due to replacement of fibre glass by carbon fibres) allowed to create the advanced composite gas cylinders rated to the pressure up to 690 bar. The weight and volume efficiency of the advanced cylinders are 80% higher than for the commercial ones, but they are more expensive. At the same time, the advanced composite cylinders have the essential market potential at the expense of reducing transportation costs.

The main characteristics of the composite gas cylinders for hydrogen storage are presented in the Table 2.

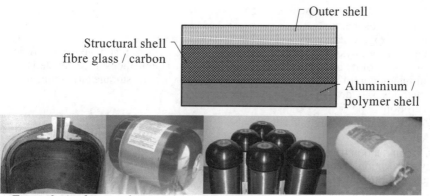

*Figure 2. A schematic layout of the wall of composite high-pressure hydrogen cylinders and general view of some of them [24–27]*

The advantage of hydrogen storage in cylinders is the simplicity in realisation and absence of power consumption to supply hydrogen. The composite cylinders allows to reach the high enough hydrogen storage weight capacity which does satisfy to DOE criterium for mobile applications. Although hydrogen compression is considerably power-consuming process (1 to 15% of hydrogen heating value), it does not sharply increase with the pressure, because the power consumption is proportional to the logarithm of compression ratio (Fig. 3).

Because of potential explosion risk for the high-pressure hydrogen, safety is the main problem of hydrogen storage in gas cylinders. Besides, hydrogen compression itself is a complicated enough engineering problem. This increases the final costs for the gas cylinder storage method. Volumetric density of hydrogen, even in the best in this respect cylinders (up to 40 g/l at 690 bar), is, nevertheless, not high enough: it is only 60% from DOE target value [7].

*2.1.2. Large-volume stationary hydrogen storage systems and pipelines*
These methods are related to the storage of compressed hydrogen. Usually hydrogen pressure therein is less than in the specially designed gas cylinders, but the volume of such systems is much more. The storage capacity of the systems can be up to several millions of cubic meters of hydrogen, so they are the single alternative for the large scale hydrogen storage [13].

For underground storage of the large volumes of hydrogen the exhausted oil and gas wells, or some other natural or artificial objects (e.g. salt rocks or cave formations) can be used [30,31]. This method is mainly applied for the storage of natural gas, but it also has been used for helium storage in Texas (~300 millions m$^3$), that has shown its feasibility to store gases with the increased leak ability [13]. The optimal way is in the usage of not open caves but volumes of the porous rock, which allow the gas to seep in. To prevent leakage, the similar porous rock, which surrounds the storage and is saturated by water or brine, is commonly used.

TABLE 2. Characteristics of the composite gas cylinders for hydrogen storage [24–27]

| | Commercial | Prototypes |
|---|---|---|
| Pressure, bar | 200–250 | 350–690 |
| Cylinder specific weight, kg/l | 0.3–0.5 | 0.15–0.20 |
| Weight hydrogen density, % | 5–7 | 10.5–13.8 |
| Volumetric hydrogen density, g/l | 14–18 | 22–40 |
| Producers | Dynetek (Canada, Germany), Quantum (USA), BOC (UK), etc. | |

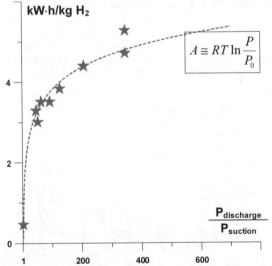

$$A \cong RT \ln \frac{P}{P_0}$$

*Figure 3. Power consumption of hydrogen compressors for filling the cylinders [28,29]*

At present large-scale underground hydrogen storage systems are absent in the world, but in the past the coal gas containing significant amount of hydrogen was stored in the rock formations both in Germany and in France.

The main disadvantage of the large scale hydrogen stores is the necessity to leave up to 50% of the gas in the system, to keep it serviceable. This so-called cushion gas is not

82

available for use, and the losses caused by it are about 25% of the total annual capital costs. A secondary loss (1 to 3% of the contents of storage system) comes from hydrogen leaks [30,31].

Apart from the possibility of creation of very large-scale underground hydrogen storage systems, there exist industrial compressed hydrogen storage facilities which are usually large spherical containers rated for the pressure 12–16 bar having volume 15000 cubic meters and more. Moreover, the distribution pipelines network (pressure 35–75 bar) can be also used as a storage system to take up peaks in supply. This storage technology is widely used in the natural gas industry and can be easy converted to hydrogen [30].

## 2.2. LIQUID HYDROGEN (LH₂)

Nowadays the technology of hydrogen liquefaction and liquid hydrogen storage is well developed. Cryogenic vessels having screen-vacuum heat isolation allow one to reach maximum weight capacity as compared to the other hydrogen storage methods. It is a reason for a preference of liquid hydrogen storage in transport (especially aerospace) applications. Significant recent developments have resulted in a creation of highly efficient cryogenic tanks, fuelling infrastructure, as well as in the improvement of safety for the all complex of systems providing liquid hydrogen storage.

Fig. 4 shows the comparative data on energy content in liquid and compressed hydrogen under different pressures. For a $LH_2$ tank system, the typical nominal operational pressures are in the range from 1 to 3.5 bar. Grading for $CGH_2$ considers 240 bar as a common pressure today and 350 bar common in the near future. Pressures up to 700 bar reflect an envisaged technical goal. As can be seen, the specific energy content of $LH_2$ is 1.6–4 times higher, even as compared with the existing prototypes of 700 bar composite pressure cylinders. In practice, this advantage is even more pronounced, due to easier up-scaling of the cryogenic $LH_2$ storage units. So, a 40 tons trailer can carry 6.4 times more hydrogen than the same trailer loaded with compressed hydrogen gas in the cylinders [27,32]. The more pronounced scale effect of cryogenic $LH_2$ storage units is in the space applications where hydrogen storage capacity of ~86 wt.% is achieved. So, the $LH_2$ tank with the propellant for the main engines of Space Shuttle contains 84800 kg of liquid hydrogen under the pressure of 2.2–2.3 bar; the total volume of the tank is 1516 $m^3$ and its dry weight is only 13150 kg [33].

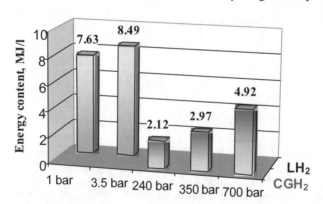

*Figure 4. Energy content in the liquid (LH₂) and the compressed gaseous (CGH₂) hydrogen at different pressures [32]*

Apart from large- and medium-scale hydrogen storage units, some test prototypes demonstrating automotive applications of liquid hydrogen storage have been successfully created. These prototypes include both LH₂-fueled cars and cryogenic filling stations. An example is the NECAR 4 car with $LH_2$ at Munich airport having the following characteristics [34]:

- Mercedes A model
- Range            450 km
- Speed            145 km/h
- Fuel             $LH_2$ 100 l = 5 kg, 9 bar; Linde AG
- Evaporation rate    1% /day
- H volume density:    50 g/l

Table 3 shows comparative information about performances of $LH_2$ and $CGH_2$ hydrogen storage methods as applied to 6.4 kg of stored hydrogen (target value for the hydrogen-fueled car). As it can be seen, the main problem of liquid hydrogen storage is in too high power consumption, up to one-third of hydrogen heating value. Fig. 5 shows our estimations of power inputs for an ideal cooling machine to cool hydrogen from the room temperature. We roughly assumed the constant specific heat capacity of hydrogen taken from the reference [13]. A comparison of these data with our estimations of power consumption for hydrogen compression (Fig. 3) shows that the value of mechanical work to cool hydrogen from 300 K down to boiling point (20 K) more than 5 times exceeds the work to compress hydrogen gas from 1 to 700 bar at room temperature. Taking into account that the efficiency of the modern hydrogen compressors [28,29] can be up to 50%, we can conclude that even at very high efficiency of hydrogen liquefaction process (about 80%, starting from the data of Table 3), power inputs for hydrogen liquefaction would exceed ones for hydrogen compression more than twice. It makes the method too expensive and competitive only in the some special cases, like aviation and space industry.

TABLE 3. Comparison of $LH_2$ and $GH_2$ on a basis of 6.4 kg of stored hydrogen [27,32]

| Performance | | $LH_2$: vacuum-insulated cryostat, P=1 bar | $CGH_2$: Commercial composite cylinder. P=200 bar |
|---|---|---|---|
| System volume, litres | | 110 | 495 |
| System weight, kg | | 86 | 250 |
| Energy consumption | MJ | **254** | 136 |
| | % of $H_2$ heating value | **28** | 15 |
| | Estimated "ideal" value, MJ | **~200** | ~45 |

Apart from extremely high power consumption, cryogenic $LH_2$ storage is characterised by evaporation losses that may occur during fuelling $LH_2$ tanks and during periods of inactivity (dormancy), due to heat penetration from environment. Hydrogen losses from cryogenic storage unit depend on a container design and volume, material of heat isolation, and now can be as low as 10% per a year for large-scale storage units (thousands of $m^3$). However, the lower-scale units are characterised by larger losses, so as the storage time of $LH_2$ in an on-board cryogenic tank on a car is limited approximately by 1 week [11]. Efficient usage of liquid hydrogen fuel by a car with 17 km / litre vehicle can be achieved only if the daily driving distance exceeds 100 km [35].

## 2.3. CRYOGENIC PRESSURE CONTAINERS

One of the recent developments, which seems to be a prospective hydrogen storage method combining advantages of both compressed and liquid $H_2$ storage, is in the development of cryogenic pressure containers [35,36]. The main idea here is in usage of commercially available pressure vessels for liquid hydrogen storage. The realisation is in combination of such a vessel (aluminium-lined glass- or carbon-fibre-wrapped) placed inside cryogenic heat insulation shell. The storage unit can be loaded either with liquid hydrogen or with compressed one.

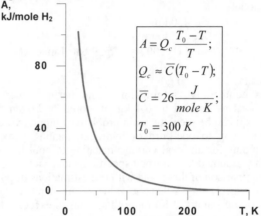

$$A = Q_c \frac{T_0 - T}{T};$$
$$Q_c \approx \overline{C}(T_0 - T);$$
$$\overline{C} = 26 \frac{J}{mole\ K};$$
$$T_0 = 300\ K$$

*Figure 5. Estimated power inputs ("ideal" mechanical work) for hydrogen cooling.*

Periodic cooling of the pressure vessel does not worsen its strength performances, moreover its burst pressure even increases, may be due in part to work hardening that took place during the cold cycling.

In our opinion, further prospectives in the development of this method is in combination of high hydrogen pressure with cooling down to liquid nitrogen temperatures. Indeed, as it can be seen from the Fig. 5, a sharp increase of power inputs for hydrogen cooling is observed at the temperatures below 50 K, so as the less-deep chilling (e.g. to the boiling point of liquid nitrogen which is produced by the industry in large quantities for moderate prices) requires much lower power inputs. From the other hand, a combined cooling and compression allow to reach rather high hydrogen volumetric densities, even more than for liquid hydrogen at P=1 bar. Our corresponding estimations based on hydrogen P – V – T reference data [13,19] are shown in the Fig. 6.

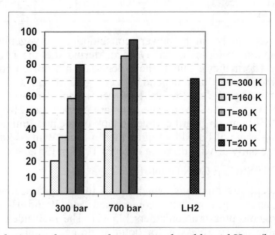

*Figure 6. Volumetric densities of compressed and liquid $H_2$, g/l.*

## 2.4. TECHNOLOGY DEVELOPMENTS IN HYDROGEN STORAGE

Here we briefly discuss some other hydrogen storage methods being now at the R&D stage. The commercial potential of some of these methods is not clear yet, the others seem not to be market-competitive and are mostly of academic interest. Nevertheless, their consideration would be useful to have a more complete presentation of the state of the art in hydrogen storage.

### 2.4.1. Encapsulated hydrogen storage (glass micro-spheres)

Hydrogen storage in glass micro-spheres appeared as a side result of R&D activities in laser-induced nuclear fusion. The developed technology allows to produce hollow glass spheres, 5–500 microns in diameter and ~1 micron in wall thickness, by using a process of spraying glass gels [7,13]. The glass is permeable for hydrogen at moderate temperatures. Because of that, glass micro-spheres can be filled with hydrogen. Hydrogen loading requires to apply temperatures of 200–400 °C and pressures 350–630 bar; loading is completed in approximately 1 hour. The weight hydrogen fraction in the micro-spheres can reach 10 wt.% which corresponds to hydrogen volume capacity of 20 $g/dm^3$. The energy consumption in this method (for hydrogen compression and heating of the micro-spheres during hydrogen charge and discharge) is higher than for the compressed $H_2$. The main disadvantage of this method is in very high losses of hydrogen (up to 50 % during 100–110 days), because of mechanical destruction of the micro-spheres (for example, during transportation) and, also, due to the hydrogen permeability through the walls, even at room temperature. As a result of the detailed cost and performance analysis, the R&D activities in this direction in the framework of DOE Hydrogen Program have recently been suspended [7].

### 2.4.2. Adsorption methods of hydrogen storage

*Zeolites and metal-organic adsorbents* As it has been shown in some experimental studies (see, for example, [37]), hydrogen adsorption by some zeolites can be applied to achieve reversible hydrogen storage in the temperature range 20–200 °C and pressures 25–100 bar. However, the maximum storage capacity (less than 10 $cm^3/g$) is still too low to exhibit a competition with other storage systems. Nevertheless, it is possible to improve hydrogen storage performances of zeolites by realisation of the low-temperature adsorption, as well as by applying modern techniques of their synthesis and modification. The latter approach also can have some perspectives concerning the creation of new synthetic hydrogen adsorbents with similar microporous structure. Metal organic frameworks (MOF) are example of such compounds. One representative of MOF is the compound $Zn_4O[O_2C-C_6H_4-CO_2]_3$, whose structure is the high-porosity cubic frame having specific surface area of 2500–3000 $m^2/g$. This compound adsorbs up to 4.5 wt.% H at 70 K and 20 bar, or up to 1 wt.% H at the room temperature and the same pressure [38]. The question about possibility to increase hydrogen sorption capacity of this class of compounds, as well as about their synthesis in large quantities and for acceptable prices remains unclear yet.

*Hydrogen cryo-adsorption on activated carbon.* This method uses the process of a low temperature adsorption of hydrogen on a low-density (380 $g/dm^3$) activated carbon. Usually, temperature range is about 65–78 K. Charging of the storage unit with hydrogen is carried out at pressures about 40 bar; hydrogen discharge takes place at the same temperature when pressure is reduced to ~2 bar. At such conditions up to 60–65 % of hydrogen adsorbed by the carbon is reversibly stored. Hydrogen sorption capacity of the adsorbent is about 7–8 wt.% (reversible capacity is equal to 4–5 wt.%). The density of stored energy is about 1.7 kW·h/kg or 0.645 kW·h/m$^3$ [13]. The energy consumption necessary for cooling the adsorbent bed and for the hydrogen compression is relatively high, however, it is much smaller compared to the figures for the storage of liquid hydrogen. Moreover, cryo-adsorption method uses a "cheap" cold, because the operating temperature is close to the boiling point of nitrogen (77 K), which is a common commercial refrigerant. The most prospective way to apply this type of storage is its combination with compressed $H_2$ storage in the cryogenic pressure containers.

In the recent studies of hydrogen adsorption on activated carbon it is mentioned that the H storage density at 77 K and 55 bar is as high as 35 g/l (10-13 wt.%) [11]. It was attributed to a condensation of liquid hydrogen in the micropores of activated carbon. This temperature, 77 K, exceeds the critical temperature for the molecular hydrogen in more than two times. However, the experimental results do not agree with recent modelling of hydrogen adsorption on active carbon providing the maximum adsorption capacity of 2.0–3.3 wt.% at 77 K [39]. Further detailed studies are necessary to reveal the potential of carbon materials in hydrogen storage

*Hydrogen storage in carbon nanomaterials.* Carbon inhabits a multitude of different nano-scale morphologies, including recently discovered fullerenes and nanotubes, as well as graphite nanofibres. All these materials are of significant interest concerning their applications for hydrogen storage [11,39,40]. Except of fullerenes whose interaction with hydrogen is a chemical process similar to the formation / decomposition of "organic hydrides", the other types of carbon nanomaterials interact with hydrogen gas either via van der Waals attractive forces (physisorption) or via a dissociative chemisorption of $H_2$ molecules [41].

The available literature data concerning hydrogen sorption capacity of carbon nanomaterials are rather conflicting [40,41]. At ambient temperatures and pressures the $H_2$ sorption capacity of similarly prepared SWNT spans the range from ~8 wt.% [42], down to < 1 wt.% H [43]. There are also available the data concerning extremely high (up to 60 wt.% H) hydrogen sorption capacity of some kinds of carbon nanofibres [44]. However, these reports have not been reproduced elsewhere. Several groups have reported a correlation between the amount of adsorbed $H_2$ in carbon nanomaterials and the specific surface area [45,46] and / or the micropore volumes [47] of the sorbents. The sorption capacities observed do not exceed the values reported for other carbon materials with high surface area (e.g. activated carbon [48]). The main conclusion from these studies is that the $H_2$ uptake is due to a physisorption process, and that a considerable room-temperature-$H_2$-storage does not take place in carbon nanotubes and nanofibres.

Apart from differences in the structure of the materials used by different authors, sample weight, degree of purity, and P–T conditions, an additional source of the disagreements in the reported storage capacities of carbon nanomaterials could lie in inappropriate techniques of the volumetric measurements. The major sources of errors are incorrect volume calibrations, first of all introduced by the contributions from a not properly determined density of the sample. For high-pressure experiments, a proper thermal management of the measurement system (avoiding temperature fluctuations and taking into consideration the exothermal effects from the introduction of pressurised gas into the reactor, and the temperature increase as a result of the adsorption processes) should be achieved as well [49].

No doubts that the research of hydrogen sorption properties of different carbon nanomaterials having well-defined structures is very interesting from the fundamental point of view, and a growing research activities in this field will result in better understanding of gas – solid interaction mechanism. However, the initial optimism concerning creation of very efficient hydrogen storage material on the basis of nanoscale carbon would seem to be premature yet.

### 2.4.3. Organic hydrides and fullerenes

An example of organic chemical system allowing to store reversibly hydrogen is a system "benzene – cyclohexane" [13]:

$$C_6H_6 + 3\,H_2 \leftrightarrow C_6H_{12} + 206.2\text{ kJ/mole}$$

The hydrogenation / dehydrogenation processes take place at 200–400 °C and 10–100 bar $H_2$ in presence of catalyst (Pt, Pd, $Mo_2O_3$). Hydrogen storage in organic hydrides is very efficient concerning weight and volume storage capacities (5–7 wt.% or 70–100 g/l H volume density). However, the energy consumption for the heating of the storage system providing the necessary reaction temperature is too large. It is higher than the one for cryogenic hydrogen storage [7,13]. So, this method can be used only in some specific cases (e.g. in chemical industry) when a waste heat source having sufficient temperature potential is available.

A similar option of hydrogen storage is a reversible catalytic hydrogenation of the double C=C bonds in fullerenes. In such a process, the $C_{60}$ fullerene can be hydrogenated up to the composition $C_{60}H_{48}$ that corresponds to 6.3 wt.% H. The hydrides of intermetallic compounds are proved to be the efficient catalysers of the reversible hydrogenation / dehydrogenation processes. However, the dehydrogenation temperature (more than 400 °C) is still too high for the practical purposes [40,50].

### 2.4.4. Sponge iron and other water-reacting metals

The process of hydrogen generation by a high-temperature reaction of water vapours with sponge iron is known for a long time. The following reaction, taking place at T=550–600 °C, is a basis of this hydrogen storage method:

$$3\,Fe + 4\,H_2O \rightarrow Fe_3O_4 + 4\,H_2$$

It is possible to make this process reversible, because the starting iron can be recovered by reducing $Fe_3O_4$, for example using carbon monoxide. By arranging the thermodynamic cycle of water decomposition using sponge iron, including an appropriate selection of the catalyst, the temperature of hydrogen generation and further recovery of iron can be reduced to some extent [13]. However, an economic and environmental analysis of this technology has resulted in a conclusion that it is not efficient enough on a competitive scale [7].

Another technology development in a similar direction is in the usage of the reaction of aluminium or silicon alloys (so-called, power-accumulating substances) with alkaline solutions for hydrogen generation. By appropriate selection of the alloying components, as well as design features and operating conditions of hydrogen-generating reactor, it is possible to reach rather fast reaction kinetics in the soft conditions and to easily control hydrogen generation. The corresponding R&D was intensively conducted in 1980th, in particular at the Institute of Mechanical Engineering Problems, Ukraine [51]. This technology efficiently generates hydrogen and is not appropriate for the reversible $H_2$ storage. But in some special cases, including the availability of peak electric power, it may be prospective in energy storage.

### 2.4.5. Ammonia and methanol

The advantage of hydrogen storage and transportation in the form of ammonia or methanol is in high hydrogen volume density (109 g/l for liquid $NH_3$ at T=15 °C, P=7.2 bar; and 99 g/l for $CH_3OH$ at room temperature). This is higher, in both cases, than for $LH_2$. Hydrogen is obtained by the following catalytical reactions [13]:

$$2\,NH_3 \rightarrow N_2 + 3\,H_2 - 92\text{ kJ}$$
$$CH_3OH + H_2O \rightarrow CO_2 + 3\,H_2 - 49\text{ kJ}$$

The dissociation of ammonia is carried out at temperatures 800–900 °C using iron as a catalyser. The process of hydrogen generation from methanol requires the temperature 300–400 °C in presence of the zinc – chromium catalyser. Both for ammonia and for methanol, the storage medium is used only once. This irreversibility, as well as too high power inputs for hydrogen generation, restricts a wider implementation of the above-mentioned hydrogen storage methods. Besides, the corrosion activity of ammonia is also a serious factor restricting its usage in hydrogen storage [39]. At the same time, if to reduce the temperature of methanol decomposition reaction, its usage as hydrogen storage medium can be competitive in a number of applications, including motor transport [11].

## 3. Metal hydrides as hydrogen storage medium and their potential in related applications

### 3.1. GENERAL CONSIDERATION

Metal hydride method of hydrogen storage is based on the process of the reversible hydrogen absorption in hydride forming metals or intermetallic compounds with the formation of metal hydrides (MH). Hydrogen in the MH is placed in interstitials of crystal structure of the matrix of the metal (intermetallide) as individual, not associated in molecules, H atoms.

Two options of hydrogen absorption by MH are available: by gas – solid reaction (1) or electrochemical charging from an aqueous electrolytic solution (2):

$$\text{M (s)} + x/2 \text{ H}_2 \text{ (g)} \leftrightarrow \text{MH}_x \text{ (s)} + Q; \tag{1}$$

$$\text{M (s)} + x \text{ H}_2\text{O (l)} + e^- \leftrightarrow \text{MH}_x \text{ (s)} + \text{OH}^- \text{ (l)}; \tag{2}$$

where $\mathbf{M}$ denotes hydride forming metal or intermetallic compound; indexes $\mathbf{s}$, $\mathbf{g}$ and $\mathbf{l}$ correspond to the solid, gas and liquid phases respectively. Both reactions are reversible, so as by the small changes of the temperature or pressure we can change direct process (charging with hydrogen) to the opposite (hydrogen discharge) and vice versa.

Hydrogen directly interacts according to mechanisms (1) or (2) with many individual metals forming binary hydrides, $\mathbf{MH_x}$. The formation of the binary hydrides is accompanied by the essential changes in the mutual location of the $\mathbf{M}$ atoms taking place in the course of expansion of the crystal lattice. The latter either remains invariant, or (in the major cases) rearranges itself (Table 4). This process requires considerable energy contribution and, as a rule, it is conditioned by a high metal – hydrogen chemical affinity, i.e. a strong chemical bond $\mathbf{M - H}$ must be formed. It takes place for the stable binary hydrides, such as hydrides of alkali and alkali-earth metals, rare-earth elements, actinides, elements of titanium subgroup, etc. For these metals, reaction (1) is reversible only at increased temperatures and, as a rule, it has too slow kinetics. It is unacceptable for the most applications. Another extreme case is the formation of unstable binary hydrides of transition metals (e.g., Ni) having weak $\mathbf{M - H}$ bond. For these metals, hydride formation takes place by hydrogen sorption from a gas phase (direct process of the reaction 1) only under high, about several kilobars, hydrogen pressures. The corresponding electrochemical reaction (2) can be realised in this case only at high overpotentials.

Apart from binary ones, the class of hydrides of intermetallic compounds (IMC) having a common formula $A_mB_nH_x$ exists. The IMCs are the compounds of two or more metals, at least one of which (A) has a strong affinity to hydrogen, e.g. forms a stable binary hydride. The component B does not interact with hydrogen at usual conditions. Hydride forming intermetallides are accepted to classify starting from the ratio of their hydride forming (A) and non hydride forming (B) components. From a great number of families of hydride forming intermetallides several ones are of practical importance, namely: $AB_5$ ($CaCu_5$ structure type), $AB_2$ (Laves phases), AB (CsCl-relative structure types), and $A_2B$ ($AlB_2$-relative structure types). In the $AB_5$ compounds rare-earth elements and / or calcium are the A-component; in the $AB_2$ and AB A is the element of titanium subgroup; and in $A_2B$ A is mainly magnesium. Transition metals (Fe, Co, Ni, V, Mn, Cr, etc.) constitute mainly the B component in all families. The formation of intermetallic hydrides is accompanied by the expansion of the crystal lattice, typical values of increasing the volume of the unit cell are varied from 10–15 to 20–25%. The symmetry of the metal matrix remains, as a rule, invariant in the course of hydride formation (Table 4). However, when the contents of the A-component increases, hydrogenation of the intermetallide results in more significant changes of the arrangement of metallic atoms that results either in the essential changing the symmetry (TiFe,

Mg$_2$Ni), or even in the degradation with the formation of the mixture of stable A-based binary hydride and B-enriched IMC (Mg$_2$Cu). Intermetallic hydrides are characterised by the excellent kinetics of reversible hydrogen sorption – desorption according to (1) or (2) mechanisms which take place in the soft conditions, close to "room" ones.

From the classes of MH under consideration, hydrides of IMC have the biggest practical importance, including hydrogen storage applications.

One more class of MH are coordination compounds where hydrogen presents as a ligand. These compounds (complex hydrides), as a rule, can be synthesized only by methods of preparative chemistry and can not be a medium for reversible hydrogen storage. However, in the systems on the basis of sodium alanate, NaAlH$_4$, the reversible hydrogen sorption – desorption is possible in the presence of catalyser. So, such systems will be also briefly discussed in this section, together with binary and intermetallic hydrides.

The majority of MH are characterised by a high hydrogen content: the ratio of number of atoms of the bounded hydrogen to the number of atoms of the parent metal varies from 0.7–1.1 for intermetallic hydrides up to 3.75 for Th$_4$H$_{15}$, and even to 4.5 for recently discovered complex hydride BaReH$_9$ [39].

Typical relations between volumetric and weight hydrogen capacities for different MH are shown in the Fig. 7 [39,52–67]. Also, the points corresponding to hydrogen storage capacities of liquid hydrogen (LH$_2$) [32] and compressed hydrogen gas in the modern composite gas cylinders (CGH$_2$) [27] are plotted here. It is seen from the figure that metal hydride hydrogen storage method can compete with conventional ones as to compactness, but is it inferior in hydrogen weight capacity. The latter is somewhere better for the hydrides of the light elements (group II in Fig. 7), but their usage as a hydrogen storage medium in most cases is problematic. So, the decomposition of some of such hydrides BeH$_2$, LiBH$_4$) is the irreversible process, and formation / decomposition of the others on mechanism (1) takes place at too high temperatures and requires significant energy consumption.

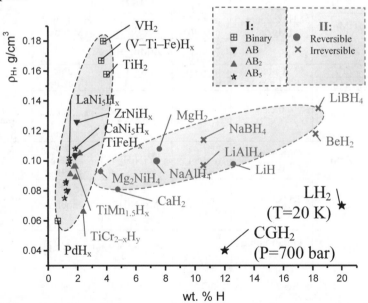

*Figure 7. Relation between weight and volume hydrogen capacities for binary and intermetallic hydrides on the basis of transition metals (I), and binary and complex hydrides of the light elements (II)*

TABLE 4. Kinds and some typical representatives of metal hydrides [39,52–67]

| kind / family | Parent metal (compound) | | | Higher hydride | | |
|---|---|---|---|---|---|---|
| | Formula | Structure (type) | Lattice periods, Å | Formula | Structure (type) | Lattice periods, Å |
| Binary | Li | b.c.c. | a=3.50 | LiH | f.c.c. (NaCl) | a=4.083 |
| | Mg | h.c.p. | a=3.20 c=5.20 | $MgH_2$ | Tetragonal (rutile) | a=4.5168 c=3.0205 |
| | Ca | c.c.p. | a=5.56 | $CaH_2$ | Ortho-rombic $(PbCl_2)$ | a=5.936 b=6.838 c=3.600 |
| | Ti | h.c.p. | a=2.95 c=4.68 | $TiH_2$ | f.c.c. $(CaF_2)$ | a=4.454 |
| | V | b.c.c. | a=3.027 | $VH_2$ | | a=4.271 |
| | La | h.c.p. ABAC | a=3.77 c=12.16 | $LaH_3$ | Cubic $(CeH_3)$ | a=5.602 |
| | Pd | c.p.p. | a=3.88 | $PdH_{0.7}$ | f.c.c. (NaCl) | a=4.085 |
| | Th | | a=5.08 | $Th_4H_{15}$ | Cubic $(Th_4H_{15})$ | a=9.110 |
| Intermetallic $AB_5$ | $CaNi_5$ | Hexagonal $(CaCu_5)$ | a=4.950 c=3.936 | $CaNi_5H_{4.5}$ | Hexagonal $(CaCu_5)$ | a=5.305 c=4.009 |
| | $LaNi_5$ | | a=5.015 c=3.987 | $LaNi_5H_6$ | | a=5.426 c=4.269 |
| | $LaNi_4Al$ | | a=5.066 c=4.070 | $LaNi_4AlH_{4.3}$ | | a=5.319 c=4.235 |
| Intermetallic $AB_2$ | $TiCr_{1.9}$ | Hexagonal $(MgZn_2)$ | a=4.927 c=7.961 | $TiCr_{1.9}H_{2.9}$ | Hexagonal $(MgZn_2)$ | a=5.171 c=8.518 |
| | $TiMn_{1.5}$ | | a=4.927 c=7.961 | $TiMn_{1.5}H_{2.5}$ | | a=5.271 c=8.579 |
| | $ZrMn_{3.2}$ | | a=4.988 c=8.200 | $ZrMn_{3.2}H_4$ | | a=5.424 c=8.682 |
| | $ZrV_2$ | Cubic $(MgCu_2)$ | a=7.416 | $ZrV_2H_{4.8}$ | Cubic $(MgCu_2)$ | a=7.956 |
| | $ZrCr_2$ | | a=7.207 | $ZrCr_2H_{3.5}$ | | a=7.697 |
| Intermetallic AB | TiFe | b.c.c. (CsCl) | a=2.975 | $TiFeH_{1.9}$ | Monoclinic | a=4.704 b=2.830 c=4.704 β=97.0° |
| | ZrNi | Orthorombic (CrB) | a=3.258 b=9.941 c=4.094 | $ZrNiH_3$ | Ortho-rombic (CrB) | a=3.53 b=10.48 c=4.30 |
| 'Intermetallic' $A_2B$ | $Mg_2Ni$ | Hexagonal $(Mg_2Ni)$ | a=5.19 c=13.22 | $Mg_2NiH_4$ | Monoclinic | a=6.497 b=6.414 c=6.601 β=93.2° |
| | $Mg_2Cu$ | Orthorombic | a=5.284 b=9.07 c=18.25 | $MgH_2$ + ½ $MgCu_2$ | | |

Generally, reaction (1) has three stages. On the first one hydrogen forms a solid solution in the matrix of the metal **M** ($\alpha$-phase); in so doing its equilibrium concentration, $C = $ H/M, is determined by the pressure, $P$, of hydrogen gas and the temperature, $T$. The relation can be approximately[*] described by the Henry – Sieverts law:

$$C = K(T)P^{\frac{1}{2}}.\tag{3}$$

After reaching a higher limit which corresponds to the hydrogen concentration in the saturated solid solution, $C = a$, the further hydrogen sorption (second stage) is accompanied by hydride ($\beta$-phase) formation with hydrogen concentration $C = b$ ($b > a$), so as the material balance of the reaction (1) can be written as:

$$1/(b\text{-}a)\ \textbf{MH}_a\ \textbf{(s)} + 1/2\ \textbf{H}_2\ \textbf{(g)} \leftrightarrow 1/(b\text{-}a)\ \textbf{MH}_b\ \textbf{(s)}.\tag{4}$$

According to the Gibbs phase rule, the process (4) is reversible at the constant hydrogen pressure, $P_D$, which corresponds to the appearance of the plateau on the pressure – composition isotherm, e.g. the dependence of the equilibrium hydrogen pressure ($P$) on hydrogen concentration ($C$) at constant temperature ($T$) – Fig. 8a. After complete transition of the solid solution to the hydride, further hydrogen sorption (third stage) is hydrogen dissolution in the hydride $\beta$-phase. In so doing, hydrogen equilibrium concentration increases with the pressure, asymptotically approaching to its upper limit, $C_{max}$ which is determined by hydrogen capacity of the metal, or the number of the available for hydrogen insertion interstitials per the number of the metal atoms.

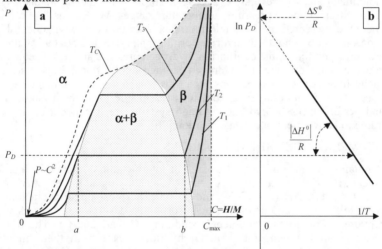

*Figure 8. A schematic representation of the idealised PCT diagram of hydrogen – metal system: a – pressure – composition isotherms at the temperatures $T_1 < T_2 < T_3 < T_C$; b – temperature dependence of plateau pressure*

Starting from the equilibrium condition in the plateau region that is the equality of hydrogen chemical potentials in the gas phase and two solid ones (solution **MH$_a$** and hydride **MH$_b$**), the known Van't Hoff relation determining the temperature dependence of plateau pressure can be derived:

$$\ln P_D = -\frac{\Delta S^0}{R} + \frac{\Delta H^0}{RT},\tag{5}$$

---

[*] The Henry – Sieverts law is valid for ideal solid solutions that is equivalent to $C \rightarrow 0$ for metal – hydrogen systems

where $\Delta H^0$ and $\Delta S^0$ are correspondingly standard enthalpy and entropy of the hydride formation reduced to 1 mole of hydrogen gas. The dependence of plateau pressure on the temperature in $\ln P_D - 1/T$ coordinates is a straight line whose slope is proportional to $\Delta H^0$ and ordinate of crossing with pressure axis (at $1/T=0$) – to the $\Delta S^0$ value (Fig. 8b).

The concentration limits (*a*, *b*) of the existence of the two-phase ($\alpha+\beta$) plateau region (miscibility gap) are temperature-dependent. They become closer when the temperature increases, and coincide when it becomes equal to a critical value, $T_C$, – this corresponds to plateau degeneration into inflection point. Above the critical temperature hydrogen in metal exists only as an $\alpha$-solution (Fig. 8a).

The above-mentioned dependence between equilibrium pressure / concentration values and the temperature (PCT diagram) is the most practically important characteristic for hydride forming metals and intermetallides. As a rule, the reference literature presents it's simplified parameters that are hydrogen capacity of the hydride and the enthalpy and the entropy of its formation determining temperature dependence of plateau pressure. The enthalpy, $\Delta H^0$, is considered to be approximately equal to the heat effect, $Q$, of the reaction (1).

Fig. 9 shows the temperature dependencies of plateau pressure for a number of MH materials having practical importance. Depending on the nature of the hydride forming metal or intermetallide, the reversible hydrogen sorption on the mechanism (1) can be realised in the extremely wide pressure / temperature ranges. More visually it can be seen in Fig. 10 where we presented the available reference data on thermodynamics of hydrogen sorption – desorption by about 100 hydride forming materials [52,56–69] as the values of temperature corresponding to the equilibrium hydrogen pressure above MH equal to 1 bar.

From the point of view of hydrogen storage applications, it is conventional to separate metal hydrides into two groups, viz. "low-" and "high-" temperature ones (Fig. 10). This separation is quite artificial, but it is reasonable to adopt it, since usually the features and performances of hydrogen storage systems are dependent on metal hydride type according to above-mentioned classification [70].

## 3.2. LOW-TEMPERATURE METAL HYDRIDES.

This type contains hydrides of several classes of intermetallic compounds, e.g. $AB_5$ (A – rare earth metal; B – Ni as a basis, Co, Al, Mn, etc. as alloying additives), $AB_2$ (A – Ti, Zr; B – Mn, Fe, Cr, etc.), AB (e.g. TiFe) and so on. The most important feature of these materials is in relatively high rate of hydrogen sorption – desorption processes at pressures and temperatures close to the "ambient" conditions (1 bar, 20 °C). The advantage of hydrogen storage in low-temperature metal hydrides, apart from their high hydrogen volume density, is a high purity of supplied hydrogen, due to the selectivity of hydrogen absorption by the hydride-forming alloys. Efficient control of hydrogen pressure at the output of MH storage unit, by controlling thermal action onto the MH layer is an additional advantage of the low-temperature MH, allowing for an extension in the performance of the hydrogen storage unit by the function of hydrogen compression. A combination of conventional hydrogen storage methods with the metal hydride H storage (e.g., using of the MH in the $H_2$ compressing and liquefying systems) is considered to be very promising.

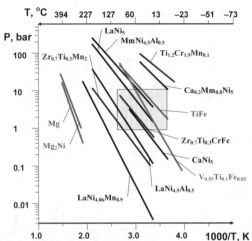

*Figure 9. Equilibrium pressure – temperature dependencies for some MH [56–69]. The shaded area (P=1–10 bar, T=0–100 °C) corresponds to the most common specifications for hydrogen storage systems*

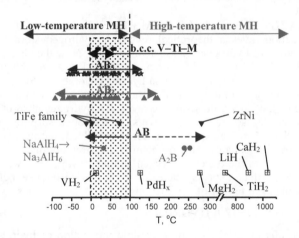

*Figure 10. Temperatures of hydrogen desorption from the different MH at P=1 bar. The shaded area (T=0–100 °C) corresponds to the most common specifications for hydrogen storage systems*

The main drawback of the low-temperature MH hydrogen storage is in a rather low weight content of H (typical value is up to 1.5 wt.% for $AB_5$ and up to 2 wt.% for $AB_2$ and AB alloys). Moreover, the real values of the weight and volume hydrogen storage capacities are even smaller in the MH hydrogen storage units (see below). Similarly, the power consumption to provide hydrogen release from the MH container is higher than the value of the corresponding heat of the decomposition of MH. Because of a low heat conductivity of the MH layer and heat losses, the thermal efficiency of the MH devices rarely exceeds 30–35% [71].

Nevertheless, the majority of MH applications including hydrogen storage units are based on the use of low-temperature MH, first of all, due to their technological flexibility.

### 3.3. HIGH-TEMPERATURE METAL HYDRIDES

This type mainly comprises magnesium-based alloys and intermetallics, which, as compared to the low-temperature hydrides, have larger weight capacity (up to 7.6 wt.% for the magnesium hydride) and a smaller price. However, hydrogen absorption – desorption kinetics in these materials is usually slow. Both absorption and desorption processes take place at increased temperatures (250–350 °C). High power consumption for the high-temperature MH hydrogen storage is caused by a necessity to maintain elevated temperatures, as well as to stimulate hydride decomposition which heat of formation exceeds the values for the low-temperature MH in more than two times. Nevertheless, this hydrogen storage method can be acceptable for the specific applications where the waste heat with corresponding temperature potential is available.

### 3.4. COMPLEX HYDRIDES (ALANATES)

Recently, some of the complex hydrides (alanates) became a subject of intensive research concerning their possible applications as hydrogen storage materials. Indeed, these compounds have rather large weight hydrogen fraction (7.4 wt.% H for $NaAlH_4$). For many years these compounds are known as nonreversible "chemical" hydrides and have not been considered as reversible MH hydrogen storage materials. However, in 1997 it was discovered [62] that in the presence of $TiCl_3$ as a catalyst the following two stage reversible reaction is possible:

$$NaAlH_4 \leftrightarrow 1/3\ Na_3AlH_6 + 2/3Al + H_2 \leftrightarrow NaH + Al + 3/2\ H_2$$

The net reaction gives a theoretical reversible hydrogen storage capacity of 5.6 wt.%, with the potential of hydrogen release about 100 °C. Now the alanate systems are the subject of intensive R&D studies. In particular, these R&D are the part of the DOE Hydrogen Program for the creation of efficient hydrogen storage systems [7,62–65].

Nevertheless, in our opinion, this direction has no large commercial potential, in particular, because of too slow kinetics of the reversible hydrogen absorption – desorption, as well as due to the degradation of the reversible hydrogen sorption properties during prolonged cycling. These problems, perhaps, could be solved, but not the other important principal drawback of the alanates. This problem is in extremely high chemical activity of the alanates with respect to the traces of oxygen and water vapours – which can even lead to the explosion. So, safety aspects make an important difference between the metal hydrides and the alanates in favour of applications of the metal hydrides.

### 3.5. ADVANTAGES OF METAL HYDRIDES

Since the end of 1960s the processes of reversible interaction of hydrogen with hydride forming metals and alloys are in the focus of the R&D activities aimed to the creation of the extremely compact and technologically flexible hydrogen storage units. First of all, it is conditioned by the record high hydrogen volumetric density in the MH. The latter usually is about 0.09–0.1 $g/cm^3$ [57] that is higher than the density of liquid hydrogen (0.07 $g/cm^3$) and can be compared with hydrogen density in such hydrogen-saturated compounds as water (0.11 $g/cm^3$). The maximum hydrogen density in the metal hydrides corresponds to 0.15 $g/cm^3$ for $TiH_2$ and 0.18 $g/cm^3$ for $VH_2$ [58], so the hydrogen packing density in the MH can 2 to 2.7 times exceed the density of liquid hydrogen. Recently discovered intermetallic hydrides with anomaly short H – H distances [72] have the local hydrogen density of 0.56 $g/cm^3$ that is 8 (!) times more than the $LH_2$ density.

A very important advantage of hydrogen storage in metal hydrides is in the improved safety. Indeed, at room temperature hydrogen gas pressure above metal hydride does seldom exceed several bars. This feature, together with multi-functionality of MH hydrogen storage systems (possibility of hydrogen purification due to high selectivity of hydrogen sorption / desorption processes with MH materials, easy pressure control at the output of MH storage unit, the increased chemical activity of supplied hydrogen, etc.) makes these systems the most preferable, for example, in laboratory applications. Small- (tens litres $H_2$) and medium-scale (several cubic meters) MH hydrogen storage systems are commercially available now, and their niche in the market increases, maybe not fast but permanently.

## 4. Metal hydride hydrogen storage units

Already three decades passed since the first demonstrations of the applications of MH. A wide implementation of this technology can be observed in the MH rechargeable batteries and, also, in MH hydrogen storage units. Dozens of companies all over the world, in USA, Japan, West and East Europe, China and other regions, are engaged into these activities. A brief summary concerning the companies and institutions engaged in the applications of MH is presented in Table 5.

Mainly, commercial MH storage units are intended for laboratory applications providing safe and compact storage of high-purity hydrogen (capacity from 30–40 litres to few m³, supply pressure 1–10 bar). The available offers of these units often propose a few typical layouts, in agreement with the specification from consumer. Technological flexibility of the MH units allows to adjust their characteristics to the specific consumer's conditions.

Apart from the small- and medium- scale metal hydride hydrogen storage units, several examples of relatively large-scale ones are also known. Usually, these units are the intermediate (buffer) storage reservoirs being component parts of hydrogen refuelling stations. The units (75 to 90 m³ $H_2$ STP) are characterised by compactness, safety and easy operation. The largest MH hydrogen storage unit (~15000 m³ $H_2$ stored in 100 tons of metal hydride) was produced by GfE Gesellschaft für Elektrometallurgie mbH [78], which now is a part of the HERA Hydrogen Storage Systems, GmbH company [79]. It was built for the German submarine U-212.

It is also feasible to use metal hydride hydrogen storage units in some special mobile applications, where weight storage capacity is not so critical, but safety and compactness are necessary (Fig. 11). One example is a zero emission technological transport, like tractor or forklift. The latter approach is of special interest, because here MH hydrogen storage unit can be mounted as a counter balance. Such a solution, in particular, was successfully implemented in the mid 80-th at the Institute of Mechanical Engineering Problems, Ukraine [70]. The similar one was also mentioned in the comprehensive review on vehicular hydrogen storage systems prepared by Volvo [11].

It should be noted that in the majority of reviews [39,57,58, etc] the values of weight and volume hydrogen sorption capacities of the metal hydride hydrogen storage method are presented as the maximum capacities of the bulk MH material. Namely such values were also presented by us in Fig. 7. In the real MH hydrogen storage units these parameters are lower. It can be illustrated by Fig. 12 where the weight and volume efficiencies of hydrogen storage for 24 laboratory and vehicular MH units on the basis of "low-temperature" $AB_5$, $AB_2$ and AB materials are presented as the function of H storage capacity of the unit.

TABLE 5. Companies and institutions engaged in metal hydride applications

| Organisation (country) | Main activities |
| --- | --- |
| Ergenics, Inc. (USA) [73] | • Hydride forming alloys<br>• Fuel cells<br>• Hy-Stor® segmented MH battery<br>• Thermal sorption compressors<br>• Heat pumps<br>• Heat storage |

| Hydrogen Components, Inc. (USA) [74] | • SOLID-H™ MH systems for hydrogen storage, compression and purification<br>• Teaching aids<br>• Alternative fuels (including hydrogen) and filling systems |
|---|---|
| ECD Ovonics (USA) [75] | • Fuel cells<br>• Ni–MH batteries<br>• Hyrdide forming alloys and hydrogen storage systems |
| The Japan Steel Works, Ltd. (Japan) [76] | • Hydride forming alloys<br>• MH hydrogen storage units<br>• Ni–MH batteries<br>• Fuel cells<br>• Heat storage, transformation and transportation<br>• Actuators |
| Japan Metals & Chemicals Co., Ltd. (Japan) [77] | • Hydride forming alloys<br>• MH hydrogen storage units<br>• MH heat pumps |
| GfE Gesellschaft für Elektrometallurgie mbH (Germany) [78] | • MH hydrogen storage units<br>• Hydride forming materials<br>• Testing equipment for MH |
| HERA Hydrogen Storage Systems, GmbH (Canada – Germany) [79] | |
| Institute for Energy Technology IFE (Norway) [80] | • Hydride forming alloys<br>• Laboratory MH hydrogen storage units<br>• Stand alone energy systems with MH units |
| LABTECH Int. Co. Ltd. (Bulgaria) [81] | • Hydride forming materials<br>• MH hydrogen storage units<br>• Negative electrodes for the MH based chemical current sources |
| Chinese Academy of Science (China) [82] | • Hydride forming alloys<br>• Thermal sorption compressors for microcryogenic systems<br>• Combined MH systems for hydrogen extraction, purification, storage and transportation |
| PA "Moscow Polymetal Plant" (Russia) [83] | • Hydride forming alloys<br>• MH hydrogen storage units |
| Institute of Mechanical Engineering Problems of National Academy of Sciences of Ukraine (Ukraine) [84] | • MH hydrogen storage units<br>• Combined MH systems for the storage, purification and controlled supply of hydrogen (deuterium)<br>• Thermal sorption compressors<br>• MH systems for evacuation and gas supply in vacuum-plasma technology |
| Institute for Problems in Materials Science of National Academy of Sciences of Ukraine (Ukraine) [85] | • Ni–MH batteries<br>• Composite hydride forming materials |

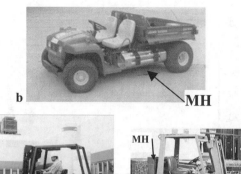

*Figure 11. Transport applications of MH hydrogen storage units:*
*a – metal hydride storage tank in minivan [86]*
*b – a utility vehicle with PEM FC and MH hydrogen storage [59]*
*c – hydrogen-powered PEM FC forklift (Siemens AG; photo from Fuel*
*Cells 2000 1999 [11])*
*d – hydrogen-powered ICE forklift with 27 Nm³ H₂ MH storage unit*
*(IPMach [70])*

The characteristics of the units produced by Ergenics [73], HERA [79], IPMach [70], IFE (our data), as well as some ones reviewed by Volvo in [11] were used as starting data. The volumetric efficiency was defined as the ratio of the real volumetric H capacity (quantity of stored hydrogen divided by volume of the unit) to the maximum volumetric hydrogen density of the MH used in the unit. Similarly, the weight efficiency was determined as the ratio of the weight H capacity (hydrogen quantity divided by the unit weight) to the maximum weight H capacity of the corresponding MH.

It is seen from Fig. 12a that the volumetric efficiency of hydrogen storage in the real MH units does not practically correlate with the scale of the unit, but as a rule, it does not exceed 40–50% of the total volumetric H density in the MH material. As to the weight efficiency, it is observed a tendency of its increase with the unit scale. Moreover, for the low-scale units (up to ~500 litres $H_2$) this effect is more pronounced. Anyway, 70–75% is the upper limit for the weight efficiency of the MH hydrogen storage systems.

There exist several reasons reducing the hydrogen storage capacity of the MH units below the capacity of the MH material. First, the useful storage capacity of the unit is determined not by the total, but by the reversible hydrogen capacity of the material, i.e. by the quantity of hydrogen which can be reversibly absorbed / desorbed in the course of the unit operation. This parameter depends both on the properties of the used MH (mainly on PCT characteristics of the corresponding system with hydrogen gas), and, also, on working pressure and temperature range. The system $TiCr_{2-X} - H_2$ (Fig. 13) is the vivid illustration; here the reversible hydrogen capacity at $T=0-120$ °C and P=10–100 bar is less than 20% of the total H capacity of the material.

In most cases the reversible capacity of "low-temperature" MH is about 60–80% of the total hydrogen capacity for AB, $AB_2$ and $AB_5$ [56], and 50–60% for pseudo-binary hydrides on the basis of b.c.c. vanadium alloys [61]. For the "high-temperature" hydrides,

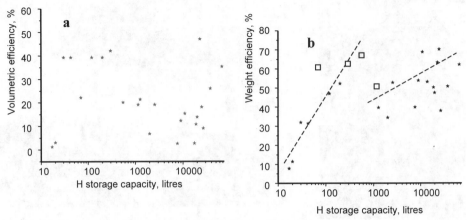

*Figure 12. Volumetric (a) and weight (b) storage efficiencies of MH hydrogen storage units*

based on pure magnesium and its intermetallides (e.g., $Mg_2Ni$), the reversible capacity derived from PCT characteristics is more than 90% of the total value [56]. However, high-alloyed magnesium alloys containing significant amounts of relatively heavy components (rare earths, Ni, etc.) added for improvement of hydrogen sorption – desorption kinetics have the real capacity about 5.5 wt.% H that is 72% of the total capacity of $MgH_2$ (7.6 wt.%) [89]. For the catalysed systems on the basis of $NaAlH_4$ the reversible hydrogen capacity, corresponding to the reaction (6) is of 5.6 wt.%, or 75% of the total hydrogen quantity in the alanate (7.4 wt.%); if only the first low-temperature stage of this reaction is taken into account, its reversible capacity is already 3.7 wt.%, i.e. 50% of total value [65].

Another factor reducing the weight hydrogen capacity of the MH storage units is material consumption of the container. This factor is also important in the large-scale units where, apart from container itself, the presence of special heat transfer elements for improvement heat transfer in the MH layer is necessary. The usage of lighter design materials allows to reduce in some extend the losses in the weight capacity of the system. However, this effect is mainly essential for the low-scale units. So, in the Fig. 12a the points denoted by opened squares correspond to the aluminium-made containers, and the stars correspond to containers made of stainless steel.

The main reason of reducing in volumetric hydrogen capacity in the real MH storage units is the upper limit of safe packing density of the MH loaded into container. It is obligatory to avoid the damage of the swelling effect of the MH material in the course of hydrogen absorption. This limit depends on the value of volume increase of the metal matrix under hydrogenation. For $AB_5$, according to the experimental data [90], the safe packing density is about 45% of the bulk density of the H storage alloy.

Taking into account the above-mentioned, it can be concluded that MH hydrogen storage units having a conventional layout (rigid container with the internal heat exchanger or without it where the MH powder is loaded) will have the volumetric hydrogen storage capacity not higher than 50% and the weight one no more than 75% of the total capacity of

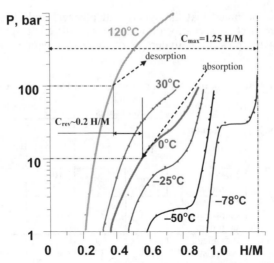

*Figure 13. Total ($C_{max}$) and reversible ($C_{rev}$) hydrogen sorption capacity for the TiCr$_{2-X}$ – H$_2$ system [87,88]*

the MH. According our estimations, it corresponds to the storage of gaseous hydrogen under the pressure 400 to 800 bar. But the real pressure of hydrogen gas inside MH hydrogen storage unit usually does not exceed several bars what is important from the reasons of safety and the absence of expenses for hydrogen compression. Moreover, the improvement of the design of MH containers (usage of light materials, non-traditional layout, etc.) can improve weight- and volumetric indices of MH hydrogen storage units. So, according to Volvo estimations [11], hydrogen storage capacity of the mobile MH H storage units can be only 10–20% lower than the hydrogen capacity of the MH.

## 5. Evaluation of suitability of different methods for different applications based on physical and chemical characteristics and particular properties

This Section summarizes the information about performance characteristics of different hydrogen storage methods considered above

The comparison of the performances of the main hydrogen storage methods is presented in Fig. 14 and Fig. 15. More detailed characteristics of the methods are given in the Table 6. We used typical characteristics based on the available reference data. We note that some of the reference data vary from source to source, especially in case of the methods being on the R&D stage. Nevertheless, these data reflect the existing trends and can be used for the comparative purposes.

From the data of Table 6 it can be concluded that all methods have advantages and disadvantages and there is no a specific hydrogen storage method which is superior to the remaining alternative ones. Cost, volume, weight and performance should be considered together in making a choice for optimal storage method suiting the specific requirements.

In the industrial-scale methods of hydrogen storage significant progress is achieved in the improvement of the weight and volume efficiencies of H storage in compressed H$_2$ (350–700 bar composite cylinders) and LH$_2$ forms with weight H capacities reaching 20 wt.% for the vehicle applications and exceeding 50 % for the large scale storage units. Metal hydrides provide the best solution in achieving safe H storage and reaching the highest level of H volume densities required for the applications in stationary and stand alone systems where the weight density is not critical.

It should also be noted that the synergy of physical and chemical methods of H storage allows a prospective scenario allowing to suit different applications. Combined systems (e.g., cryogenic pressure cylinders or pressure cylinders + MH compressors) are the most prospective.

TABLE 6. Comparison of hydrogen storage methods

| Group | Subgroup | Method | Storage conditions | | Storage performances | | |
|-------|----------|--------|----------|--------|----------|----------|----------|
| | | | P, bar | T, °C | Weight % H | H Volume density, g/dm$^3$ | Energy consumption, %[2] |
| Physical | Compressed gas storage | Steel cylinders | 200 | 20 | $\frac{100^{(3)}}{2.5}$ | 17.8 | 9 |
| | | Commercial composite cylinders | 250 | 20 | $\frac{100}{6}$ | 22.3 | 10 |
| | | Advanced composite cylinders | 690 | 20 | $\frac{100}{8}$ | 29.7 | 12.5 |
| | | Glass micro-spheres | 350–630 | 200–400 | $\frac{100}{10}$ | 20 | 25 |
| | Cryogenic (LH$_2$) | | 1 | –252 | $\frac{100}{15-20}$ | 71 | 27.9 |
| Chemical | Cryo-adsorption | | 2–40 | –208... ...–195 | $\frac{6.0-8.2}{3.8-5.2}$ | 15–30 | 8.1 |
| | Metal hydrides | "Low-Temperature" (20–100°C) | 0.01–20 | 20–100 | $\frac{1.5-1.8}{0.8-1.3}$ | $\frac{90-100}{60-70}$ | 10.4 |
| | | "High-Temperature" (250–400°C) | 1–20 | 250–350 | $\frac{3.5-7.6}{2.5-5.6}$ | $\frac{90-100}{60-70}$ | 20.6 |
| | | Complex hydrides (alanates) | 1–200 | 125–165 | $\frac{4.0-5.6}{3.0-4.0}$ | 30 | 13.4 |
| | Organic hydrides | | 10–100 | 300–400 | $\frac{5.0-7.0}{3.5-4.7}$ | 70–100 | 28 |
| Gasoline[4] | | | 1 | 20 | $\frac{41}{28}$ | 29 | – |

---

[2] As compared with hydrogen heating value 39.413 kW·h/kg
[3] Numerator – "ideal" value; denominator – real value (due to limited efficiency of the storage system)
[4] In energy equivalents

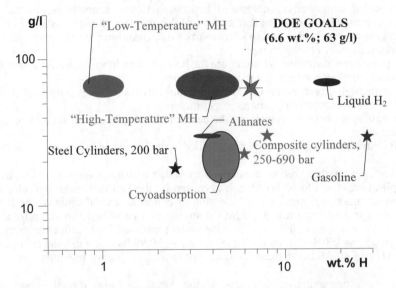

*Figure 14. Weight- and volume capacities of hydrogen storage methods*

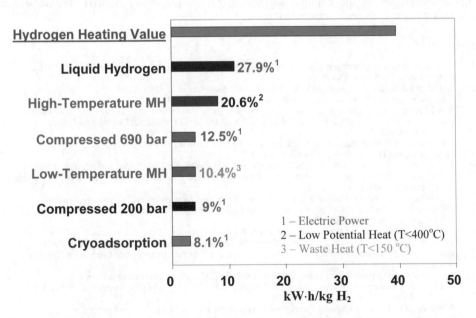

*Figure 15. Energy consumption of hydrogen storage methods*

A detailed comparative analysis of hydrogen storage alternatives resulted in the following recommendations for the applications of the different storage methods [11]:

> **Underground reservoirs and moderate-pressure chambers** are recommended for seasonal stationary energy storage;
> **High-pressure composite cylinders** are the best solution in the transport applications with large carrying capacity (buses, trucks, ferries);
> **Cryogenic high-pressure vessels and methanol reformers** are recommended in the smaller-scale transport applications (cars, tractors);
> **Cryogenic low-pressure vessels** are the best solution for the use in aviation and space applications;
> **Metal hydrides** are recommended for the use in the low-scale hydrogen storage units.

In our opinion, these recommendations are rather realistic, except for metal hydrides whose applications seem to be broader. It is caused by the fact that metal hydrides form a basis for multi-functional applications allowing to combine several options, for example, hydrogen storage, purification and controlled supply to a consumer, provided by a single device [70]. As an example, the metal hydride system intended for hydrogen recovery from ammonia purge gas (50% $H_2$), its purification (purity 99.999%), storage and transportation (200 $Nm^3$ $H_2$ in four MH containers) and supply to a glass factory in China can be given [91].

The most promising applications of the metal hydrides directly connected to alternative hydrogen storage methods involve MH-based hydrogen thermal sorption compressors as auxiliary facilities for filling gas cylinders, as well as for re-liquefaction systems in $LH_2$ hydrogen storage [92]. The main R&D activities in the implementation of this approach are in the detailed adjustment of the properties of MH, system design according to the consumer's specification.

## References

1. F. Barbir, *Review of Hydrogen Conversion Technologies*, Clean Energy Research Inst., Univ. of Miami, Coral Gables, FL 33124, U.S.A.
2. J.O'M. Bockris, and T.N. Veziroglu, *Environmental Conservation* 12 (1985) 105
3. T.N. Veziroglu and F. Barbir, *Int. J. Hydrogen Energy* 17 (1992) 391
4. T.N.Veziroglu, in: Y.Yürüm (Ed), *Hydrogen Energy System. Production and Utilization of Hydrogen and Future Aspects (NATO ASI, Series E, Vol.295)*, Kluwer Academic Publishers, Dordrecht – Boston – London,1994, p.1
5. E.E.Spilrain, S.P.Malyshenko and G.G.Kuleshov, *An Introduction to Hydrogen Energy*, "EnergoAtomIzdat", Moscow, 1984 (in Russian)
6. J.O'M. Bockris and J.C.Wass, in: T.N. Veziroglu and A.N. Protsenko (Eds.), *Hydrogen Energy Progress VII. Proc. 7th World Hydrogen Energy Conf., Moscow, U.S.S.R, 25-29 Sept. 1988, Vol.1*, Pergamon Press, New York e.a., 1988, p. 101
7. *A Multiyear Plan for the Hydrogen R&D Program. Rationale, Structure, and Technology Roadmaps*, Office of Power Delivery, Office of Power Technologies, Energy Efficiency and Renewable Energy, U.S. Department of Energy, August 1999
8. *Report to Congress on the Status and Progress of the DOE Hydrogen Program*, February 4, 1999
9. *International Energy Agency. Experience Curves for Energy Technology Policy*, OECD/IEA, 2000
10. *List of Task 12 Publications and Presentations. IEA Task 12: Metal Hydrides and Carbon for Hydrogen Storage*, 2001
11. J.Pettersson and O.Hjortsberg (Volvo Teknisk Utveckling AB), *Hydrogen storage alternatives – a technological and economic assessment*, KFB-Meddelande 1999:27, published December 1999, KFBs DNR 1998-0047
12. R.Wurster, Overview of Existing Experience in Hydrogen and Fuel Cells in Germany and Europe, *Recherches et perspectives industrielles sur la pile à combustible et l'hydrogène Paris, 13 décembre 2001*
13. D.Yu.Hamburg and N.F.Dubovkin (Eds.), *Hydrogen: Properties, Production, Storage, Transportation, Applications. Reference Book*, "Khimia", Moscow, 1989 (in Russian)
14. *Properties of Fuels*, Alternative Fuels Data Center (AFDC); US Department of Energy, Office of Transportation Technologies, http://www.afdc.doe.gov/pdfs/fueltable.pdf

15. *Hydrogen Energy Equivalents*, H–ION SOLAR, Inc, http://www.hionsolar.com/n-heq1.html
16. *Hydrogen Fuel*, http://www.dynein.com/Hydrogen/hydrogen.htm
17. G.Nicoletti, *Int. J. Hydrogen Energy*, 20 (1995) 759
18. W.Zittel and R.Wurster, *Hydrogen in the Energy Sector*, Ludwig-Bölkow-Systemtechnik GmbH, – Issue : 8.7.1996, http://www.hyweb.de/Knowledge/w-i-energiew-eng.html
19. M.P.Malkov (Ed.), *Reference Book on Physical-Technical Basics of Cryogenics*, "EnergoAtomIzdat", Moscow, 1985 (in Russian)
20. L.Schlapbach, *Materials Research Bulletin*, 27 (2002) 675
21. L.Browning, *Projected Automotive Fuel Cell Use in California*, P600-01-022, Consultant Report – Prepared for California Energy Commission, October 2001
22. R.Ramachandran and R.K.Menon, *Int. J. Hydrogen Energy*, 23 (1998) 593
23. V.V.Solovey (Ed.), *Development of the concept of increasing the efficiency of alternative energy carriers and non-conventional power sources in the production complex of the Ukraine*, Report by A.N.Podgorny Institute of Mechanical Engineering Problems of National Academy of Sciences of Ukraine, Kharkov, 2000.
24. *Composite Cylinders Latest Developments*, by Dynetek Advanced Lightweight Fuel Storage Systems TM; Asia- Pacific Natural Gas Vehicles Summit, Brisbane, Australia; April 10, 2001; Rene Rutz, VP Marketing & Business Development
25. *Quantum Technologies, Inc.*, http://www.qtww.com
26. *Hydrogen composite tank program*, Proc. 2002 U.S. DOE Hydrogen Program Review NREL/CP-610-32405
27. R.S.Irani, *Materials Research Bulletin*, 27 (2002) 680
28. *PDC Machines, Inc.* (www.pdcmachines.com), compressors data.
29. *RIX Industries* (www.rixindustries.com), RIX 3KX series compressors
30. D.Hart, *Hydrogen Power: The Commercial Future of the "Ultimate Fuel"*, Financial Times Energy Publishing, a Division of Pearson Professional Limited, 1997
31. R.D.Venter and G.Pucher, *Int. J. Hydrogen Energy* 22 (1997) 791
32. J.Wolf, *Materials Research Bulletin* 27 (2002) 684
33. *Space Shuttle Projects Office*, http://shuttle.msfc.nasa.gov
34. *KNOW HOW*, 2/2000, LINDE TECHNISCHE GASE GMBH.
35. S.M.Aceves, J.Martinez-Friaz and O.Garsia-Villazana. Low temperature and high pressure evaluation of insulated pressure vessels for cryogenic hydrogen storage, *Proc. 2000 Hydrogen Program Review*, NREL/CP – 570 - 28890
36. S.M.Aceves, J.Martinez-Frias, O.Garsia-Villazana and F.Espinosa-Loza, Performance and certification testing of insulated pressure vessels for vehicular hydrogen storage, *Proc. 2001 DOE Hydrogen Program Review*, NREL/CP – 570 - 30535
37. J.Weitkamp, M.Fritz and S.Ernst, *Int. J. Hydrogen Energy* 20 (1995) 967
38. N.L.Rosi, e.a. *Science* 300 (2003) 1127
39. L.Schlapbach and A.Züttel, *Nature* 414 (2001) 353
40. V.I.Trefilov, D.V.Schur e.a., *Fullerenes*, "ADEF Ukraine", Kiev, 2001 (in Russian)
41. A.Züttel and S.-I.Orimo, *Materials Research Bulletin* 27 (2002) 705
42. A.C. Dillon, e.a., Hydrogen storage in carbon single-wall nanotubes, *Proc. 2002 DOE hydrogen program review*, 2002, National renewable energy laboratory, Golden, CO 80401-3393.
43. M. Hirscher, e.a., *Applied Physics A-Materials Science & Processing*, 72 (2001) 129
44. A.Chambers, C.Park, R.T.K.Baker and N.M.Rodriges, *J. Phys. Chem.* B102 (1998) 4253
45. A. Züttel, e.a., *J. Alloys and Compounds* 330 (2002) 676
46. R. Strobel, e.a., *J. Power Sources*, 84 (1999) 221
47. M.G. Nijkamp, e.a., *Applied Physics A-Materials Science & Processing* 72 (2001) 619
48. R. Chahine and T.K. Bose, *Int. J. Hydrogen Energy* 19 (1994) 161
49. B.P.Tarasov, J.P.Maehlen, e.a., *J. Alloys and Compounds* 356–357C (2003) 510
50. D.V.Schur, B.P. Tarasov, e.a., in: T.N.Veziroglu et al (Eds.), *Hydrogen Materials Science and Chemistry of Metal Hydrides* (NATO Science Series II, Vol.82), Kluwer Academic Publishers, 2002, p.1
51. B.A.Troshenkin, *Circulation and film evaporators and hydrogen reactors*, "Naukova Dumka", Kiev, 1985 (in Russian)
52. M.Mueller, J.P.Blackledge and G.G.Libovitz (Eds.), *Metal Hydrides*, Academic Press, New York & London, 1968
53. V.A.Yartys, V.V.Burnasheva and K.N.Semenenko, *Uspekhi Khimii* 52 (1983) 529 (in Russian)
54. W.B.Pearson. *The crystal Chemistry and Physics of Metals and Alloys*, Russian Translation (in 2 volumes), "Mir", Moscow, 1977
55. P.Villars and L.D.Calvert. *Pearson's Handbook of Crystallographic Data for Intermetallic Phases*, Second Edition. Volume 4.– ASM International, The Materials Information Society, Materials Park, OH 44073.
56. G.Sandrock, *J. Alloys and Compounds* 293-295 (1999) 877
57. P.Dantzer, in: H.Wipf (Ed.), *Hydrogen in Metals. III. Properties and Applications (Topics in Applied Physics, Vol. 73)*, Springer-Verlag, Berlin – Heidelberg, 1997, p.279

104

58.  G.Sandrock, S.Suda and L.Schlapbach, in: L.Schlapbach (Ed.), *Hydrogen in Intermetallic Compounds. II. Surface and Dynamic Properties, Applications*, Springer-Verlag, 1992, p.197
59.  R.C.Bowman, Jr. and B.Fultz, *Materials Research Bulletin* 27 (2002) 688
60.  J.-M.Joubert, M.Latroche and A.Percheron-Guegan, *Ibid*. 694
61.  E.Akiba and M.Okada, *Ibid*. 699
62.  B.Bogdanovic and G.Sandrock, *Ibid*. 712
63.  K.J.Gross, G.J.Thomas, G.Sandrock, Hydride Development for Hydrogen Storage, *Proc. 2000 Hydrogen Program Review*, NREL/CP–570–28890.
64.  J.Tolle, *Ti- oder Ti- und Fe-dotierte Natriumalanate als nue reversible Wasserstoffspeichermaterialien*, Dissertation aus dem Max-Planck-Institut fur Kohlenforschung, Mulheim am Rhur, 1998, Bd.4, S.1–144.
65.  B.Bogdanovic, e.a., *J. Alloys and Compounds* 302 (2000) 36
66.  P.Dantzer, F.Meuner, *Materials Science Forum* 31 (1988) 1
67.  B.A.Kolachev, R.E.Shalin, A.A.Il'in. Alloys *for Hydrogen Storage: Reference Book*, "Metallurgy", Moscow, 1995 (in Russian)
68.  E.L.Huston, G.D.Sandrok, *J. Less-Common Metals* 74 (1985) 435
69.  G.G.Libowitz, K.T.Feldman Jr., C.Stein, *J. Alloys and Compounds* 253–254 (1997) 673
70.  V.V.Solovey, Yu.F.Shmal'ko and M.V.Lototsky, *Problemy Mashinostroeniya (J. of Mechanical Engineering)* 1 (1998) 115 (in Russian).
71.  Yu.F.Shmal'ko, A.I.Ivanovsky, e.a., in: T.N.Veziroglu, e.a. (Eds), *Hydrogen Energy Progress. XI. Proc. 11-th World Hydrogen Energy Conf., Stuttgart, Germany, 23–28 June, 1996*, Int. Association for Hydrogen Energy, 1996, vol. 2, p.1335
72.  V. A. Yartys, e.a., *J.Alloys and Compounds* 330-332 (2002) 132
73.  *Ergenics, Inc.* 373 Margaret King Ave. Ringwood, NJ 07456, USA; http://www.hydrides.com/
74.  *Hydrogen Components, Inc.* 12420 North Dumont Way, Littleton, CO 80125, USA; http://www.hydrogencomponents.com/
75.  *ECD Ovonics*. 2956 Waterview Drive Rochester Hills, Michigan 48309, USA; http://www.ovonic.com/
76.  *The Japan Steel Works, Ltd.* Hydrogen Energy Center. Hibiya Mitsui Bldg., 1-2, Yurakucho 1-Chome, Chioda-ku Tokyo 100-0006, Japan; http://www.jsw.co.jp/
77.  *Japan Metals & Chemicals Co., Ltd. Department of Metal Hydride Alloy Business.* 8-4 Koami-Cho, Nihonbashi, Chuo-ku, Tokyo 103-8531, Japan; http://www.jmc.co.jp
78.  *GfE Gesellschaft für Elektrometallurgie mbH.* Hoefener Strasse 45 D-90431 Nuremberg, Germany; http://www.gfe-online.de
79.  *HERA Hydrogen Storage Systems, GmbH.* 577 Le Breton Longueuil, Québec, Canada J4G 1R9; Hoefener Strasse 45, 90431, Nuremberg, Germany; http://www.herahydrogen.com/
80.  *Institute for Energy Technology*, P.O.Box 40, Kjeller, N 2027, Norway; http://www.ife.no
81.  *LABTECH Int. Co. Ltd. Alloys Research and Manufacturing.* Mladost-1, bl.25/A, Sofia-1784, BULGARIA; http://labtech.solo.bg/
82.  *Chinese Academy of Sciences.* Zhong guan kun, Beijing 100080, People's Republic of China.
83.  *PA "Moscow Polymetal Plant"*, 49 Kashirskoe Road, Moscow 116409, Russia
84.  *A.N.Podgorny Institute of Mechanical Engineering Problems of National Academy of Sciences of Ukraine*, 2/10 Pozharsky Str., Kharkov 61046, Ukraine; http://ipmach.kharkov.ua
85.  *Institute for Problems in Materials Science of National Academy of Sciences of Ukraine*, 3 Krzhyzhanovsky Str., Kiev 03180, Ukraine; http://hydro.ipms.kiev.ua
86.  T.Schucan, Case Studies of Integrated Hydrogen Energy Systems, *International Energy Agency. Hydrogen Implementing Agreement. Task 11: Integrated Systems Final report of Subtask A*, Chapter 2, IEA/H2/T11/FR1-2000
87.  J.R.Johnson, *J.Less-Common Met.*, 73 (1980) 345
88.  O.Beeri, e.a., *J. Alloys and Compounds*, 267 (1998) 113
89.  B.P.Tarasov, e.a., in: D.V.Schur, S.Yu.Zaginaichenko and T.N.Veziroglu (Eds.), *Hydrogen Materials Science and Chemistry of Carbon Nanomaterials (ICHMS'2003) VIII International Conference, Sudak – Crimea – Ukraine, September 14–20, 2003*, Kiev: 2003, p.298–301.
90.  K. Nasako, Y. Ito, N. Hiro, M. Osumi, *J. Alloys and Compounds* 264 (1998) 271
91.  M.Au, e.a., Int. J. Hydrogen Energy 21 (1996) 33
92.  V.A.Yartys and M.V.Lototsky. Metal Hydride High Pressure Hydrogen Compression: Present Status. – Review, Institute for Energy Technology, Department of Energy Systems, 2002

# ALUMO- AND BOROHYDRIDES OF METALS: HISTORY, PROPERTIES, TECHNOLOGY, APPLICATION

B.M. BULYCHEV
*M.V. Lomonosov Moscow State University, Chemistry Department,*
*119992, Moscow, Leninskie Gory, Russia*
E-mail: b.bulychev@highp.chem.msu.ru

## Summary

The paper considers a number of particular questions arising in the course of solution of the hydrogen energy problem. The issue of hydrogen storage in the ternary hydrides of light metals and aluminium (tetrahydridealuminates) is addressed in more details.

## 1. Introduction

As it follows from the analysis of literature data, the hydrogen energy problem, despite its seeming simplicity and obviousness, raises much more particular issues, extremely complicated in implementation, than all other problems connected to so-called renewable energy sources. It is caused by many reasons and, first of all, by the fact that hydrogen is the secondary energy source requiring for its production the usage of primary sources. Besides, it has such "inconvenient" for applications physical and chemical properties, as low boiling point, relative inertness of the pure gas and formation of explosive mixtures with oxygen, high enough strength of the H–H covalent bond and weakly pronounced polarization ability, etc.

The solution of the problem to use hydrogen as a fuel, in fact, can be divided into solutions of three great problems.

The **first** and the main problem is in the development of an effective method of hydrogen production, whose implementation would allow one to sell hydrogen for the price which is comparable or is not much higher than the prices of conventional energy carriers. Now this problem has not been solved. It can not be solved in the framework of traditional directions and approaches (methane pyrolysis, water electrolysis, catalytic decomposition of hydrocarbons), because in this case hydrogen never will be even comparable on its price with conventional fuels. Certainly, there can arise the circumstances forcing one to use the more expensive fuel because of serious environmental problems in a specific region. But at the same time in the more safe hitherto regions, where because of these circumstances hydrogen production will be developed, the environment will be damaged. The notion about safety of hydrogen as an energy carrier is grounded on the consideration of the single harmless reaction of its oxidation. Unfortunately, the system approach to the problem under consideration is absolutely absent, and nobody takes into account the dozens and hundreds of processes which should be realised to obtain hydrogen and to recycle wastes after its consumption.

*T.N. Veziroglu et al. (eds.),*
*Hydrogen Materials Science and Chemistry of Carbon Nanomaterials,* 105-114.
© 2004 *Kluwer Academic Publishers. Printed in the Netherlands.*

Now we can see only two methods of production of the cheap hydrogen. Both methods are fantastic enough. The first is connected to the development and usage of the thermonuclear energy which is needed to be produced yet, and its safety and ecological compatibility should be estimated and argued. The second way seems to be more "natural", and it is connected with the creation of the effective hydrogen-generating microbes, or hydrogenase-like enzymes, i.e. with bio-technology. However, even here, after decades of activities of a lot of R&D groups, a break-through is not observed. So, we guess that if the cheap hydrogen sources will be absent, hydrogen applications will be limited only by advertising prototypes and special programs.

The **third** problem connected to hydrogen burning in fuel cells seems to be the most clear up to date. However, in the course of its technical implementation the quite serious problems can arise. We are not the experts in this field, so we can not adequately estimate the complexity of these problems and perspectives of their solution in the time and engineering ranges. But the existence of these problems follows at least from the absence of a general line in the selection of layout of a fuel cell for mobile and stationary power plants. It seems that all agree that the usage of alkali electrolyte in fuel cells, requiring to burn the extra-pure gases, is a dead-end way, in spite of its highest efficiency. The solid oxide electrolytes can operate only at the temperatures more than 600 °C. The most effective way, apparently, would be in the application of a polymer electrolyte for mobile power plants. But up to now such an electrolyte is absent in the choice of electrochemists. The known Nafion unlikely will have a perspective for its large-scale use, because it is very difficult to expect the operating stability of the substance having a functional group, in contact with platinum-group metal; i.e. at the simultaneous presence of very powerful hydriding catalyst and hydrogen gas in the system. We can find more weak points in this problem, but let leave it to the experts.

The **second** problem is connected with hydrogen transportation and storage in *gaseous*, *liquefied* or *bounded* state. It is quite obvious that the first two methods, although being not very convenient, are nevertheless acceptable in the laboratory conditions and in the low-power stationary or even autonomous installations. It is conditioned by the creation of the light gas cylinders or enhanced Dewar flasks. But all the similar solutions are not acceptable in the course of realisation of **hydrogen energy** program as a whole, because this program is aimed at first to hydrogen utilisation in the mobile and autonomous devices. Therefore, the **bounded** hydrogen (the forces – chemical or physical – are not important here) is apparently the most prospective for a global solution of the second problem. Mostly, we devote the paper to the consideration of one way of such a solution.

## 2. Hydrogen storage materials: an overview

Conditionally, all the materials which are able to store bounded hydrogen can be divided starting from the bond energy of hydrogen atom or molecule either with the matrix (host material) or with the other atoms in a molecule. It is **physically** ad- or absorbed hydrogen having the weakest bond, **chemisorbed** hydrogen, and **valence** bounded hydrogen in the molecular or ionic, binary or ternary metal hydrides.

It is quite logical that the materials binding hydrogen **physically** should contain the minimal quantity of the gas and easy return it under normal conditions. Nevertheless, some agiotage arisen after the publications by N. Rodriguez et al., concerning the properties of developed in her group "herringbone" type carbon material sorbing up to 60 wt.% of molecular hydrogen [1], has shaken this assurance in some extent. However, the years passed after the first publication turned the scale at the well-known ~1–1.5 % $H_2$ which is the characteristic sorption capacity of active birchen charcoal. This fact is well explained in the framework of existent ideas about the mechanisms of sorption of gases on the neutral surfaces. The theoretical calculations carried out under the supervision of M. Dresselhaus [2] and by other authors [3] also did not bring the optimism to the scientists working in this direction. We would not like to discuss here the indistinct mechanism and reasons of unjustified high capacity of the sorption of covalent-bounded and weakly polarized gas by carbon materials which is known in the literature as "capillary condensation" [4,5]. It supposes the hydrogen liquefaction – but it is not clear, what is the reason? Let only note that at the monolayer filling of the internal surface of the carbon tube with hydrogen molecules, it can be expected the sorption of not more than 4 wt.% of the gas [2,3]. In the course of multilayer placement of the molecules, both inside and outside the carbon-contained layer, at 77 K and relatively high hydrogen pressure, this value can reach 8–14 wt.% [2,3], although it is quite not clear what can force hydrogen molecules to form two layers and to keep themselves therein. So, even if do not take into account very low values of the packing density of carbon materials[*] considered in the literature as hydrogen sorbents, the modern state of the art of the R&D in this field does not allow us to make even moderately optimistic forecast concerning the feasibility of hydrogen storage applications based on these materials.

The further progress in this direction without changing the ideology, i.e. without usage of the realistic models describing the sorption of the relatively inert gas by the inert matrix seems to have no prospects. It can be supposed that the application of active matrices obtained, for example, by modification of the carbon framework with electron-deficit atoms, and / or the preliminary activation of the gas, i.e. transfer from the physical sorption to **chemisorption** mechanism, will be more fruitful. The striking example of successful implementation of the latter approach is known from the chemistry of intermetallic hydrides. Note that the disadvantage of the IMC hydrides hampering their usage in the mobile hydrogen storage applications is connected with the low weight storage capacity, at very high density of the IMC itself. So, the storage of 1 kg $H_2$ requires to use about 50 kg of IMC, without weight of high-pressure container storing the metal and hydrogen. Nevertheless, we consider that the final word in this R&D direction has not said yet, and it can be quite favourable. So, in [6] it is shown that the composites {$MgH_2 + Mg_2NiH_4$}, obtained by ball milling and containing up to 5 wt.% H, have the properties radically differing from the properties of their components. For example, they are characterised by the one-stage hydrogen desorption at 220 °C. The results obtained in the course of studies of metal hydrides formation in the presence of carbon (graphite) [7,8] are also quite prospective, but they have not been explained yet.

**Valence** bounded hydrogen is known in the molecular or ionic, binary or ternary metal hydrides. We do not stop on the binary hydrides decomposing at high enough and sometimes hardly available temperatures. We also do not consider the reactions of

---

[*] This factor is very important for the technical implementation of the process, because even if the carbon material absorbs 10 wt.% H, then, at its packing density of 0.1 $g/cm^3$, the volume of a tank containing 1 kg of hydrogen must be not less than 100 litres.

synthesis of borohydrides, because these substances, even despite of the advertising by "Millenium Cell" Company, are of the less interest, as to their applications in the autonomous hydrogen sources, than alumohydrides. It is caused by a number of not very "convenient" properties, for example, by the extremely high toxicity of diborane which is always formed in the small quantities during reactions of hydrolysis and thermolysis of borohydrides.

## 3. Alumo- and borohydrides: state of the art

### 3.1. BRIEF HISTORICAL INFORMATION

The problem of application of the ternary metal alumo- and borohydrides as reversible hydrogen accumulators has arisen in the scientific community quite recently. This is the specific class of compounds whose studies had their peak of popularity in 1960th – 1970th. But the chemistry of complex hydrides of the IIIA-group metals has already 80 years' history. It was started in Germany in 1920th, by A. Stock's works. Later, in 1940th, these activities were developed in the works of H.I. Schlesinger' group, USA. Namely these authors have proposed all main methods of synthesis of boro- and alumohydrides of alkali metals. Their works have not lost their value and technological importance until now. In the USSR these activities were picked up in 1950th–1960th by the laboratories headed by V.I.Mikheeva (Institute of General and Inorganic Chemistry), B.M.Mikhailov (Institute of Organic Chemistry), L.I.Zakharkin (Institute of Hetero-Organic Compounds), K.N.Semenenko (Moscow State University and Institute of New Chemical Problems) and, more recently, by U. Mirsaidov (Institute of Chemistry, Tajikistan). Mainly these works were supported by military departments intended to use these substances as the components of rocket propellants. Apparently, the same motivation took place outside USSR. The plans of militaries were partially realised in 1980th, whereupon the studies on this direction had been quietly ended. The most significant results on this direction were obtained outside USSR by E.C. Ashby (USA), E. Wiberg and later, by his follower H. Noth (Germany) – he is still to work now and can be considered as a patriarchy of this field of chemistry. J. Plešek and S. Heřmanek (Czechoslovakia), B. Bonnetot, J.-P Bastide and P. Claudy (France) made significant contribution into the subject as well.

### 3.2. SYNTHESIS AND SOME SPECIFIC PROPERTIES

The main reactions of synthesis of alumohydrides are presented below. As a rule, the reactions are carried out in the medium of a donor-type organic solvent, usually in diethyl ether, tetrahydrofurane or in alcohols. The methods of synthesis of boro- and alumohydrides, as well as their properties, are described in details in numerous reviews and monographs.

Industrial technology of lithium and sodium alumohydrides which are in the most demand, as well as all further technologies of their application, are based on the reaction (eqn.1) discovered by H.I. Schlesinger et al. in 1947 [9]:

$$4MH + AlCl_3 \xrightarrow[\text{Solv}]{20\text{-}30\ ^\circ C} MAlH_4 + 3MCl \tag{1}$$

$$M - Li, Na$$

The most difficulties in the course of realisation of this reaction arise in the case of synthesis of lithium alumohydride which is now the main large-tonnage product among similar compounds. This substance is kinetically unstable, so its synthesis requires using the extra-pure raw materials. The most rigid requirements are stated for the contents of group IV and VIII transition metals therein. If these requirements are not satisfied, the alumohydride catalytically decomposes already during the synthesis.

The sodium alumohydride is stable both kinetically and thermodynamically. So, the requirements to the purity of the raw materials for its synthesis are less rigid. The direct synthesis of the alumohydrides from elements, or starting from binary hydrides, was developed by E.C. Ashby and V.V. Gavrilenko at the beginning of 1960th [10,11]:

$$Al + 2H_2 + Na \xrightarrow[\text{Ti, AOC, (Solv)}]{P,T} NaAlH_4 \tag{2}$$

$$MH + Al + 3/2 H_2 \xrightarrow[\text{Ti, AOC, (Solv)}]{P,T} MAlH_4 \tag{3}$$

$$M - Li, Na$$

This process is more appropriate economically for the preparation of $NaAlH_4$. It is carried out in the medium of an organic solvent, at $P_{(H2)}=30\text{–}300$ atm and $T=100\text{–}200$ $^\circ C$, at the presence of activators ($AlR_{3-n}X_n$) and catalysers (titanium salts). The synthesis of $NaAlH_4$ from elements in the melt (220–240 $^\circ C$, 250 atm), but also at the presence of titanium, was realised by T.N.Dymova et al. [12].

Alumo- and borohydrides of alkali-earth metals, except for magnesium alumohydride becoming recently much more studied including its application for hydrogen storage [13–15], are less available and poorly investigated [16], although some of them have high enough hydrogen sorption capacity (Table 1). It is, more apparently, caused by the difficulties of their obtaining as pure compounds, because the heterogenic reaction with the participation of the lithium alumohydride, which is the basis of their synthesis, in fact results in the formation of double salts:

$$3\ MAlH_4 + M'X_2 \xrightarrow{Et_2O} MAlH_4 \cdot M'(AlH_4)_2 + 2\ MCl \tag{4}$$

$$\text{where } M - Li, Na; M' - Mg, Ca$$

At the same time, the homogeneous reaction with the participation of $NaAlH_4$, but carried out in tetrahydrofurane, yields the strongly bounded solvate which does not release the tetrahydrofurane molecule at heating as far, as the hydride is decomposed. However, these compounds can be obtained in the non-solvated state, according to the solid state reaction under conditions of thermobaric synthesis:

$$MH_2 + 2AlH_3 \xrightarrow{P,T} M(AlH_4)_2 \qquad (5)$$
$$M-Mg,Ca$$

Also, the non-solvated alumohydrides of alkali-earth metals can be obtained by the heterogeneous exchange reaction of $NaAlH_4$ with anhydrous chlorides in diethyl ether with the consequent elimination of the ether by vacuum heating:

$$3NaAlH_4 + MX_2 \xrightarrow{Et_2O} M(AlH_4)_2 + 2NaCl \qquad (6)$$

Unfortunately, because of the extremely low solubility of $Mg(AlH_4)_2$ in the solvent, this method (eqn.6) can not be considered as a preparative one.

TABLE 1. Some properties of ternary hydrides of alkali and alkali-earth metals in decomposition reactions

| Hydride | Wt. % $H_2$ | $\rho$ g/cm$^3$ | Decomposition process | Decomposition temperature, $^\circ$C | Decomposition conditions | $H_2$ volume, m$^3$/kg |
|---|---|---|---|---|---|---|
| $NaBH_4$ | 8.3 | 1.074 | Hydrolysis | 25-30 | pH<6, catalyst | 2.48 |
| $LiAlH_4$ | 10.5 | 0.92 | Thermolysis | 25-110 | Catalyst, Ti,Fe | 0.88 |
| $NaAlH_4$ | 7.4 | 1.28 | Thermolysis | 80-150 | Catalyst, Ti,Fe | 0.62 |
| $KAlH_4$ | 5.7 | 1.33 | Thermolysis | 250-280 | | 0.48 |
| $Mg(AlH_4)_2$ | 9.4 | - | Thermolysis | 120-150 | - | 0.82 |
| $Ca(AlH_4)_2$ | 7.8 | - | Thermolysis | 130-160 | - | 0.65(?) |
| $LaNi_5H_6$ | 1.4 | ~6 | Thermolysis | <40 | | 0.15 |

All the alumohydrides, as well as $LiBH_4$, are the extremely fire-dangerous substances (A category), so any operation with them requires the special safety measures. On the opened air they are quickly and completely oxidised and hydrolysed. But in the inert medium, or in vacuum, they, except for $LiAlH_4$, can be stored for years, without change of their properties. Being contacted with liquid water, the alumohydrides ignite, often with an explosion.

3.3. STRUCTURE AND SOME PHYSICAL PROPERTIES

$LiBH_4$ is the most covalent among the alkali borohydrides. It melts at 280 $^\circ$C with decomposition evolving 12–13 % $H_2$ of its total quantity which is equal to 18.2 %. Hydrogen residue presents in the form of lithium hydride. The ionic $NaBH_4$ (10.5 % $H_2$) is melted with decomposition at 505 $^\circ$C.

In the monoclinic $LiAlH_4$ having the density of 0.9 g/cm$^3$ [17], the bond between cations $Li^+$ and anions $[AlH_4]^-$ is essentially covalent. This substance, even being extremely pure, decomposes according the reaction:

20-110°C          140-200°C

$$3LiAlH_4 \rightarrow \{[Li_3AlH_6]+3H_2\} \rightarrow 3LiH+3Al+ 9/2H_2 \quad (\Delta H^0 = -6.8 \text{ kcal/mole}) \qquad (7)$$

Reaction (7) slowly takes place already at room temperature; at T>100 °C it proceeds rapidly and more completely evolving ~7.8 % $H_2$ from the 10.5 % of the total contents [18]. Activation energy of the decomposition reaction changes from 18 to 25 kcal/mole depending on dispersity of the reactants, and quantity and quality of the impurities.

The tetragonal $NaAlH_4$ has the density of 1.28 $g/cm^3$ [19]. It is mainly ionic, stable compound. The substance melts with decomposition at 180 °C evolving ~5.6 % $H_2$ from the total 7.4 %. $Na_3AlH_6$ is formed during decomposition as an intermediate substance [10,20]. The standard decomposition enthalpy for $NaAlH_4$ is equal to −13.5 kcal/mole. The influence of the salts of transition metals on the thermal stability of sodium alumohydride is less pronounced than for $LiAlH_4$, but their introduction results in reduction of the activation energy as well.

Recently completed repeated structure investigations, as a whole, did not contribute some important changes into the knowledge about complex boro- and alumohydrides. To our regret, in a number of cases these works do not contain the references to their precursors.

## 3.4. APPLICATIONS

Both boro- and alumohydrides are the strong reducing agents. The main field of their application is organic chemistry and pharmaceutics; in inorganic synthesis they are used as starting materials for preparation of hydrides of non-transition and some transition metals. Recently they are considered as prospective materials for hydrogen storage application in autonomous power plants. As it is seen from the Table 1, in this aspect they have the more favourable physical-chemical properties than other metal hydrides, first of all, hydrogen capacities, densities and decomposition temperatures.

Two schemes of hydrogen take-off based on hydrolysis and thermolysis are available. The first scheme is irreversible; it is partially realised by "Millenium Cell", by hydrolysis of $NaBH_4$ evolving under this operation up to 20 % $H_2$. However, if to take into account the necessity of water excess in 10–15 times, the hydrogen yield per weight unit will be already not so impressive, rather vice versa. Properly speaking, this remark concerns all the cases when hydrolysis of hydrides is used for hydrogen generation. The second principal remark is connected with the fact that the problem of regeneration of the products of hydrolysis of binary or ternary hydrides, independently on their nature, can not be solved within economically acceptable framework.

In the scheme of thermal decomposition of the alumohydrides the take-off of 75 % of bounded hydrogen is planned to be carried out as a result of the "soft" themolysis reaction. To improve kinetics of hydrogen evolution and to reduce the temperature of the thermolysis, the method of alumohydrides modification by the cations of transition metals is used. This method has been proposed at the end of 1990th by B. Bogdanovich et al. [21]. The modifiers simultaneously are the catalysts for the hydriding of aluminium, in so doing they promote the closed "sorption – desorption" cycle at technologically acceptable P, T parameters. This solution really has the innovation features, and this innovation has the opposite sign as compared to the methods known in the chemistry of alumohydrides earlier – the latter, *viz.* the synthesis of these substances in organic solvents, requires very pure parent materials, especially as to the impurities of transition metals.

The novelty of B. Bogdanovich' approach and his merit is that he broke the usual stereotype and doped already finished form of sodium alumohydride by the compounds of IVB-group metals. Quite recently it was shown that the atoms of transition metal, under this way of modification, do not form their own phase, but enter into the lattice of the alumohydride creating defects, destabilising the lattice and changing the decomposition kinetics [22,23]. All this does not concern the potassium alumohydride whose "synthesis↔decomposition" reaction can be reversibly realised without modification, but, unfortunately, at the temperatures above 250 °C [24].

The decomposition of alumohydrides takes place in three stages. Only two fist stages from these three, where hexahydridoaluminate is formed and decomposed, have the temperatures acceptable for applications. The figures in the Table 1 showing the quantity of hydrogen generated by the alumohydride belong to namely these two stages. It follows from there that ¼ of hydrogen is left in the bounded state, in the form of binary hydride decomposing above 450 °C:

$$3NaAlH_4^* \xrightarrow{\;80\text{-}100^\circ C\;} Na_3AlH_6^* + 2Al + 3H_2 \xrightarrow{\;120\text{-}150^\circ C\;}$$

$$\xleftrightarrow{} 3MH^* + Al + 3/2H_2 \xrightarrow{\;>450^\circ C\;} M/Al^* + 3/2H_2$$

The most unpleasant property of alumohydrides is their extremely high chemical activity. The modification makes alumohydrides more dangerous in this respect. So, if kinetically unstable but pure lithium alumohydride is decomposed to hexahydridoaluminate at room temperature in several years, then the doped one – in several weeks or even days. Pure sodium alumohydride is stable both kinetically and thermodynamically. But in the presence of the doping agent the activation energy of its decomposition is sharply reduced. The temperatures when the decomposition becomes visible are reduced correspondingly. Also, it is natural that the regeneration of the alumohydrides from {MH + Al} mixtures being the products of their decomposition obligatory requires the usage of extra-pure hydrogen, which has passed the purification on palladium membrane or intermetallic hydrides. It is caused by extremely high sensitivity of the residue to oxygen and water.

Nevertheless, this direction seems to be the most prospective among all known variants using the bounded hydrogen. However, taking into account chemical properties of alumohydrides, it is difficult to imagine in a car an artificial "bomb" made of the chemical substance where the bonds are specially weaken by modifiers. The requirement to apply the extra-pure hydrogen for charging the alumohydride storage unit can much more complicate the practical implementation of the idea. The usage of this material in a car by a man, who has not learnt yet to use gas cylinders, seems to be unbelievable as well. But in some special applications the activated alumohydrides, undoubtedly, can be successfully used as reversible hydrogen storage materials.

## 4. Concluding remark

Finally, I would like to return again to the problem of physically adsorbed gases. Note one more, that I have the great pessimism as to a possibility of using carbon materials for these purposes. This way can lead to success only by pure accident, or, as it was mentioned above, under reasonable and scientifically grounded **transfer from the mechanism of physical sorption to the mechanism of chemisorption**. In fact, it does mean to start the new direction from scratch. Therefore, we consider as more perspective the development of recently appeared activities on purposive synthesis of low-density framework compounds. Figs. 1, 2 show the structure of two representatives of such compounds. As it is seen, these substances are the ideal objects of supramolecular chemistry. Benzyle-substituted bis-carboxylated zinc complex [25] (Fig. 1) is able to absorb 4.5 wt.% of hydrogen at 77 K, or 17.2 hydrogen molecules per formula unit. At room temperature and pressure of 20 atm the sorption capacity is about 1 wt. %, or ~2 molecules per formula unit. The sphere inside cubic framework is the Van Der Waals void whereto hydrogen is included. Increasing the void by the increase of the substituent volume (e.g. replacing the benzene with naphthalene), it is possible to increase hydrogen capacity up to 9 $H_2$ molecules per formula unit at room temperature and 20 atm.

Dipiridine complex of copper hexafluorosilane [26] (Fig. 2) is able to absorb at 36 atm and 20 °C 6.5 mmole of methane per 1 g of the matrix, against 3.7 mmole for the 5A zeolite; the density of sorbed methane is of 0.21 g/ml against 0.16 g/ml for compressed methane (280 atm, 27 °C). There is no information about hydrogen sorption by this substance, although the problem of methane sorption today is not less and, maybe, more practically important than that for hydrogen.

In the considered examples, in contrast to questionable works on hydrogen sorption in carbon materials, all is definite and transparent. Their authors do not break any law of physical sorption of gases, it is not necessary to operate by the farfetched explanations of the high sorption capacity, like "capillary condensation"; and mainly, there appears the perspective of really purposeful synthesis of matrices which is based on direct structure investigations.

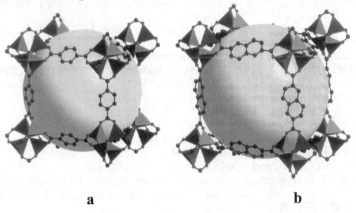

a         b

*Figure 1. Structures of (a) $Zn_4O(1,4$-benzenedicarboxilate)*
*and (b) $Zn_4O(1,4$-naphtalededicarboxilate)*

114

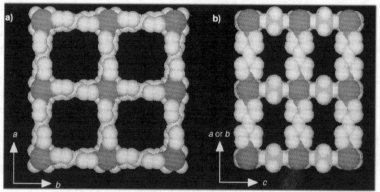

*Figure 2. Structure of [{CuSiF₆ (4,4'-bipyridine)₂}ₙ]*

## References

1.  A.Chambers, C.Park., R.T.K. Baker, N.M. Rodriguez, *Phys.Chem. B.*, **1998**, *102*, 4253.
2.  M.S. Dresselhaus, K.A. Williams, P.C. Eklud, *MRS Bulletin*, **1999**, *24*, 45.
3.  S.M. Le, Y.C. Choi, Y.S. Park, J.M. Bok, D.J. Bae, K.S. Nahm, Y.G. Choi, S.C. Yu, N.-g. Kim, T. Frauenheim, Y.H. Lee, *Synth.Metals*, **2000**, *113*, 209.
4.  M.R. Pederson, J.Q. Broughton, *Rhys.Rev.Lett.*, **1992**, *69*, 2689.
5.  A.C. Dillon, K.M. Jones, T.A. Bekkedahl, C.H. Kiang, D.S. Bethune, M.J. Heben, *Nature*, **1997**, *386*, 377.
6.  A. Zaluska, L. Zaliski, J.O. Ström-Olsen, *Appl.Phys. A.*, **2001**, *72*, 157
7.  A. Zaluska, L. Zaliski, J.O. Ström-Olsen, *J.Alloys and Comp.*, **2000**, *298*, 125.
8.  H. Imamura, S.Tabata, N. Shigetomi, Y. Takesue, Y. Sakata, *J.Alloys and Comp.*, **2002**, *330-332*, 579.
9.  A.E. Finholt, A.C. Bond, H.I. Schlesinger, *J.Am.Chem.Soc.*, **1947**, *69*, 1199.
10. L.I.Zakharkin and V.V.Gavrilenko, *Doklady AN SSSR (Reports of the USSR Acad. Sci.)*, **1962**, *145*, 793.
11. E.C. Ashby, G.J. Brendel, H.E. Redman, *Inorg.Chem.*, **1963**, *2*, 499.
12. T.N.Dymova, N.G.Eliseeva, S.I.Bakum, Yu.M.Dergachev, *Doklady AN SSSR (Reports of the USSR Acad. Sci.)* **1974**, *215*,1369.
13. K.N.Semenenko, B.M.Bulychev, K.B.Bitsoev, *Vestnik MGU (Bull. of Moscow State University, Series Chemistry)*, **1974**, 185
14. M. Fichtner, O. Fuhr, O. Kircher, *J.Alloys and Comp.*, **2003**, *356-357*, 418.
15. M. Fichtner, J. Engel, O. Fuhr, A. Glöss, O. Rubner, et. al. *Inorg.Chem.*,**2003**,*42*, 7060.
16. B.M.Bulychev, V.K.Belsky, A.V.Golubeva, P.A.Storozhenko, *J. Neorgan. Khimii (Russ. J. Inorg. Chemistry)*, **1983**, *28*, 1131.
17. N. Sklar, B. Post, *Inorg.Chem.*, **1967**, *6*, 669.
18. E.C. Ashby, *Advances in Inorganic and Radiochemistry*, **1966**, 8, 283.
19. J. Lauher, W. Dougherty, P.J. Herley, *Acta Crystallogr.*, **1979**, *B35*, 1454.
20. E.C. Ashby, P. Kobetz, *Inorg. Chem.*, **1966**, *5*, 1615.
21. B.Bogdanović, G. Sandrock., *MRS Bulletin*, sept. **2002**, 712
22. H.W.Brinks, B.C.Hauback, P. Norby, H. Fjellvåg, *J.Alloys and Comp.*, **2003**, *351*, 222.
23. D. Sun, T. Kiyobayashi, H.T. Takeshita, N. Kuriyama, C.M. Jensen, *J.Alloys and Comp.*, **2002**, *337*, L8.
24. H.Morioka, K.Kakizaki, S.-C. Chang, A. Yamada, *J.Alloys and Comp.*, **2003**, *353*, 310.
25. M.L. Rosi, J. Eckert, M. Eddaoudi, D.T. Vodak, J. Kim., M. O'Keeffe, O.M. Yaghi, *Science*, **2003**, *300*, 1127
26. S.-I. Noro, S. Kitagava, M. Kondo, K. Seki, *Angew.Chem.Int.Ed.*, **2000**, *39*, 2082.

# STRUCTURAL – PHASE TRANSFORMATIONS
# IN TITANIUM – FULLERENE FILMS AT IMPLANTATION OF BORON IONS

BARAN L.V.[(1)], SHPILEVSKY E.M.[(2)], OKATOVA G.P.[(3)]
[(1)]Belarusian State University, 4 F. Skorina Av., Minsk 220050, Belarus,
e-mail: brlv@mail.ru
[(2)]Institute of Heat and Mass Transfer, 15 P. Brovki Str., Minsk 220072, Belarus
[(3)]NII Powder Metallurgy, 41 Platonov Str., Minsk 220071, Belarus

## Abstract

Three-layered titanium-fullerene-titanium films are obtained by thermal evaporation method in vacuum at residual vapor pressure of $1{,}3 \cdot 10^{-3}$ Pa in the chamber. Titanium – fullerene films were implanted with boron ions of 80 keV energy, at this the doze of implantation was $1 \cdot 10^{16}$ ion/sm$^2$. Structural - phase transformations in films were studied by means of X-ray diffraction, raster electronic microscopy, Auger spectroscopy and X-ray spectrum microanalysis. It is established, that during condensation of a fullerite layer on spreading titanium layer, and then a titanium layer on a fullerite layer there is an intensive diffusion as titanium in a layer of fullerite, and fullerene molecules in a layer of the titanium. Boron ion implantation (E = 80 keV, D = $1 \cdot 10^{16}$ ion/sm$^2$) of three-layered titanium – fullerenes – titanium films results in formation of new phase of $Ti_xO_yC_{60}$, mixing titanium and fullerene layers, increase oxygen atom part in the films as well.

*Keywords:* A. Fullerene; B. Coating, implantation; C. Scanning electronic microscopy, X-ray diffraction; D. Microstructure.

## 1. Introduction

Studying the titanium – fullerene films causes considerable practical interest from the point of view of formation of new compounds with unique physical and chemical properties. High characteristics of physical and chemical properties of titanium carbide allow coming out with the assumption of uniqueness of properties of compounds of titanium with fullerenes. The studies carried out in the works [1-3] have shown that the formation of new phases in the system Ti – $C_{60}$ occurs both at the condensation stage of films and in the process of the subsequent annealing, in this case large mechanical stresses, which lead to the disturbance of the adhesion of backed films can appear. So, while single-layered titanium – fullerene films [1] were obtained from the combined atom-molecular stream in ultrahigh vacuum, it was established that the phase $Ti_{5,3}C_{60}$ is formed. However, after endurance in the air the films were oxidized, and $Ti_xO_yC_{60}$ was formed.

One of the widespread methods of synthesis of new phases in thin-film structures is ion implantation. Investigations of interaction of the accelerated ions with fullerene films, carried out in works [4-10], have shown, that even at low energy of implanted ions

115

*T.N. Veziroglu et al. (eds.),*
*Hydrogen Materials Science and Chemistry of Carbon Nanomaterials,* 115-122.

(< 100 keV) at the certain parameters of collision with molecules of a matrix there is destruction of $C_{60}$, and the high factor of dispersion is observed. The problem of interaction in layered metal – fullerene structure at ionic influence, when a metal layer protects a fullerene film is of great scientific interest.

The purpose of the present work is studying structural – phase transformations in the titanium – fullerene films at implantation by boron ions.

## 2. Technique of experiment

Titanium – fullerene films were prepared using thermal vacuum evaporation plant "VUP–5M". The layers of metal and fullerene were consecutively deposited onto the oxidized single-crystal silicon plate at substrate temperature 293 K. Fullerene sublimation took place from the effusion

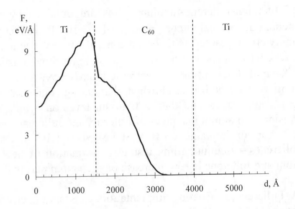

Figure 1. Profile of distribution of the energy allocated by elastic collisions by boron ions in films
Ti (d = 120 nm) – $C_{60}$ (d = 250 nm) – Ti (d = 150 nm)

cell at the temperature 773 K. Three-layer films Ti (d = 120 nm) – $C_{60}$ (d = 250 nm) – Ti (d = 150 nm) were obtained. Titanium – fullerene – titanium films were implanted by boron ions of 80 keV energy. The implantation dose for the films was $1 \cdot 10^{16}$ ion/sm$^2$ at the ion current density 3,5 µkA/sm$^2$. Thickness of the layer was selected in such a way that the maximum of the defects, created by implantation, would fall to the interface of the layers of titanium and fullerite from the side of implantation, and the distant interface $C_{60}$ – Ti was not reached by the ions. At such an experiment scheme, the boundary surface can serve as a reference point for the analysis of diffusion processes, occurring on near border of the unit as a result of the cascade of collisions, caused by the accelerated ions at penetration into a film. Results of calculation of defect distribution profile in implanted layered structure with the help of program TRIM are submitted in the Fig. 1. Rentgenphase studies were carried out on the diffractometer DRON - 3.0 in copper $K_\alpha$ – radiation with the application of automation system on the base of personal computer, which

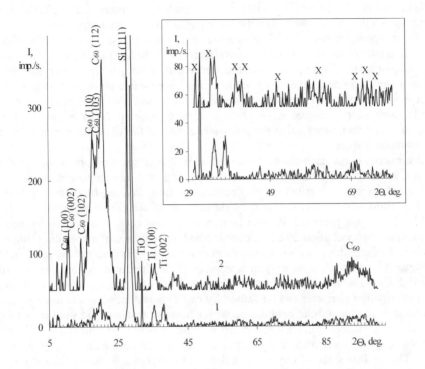

Figure 2. X-ray diffraction patterns of three-layer film Ti – C$_{60}$ – Ti, applied on oxidized Si:
1 — unimplanted film; 2 — implanted by boron ions film

scanning electronic microscope LEO 1455 VP firm Carl Zeiss, the analysis of element structure was using X-ray spectrum michroanalyser by the firm RÖNTEC and Auger spectrometer PHI – 660 by the firm Perkin Elmer.

## 3. Results and discussion

By the method of X-ray diffraction it is established that the fullerite structure is formed during the condensation on the base layer, identified in the hexagonal syngony (Fig. 2), and the subcrystalline structure of titanium α – modification. The intensive line of titanium mono-oxide is presented too in the diffractogram. Presence the titanium oxide in just prepared films is caused by sorption properties of the titanium and low vacuum in the chamber (pressure residual vapor gas at reception of films was $1,3 \cdot 10^{-3}$ Pa).

Implantation of titanium – fullerene films by boron ions leads to the essential structural-phase changes. Apparently from Fig. 2, the intensity of the X-ray reflections of fullerite phase on the survey spectrum of X-ray diffraction increases by almost 10 times. There is a number of new lines of the small intensity designated by the letter X in the diffractogram. The sharp increase in intensity of X-ray reflexes can testify to course recrystallization processes in fullerite phase, as well as the occurrence of new diffraction maxima may indicate a new phase formation. The intensity of Ti line decreases, what can be explained by the radiation- accelerated diffusion of titanium atoms into the fullerite matrix. The line of TiO (102) phase, which was presented in the diffractogram of titanium-fullerene sample before the ionic implantation, also disappears.

As a result, during ion implantation both amorphous clusters (the disorder formations), and clusters with correct packing (crystal clusters) are formed. The increase in a doze of an irradiation, strengthening amorphism, results in that the crystal clusters, arising in an amorphous layer as dissemination of more stable phase, start to grow due to amorphous clusters. The parity of phases in an implanted layer is formed as a result of development of two processes — process of atom hashing at an initial stage of an irradiation and their segregation process under action of forces of the chemical nature in the subsequent stage.

As investigations have shown by a method of Auger spectroscopy, concentration of oxygen in implanted films increases in comparison with not implanted ones. Typical profiles of elements distribution on thickness of three-layered titanium – fullerenes – titanium films with condensed on the oxidized silicon are presented in the Fig. 3. It is visible, that in just prepared films on borders of the unit there is a peak of oxygen with the maximal concentration 20 at. %, chemisorbed on a surface of a spreading film of the titanium. In the bottom titanium film concentration of oxygen does not exceed 7 at. %. In top layer Ti the contents of oxygen is appreciably higher, and monotonous growth as approaching the surface takes place: from 6 % in depth of a film up to 60 % on a surface. Such distribution character can be caused by two circumstances: in the lowered density of a layer of the titanium in comparison with a massive material and storage of samples on air. The increased contents of oxygen on distant border of unit $C_{60}$ – Ti is caused sorption, oxygen microvoids at contact to an atmosphere before drawing of the second layer. The nuclear share of oxygen in fullerite film makes less than 1 %, and the same quantity is revealed also the titanium. This fact allows asserting that titanium single-oxide is dissolved in fullerite matrix.

Distribution of carbon in just prepared films has the trapezoidal form with the maximal concentration at a level of 98 %. The small part of carbon lies on border of the unit between films, creating a transitive layer that testifies to hashing atoms of the titanium and fullerene molecules already at a stage of reception of samples. It is established by the method of Auger spectroscopy and scanning electron microscopy that after ion implantation there is an increase of the atomic fraction of carbon in the upper film of titanium occurs (Fig. 3, 4). In the electronic photograph of the structure of the implanted film (Fig. 4) are visible the bright formations with the dimension of 100...250 nm, some of them reach 500...700 nm. It is established by the method of X-ray spectral microanalysis that these particles consist of Ti, C and O. An increase of the atomic fraction of carbon phase in the titanium layer, it can be explained by the radiation- accelerated diffusion and hashing of layers during implantation.

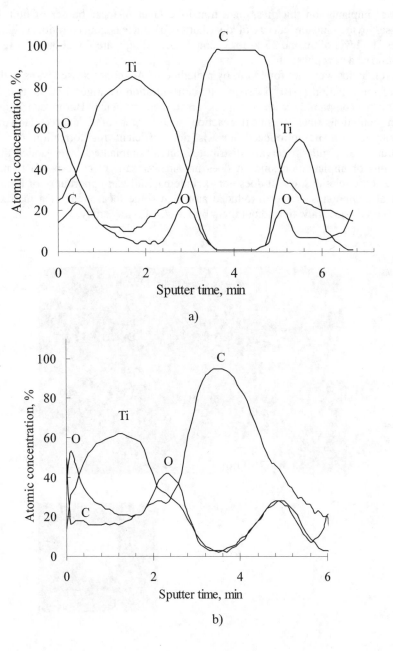

Figure 3.  AES depth profile of Ti (d = 120 nm) – $C_{60}$ (d = 250 nm) – Ti (d = 150 nm) thin film:
a — unimplanted film;
b — implanted by boron ions film

After implantation thickness of a transitive layer on near border of unit Ti – $C_{60}$ increases twice, and on curves of distribution of nuclear concentration (Fig. 3) half-glasses at a level of 30 and 35 % for C and Ti accordingly are found out, that testifies to formation of a new phase.

Boron atoms were not found out by a method of Auger spectroscopy, as at the given doze of implantation ($1 \cdot 10^{16}$ ions/sm$^2$) its concentration averages 1 at. %, that for easy elements is comparable with a margin error measurements. Boron can enter under certain conditions into chemical reaction with atoms of a matrix. However, the concentration of the introduced ions is not sufficient for formation of chemical compounds in regular intervals distributed on all implanted layer under the given conditions of implantation since it does not answer satisfy (correspond) to necessary stochiometric relationships. It does not exclude basically an opportunity of formation of chemical compounds of the introduced atoms in some local areas of a matrix, which small sizes do not allow to find out them by means of X-ray study.

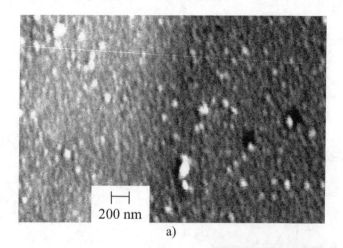

200 nm

a)

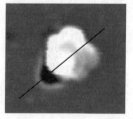

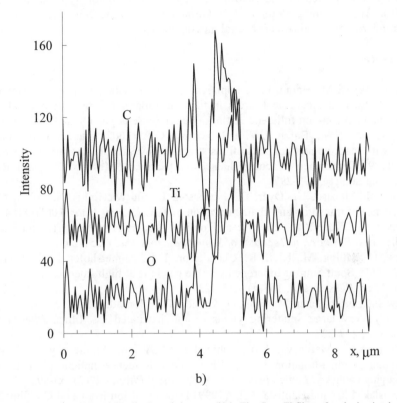

b)

Figure 4. Microstructure (a) and distribution of elements (b) in Ti – $C_{60}$ – Ti films after the ion implantation.

After the ion implantation, the atomic fraction of oxygen increases throughout the entire depth of upper film twice that is caused by diffusion of oxygen from an atmosphere (Fig. 3). Thus in a zone of border of the unit where the maximum of radiating defects is located, ascending diffusion is observed. The atomic fraction of oxygen and titanium in fullerite film increases and consists 3…4 %. Probably, the disappearance of the lines of titanium oxide in difractogram is related to the formation of the new thermodynamic more advantageous phase: $Ti_xO_yC_{60}$.

## 4. Conclusions

1.   In the process of the condensation of fullerite layer to the underlying titanium layer, and then titanium layer to the fullerite layer, the intensive diffusion occurs for both titanium into the fullerite layer and fullerene molecules into the titanium layer.
2. Boron ion implantation (E = 80 keV, D = $1 \cdot 10^{16}$ ion/sm$^2$) of three-layered titanium – fullerenes – titanium films results in mixing titanium and fullerene layer and formation of new phase $Ti_xO_yC_{60}$.
3. Redistribution of composition in implanted layered structures Ti – $C_{60}$ – Ti, taking place at their storage on air, is caused by diffusion of oxygen from an atmosphere.

122

This work is supported by the Belorussian republican fund for basic research (Grant P01-116). Authors are grateful to S.V. Gusakova and V.A. Ukhov for carrying out electron-microscopic and Auger spectral measurements.

## 5. References

1. Shpilevsky, E.M., Baran, L.V., Shpilevsky, M.E., Chekan, V.A. and Okatova, G.P. (2002) Internal mechanical stress in titanium-fullerene films. Extended abstracts, 2nd biennial conf. on fullerenes. Minsk (Belarus): Belarusian State University, 38.
2. Shpilevsky, E.M., Baran, L.V. and Okatova, G.P. (2002) Interaction of fullerenes $C_{60}$ with titanium and copper in films structures by vacuum annealing. Proceedings of the 9$^{th}$ Conference «Vacuum shience and technics». Moscow (Russian): Russian Vacuum Society, 293-297.
3. Norin, L., Jansso, U., Dyer, C., Jacobsson, P. and McGinnis, S. (1998) On the existence of transition and characterization of $Ti_xC_{60}$, *Chem. Mater* **10**(4), 1184-1186.
4. Makarova, T.L. (2001) Electric and optical properties monomeric and hardened fullerenes, *Physics and Technics of Semiconductors* **35**(3), 257-293.
5. Fink, D., Müller, M., Kleff, R., Chadderfon, L.T., Palmetshofer, L., Kastner, J., *et al.* (1995) Sputtering of fullerene by noble gas ions at high fluences, *Nucl. Instrum. and Meth. Phys. Res. B.* **103**(4), 415-422.
6. Zawislak, F.C., Behar, M., Fink, D., Grande, P.L., Jornada, J.A.H. and Kaschny, J.R. (1997) Very large sputtering yields of ion irradiated $C_{60}$ films, *Phys. Lett. A.* **226**(3-4), 217-222.
7. Vasin, A.V., Matveeva, L.A., Juhimchuk, V.A. and Shpilevsky, E.M. (2001) Structural transformation in $C_{60}$ films under influence helium plasmas of the decaying category, *Letters in Journal of Technical Physics* **27**(2), 65-69.
8. Zawislak, F.C. and Baptista, D.L. (1999) Damage of ion irradiated $C_{60}$ films, *Nucl. Instrum and Meth. Phys. Res. B* **149**(3), 336-342.
9. Jin, J., Khemliche, H., Prior, M.H. and Xie, Z. (1996) New highly charged fullerene ions: Production and fragmentation by slow ion impact, *Physical Review. A.* **53**(1), 615-618.
10. Foester, C.E. and Serbena, F.C. (1999) The effect of fluence on hardening of $C_{60}$ films irradiated with He and N ions, *Nucl. Instrum and Meth. Phys. Res. B.* **148**(1-4), 634-638.

# FIELD EMISSION INVESTIGATION OF CARBON NANOTUBES DOPED BY DIFFERENT METALS

NIKOLSKI K.N.[1], BATURIN A.S.[1], BORMASHOV V.S.[1], ERSHOV A.S. [1], KVACHEVA L.D.[2], KURNOSOV D.A.[1], MURADYAN V.E.[3], ROGOZINSKIY A.A.[4], SCHUR D.V.[4], SHESHIN E.P.[1], SIMANOVSKIY A.P.[4], SHULGA Yu.M.[3], TCHESOV R.G.[1], ZAGINAICHENKO S.Yu.[4]

[1]  Moscow Institute of Physics and Technology,
Institutskij per., 9, Dolgoprudny, 141700, Russia
[2]  A.N. Nesmeyanov Institute of Organoelement Compaunds of RAS,
28 Vavilov St., Moscow, 111991, Russia
[3]  Institute of Problems of Chemical Physics RAS,
Chernogolovka, 142432, Russia
[4]  Institute for Problems of Materials Science of NAS of Ukraine,
Krzhizhanovsky str., 3, Kiev-142, 03680, Ukraine

FAX: +7 095 409 95 43;  e-mail: nkn@lafeet.mipt.ru

## Abstract

Carbon nanotubes (CNT) possess very promising properties for field emission applications. At this paper the two methods of field emission properties improving of multi-wall nanotubes (MWNT) are described. The first is filling nanotubes by barium with aim to decrease work function of CNT cathodes. The nanotubes synthesized by arc discharge technique were selected for the tests. CNTs were boiled in the nitric acid solution containing the barium nitrate. Another method is doping the CNTs during arc discharge synthesis. By this technique nanotubes were doped with the following metals: Zr, Fe, Ni, Mn, Ti, Cu, Co.

The cathodes for field emission tests were produced by the screen-printing technique. Field emission tests showed that filling nanotubes by barium allows reduction cathode's operating voltage by 30%.

Keywords: A. Carbon nanotubes; B. Arc discharge, Doping; D. Field emission

## 1. Introduction

There are lot of areas of field emission cathode (FEC) applications: flat displays with high brightness, large view angle and high drawing speed; high efficiency light sources; compact x-ray sources; electron gun and over vacuum electronic devices. The key problem for custom application of FECs is they high operating voltage. There are two ways of cathodes operating voltage reduction: to increase the field amplification factor by cathode geometry optimization and reduction of cathode work function.

T.N. Veziroglu et al. (eds.),
Hydrogen Materials Science and Chemistry of Carbon Nanomaterials, 123-130.
© 2004 Kluwer Academic Publishers. Printed in the Netherlands.

From 1991 than Iijima [1] discover the nanotubes in they are became are popular materials in different field of science. Carbon nanotubes (CNTs) possess very promising properties for field emission applications. [2 and references therein]. The first papers described nanotubes as field emission source appears in 1993–1994 [3–6]. Due to CNTs specific shape CNTs provide essential amplification of electric field, which cause reduction of operating voltage for CNT–based field emission cathodes (FEC) in comparison with other carbon cathodes.

At present time there are two main techniques of CNT synthesis: arc discharge method [7] and chemical–vapor deposition (CVD) method [8]. The CVD method allows growing nanotubes according to the stipulated pattern directly on the base, which further can be used as a field emission cathode (see Fig. 1). Initially researchers connect a lot of prospects to this method. Nevertheless it has one substantial disadvantage — high deposition temperatures (≈900 °C), which make impossible the use of glass bases and increase of cathode cost.

Fig. 1. SEM images of CNTs FEC made by CVD technique.
(Image from Chemical Physics Letters 312, 1 1999 461-469
Field emission investigation of carbon nanotubes doped by different metal

In case of the arc discharge method the first step is a production of nanotubes. Then they are deposited to a base with the help of different methods (electrophoresis, printing process). FECs making by printing process is generally used. This method allows getting cathodes with different patterns and practically on any base. Besides, the printing technology is quite developed and very cheap.

At this work we try to make low field emission cathode from arc discharge producing multi-wall nanotubes (MWNTs). At thermal cathode producing the core covering by active layer is frequently used [9]. The barium often applies as activators. The dipole layer formation on the core surface leads to the potential barrier reduction and as a result to the reduction of work function. The minimum work function is achieved at monolayer activator.

The method of improving carbon materials emission properties by the metals having low work function, which allows considerable improvement of their field emission properties, was described at [10]. By now the doping was tried for FPG-6 graphite cathodes and carbon fibers cathodes [10]. In all cases the reduction of cathode's operating voltage was observed. Also results of field emission investigation of CNTs doped by different metal are represented at this paper. The nanotubes were prepared in arc discharge. In the synthesis process they were doped by Zr, Fe, Ni, Mn, Ti, Cu, Co.

## 2. Experimental

### 2.1. Nanotubes doped after synthesis

We represent the results of studying of FECs based upon CNTs prepared by arc discharge method. MWNTs columnar structure have been chosen for investigation. Nanotubes were synthesized as follows: 8 mm of the graphite rod was evaporated at the current 90 A in the helium atmosphere under the reactor's pressure 500 Torr. The used method described in detail at [11]. The typical size of MWNTs is 5-40 nm in diameter and about 5 μm length.

After CNT production the special method was used for cathode emission properties improving. The powder containing MWNTs was boiled during 24 hours in the nitric acid solution containing the barium nitrate. After boiling the solution was filtrated and the supported nanotubes was dried at the temperature of 110°C in air atmosphere. The barium work function is 2.49 eV in comparison with 4.7 eV for carbon and it have to cause the decreasing of CNT cathode work function. The TEM-image supported barium MWNTs shown in the Fig. 2.

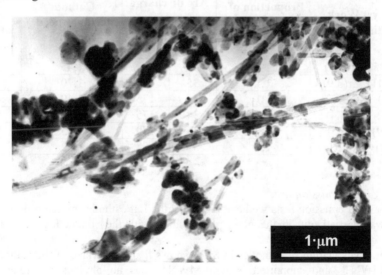

Fig 2. TEM-image of nanotubes powder: black particles — barium.

## 2.2. Nanotubes doped in the process of their synthesis

Various metals are often used as catalysts in the process of producing nanotubes by the arc discharge method. The catalyst's choice strongly influences the properties of produced nanotubes. One of the reasons of this phenomenon is doping nanotubes with the catalyst's material. To study the influence of doping on the carbon nanotubes' field emission properties we have fulfilled the experiments with carbon nanotubes doped with various metals during the synthesis process.

Carbon nanotubes were synthesized in the arc discharge. To get the doped nanotubes we have inserted a catalyst in the reactor. For this purpose we have drilled the hole of 3 mm diameter and 100 mm depth in the graphite anode along its axis. Then we prepared the mixture of the graphite and the introduced catalyst. The catalyst's quantity was calculated so that its mass would be 2–4% of the total mass of the evaporated anode. This mixture was being milled in the ball mill during 1.5 hours, which conduces to getting more uniform mass. Then this mass was inserted into the drilled hole and compacted. Further the anode was mounted in the reactor. After pumping out and washing with helium the reactor's chamber, we heated the anode during 1 minute in the short circuit mode (the cathode and the anode in the contact position). Then the anode was disconnected from the cathode and evaporated at the helium pressure of 90 Torr in the chamber. The temperature of the cooled surface of the reactor did not exceed 40°C. Further the reactor was unloaded and the produced cathode deposit was milled in the ball mill during 1.5 hours. The table 1 represents the technological parameters of the doped carbon nanotubes' synthesis.

TABLE 1. Conditions of the synthesis of nanotubes doped with different metals

| Catalyst Work function, (eV) | Proportion of the catalyst components | Arc discharge parameters | | Cathode feeding speed, mm/min | Deposit of the anode's mass, % |
|---|---|---|---|---|---|
| | | U, V | I, A | | |
| Ti(3.95)+Zr(3.9) | 1:1 | 35 | 180 | 1.45 | 50.0 |
| Co(4.41)+Zr(3.9) | 1:1 | 35 | 180 | 1.70 | 40.1 |
| Zr(3.9)+Mn(3.63) | 1:1 | 36 | 175 | 1.49 | 49.2 |
| Zr(3.9)+Cu(4.4) | 1:1 | 34 | 175 | 1.73 | 55.1 |
| Zr(3.9)+Ni(4.5) | 4:1 | 35 | 170 | 1.65 | 63.9 |
| Zr(3.9)+Fe(4.31) | 1:1 | 33 | 185 | 1.72 | 14.7 |

## 2.3. Field emission tests

For field emission tests we have produced the flat cathodes by the printing process. Printing was carried out on the glass plate coated with the conducting ITO layer. The thickness of the layer was 20 μm.

The investigation of cathodes in SEM shows that the surface of the cathodes made of pure MWNTs and supported barium MWNTs has no obvious difference in they geometry. Lets consider in future that cathode have the same geometry and due to same field enchantment factor. From the SEM images of the FEC made of the powder containing the carbon nanotubes doped during the synthesis follows that nanotubes are practically absent at the cathode's surface.

The FECs was installed into diode testers. The anode–cathode distance was fixed by the glass spacers and was equal to 300 µm. The glass plate coated with the ITO conducting layer and the phosphor layer was used as an anode. The construction of the tester showed on the Fig. 3. Each tester consists of two pair of anode and cathode for control of cathodes reproducible. Field emission properties were measured in vacuum chamber with residual gas pressure lower than $5\times10^{-7}$ Torr.

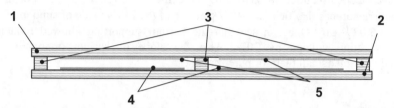

Fig. 3. The construction of the tester: anode plate — 1; cathode plate — 2; glass spacers — 3; cathode coating — 4; phosphor — 5.

The schematic of the experimental set-up is shown in Figure 4. The voltage was measured via a high voltage divider ($R_1 = 500$ MOhm, $R_2 = 500$ kOhm), and the emission current was measured as a voltage drop on resistor ($R = 51$ kOhm) in cathode circuit. Voltage and current signal were connected to the ADC channels. The precision (one ADC reading) of current measurements was $\delta I = 0.46$ µA and of voltage measurements was $\delta V = 2.62$ V.

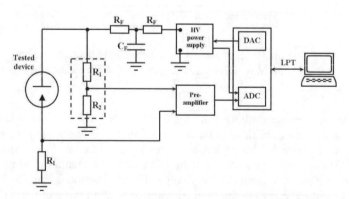

Fig. 4. Scheme of experimental set-up. ADC – analog-to-digital converter and DAC – digital-to-analog converter.

128

I-V characteristic measurements were carried out in the "fast" regime. To perform such measurements the computer sends the switch off control signal to the high voltage power supply. Then a set of 4000 current and voltage values is measured with time period 100 μs. Measurement of one characteristic takes 400 ms.

## 3. Results and discussion

The Fig. 5 represents the current–voltage characteristics of the FECs made of pure nanotubes and supported barium MWNTs. As criteria of cathodes comparison we used the value of electric field required for achieving the emission current of 50 μA. For the FECs made of pure nanotubes this field is 4.45 V/μm, for of the FECs made of supported barium MWNTs — 3.02 V/μm. One can see that the barium supporting allowed reducing the cathode's operating voltage by 30%. Also the slope of barium improved cathode characteristic is rather small than for pure MWNTs.

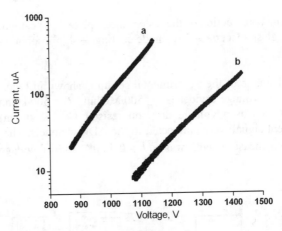

Fig 5. Current–voltage characteristics of the cathode made of the nanotubes filled by barium MWNTs — a, and made of the pure nanotubes — b.

As we mentioned above from the SEM images one can conclude that the cathodes have the same field amplification factor. Therefore the difference in the emission properties will be caused only by the presence of the doping material. In connection with the FN theory [12] one could estimate work function from the relation of slopes. The cathode's work function one can found that the change of work function is equal to 0.77 eV.

We did not carry out the field emission tests of all types of nanotubes doped during synthesis process. They would not allow unambiguous determining the influence of doping on the emission properties of nanotubes because of their small content. The absence of nanotubes in the powder can be explained by the long milling of the carbon material, produced in the reactor, in the ball mill. We can test only two types of synthesis doped CNTs: doped with Zr + Fe and doped with Zr + Ni. For the comparison of all investigated cathodes we draw on the graph (see Fig. 6) essential electric field for 10 uA emission from different CNTs FEC. The emission properties of cathode made of these CNTs are not very different from FEC made of pure nanotubes.

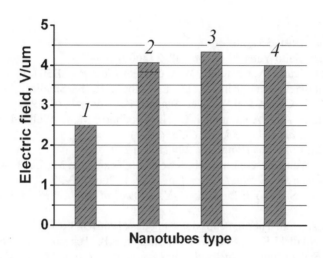

Fig. 6. Comparative graph of essential electric field for 10 uA emission from different CNTs FEC: 1 — CNT filled by Ba; 2 — pure CNT; 3 — CNT doped with Zr + Fe in synthesis process; 4 — CNT doped with Zr + Ni in synthesis process.

## 4. Conclusions

Thus, it was shown, that supporting the barium into CNTs could decrease value of work function of cathode. We have tried the doping technique at the FECs made of carbon nanotubes. The fulfilled experiments showed that applying the doping technique to carbon nanotubes allows reducing cathode's operating voltage by 30%. It was shown, that difference in the emission properties caused only by work function changing not the geometrical changes of cathodes. We assume that one more reason of operating voltage reduction of CNTs FEC filled by Ba exists. During the filling process the bundles of nanotubes separates to the single nanotubes. And because of that quantity of single nanotubes on the cathode surface are increases. Due to this fact the lot of emission sites appears on the cathode. In future we plan to fulfill the repeated experiments with doping in synthesis process without long milling.

## 5. Acknowledgements

Financial support from the RFBR (project 02-03-33226) is gratefully acknowledged by Muradyan V.E.

## 6. References

1. Iijima S. *Nature* 1991, **354**, 56.
2. Sashiro Uemura, Saito Yahachi *Carbon* **38**, 2 (2000) 169-183
3. W.S. Baska, D. Ugarte, A. Châtelain, W.A. de Heer *Phys. Rev.* **B50** (1994) 15473
4. T.Guo, P. Nikolaev, A.G. Rinzler, D. T. Colbert, R. E. Smalley, D. W. Owens, P.G. Kotula, C. B. Carter, J. H. Weaver, *Science, Vol. 266, p1218, 1994*
5. S. R. P. Silva, G. A. J. Amaratunga, C. N. Woodburn, M. E. Welland, and S. Hang, *Appl. Phys. Lett.,Part 1, Vol. 33, No. 6458, (1994)*
6. Chernozatonskii L.A. *ChemPhys.Lett.* **209** (1993) 229
7. Chuang F.-Yu, Wang W.-C., Lin I.-N., Kwo J.-L., Yokoyama M., Lee C.-C., Tsou C.C. *Journal of Vacuum Science and Technology B* **19**, 1 (2001) 23-27
8. Han J., Yang W.-S., Yoo Ji-B., Park C.-Y. *Journal of Applied Physics* **88**, 12 (2000) 7363-7366
9. Kudintseva G.A. et.al. Thermal emission cathodes (1966) 368 p.
10. Baturin A.S., Nikolski K.N., Sharov V.B., Sheshin E.P., Tchesov R.G. *Abstract of 11$^{th}$ Intercalation Symposium On Intercalation Compounds (ISIC), Moscow Russian*, 2001, p.77.
11. Muradyan V.E., Tarasov B.P., Shul'ga Yu.M., Ryabenko A.G., Fursikov P.V., Kuyunko N.S. et al *Abst. VII$^{th}$ Int. Conf. "Hydrogen Materials Science and Chemistry of Metal Hydrides" (ICHMS'2001)*, Alushta-Cremia-Ukraine, 2001, p.548-551.
12. L. W. Nordheim Proc. Roy. Soc. **A121** (1928) 626

# DEVELOPMENT OF METHODS OF DEPOSITION OF DISCONTINUOUS NICKEL COATINGS ON POWDERS OF AB$_5$ TYPE ALLOYS

I. SLYS, L. SHCHERBAKOVA, A. ROGOZINSKAYA,
D. SCHUR, AND A.ROGOZINSKII

*Frantsevich Institute for Problems of Materials Science of the NAS of Ukraine,*

*03142, 3, Krzhizhanovskii St., Keiv, Ukraine,*

## Abstract

In the present work, the possibility to modify the surface of the intermetallic compound MmNi$_5$ (Mm - La$_{0.815}$Ce$_{0.185}$) by nickel during decomposition of nickel oxalate (C$_2$O$_4$Ni) either during reducing heating in hydrogen or during intensive grinding in a planetary ball mill was investigated. The phase composition and transformation of the alloy lattice depending on conditions were controlled by X-ray diffractometry. Reducing heating in hydrogen products and grinding products of mixtures of the alloy with nickel oxalate contain the second phase, namely, elemental Ni. In the case of milling with nickel oxalate, particles and clusters of the Ni phase were smaller, uniformly distributed on the surface of alloy particles, and did not form a continuous coating. The lattice volume remains unchanged for powders ground in vacuum or air and increases by 1% and 5% for those ground in Ar and H$_2$, respectively. It was made the assumption of the preferred location of hydrogen atoms between hexagonal planes along the axis *c*. After mechanical treatment, the discharge capacity of the MmNi$_5$ alloy increases by 20%, and the combination of grinding and coating by Ni does not enhance this effect. The catalytic activity of the alloy surface modified by Ni during grinding in Ar is higher than that after grinding in air by 10-15%.
Keywords:ball milling, hydrogen, nickel oxalate, discharge capacity

## 1. Introduction

From the viewpoint of hydrogen storage, lanthanum nickelide LaNi$_5$ and its alloys are considered to be most promising. To improve service properties of LaNi$_5$-based electrode materials, microencapsulating of particles of LaNi$_5$-based alloys by different metals, which enables one to decrease significantly the oxidation and disintegration of electrodes during hydrogen adsorption/desorption, is used [1]. Deposition of coatings on alloy powders is traditionally performed by chemical methods in aqueous solutions with using reducers. A ball-milled technique to affect surface alloying [2] has been introduces to modify their electrochemical behavior. The ball-milled TiMn$_2$-type alloys in combination with nickel powder exhibit a considerable discharge capacity [3].

In the present work, we present results of a study of the possibility of using two alternative methods of deposition of a nickel coating on the surface of powders that are

*T.N. Veziroglu et al. (eds.),*

*Hydrogen Materials Science and Chemistry of Carbon Nanomaterials,* 131-136.

hydrogen sorbents as a result of decomposition of nickel oxalate either during reducing heating in hydrogen or during intensive grinding in a planetary ball mill by the reaction

$$NiC_2O_4 = Ni + CO_2$$

## 2. Experimental details

The intermetallic compound $MmNi_5$ ($Mn = La_{0.84}Ce_{0.16}$) was obtained by induction melting in argon followed by grinding of the ingot and cycling in a hydrogen atmosphere with the aim of its further comminution. The reducing heating of an alloy-nickel oxalate mixture was carried out in a hydrogen flow at temperatures of 250-450°C for 2.5 h. The mechanochemical modification of the surface was performed in a planetary mill in different gas atmospheres (air, argon, hydrogen, and vacuum) for 6 h under conditions of intensive grinding of the alloy-nickel acetate mixture.

Electrochemical measurements (plotting of voltammograms and charging curves) were carried out in a three-electrode cell, with an anodic and a cathodic space being separated , with using a P-5848 potentiostat. The investigated electrode made in the form of a pellet 1 cm in diameter was pressed into a nickel holder-current tap. As a reference electrode, a Hg/HgO electrode was used, and a platinum plate served as a counter electrode.

Changes in the morphology of the surface and phase composition, as well as transformation of the crystal lattice of the investigated objects depending on the conditions of deposition of coatings were controlled by methods of X-ray diffractometry (Model DRON-3M) in $2\theta$ geometry with Cu K$\alpha$ radiation, local X-ray local microanalysis and scanning Auger spectrometry.

## 3. Results and discussion

3.1 Decomposition of nickel oxalate during reducing heating or grinding.
Initial nickel oxalate powder and its behavior during annealing (for 2.5 h) in hydrogen at temperatures of 300, 400, and 500 °C and during grinding for 2, 5, 7, 9, and 18 h were investigated.

Nickel oxalate ($C_2O_4Ni$) had a monoclinic lattice with the following parameters: $a = 11.775$ Å, $b = 5.333$ Å, and $c = 9.33$ Å. During annealing in hydrogen, the reduction to metallic nickel occurred; the intensity and sharpness of its lines (111) increased with temperature, the lattice constant being equal to the value available in the literature ($a = 3.524$ Å).

An analysis of X-ray diffraction patterns of ground nickel oxalate powders showed that they changed with the grinding time. The grinding for 2 h led to the strain hardening of the powder, which manifested itself in the broadening of the most intensive ($I = 100\%$) line (221) at $2\theta = 47.75°$ ($d = 2.067$ Å). After grinding for 5 and 7 h, nickel oxalate partially decomposed with the formation of metallic nickel. In the X-ray diffraction patterns, at an angle $2\theta = 44.5°$ ($d = 2.03$ Å), the line (111) of nickel, which caused the pronounced asymmetry of the line (221) of nickel oxalate, appeared. As the grinding time was increased up to 18 h, the severe deformation (strain hardening) of the powders leading to the smearing of the diffraction lines and "smearing" of the effect of formation of metallic nickel, was observed.

It was established that the amount of nickel oxalate introduced into the alloy powder influences essentially the structure, phase composition, and electrochemical characteristics of the heat treated powder. On addition of about 10 mass % (in terms of Ni) of oxalate, in X-ray diffraction patterns of the heat treated powder, significant decreases in the intensities of lines characteristic for the $MmNi_5$ alloy are observed, and the intensive lines of the free nickel phase are detected. Alloy particles are covered with a nickel layer 5 μm in thickness, the coating is dense, which leads to an abrupt decrease in the surface area accessible for proceeding volume (for hydrogen) reactions.

The electrochemical discharge capacity decreases from 190 mA*h/g for the initial alloy to zero, and the hydrogen capacity in the gas atmosphere decreases from 165 ml/g to 70 ml/g. In Auger spectra obtained for mixtures of alloy powder with less than 1 mass % of nickel oxalate (in terms of nickel) after heat treatment at 400°C for 2.5 h, lines of the Ni phase are present. On local scanning of the surface, the intensities of the lines of this phase change by several orders of magnitude, which indicates the nonuniform distribution of conglomerates and individual Ni nanoparticles on the surface of alloy particles. The hydrogen capacity of the powders modified by nickel is equal to that of the initial alloy. It is likely that the electrochemical capacity on discharge falls by 40-50% as a result of the partial contamination of the surface of the $MmNi_5$ alloy by products of thermal decomposition of nickel oxalate (CO, $CO_2$, and $H_2O$).

TABLE 1. Influence of the atmosphere of milling on the parameters of the crystal lattice of the alloy $La_{0.84}Ce_{0.16}Ni_5$

The alloy $La_{0.84}Ce_{0.16}Ni_5$ was single-phase and had a hexagonal structure characteristic of $LaNi_5$, after milling the alloy for 6 h, changes in characteristic peaks in

| Subject of inquiry | Atmosphere of milling | Lattice parameters, Å | | Lattice volume, V, $Å^3$ | $\Delta V$, % |
|---|---|---|---|---|---|
| | | $A$ | $B$ | | |
| $La_{0.84}Ce_{0.16}Ni_5$ | Air | 4.9754 | 3.9756 | 85.220 | 0 |
| $La_{0.84}Ce_{0.16}Ni_5$ + $C_2O_4Ni$ (3 mass.% Ni) | Air | 4.9756 | 3.9757 | 82.238 | 0.02 |
| | Vacuum | 4.9755 | 3.9757 | 85.230 | 0.01 |
| | Argon | 4.9802 | 4.0123 | 86.168 | 1.00 |
| | Hydrogen | 4.9865 | 4.1807 | 90.023 | 5.60 |

X-ray diffraction patterns were not observed (see Figure 1a). In diffraction patterns of the products of milling of the powder mixture alloy - nickel oxalate (3 mass % in terms of Ni), lines characteristic of $C_2O_4Ni$ were not detected, and in the region of small angles, at the location of the most intensive (I = 100%) line of Ni (111) ($2\theta = 44.5°$), a shoulder is observed on the line of the main phase $LaNi_5$ (002). Another line of Ni

(200) (I = 50%) appeared in the X-ray diffraction patterns in the region 2θ = 52° (see Figure 1,b). The analogous appearance of two lines of Ni was observed in diffraction patterns of the products of milling of the powder mixture alloy - 3 mass.% of metallic Ni. In this case, however, the intensities of both lines were higher than those for the mixture with nickel oxalate (see Figure 1,c).

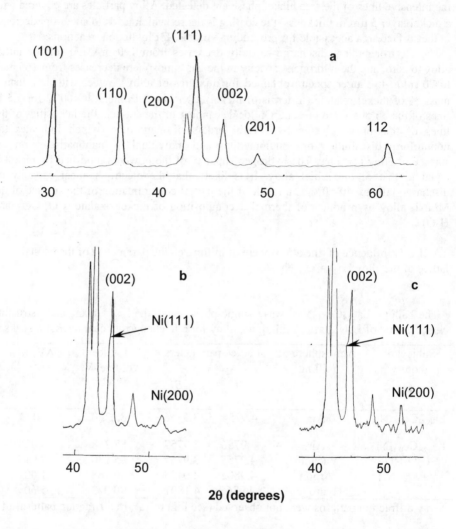

*Figure 1.* A fragment of an X-ray diffraction pattern of the alloy $La_{0.84}Ce_{0.16}Ni_5$ after mechanical treatment at room temperature for 6 h in air: a -alloy powder; b - powder mixture alloy - nickel oxalate (3 mass % in terms of nickel); c - powder mixture alloy - 3 mass % nickel.

Changes in the parameters and volume of the crystal lattice of the alloy (V) depending on the atmosphere of milling is presented in Table 1. Pronounced changes in the lattice parameters occurred only during milling in hydrogen. A significant change of the lattice size along the axis $c$ enables us to assume that hydrogen atoms enter actively the lattice of LaNi$_5$ during milling and arrange between its hexagonal planes along the axis c, moving apart these planes.

A study of the distribution of Ni particles on the surface of an alloy with a high (80 at.%) content of this element was a difficult methodical problem. An examination of the surface of the product of milling of the powder mixture alloy - nickel oxalate by electron microscopy analysis in characteristic La radiation showed a significant decrease in the intensity of La radiation at locations of the nickel phase on the surface. The obtained images of regions of the surface show a relatively uniform local distribution of individual Ni particle and their conglomerates on the surface of alloy particles. In the case of milling of the alloy with 3 mass % of Ni powder, the character of distribution of the Ni phase on the surface is the same, but formed Ni particles and their conglomerates are larger that those after milling with $C_2O_4Ni$.

Thus, when Ni oxalate particles collide with alloy particles under the action of ceramic balls at high speed, a high kinetic energy can cause oxalate decomposition and the formation of nickel at the points of impact. It is still unclear if formed Ni particles are isolated particles of pure nickel or combine partially with alloy particles

3.2 Electrochemical measurements

For all as-prepared electrodes, rather insignificant discharge capacities (170-190 mA*h/g) were obtained after their single saturation by hydrogen, which can be connected with the partial oxidation of the surface of powders in air during making the electrodes. However, even in the second cycle, values of the discharge capacities increased by 30-50%, which testifies to the self-activation of the surface during charging. In Table 2 are presented, sorption parameters calculated on discharging curves of electrodes in the second cycle base, namely the discharge capacity and the amount of electrochemically extracted hydrogen in terms of a mol of the alloy are summarized. Irrespective of the technology of application of nickel coatings, discharging curves are characterized by stable plateaus of the $\alpha+\beta \rightarrow \alpha$ transition. Compared to the technology of comminution of powder by cycling in hydrogen, the technology considered provides 20% increase in the discharge capacity of the alloy after mechanical milling (Table 2), which is connected with the formation of a highly developed surface active as to hydrogen exchange reactions. The grinding of the powder mixtures in air results only in the partial oxidation of the surface and in a fall of the discharge capacity almost by 20% in comparison with that of powders ground in vacuum. When the Ni content in the mixture is increased up to 3 mass %, the catalytic activity of the surface rises. On keeping in air for 1.5 months, $La_{0.84}Ce_{0.16}Ni_5$ alloy powders modified by nickel during milling retained entirely their service characteristics.

TABLE 2. Electrochemical capacities and hydrogen contents in the alloy $La_{0.84}Ce_{0.16}Ni_5$

| Subject of inquiry | Atmosphere of milling | Discharge capacity, mA·h/g | Hydrogen content in hydride | |
|---|---|---|---|---|
| | | | mass % | H/M |
| $La_{0.84}Ce_{0.16}Ni_5$ | - | 215 | 0.79 | 3.2 |
| $La_{0.84}Ce_{0.16}Ni_5$ | Air | 262 | 0.97 | 4.2 |
| $La_{0.84}Ce_{0.16}Ni_5$ + | Air | 218 | 0.81 | 3.4 |
| $C_2O_4Ni$ (3 mass.% | Argon | 242 | 0.90 | 3.9 |
| Ni) | Vacuum | 264 | 0.97 | 4.2 |
| | Hydrogen | 240 | 0.89 | 3.9 |
| $La_{0.84}Ce_{0.16}Ni_5$ + 1.5 mass.% Ni powder | Air | 237 | 0.88 | 3.8 |
| $La_{0.84}Ce_{0.16}Ni_5$ + 3.0 mass.% Ni powder | Air | 270 | 1.00 | 4.3 |

## 4. Conclusions

On thermal decomposition of nickel oxalate in hydrogen at 350°C, either continuous or patch-like nickel coatings form. During deposition of coatings, the discharge capacity of the alloy powders decreases due to the contamination of the surface by the decomposition products of the salt. During mechanical treatment of the alloy in the presence on nickel oxalate, isolated nickel particles and their conglomerates that are uniformly distributed over the surface form. As compared to other methods of grinding of the alloy, the discharge capacity increases almost by 30% and varies within 20% depending on atmosphere used in grinding.

## 5. References

1. Arantes D.R., Santos M.B., Krichheim R., *11<sup>th</sup> World Hydrogen Energy Conference*, Stuttgard, Germany, 1996, **3.**
2  J.H.Harris, S.Heights, M.A.Tuenhover and R.S.Heuderson, *US Patent 4 859 413*.
3. Ming Au, F.Pourarian, S.Simisu, S.G.Sankar, Lian Zhang, (1995) *J.Alloys Comp., 223*, 1.

# XRD PATTERNS OF CATHODE DEPOSITS FORMED IN ELECTRIC ARC SPUTTERING Zr-Me-GRAPHITE ELECTRODES

Yu.M. SHULGA, D.V. SCHUR, S.A. BASKAKOV, A.P. SIMANOVSKIY,
A.A. ROGOZINSKAYA, A.A. ROGOZINSKIY, A.P. MUKHACHEV

*Institute of Problems of Chemical Physics RAS, Chernogolovka,
Moscow region, 142432 Russian Federation
E-mail: shulga@icp.ac.ru
Institute for Problems of Materials Science of NAS of Ukraine,
3 Krzhizhanovsky St., Kiev, 03142, Ukraine
State Scientific Industrial Enterprise "Zirconium", Dneprodzerjinsk, Ukraine*

## Abstract

XRD X-ray diffractogram patterns have been obtained for the deposits formed in electric arc sputtering of Zr-Me-graphite electrodes, where Me is Ti, Mn, Fe, Co, Ni, Cu. The interlayer distances in the multi-wall carbon nanotubes, as the main component of the deposit, slightly depend on the presence of metal Me. It has been shown that zirconium is present in the deposit as ZrC carbide. The sputtering of Zr-Ti-graphite electrode gives rise to two carbides ZrC and Ti(Zr)C.

## 1. Introduction

Currently carbon materials have been intensively studied due to their supposed high hydrogen sorption capacitance. One of the most important parameters, characterizing the properties of layer materials, is the interlayer distance. It has been known that the distance between layers (002) in an ideal crystal of graphite is equal to 0.3354 nm [1]. In the case of multi-wall carbon nanotubes (MWNTs) the distance between neighboring layers is considerably larger. It is assumed that this distance is close to 0.34 nm [2]. However, there exist a number of factors which can distort the ideal cylindrical structure of MWNTs. Among these factors are the defects within the layer (five- and seven terms cycles), the availability of atoms (ions) of another elements between layers, the interaction of nanotubes with each other that can manifests itself at their uniting in bundles, and others. It would appear natural that addition of metal atoms into the interlayer space or into the tubes can have influence on their sorption characteristics.

In the present work we pursue the objective to study XRD patterns of the deposits formed in electric arc sputtering Zr-Me-graphite electrodes, where Me is metal of group 3B (or 3d metal). The cylindrical growth on the cathode is called simply deposit. Deposit is known to consist of MWNTs mainly. Addition of such metal as zirconium into the anode composition was caused by this metal ability to form carbide which due to high fusion temperature ($3420°C$ [3, 4]) can modify composition of the deposit formed at sufficiently high temperature (more than $2000°C$ [4]). Addition of the second metal component, according to our initial intention, was aimed to test the possibility of softer doping of deposit and, consequently, MWNTs. The principle parameter, in which we were interested first of all, is the interlayer distance in MWNT, easily calculated from

137

*T.N. Veziroglu et al. (eds.),*
*Hydrogen Materials Science and Chemistry of Carbon Nanomaterials,* 137-142.
© 2004 *Kluwer Academic Publishers. Printed in the Netherlands.*

XRD patterns [5,6]. It was also interested to gain information about the structure of metal components in the deposit.

## 2. Experimental features

The reactor, similar to that for fullerene production and described in [6], was used for electric arc sputtering of graphite and composition electrodes.

The graphite anode 6 mm in diameter was drilled along its axis with a drill 3 mm in diameter. Small portions of catalyst were filled in this hole using a special funnel then catalyst was rammed. The catalyst comprises preliminary milled powder of metals and graphite mixture. This powder was prepared by combined milling metal alloys and powder graphite in a ball mill for 5 h. Before sputtering, the anode was placed in the reactor, the system was evacuated down to 1 Pa than helium was let in up to 0.05 MPa, the anode was annealed for 1 min in the short circuit regime. After annealing the system was evacuated again and helium was let in up to $5 \times 10^4$ Pa. The temperature of the cooled external surface of the reactor did not exceed 40°C. After completing sputtering and cooling of all system down to the room temperature, at first the cathode with deposit was removed from the reactor what, to our opinion, kept clear the cathode of soot from the walls. Features of sputtering regimes, catalysts composition and volume of produced deposits are given in Table I.

TABLE I. Parameters of synthesis

| № | Catalyst | Proportion of catalyst components | Catalyst % | Regime | | Rate of cathode feed mm/min | Deposit, % of mass of sputtering anode |
|---|---|---|---|---|---|---|---|
| | | | | U, V | I, A | | |
| 1. | Zr + Ti | 1:1 | 2÷4 | 35 | 180 | 1,45 | 50,0 |
| 2. | Zr + Mn | 1:1 | 2÷4 | 36 | 175 | 1,49 | 49,2 |
| 3. | Zr + Fe | 1:1 | 2÷4 | 33 | 185 | 1,72 | 14,7 |
| 4. | Zr + Co | 1:1 | 2÷4 | 35 | 180 | 1,70 | 40,1 |
| 5. | Zr + Ni | 4:1 | 2÷4 | 35 | 170 | 1,65 | 63,9 |
| 6. | Zr + Cu | 1:1 | 2÷4 | 34 | 175 | 1,73 | 55,1 |

The deposit was parted from the cathode and milled in the ball mill to receive XRD patterns. XRD patterns were recorded using DRON ADP-1 diffractometer (Cr $K_\alpha$- or Cu $K_\alpha$-radiation).

## 3. Experimental results and discussion

Fig. 1 shows XRD pattern for one of the studied sample. The most intensive peak is that result from (002) reflection of MWNTs according to the existing literature concepts (for example, [5]). This peak is significantly broader than that of graphite [7] and shifted towards the lower angles (Fig.1, inset). Fig.1 also shows other less intensive peaks of MWNTs present in the sample. The identification of these reflections was made

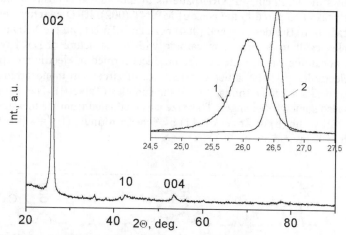

*Figure 1.* XRD pattern of the deposit formed in sputtering of Zr-Cu-graphite electrode. The inset at right of illustration shows imposition of XRD patterns for graphite and deposit in the region of the most intensive peak (002). The radiation is Cu $K_\alpha$

according to the data presented in papers [8]. The values of the interlayer distances (002) in carbon nanotubes of the studied samples are given in Table II. As it was expected, the values of $d_{002}{}^m$ and $d_{002}{}^{cg}$ for deposits significantly exceed those for graphite (according to our measurements for graphite $d_{002}{}^m = d_{002}{}^{cg} = 0.3358$ nm). Note also that the interplanar distances calculated by the positions of the maximum and the center of gravity (002) is significantly different for some specimens. In this case the difference $\Delta = d_{002}{}^m - d_{002}{}^{cg}$ can be both positive and negative. In transition from one deposit to another the change of the $d_{002}{}^m$ value fits in the rather narrow region from 0.3421 to 0.3436 nm. Any regularity in changing $d_{002}{}^m$ was not found in a number of modifying additives studied.

TABLE II. Interlayer distances $d_{002}$, average dimensions D002 of MWNT, lattice constant for zirconium carbide $a_o$, size $D_{111}$ of ZrC particles and fusion temperature of 3d metal. The $d_{002}$ values with index **m** were determined from the position of park (002) maximum and that with index **cg** – by the position of gravity centre of the peak.

| N⁰ | Catalyst | $d_{002}{}^m$, nm | $d_{002}{}^{cg}$, nm | $D_{002}$, nm | Lattice constant for zirconium carbide, nm | Size of ZrC particles, nm | Fusion tempe-rature of 3d-metal, °C |
|---|---|---|---|---|---|---|---|
| 1 | Zr + Ti | 0.3427 | 0.3431 | 15.7 | 0.4709 | 27 | 1668 |
| 2 | Zr + Mn | 0.3423 | 0.3424 | 12.8 | 0.4702 | 85 | 1245 |
| 3 | Zr + Fe | 0.3436 | 0.3429 | 15.1 | 0.4703 | 56 | 1539 |
| 4 | Zr + Co | 0.3431 | 0.3427 | 15.8 | 0.4703 | 57 | 1492 |
| 5 | Zr + Ni | 0.3427 | 0.3417 | 16.8 | 0.4702 | 61 | 1452 |
| 6 | Zr + Cu | 0.3421 | 0.3424 | 16.9 | 0.4701 | 51 | 1083 |

140

Fig. 2 shows the fragments of XRD patterns of the studied deposits in the region of less intensive peaks compared to the peak of MWNT (002). XRD patterns were obtained using Cr $K_\alpha$-radiation. Besides (10) and (004) peaks of MWNT, all the XRD patterns also show the peaks result from zirconium carbide with the structure of NaCl type. Earlier zirconium monocarbide was detected in the products formed in electric arc sputtering of Zr-graphite electrodes [9]. The lattice constant, $a_o$, of zirconium monocarbide evaluated by three noted peaks is rather similar for all the samples (Table II). The average size of ZrC particles also somewhat changes. This size was evaluated from the measured position (2Θ) and the half width (β) of Zr peak (111) by Scherer formula $D_{111} = \lambda/\beta\cos\Theta$, where λ is wavelength of X-ray radiation.

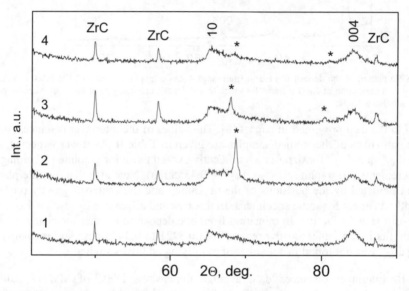

*Figure 2.* XRD patterns of the deposits formed in electric arc sputtering of Zr-Me-graphite electrodes, where Me=Mn (1), Fe(2), Co(3) or Ni(4). The radiation is Cu $K_\alpha$

The peaks marked with (*) in Fig. 2 are resulted from metal Me presence in the studied samples. However, different metals manifest variously themselves in studied samples. So the presence of manganese or its compounds (alloys) were not detected in the deposit of the Zr-Mn-graphite system by its diffractogram. It is notable that the the superposition of diffraction peaks on the peaks of multilayered nanotubes inhibits the revealing the manganese carbide $Mn_3C_2$. Iron appears on diffractogram as the relatively intensive peak at 2Θ=68.65 degree (d=0.2031 nm) what corresponds to α-Fe (*bcc* lattice). Cobalt is present in the deposit of Zr-Co-C system as metal small particles with the *fcc* lattice while at room temperature the *bcc* lattice is more stable for cobalt. It seems to be related to the small size of cobalt particles in deposit. The formation of intermetallic compound ZrCo must not be ruled out [10]. Nickel can be also detected in the deposit using XRD patterns in spite of the low nickel concentration in the electrode sputtered. Nickel is present in the deposit as very small-sized particles which give the broad peaks of low intensity with d=0.2037 and 0.1764 nm. When studied XRD pattern of the sample from Zr-Cu-C system, the peak with low intensity can be distinguished on the background of the broad peak (10) from carbon nanotubes. By its position this peak can be assigned to that of Cu (110). However, another peak of copper (Cu(200)) with somewhat lower intensity is not already observed. For this reason the presence of copper in the deposit is still open to question.

Titanium is present in the deposit as titanium carbide, in which lattice the part of Ti atoms are substituted for Zr atoms, i.e. Ti(Zr)C. The Ti(Zr)C lattice constant is equal to 0.4361 nm, ($a$ = 0,4324 nm for TiC). In consequence of leaving of Zr atoms to solid solution Ti(Zr)C the lines intensity of ZrC carbide is diminished in comparison with the lines of previous samples. This points to the fact that the amount of zirconium carbide in Zr-Ti-C deposit is decreased compared to the above specimens.

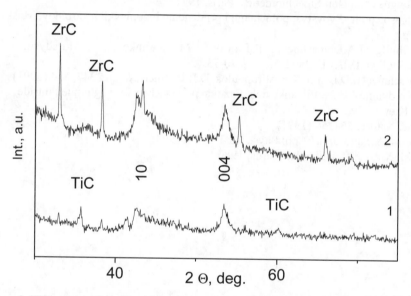

*Figure 3.* XRD patterns of the deposits formed in electric arc sputtering of Zr-Me-graphite electrodes, where Me=Ti(1) or Cu(2). The radiation is Cu K$_\alpha$.

## 4. Conclusions

The most part of the deposit is formed at high temperature. However, if the deposit size is sufficiently large, temperature along its outer cylindrical surface may differ considerably. The part of the deposit close to the space between the electrodes has maximum temperature (2000-3000 °C), and the part of the deposit formed at the beginning of the process has minimum temperature. The temperature in this part does not exceed 600°C for deposits 5 cm long and more. This conclusion can be made because this part is of dark colour (does not glow) in the course of sputtering. According to these observations, the inner part of the deposit is formed from refractory material while more fusible materials can deposit on the gradually cooled outer surface.

This point of view is confirmed by the fact that we did not detect the presence of low-melting copper and manganese in XRD patterns. At the same time, the rest of metals with $t_{fus} \geq 1450°C$ are present in the deposit as small metal particles.

It should be noted that metals present in the sputtering electrode and not forming refractory carbides lead to the growth of larger zirconium carbide particles in the deposit.

Finally, the composition variations of the sputtered anode described in this paper has a weak effect on the interlayer distance in MWNTs present in the deposit.

The work has been carried out under financial support of RFBI (project N 03-03-32796).

## 5. References

1. Y.Saito, T.Yoshikawa, S.Bandow, M.Tomita, T.Hayashi, Phys.Rev., **B48**, 1907 (1993)
2. A.V.Eletskiy, UFN, 167, 945 (1997)
3. R.Collougues "La Non-Stoechiometie", Paris, 1971
4. T.Ya. Kosolapova, Svoystva, poluchenie i primenenie tugoplavkikh karbidov, Moskva, 1986 g.
5. Yu.M.Shul'ga, I.A.Domashnev, B.P.Tarasov, A.M.Kolesnikova, V.E.Muradyan, N.Yu.Shul'ga, ISJAEE, 2002, No 1, p.70-73.
6. W.Kratschmer, L.D.Lamb, K.Fostiropoulos, D.R.Huffner, Nature, **347**, 354 (1991)
7. A.V. Kurdyumov, A.N. Pilyankevich, Fazovie prevrascheniya v uglerode i nitride bora, Kiev, 1979.
8. S.Iijima, Nature, **354**, 56 (1991).
9. S.Bandow, Y.Saito, Jpn.J.Appl.Phys., 32, L1677 (1993).
10. ASTM, X-ray diffraction date cards, 1980.

# METAL HYDRIDE ACCUMULATORS OF HYDROGEN ON THE BASIS OF ALLOYS OF MAGNESIUM AND RARE-EARTH METALS WITH NICKEL

B.P. TARASOV, S.N. KLYAMKIN, V.N. FOKIN, D.N. BORISOV, D.V. SCHUR, V.A. YARTYS
*Institute of Problems of Chemical Physics of Russian Academy of Science, 142432 Chernogolovka, Moscow Region, Russia*
*Lomonosov Moscow State University, 117000 Moscow, Russia*
*Institute for Problems of Material Science of National Academy of Sciences of Ukraine, 03142 Kiev, Ukraine*
*Institute for Energy Technology, N-2077 Kjeller, Norway*

## Abstract

In present work we propose the compositions obtained by mechano-chemical treatment of the mixture of 90 wt.% of the powders of hydride phases of Mg–Ni–Mm eutectic alloys and 10 wt.% of $(La,Mm)Ni_5$ based hydrogen activators as a working material for the metal hydride hydrogen accumulators. Mm denotes michmetal that is the commercial mixture of rare-earth metals.

*Keywords:* powdered hydride phase, hydrogen accumulator, isotherms of sorption and desorption, phase transition, Mg-Ni

## 1. Introduction

The reversible hydrogenation of the intermetallic compounds (IMC) and alloys on the basis of hydride forming metals is one of widely used solutions for the problem of hydrogen storage [1,2].

As to their high hydrogen capacity, availability and cost, magnesium alloys can meet the modern requirements for the metal hydride systems. Among them two-phase alloys Mg–Ni and three-phase alloys Mg–(La,Ce)–Ni in the region of ternary eutectic $Mg–Mg_2Ni–(La,Ce)_2Mg_{17}$ are the most attractive materials for metal hydride hydrogen accumulators. Those alloys have hydrogen sorption capacity exceeding 5 wt.% [3-5].

The use of magnesium alloys for hydrogen storage is often connected with the difficulties conditioned by the low heat conductivity of the powdered hydride phases. It causes the local overheating and sintering that results in the losses of significant amount of the material from the working cycle.

## 2. Experimental

According to the data of X-ray and chemical analysis, the $La_{0.67}Mm_{0.33}Ni_5$ alloy is crystallyzed in the hexahonal structure type with lattice parameters a=0.5015 nm and

143

*T.N. Veziroglu et al. (eds.),*
*Hydrogen Materials Science and Chemistry of Carbon Nanomaterials, 143-146.*
© *2004 Kluwer Academic Publishers. Printed in the Netherlands.*

c=0.3987 nm; the alloy 72% Mg– 8% Mm– 20% Ni consists of the ternary eutectic containing Mg, $Mg_2Ni$ and $Mm_2Mg_{17}$ phases.

The microstructure analysis of the last alloy testify to its high homogeneity and dispersivity.

The characteristics of the $La_{0.67}Mm_{0.33}Ni_5 - H_2$ system are presented in the Fig.1. It can be concluded that the temperature dependence of hydrogen equilibrium pressure in the region of the $\alpha \leftrightarrow \beta$ phase transition on temperature follows the equation $lgP_{H2}=5.95-1690/T$, and the heat of formation is equal to $\Delta H= -30$ kJ/mole $H_2$.

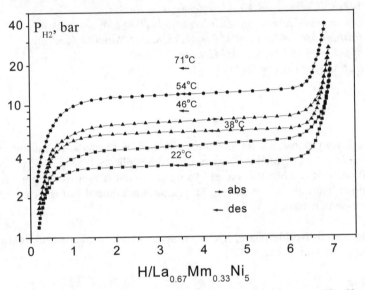

Figure 1. Isotherms of sorption ($\rightarrow$) and desorption ($\leftarrow$) in the $La_{0.67}Mm_{0.33}Ni_5 - H_2$ system

The disperse analysis of $La_{0.67}Mm_{0.33}Ni_5$ powders after 5 $H_2$ absorption $\leftrightarrow$ desorption cycles testifies that the cross diameter of 90 wt.% of the particles is within the interval of 4 to 15 μm.

Hydrogen interaction with initially powdered (particle size < 2 mm) Mg–Mm–Ni alloys in the region of ternary eutectic at the temperature of 250–300°C and the pressure of 3–5 MPa, results in the formation of the mixture of three hydrides:

$$(Mg-Mg_2Ni-Mm_2Mg_{17}) + xH_2 \rightarrow$$
$$MmH_3 + MgH_2 + Mg_2NiH_4$$

The maximum hydrogen sorption capacity of such a composition is 5.5 wt.%. The presence of the disperse eutectic structure and synergetic influence of rare-earth metals and Ni allow one to reduce the hydrogenation temperature and thus to avoid the negative effect of sintering process which is an distinguishing feature of Mg- based powders.

Fig.2 shows the hydrogen absorption and desorption isotherms at 573–623 K where two plateaus corresponding to the phase transitions $Mg_2Ni-H_2$ ($\Delta H= -65$ kJ/mole $H_2$) and $Mg-H_2$ ($\Delta H= -75$ kJ/mole $H_2$) are observed.

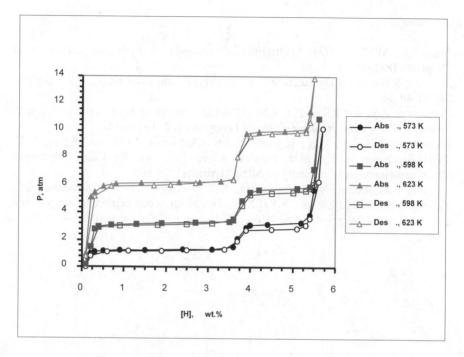

Figure 2. Sorption and desorption isotherms in the {Mg-Mm-Ni} – $H_2$ system

The disperse analysis after replicates of experiment from 5 hydrogen sorption ↔ desorption cycles testifies that 90 wt.% of the powder contains the particles 20–200 μm in diameter.

By planetary-mill treatment of the mixtures of 90 wt.% of the hydrogenated magnesium alloy with 10 wt.% of $La_{0.67}Mm_{0.33}Ni_5H_x$ the homogeneous high-disperse compositions are obtained. These compositions after dehydriding quickly interact with hydrogen.

On the basis of $La_{0.67}Mm_{0.33}Ni_5$ and Mg–Mm–Ni several types of the metal-hydride hydrogen accumulators (Table 1) have been produced.

TABLE 1. The technico-operating characteristics of hydrogen accumulators

| Alloy composition | Mg–Mm–Ni | $(La,Mm)Ni_5$ |
|---|---|---|
| Hydrogen amount, $l$ | 50-250 | 150-2500 |
| Alloy mass, kg | 0.1-0.4 | 1-15 |
| Working pressure range, MPa | 0.3-5 | 0.1-0.3 |
| Working temperature range, °C | 250-300 | 10-100 |
| Heating–cooling system | Electric | Water |

This work has been supported by Russian Foundation of Basic Research, Russian Academy of Sciences and and Nordisk Energiforskning.

## 3. References

1. Antonova, M.M. (1993) Magnesium compounds – hydrogen accumulators (Preprint), IPMS, Kiev, 41 p.
2. Schwarz, R.B. (1999) Hydrogen storage in magnesium-based alloys, *MRS Bulletin* **24**(11), 40-44.
3. Verbetsky, V.N. and Klyamkin, S.N. (1988) Interaction of magnesium alloys with hydrogen, *Pergamon: Hydrogen Energy Progress VII* **2**, 1319-1342.
4. Tarasov, B.P., Fokin, V.N., Klyamkin, S.N., Antonova, M.M. and Schur, D.V. (2001) Hydrogen storage in magnesium alloys. Proc. VII Int. Conf. "Hydrogen Materials Science and Chemistry of Metal Hydrides", 98-101.
5. Tarasov, B.P., Fokin, V.N., Klyamkin, S.N., Borisov, D.N., Gusachenko, E.I., Yakovleva, N.A. and Shilkin, S.P. (2004) Hydrogen accumulation by magnesium and rare earth alloys with nickel, *Int. J. Alternative Energy and Ecology* No.1, 58-63.

# SYNTHESIS OF NANOTUBES IN THE LIQUID PHASE

SCHUR D.V. DUBOVOY A.G., LYSENKO E.A., GOLOVCHENKO T.N.,
ZAGINAICHENKO S.Yu., SAVENKO A.F., ADEEV V.M.,
KAVERINA S.N.
*Institute for Problems of Materials Science of NAS of Ukraine, lab. # 67,
3, Krzhizhanovsky str., Kiev, 03142 Ukraine*

**Keywords:** nickel microparticle, sputtering of graphite, carbon nanotubes, liquid hydrocarbons

## 1. Introduction

In connection with the discovery of fullerenes, the research devoted to the production of ultradispersive powders of metals and started in 80th of the last century by A.G. Dubovoy and colleagues [1, 2], has an original continuation.

In the first works it was supposed that the ultradispersive particles produced by the arc sputtering of metal can change their properties depending on the synthesis conditions.

In their works [3, 4] the authors sum up investigations performed in the course of ten years in Institute for Metals Physics of National Academy of Sciences of Ukraine. It was noted that the arc parameters and the nature of the medium for synthesis affected the structure and properties of particles. By changing the conditions for synthesis, one can change properties of the product in the wide range.

The present work is aimed to test the hypothesis for possibility of synthesis of carbon nanotubes in the liquid medium which is a source for carbon.

## 2. Experiment

The technique of electric-spark dispersion of metals (ESD) mentioned above has been used to verify this hypothesis. The original apparatus designed in laboratory 67 of Institute for Problems of Materials Science of National Academy of Sciences of Ukraine has been used.

According to [4], the main positive moments of the method employed are:
1. high temperature in the arc zone is ~4000 °C,
2. high cooling rate of sputtered products is $10^9$-$10^{14}$ °C/s,
3. high dispersion level. The size of the particles produced is 1-100 nm.
4. high rate of nucleation at the low rate of the particles growth.

All these conditions are in a good agreement with the conditions of fullerenes and nanotubes synthesis by the arc sputtering of graphite. Experiments have been performed in the medium of alcohol $C_2H_5OH$, benzene, toluene and hexane.

The reaction products have been analyzed using the scanning and transmission microscopes.

## 3. Results and discussion

On the basis of commonly accepted concepts of mechanisms of the carbon nanotubes growth, as expected, the sputtered nano-dispersed nickel must catalyze the growth of

*T.N. Veziroglu et al. (eds.),*
*Hydrogen Materials Science and Chemistry of Carbon Nanomaterials,* 147-151.
© 2004 *Kluwer Academic Publishers. Printed in the Netherlands.*

148

carbon nanotubes. The source for carbon should be carbon from the hydrocarbon that transforms into vapor in the arc zone. It was supposed to prepare the single-wall nanotubes on the nickel particles of size 1-10 nm, and to produce the layer of nanotubes up to 1 μm thick on the larger nickel particles.

Electron-microscopic studies have indicated that carbon nanotubes do not form on the nickel particles in the media chosen (Fig.1, a,b - alcohol, c,d,e,f - toluene, g,h - hexane).

However, at mixing of hydrocarbons, we succeeded in production of carbon nanotubes up to 100 nm in diameter on the surface of nickel microparticles. Nanotubes are not perpendicular to the surface (after pyrolysis), but they are arranged parallel to it. Tubes on the surface of particles form the continuous net (Fig.2,3).

In these conditions the carbon cores sputtering has resulted in the solutions which resemble fullerene extracts by colour. We have failed to extract fullerenes from the prepared mixture chromatographically.

## 4. Conclusions

All obtained results are of scientific and practical interest. The materials prepared require further investigations. The proposed method can be one of the most effective method to synthesize fullerenes and nanotubes.

## 5. References

1. A.G. Dubovoy, A.E. Perekos, K.V. Chuistov. Metallofizika 1984, V. 6, No. 5, P. 29-131 (*in Russian*).
2. A.G. Dubovoy, V.P. Zalutskiy, I.Yu. Ignat'ev. Metallofizika, 1986, V. 8, No. 4, P. 101-103 (*in Russian*).
3. K.V. Chuistov, A.E. Perekos, V.P. Zalutsky et al. Metallofizika i noveyshie tekhnologii, 1996, V. 18, No. 8 P. 18-25 (*in Russian*).
4. K.V. Chuistov, A.E. Perekos. Metallofizika i noveyshie tekhnologii, 1997, V. 19, No. 1, P. 36-53 (*in Russian*).

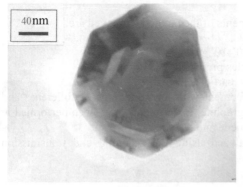
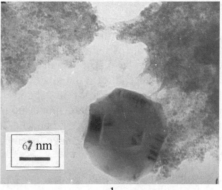

a                                    b

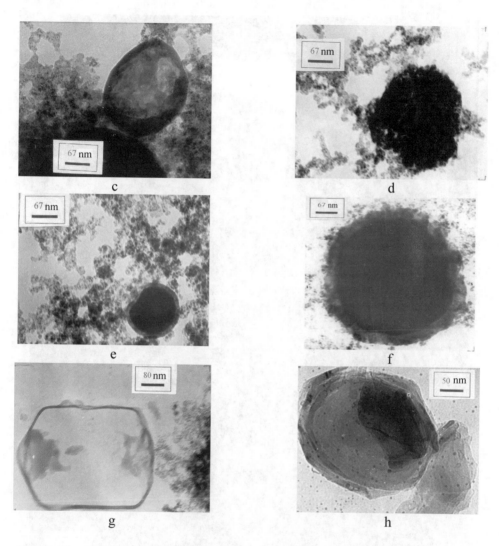

Fig. 1. The nickel particles with nanocarbon produced in different liquid medium
a, b – alcohol, c, d, e, f – toluene, g, h – hexane

150

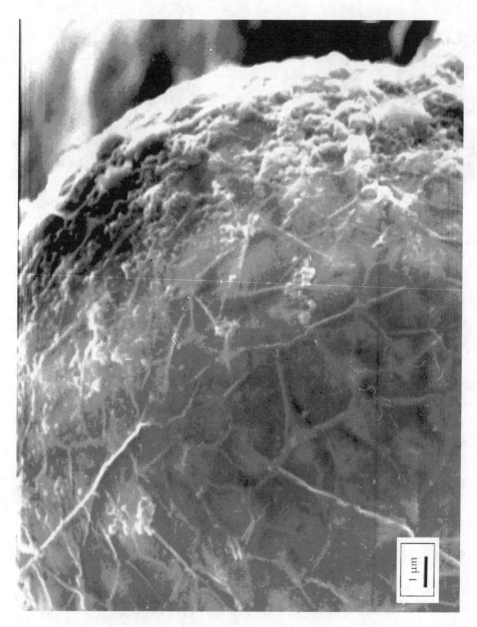

Fig. 2. The nickel particle with carbon nanotubes on its surface

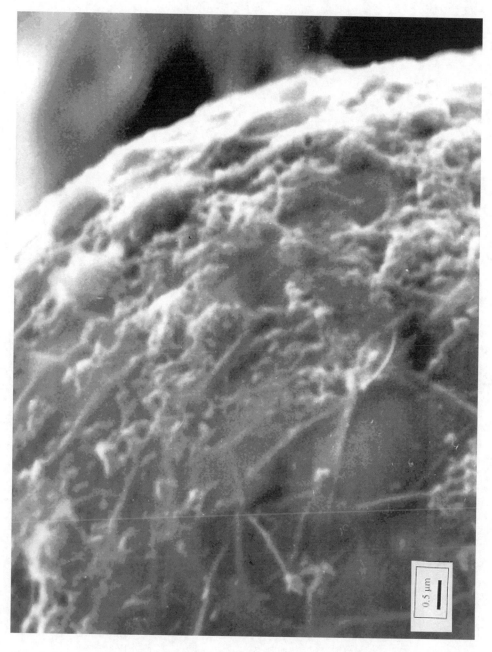

Fig. 3. Fragment of nickel particle with carbon nanotubes on the surface of particle

# MOSSBAUER STUDY OF CARBON NANOSTRUCTURES OBTAINED ON Fe-Ni CATALYST

T.Yu. KISELEVA, A.A. NOVAKOVA, B.P. TARASOV, V.E. MURADYAN
*Moscow M.V.Lomonosov State University, Department of Physics,*
*Vorobiovy Gory, 119992, Moscow, Russia*
*Institute for New Chemical Problems RAS , Chernogolovka, 142432, Russia*

## Abstract

A microstructure of carbon containing iron and nickel compounds, obtained during arc-disharge synthesis have been studied by electron microscopy, Mossbauer spectroscopy, X-ray diffraction and thermogravimetry. The optimal concentration of bimetal Fe:Ni catalyst determines quantitative ratio between formed large inert carbon encapsulated metal particles and small-sized metal nanoparticles being catalytic centers of SWNT origin.

*Keywords*: Carbon nanotubes, arc discharge, Fe-Ni catalyst, nanostructures,
Mössbauer spectroscopy

## 1. Introduction

It is known, that catalytic properties of the nanosystem are determined by a surface conditions of a catalytic cluster, its size and interaction activity with a matrix [1]. The last experimental data testify, that the efficiency of the catalyst for single wall carbon nanotubes (SWNT) formation is caused by its ability for carbon graphitization, its low solubility in carbon and possibility of its stable crystallographic orientation on graphite. In spite of enormous progress in the synthesis, the basic SWNT growth mechanisms and interaction of transition metal atoms with graphite or graphene sheet has been the subject of extensive investigations and discussion. Nevertheless, it is already established, that SWNT formation happens most heavily when bimetallic catalytic mixtures are applied [2-4].

Considering the importance of alloying effects in heterogeneous catalysis the aim of this study was to investigate the effects associating with Fe and Ni interactions in the carbon nanotubes arc-synthesis. The use of iron as a component of a binary catalytic mixture allows to use nanostructure sensitive Mossbauer spectroscopy in accordance to electron microscopy, x-ray diffraction and thermal analysis to reveal the chemical nature and structure of metal particles – containing centers of SWNT growth.

## 2. Experimental

In our work we used conventional arc-discharge method with bimetallic catalyst powders to obtain single-wall carbon nanotubes [5-6]. The discharge experiment was

153

*T.N. Veziroglu et al. (eds.),*
*Hydrogen Materials Science and Chemistry of Carbon Nanomaterials,* 153-158.
© 2004 *Kluwer Academic Publishers. Printed in the Netherlands.*

carried out using standard arc chamber with a pair of electrodes. The reaction chamber was maintained at 650 Torr helium atmosphere. The stabilized voltage set was 100 A and 25-28 V. The carbon rod (anode) with sizes of 8.2x79 mm was drilled (the hole was of 3.3x71mm) and filled with metall powder. The distance between the carbon electrodes was 1.5-2.5 mm. The portion of catalyst in evaporated part of the electrod was 35wt%. The SWNT synthesis was carried out on a catalytic mixture Fe:Ni with different mutual ratio of components Fe: Ni - 1:0; 3:1; 1:1.

A largest amount of SWNTs were found in the soot collected from the "collar" formed on graphite electrode. This metal-containing carbon nanomaterial have been analyzed by means of electron microscopy, thermogravimetry, x-ray diffraction and Mossbauer spectroscopy.

Investigations of microstructures and evaluation of linear sizes of obtained nanostructures were carried out on JEM 200CX and JEM 2000FX electron microscopes. An estimation of SWNT content in the soot was performed by means of oxidizing thermogravimetry. The thermal oxidation of the soot has been studied with a Q-1000 derivatograph and DuPont 2100 thermal analyzer through linear heating of a weighed portion of the sample in air at a rate of $5°/min$. X-ray diffraction measurements were performed at Rigaku D/Max $\theta$-$2\theta$ diffractometer with Cu K$\alpha$ radiation. Mossbauer spectra were obtained at room temperature using constant acceleration spectrometer with Co57(Rh) source and fitted by means of the UNIVEM(Mostek) software. Mossbauer spectroscopy is very sensitive to the chemical and structural environment of iron atoms on a nearest-neighbor length scale and therefore allows to analyse the chemical and magnetic structure of small iron particles as well as their size and volume.

### 3. Results and discussion

Fig1. shows electron microphotographs of the typical "collar" soot, obtained on Fe-Ni (1:1) catalyst. Based on samples morphology exhibited by this TEM electron microscopy it is clear that the soot consists of different nanostructures. In this fraction SWNTweb penetrates through the different sized metal nonuniform particles. Their sizes ranges from 2 to 25 nm. Thin threads of the web have some tens micron length and up to 2 nm diameters. The biggest particles are covered with a carbon shell of a various nature identifying the core-in-shell structure. The small particles (about 1-2 nm) are situated at the bottom or inside of SWNT.

Fig 2 (b-c) show X-ray diffraction patterns for collar soot samples, obtained on the catalysts with different mutual Fe-Ni concentrations.It is clear that these XRD patterns containt only structural maxima of fcc-FeNi with average concentration close to $Fe_{64}Ni_{36}$ in addition to the graphite ones. Contrary, the XRD-pattern of collar soot obtained on pure Fe catalyst (fig2 a) consists of Fe(C) phases reflexes such as $\alpha$-Fe(C), $\gamma$-Fe(C) and $Fe_3C$ and graphite [7]. Metal particles participating in the graphite electrode evaporation interact together and with carbon atoms. Addition of Ni to Fe catalyst obviously influences phase constitution and microstructure of formed binary particles. In Fe-Ni nanostructures two phases of different structure usually may coexist [8-9]. Both nickel and carbon atoms are beneficial to form solid solution with fcc-structure. As for iron it favors to solubilize carbon atoms in Fe-Ni(C) solution. The slight increase of this phase lattice parameters is a result of small carbon dissolution. The

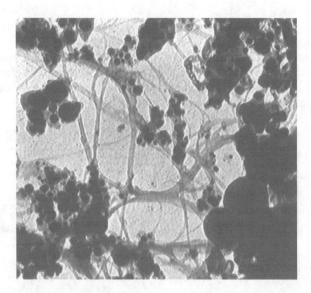

Figure 1. TEM photograph of the soot formed by electric arc evaporation of graphite in the presence of the Fe:Ni(1:1) catalyst

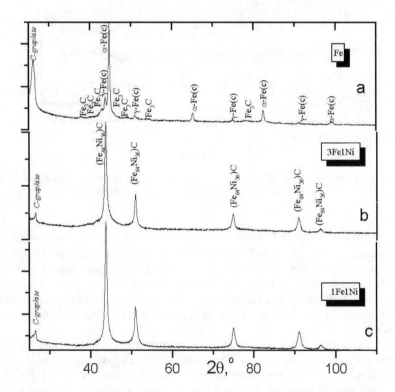

Figure 2. X-ray diffraction patterns of SWNT containing collar soot obtained by arc discharge synthesis with Fe:Ni catalyst of different concentration

evaluation of metal $Fe_{64}Ni_{36}$ particles size by Hall method gives an average value of their diameter about 15 nm.

Mossbauer spectra analysis (Fig.3) of collar soot obtained in the presence of different concentration Fe:Ni catalyst shows that the structure of formed metal particles depends on the mutual concentrations of Fe and Ni in catalytic mixture and resulted from the high temperatures achieve in arc discharge.

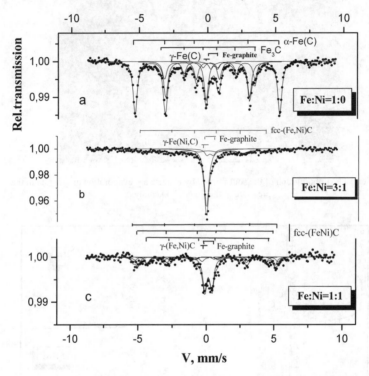

Figure 3. Mossbauer spectra of SWNT containing collar soot obtained by arc diasharge synthesis with Fe:Ni catalyst of different concentration

The resolved Mossbauer spectra components with narrow intensive lines and corresponding to large $\gamma$-Fe(Ni,C) particles which are predominant in the samples obtained with the greater concentration of iron. Besides that these spectra contain components of ultradispersed iron carbide (Fig3a) and iron-graphite complex being a transitional stratum between a metal particle and a carbon nanotube. This Mossbauer spectrum component may be served as indicator of SWNT formation in the synthesis products [3].

For higher concentration of a nickel in the catalyst (Fig.3 b, c) the Mossbauer spectra testify the preferrable formation of ultra disperse particles of fcc-(Fe,Ni)C structure with the presence of the iron-nickel-graphite complex too. As nickel concentration increase the magnification of corresponding to this complex component intensity is observed (Fig.4). The obtained from the Mossbauer spectra results are in agree with data obtained from the oxidizing thermogravimetry applied to evaluate the quantity of SWNT (Fig.4 right axis). According these data SWNT yield increased up to 10% when nickel concentration in a catalytic mixture magnifying in following sequence Fe < Fe3Ni < FeNi.

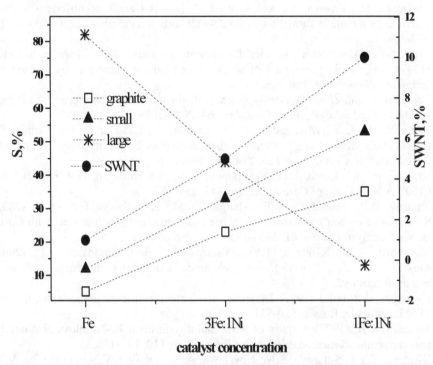

Figure 4. Large and small metall particles content in collar soot together with iron-graphite complex and SWNT yeld for different concentration of used Fe:Ni catalyst

Experimental data, obtained in our present work, testify that high-temperature interaction of Fe and Ni catalyst particles in an arc results in their alloying and dispersion with formation more small-sized, than in case of pure iron, particles. Such anomalous dispersion at certain mutual concentration of transition elements in binary systems have been observed in our previous work deals with Fe-Ni, Fe-Mo, Fe-W, Fe-N nanostructured compounds [10-12]. The average size (~ 15 нм) of fcc-(FeNi) metal nanoparticles is close to an optimum size of catalytic particle for SWNT formation. The increase of such particles amount with Ni-concentration rise (Fig.4) causes optimal conditions for high yield SWNT synthesis.

## 4. Conclusions

Our structural study of SWNT containing arc synthesis fractions have shown, that their structure is determined by composition of the used binary catalyst. The initial mutual concentration of Fe:Ni catalytic mixture in an electrode determines quantitative ratio between formed large inert carbon encapsulated metal particles and small-sized metal nanoparticles being catalytic centers of SWNT origin thereby causing the optimum conditions for high SWNT yield.

## 5. References

1.  Yudasaka, M., Kasuya, Y., Kokai, F. et al. (2002) Causes of different catalytic activities of metals in formation of single-wall carbon nanotubes, *Appl. Phys. A.* **74**, 377-385.
2.  Journet, C., Maser, W.K., Bernier, P., Loiseau, A., Lamy de la Chapelle, M., et al (1997) Large scale production of single-walled carbon nanotubes by electric arc technique, *Nature* **388**, 756-768.
3.  Seraphin, S. and Zhou, D. (1994) Single-walled carbon nanotubes produced at high yield by mixed catalyst, *Appl. Phys. Lett.* **64**, 2087-2089.
4.  Saito, Y., Tani, Y., Miyagawa, N., Mitsushima, K., Kasuya, A. and Nishina, Y. (1998) High yield of single-walled carbon nanotubes by arc disharge using Rh-Pt mixed catalyst, *Chem. Phys. Lett.* **294**, 593-598.
5.  Kratschmer, W., Lamb, L.D., Fostipopoulos, K. and Huffman, D.R. (1990) Solid C(60): A New Form of Carbon, *Nature* **347**, 354-357.
6.  Tarasov, B.P., Muradyan, V.E., Shulga, Yu.M., Krinishnaya E.P. et al. (2003) Synthesis of carbon nanostructures by arc evaporation of graphite rods with Co-Ni an Yni2 catalyst, *Carbon* **41**, 1357-1364.
7.  Novakova, A.A., Kiseleva, T.Yu., Tarasov B.P. and Muradyan, V.E. (2003) Mossbauer study of iron-graphite electrode evaporation products, *Advanced materials* **6**, 86-92.
8.  Lyakishev, I.P. (eds) (1997) Diagrams of binary metallic systems, V.2, Moscow, Mashinostroenie, Russia: 510-521.
9.  Scorzelli, R.B. (1997) A study of phase stability in invar Fe-Ni alloys obtained by non-conventional methods, *Hyperfine Interactions* **110**, 143-150.
10. Kiseleva, T.Yu., Sidorova, E.N., Novakova, A.A., Sokolov, V.N. and Levina, V.V. (2003) Surface investigations. Study of intermediate compounds during nanocrystalline Fe-Ni compounds prepare **3**, 61-66.
11. Kiseleva, T.Yu., Novakova, A.A. and Hait, E.I. (2002) NATO ASI Series, Hydrogen Materials Science and Chemistry of Metal Hydrides, *ed.* T. N. Veziroglu, *Kluwer Acad. Publ.* 82:213-220.
12. Novakova, A.A., Kiseleva, T.Yu., Levina, V.V., Dzidziguri, E.L., Kuznetsov, D.V. (2001) Low temperature formation of nanocrystalline Fe-Mo and Fe-W compounds, *Journal of Alloys and Compounds* **317-318**: 423-427.

# DETERMINATION OF AN OPTIMUM PERFORMANCE OF A PEM FUEL CELL BASED ON ITS LIMITING CURRENT DENSITY

AYOUB KAZIM
United Arab Emirates University Department of Mechanical Engineering
P.O. Box 17555 Al-Ain, U.A.E.
Fax +971-3-7623-158 E-mail: akazim@uaeu.ac.ae

## Abstract

In this paper, determination of optimum current density and cell voltage of a PEM fuel cell based on its limiting current density is carried out. The optimum values are calculated in terms of A and B, which represent the coefficients of the activation and concentration overpotentials, and a factor K, which relates the two overpotentials in the standard cell voltage equation. The analysis is conducted over a wide range of a fuel cell limiting current density, ohmic resistance and K. The results showed that higher overpotentials and optimum current densities are achieved with higher values of A, B, ohmic resistance or limiting current density. Furthermore, higher activation overpotential and optimum cell voltage can be achieved with an increase in K.

**Keywords:** PEM fuel cell, cell overpotential, optimum current density, optimum cell voltage.

## Nomenclature

| | |
|---|---|
| A | Activation overpotential coefficient, V |
| B | Concentration overpotential coefficient, V |
| K | factor, $10^3$ to $10^5$ |
| F | Faraday's constant, 96487 C per equivalent |
| i | Average current density, $A/cm^2$ |
| $i_o$ | Exchange current density, $A/cm^2$ |
| $i_L$ | Limiting current density, $A/cm^2$ |
| n | Number of electrons involved in the reaction |
| $\Re$ | Gas constant, 82.06 $cm^3$.atm/mol K |
| R | Membrane ohmic resistance, $\Omega \cdot cm^2$ |
| T | Fuel cell operating temperature, $^\circ C$ |
| $V_{FC}$ | Fuel cell voltage, V |
| $V_{oc}$ | Open circuit voltage, V |

### Greek symbols

| | |
|---|---|
| $\eta$ | Electrode over-potential, V |
| $\alpha$ | Transfer coefficient |
| $\varepsilon$ | Ratio of the current density to the limiting current density |

### Subscripts and superscripts

| | |
|---|---|
| act | activation |
| con | Concentration |
| ohm | Ohmic |

T.N. Veziroglu et al. (eds.),
Hydrogen Materials Science and Chemistry of Carbon Nanomaterials, 159-166.
© 2004 Kluwer Academic Publishers. Printed in the Netherlands.

## 1. Introduction

Determination of irreversible losses of a proton exchange membrane (PEM) fuel cell is considered to be extremely essential to assess its performance in terms cell voltage, limiting current density and power density. There are several sources that contribute to irreversible losses in a fuel cell during the operation. These losses are often called overpotentials or polarizations that could be originated primarily from three sources namely activation overpotential, ohmic overpotential and concentration overpotential. Activation overpotential is associated mainly with the slowness of electrochemical reaction in the fuel cell. Ohmic overpotential is associated with the resistance of the membrane to the flow of migrating ions during an electrochemical process. And the concentration overpotential occurs due to concentration gradients established as a result of rapid consumption of the reactant (oxygen) in the electrode during the electrochemical reaction.

In the past decades, several studies were performed on modelling the polarization curve by simple mathematical equations [1, 2, 3]. Some studies carried out calculating the performance of the fuel cell based on formulation of several parametric equations at various operating conditions such as temperature and pressure [4]. Others set more complex mathematical models to determine the performance of a fuel cell [5,6]. However, these studies did not address the calculation of an optimum performance of a fuel cell based on its limiting current density. Moreover, their mathematical models were considered to be complex and inflexible although most of these models were valid at various operational conditions.

The objective of the current study is to determine the optimum current density and cell voltage of a PEM fuel cell based on its given limiting current density. The optimum values are to be calculated over a wide range of a fuel cell limiting current density, ohmic resistance and activation and concentration overpotentials. In addition, the analysis should be simple, flexible and valid at various operational conditions of a PEM fuel cell.

## 2. Fundamental equations

At a certain operational condition, voltage of a PEM fuel cell $V_{FC}$, can be determined through the following general equation [7]:

$$V_{FC} = V_{oc} - \eta_{act} - \eta_{con} - \eta_{ohm} \qquad (1)$$

where, $V_{oc}$, $\eta_{act}$, $\eta_{con}$, $\eta_{ohm}$ are open–circuit voltage, activation overpotential, concentration overpotential and ohmic overpotential, respectively. The open circuit voltage of a fuel cell depends on the cell temperature described by Parthasarathy *et al.* [8]. However in the current study, the open-circuit voltage is set to take a constant value of 1.1V [9].

The activation overpotential is expressed in terms of gas constant $\mathfrak{R}$, fuel cell operating temperature T, transfer coefficient $\alpha$, number of electrons involved n and Faraday's constant F, current density i, and exchange current density $i_o$ [7]:

$$\eta_{act} = \frac{\Re T}{\alpha nF} \ln\left(\frac{i}{i_o}\right) \tag{2}$$

Similarly, the concentration overpotential $\eta_{con}$, is expressed as [7]:

$$\eta_{con} = \frac{\Re T}{nF} \ln\left(1 - \frac{i}{i_L}\right) \tag{3}$$

where, $i_L$ is the limiting current density of the fuel cell. The ohmic overpotential $\eta_{ohm}$ is described in terms of operating current density of the fuel cell and the ohmic resistance R, which is related to the ionic conductivity and discussed thoroughly by Springer *et al.* [10]:

$$\eta_{ohm} = i R \tag{4}$$

Substituting the above-described overpotentials into equation (1) to yield a more elaborative expression of the fuel cell's voltage:

$$V_{FC} = V_{oc} - \left\{\frac{\Re T}{\alpha nF} \ln\left(\frac{i}{i_o}\right)\right\} - \left\{\frac{\Re T}{nF} \ln\left(1 - \frac{i}{i_L}\right)\right\} - i R \tag{5}$$

## 3. Determination of the optimum values

In order to determine the optimum current density and voltage of a PEM fuel cell, the following steps are going to be taken:

a. Let a variable $\varepsilon$, to represent the ratio of the current density i, to the limiting current density $i_L$ :

$$\varepsilon = \left(\frac{i}{i_L}\right) \tag{6}$$

b. Develop a relation between the ratios $\left(\frac{i}{i_o}\right)$ and $\left(\frac{i}{i_L}\right)$ of both activation and concentration overpotentials, through the following relation:

$$\frac{i}{i_o} = K\left(\frac{i}{i_L}\right) = K\varepsilon \tag{7}$$

where, factor K is set to vary from $10^3$ to $10^5$. This relation is considered to be valid since $\left(\frac{i}{i_o}\right)$ is greater than $\left(\frac{i}{i_L}\right)$ by order of magnitude of at least 3.

c. Set A and B to represent the coefficients of activation and concentration overpotentials such that:

$$A = \frac{\Re T}{\alpha nF} \tag{8}$$

$$B = \frac{\Re T}{nF} \qquad (9)$$

d. The activation and concentration overpotentials can now be expressed in simplified forms in terms of A, B, K and $\varepsilon$ :

$$\eta_{act} = A \ln(K\varepsilon) \qquad (10)$$

$$\eta_{con} = B \ln(1-\varepsilon) \qquad (11)$$

By the same token, the ohmic overpotential of equation (4) can be described as:

$$\eta_{ohm} = iR = \varepsilon i_L R \qquad (12)$$

e. Substitute equations (10) through (12) into equation (5) to obtain a more simplified equation of a fuel cell voltage $V_{FC}$ :

$$V_{FC} = V_{oc} - A \ln(K\varepsilon) - B \ln(1-\varepsilon) - \varepsilon i_L R \qquad (13)$$

f. Differentiate equation (13) with respect to $\varepsilon$ to yield:

$$\frac{dV_{FC}}{d\varepsilon} = \frac{-A}{\varepsilon} + \frac{B}{1-\varepsilon} - i_L R \qquad (14)$$

By setting $\dfrac{dV_{FC}}{d\varepsilon} = 0$ , the above second-order equation is then solved with respect to $\varepsilon$ in order to obtain the optimum values (roots). Subsequently, the optimum current density $i_{opt}$ , is calculated at a given limiting current density and then all the related overpotentials and the optimum fuel cell voltage $V_{opt}$ are determined.

## 4. Results and discussions

Calculations of $i/i_L$, optimum operating current density $i_{opt}$, and optimum cell voltage $V_{opt}$ were carried out at a given limiting current density ranging from $i_L = 0.5$ A/cm$^2$ to 1.5 A/cm$^2$, ohmic resistance of R=0.10 $\Omega \cdot$ cm$^2$ and R=0.20 $\Omega \cdot$ cm$^2$ and K ranging from $10^3$ to $10^5$ as presented in tables 1 to 5. At overpotential coefficients A and B of 0.05 V and ohmic resistance of R=0.10 $\Omega \cdot$ cm$^2$, the optimum values of current density, overpotentials and cell voltage are calculated and presented in table 1, which represents the base case. A drastic increase in the optimum current density and cell overpotentials occur when the limiting current density of the fuel cell is increased. For instance, the optimum fuel cell operating current density increases from i=0.31 A/cm$^2$ to 1.15 A/ cm$^2$ as the limiting current density increases from $i_L = 0.5$ A/ cm$^2$ to 1.5 A/ cm$^2$. Conversely, the optimum cell voltage decreases with increasing the optimum current density and as a result of increase in the activation, concentration and ohmic overpotentials deduced from equation (13).

TABLE 1: Optimum values of a PEM fuel cell operating current density, cell overpotentials and cell voltage at A=B=0.05V and R= 0.10 ohm. cm$^2$.

| $i_L$ (A/cm$^2$) | Current Density, $i_{opt}$ (A/cm$^2$) | $\eta_{con}$ (V) | $\eta_{ohm}$ (V) | $\eta_{act}$ (V) | | | Fuel Cell Voltage, $V_{opt}$ (V) | | |
|---|---|---|---|---|---|---|---|---|---|
| | | | | K= 10$^3$ | K= 10$^4$ | K= 10$^5$ | K= 10$^3$ | K= 10$^4$ | K= 10$^5$ |
| 0.5 | 0.31 | -0.048 | 0.031 | 0.321 | 0.436 | 0.552 | 0.796 | 0.681 | 0.566 |
| 0.75 | 0.5 | -0.055 | 0.05 | 0.325 | 0.440 | 0.555 | 0.780 | 0.665 | 0.550 |
| 1.0 | 0.71 | -0.061 | 0.071 | 0.328 | 0.443 | 0.558 | 0.763 | 0.648 | 0.532 |
| 1.25 | 0.92 | -0.067 | 0.092 | 0.330 | 0.445 | 0.560 | 0.744 | 0.629 | 0.514 |
| 1.5 | 1.15 | -0.073 | 0.115 | 0.332 | 0.447 | 0.562 | 0.726 | 0.610 | 0.495 |

Obviously, the activation overpotential would be greater than the concentration overpotential if the coefficient A is set be higher than B and keeping R=0.10 $\Omega \cdot cm^2$, as indicated in table 2. A general comment can be made about this case is that the optimum current density, concentration overpotential, activation overpotential and ohmic overpotential, are at least 5% greater than those of the base case where A=B. This is mainly attributed to a higher coefficient A in the second order equation leading the results to be slightly higher than the base case. Change in the activation overpotential $\eta_{act}$, is considered to be the highest among all of the parameters due to 50% increase in the coefficient A. Similar to the base case, a lower optimum cell voltage is calculated as a result of higher operating current density and cell overpotentials.

TABLE 2: Optimum values of a PEM fuel cell operating current density, cell overpotentials and cell voltage at A= 1.5B= 0.075V and R= 0.10 ohm. cm$^2$.

| $i_L$ (A/cm$^2$) | Current Density, $i_{opt}$ (A/cm$^2$) | $\eta_{con}$ (V) | $\eta_{ohm}$ (V) | $\eta_{act}$ (V) | | | Fuel Cell Voltage, $V_{opt}$ (V) | | |
|---|---|---|---|---|---|---|---|---|---|
| | | | | K= 10$^3$ | K= 10$^4$ | K= 10$^5$ | K= 10$^3$ | K= 10$^4$ | K= 10$^5$ |
| 0.5 | 0.34 | -0.058 | 0.034 | 0.490 | 0.662 | 0.835 | 0.634 | 0.461 | 0.288 |
| 0.75 | 0.54 | -0.064 | 0.054 | 0.494 | 0.666 | 0.839 | 0.616 | 0.443 | 0.271 |
| 1.0 | 0.75 | -0.069 | 0.075 | 0.496 | 0.669 | 0.842 | 0.598 | 0.425 | 0.252 |
| 1.25 | 0.97 | -0.074 | 0.097 | 0.499 | 0.672 | 0.844 | 0.579 | 0.406 | 0.233 |
| 1.5 | 1.20 | -0.079 | 0.119 | 0.501 | 0.674 | 0.846 | 0.559 | 0.386 | 0.214 |

Now, if the ohmic resistance is doubled to R=0.20 $\Omega \cdot cm^2$, the ohmic overpotential will increase by two-fold as demonstrated in table 3. This is clearly a result of a linear relation between the ohmic overpotential and fuel cell's membrane resistance. In addition, the optimum cell voltage is considered to be less than the optimum voltage calculated at the base case, due to increase in the ohmic overpotential. Through increasing the coefficient A by 50% and letting the ohmic resistance unchanged, all overpotentials will increase especially activation overpotential, as shown in table 4. For instance, at $i_L$=0.5 A/m$^2$ and at K=10$^3$, the activation, concentration and ohmic overpotentials would increase by 51%, 13% and 6%, respectively. Clearly, activation overpotential is mostly influenced by this increase due to increase in A.

TABLE 3: Optimum values of a PEM fuel cell operating current density, cell overpotentials and cell voltage at A=B= 0.05V and R= 0.20 ohm. cm$^2$.

| $i_L$ (A/cm$^2$) | Current Density, $i_{opt}$ (A/cm$^2$) | $\eta_{con}$ (V) | $\eta_{ohm}$ (V) | $\eta_{act}$ (V) | | | Fuel Cell Voltage, $V_{opt}$ (V) | | |
|---|---|---|---|---|---|---|---|---|---|
| | | | | K=10$^3$ | K=10$^4$ | K=10$^5$ | K=10$^3$ | K=10$^4$ | K=10$^5$ |
| 0.5 | 0.35 | -0.061 | 0.071 | 0.328 | 0.443 | 0.558 | 0.763 | 0.647 | 0.532 |
| 0.75 | 0.58 | -0.073 | 0.115 | 0.332 | 0.447 | 0.562 | 0.726 | 0.610 | 0.495 |
| 1.0 | 0.81 | -0.083 | 0.162 | 0.335 | 0.450 | 0.565 | 0.686 | 0.571 | 0.456 |
| 1.25 | 1.05 | -0.091 | 0.210 | 0.337 | 0.452 | 0.567 | 0.645 | 0.530 | 0.415 |
| 1.5 | 1.29 | -0.098 | 0.258 | 0.338 | 0.453 | 0.568 | 0.602 | 0.487 | 0.372 |

TABLE 4: Optimum values of a PEM fuel cell operating current density, cell overpotentials and cell voltage at A=1.5B= 0.075V and R= 0.20 ohm. cm$^2$.

| $i_L$ (A/cm$^2$) | Current Density, $i_{opt}$ (A/cm$^2$) | $\eta_{con}$ (V) | $\eta_{ohm}$ (V) | $\eta_{act}$ (V) | | | Fuel Cell Voltage, $V_{opt}$ (V) | | |
|---|---|---|---|---|---|---|---|---|---|
| | | | | K=10$^3$ | K=10$^4$ | K=10$^5$ | K=10$^3$ | K=10$^4$ | K=10$^5$ |
| 0.5 | 0.38 | -0.069 | 0.075 | 0.496 | 0.669 | 0.842 | 0.598 | 0.425 | 0.252 |
| 0.75 | 0.60 | -0.079 | 0.119 | 0.501 | 0.674 | 0.846 | 0.559 | 0.386 | 0.214 |
| 1.0 | 0.83 | -0.088 | 0.166 | 0.504 | 0.677 | 0.849 | 0.518 | 0.346 | 0.173 |
| 1.25 | 1.06 | -0.096 | 0.213 | 0.506 | 0.679 | 0.851 | 0.476 | 0.304 | 0.131 |
| 1.5 | 1.31 | -0.102 | 0.261 | 0.508 | 0.680 | 0.853 | 0.433 | 0.261 | 0.088 |

If we compare the base case of table 1 with the case where both coefficients A and B are increased by 50% presented in table 5, we notice a major increase in both concentration and activation overpotentials. On the other hand, the ohmic overpotential decreases by at least 6%. By the same token, the optimum current density and cell voltage at K=10$^5$ are decreased by at least 6% and 45%, respectively.

TABLE 5: Optimum values of a PEM fuel cell operating current density, cell overpotentials and cell voltage at A=B= 0.075V and R= 0.10 ohm. cm$^2$.

| $i_L$ (A/cm$^2$) | Current Density, $i_{opt}$ (A/cm$^2$) | $\eta_{con}$ (V) | $\eta_{ohm}$ (V) | $\eta_{act}$ (V) | | | Fuel Cell Voltage, $V_{opt}$ (V) | | |
|---|---|---|---|---|---|---|---|---|---|
| | | | | K=10$^3$ | K=10$^4$ | K=10$^5$ | K=10$^3$ | K=10$^4$ | K=10$^5$ |
| 0.5 | 0.29 | -0.065 | 0.029 | 0.477 | 0.650 | 0.823 | 0.659 | 0.486 | 0.313 |
| 0.75 | 0.46 | -0.072 | 0.046 | 0.482 | 0.655 | 0.827 | 0.644 | 0.471 | 0.298 |
| 1.0 | 0.65 | -0.079 | 0.065 | 0.486 | 0.659 | 0.831 | 0.628 | 0.455 | 0.283 |
| 1.25 | 0.85 | -0.086 | 0.085 | 0.489 | 0.662 | 0.835 | 0.611 | 0.439 | 0.266 |
| 1.5 | 1.06 | -0.092 | 0.106 | 0.492 | 0.665 | 0.837 | 0.594 | 0.421 | 0.249 |

It should be emphasized that the above analysis could be valid over wide ranges of operational conditions through interpolation and extrapolations. For instance, at A=0.06V, B=0.05V, limiting current density of $i_L$=1.5 A/cm$^2$, ohmic resistance of R=0.10 $\Omega \cdot cm^2$ and K=10$^3$, the optimum current density and cell voltage can be calculated after interpolation to be 1.17 A/cm$^2$ and 0.659V, respectively. Furthermore, at coefficients A=0.06V and B=0.06V, limiting current density of $i_L$=1.0 A/cm$^2$, resistance R=0.25 $\Omega \cdot cm^2$ and K=10$^3$, the optimum current density and cell voltage are calculated after extrapolation to be 0.84 A/cm$^2$ and 0.601V, respectively.

Finally, a general remark can be made regarding the above results, is that the calculated values of concentration overpotential are negative if compared with ohmic and activation overpotentials. This is mainly attributed to the value inside the logarithmic function to be less than unity. Moreover, values of activation overpotentials and optimum cell voltages increase with increasing K, which is restricted between the limits 10$^3$ and 10$^5$. Value of K outside these limits would be considered unrealistic in most cases specially in determining the fuel cell activation overpotential [11]. By the same token, values of coefficients A and B are posed to be typical for concentration and activation overpotentials of a PEM fuel cell during the operation [12].

## 5. Conclusions

Determination of optimum current density and cell voltage of a PEM fuel cell based on its limiting current density was carried out. The conclusions are summarized as follows:

- Higher overpotentials and optimum current density are achieved with higher limiting current density. Conversely, a lower optimum cell voltage can be detected with higher $i_L$.
- Higher activation overpotential and optimum cell voltage can be obtained with an increase in K.
- Higher overpotentials, optimum current density are achieved with increase in the ohmic resistance R. On the other hand, a slight decrease in the optimum cell voltage can be obtained with increasing R.
- Higher overpotentials and optimum current densities are achieved with higher values of A, B, ohmic resistance or limiting current density. On the other hand, a decrease in the optimum cell voltage can be obtained with increasing any of the above coefficients.
- Values of coefficients A, B and K used in the study, were considered to be typical for a normal operational conditions of a PEM fuel cell.
- The analysis is considered to be simple, flexible and valid over wide ranges of the operational conditions of a PEM fuel cell.

## 6. References

1. Kim J, Seong L, Srinivasan S. Modeling of proton exchange membrane fuel cell performance with an empirical equation. J Electrochem Soc. 1995; 142: 2670-2674.
2. Bernardi D M, Verbrugge M W. A Mathematical Model of the Solid-Polymer-Electrolyte Fuel Cell. J Electrochem Soc. 1992; 140: 2767-2772.

3. Bevers D, Wohr M, Yasuda K, Oguro K, Simulation of a Polymer Electrolyte Fuel Cell Electrode. J Applied Electrochem. 1997; 27: 1254-1264.

4. Amphlett J C, Baumert R M, Mann R F, Peppley B A, and Roberge P R, Performance Modeling of the Ballard Mark IV Solid Polymer Electrolyte Fuel Cell. J Electrochem Soc. 1995; 142: 1-8.

5. Gurau V, Liu H and Kakaç S, Two Dimensional Model for Proton Exchange Membrane Fuel Cells. AICHE Journal. 1998; 44: 2410-2422.

6. Yi J S, Nguyen T V, An Along-the-Channel Model for Proton Exchange Membrane Fuel Cells. J Electrochem Soc. 1998; 145: 1149-1159.

7. Kinoshita K, McLarnon F. and Cairns E. Fuel cells a handbook. U.S. Department of Energy, 1988.

8. Parathasarathy A., Srinivasan S. and Appleby A. Temperature dependence of the electrode kinetics of oxygen reduction at the Platinum/Nafion interface- A microelectrode investigation. J. Electrochem Soc. 1992; 139: 2530.

9. Kazim A, Liu H and Forges P. Modeling of performance of PEM fuel cells with conventional and interdigitated flow fields. J Applied Electrochem 1999; 29 (12): 1409-1416.

10. Springer T. E., Zawodzinski T. A. and Gottesfeld S. Polymer electrolyte fuel cell model. J Electrochem Soc. 1991; 138: 2334-2342.

11. Newman J.S. Electrochemical systems. Prentice Hall, New Jersey, 1991.

12. Larminie, J. and Dicks, A. Fuel cell systems explained. John Wiley & Sons, LTD. 2001.

# PHOTOINDUCED MODIFICATIONS OF THE STRUCTURE AND MICROHARDNESS OF FULLERITE C60

I.MANIKA,[A] J.MANIKS[A], J.KALNACS[B]
*[a]Institute of Solid State Physics, University of Latvia, 8 Kengaraga St., Riga LV-1063, Latvia*
*[b]Institute of Physical Energetics, Latvian Academy of Sciences, 21 Aizkraukles St., Riga LV-1006, Latvia*
*FAX: +371 7132778    E-mail: manik@latnet.lv*

## Abstract

The wavelength dependence, temperature limits, time and depth evolution of the photoinduced hardening of $C_{60}$ crystals in air have been investigated by microindentation and dislocation mobility methods. Two photopolymerized phases, which differ in the hardness and thermal stability, are found to appear. We suggest that formation of fullerene dimers (ie. $C_{120}$) in pristine fullerite and growth of $C_{120}O$ phase in oxygenated fullerite *via* [2+2] photoaddition reaction is responsible for it. It has been found from depth distribution data of the hardness that the $C_{120}O$ phase is located in oxygen-contaminated subsurface layer of 0.8- 1 µm and appears under illumination at 290-330 K.
**Key Words** – $C_{60}$, photopolymerization, indentation hardness

## 1. Introduction

During the recent years the polymeric phases of $C_{60}$ have attracted much attention. The photopolymerization is of special interest suggesting possible photolithographic and other applications of fullerite. In this photoinduced reaction some of the double bonds on the fullerene molecules break up and by a [2+2] cycloaddition form linking bonds to neighboring molecules [1]. Phototransformed fullerite is appreciated as a one-dimensional polymer, which consists mainly of covalently bonded fullerene dimers or short chains.

Because $C_{60}$ absorbs light very efficiently, the photopolymerization occurs in a thin near-surface layer, and the limited sample volumes available have made a full characterization of phototransformed $C_{60}$ rather difficult. Despite the large number of studies devoted to $C_{60}$ films there remain unanswered questions regarding photopolymerization processes in single crystals.

Considering different possible applications, there is of interest to investigate photopolymerization of fullerite in air atmosphere. Different possible mechanisms for the photochemical transformations under such conditions are offered, however, oxygen effects on the structural phase transitions still remain unclear. Contradictions arise also with interpretation of polymerized structures formed at the saturation stage of photopolymerization.

*T.N. Veziroglu et al. (eds.),*
*Hydrogen Materials Science and Chemistry of Carbon Nanomaterials,* 167-176.
© 2004 *Kluwer Academic Publishers. Printed in the Netherlands.*

In the present report, the behavior of photoinduced polymerization of $C_{60}$ single crystals and polycrystalline films in air has been investigated by microindentation and dislocation mobility methods. Effects of the wavelength, illumination time, air exposure and temperature are studied.

## 2. Experimental

$C_{60}$ single-crystals were grown from the vapor phase by sublimation of a twice-sublimed $C_{60}$ powder (99.95% $C_{60}$). The typical size of crystals was about $2x0.5x12mm^3$ and the density of "grown-in" dislocations was $10^3$ $cm^{-2}$. Investigations were performed on the (111) face of as-grown crystals. The quality of crystals was characterized by X-ray diffraction [2] and FTIR Raman spectra (Fig.1). Fullerite films with a thickness of 0.5-1μm and grain size of 100-300 nm were prepared by sublimation of $C_{60}$ powder at $10^{-4}$ Pa on glass substrates.

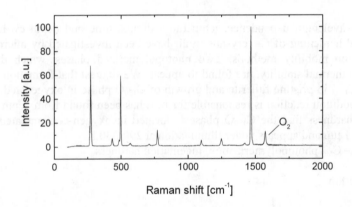

Fig.1. FTIR Raman spectrum of $C_{60}$ crystal. Excitation was provided by Nd-YAG laser (1064 nm).

Experiments were performed in the wavelength range of 350-900 nm. As the light sources we used a monochromator SPM-2 (Carl Zeiss) for the wavelength 500-900 nm and a 125 W mercury lamp with the filters for 365 nm UV light. The wavelength dependence was investigated at a constant power density of 2 $mW/cm^2$. Investigations of the time evolution of polymerization were conducted under irradiation of a low fluency He-Ne laser (632.8 nm) at power density of 1.5 $mW/mm^2$.

For characterization of the extent of polymerization, we used indentation hardness technique, which is a local and structure sensitive method with a good response to change in bonding. The microhardness measurements were conducted at light loads (3.2—10 mN) using a vibration-insensitive loading system for ensuring accurate measurements. The scatter of the hardness measurements was typically about 12 %. The dislocation structure around indents was visualized by selective etching the crystals in toluene. A six-armed dislocation etch-pit pattern was obtained, which consists of the rows of edge dislocations with arms parallel to <110> directions. The parameter $\Delta l/l_0$ (where $l_0$ and $l$ is the dislocation arm length around indents before and after light-irradiation, respectively, and $\Delta l$ is the difference between them) was used as a characteristic of photoinduced change in dislocation mobility. The length $l$ of the dislocation rosette arms was measured with an accuracy of 3.5 %.

Raman spectra were measured with a Nicolet FTIR spectrometer. Excitation was provided by Nd-YAG laser (1064 nm). An UV-VIS spectrometer Specord M40 was used for measurements of optical absorption spectra.

## 3. Results and discussion

The residual hardening and reduction of dislocation mobility in $C_{60}$ crystals after illumination with visible and ultraviolet light was observed confirming earlier results [3-8]. At the saturation stage, the hardness of illuminated crystals by a factor of three or more exceeds the hardness before illumination. The hardening effect is related to photopolymerization of fullerite and measurements of Raman spectra give evidence for this [4]. The hardening is typical for all fullerite polymers and the indentation hardness technique is successfully used for investigations of the behavior of polymerization.

In the first series of experiments, the optimum conditions for illumination and indentation tests were cleared up. With this aim, the wavelength dependence of photoinduced effects in hardness and dislocation mobility was investigated. The results are shown in Fig. 2 where the optical absorbance, photoinduced change in hardness and dislocation mobility is plotted as a function of incident photon energy.

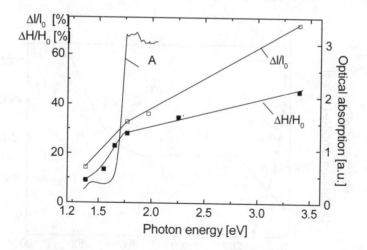

Fig.2. The wavelength dependence of the optical absorption A, photoinduced hardening ($\Delta H/H_0$) and reduction of the dislocation mobility ($\Delta l/l_0$) of $C_{60}$ crystal. Illuminated at 2 mW/cm$^2$ for 80 min ($\Delta H/H_0$) and 20 min ($\Delta l/l_0$) at 290 K. The load was 3.2 mN.

The photoinduced effects clearly appear at photon energies above the bandgap. The magnitude of the effects linearly increases with increasing the photon energy. For photon energies below the bandgap the effects rapidly decrease. The effects disappear at the threshold energy (1.5-1.6 eV), which may originate from exciton transitions [9]. It follows from the obtained results that the polymerization occurs under the excitation with visible and UV light in the fundamental absorption region and adjacent tail zone of optical absorption. The results obtained by microhardness and dislocation mobility methods are similar. However, at comparable illumination conditions the photoinduced effect in dislocation mobility is more marked because the dislocation mobility is sensitive not only to formation of strong obstacles for dislocations as in case of hardness, but also to weak barriers, such as absorbed impurities and point defects. Moreover, significant role in formation of dislocation rosette play surface effects.

From the methodical point of view, the optimum conditions for photopolymerization procedure and indentation tests can be ensured at the wavelength of about 630 –700 nm where the photoinduced effects reach stabilized values and the light penetration depth (2-3 μm) still remains sufficiently high for performing correct measurements of the microhardness in the illuminated layer. Because of this, a low fluency 632.8 nm He-Ne laser was used as the light source for further detailed investigations.

Investigations of the illumination time evolution of photoinduced hardening showed a stepwise behavior (Fig.3). After a short initial stage, a plateau of the hardness at about 450 MPa was reached indicating to formation of a photopolymerized phase with a constant hardness. This phase is considerably harder than pristine fullerite (165 MPa). During further light irradiation, the increase in hardness continued up to 1 GPa giving evidence for formation of another considerably harder phototransformed product.

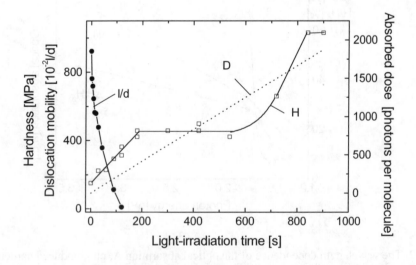

Fig.3. Photoinduced change in hardness *H*, dislocation mobility *l/d* and photon dosage *D* as a function of light-irradiation time. The load was 3.2 mN.

The photoinduced effect in dislocation mobility was more marked (Fig.3). A rapid reduction of the dislocation arm length around indents was observed and after about 180 s the values of $l/d$ were reduced almost to zero confirming formation of a toluene-insoluble phototransformed layer on the irradiated surface. However, its thickness at this initial stage is too low to cause the change of hardness. The minimum thickness of the hardened layer, which formation we were able to detect by the microhardness method was about 0.2 μm. The illumination-time evolution of the dislocation mobility confirms the typical for diffusion controlled processes parabolic law (Fig.4). As the intensity of self-diffusion processes in fullerite at room temperature is known to be small, we suggested that the restriction of dislocation mobility under light-irradiation might be controlled by oxygen penetration in fullerite from air atmosphere and its reaction with fullerite.

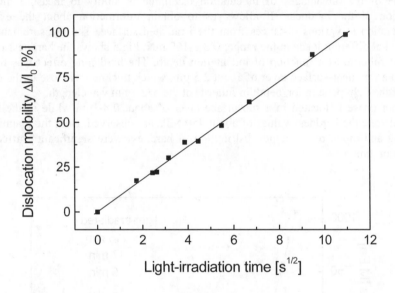

Fig.4. Dislocation mobility parameter ($\Delta l/l_0$) as a function of $t^{1/2}$.

It was of interest to estimate the photon dosage for different stages of photopolymerization. The absorbed dose of irradiation (photons per fullerite molecule) was estimated as a ratio of the number of photons $N$ at a given light exposure and molecular density $d$ of fullerite: $D=N/d$. The number of photons was calculated from $N=I\alpha t/\hbar\omega$, where $I$ is the light intensity, $\alpha$ is the light absorption coefficient (for the 632.8 nm wavelength we used $\alpha=9\times10^3$ cm$^{-1}$), $t$ is the illumination time, $\hbar\omega$ is the photon energy. The molecular density was found as $d=4/a^3$, where $a$ is a lattice constant of fullerite. The obtained values of irradiation dose are presented in Fig.3 by a dotted line. Detectable change in the dislocation mobility is found to appear at a dose of about 6 photons per molecule while about 250 photons per molecule are needed for full immobilization of dislocations and formation of toluene insoluble polymerized surface layer. The photoinduced hardening is initiated at the absorbed dose of about 80 photons per molecule while the hardness level of 450 MPa is reached at about 380 photons per fullerite molecule.

The multiphoton origin of photopolymerization follows from the condition that two adjacent fullerene molecules need to be located close to each other, simultaneously activated and oriented in a favorable position. Obviously, there is a small probability for fulfilling of this condition under single photon excitation. This is also the reason why the [2+2] photoaddition mechanism works only in the face-centered-cubic phase (above 260 K) where the fullerene molecules are able to rotate freely [1].

We observed photoinduced hardening also in fullerite films. On contrary to obtained results for single crystals, the photopolymerization in fullerite films occurs at considerably higher photon dosage $(10^4$-$10^5$ photons per fullerene molecule) [10]. Obviously, the structural defects (grain boundaries, dislocations, stacking faults and vacancies) significantly reduce the efficiency of polymerization.

One of the advantages of indentation technique is ability to measure the depth distribution of the hardness. It allows us to obtain information about the extent of polymerization at various distances from the irradiated surface. The measurements were performed at different loads in the range of 3.2-100 mN. Fig.5 shows the hardness of light-irradiated fullerite as a function of indentation depth. The hardening effect was found to appear in a thin near-surface layer of about 2.5 μm, which thickness is close to the referred values of the light penetration depth in fullerite at the 632.8 nm wavelength.

The harder phase is located in a subsurface layer of about 0.8-1μm while in the bulk of irradiated zone the hardness values of about 450 MPa are observed. Both the magnitude of hardening and mode of the depth distribution of hardness were significantly affected by illumination time.

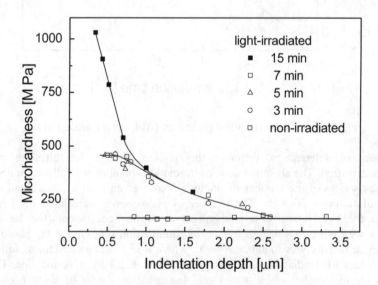

Fig.5. Hardness of pristine and laser-irradiated $C_{60}$ as a function of indentation depth.

It follows from Fig.3 and Fig.5 that two different polymers of fullerite: soft (HV=450 MPa) and hard (0.65-1 GPa) can be created under laser irradiation. Our further experiments showed that the laser-irradiation of $C_{60}$ crystals at T> 340 K produces only the soft polymer while at 290-330 K both the soft and hard phase can be obtained depending on light exposure. Recent atomic force microscopy (AFM) [11] and Raman studies [12] also suggested that two different photopolymerized states might exist. AFM investigations of the surface structure showed that photopolymerization at temperatures above 350 K produce mainly fullerene dimers, while material polymerized at 320 K and below contains larger chains typically six molecules long [11].

An essential characteristic of fullerite polymers is the depolymerization temperature. It was found that abrupt decrease of the hardness of the softer phase occurs at 470-480 K. After such thermal treatment the hardness reached values, which are typical for non-irradiated samples. The observed softening temperature coincides with the depolimerization temperature for the fullerite photopolymer consisting mainly of fullerene dimers [13]. This result allows us to recognize the softer phase as the well-known dimerized fullerite photopolymer.

The depolymerization behavior of the harder polymer is shown in Fig.6. A two-stage irreversible softening on heating the samples at the rate of 3 K/min was observed. At first stage, a rapid reduction of the hardness of crystals down to 430 MPa on heating to 320 K was found. At the final stage, softening down to 185 MPa on heating to about 480 K was reached. Similar change of the hardness on heating was observed for polymerized fullerite films. The obtained data were confirmed also by annealing experiments. The hardness of photopolymerized crystals was reduced from 0.85-1 GPa to 390-440 MPa after annealing the samples in air 340 K for 10 min. Both hard and soft photopolymer return to non-polymerized state after annealing at 480 K for 10 min. The hardness values (about 185 MPa) after annealing coincide with those for the oxygenated fullerite monomer [14, 15].

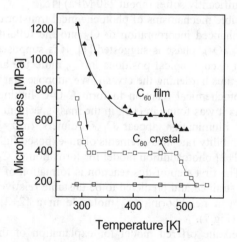

Fig.6. Hardness of phototransformed $C_{60}$ single crystal and polycrystalline film as a function of temperature. The load was 3.2 mN and heating rate was 3 K/min.

It follows from the comparison of different $C_{60}$ polymers that their thermal stability increases if the extent of polymerization is increased [13]. Consequently, a higher thermal stability of the harder photopolymer is expected. However, our experiments show that the softer phase depolymerizes at 480 K while the harder phase is transformed at 320-340 K. This transformation temperature is lower than that, at which the covalent carbon bonds between fullerene molecules can be thermally broken and fullerite photopolymer returns back to monomeric $C_{60}$. Because of this, the hard phase of photopolymer cannot be considered as highly polymerized all-carbon structure.

On the other hand, our experiments are performed in air atmosphere and the obtained results are indicative of the participation of oxygen in the phototransformation. The investigations of dislocation mobility and hardness indicate that the photoinduced structural change is initiated in the oxygen-contaminated subsurface layer of 0.8-1 μm and here the formation of hard phase is observed (see Fig.5). It is well established that oxygen molecules can be physically absorbed in fullerite solid occupying the octahedral interstitial states. The FTIR Raman spectra of our crystals (see Fig.1) show the presence of a maximum at 1575 $cm^{-1}$, which is typical for oxygenated fullerite and is not observed for oxygen-free fullerite [16]. Besides, the light-irradiation enhances incorporation of $O_2$ into the fullerite lattice as it was observed in [1]. The physically absorbed molecules can act as obstacles for dislocations and thus harden the crystal. Our previous investigations show that only a slight increase of the hardness (up to about 15%) appears during long-time exposure the samples to air in the dark [14], however, the long-time holding of fullerite crystals in air facilitates their photopolymerization [4,17]. Thus, the contribution of physically absorbed oxygen in the photo-induced hardening turns out to be small. Obviously, the observed hardening in subsurface layer by more than 500% is mainly due to photochemical reactions in oxygenated fullerite.

The properties of hard phase significantly differ from those of oxide film. Unlike oxide, the hard phase reverts to pristine fullerite on heating in air. Besides, the oxide film was found to be significantly softer (about 280 MPa) [15].

Different possible mechanisms of photochemical transformation of fullerite in air are offered. A photo-enhanced incorporation of $O_2$ into the fullerite lattice and formation of toluene insoluble $C_{60}(O_2)_x$ phase is suggested [1]. It is supposed in [7] that oxygen under light-irradiation can occupy apical positions in $C_{60}$ molecule and form covalent bonds to C - neighbor atoms thus hardening the crystal. We suppose that the absorbed oxygen could be involved in photochemical reaction to form C-O-C linking bonds between fullerene molecules [8,16]. As it was found in [18], in the mass spectrum of fullerite after holding in air under daylight illumination appear $C_{120}O$ dimers (m/z=1475). Similar conclusion follows from the solubility rate measurements of air-exposed fullerite [19].

A modified [2+2] photoaddition model for formation of $C_{120}O$ phase in aged fullerite is offered in [19]. The first step of this reaction is formation of fullerite mono-oxide $C_{60}O$. The mono-oxide is strained and predicted to be unstable relative to $C_{120}O$. Due to this, it is quickly captured by a neighboring $C_{60}$ molecule in a [2+2] reaction, which leads to formation of $C_{120}O$ (Fig.7).

We adopted the offered model for explanation of the photochemical reaction, which leads to formation of the hard phase at the saturation stage of photopolymerization. In such way, all obtained results can be explained by [2+2] photoaddition reaction. In pristine fullerite this reaction leads to formation of all-carbon polymer, consisting mainly of

$C_{60}$ dimers (ie. $C_{120}$) where two fullerene molecules are bonded by four- member ring of carbon atoms. We suggested that oxygenated fullerite polymerizes *via* formation of $C_{120}O$ phase, in which the bonding of fullerene molecules is ensured by five-member ring where C-O-C linking bonds are involved as shown in Fig.7 c.

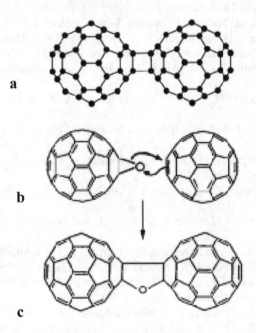

a

b

c

Fig.7. Photoinduced formation of $C_{120}$ and $C_{120}O$ fullerene dimers *via* [2+2] cycloaddition reaction.

## 4. Conclusions

The behavior of photopolymerization of $C_{60}$ crystals in air has been investigated by microindentation and dislocation mobility methods. It has been found that the photoinduced increase in hardness and the reduction of dislocation mobility occur under the excitation with visible and UV light in the fundamental absorption region and linearly increases with increasing the photon energy. Two photopolymerized phases, which differ in hardness and thermal stability, are found to appear. The depth distribution of the hardness was investigated. It was found from the depth distribution measurements of the hardness that the harder polymerized phase is located in the oxygen-contaminated subsurface layer of 0.8- 1 μm. We suggest that besides the well-known all-carbon fullerite photopolymer consisting mainly of $C_{60}$ dimers (ie. $C_{120}$), formation of $C_{120}O$ phase in oxygenated subsurface layer by means of [2+2] photoaddition reaction might occur. This phase appears under illumination at 290-330 K reverts to $C_{120}$ on heating above 320 K.

176

## 5. References

1. Eklund, P.C., Rao, A.M., Zhou, P., Wang Y. and Holden, J.M., *Thin Solid Films*, 1995, **257**, 185.
2. Shebanovs L., Maniks J.and Kalnacs J., *J.Cryst.Growth*, 2002, **234**, 202.
3. Manika I., Maniks J.and Kalnacs J., *Latvian J.Phys.Techn.Sci.*,1997, 3,12.
4. Tachibana M., Sakuma H. and Kojima K., *J.Appl.Phys.*, 1997, **82**, 4253.
5. Manika I., Maniks J.and Kalnacs J., *Phil. Mag. Lett.*,1998, **77**, 321.
6. Tachibana M., Kojima K., Sakuma H., Komatsu T. and Sunakawa T., *J.Appl.Phys.*, 1998, **84**, 1.
7. Haluska M., Zehetbauer M., Hulman M. and Kuzmany H., *Mat.Sci.Forum*,1996, **210-213**, 267.
8. Manika I., Maniks J., Pokulis R., Kalnacs J. and Erts, D., *Phys.Stat.Sol.(a)*, 2001, **188**, 989.
9. Licas A., Gensterblum G., Pircaux J.J., Thiry P.A., Caudano R. and Vigneron, J.P., *Phys. Rev.B*, 1992, **45**, 12694.
10. Makarova T.L., Saharov V.I., Serenkov I.T. and Vul A.J., *Fizika Tverdovo Tela*, 1999, **41**, 554.
11. Hassanien A., Gasperic J., Demsar J., Muslevic I. and Mihailovic D., *Appl. Phys. Lett.* 1997,**70**, 417.
12. Vagberg T., Jacobson P. and Sundqvist B., *Phys.Rev.B*, 1999, **60**, 4535.
13. Nagel P., Pasler V., Lebedkin S., Soldatov A., Meingast C., Sundquist B., Persson P.A., Tanaka T., Komatsu K., Buga S. and Inaba A., *Phys.Rev.B*, 1999, **60**, 16920.
14. Manika I, Maniks J. and Kalnacs J., *Carbon*, 1998, **36**, 641.
15. Manika, I., Maniks J. and KalnacsJ., *Fullerene Science &Technology*, 1999,**7**, 825.
16. Matsuishi K., Tada K., Onari S. and Arai T., *Phil.Mag.B*, 1994,**70**,795.
17. Manika I., Maniks J., Kalnacs J., *Latvian J.Phys.Tech. Sci.*, 2000, **4**, 48.
18. Lebedkin S., Ballenweg S., Cross J.,Taylor R. and KratchmerW.,*Tetrahedron Lett.*, 1995, 4571.
19. Taylor R., Barrow P.M. and Drewello T., *Chem.Commun.*, 1998, 2497.

# INVESTIGATIONS ON THE CARBON SPECIAL FORM GRAPHITIC NANOFIBRES AS A HYDROGEN STORAGE MATERIALS

BIPIN KUMAR GUPTA and O.N.SRIVASTAVA
*Department of Physics, Banaras Hindu University,*
*Varanasi-221005 , INDIA*
*Email: bipinbhu@yahoo.com*

## Abstract

It is time to invent new nanomaterials and in this regards graphitic nanofibres is one of the new break through in the field of hydrogen storage materials. The unique properties of graphitic nanofibres (GNFs) have generated intense interest in the application of these new carbon materials toward a number of applications including selective absorption, hydrogen storage, polymer reinforcement and catalytic support.

The present study is aimed at synthesis, microstructural characterization and hydrogenation behaviour of graphitic nanofibres (GNF) prepared by ethylene gas. The GNF were prepared by thermal cracking of ethylene gas at a temperature of ~600°C for time durations of two hrs. The structural characterizations have been evaluated by X-ray diffraction and transmission electron microscopy. It was found that the typical length of GNF was in the range of ~1 to ~6μm and most of GNF sample exhibits coil like configuration. The microstructural characteristics as studied by transmission electron microscopy (TEM) techniques after hydrogenation/dehydrogenation were found to consist of curved voids and were significantly different as compared to as grown GNF. The graphitic nanofibres were hydrogenated at 120atm. for 24 hrs. and then dehydrogenated upto 1 atm. The P-C-T curves drawn based on several dehydrogenation runs have revealed the hydrogen storage capacity of present GNF as ~15wt.%. In addition to hydrogen storage capacity, we have also investigated the desorption kinetics of GNF samples, the hydrogen was desorbed at the rate of ~57ml$^3$/min.

**Key Words** : Graphitic Nanofibres,Synthesis, Microstructural Characterization, Hydrogen Storage

## 1. Introduction

The depleting and polluting fossil fuel makes it imperative to find renewable and clean 21$^{st}$ century fuel. Decades of R&D efforts have revealed that Hydrogen is indeed such a fuel. The harnessing of hydrogen as a 'Green Fuel' on a massive scale is held up largely due to lack of a viable storage system. The state of the art storage media, the hydrides are too heavy and have a storage capacity at ambient conditions of only about ~1.5wt.% [1]. Therefore, there is an ever on going search for new storage materials. In this search recently a new material in the form of graphitic nanofibres with very high hydrogen storage capacity has been found [2].

177

*T.N. Veziroglu et al. (eds.),*
*Hydrogen Materials Science and Chemistry of Carbon Nanomaterials,* 177-184.

The exotic hydrogen storage characteristics of graphitic nanofibres prepared by catalysed decomposition of ethylene over Ni:Cu catalysts was first reported by A. Chambers et al. (1998). They claimed storage of about 23 litres or 2 g of hydrogen per gram of carbon corresponding to a storage capacity of ~67wt% [2]. This hydrogen storage capacity is an order of magnitude higher than that of the conventional hydrogen storage material i.e. metal hydrides. However, later in a similar effort Ahn C.C. et al. (1998) did not get as high storage capacity from GNF. They in fact obtained a storage capacity of only ~2.7wt.% [3] and concluded that the earlier results of Chambers et al. may not be valid. In their further work the Boston group (C. Park et al., 1999) suggested that careful pretreatment of GNF to remove chemisorbed gases is a critical step for obtaining high storage capacity of GNF [4]. In view of these two early contradictory results, there has been interest to find out the real position regarding the hydrogen storage capacity of GNF. In another recent study of GNF, hydrogen storage capacity of ~6.5wt% has been found [5].

We have carried out studies on GNF grown in a manner similar to that of the above said earlier studies (2000). In our studies we have found reproducible storage capacity of ~10wt.% [6]. The aim of the present paper is to describe and discuss further studies on the synthesis characterization and hydrogenation behaviour of GNF prepared by thermal cracking of ethylene. Several new results embodying (a) higher reproducible hydrogen storage capacity of ~15wt.% as compared to our previously reported capacity of ~10wt.%, (b) curious microstructural characteristics of the hydrogenated GNF and its comparison with the microstructure of the as synthesized samples and (c) kinetics of desorption, will be described and discussed.

## 2. Experimental Technique

The growth of graphitic nanofibres, in the present investigation, have been achieved through thermal cracking of carbon rich gases. The gas employed was ethylene (C2H4). The thermal decomposition/cracking was done by taking the gas in a silica tube (75 cms long, 2cms diameters). The tube contained inlet ports for hydrogen and helium gases. For efficient thermal cracking, in keeping with known results, fine Ni and Cu (98% Ni, 2% Cu, total weight 100 mg) were taken in a ceramic boat and were passivated by keeping them in the growth tube which was evacuated [~$1.31 \times 10^{-3}$ mbar($10^{-3}$ atm.)] and then filled with $H_2$ and He [10% each and at a pressure of ~52.63 mbar($10^{-3}$ atm.)]. The size of the employed catalyst was ~0.4µm. The essential role of catalyst in growth of GNF is in the formation of solid solution of carbon in the metal catalyst followed by diffusion of carbon and formation of GNF [6]. The cracking was carried out in a stationary mode. The thermal cracking of the gas was achieved by heating in the presence of Ni+Cu (98% & 2%) catalyst powders at a temperature $600^0$C for 2hrs. in a resistance heated furnace. After this the tube was allowed to cool at a rate of $5^0$C/min. At this stage the carbon deposits near the tube regions where catalyst powders were located could be seen. Before taking out the carbon deposits, the grown material was flushed with 2% He and Air. The total quantity ~ 1.75 gms of GNF was collected through several (five or six) runs.

The hydrogenation of the as synthesized GNF was carried out in a Sievert's equipments fabricated in our laboratory [7]. Before subjecting to hydrogenation through exposure to $H_2$ (purity 99.9% and ~120atm. pressure), the GNFs' were activated by annealing upto a temperature of $150^0C$ under a vacuum $10^{-5}$ torr. The dehydrogenation was done by volumetric displacemental method through incremental decrease of pressure.

The structural characterization of the as-synthesized and hydrogenated/ dehydrogenated GNF was carried through X-ray diffraction technique employing Phillips PW-1710 X-ray diffractometer equipped with graphite monochromator. The microstructural characterizations were carried out through transmission electron microscopy (EM-CM 12) in diffraction and imaging modes.

## 3. Results and discussion

In the studies carried out so far by other workers on hydrogenation of GNF, it has been found that the amount of hydrogen desorbed is always less than that which is absorbed. It has been conjectured that some hydrogen even after discharge is still left within the GNF. The feasible mechanism of hydrogen storage corresponds to adsorption of hydrogen molecules between graphitic sheets. It has been suggested that as a result of expansion, multilayer coverage takes place within the mobile pore walls formed from graphitic sheets. It is further suggested that under these conditions mobility is suppressed and hydrogen molecules get agglomerated in a liquid like configuration. In order to visualize whether lateral expansion of graphitic sheet takes places, we have carried out detailed microstructural characteristics of the as synthesized as well as hydrogen desorbed GNF employing the technique of transmission electron microscopy.

X-ray diffraction studies of the as-synthesized, hydrogenated and dehydrogenated GNFs' samples revealed some curious characteristics. These are listed in the following :

a)      The (002) peak which manifests the graphitic layers was invariably poorly defined in the XRD patterns of the as synthesized GNFs'. On the other hand, the (101) peak and other non (00l) peaks were found to be sharp and of high intensity. These features are brought out by the representative XRD pattern as shown in Fig. 1(a).

b)      The (002) peak of the hydrogenated samples (upto ~120 atm. of $H_2$) which were taken out from the reactor after hydrogenation for 24 hrs. and then dehydrogenated is better defined than that of the as grown GNF in Fig.1(b).

It may be pointed out that for GNFs' even on repeated dehydrogenation some amount of hydrogen is always present. The hydrogenation leads to lateral expansion of graphene sandwich layers and hence an increase of the interlayer spacing. This is clearly borne out by the estimates of $d_{002}$ which is 3.46Å for the as synthesized and the average $d_{002}$ on hydrogenation/dehydrogenation.

For graphitic nanofibres, the (00l) layers are of finite extent (100 to ~1000Å) arranged nearly perpendicular or inclined to the fibre axis. In all the firbres which will be arranged randomly the (00l) planes in GNF have finite extents. Therefore, the XRD peak profile for (00l) planes including (002) will be broadened in Fig. 1(a). The other planes like (h0l) e.g. (101) or (hk0) e.g. (110) do not have finite extents like (00l) planes and therefore, the peaks corresponding to these will not be broadened but comparatively sharp. This is what is actually observed in Fig. 1(b).

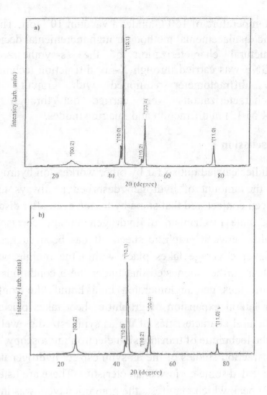

Fig. 1 (a,b)    X-ray diffractogram from the GNF prepared from thermal cracking of ethylene gas, Fig. 1(a) and Fig. 1(b). It can be noticed that the (002) peaks was improved after desorption of hydrogen from GNF sample (see text for details).

Now when hydrogen goes in between the graphene planes, lateral expansion of graphene planes will take place. Transmission electron microscopic investigations reveal that the hydrogenated/dehydrogenated GNF samples show expansion of the graphene planes through the existence of gaps between them.

Figs. 2(a), 2(b) and 2(c) show the TEM micrograph of the as synthesized and hydrogenated/dehydrogenated GNF, respectively. As is easily discernible, hydrogenation has led to lateral expansion of the graphene planes as brought out by the void like spaces {as marked by arrows in Fig. 2(b)}. The TEM micrograph shown in Fig. 2(b) was taken after 10 minutes of hydrogenation.

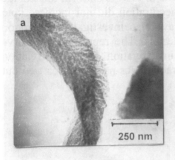
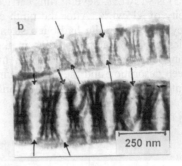
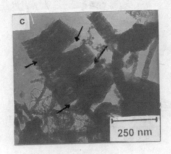

Fig. 2 (a,b,c) Transmission electron micrograph of graphitic nanofibres, Fig. 2(a) as synthesized GNF sample Fig. 2(b) immediately hydrogenated samples. Fig. 2(c) dehydrogenated samples. It can be noted that after hydrogen desorption several interlayer voids are produced in GNF sample.

The expansion is irreversible since on dehydrogenation carried out after 14 hrs. of hydrogenation also the void spaces are still present in Fig. 2(c). The TEM observations suggest that the graphene planes get distorted and become somewhat curved in Figs. 2(b) and 2(c). Under these conditions, the extents of the (002) planes will increase. Therefore, the (00l) XRD peaks typified by (002) peak will improve in the hydrogenated/ dehydrogenated GNF samples. This is as per the observed results as typified by Fig. 1(b) which brings out XRD pattern of the GNF sample hydrogenated for 24 hrs. and then dehydrogenated. In passing it should be mentioned that some hydrogen always remains inside even in the best dehydrogenated GNF samples. Based on the analysis of FWHM in Figs. 2(a) and 2(b), the size of (002) planes (particle size) has been found to be 195nm in Fig. 2(a) in the hydrogenated/ dehydrogenated GNF's as compared to 130nm in the as synthesized samples.

The hydrogenation characteristics of the GNF (~1.75 g) are brought out by the representative P-C-T desorption curves shown in Fig. 3. The P-C-T desorption curves were evaluated for several hydrogenation/dehydrogenation cycles. The P-C-T desorption curves shown in Fig. 3 correspond to the first, third, seventh, ninth and fifteenth cycle. The highest hydrogen storage capacity as brought out by the P-C-T desorption curves from the 9th cycle

182

(Fig. 3) is ~15wt.%. Several runs carried out on different lots of GNF's confirm the storage capacity of ~15wt.%. After the ninth cycle hydrogen concentration did not increase upto 15th cycle. Besides the hydrogen storage capacity, we have also investigated desorption kinetics of GNF sample studies in the present investigations. The representative curve bringing out the kinetics is shown in Fig. 4. Further investigation on the correlation between desorption and the structural/microstructural characteristics have been carried out presently and results will be forthcoming.

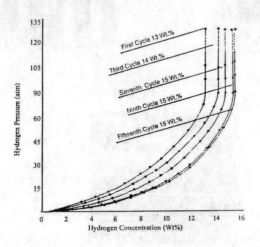

Fig. 3    Representative P-C-T desorption curves revealing hydrogen storage capacity of GNF as prepared from ethylene gas. The desorption curves bring out the maximum storage capacity of ~15wt.% at room temperature (~27°C).

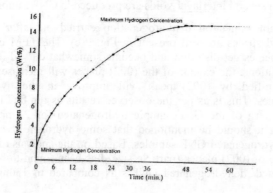

Fig. 4    Desorption kinetics of GNF samples (hydrogen concentration vs time) at 27°C.

## 4. Conclusions

The present investigation forms furtherance of our earlier result on hydrogenation behaviour of graphitic nanofibres. Some results coming out of the present studies are enumerated in the following :

(i)     The actual hydrogen storage capacity will depend on the treatments imparted to GNF. In the present case through suitable activation, it was possible to increase, the storage capacity from 10 to 15wt.%. The storage capacity may get raised to still higher values.

(ii)    The small extents and dispositions of (00l) planes makes (00l) X-ray diffraction peaks e.g. (002) broadened unlike the (h0l), (hk0) peaks which are comparatively sharp.

(iii)   The XRD and TEM investigations suggest that hydrogen enters between the (00l) graphene layers, leads to lateral expansion of these layers creating void like space and turning the (00l) planes curved and lengthened. The periphery of the void space may be favoured sites of adsorption of hydrogen molecules in further hydrogenation cycles.

(iv)    The present investigation on desorption kinetics of GNF sample shows that the average rate of hydrogen desorption $\sim57ml^3/min$. The highest desorption kinetics of GNF sample corresponding to $\sim90ml^3/min$ was obtained for hydrogen concentrations ranging between $\sim6.0$ to 8.0wt%.

## Acknowledgements

The author are thankful to Professor A.R. Verma, Professor C.N.R. Rao, Dr. R. Chidambaram and Professor Y.C. Simhadri for their encouragement. Helpful discussions with Dr. R.S. Tiwari, Dr. A.K. Singh and Sri Vijay Kumar (Scientific Officer) are gratefully acknowledged. Financial assistance from MNES, UGC and CSIR is also gratefully acknowledged.

## 5. References

1.     S. Iijima, Helical microtubules of graphitic carbon, Nature  354(1991)56-58.

2.     A. Chambers, C. Park, R.T.K. Baker, N.M. Rodriguez, Hydrogen storage in graphitic nanofibres, J Phys, Chem B 102(1998)4253-56.

3.     C.C. Ahn, Y. Ye, B.V. Ratnakumar, C. Witham, R.C. Bowman, B. Fultz, Hydrogen desorption and adsorption measurements on graphite nanofibres, Appl Phys Letters 73(1998)3378-80.

4.     C. Park, P.E. Anderson, A. Chambers, C.D. Tan, R. Hidalgo, N.M. Rodriguez, Further studies of the interaction of hydrogen with graphite nanofibres, J Phys. Chem B. 103(1999)10572-10581.

5      D.J. Browning, M.L. Gerrard, J.B. Lakeman, I.M. Mellor, R.J. Mortimer, M.C. Turpin, Investigation of the hydrogen storage capacities of carbon nanofibres prepared from an ethylene precursor, Proc. of the 13th World Hydrogen Energy conference, Beijing, China, June 2000: 554-559.

184

6.     Bipin Kumar Gupta, O.N. Srivastava. Synthesis and hydrogenation behaviour of graphitic nanofibres, Int. J of Hydrogen Energy 25(2000)825-830.

7.     K. Ramakrishna, S.K. Singh, A.K. Singh, O.N. Srivastava, R.P. Dhiya editors. Solid state materials for hydrogen storage, Progress in hydrogen energy, Boston, MA: Reidel, 7(1987)81.

# HETEROMETALLIC FULLERIDES OF TRANSITION METALS WITH THE COMPOSITION $K_2MC_{60}$

V.A. KULBACHINSKII, B.M. BULYCHEV[A], R.A. LUNIN, A.V. KRECHETOV, V.G. KYTIN, K.V. POHOLOK[A], K. LIPS[B], J. RAPPICH[B]
*Moscow State University, Department of Physics; [a]Department of Chemistry; 119992, GSP-2, Moscow, Russia*
*fax: +7 (095) 932-8876,*
*e-mail: kulb@mig.phys.msu.ru*
*[b]Abteilung Silizium-Photovoltaik, Hahn-Meitner-Institute, D-12489, Berlin, Germany*

**Abstract**
Heterometallic fullerides with composition $K_2MC_{60}$, synthesized by exchange chemical reaction of $K_5C_{60}$ or $K_4C_{60}$ with chlorides of metals Fe and Cu groups have been investigated by X-ray diffraction, magnetic resonance, Raman and Mössbauer spectroscopy. Magnetization and susceptibility measurements also have been carried out. Metal chlorides from Fe and Cu groups enable to cover the whole range of electronic configuration of metal from $d^1$ to $d^{10}$. Heterometallic fullerides with M=$Cu^{+2}$, $Fe^{+2}$, $Fe^{+3}$, $V^{+2}$, $Cr^{+3}$ and $Ni^{+2}$ appeared to be superconductors with $T_c$=13.9-16.5K.

*Keyword*: Fullerene, intercalation, Mössbauer and Raman spectroscopy, superconductivity, ferromagnetism

## 1. Introduction

Since the discovery of the fullerene $C_{60}$ [1], a lot of its compounds revealed a wide variety of interesting physical properties. The most fascinating of them was the discovery of superconductivity in alkali doped fullerite [2]. The superconducting properties of fullerides detected for homo and heteronuclear compounds of alkali metals [2-4] subsequently were discovered for alkali-earth and some rear-earth metals [5]. The intercalation of multivalent metallic atoms in fullerite lattice voids causes the consecutive filling by electrons of $t_{1u}$ sublevel and higher-lying $t_{1g}$ sublevel of a fullerene molecule.

Trending efforts on synthesis of heterometallic fullerides with composition $K_2MC_{60}$ we chose metal chlorides from Fe and Cu groups as an objects of further investigations. These chlorides allow covering the whole range of electronic configuration of metal from $d^1$ to $d^{10}$.

Here we present measurements of X-ray diffraction, electron paramagnetic resonance (EPR), Raman and Mössbauer spectroscopy in the new heterometallic fullerides. In order to determine critical temperature of the superconducting transition the ac susceptibility measurements were carried out. The new method of synthesis was developed, based on homogeneous-heterogeneous oxidation-reduction reactions of potassium fullerides with $Fe^{+2}$, $Fe^{+3}$, $Co^{+2}$, $Ni^{+2}$, $Cu^{+1}$, $Cu^{+2}$, $V^{+2}$, $Cr^{+3}$ and $Ag^{+1}$ chlorides in organic solvents at temperature $T < 340$ K. For the samples with the assumed composition $K_2FeC_{60}$ our Mössbauer spectroscopy studies indicate the presence of Fe atoms in the crystal lattice of fulleride.

*T.N. Veziroglu et al. (eds.),*
*Hydrogen Materials Science and Chemistry of Carbon Nanomaterials,* 185-192.
© 2004 *Kluwer Academic Publishers. Printed in the Netherlands.*

## 2. Experimental

The potassium fulleride was prepared by reaction of fullerite with the calculated quantity of metal in the environment of toluene at 120-130 $^{\circ}$C [6]. The synthesis of initial alkali metal fullerides, the elimination of toluene, the mixing of fulleride with a suspension of metal halogenide and the realization of an exchange reaction, the drying and prepacking of heterometallic fullerides in ampoules for different physical and spectroscopic investigations were conducted at vacuum in all-soldered glass facility. As an example the $K_2Fe^{+3}C_{60}$, $KFe_2^{+2}C_{60}$ and $K_2Ni^{+2}C_{60}$ synthesis methods are described:

$$K_5C_{60} + Fe^{+3}Cl_3 \rightarrow K_2Fe^{+3}C_{60} + 3KCl$$
$$K_5C_{60} + 2Fe^{+2}Cl_2 \rightarrow KFe_2^{+2}C_{60} + 4KCl$$
$$K_4C_{60} + Ni^{+2}Cl_2 \rightarrow K_2Ni^{+2}C_{60} + 2KCl$$

The mixture of potassium fulleride and metal chloride is placed for 15 day in the oven with temperature 65-75$^{\circ}$ C and periodically intermixed by shaking. After the end of reaction the solvent is removed from the reactor, the precipitate is dried in vacuum. For different measurements the dry powdered agent was placed in ampoule, which was soldered after filling by helium at pressure 350-400 millimeters of mercury. The reactions were controlled by X-ray method. The X-ray data of fullerides had been recorded by photomethod with Ginie chamber (radiation $\lambda CuK_{\alpha}$) and listed in table 1. As the reference the data for $K_3C_{60}$ synthesized by method [6] also is shown. According to the X-ray data in all obtained precipitations the phases of metals, free fullerite and initial substances are absent. As the main phases the one of two phases of potassium halogenides (KCl or KI) are registered. This fact indicates an execution of an oxidation-reduction reaction. As it is seen in table 1 parameter $a$ of fcc lattice of heterofullerides less than the value of $a$ in $K_3C_{60}$, synthesized under the same conditions. It seems to be reasonable because the dimensions of ions for metals of Fe and Cu groups are less than K, hence metals were intercalated to fulleride lattice.

Laser induced ion mass spectrometry (LIMS) analysis was made by LAMMA-1000 device (Germany, Leybold AG) using Q-switched Nd-YAG laser ($\lambda= 266$ nm, $\tau=8$-10 ns) with power density on the sample surface $10^8$-$10^{11}$ W/cm$^2$. According to these data the structure of $C_{60}$ molecule in heterofulleride is the same as in the host potassium fulleride.
In the present work we used the conventional low frequency ac susceptibility measurements to determine the critical temperature of the superconducting transition. The magnetic susceptibility $\chi$ was measured in the temperature range 4.2 $<T< 50$ K.

Electron paramagnetic resonance was studied in X-Band (9 GHz) Bruker Elexsys 580 ESR Spectrometer. A gas flow Oxford Cryostat was used to control the temperature in the range 5 – 300 K.

Raman spectra were measured by Perkin Elmer Raman Spectrometer at room temperature. For excitation 632.7 nm red line of He-Ne laser was used.

Magnetization curves were recorded by vibrating sample magnetometer EG&G Princeton applied research (USA), model 155, equipped by gas-flow cryostat.

Page 187

**TABLE 1** Some parameters of synthesized samples

| Composition of the initial components | Assumed composition of the fulleride | $T_c$ (K) | Electron configuration of metal ions | Color | Lattice parameter $a$ (Å) | $d\,M^{+n}$ (Å) |
|---|---|---|---|---|---|---|
| | $K_3C_{60}$ | 18.5 | | | 14.3110(5) | |
| $5K+Ti^{+3}$ | $K_2TiC_{60}$ | - | $3d^1$ | Black | | 0.72 |
| $6K+Zr^{+4}$ | $K_2ZrC_{60}$ | - | $4d^0$ | Black | | 0.80 |
| $5K+V^{+3}$ | $K_2V^{+3}C_{60}$ | - | $3d^2$ | Black | | 0.74 |
| $4K+V^{+2}$ | $K_2V^{+2}C_{60}$ | 12 | $3d^3$ | Black | | 0.82 |
| $5K+Cr^{+3}$ | $K_2CrC_{60}$ | 14 | $3d^3$ | Black | | 0.69 |
| $5K+Fe^{+3}$ | $K_2Fe^{+3}C_{60}$ | 15 | $3d^5$ | Black | 14.23(2) | 0.64 |
| $4K+Fe^{+2}$ | $K_2Fe^{+2}C_{60}$ | 16.5 | $3d^6$ | Black | 14.24(6) | 0.76 |
| $5K+2Fe^{+2}$ | $KFe_2^{+2}C_{60}$ | - | $3d^6$ | Black | 14.28(2) | 0.76 |
| $4K+Co^{+2}$ | $K_2CoC_{60}$ | - | $3d^7$ | Black | 14.264(4) | 0.74 |
| $5K+2Co^{+2}$ | $KCo_2C_{60}$ | - | $3d^7$ | Black | - | 0.74 |
| $4K+Ni^{+2}$ | $K_2NiC_{60}$ | 13.9 | $3d^8$ | Black | 14.244(3) | 0.72 |
| $5K+2Ni^{+2}$ | $KNi_2C_{60}$ | - | $3d^8$ | Black | - | 0.72 |
| $4K+Cu^{+2}$ | $K_2Cu^{+2}C_{60}$ | 14.5 | $3d^9$ | Black | - | 0.69 |
| $3K+Cu^{+1}$ | $K_2Cu^{+1}C_{60}$ | - | $3d^{10}$ | Red | - | 0.96 |
| $3K+Ag^{+1}$ | $K_2Ag^{+1}C_{60}$ | - | $4d^{10}$ | Brown | - | 1.26 |

## 3. Results and discussion

### 3.1 Ac-susceptibility

The temperature of superconducting transition is determined as the onset of the transition. **Fig. 1** shows the dependence of the magnetic susceptibility on temperature for some superconducting fullerides with the composition $K_2MC_{60}$ (M=Fe, Ni, Cu). As the reference sample $K_3C_{60}$ also is shown. Critical temperatures $T_C$ are listed in table 1. Heterometallic fullerides with composition $K_2MC_{60}$ (M=$Ag^{+1}$, $Cu^{+1}$, $Co^{+2}$) and $KM_2^{+2}C_{60}$ (M=Co, Ni, Fe) are not superconductors.

Magnetic susceptibility measurements showed that exchange reaction of $K_3C_{60}$ with $Cr^{+1}Cl$ and $Ag^{+1}Cl$ chloride did not produce superconducting heterometallic fullerides. This reaction reduces metal down to zero-valence state (see table 1) and produces its own phase. For example, in the reaction of $K_4C_{60}$ with $CuCl_2$ the fulleride $K_2Cu^{+2}C_{60}$ was synthesized. Despite the small size of diameter $d$ of the copper ion $Cu^{+2}$ ($d$=0.69 Å), $K_2Cu^{+2}C_{60}$ is the superconductor with the critical temperature $T_c$=14.5 K. Thus, not filled $d$-shell in intercalated metal leads to superconductivity of fullerides in contrast to f-elements, for which superconductivity was observed in fullerides with filled $f$-shell. Indeed in the reactions of potassium fullerides with $Fe^{+3}$ chloride ($d$=0.64 Å), $Fe^{+2}$ chloride ($d$=0.76 Å), and $Ni^{+2}$ chloride ($d$=0.72 Å) superconducting materials with composition $K_2MC_{60}$ were synthesized with $T_c$=13.9-16.5 K.

### 3.2 Mössbauer spectroscopy

Mössbauer spectroscopy is a powerful tool in the determination of the valence of Fe and its positioning in the lattice [7].

188

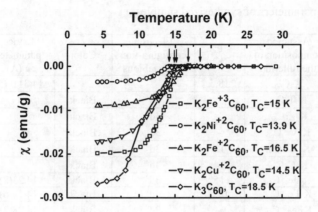

Fig. 1. Temperature dependence of the magnetic susceptibility of samples with the composition $K_2MC_{60}$ (M=Fe, Ni, Cu). $K_3C_{60}$ is shown as the reference

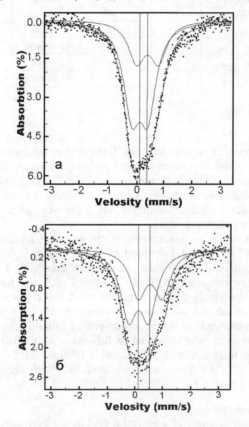

Fig. 2. Mössbauer spectra of $K_2Fe^{+3}$ (a) and $K_2Fe^{+2}C_{60}$ (b). Solid lines are Lorentzian fitting to the experimental data. Vertical lines indicate positions of two peaks of different signals

Mösbauer spectrum for $K_2Fe^{+3}C_{60}$ and $K_2Fe^{+2}C_{60}$ is shown in Fig. 2. For $K_2Fe^{+3}C_{60}$ the presence of two significant peaks (marked in fig. 2a by two vertical solid lines), and the width of the signal $\Gamma=1.5$ mm/s means that the spectrum consists of two different signals. The first signal has IS=0.09 mm/s and $\Gamma=0.7$ mm/s, and the second's IS=0.5 mm/s and $\Gamma=0.7$ mm/s. Large width of each line means that there is quadrupole splitting of the signals ($\Delta=0.71$ mm/s for the first signal and $\Delta=0.86$ mm/s for the second). Quadrupole splitting can be attributed to the presence of Fe in fullerides lattice at minimum of two inequivalent positions. The resulting fitting lines are plotted in fig. 2 by two solid Lorentzian curves [7]. The isomer shift of the first signal is typical for $Fe^0$ and IS of the second one is very close to $Fe^{+3}$. For $K_2Fe^{+2}C_{60}$ fulleride the spectrum is very similar, that is corresponds to iron in $Fe^0$ and $Fe^{+3}$ states. So we suppose that iron intercalates potassium fulleride with valences 0 or +3.

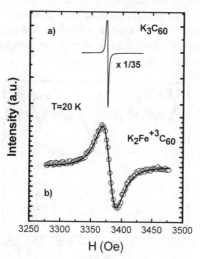

Fig. 3. ESR spectrum at T=20 K of $K_3C_{60}$ (a) and $K_2Fe^{+3}C_{60}$ (b) (points are experimental data, solid line is the best fitting with Lorentzian line)

### 3.3 *ESR spectroscopy*

The typical electron paramagnetic resonance spectra at 20 K for $K_3C_{60}$, is shown in Fig. 3a. Electron paramagnetic resonance in $K_3C_{60}$ is electron spin resonance (ESR) [8]. ESR signal is due to conduction electrons, $C_{60}^{3-}$, $C_{60}^{3-}$-O-$C_{60}^{3-}$ and other anions [9,10]. The localized anions (mainly $C_{60}^{3-}$-O-$C_{60}^{3-}$ in $K_3C_{60}$) give narrow line with g-value about nearly 2 [10]. The spectrum is asymmetric and can be fitted with two Lorentzian lines. The width of lines increases with the increase of temperature, while the amplitude of the lines decreases. The g-values of both lines are almost temperature independent.

Typical spectrum of electron paramagnetic resonance for $K_2FeC_{60}$ is shown in **Fig.3b**. It is reasonable to associate broad line observed in our measurements with the signal from conducting electrons, although a small contribution from localized $C_{60}^{3-}$ anions cannot be excluded. The width of electron paramagnetic resonance in $K_2FeC_{60}$ fullerides is significantly larger, than the width of ESR curves in $K_3C_{60}$, while the amplitude of the signal is essentially smaller.

The double integrated ESR intensity is proportional to the paramagnetic magnetization which is proportional to the paramagnetic susceptibility of the sample [8]. In $K_3C_{60}$ conducting electrons obey Fermi statistics, while localized $C_{60}$ anions obey Boltzmann statistics. Therefore contribution to the magnetization from conducting electrons is almost temperature independent, while the contribution from $C_{60}$-anions follows Curie-Weis law. The model function (Curie-Weis law + const) describes an experimental dependence with a good agreement. This gives additional evidence that ESR signal comes from conducting electrons and localized anions.

ESR spectra are shown in Fig. 4 for $K_2FeC_{60}$ (a) and $K_2NiC_{60}$ (b) fullerides. The resonance curves have almost ideal Lorentzian shape and are essentially broader in Fe, Ni-containing compound than in $K_3C_{60}$. The amplitude of curve decreases with the increase of temperature, while the width of the resonance increases (see fig. 4). The position of the resonance is sensitive to the sweep range and sweep direction of magnetic field. The fluctuations of the background signal were observed in $K_2FeC_{60}$ as in the case of $K_3C_{60}$. These fluctuations disappear at temperatures above the temperature of superconducting transition determined from magnetization measurements.

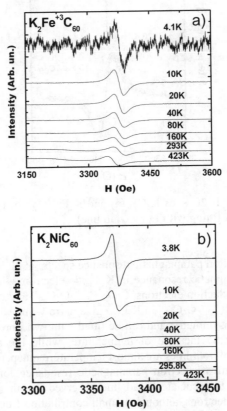

Fig. 4. ESR spectra of $K_2Fe^{+3}C_{60}$ (a) and $K_2NiC_{60}$ (b) at different temperatures

All resonance curves measured on heterofullerides containing Fe or Ni can be very well fitted by one Lorentzian line. The reasonable interpretation of this fact is that the spins in fulleride are ordered and observed electron paramagnetic resonance in $K_2MC_{60}$ (M=Fe, Ni) is ferromagnetic resonance. The g-factor for the ferromagnetic resonance is often close to 2 (see for example [11]), but sensitive to the shape of sample. It is worth to note that g-factor in $K_2NiC_{60}$ and $K_2CoC_{60}$ at T<20 K is equal to 1.9996±0.0003 and 1.9986±0.0003 respectively. g-фактор of $K_2Fe^{+3}C_{60}$ under the same temperature is equal to 1,9965±0,0001. These values are less than g-factor in $K_3C_{60}$. The difference in g-factors measured in different direction of the field sweeping can be explained by hysteresis of magnetization as well as by redistribution and reorientation of particles in the powdered sample. The width of resonance curve is determined by relaxation of total magnetic moment. In our case the width is larger for the sample, which was obtained from the $Fe^{+2}$, then for the sample obtained from $Fe^{+3}$.

### 3.4 Magnetization

Typical dependences of magnetic moment on applied magnetic field are shown in Fig. 5 for $K_2Fe^{+3}C_{60}$ fulleride in small applied field. At small variation of magnetic field magnetic moment of the sample decreases with an increase of applied field and vice versa. Observed magnetization curve is typical for superconductor except non-zero magnetic moment at zero applied fields. The latter is the manifestation of magnetic ordering. In high magnetic field a magnetization curves typical for ferromagnetic was observed. Magnetization increases strongly with an increase of applied field. A residual magnetization remains in the sample. Hence, ferromagnetic ordering and superconductivity coexist in investigated fullerides.

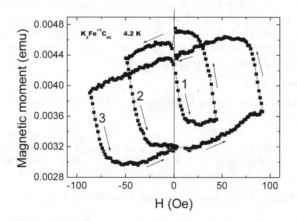

Fig. 5. Magnetization curve for $K_2Fe^{+3}C_{60}$ at T=4.2 K in low magnetic field; 1, 2, 3 shows the sequence of magnetization loops

### 3.5 Raman Scattering

Raman spectra measured on $K_3C_{60}$ exhibits peaks at the positions close to the positions known from literature [12]. In alkali-metal doped fulleride $C_{60}$ ions line $A_g(2)$ can be used for determination of charge state of intercalated compound [9]. A part of the

192

Raman spectrum near $A_g(2)$ line of $K_3C_{60}$, $K_2Fe^{+2}C_{60}$, $K_2Fe^{+3}C_{60}$ and $KFe_2^{+2}C_{60}$ is shown in Fig 6. The position and shape of $A_g(2)$ line in $K_2Fe^{+3}C_{60}$ is almost the same as in $K_3C_{60}$, while in $K_2Fe^{+2}C_{60}$ and $KFe_2^{+2}C_{60}$ peak is shifted towards larger values of Raman shift and the line is asymmetrically broadened. Moreover the shape of the line in $K_2Fe^{+2}C_{60}$ and $KFe_2^{+2}C_{60}$ is very similar.

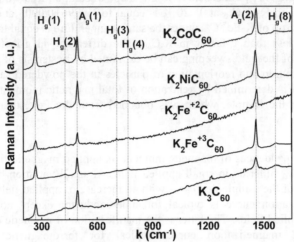

Fig. 6. Raman spectra of fullerides at room temperature

## 4. Conclusions

A new method of synthesis of heterometallic fullerides has been developed. The obtained samples have fcc lattice. Mössbauer spectroscopy and other data confirm the intercalation of metal atoms in fullerides lattice. Ac susceptibility data show that fullerides with composition $K_2MC_{60}$ (M=$Fe^{+2}$, $Fe^{+3}$, $Ni^{+2}$, $Cu^{+2}$) are superconductors with $T_C$=13.9 - 16.5 K, this is also confirmed by dc-magnetization.

## Acknowledgements

The work was supported by RFBR (grants 02-03-32575 and 03-03-06354).

## 5. References

1. Kroto H. W., Heath J. R., Obrien S. C., Curl R. F., Smalley R. E., Nature 1985; **318**: 162-163.
2. Hebard A. F., Rosseinsky M. J., Haddon R. C., Murphy D. W., Glarum S. H., Nature 1991; **350**: 600-601.
3. Rosseinsky M. J., Ramirez A. P., Glarum S. H., Murphy D. W., Haddon R. C., Phys. Rev. Lett. 1991; **66**: 2830- 2832.
4. K. Holczer, Karoly, O. Klein, Olivier, Huang, Shiou-Mei, Kaner, Richard B., Fu, Ke-Jian, Science 1991; **252**: 1154-1157.
5. Kortan A. R., Kopylov N., Glarum S., Gyorgy E. M., Ramirez A. P., Fleming R. M., Thiel F. A., Haddon R. C., Nature 1992; **355**: 529-532.
6. Bulychev B.M., Privalov V.I., Dityat'ev A.A., Zournal Neorg. Khimii. 2000, **45**, 931-938.
7. Goldanskii V.I., Herber R.H., Chemical Applications of Mössbauer Spectroscopy, 1968: Ch. 1, 503 pages (1968).
8. see for example Robert J., Petit P., Yildirim T., Fisher J. E., Phys. Rev. B 1998, **57**, 1226-1230.
9. Reed C. A., Bolskar R. D., Chem. Rev. 2000, **100**, 1075-1120.
10. Eaton S.S., Kee A., Konda R., Eaton G.R., Trulove P.C., Garlin R.T., J. Phys. Chem. 1996; **100**, 6910-6919.
11. Gurevich A. G., Magnetic resonance in ferrites and antiferromagnetics, "Nauka", 591 pages, Moscow 1973.
12. Zhou P., Wang K.-A., Eklund P. C., Dresselhaus G., Dresselhaus M. S., Phys. Rev. B 1993; **48**: 8412-8417.

# CHARACTERIZATION OF NANOPARTICLES PROCESSED BY ARC - DISCHARGE BETWEEN CARBON ELECTRODES CONTAINING Ni$_2$Y CATALYST

LEONOWICZ M.[(1)], SHULGA Yu.M.[(2)], MURADYAN V.E. [(2)], WOZNIAK M. [(2)], WEI XIE[(2)]

[(1)] *Faculty of Materials Science and Engineering, Warsaw University of Technology, Woloska 141, 02-507 Warszawa, Poland*
[(2)] *Institute of Problems of Chemical Physics RAS, 142432 Chernogolovka, Moscow Region, Russia*
*mkl@inmat.pw.edu.pl, fax: (48 22) 6608725.*

## Abstract

Chemical composition, structure and magnetic properties of encapsulated in carbon powder particles obtained by arc discharge between carbon electrodes containing Ni$_2$Y catalyst were investigated. It has been found that the composition and magnetic properties of the encapsulates obtained in this process substantially differ from those of the starting intermetallic compound Ni$_2$Y. Decomposition mode for the Ni$_2$Y compound, undergoing in the course of arch discharge, has been proposed. Mean particle size of the encapsulated metal has been found to vary within a range of about 5 - 42 nm. The evidence for the variation of the particles lattice parameters with the change of their size is also presented.

**Key words**: carbon nanotubes, carbon encapsulates, ferromagnetic nanoparticles.

## 1. Introduction

Synthesis of carbon nanotubes (CNT) using arch discharge between graphite rods requires the presence of a catalyst. Very effective catalysts can be made from mixtures of two metals from iron group; however, search for the most appropriate catalysts is still under way. Processing of CNT is accompanied by formation of other by-products such as carbon rods, ribbons, encapsulates and onion-like structures. One of the most prospective catalysts for the synthesis of CNT appears to be Ni$_2$Y [1]. Proper control of the process requires knowledge of the amount, structure, composition and properties of the discharge products. Another problem to be solved is separation of the particular product from other forms.

In this study the structure and magnetic properties of the products obtained by arch discharge between carbon electrodes containing Ni$_2$Y catalyst have been investigated. Particular focus has been put on the ferromagnetic properties of the carbon encapsulates.

*T.N. Veziroglu et al. (eds.),*
*Hydrogen Materials Science and Chemistry of Carbon Nanomaterials,* 193-202.

## 2. Experimental

The experimental method applied in this study is commonly used for the processing of fullerens and carbon nanotubes [1]. The cathode was made from bulk graphite. A hollow graphite rod, filled with the catalyst was used as an anode. Finely milled powder of $Ni_2Y$ intermetallic phase was applied as the catalyst (specimen 1). Before milling the powder was charged with hydrogen and subsequently pressed into the hole in the graphite anode. The fraction of the metallic catalyst in the graphite rod was 14 wt.%. Before the discharge the electrodes were annealed in vacuum of 0.15 Pa, at 900 °C.

The discharge was performed in helium gas atmosphere (650 hPa). The discharge current and voltage were 90 A and 28-30 V, respectively. The distance between electrodes during the discharge was kept constant around 2 mm. The distance between electrodes and the water-cooled chamber walls was 70 mm.

The final carbon products, obtained in the course of discharge, were divided into four parts, depending on the collecting area. The first part, (*soot*), most plentiful, was collected from the chamber walls. The second part, (*collar*), grew up during the discharge around the cathode. The latter material was friable and, in a contrast to the soot, exhibited some elasticity. On the cathode grew also the third part - dense coating (*deposit*). At the bottom of the chamber accumulated material consisting of pieces of graphite, which had ripped out of the electrodes and some soot, which for some reasons, did not deposit on the chamber walls nor as the collar. In these investigations the soot (sample 2) and collar (sample 3) were studied in details. TEM studies were done for the collar and deposit. The magnetic properties were measured using a vibrating sample magnetometer EG&G PARC M4500. For those measurements the specimens were placed into a thin paramagnetic nylon capsule. The magnetic signal from the capsule was subsequently subtracted from the total signal. Curie temperature was measured using a Faraday magnetic balance. TEM studies were performed with the application of the HRTEM Jeol 3010. X-ray analysis was done with the application of the ADP-1 diffractometer using CrKα radiation. Assessment of the metallic particle size was done on a basis of the Scherrer method, using the formula $D_{111}= \lambda/\beta\cos\Theta$, where $\lambda$ is the wavelength of the x-ray radiation, $\beta$ represents peak width in the half of its length and $\Theta$ is the angle of the peak position.

## 3. Results and discussion

It is well established that nickel ions in the $Ni_2Y$ intermetallic do not contribute to the net magnetic moment [e.g. 1]. However, in the course of milling, which follows the hydrogen decrepitation, partial decomposition of the $Ni_2Y$ and formation of ferromagnetic and probably also superparamagnetic particles of metallic Ni may occur. The magnetic properties of the catalyst material are represented in Fig. 1 by the loop 1. The magnetisation of this powder does not achieve saturation, which is better visible in Fig. 1B (loop 1). In this figure the magnetisation for each loop is reduced to the value obtained for 10 kOe (800 kA/m) of the magnetising field.

For the products of arch discharge (collar and soot) the magnetisation values close to saturation can be achieved in much lower field of 5 kOe (400 kA/m) (Fig. 1). The magnetisation in the external field of 10 kOe (800 kA/m) for discharge products is much higher than those for the initial catalyst (Table 1).

TABLE 1. Magnetisation $\sigma$ and coercivity $H_c$ for the specimens investigated.

| Specimen | $\sigma$ (emu/g) | $H_c$ (Oe / kA/m) |
|---|---|---|
| Starting Ni$_2$Y (specimen1) | 0.08 | 114 / 9.12 |
| *Soot* (specimen 2) | 0.76 | 87 / 6.96 |
| *Collar* (specimen 3) | 3.06 | 111 / 8.88 |

The assessment of Ni content on the basis of magnetisation level gives values 2 wt% and 8 wt% in the soot and collar, respectively. If this assessment is done on a basis of the absorption spectroscopy the respective contents are 11 and 14 wt.% of Ni, respectively. We suppose that not all Ni particles existing in the product are ferromagnetic. Some of them can either be superparamagnetic or in a form of a diamagnetic compound. However, the shape of the hysteresis loops do not show clear enough evidence for the superparamagnetic contribution, which would enable to separate both signals.

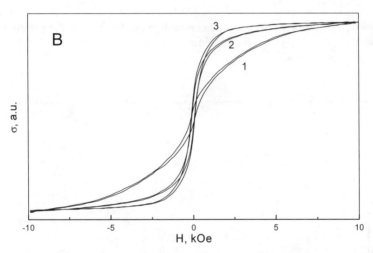

Fig. 1. Hysteresis loops for the specimens investigated - A. Reduced magnetisation - B; (for each specimen the magnetisation is divided by the value obtained for external field H=10 kOe (800 kA/m)). Hysteresis loops for the specimens investigated; catalyst -1, soot - 2, collar -3.

X-ray patterns for the discharge products (samples 2 and 3) are shown in Fig. 2. Two diffraction peaks having $2\Theta$=68.4 and 80.9° represent (111) and (200) crystallographic planes for the FCC Ni lattice, respectively. For the collar the 68.4° peak represents superposition of two reflections: a narrow one and a wide one (Fig. 3). The presence of the narrow and wide peaks gives an evidence for the existence in the products of broad distribution of the particle size. For the collar, the calculations of the particle size, $D_{111}$, for both peaks, made on a basis of the Scherrer method, give values 42.4 nm and 5.4 nm for the large and small one, respectively. The intensity ratio of the narrow to wide peaks is 3:2. In the soot the particle size is much more homogeneous with a mean value of about 5 nm.

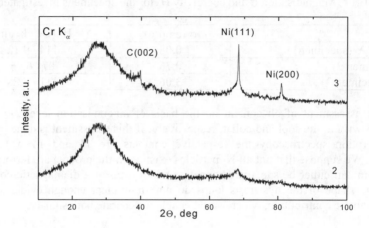

Fig. 2. Diffraction patterns for the specimens investigated. Number of the pattern represents the specimen number.

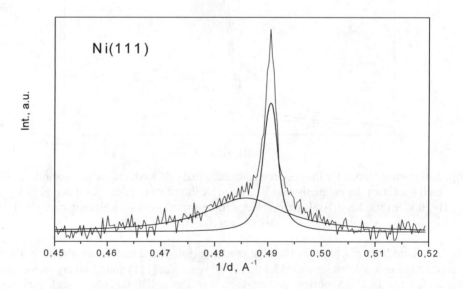

Fig. 3. Partition of the Ni (111) diffraction line for the contributions from small and large particles, respectively.

The lattice constant for the plane (111), for all the specimens studied, exceeded those for pure nickel (2.034 Å [4], Table 2). This difference can be explained by the dissolution of some carbon in Ni. Liquid Ni can dissolve substantial amounts of carbon, which in the course of cooling precipitates as graphite. Maximum equilibrium solubility of carbon in solid Ni metal is 0.55 wt.%, at 1318 °C [4]. Assuming that the increase of lattice parameters is fully related to the carbon dissolution, we can estimate the carbon content, for smaller particles in the collar, to be 2 wt.% for the $d_{111}$=2.058 Å. Moreover, the smaller the particles the larger the lattice constants and carbon content.

TABLE 2. Lattice parameters $d_{111}$ for the Ni (111) planes and Ni particle size $D_{111}$ for the specimens investigated.

| Specimen | $d_{111}$ (Å) | $D_{111}$ (nm) |
|---|---|---|
| *Soot* (specimen 2) | 2.048 | 5.0 |
| *Collar* (specimen 3) | 2.039 | 42.4 |
| | 2.058 | 5.4 |

The supposition of the carbon dissolution can further be confirmed by the fact that the increase of the coercivity for the particles follows the change of the lattice parameters (compare $H_c$ for specimens 2 and 3 Table 1).. Moreover, the measurements of the Curie temperature gives, for the collar, a value of 345 °C, which is somewhat lower than those for pure Ni (354.4 °C), pointing to change of the distance between Ni atoms in the lattice.

TEM studies revealed that that in the collar nanotubes are not present. Carbon ribbons and encapsulates are the only structures visible (Fig. 4). The electron diffraction analysis proved that the encapsulates contain mainly nickel (Fig. 5). This means that the starting catalyst $Ni_2Y$ decomposes in the course of arc-discharge and Ni goes mainly to the collar and in smaller amount to the soot (from magnetic measurements). The encapsulates, which are particularly of interest in these investigations, have size within a range of about 5 to 40 nm which generally agrees with the calculations on the basis of the x-ray studies. Somewhat different products are in the deposit, which consists of carbon ribbons, nanotubes, encapsulates (onions) and particles of the catalyst. Some proportion of amorphous carbon is also visible (Fig. 6a). Multiwall nanotubes have mean internal and external diameters 2 nm and 15 nm, respectively, and are either empty (Fig. 6b) or contain some amount of the catalyst (Fig. 4c). The catalyst is located also in onion-like capsules and individual free particles without any cover (Fig. 6d-f). The electron diffraction revealed that the particles are in fact yttrium oxide without nickel (Fig. 7).

Summarising, it has been found that the arch-discharge between the composite electrodes graphite - $Ni_2Y$ results in decomposition of the intermetallic $Ni_2Y$ phase and formation of Ni ferromagnetic nanoparticles, encapsulated in the carbon shells, in the collar and soot, and yttrium oxide in the deposit. The Ni particles exhibit increased lattice parameter, when compared with pure metal, which we suppose to be due to the carbon dissolution. The difference in the lattice constants increases with decreasing particle size. The carbon concentration in smaller crystallites exceeds those in the larger ones.

198

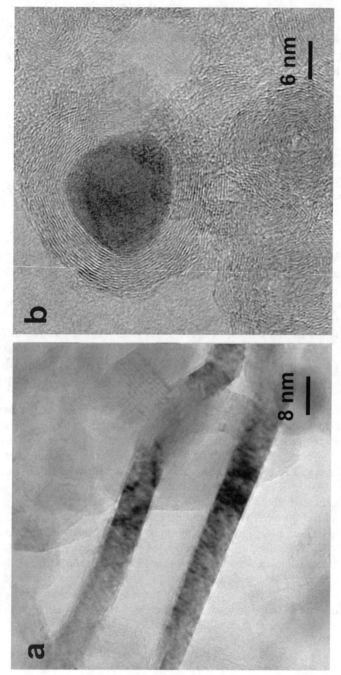

Fig. 4. Various forms of carbon in the collar. Carbon ribbon - a; Ni encapsulate - b.

199

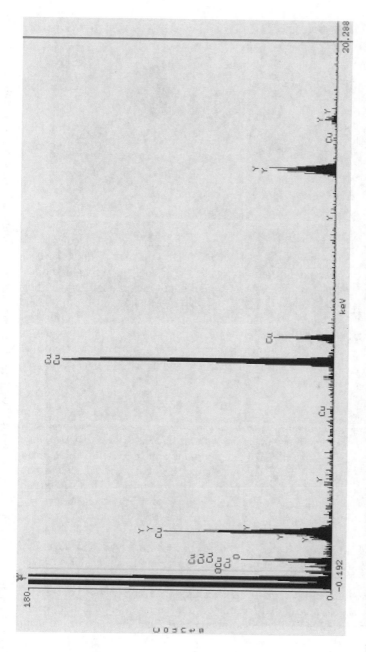

Fig. 5. Electron diffraction pattern of the encapsulated particles in the collar. (Signal from copper comes from the Cu grid)

200

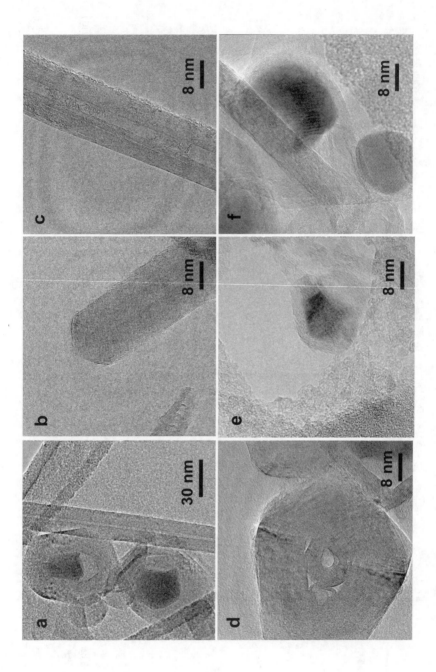

Fig. 6. Various forms of carbon in the deposit. Mixture of different forms -a; multiwall nanotube - b; nanotube filled with the catalyst - c; encapsulates - d,e; free catalyst - f.

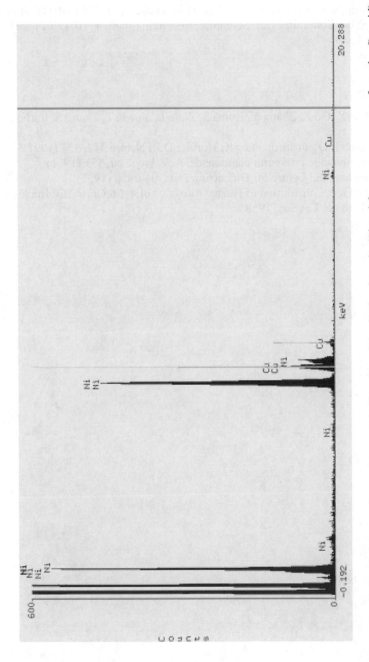

Fig. 7. Electron diffraction pattern of the encapsulated particles in the deposit. (Signal from copper comes from the Cu grid).

202

## 4. Acknowledgements

Financial support from the RFFU (projects 00-03-32106 and 02-03-33226), MHTC (project 1580) and Polish State Committee for Scientific Research (grant 4 T08D 021 23) are gratefully acknowledged.

## 5. References

1. Shi Z, Lian Y, Zhou X, Gu Z, Zhang Y, Iijima S, Zhou L, Yue K.T, Zhang S. Carbon **37**, 1449 (1999).
2. Kratschmer W, Lamb L.D., Fostiropoulos K, Huffner D.R, Nature **347**, 354 (1991).
3. Taylor K.N.R "Intermetallic rare-earth compounds, Adv. Phys. **20**, 551 (1971).
4. Files JCPDS – International Centre for Diffraction Data, 04-0850 (1995).
5. Hansen M, Anderko K, "Constitution of Binary Alloys", vol. I, McGraw-Hill Book Comp., Inc., New-York, Toronto, London, 1958.

# PROTECTION OF SECURITIES BY THE APPLICATION OF FULLERENES

D.V. SCHUR, N.S. ASTRATOV, A.P. POMYTKIN,
A.D. ZOLOTARENKO, T.I. SHAPOSHNIKOVA

*Institute for Problems of Materials Science of NAS of Ukraine, lab. # 67,
3, Krzhizhanovsky str., Kiev, 03142 Ukraine
E-mail: shurzag@materials.kiev.ua*

## 1. Introduction

After discovery of fullerenes in 1985 many scientists participating in their investigations asked themselves about possible applications of these expensive objects. When the arc synthesis of fullerenes was discovered by W. Kratschmer, there appeared especially many results of research on the use of fullerenes in different fields of science and technique. This method allowed the decrease in the cost of fullerenes in hundreds times what made them available for many investigators. At present it is impossible to describe in one article the fields in which many enthusiasts try to apply fullerenes.

Authors of the present work have attempted to introduce fullerenes into the composition of securities for their protection.

## 2. Experimental

The fullerenes used have been prepared in laboratory 67 of Institute for Problems of Materials Science of National Academy of Sciences of Ukraine. Fullerenes have been prepared by electric arc graphite sputtering in helium. Operation mixtures have been treated on UZDN-1 Y4.2 ultrasonic apparatus at 22 kHz. Samples have been analyzed on a scanning electron microscope.

The trial paper mouldings have been prepared on the experimental apparatus at the department of ecology and paper of National Technic University of Ukraine "Kiev Polytechnic Institute". Sulphate white cellulose from softwood ($l$=3-5 mm), hardwood ($l$=0.7-1.2 mm) and cotton pulp ($l$=5-7 mm) have been used in experiments.

## 3. Results and discussion

We have considered two types of fullerene-containing cellulose:
1) exofullerized cellulose ($C_{exo}$);
2) endofullerized cellulose ($C_{endo}$).

The first variant provides the precipitation of fullerites in nanodispersed state on the surface of cellulose fibers.

*T.N. Veziroglu et al. (eds.),*
*Hydrogen Materials Science and Chemistry of Carbon Nanomaterials,* 203-206.
© 2004 *Kluwer Academic Publishers. Printed in the Netherlands.*

The second method provides the introduction of fullerenes into cellulose fibers followed by removing the solvent and transition of fullerenes to the solid state in the fiber bulk.

Moreover, we have considered two techniques for fullerene introduction into the paper as a final product. The first one involved in synthesis of fullerized cellulose and its introduction into the initial stock. The second one consisted in direct introduction of fullerenes into the initial stock during the technological process.

The first technique provides preliminary synthesis of fullerene-containing cellulose with the certain parameters specified beforehand. In the second case it is necessary to convert fullerenes into the water-soluble state, and then, when paper was formed, to convert them into the initial state.

Fig. 1 ($a,b,c,d$) show the microphotographs of paper containing endofullerized fibers of cellulose. Fullerite crystals are absent on the external surface of fibers. Fig. 1($e$) shows the photograph taken by an optical microscope on a color film. The fullerite microcrystals are clearly seen in the photograph. In this case the exofullerized cellulose fibers prepared by introduction of fullerite in water-soluble state into the initial suspension are seen in the photograph.

At present laboratory 67 of Institute for Problems of Material Science of National Academy of Sciences of Ukraine and the department of ecology and paper of National Technic University of Ukraine "Kiev Polytechnic Institute" perform investigations of physical and chemical properties of cellulose-fullerene composites, features of their synthesis, structure, construction both at micro- and nanolevels.

## 4. Conclusions

In our opinion, the protection of securities with the use of fullerenes can be protection at the highest level. As fullerenes are new inaccessible materials, nobody owns the method for their detection in paper.

Fullerenes introduction into the paper is the difficult process that requires special skill and knowledge. Moreover, the formation of solid substitution solutions in the $C_{60}$ crystal lattice (with $C_{70}$ and other fullerenes) gives possibility to combine fullerenes in the wide range of concentrations. This allows the creation of compositions which decoding is very labor-consuming and requires much time and special equipment. All mentioned above makes counterfeit of the paper encoded by fullerenes practically impossible. The pink color of paper, caused by introduced fullerenes, will allow application of fullerenes for protection of pink dollars.

Our scientific group has a number of patents of Ukraine on chemical and technological methods for fullerenes conversion into the water-soluble state and inversely and also on methods and techniques for introduction of fullerenes and fullerene-containing products into the paper composition.

Fig.1. (a,b,c,d)

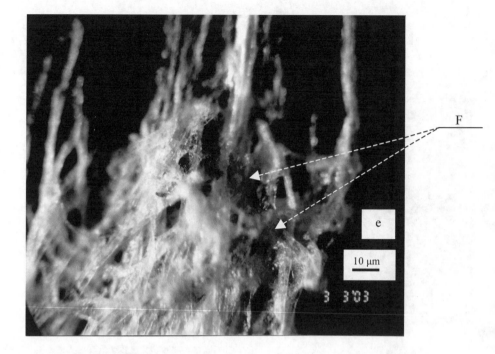

Fig.1. (a,b,c,d) are the fibers of cellulose containing fullerite in their volume, (e) is cellulose fibers containing the fullerite crystals on the surface of fibers.

# SPECTROPHOTOMETRIC ANALYSIS OF $C_{60}$ AND $C_{70}$ FULLERENES IN THE TOLUENE SOLUTIONS

N.S.ANIKINA, S.Yu.ZAGINAICHENKO, M.I.MAISTRENKO,
A.D.ZOLOTARENKO, G.A.SIVAK, D.V.SCHUR, L.O.TESLENKO
*Institute for Problems of Materials Science of Ukrainian Academy of Sciences, 3, Krzhizhanovsky str., Kiev, 03142 Ukraine*

## Abstract

The method for determination of the molar absorption coefficient (MAC) of $C_{60}$ and $C_{70}$ fullerenes has been developed. The values of MAC obtained by the graphic method are refined there after by the mathematical method of successive approximation.

The levels of analytical wavelengths, the ranges for concentrations of fullerene mixtures, solutions optimum for the qualitative analysis have been established.

The spectra of $C_{60}$ solutions and solutions of its mixtures with $C_{70}$ containing $C_{60} \geq 0.5 \cdot 10^{-4}$ mol/$l$ show the bathochromic shift of the absorption band with the maximum at $\lambda = 335$ nm.

**Key words:** fullerenes $C_{60}$, $C_{70}$, absorption spectroscopy, toluene, crystals, absorption factor.

## 1. Introduction

The arc synthesis of fullerenes is the most accessible and prevalent method for production of fullerene-containing soot. The percentage of fullerenes in this soot varies from 6 to 24%. The fullerene yield is significantly affect

At present ed by: practically all technological parameters, chemical purity of the graphite evaporated as well as the size and the shape of graphite electrodes etc.

The optimization of technology for preparation, extraction and separation of fullerenes largely depends on the analytical provision. Being rather precise, fast and accessible method for the qualitative analysis, the absorption spectrophotometric method was not practiced widely mainly owing to the absence of the accurate values for MACs of C60, C70 solutions. The available literature data on MACs are ambiguous [1-4]. Besides these reasons, there exist a number of factors affecting on the method precision. Some of them require a special attention:

1. As fullerenes significantly absorb in the UV region, one has to work with highly dilute solutions what contributes to the analytical errors, complicates and extends the procedure of the solution preparation.

2. In contrast to the common organic substances, fullerenes possess property to entrain solvent molecules during crystallization. Impurities are absorbed by the surface as well as by the substance bulk [5-7]. In the course of thermal treatment the solvent molecules can be fixed in the spaces of the crystals sintered what complicates purification of fullerenes by their holding under vacuum even with simultaneous heating. The strongest occlusion of impurity occurs in fast crystallization accompanied by sorption of the solvent on the developed crystal surface [8]. The solvent content in the fullerite crystals amounts up to 5-6% [2].

*T.N. Veziroglu et al. (eds.),*
*Hydrogen Materials Science and Chemistry of Carbon Nanomaterials,* 207-216.
© 2004 *Kluwer Academic Publishers. Printed in the Netherlands.*

3. Method of spectrophotometric analysis is confined to upper and lower limits of the optical density of the studied solutions. At high concentrations of the absorbing substance and due to the weak energy of passed radiation, the measurement error results from insufficient sensitivity of the photocell. At low concentrations the instrument error becomes comparable with the measured value. It follows from the above that there exists the "optimum" concentration at which measurement accuracy is maximum.

As it was found empirically in the first works on photometric analysis, the most accurate results were obtained using the solution which transmission was equal to 37% (corresponding "optimum" optical density, $A_{od}$, equals to 0.434 rel.un.) [9]. If the studied system adheres to Lambert-Beer's law of absorption, the same value may be attained mathematically from the expression for the optical density A:

$$A = \lg P_0 - \lg P \ , \tag{1}$$

where $P_0$ and P are intensities of incoming and outgoing light fluxes, respectively.

The maximum accuracy of optical density A measurement will be achieved, when the relative error $\Delta A / A$ becomes minimum. To find the value of $A_{op}$, one should differentiate Eq.(1) twice and equate the second derivative to zero:

$$dA = 0 - (\lg e)\frac{1}{P}dP = 0 - 0{,}434\frac{1}{P}dP \ . \tag{2}$$

Substituting P for the equal expression $P_0 \cdot 10^{-A}$ and dividing both parts by A, we have:

$$\frac{1}{A}dA = -\frac{0{,}434}{A \cdot P_0 \cdot 10^{-A}}dP \ . \tag{3}$$

Substituting differentials for finite increments, we find:

$$\frac{\Delta A}{A} = -\frac{0{,}434 \cdot \Delta P}{P_0}\left(\frac{1}{A \cdot 10^{-A}}\right) \ . \tag{4}$$

Noise resistance limits sensitivity of antimony-cesium cells. This sort of noises is constant and not related to the intensity of the measured radiation. At constant $\Delta P$ the second differentiation gives:

$$\frac{d\,(\Delta A/A)}{dA} = -\frac{0{,}434\Delta P}{P_0}\left(\frac{10^A \cdot \ln 10}{A} - \frac{10^A}{A^2}\right) \ . \tag{5}$$

The value of $\Delta A / A$ is minimum provided that the expression in brackets equals to zero.

$$10^A \ln 10 / A = 10^A / A^2 \ . \tag{6}$$

Then

$$A_{od} = 1/\ln 10 = 0{,}434 \ . \tag{7}$$

The most prevalent spectrophotometric technique is the method of calibration plots, when the solution subjected to analysis contains only one absorbing substance. In this case Lambert-Beer's law is not strictly realized.

Quantitative analysis of multicomponent systems is based on additivity of optical density:

$$A = \sum_i \varepsilon_i \cdot c_i \cdot b \ , \tag{8}$$

where $c_i$ is concentration of i-component, $\varepsilon_i$ are molar coefficients of the mixture components absorption, b is thickness of the absorbing layer, cm.

Investigating solutions of the fullerenes mixture, amount of $C_{60}$ and $C_{70}$ has been determined by solving two linear equations:

$$A_1 = \varepsilon_1^{60} \cdot X + \varepsilon_1^{70} \cdot Y \ , \tag{9}$$

$$A_2 = \varepsilon_2^{60} \cdot X + \varepsilon_2^{70} \cdot Y \ , \tag{10}$$

X and Y are molar concentrations of $C_{60}$ and $C_{70}$, respectively; indexes 1 and 2 show that the given value was determined for $\lambda_1$ or $\lambda_2$ wavelength; the thickness of the absorbing layer $b$ equals to 1 cm.

The main condition of the calculation method is that absorption spectra of the mixture components are not overlapped. This can be qualitatively characterized by inequalities:

$$\varepsilon_1^{60} > \varepsilon_2^{60} \ , \quad \varepsilon_1^{70} < \varepsilon_2^{70} \ , \quad A_1 > A_2 \tag{11}$$

## 2. Experiment

In this paper the fullerenes extracted by toluene from the fullerene-containing soot have been studied. The soot was produced by the electric arc method.

The extract was separated into $C_{60}$ and $C_{70}$ in several cycles on the chromatographic column with graphite filling.

Special experiments have shown that purity of the prepared fullerites depends on the temperature regimes of fullerenes extraction from the fullerene-containing soot and fullerite crystallization. Four sorts of fullerene crystals have been prepared:

Crystals 1: a) extraction temperature 12 °C; b) toluene evaporation in a rotary evaporator under vacuum at 35-40 °C;

Crystals II: a) extraction at the boiling temperature in a flask with a reflux condenser; b) toluene evaporation at 80-90 °C;

Crystals III: a) extraction in Soxhlet apparatus; b) evaporation at 80-90 °C;

Crystals IV: a) extraction in Soxhlet apparatus; b) evaporation in a rotary evaporator under vacuum at 35-40 °C.

In every case the crystals have been washed with diethyl ether three times, kepted under vacuum and periodically weighted (in 24-36 h) until the weight became constant. The purity of fullerenes has been evaluated according to the obtained results of X-ray analysis using DRON-3 X-ray apparatus.

The mass fraction of solvent has been determined by the derivatographic method. Derivatograms have been recorded using "Derivatograph Q-1500D" apparatus.

Crystals 1 are the purest and contain ≤ 1% of impurity. Crystals II contain the largest amount of impurity (up to 5÷6%). Crystals IV contain more impurity (2-3%) than crystals I.

For spectrum analysis the solutions have been prepared from the crystals of the first sort which concentration has been calculated with regard to the residual amount of toluene in the crystal solvates.

The fullerene solutions have been protected from the light.

The absorption spectra have been measured in the 335-600 nm region in the 10 mm optical cells using SF-26 spectrophotometer with the digital data output and the wavelength ranging 2 to 5 nm.

Positions of the maximum of absorption peaks do not differ (in the range of the measurement error (0.1%)) from the literature data [10].

## 3. Evaluation of molar absorption coefficients of toluene solutions of $C_{60}$ and $C_{70}$ fullerenes

As shown above, the least error of spectrophotometric measurements will be achieved when intensity, $P$, of the light flow outgoing from the solution equals to 37 % and corresponding optimum optical density $A_{od}$ equals to 0.434 rel.un. As MAC is equal to the relation

$$\varepsilon = \frac{A_{od}}{c_{od}} = \frac{0,434}{c} , \qquad (12)$$

the problem on its evaluation is to find the concentration of the solution absorbing the substance, which optical density equals to 0.434 rel.un. for the selected wavelength.

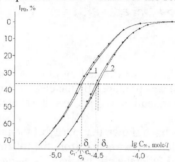

Figure 1. The intensity of the passed beam $I_p$ (%) as a function of logarithm of the molar concentration for the $C_{70}$ toluene solution at $\lambda_1 = 407$ nm, $\lambda_2 = 472.8$ nm; $\delta_1$ and $\delta_2$ are concentration ranges used in the method of mathematical successive approximations

In this paper the optimum concentration has been evaluated by the curve plot for the relationship between intensity of the light flow $P$ outgoing from the solution and logarithm of the concentration $c$. The curves $P(lgc)$ (Fig.1) have been plotted using absorption spectra of $C_{60}$ and $C_{70}$ fullerene solutions (Fig 2).

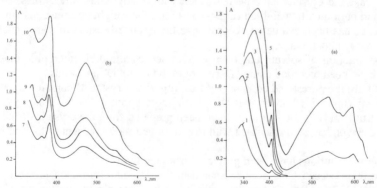

Fig.2. Spectra of $C_{60}$(a) and $C_{70}$(b) in toluene solutions, concentration of solutions in mol/$\lambda$ x10$^{-4}$: 1 – 0.012; 2 – 0.033; 3 – 0.100; 4 – 0.164; 5 – 0.300; 6 – 0.981; 7 – 0.016; 8 – 0.026; 9 – 0.032; 10 – 0.640

If the wide range of concentrations is studied, the curve $P(lgc)$ has S-like shape. In the point of inflection, in which the tangent line to the curve of $P(lgc)$ function has the largest slope to the x-coordinate, the relation

$$\frac{dP}{d\,lgc} = \frac{dP}{\dfrac{dc}{c}} \qquad (13)$$

takes the maximum value, and the value of the relative error in concentration, dc/c, becomes minimum. If the system adheres to Lambert-Beer's adsorption law, the ordinate position of the inflection point does not depend on the wavelength and equals to ~37%, and the concentration corresponding to the point of inflection, $c_{pi}$, is required.

However, because in the ranges of high and low concentrations the instrumental errors of spectrophotometric measurements are sharply increased, there exists inversion in abscissas of the inflection point that are in the range of $\delta$ concentrations (Fig.1). The mathematical method of successive approximations has been developed to evaluate the concentration $c_{pi}$ within the highest accuracy. Several concentrations have been chosen from the $\delta$ range (in the considered example we have taken three concentrations $c_1$, $c_2$, $c_3$: two extreme and one middle (Fig.1)) and three values of MAC have been calculated for each fullerene, $C_{60}$ and $C_{70}$, for $\lambda_1$ and $\lambda_2$ wavelength by Eq.(12). Then solving Eqs.(9) and (10) and varying 12 values of calculated MACs, we have evaluated the "calculated" concentration, $c_c$, of the "control" solution. Taking the "analytical" concentration, $c_a$, of the control solution as 100%, we have evaluated the error of the "calculated" concentration, $N$, by the equation:

$$N=100\left(1-\frac{c_c}{c_a}\right) . \qquad (14)$$

The values of $\varepsilon^1_{60}$, $\varepsilon^2_{60}$, $\varepsilon^1_{70}$ and $\varepsilon^2_{70}$ are the best for the studied system, the use of which gives the full agreement between the concentration of the "calibration" solution and numeral approximations.

The obtained values of MACs are given in the Table 1.

TABLE 1. Molecular absorption spectra of $C_{60}$ and $C_{70}$ in toluene.

| $\lambda$, nm | $\varepsilon_{60}$ | $\varepsilon_{70}$ |
|---|---|---|
| 356,3 | 22894 | |
| 365,2 | 15630 | |
| 372,0 | 11722 | |
| 400,0 | 3293 | 16134 |
| 407,0 | 3136 | 14371 |
| 410,0 | 2597 | 12802 |
| 418,4 | 1054 | 13108 |
| 439,0 | 334 | 15049 |
| 472,8 | 613 | 21254 |
| 540,0 | 933 | 9604 |

Equations have been solved using the developed computer program. To alleviate the data entry, the matrix has been developed (Fig.3) for each wavelength, $\lambda_1$ and $\lambda_2$. The matrix elements are combinations of $\varepsilon^{60}$ and $\varepsilon^{70}$ calculated by Eq.(12). 81 solutions have been fulfilled with the use of matrix elements for every chosen pair of wavelengths.

$$\lambda_1$$

| | | |
|---|---|---|
| $(\varepsilon_{60})_1^1$ | $(\varepsilon_{60})_1^2$ | $(\varepsilon_{60})_1^3$ |
| $(\varepsilon_{70})_1^1$ | $(\varepsilon_{70})_1^1$ | $(\varepsilon_{70})_1^1$ |
| $(\varepsilon_{60})_1^1$ | $(\varepsilon_{60})_1^2$ | $(\varepsilon_{60})_1^3$ |
| $(\varepsilon_{70})_1^2$ | $(\varepsilon_{70})_1^2$ | $(\varepsilon_{70})_1^2$ |
| $(\varepsilon_{60})_1^3$ | $(\varepsilon_{60})_1^2$ | $(\varepsilon_{60})_1^3$ |
| $(\varepsilon_{70})_1^2$ | $(\varepsilon_{70})_1^3$ | $(\varepsilon_{70})_1^3$ |

$$\lambda_2$$

| | | |
|---|---|---|
| $(\varepsilon_{60})_2^1$ | $(\varepsilon_{60})_2^2$ | $(\varepsilon_{60})_2^3$ |
| $(\varepsilon_{70})_2^1$ | $(\varepsilon_{70})_2^1$ | $(\varepsilon_{70})_2^1$ |
| $(\varepsilon_{60})_2^1$ | $(\varepsilon_{60})_2^2$ | $(\varepsilon_{60})_2^3$ |
| $(\varepsilon_{70})_2^2$ | $(\varepsilon_{70})_2^2$ | $(\varepsilon_{70})_2^2$ |
| $(\varepsilon_{60})_2^1$ | $(\varepsilon_{60})_2^2$ | $(\varepsilon_{60})_2^3$ |
| $(\varepsilon_{70})_2^3$ | $(\varepsilon_{70})_2^3$ | $(\varepsilon_{70})_2^3$ |

Figure 3. Matrixes for $\lambda_1$ and $\lambda_2$ which elements have been used to solve the system equations (9) and (10)

The superscript points that MAC has been calculated by Eq.(12) using the concentrations $c_1$, $c_2$ or $c_3$, respectively (Fig.1). The subscript points that the given value has been evaluated for $\lambda_1$ or $\lambda_2$, respectively

TABLE 2. Data input and attained results of solutions for $\lambda_1$=400 nm, $\lambda_2$=472.8 nm

| № calculation | $\lambda_1$=400 nm | | $\lambda_2$=472,8 nm | | $c_c$, g/$l$ | Error N, % |
|---|---|---|---|---|---|---|
| | $\varepsilon_{60}^1$ | $\varepsilon_{70}^1$ | $\varepsilon_{60}^2$ | $\varepsilon_{70}^2$ | | |
| 2 | 3293 | 16134 | 613 | 21254 | 0,10003 | +0,034 |
| 20 | 3347 | 16134 | 613 | 21254 | 0,09843 | -1,570 |
| 11 | 3370 | 16134 | 613 | 21254 | 0,09776 | -2,240 |
| | | | | | | |
| 2 | 3293 | 16134 | 613 | 21254 | 0,10003 | +0,034 |
| 29 | 3293 | 16502 | 613 | 21254 | 0,09882 | -1,180 |
| 56 | 3293 | 16887 | 613 | 21254 | 0,09754 | -2,464 |
| | | | | | | |
| 1 | 3293 | 16134 | 599 | 21254 | 0,09978 | -0,219 |
| 2 | 3293 | 16134 | 613 | 21254 | 0,10003 | +0,034 |
| 3 | 3293 | 16134 | 627 | 21254 | 0,10029 | +0,289 |
| | | | | | | |
| 2 | 3293 | 16134 | 613 | 21254 | 0,10003 | +0,034 |
| 5 | 3293 | 16134 | 613 | 21754 | 0,1009 | +0,993 |
| 8 | 3293 | 16134 | 613 | 22256 | 0,10191 | +1,906 |

Tables 2 and 3 show the data input on MACs values calculated by Eq.(12), the attained "calculated" concentrations $c_c$ and the values of errors $N$ for two pairs of wavelengths: $\lambda_1 = 400$ nm,. $\lambda_2 = 472.8$ nm (Table 2) and $\lambda_1 = 410$ nm, $\lambda_2 = 472.8$ nm (Table3).

TABLE 3. Data input and attained results of solutions for $\lambda_1 = 410$ nm and $\lambda_2 = 472.8$ nm

| № calculation | $\lambda_1 = 410$ nm | | $\lambda_2 = 472{,}8$ nm | | $c_c$, g/$l$ | Error N, % |
|---|---|---|---|---|---|---|
| | $\varepsilon_{60}^1$ | $\varepsilon_{70}^1$ | $\varepsilon_{60}^2$ | $\varepsilon_{70}^2$ | | |
| 2 | 2597 | 12802 | 613 | 21254 | 0,10002 | +0,0167 |
| 11 | 2614 | 12802 | 613 | 21254 | 0,09937 | -0,63 |
| 20 | 2676 | 12802 | 613 | 21254 | 0,09708 | 2,920 |
| | | | | | | |
| 2 | 2597 | 12802 | 613 | 21254 | 0,10002 | +0,0167 |
| 29 | 2597 | 13112 | 613 | 21254 | 0,09871 | -1,289 |
| 56 | 2597 | 13395 | 613 | 21254 | 0,09751 | -2,490 |
| | | | | | | |
| 1 | 2597 | 12802 | 599 | 21254 | 0,09976 | -0,239 |
| 2 | 2597 | 12802 | 613 | 21254 | 0,10002 | +0,0167 |
| 3 | 2597 | 12802 | 627 | 21254 | 0,10027 | +0,274 |
| | | | | | | |
| 2 | 2597 | 12802 | 613 | 21254 | 0,10002 | +0,0167 |
| 5 | 2597 | 12802 | 613 | 21754 | 0,10099 | +0,988 |
| 8 | 2597 | 12802 | 613 | 22256 | 0,10191 | +1,912 |

Concentration of the control solution is 0.1000 g/$l$. The optical densities of the "control" solution are 0.622, 0.492 and 0.360 for the wavelengths 400, 410 and 472.8 nm, respectively.

Analysis of the obtained data has shown that the smallest errors of the "calculated" concentration for the "control" solution have been obtained at combinations of MACs used in calculation No2, Table 2 and No2, Table 3. The values of $\varepsilon_{60}^2$ and $\varepsilon_{70}^2$ coincide in the given calculations.

This confirms the high resultativity of the method proposed for evaluation of absorption coefficients for fullerene solutions.

## 4. Spectra of $C_{60}$ and $C_{70}$ fullerene mixture solutions

The accuracy of determination of $C_{60}$ and $C_{70}$ fullerenes concentration in their mixture solutions by the calculating method depends on accuracy of MACs values as well as on the chosen wavelength and concentration of the studied solution.

The wavelengths, $\lambda_1$ and $\lambda_2$, have been chosen so that $A_1$ and $A_2$ values (corresponding to them) of the analyzed solution of fullerenes mixture are in the range from 0.3 to ~ 0.7 rel.un. Numerous calculations of concentrations for the control mixture solutions have confirmed this. The $A_1$ and $A_2$ values have been varied according to the chosen values of $\lambda_1$ and $\lambda_2$. When $A_1$ and $A_2$ does not go out the given range, the estimated error $N$ does not exceed some hundredth of percent.

Fig.4 shows absorption spectra of $C_{60}$ and $C_{70}$ mixture solutions. The thickened lines mark the sections of spectra embedded the region of analytical wavelengths. The concentration ranges (corresponding to these wavelengths) of the fullerene mixture solutions are given in Table 4. The long-wave region of analytical wavelengths is the most acceptable for calculations. Two regions, 400-410 nm and 460-480 nm, are sufficient for this purpose.

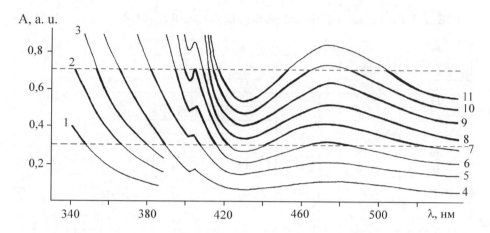

Figure 4. Spectra for solutions of $C_{60}$ and $C_{70}$ fullerene mixtures; the values of concentrations for spectral solutions and corresponding regions of analytical wavelength are given in TABLE 3

In the short-wave region the $C_{60}$ and $C_{70}$ spectra have combined partially what violates requirement (11) and reduces the accuracy of analysis.

From theoretical and practical viewpoints information on the qualitative ratio of $C_{60}$, $C_{70}$ fullerenes in the fullerene-containing soot and in the fullerene mixture solutions, used to extract and separate the fullerene mixtures into components, is of great importance.

In this paper the ratio is evaluated by the following equation:

$$\frac{X}{Y} = \frac{\varepsilon_{70}^2 - \varepsilon_{70}^1}{\varepsilon_{60}^1 - \varepsilon_{60}^2} .$$

(15)

This method has been developed as the alternative one. For solution of this equation, the values of MACs have been taken for the $\lambda_1$ and $\lambda_2$ wavelengths for which the optical densities of the studied solution are equal.

Spectra of $C_{60}$ solutions and solutions of its mixtures with $C_{70}$ fullerene containing $C_{60} \geq 0.5 \cdot 10^{-4}$ mol/$l$ show the bathochromic shift of the absorption band with the 335 nm maximum. We have reasoned that this shift is caused by the formation of aggregates consisting of some number of fullerenes $(C_{60})n$ (Figs.2 a). The possibility of forming $(C_{60})n$ clusters has been discussed in papers [11,12] in connection with the experimental character of isotherms of $C_{60}$ fullerene dissolution in some organic solvents including toluene.

TABLE 4. Optimum concentrations of fullerenes and corresponding regions of wavelengths.

| № spectrum | c, g/$l$ | Region of analytical wave lengths, nm |
|---|---|---|
| 1 | 0,0065 | 340-350 |
| 2 | 0,0129 | 340-370 |
| 3 | 0,0194 | 350-380 |
| 4 | 0,324 | 365-390 |
| 5 | 0,0647 | 385-410 |
| 6 | 0,0971 | 395-416 450-495 |
| 7 | 0,1294 | 400-540 |
| 8 | 0,1618 | 410-550 |
| 9 | 0,1942 | 410-570 |
| 10 | 0,2265 | 415-560 |
| 11 | 0,2589 | 418-450 510-600 |

## 5. Conclusions

The method allowing to determine, to a high degree of accuracy, the values of absorption factors for the $C_{60}$ and $C_{70}$ fullerenes in toluene solutions has been developed. This opens the possibility to define the quantitative contents of $C_{60}$ and $C_{70}$ fullerenes in the solutions of their mixtures by the spectrophotometrie method with the accuracy of the hundredth of percent.

## 6. References

1. R.V. Bensasson, E. Bienvenue, M. Dellinger, S. Leach, P. Seta. *J. Phys. Chem.*, *1994, 98, 3492.*
2. Moravsky A.P., Fursikov P.V., Kachapina L.M., Khramov A.V., Kiryakov N.V. Fullerenes. Recent Advances in the Chemistry and Physics of Fullerenes and Related Materials / Eds. Kadish K.M., Ruoff R.S. The Electrochemical Society: Pennington, N.J. 1995. v.2. P.156.
3. Rűdiger Bauernschmitt, Reinhart Ahlrichs, Frank H. Hennrich, and Manfred M. Kappes, *J. Am. Chem. Soc. 1998, 120, 5052-5059.*
4. Tanigaki K.; Ebbesen T; Kuroshima S. *Chem. Phys. Lett. 1991, 185, 189.*
5. H. Werner, D. Bublak, U. Gőbel, B. Henschke, W. Bensch, R. Schlőgl. Angew. *Chem., 104, 909. (1992)*
6. E.V. Skokan, V.I. Privalov, I.V. Arkhangelsky, V.Ya. Davydov, N.B. Tamm. *J. Phys. Chem. B, 103, 2050 (1999).*
7. V. V. Dikiy, G. Ya. Kabo. Uspekhi khimii. *69 (2) 2000.*

216

8.     V.P. Kolesov, S.M. Pimenova, V.K. Pavlovich, N.B. Tamm, A.A. Kurskaya. *J. Chem. Thermodyn.*, *28, 1121 (1996)*

9.     Gilbert H. Ayres. *Analytical Chemistry. V. 21, № 6, June 1949.*

10.    A. Weston and M. Murthy. *Carbon* vol. *34, № 10, pp. 1267-1274, 1996.*

11.    Besmelnitsin V.N., Eletskii A.V., Stepanov E.V. *J. Phys. Chem.*, 98, 6665 (1994)

12.    Besmelnitsin V. N., Eletskii A.V., Stepanov E.V. *J. fiz. khimii 69, 735 (1995)*

# EFFECT OF THE NATURE OF THE REACTOR WALL MATERIAL ON MORPHOLOGY AND STRUCTURE OF PRODUCTS RESULTED FROM ARC GRAPHITE SPUTTERING

A.D. ZOLOTARENKO, A.F. SAVENKO, A.N. ANTROPOV,
M.I. MAYSTRENKO, A.YU. VLASENKO, V.K. PISHUK,
V.V. SKOROHOD, D.V. SCHUR, A.N. STEPANCHUK[1], P.A. BOYKO[1]
*Institute for Problems of Materials Science of NAS of Ukraine, lab. # 67,
3 Krzhizhanovsky str., Kiev, 03142 Ukraine*
[1] *National Technical University of Ukraine "Kiev Polytechnical Institute",
Pobedy avenue 37, Kiev, Ukraine*

## Abstract

Morphology and structure of the forming layers have been studied. The films from the nanostructured synthesized carbon materials have been produced on the surface of Ti, Fe, Cu, Al, W, manganin, stainless steel, Ni, Mo substrates. The effect of nature of the metal substrate on morphology and structure of the prepared carbon nanomaterials have been studied in this paper.

It has been established that the nature of the metal substrate affects both morphology of the forming nanostructured carbon films and the shape and the structure of the carbon nanotubes entering into the composition of these films.

*Keywords:* fullerenes, reactor, carbon nanostructures, structure, morphology, substrate

## 1. Introduction

Fullerenes and nanotubes may be considered as one of the modifications of nanostructured carbon. It has been known that they are synthesized during graphite evaporation in the electric arc or under the action of a powerful laser pulse at very sharp cooling of carbon vapor. After spontaneous condensation of derivatives resulted from thermal decomposition of hydrocarbons, fullerenes are present in any smoking flame. It is thought that the flame of the common paraffin or stearin candle can be used to produce a sufficient amount of fullerenes.

A knowledge of mechanisms of the nanostructured carbon formation and the structure of soot particles is of great importance. Nature of these processes is still investigated intensively in the context of physics and chemistry of different modifications of solid carbon as well as to account for a number of the fundamental occurrences proceeding in the condensed media with a different extent of structural ordering.

In spite of the great attention focused by scientists [1-4] on research of the processes occurring in the reactor for fullerene synthesis, many problems have not been solved yet.

*T.N. Veziroglu et al. (eds.),*
*Hydrogen Materials Science and Chemistry of Carbon Nanomaterials,* 217-223.
© 2004 *Kluwer Academic Publishers. Printed in the Netherlands.*

Every experimental work, called to answer one or another question, induces more questions than gives answers.

The investigation presented is brought about by the consideration of some processes occurring in the gas phase and on the surface of the solid being in the reaction zone.

Experiments performed are called to answer the question about the effect that the nature of solid has on the structure and the morphology of products formed on its surface.

## 2. Experimental

The arc vacuum plasmo-chemical apparatus has been used in the experiment. The reactor diameter is 150 mm. The cores from MPG-7 graphite has been used as a source for carbon vapor. The cores are 800 mm in length and have $9 \times 9$ mm cross-section. Graphite has been evaporated under He pressure about 0.5 atm. and at the following arc parameters: current of 180 A and voltage of 26 V.

The temperature of the external reactor wall has been kept constant (about 25-30 °C) due to the temperature-controlling water-cooled jacket. The temperature of the internal wall and the experimental cage ranges around 500-600 °C.

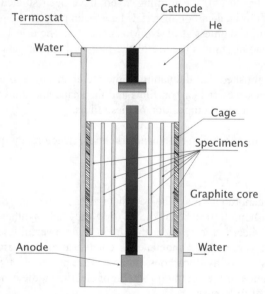

Figure 1. Scheme of the reactor for electric arc synthesis of carbon nanomaterials

As indicated earlier, hypotheses for the structure of carbon nanostructured materials depends on the material of the substrate on which they are formed needs to be verified experimentally. In this connection the influence of the nature of some substrates (Ti, Fe, Cu, Al, W, manganin, stainless steel, Ni, Mo) has been investigated in this paper. A special cylindrical cage has been designed to ensure identical conditions for formation of carbon nanomaterials on the different substrates. The above substrates are placed on the internal surface of this cage (Fig. 1). The graphite core sputtered is placed along the axis of the cage what makes possible the equidistance between the substrates and the source of carbon vapor and equal temperature conditions for their heating. The scheme for synthesis of carbon nanomaterials used in the paper is given in Fig. 2.

As indicated earlier, formation of nanostructured materials occurs from the gas phase. In this connection the plasma of electric arc in the helium medium has been used to transform carbon into the gas phase in the presented apparatus.

The operating condition of the apparatus depends on the type and the diameter of the initial material evaporated at the given moment.

Time of the evaporation cycle for one core was 3 h.

Thereupon the cage was removed from the reactor. The films formed on the surface of the specimens have been studied using TEM and SEM.

## 3. Results and discussion

Upon completion of the process, the structure of the films from the carbon materials formed on the surface has been studied using SEM T-20 and TEM microscopes.

It has been known from various scientific publications [1-4] that in the course of synthesis the out-growth called deposit forms on the permanent electrode. It consists of multi-wall carbon nanotubes and some amount of graphitized mass. This product is formed from charged particles which move under action of the electromagnetic field created by electrodes (Fig.2, a). As the deposit grows and remains conductor, its structure forms in such a way that its ohmic resistance remains minimum.

In the course of the arc graphite evaporation the gas phase in the plasma state with T>4000K shifts from the arc zone at a velocity about 20-25 m/s and reaches the reactor wall for $3.5 \cdot 10^{-3}$s. Within this time a number of reactions may occur; mechanisms for two of them are given in the scheme (Fig. 2, b). This is formation of fullerenes and onions. When the gas phase contains catalyst, the single-wall carbon nanotubes may be formed.

When the vapor phase reaches the wall, the nanostructures like tubes may also grow in the carbon film. One of the supposed mechanisms of the reaction is given in Fig. 2, c.

The processes occurring on the reactor wall are considered in this paper. To date these problems have not been received serious attention as all researchers use reactors from stainless steel.

We presume that this process is especially effected by the boundary layer "gas-solid" that forms over the surface of the used metal at T=500-600 °C. Depending on the electron structure of the metal and, consequently, on the melting point, the chemical composition of 1-2 monolayers on the surface of each metal has different defects what enables atoms to transfer to the gas phase. This is one of the factors of the catalytic effect of solid on the carbon vapor that forms the structure and the morphology of the products. This situation is analogous to the process of hydrocarbon pyrolysis at 500-600 °C when carbon nanotubes form on the catalyst and the source of carbon is different hydrocarbons.

All the above processes were studied insufficiently and require additional investigations.

Research into the films of the product has revealed that the layers of the product on the surface of different metals are distinguished by their morphology as well as the structure in the depth of the film: Ti, Cu, Fe (Fig. 3, a, b, c). Morphology in the depth of the films is seen to be different. The sponge film forms on iron. The slightly marked double-stage film (some particles going through the upper sponge layer are thickened in the second one) is observed on copper. The brightly marked two-layer film forms on titanium (in this case particles also going through the upper sponge layer are thickened in the second one to form monolith of graphitized carbon). As indicated earlier, these processes require further investigations.

Nanostructural investigations of the synthesized products which are part of the formed films have been performed using transmission electron microscopy. It has been revealed that, besides different carbon nanoclusters of an indefinite form, the films contain nanotubes with different geometry.

Within each specimen the tubes show an equal geometric structure, but they can be distinguished by size (Fig. 4, a, b, c). However, in each metal the film contains the tubes which structure strongly differ from that usually observed in the stainless steel reactors after synthesis of these materials.

## 4. Conclusions

The construction of the experimental reactor has been designed in the course of the work. The manufactured reactor admits synthesis of nanostructured carbon materials in the equal thermodynamic conditions on the different metal substrates (up to 9 simultaneously).

The morphology and the structure of the forming layers has been studied. The films form the nanostructured synthesized carbon materials have been produced on the surface of Ti, Fe, Cu, Al, W, manganin, stainless steel, Ni, Mo substrates.

The influence that the nature of the metal substrate exerts on morphology and structure of the produced carbon materials has been studied in this paper.

It has been established that the nature of the substrate metal affects both morphology of the forming nanostructured carbon film and the form and the structure of the carbon nanotubes which are part of these films.

Interaction of carbon with metal over the metal surface shows a specific behavior and is mainly determined by the electron structure of the metal used.

In our opinion, further research into these processes may provide the means for tackling a number of questions related to the catalysis processes and kinetics of the reactions in which carbon nanostructured forms arise.

## 5. References

1. Alexeev N.I. and Duzhev G.A. Statisticheskaya model' obrazovaniya fullerenov //HTF. 2001. Vol. 71(5). P. 67-77 (*in Russian*).
2. Alekseyev N.I. and Dyuzhev G.A. Fullerene formation in an arc discharge //Carbon. 2003. Vol. 41(7). P. 1343-1348.
3. Churikov G.N., Fedorov A.S. and Novikov P.V. Obrazovanie fullerenov C60 v chastichno ionizirovannom uglerodnom pare //*Pisma v* HTF. 2002. Vol. 76(8). P. 604-608 (*in Russian*).
4. Novikov P.V. Issledovanie zavisimosty protsessa obrazovaniya fullerenov i metallofullerenov ot parametrov uglerodsoderzhaschey plazmy, Avtoreferat disertatsii, 2003 Krasnoyarsk, Institut fiziki SO RAN (*in Russian*).

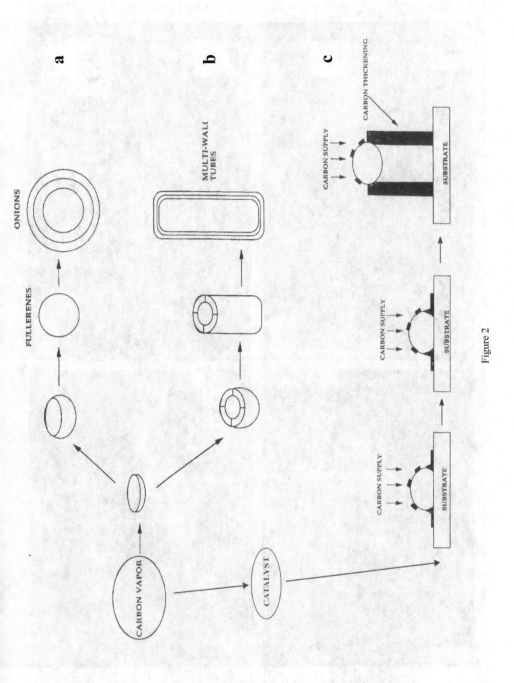

221

Figure 2

222

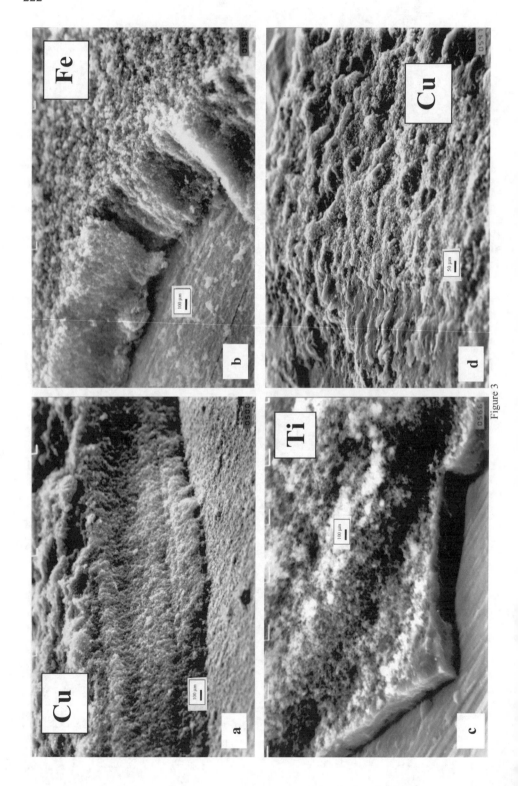

Figure 3

223

Figure 4

# STUDY OF THERMODYNAMIC PARAMETERS OF HYDROGEN GAS BY GRAPHO-ANALYTIC METHOD

B. IBRAHIMOGLU, T.N. VEZIROGLU, A. HUSEYNOV, D. SCHUR

*Mechanical Engineering department, Faculty of Architecture and Engineering, Gazi University, Celal Bayar Bulvari, Maltepe, 2317400 (Pbx) Ankara, Turkey*
*Clean Energy Research Institute, Mechanical Engineering, University of Miami, P.O.Box 248294, Coral Cables, FL33124 – 0620, USA*
*Computer Engineering Dept., Baskent University, Eskisehir Yolu 20 km, 06530, Ankara, Turkey*
*Institute for Problems of Materials Science of NAS of Ukraine, P.O. Box 195, Kiev-150, Ukraine*

## Abstract

Grapho-analythical method suggested to guarantee more precise outcomes when analyze $V - T$, $P = $ const, $1/V - P$, $T = $ const diagrams for gases. Results for hydrogen gas ($H_2$) give outcomes principially distinguished from similar researches for carbon dioxide ($CO_2$), oxygen ($O_2$), argon (Ar), helium (He), neon (Ne) and other gases.

## 1. Introduction

The thermodynamical parameters are determined by theoretical or experimental methods. Determined by theoretical method, the values of parameters hardly differ from actual values, therefore the theoretical methods have not received any preference in an industry. Now, these parameters are evaluated using the experimental methods. However, the application of such experiments is connected to technical difficulties and major financial expenses. In conditions of accelerated technical and scientific advance more effective and fast methods of testing find a use. One of such adopted methods is the grapho-analytical method [1, 2, 3].

In the given article the application of a grapho-analytical method for an investigation of the diagrams $(V - T)$, $P = $ const., and $(1/V - P)$, $T = $ const., constructed on the basis of experimental data for different gases is offered.

With the application of a grapho-analythical method to thermodynamic parameters of matters in the solid, liquid, and gaseous states it is possible to determine the system's equilibrium state, phase transitions, state diagram, critical parameters ($T_{cr.}$, $P_{cr}$), temperature of melting, boiling and freezing ($T_{th.}$, $T_{b.}$, $T_{f.}$) and other essential arguments more exactly.

For a graduation of thermometer the standard thermometer is used. As standard substance hydrogen $H_2$ is accepted [1]. The standard gas thermometers have found broad application. An attribute on which one judges temperature, is a gas pressure.

225

*T.N. Veziroglu et al. (eds.),*
*Hydrogen Materials Science and Chemistry of Carbon Nanomaterials*, 225-232.
© 2004 *Kluwer Academic Publishers. Printed in the Netherlands.*

A body's temperature is proportional to a hydrogen pressure in the gas thermometer at a constant volume taken by the hydrogen gas. Melting (0° C) and boiling (100° C) temperatures of water are accepted as reference points. These points are indicated on the diagram (Fig. 1), where $P_0$ is a pressure corresponding to the temperature of ice melting (0° C), $P_{100}$ is a pressure corresponding to the temperature of water boiling (100° C), both temperatures at pressure of 760 mm of mercurial column.

Under the laws of geometry, through two points it is possible to draw only one straight line. The line through the points $P_0$ and $P_{100}$ is drawn until an intersection with an axis t.

Equation of this line is

$$T = \frac{P - P_0}{P - P_{100}} \cdot 100 .$$ (1)

Point of intersection of line, passing through points $P_0$ and $P_{100}$, with the axis t makes the value of −273.1 °C and is called as an absolute zero [1].

We and other authors conducted research of properties of many gaseous matters by the grapho-analythic method [2, 3, 8, 9].

In a Fig. 2 the diagrams for gases of helium (He) and neon (Ne) are adduced. As a result of our studies all isobars ( P = const ) intersect in one point on an axis t and this point is a point of an absolute zero. However, at application of the grapho-analytical method to a research of the properties of hydrogen gas, the outcome received is distinct from the one above-mentioned.

## 2. Analysis of hydrogen properties by a grapho-analytical method

Hydrogen is considered as energy of the 21-st century [6], besides it is a standard gas, and no doubt that hydrogen will be used in different industries in more broad and diverse manner. From this point of view, the detail examination of its properties is represented [13].

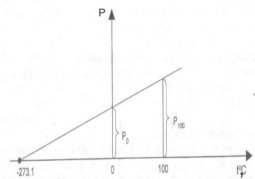

Figure 1. Determination of temperature of absolute zero by graphical method

It is clearly observed from a Fig. 3, that $H_2$ does not behave like an ideal gas at low temperature. Data points on iso-bars of hydrogen (P = const), unlike another gases (He and Ne), decline from the linear dependence to parabolic one at transition from higher temperatures to lower temperatures and approaching to origin of coordinates.

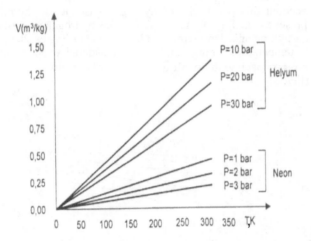

Figure 2. V-T diagram of gases He and Ne for P = const

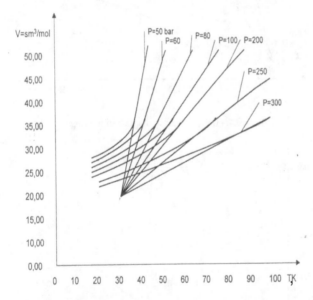

Figure 3. (V – T) P = const. diagram of hydrogen

In this case, the linear extrapolation of hydrogen curve as well as of other gases, can be applied only for definite ranges of temperature values.

For example, linear extrapolation is possible for iso-bar curve P = 50 only in the range between higher temperatures down to T = 44 K, and for iso-bar curve P = 60 –down to T=48K. The curve becomes a parabola when temperature is lower than the interval T=44 K to T=16 K. At the same time, this parabolic curve also depends on pressure.

Constructed on experimental data [5], (V – T) diagrams of hydrogen, unlike other gases, contain iso-bars P = const., which linear extrapolation does not result in an origin

228

(see Fig. 3 ), and crosspoint is placed in the range of T = 33.5K, V = 20.08 sm³/mol, T = 32K, V = 22.00 sm³/mol.

Taking into account this peculiarity of hydrogen gas, the iso-therms in (1/V –P) coordinates are presented in Fig. 4. At T < 20 K the hydrogen consists only of one modification "para-hydrogen". Iso-therms in (1/V – P) coordinates demonstrate linear character for this temperature's range under the considered pressure interval. When T→0 the iso-therms approach to the small finite values of density (1/V) that are different from each other (Fig. 5).

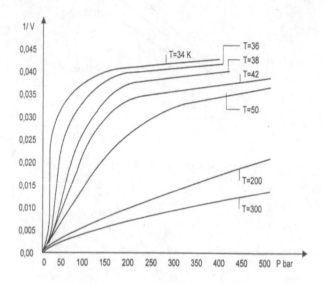

Figure 4. (1/V), T = const. diagram of hydrogen

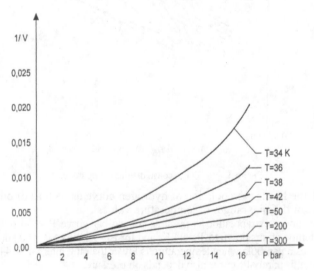

Figure 5. (1/V), T = const. diagram of hydrogen

The increase of temperature changes the iso-terms character on diagram (1/V – P). At these temperatures beginning from T > 20 K, the iso-terms are located in some levels. At low and high pressures iso-terms on diagram (1/V – P) have in a major degree linear character and between these "the linear ranges" include a transition segment in the interval 12 bar < P < 30 bar.

Mentioned multi-level structure and the modification of transition range of pressure in a temperature dependence is determined by influence microscopic interactions between "para- "and "ortho-hydrogen" moleculas.

As is known from the literature [10], with increase of temperature some moleculas of "para-hydrogen" are transformed to molecules of "ortho-hydrogen" and the gas turns to a mixture of two different modifications of hydrogen.

The existence of two modifications of molecules of hydrogen is connected to different relative orientation of nuclear spins of atoms and therefore with different values of rotational quantum numbers. In molecules of para-hydrogen the nuclear spins are antiparallel, also rotational quantum numbers even. The ortho-hydrogen has parallel spins and odd quantum numbers. The nuclear spin isomerism is the initial cause of different magnetic, spectral and thermal properties of both modifications. A para- and ortho- modifications of hydrogen have different quantity of a rotational energy and consequently have a little bit distinguishing of thermal capacity, heat conduction, pressure of saturated steams, melting point. On chemical conduct a para- and ortho-hydrogen are identical. The para- hydrogen in a pure state is stable only at temperatures, close to an absolute zero. At usual (room) temperature the isotopes of hydrogen have following equilibrium composition: protium and tritium contain upto 75 % of the para - form and 25 % of the ortho - form, deuterium correspondingly 66.67 % and 33.37 %. The isotopes of such structure are considered normal and denoted as normal hydrogen (n- $H_2$), deuterium (n- $D_2$), tritium (n- $T_2$). Ratios of ortho- and para- forms do not vary, if temperature is higher than 25°C for $H_2$, 70°C for $D_2$, and 100°C for T2 [11,12].

At the temperature of 20.4 K in equilibrium hydrogen ($H_2$), is contained about 21 % o-$H_2$, such hydrogen is identified with para-hydrogen. Equilibrium ortho - para-structure of isotopes of hydrogen in a temperature dependence is shown in a Fig. 6

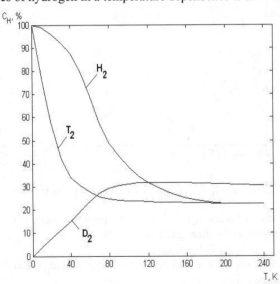

Figure 6. Dependence of equilibrium ortho- para structure of hydrogen (protium, deuterium, tritium) $C_H$ on temperature T (the curves fall into para- modification [14])

The transition of hydrogen from one modification to other also occurs at very low temperatures, but process goes with a negligible speed. It can be speeded up by a pressure buildup. The ratio of number of moleculas of these modifications in a mixture depends on temperature and does not depend on pressure [5].

Defined for different iso-therms the curves pressure-density are interpolated by the fifth-order polynomial, using the least-square method.

$$g(p) = \sum a_i p^i . \qquad (2)$$

The applied interpolation formula allows us to determine the value of a density with error 2% for arbitrary values of pressure P (Fig. 7).

## 3. Discussion

It is seen from Fig. 3 that diagrams (V – T), P = const. vary linearly in the ranges of high temperatures, where the classical approach is valid. Nevertheless, according to the laws of quantum physics for the gas moleculas, iso-bar curves decline from linear dependence with decreasing temperature values. The limiting temperature variation T→0 doesn't affect considerably a volume of gas. In limiting condition an expansion ratio $(dV/dT)_p$ approaches zero according to necessary conditions of the Nernst theorem [10].

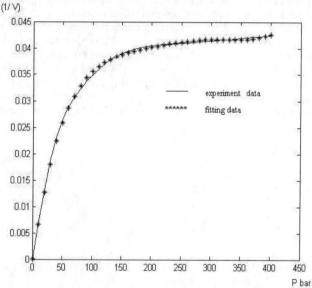

Figure 7. An interpolation diagram at T = 36 K

At an extrapolation of a linear part V – T iso-bars for values T >> 0 K, instead of the formula conforming to the ideal gas law

$$V = ( R/P ) T \qquad (3)$$

more general formula of linear dependence is demonstrated, when the line is passing through the coordinate origin

$$V_{extrapolat} = aT + b. \qquad (4)$$

Here R is a gas constant. If we do not consider quantum effect, the argument $b$ can be interpreted as a volume taken by gas at different values of pressure and T = 0 K. The constant $a$, describing chart inclination in relation to the axis, does not coincide with the value R / P (P = const.) in the ideal gas equation.

Important point is the fact, that the extrapolation of values of V – T curve from ranges of high temperatures in range of low temperatures results in intersection of a set of linear sections in one point $(T^*, V^*)$ with definite error (Fig. 8).

The point $(T^*)$, defined by means of an experiment series in ranges of high temperatures and consequent interpolation, allows to pronounce preliminary judgment on a critical point of gas. Really, if we neglect quantum effect influencing a gas law

$$F ( P, V, T) = \text{const.} \tag{5}$$

it is possible to suspect, that iso-bars should drive through a critical point $T_{cr.}$ in a phase space. Thus, determination of a point $V_{cr.}$ on the arbitrary curve of an iso-bar which is satisfying an equation

$$f (P, V^*, T^*) = \text{const.} \tag{6}$$

may be executed by taking a derivative on pressure from the previous expression

$$(df/ dP )_{T^*, V^*} = 0 \tag{7}$$

The last operation is mathematical expression of definition of extreme point from a state equation of gas.

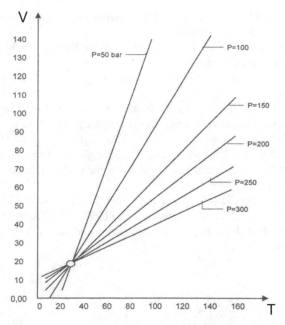

Figure 8. Intersection of hydrogen iso-bars on the diagram 1/V – T

## 4. Conclusions

1. It is established that the density $\rho$ of hydrogen in ranges P=12-20 bars and T=18-40K varies spasmodically.
2. By a grapho-analytic method it is revealed, that the iso-bars of hydrogen are not intersected

in a point of an absolute zero.

3. Different critical points correspond to "ortho-" and "para-" modifications of hydrogen. The iso-bars are intersected in one range, which one corresponds to different critical points ($T_{cr.}$) of "ortho-" and "para-" modifications of hydrogen.

4. The hydrogen can be utilized as standard matter only within the limits of definite temperatures.

5. More suitable solution is usage of mono-atomic gases of helium or neon as standard matter.

## 5. References

1. Kitaygorodskiy, A.I. (1973) Introduction in physics, Moscow: *Science*. (in Russian).
2. Farzaliev, B.I. and Ragimov, A.M. (1984) Research of phase changes in fluids, *Izvestiya of Academy of Science of Azerbaijan*, Baku 12. (in Russian).
3. Gibbs, G.V. (1950) Thermodynamic work. *Gostekhizdat*, Moscow, 422. (in Russian).
4. Farzaliev, B.I. and Aliev, N.F. (1992) Determination of melting line of gases by graphical analytic method, *Proceedings of a Scientific Conference AzTU*, Baku; 21. (in Russian).
5. Vargaftik, N.B. (1972) Reference book on thermo-physical properties of gases and liquids, Moscow, 720. (in Russian).
6. Veziroglu, T.N. and Barbir, F. (1998) Hydrogen Energy Technologies, United Nations Industrial Development Organization, Vienna.
7. Molozevskiy, A.V. (1969) A geometrical thermodynamics, Trudi MGU, Moscow, 91. (in Russian).
8. İbrahimoğlu, B.İ. (1994) Gazlarda Kritik Basıncının Grafoanalitik Yöntemle Bulunması, Türk İşi Bilmi ve Tekniği Dergisi, Ankara, 45(17x12).
9. Ibrahimoglu, B.I. and Ataer, O.E. (1997) Erime Egrisi Uzerinde Bir Uç Noktanin Belirlenmesi, Ulusal Isı Bilim ve Teknigi Kongresinin Bildirimleri, Ankara, 33(11).
10. Klimatovich, Y.L. (1982) Statistical physics, Moscow. (in Russian).
11. Zabetakis, M.G. (1961) Adv. Cryog. Eng. 6(2), 185-191.
12. Gelperin, I.I. and others (1980) Liquid hydrogen, M. Chemistry, 225. (in Russian).
13. Rober, A.M., Weber, L.A. and Goodwin, R.D. (1965) NBS Mongraf, 64(25).
14. Farkas, A. (1936) Orthohydrogen, parahydrogen and heavy hydrogen, M. ONTI, 250. (in Russian).

# SIMULATION OF OPERATION HEAT OR COLD-MAKING UNIT WITH HYDRIDE HEAT PUMP

SHANIN YU.I. *
*FSUE SRI SIA "Luch"*
*Zheleznodorozhnaya 24, Moscow area, Podolsk, Russia, 142100*

## Abstract

The developed earlier mathematical model is applied for calculations of processes in the hydride heat pump (HHP) intended for increase or reduction of a level of temperature. The model set of equations actuates non-stationary one-dimensional equations of thermal balance in hydride sorbers with allowance for of thermal effects at an absorption/desorption of hydrogen. In computer calculations the experimental data on equilibrium isotherms in systems a metal alloy hydrogen are used. As a high-temperature hydride $ZrCrFe_{1.2}$, low-temperature hydride $LaNi_5$ is used. In calculations the time cyclogramme of activity a HHP is determined: distribution of temperature of design elements on a radius of a cylindrical hydride bed, contents of hydrogen in sorbers, pressure of hydrogen, energy and thermal power. The alternative calculations are carried out for an actual module of the heat pump. The research of influence of main parameters of a mode of the heat pump (hydrogen pressure of charge, temperatures and consumptions of the heat-carrier, time of a cycle) on it the power characteristics is carried out. Is shown that at the temperature of heat source 80-95°C can be obtained an overheated water at the temperature of 110°C and the cooled water with temperature in region 0°C. It is shown that the computational data satisfactorily coincide experimental results. It is established that the developed model the HHP is possible to use for qualitative and quantitative study of influence of various parameters on operation.

## Nomenclature

| | |
|---|---|
| $\Delta H_f$ | Hydrogen heat of adsorption [J kg$^{-1}$] |
| $C_p$ | Heat capacity of sorber [J kg$^{-1}$ K$^{-1}$] |
| $G$ | Heat carrier rate [kg s$^{-1}$] |
| $N$ | Quantity of hydrogen in a sorber [kg] |
| $N_{H2}$ | Quantity of hydrogen in a gas phase in appropriate sorbers [kg] |
| $N_{hydr}$ | Quantity of hydrogen in a hydride [kg] |
| $N_{Me}$ | Density of a metal sublattice in a hydride [kg m$^{-3}$] |
| $P$ | Pressure [Pa] |
| $Q$ | Power [W] |
| $Q$ | Energy [J, kJ] |

---

* Corresponding author e-mail: syi@luch.podolsk.ru; fax: 007(096)634582

*T.N. Veziroglu et al. (eds.),*
*Hydrogen Materials Science and Chemistry of Carbon Nanomaterials, 233-242.*
© 2004 *Kluwer Academic Publishers. Printed in the Netherlands.*

| | |
|---|---|
| $Q_v$ | Heat flow per unit of sorber volume [W m$^{-3}$] |
| $R$ | Radius [m] |
| $t$ | Time [s, min] |
| $T, T_c, T_w$ | Temperature of a sorber, heat-carrier and wall [°C, K] |
| $V$ | Volume of a hydride bed (m$^3$) |
| $X$ | Concentration of hydrogen [g-atom H$_2$ (mole of alloy)$^{-1}$] |
| HT(-H), LT(-H) | High-temperature and low-temperature hydrides |
| HHP | Hydride heat pump |
| COP | Efficiency a HHP (relation of energy of an obtained cold to energy loiter on heating) |

*Greek symbols*

| | |
|---|---|
| $\alpha$ | Coefficient of a heat transfer (W m$^{-2}$ K$^{-1}$) |
| $\lambda_w$ | Heat conduction of a material of a wall (W m$^{-1}$ K$^{-1}$) |
| $\rho$ | Density (kg m$^{-3}$) |

*Subscripts*

| | |
|---|---|
| h, c | Concerns to a HTH and LTH accordingly |
| i | Sorber index (i=1,2) |
| hydr | Hydride |
| H$_2$ | Hydrogen |
| W | Wall, surface |
| eq | Equilibrium |
| h, m, l | High, mean, low level of temperatures |

**Key words**: Hydride, hydrogen, heat pump, simulation, calculation, experiment.

## 1. Introduction

The experience of development enhancement heat-transformers using two-stage heat pumps is known [1]. They were used for production of overheated fluid (water) or vapour. Such HHP as a rule have two stages and in them three various kinds of a metal hydride are used. Application the single-stage HHP for increase of temperature or generation of a cold practically is not known within the framework of one pump and couple of hydrides.

The computational substantiation of a material structure, design and operational modes is a necessary part of researches at creation a HHP [2]. Mathematical models earlier were developed, the computer simulation of working processes in a HHP and comparison with experiment with reference to refrigerating devices of the automobile [3] is carried out. Results of calculations with allowance for of bounded heat conduction of hydride beds here are considered. The calculations are carried for single modules a conditional HHP. The results of calculations are compared to experimental results for the same modules an actual HHP which included by components in the installation imitating production of a heat as an overheated water or a cold [4]. As a heat source for operation the HHP was used warmly water with a level of maximum temperature 80-100°C.

## 2. Mathematical model

Development of mathematical model and it a program realization are realized Fedorov E.M. [2]. At the first stage of application the developed model allows to make preliminary selection of couples of hydrides on specific intervals of operation temperatures and pressure of hydrogen with allowance for of design features of hydride sorbers [4]. At this stage purely is decided whether the HHP on selected hydrides (i.e. will work whether the thermodynamic cycle) in indicated ranges of working parameters (temperature, pressure) is feasible.

The second stage will realize mathematical model of working process in hydride sorbers and submits a capability of optimization of the geometric sizes and operational modes a HHP.

The mathematical model takes into account main factors of a design of hydride sorbers and processes happening in it: a method of heating and heat pick up, heat conduction in hydride sorbers, transfer of hydrogen with appropriate allocation and absorption of heat.

The mathematical model is developed for one-dimensional non-stationary case and in more detail is explained in [2,3]. In it the actual distribution physical variable - fields of temperatures, pressure, concentration of hydrogen on volume of a hydride bed a HHP and their time history is taken into account. Within the framework of one-dimensional model the change of physical sizes on one only to radial coordinate of cylindrical sorbers is taken into account (Fig.1).

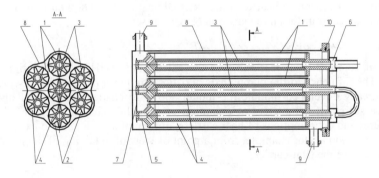

Fig. 1. The scheme of HHP modular sorber design: 1- tubular case hydride module; 2- insert-corrugation for thermal conductivity; 3- hydrogen ceramic collector-filter; 4- metal hydride; 5-tip of metal hydride filling; 6- hydrogen collector; 7- plate-spacer; 8- heat exchanger casing; 9- union; 10- flange cover of heat exchanger. Heat exchanger thermal insulation conditionally is not shown.

The set of equations describing a heat mode a HHP consisting from two sorbers containing high-temperature (HTH) and low-temperature (LTH) hydrides can be recorded as follows

$$\overline{\rho C_p}\left(\frac{\partial T}{\partial t}+\frac{\rho}{V}\,gradT\right) = div\ \overline{\lambda}\ gradT + \overline{Q}_v, \tag{1}$$

$$x=x_{eq}(P,T), \tag{2}$$

$$N_h^0=N_{hydr}^1+N_{hydr}^2+2N_{H_2}^1+2N_{H_2}^2. \tag{3}$$

Here first records a heat conduction equation (which is decided in each from hydride beds and shells of bodies). The second term in a right member takes into account allocation or absorption of heat at a sorption of hydrogen. The contents of hydrides in sorbers is calculated pursuant to equilibrium isotherms on values of pressure of hydrogen and temperature of hydrides.

The equation (1) is recorded in the supposition about absence of a jump of temperatures between gas and solid phases in a hydride bed. The equation (2) indicates that the concentration of hydrogen in each point of computational area is determined from an equilibrium condition. The equation (3) expresses a condition of a constancy of quantity of hydrogen in the heat pump. In it the superscript 0 concerns to the whole system, indexes 1 and 2 to various sorbers.

As the initial data the initial pressure of hydrogen in a system and temperature of hydrides is given. The contents of hydrogen in sorbers in an initial moment is from an equilibrium condition. During activity of pressure in sorbers it was supposed identical.

The boundary conditions are determined by character of heat exchange. In our case the hydride sorbers exchange a heat with the external heat-carrier. On their outside surface the condition of a heat transfer on the law of a Newton is set

$$-\lambda_w\ grad\ T = \alpha\,(T_w-T_c). \tag{4}$$

Problem of computational research was the calculation a HHP which design is already selected. The influence to the pump characteristics of control parameters was investigated. For the already developed module the model allows to change modes it of operation by exhaustive search of parameters of operation of a module (initial pressure of hydrogen, temperatures and consumptions of heat-carriers, of time of a cycle) and to execute selection of an optimum mode (for example, from a point of view of obtaining of the greatest power or temperature of the heat-carrier).

## 3. Results and discussion

### 3.1. Example of calculation

Let's put by way of illustration results of calculations of a design a HHP constructed because of modules. The design a HHP explicitly is described in paper [4], the scheme of the installation is shown on Fig.2 (experimental results on this installation are obtained Astakhov B.A.). The HHP works both in a mode heat-transformer, and in a mode of a refrigerator because of hydride pair $ZrCrFe_{1.2}$ - $LaNi_5$.

The module is made from sections of a pipe of a stainless steel with external diameter R=25 mm and thickness of a wall 1 mm. Length of a module containing the high-temperature hydride ($ZrCrFe_{1.2}$) - 655 mm, weight of a hydride in it - 0.57 kg. Length of a module containing the low-temperature hydride (LaNi$_5$) - 850 mm, weight of a hydride - 0.84 kg. For increase of efficiency main heat generating cycle weight a LTH in 1.5 times exceeded weight a HTH. For increase of heat conduction of a hydride bed in pipes the transcalent insertion from an aluminium foil is located. The central part of a module is taken by a collector of the collecting and distribution of hydrogen which also executes a role of the filter.

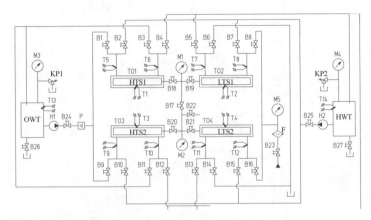

Fig. 2. The scheme of installation: HWT- constant-temperature tank of hot water; OWT- tank with overheated liquid; HTS1, HTS2 - high-temperature sorber; LTS1, LTS 2 - low-temperature sorber; T01-T04- sorber heat exchanger; H1, H2 - pump; F- filter; B1-B27- manually operated valve; M1-M5-манометр; KP1, KP2- safety valve; P- flowmeter; T1-T14 – thermocouple.

The isotherms for the data of materials are obtained experimentally and were used in calculations after their computer processing.
At increase of temperature there were following heat sources: the hot heat-carrier (high temperature level $T_h$) - water with temperature 95-110°C, heat-carrier of an average level $T_m$ - water with temperature 75-95°C and heat-carrier of a low level $T_l$ - water with temperature 8-15°C.
Heat sources at production of a cold (conversion of a cycle): the hot heat-carrier (high temperature level $T_h$) - water with temperature in range from 80-100°C, heat-carrier of an average level $T_m$ - water with temperature 8-20°C and heat-carrier of a low level $T_l$ - water with temperature 5-10°C.
The characteristic curve changes mean-integral (on a radius of a sorber pipe) temperature of hydride beds at cycling a HHP are shown in a Fig. 3.

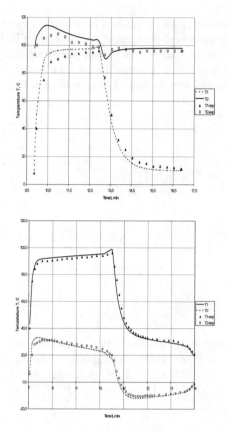

Fig. 3. The cyclogramme change of parameters of a module in a cycle: a) temperature increase; b) cold production. Points - experimental data. T1 - high-temperature sorber, T2 - low-temperature sorber.

The calculations have shown that at initial stages of non-stationary processes of heating and the cooling of sorbers arise gradients of temperatures and concentration of hydrogen causing to small heterogeneities of a temperature field on a radius of a hydride bed. It testifies that the sufficient effective heat conduction of a hydride bed is structurally reached. The registration of a hysteresis the curve pressure - concentration (at absorption $p_a$ and desorption $p_d$ of hydrogen) essentially increases complexity of calculations. Therefore influence of a hysteresis was appreciated only on limited number of versions. Generalizing results of these assessments it is possible to tell that the hysteresis (at the relation of pressure of a sorption and desorption at a level 1.1-1.3) reduces an overall performance a HHP approximately on 25-40%.

*3.2. Research of influence of parameters on the characteristics of a module a HHP*
1. **Initial pressure of hydrogen in sorbers.** The pressure variation of charge essentially has an effect for operation of a module. In a Fig. 4 the relations mean on time of

powers of heating $q_h$ and cooling $q_c$ are adduced depending on a level of initial pressure above a hydride (at the temperature of 10°C).

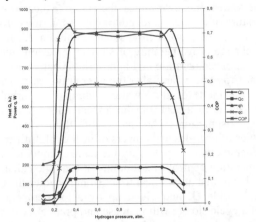

Fig. 4. Influence of initial pressure of hydrogen to the power characteristics of a module: $q_h$ - mean thermal power proceeden in a HTH, $q_c$ - mean power of a cold allotted from a LTH, COP - efficiency of a module (relation of energy allotted from the LTH $Q_c$ to energy bringed to HTH $Q_h$).

At pressure is lower $0.2 \cdot 10^5$ Pa the contents of hydrogen in a hydride corresponds to a linear segment of the phase diagram. As it is not enough of hydrogen in of an alloy the powers are small. In range from 0.4 up to $1.2 \cdot 10^5$ Pa these characteristics are maximum and practically do not depend on pressure. At further increase of pressure of charge the quantity of hydrogen transferred between sorbers and accordingly power characteristics is reduced.

2. **Temperature of the heat-carrier.** In case temperature of a raising cycle the increase of temperature of a water on an average level ($T_m$) results in more temperature of a hot water ($T_h$). Increase of temperature of a water on a low level ($T_l$) capable to worsen pump parameters since certain temperature $T_{lmax}$. At levels of temperatures $T_m$=85-95°C, $T_l$=10-15°C mean temperature in a hydride bed will be realized at a level 115-125°C that allows to calculate on obtaining of a hot water with temperature 105-110°C. The relation of useful energy obtained at heating to expended energy (efficiency or CHT HHP in a cycle of increase of temperature), edges on a level of temperature at change last in range from 80 up to 100°C and approximately CHT=0.4. Calculated power of a module is 140 W.

In case of a refrigerating cycle the increase of water temperature on a high level ($T_h$) results in lower temperature on a low level ($T_l$). The increase of water temperature on an average level ($T_m$) worsens parameters of the pump and increases temperature $T_l$. At a level of temperatures $T_h$=85-95°C, $T_m$=10-20°C mean temperature in a refrigerating cycle in a hydride bed will be realized at a level -4...-14°C that allows to calculate obtaining a cold water with temperature 0-7°C. The relation of energy obtained at cooling to energy expended on heating of a module (efficiency or COP a HHP in a refrigerating cycle) edges on a level of temperature at change last in range from 80 up to 100°C and approximately COP= 0.7. Calculated power of a module is of 100 W.

3. **Water rate.** The water rate determines factors of a heat transfer in channels and change of temperature of a water on length of sorbers. In the considered design concept the longitudinal current of a water in channels is organized. It the heating on a flow course results in change of temperature of sorbers in various their parts why there are axial misalignments of temperature and appropriate characteristics of sorbers. As well as it was necessary to expect (and also the alternative calculations have shown) that most essential is the mode of flow water - laminar or turbulent. In the stabilized laminar flow the factors of a heat transfer appear rather small and practically do not depend on the consumption. The mode of fluid flow essentially influences instantaneous values of energies and powers of heat/cool. Influence of the rate on mean (on time) of the characteristic of power for the realized module did not exceed 20 %.

4. **Cycle time.** At operation control of a module on time the calculations (with reference to cooling) have shown that at identical times of half-cycles the energy (proceeden in a high-temperature hydride $Q_h$ and given up by a low-temperature hydride $Q_c$) with growth of time of a cycle slowly will increase and practically does not vary at t>4 min. (Fig. 5). While the mean power of heating $q_h$ constantly drops the mean power of cooling $q_c$ has a slanting maximum at t≈3.5 min. The relation of energies LTH/HTH (the $COP=Q_c/Q_h$) will increase a little at growth of time from 2 up to 5 minutes and practically remains to constants at further increase of time of a half-cycle. For obtaining maximum refrigerating power the time of a cycle 7-10 min. is recommended.

*3.3. Comparison with experimental results*

The comparison of results of calculation and experiment [4] has shown the following:
1. The model is qualitatively correct describes all processes in an actual HHP: changes of temperatures, concentration and pressure of hydrogen depending on time (see Fig.3).
   The conformity of computational and experimental values is observed and for processes at change of managing parameters (initial pressure of hydrogen, temperatures and consumptions of the heat-carrier, time of a cycle). The availability of local computational maxima on curve COP (see Fig. 5) is connected apparently to inaccuracy of the computer description of relation of hydrogen pressure from it concentration for hydrides in transient phase condition of a hydride.
   At calculations was observed that the physical processes pass little bit faster on time than in an actual module. It can be associated to overestimate of computational value of effective heat conduction of hydride beds. There was an excess, obvious 30-50 %, of a computational overall performance a HHP (COP, CHT) above really reached.
2. Values of computational parameters exceeded experimental results from rather small sizes (up to 20 %) up to rather significant (about 60 %). In main it concerned initial stages of transients. A kind of computational relations more precise and expressed than at obtained experimental results (curve temperatures, pressure).

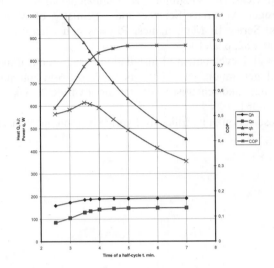

Fig. 5. Influence of time of a half-cycle of operation to the power characteristics of a module (see symbols in a Fig. 4).

The large divergences of the experimental and computational data can be referred to account inaccuracies of model (we shall remind that in majority of calculations the hysteresis was not taken into account, and for a HTH $ZrCrFe_{1.2}$ it achieved $p_d/p_d$=1.5, for the LTH $LaNi_5$ - $p_d/p_d$ =1.25, i.e. took place a rather essential hysteresis), inaccuracy of definition of the thermophysical characteristics of hydride beds.

The more careful registration of influence of a hysteresis has shown that the data obtained because of experiments and calculations fall in satisfactorily.

## Conclusions

The comparison of computational and experimental results has shown that the developed one-dimensional model does not allow to not take into account yet all actual design features a HHP. But it can use for qualitative and quantitative study of influence of various parameters on processes in a HHP.

## References

1. Suda S, Komazaki Y, Narasaki H, Uchida M. Development of a double-stage heat pump: experimental and analitical surveys. J. Lesson-Common Metals 1991; 172-174: 1092-1110.
2. Fedorov EM, Shanin YI, Izhvanov LA. Simulation of hydride heat pump operation. Int. J Hydrogen Energy 1999; 24: 1027-1032.

3.  Shanin YI. Simulation of hydride heat pump operation with reference to vehicle refrigerating devices. In: Veziroglu TN, Zaginaichenko SY, Schur DV, Trefilov VI editors. Hydrogen Materials Science and Chemistry of Metal Hydrides. NATO science series. Series II: Mathematics, Physics and Chemistry. Kluwer Academic Publishers, 2002. 82. p.97-106.
4.  Astakhov BA et al. Development of installation based on metal hydride heat pump for heat and cold generation. In: Abstracts book by 6-th intern. confer. "Hydrogen materials science and chemistry of metals hydrides". Ukraine, Yalta, (02-08) .09.1999. p.360-361.
5.  Shanin YuI. Preselection of hydrides for hydride heat pumps. (there), p.304-305.

# QUANTUM-CHEMICAL INVESTIGATIONS OF SINGLE WALL CARBON NANOTUBE HYDROGENATION PROCESSES

LEBEDEV N.G. [(1)] ZAPOROTSKOVA I.V. [(1)], CHERNOZATONSKII L.A.[(2)]
[(1)] *Volgograd State University, ul. 2-ya Prodolnaya, 30, Volgograg, 400062, Russia E-mail: nikolay.lebedev@volsu.ru*
[(2)] *Emanuel Institute of Biochemical Physics, Russian Academy of Sciences, ul. Kosygina 4, Moscow, 117334 Russia, E-mail: cherno@sky.chph.ras.ru*

## Abstract

Calculations of electron energetic characteristics of atomic hydrogen adsorption processes on external and internal surfaces of (6,6) and (10, 0) single-walled carbon nanotubes (SWNTs) having cylindrical symmetry have been carried out. Ionic-embedded covalent-cyclic cluster (IECCC) and molecular cluster models within the framework of semi-empirical quantum-chemical scheme MNDO well shown in the theoretical researches of electronic molecular and periodic solid-state structures have been used. The electronic and energy characteristics of the hydrogenation processes have been analyzed, and the most energetically favorable SWNT hydride structures have been determined. The mechanisms of SWNT hydrogenation processes have been investigated.

**Keywords**: A, Carbon nanotubes; D, Absorption, Activation energy, Adsorption properties, Electronic structure.

## 1. Introduction

In recent years, considerable attention has been focused on investigating the electronic band structure of single-walled carbon nanotubes [1 - 3] modified by different methods. In particular, it was revealed that the properties of nanotubes can change depending on the particular modifying or doping technique, the choice of incorporated or adsorbed elements, and treatment conditions. Due to the strongly curved surface the single-walled carbon nanotube causes the great interest as effective adsorbent of easy atoms and molecules

*T.N. Veziroglu et al. (eds.),*
*Hydrogen Materials Science and Chemistry of Carbon Nanomaterials,* 243-258.
© 2004 *Kluwer Academic Publishers. Printed in the Netherlands.*

[4 - 22]. The particular care is taken to investigating the adsorption properties of SWNTs for experimental synthesis of nanotube hydrides. However along with the existing theoretical quantum chemical investigations we consider the question of adsorption process mechanism and nanotube surface hydrogen saturation ways still opened.

In the paper we have studied the energy and geometry properties of (6,6) and (10,0) SWNT hydrides with the use of solid cluster model [23] within the framework of semi-empirical quantum chemical scheme MNDO [24, 25], shown in detail as a means in the theoretical researches of electronic molecular structure.

There are some ways of H atoms SWNT surface saturation but we propose three variants for its fulfillment: 1) hydrogen atoms are adsorbed on external surface of nanotube; 2) H atoms are adsorbed on internal surface (so-called capillary way of saturation); 3) H atoms are adsorbed on both external and internal surface simultaneously. Also we have studied regular adsorption of atomic hydrogen [21] on (6, 6) SWNT using ionic-embedded covalent-cyclic cluster (IECCC) model [26].

## 2. Atomic hydrogen adsorption on external (6,6) SWNT surface

The quantum chemical MNDO calculations within the molecular cluster model of (6, 6) SWNT have been carried out to study the energy and geometry properties of hydrogen atoms adsorbing processes on external surface. The extended unit cell (EUC) of the cluster consisted of 96 carbon atoms (cluster $C_{96}$) and contained four layers of carbon hexagons in each. The cluster boundary chemical bonds saturated by hydrogen atoms. First we have studied the single atom adsorption on the external site of nanotube surface. The hydrogen attack of (6,6) surface has been simulated by step-by-step approach of H to the carbon atom along a perpendicular to tube surface through carbon atom (fig. 1).

Thus the H atom (adatom) has got two degrees of freedom and can freely deviate the perpendicular in two mutually orthogonal directions. While calculating the surface center C has got the degrees of freedom allowing it to output from the tube surface. The surface C atom has been selected in middle of the cluster in order to exclude the influence of the boundary effects. The hydrogen adsorption on C-C bond and carbon hexagon center has been studied in [7 - 10]. The quantum-chemical calculations have shown that the H adsorption on the surface center is appeared energetically most favorable.

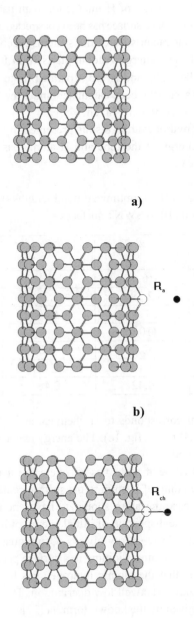

a)

b)

c)

Fig. 1. The simulation of hydrogen atom (black circle) interaction process with (6, 6) SWNT: a) H being on a large distance, b) H being on transition state distance $R_a$, c) H binding chemically with surface carbon atom (light circle).

The dependence of potential energy of H interaction with nanotube as a function of the distance from attacking atom to the tube surface has been constructed and shown in fig. 2.

The geometry and energy characteristics of the systems received as a result of attack have presented in the table 1. The energy readout starts from 0.0 eV, which corresponds to energy of the system when tube and H are separated up to ∞. Adsorption ($E_{ch}$) and activation ($E_a$) energies of all cases have been calculated under the following formula:

$$E_i = E(R_i) - E(\infty), \quad \text{where } i = a, ch. \tag{1}$$

The analysis of (6,6) tube hydrogenation results presented in fig. 2 and in the table 1 has shown that the interaction energy of the tube and H atom has got minima on distances $R_{ch} = 1.12$ Å and $E_{ch} = -1.7$ eV.

**Table 1.** Adsorption energies ($E_{ad}$), optimized bond lengths ($r_{c-H}$, Å) и active center distances ($r_{c-tub}$, Å) from (6,6) and (10,0) SWNT surfaces.

| Number of H atoms | 1 | 2 | 3 | 4 |
|---|---|---|---|---|
| (6,6). | | | | |
| $E_{ad}(H)$, eV | 16.27 | 2.29 | 3.34 | 0.13 |
| $r_{c-H}$, Å | 1.13 | 1.4 | 1.49 | 1.52 |
| (10,0). | | | | |
| $E_{ad}(H)$, eV | -2.0 | -2.6 | -2.0 | -1.6 |
| $r_{c-H}$, Å | 1.12 | 1.12 | 1.12 | 1.12 |
| $r_{c-tub}$, Å | 0.33 | 0.42 | 0.48 | 0.48 |

On our sight the minimum corresponds to H chemical adsorption on the surface that is the formation of H-C chemical bond (fig. 1c). The energy negative value means that the system stable state has been formed.

To form chemical bond hydrogen atom has to overcame a potential barrier of height $E_a = 2.2$ eV. Obviously the binding of H atom to the surface center can take place by classical and tunnel ways. In first case the atom should to have energy exceeding $E_a$ to overcome a potential barrier. Due to the dispersion of initial energy of hydrogen atom there are attacking particles with the rather large energy always. Using quasi-classical approximation it is possible to estimate a share of atoms joining to a surface per second.

For definiteness we assume that the atomic hydrogen gas has got a temperature about 1000 K and the atom speeds have Maxwell low distribution. Then the share of particles having energy exceeding $E_a$ is given by the known formula [27]:

$$\alpha = \exp\left(-\frac{E_a}{kT}\right), \tag{2}$$

where k is Boltsman's constant, T - absolute temperature. The estimation of the parameter has given $\alpha \sim 10^{-12}$. Then the particle number overcame a barrier $E_a$ and connected to

SWNT surface unit per second (reaction rate) has given by the following expression (in approaching that each collision leads to positive result) [27]:

$$V_s = \left(\frac{kT}{2\pi m}\right)^{1/2} n\alpha,$$  (3)

where n is the atomic hydrogen concentration, m – H atom mass (m=1.66 $10^{-27}$ kg). According to above described approaches the reaction rate in order of size is determined by the concentration as the function $V_s \sim 10^{-9}$ n c$^{-1}$m$^{-2}$.

Using experimental value of pressure p of molecular hydrogen (300 торр), submitted in work [4], it is possible to estimate the concentration of atomic hydrogen, which is formed in experiment, with the help of the ideal gas pressure formula n = p/kT. It appears equal n $\sim 10^{19}$b см$^{-3}$, where b - share of hydrogen atoms of received at dissociation of molecular hydrogen. The binding energy of a molecule $H_2$ makes 4.75 eV [28]. Using the formula (2) we receive b $\sim 10^{-24}$. In result the concentration of atomic hydrogen n $\sim 10^{-5}$ см$^{-3}$, and reaction rate $V_s \sim 10^{-11}$ c$^{-1}$ см$^{-2}$. The date indicates that atomic hydrogen adsorption process is quite slow. Most likely, the competing process of molecular hydrogen adsorption is realized in practice, which is a subject of a separate research.

The second way of a barrier overcoming for the particles having average energy at given temperature - tunnel. The share of atomic hydrogen makes value b $\sim 10^{-24}$. It is easy to calculate the probability of tunneling, having taken the quasi-classical approximation formula and approximating a potential barrier with peak in a point Ra (fig. 2) as square-law potential:

$$E(R) = E_a - \frac{K(R - R_a)^2}{2},$$  (4)

where K = $2(E_a - E(H))/d^2$ – a coefficient, which is defined from boundary conditions E (R) = E (H), d - characteristic semi-width of potential barrier, E (H) = 3kT/2 - kinetic the energy of attacking hydrogen atom (for given above temperature is equal $\sim 0.1$ eV). Then the probability of passage of a particle of weight m through a square-law potential barrier of height $E_a$ and characteristic semi-width d will be determined by the formula [29]:

$$w \sim \exp\left(-\frac{\pi d(E_a - 1.5kT)}{\hbar}\sqrt{\frac{2m}{E_a}}\right).$$  (5)

Choosing characteristic semi-width of a barrier d = 0.5 Å (fig. 2), in tunnel case the probability of barrier tunneling of the particle w $\sim 10^{-21}$ c$^{-1}$. Then the probability of passage of initial number of atoms H through a barrier will be equal wb $\sim 10^{-45}$ c$^{-1}$. As well as in a classical case, this value is very small for practical realization of tunnel adsorption process of atomic hydrogen.

The analysis of geometry optimization results has found out that during the attack of SWNT by atom H the surface atom C went deep approximately on 0.1 Å inside tube at first and, after the C-H distance became less than 2, has risen on 0.5 Å above a tube surface. This moment is well reflected by an inclined site on curve of potential energy (fig. 2) in a vicinity of a point 2 Å.

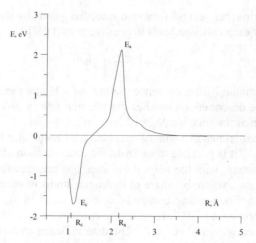

Fig. 2. The curve of potential energy of hydrogen atom interaction with (6, 6) SWNT surface as a function of the distance from hydrogen to the tube surface. The sum of reactant energies is determined as 0 eV.

In result of hydrogen adsorption on SWNT surface three C-C bonds of carbon, on which H attacking has taken place, were elongated in comparison with the initial meanings and were equal 1.47 Å. Thus, the adsorption results in a surface nanotube deformation. Moreover carbon atom, on which hydrogen adsorbing, as though rises above a tube surface and the neighbor to it displaces inside, forming an additional active center inside SWNT thus. So at $R_C$ point (fig. 2) the carbon atom leaves the tube on ~ 0.3 Å. The nearest atoms also changed the situation in space - were displaced inside the tube on distance $\delta R \sim 0.1$ Å.

Further second hydrogen atom adsorption has been investigated. As adsorption centre the neighbor carbon atom (fig. 3) has been used.

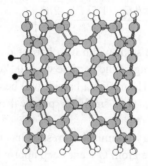

**Fig. 3.** The carbon (6,6) nanotube cluster with two adsorbed hydrogen atoms (black circle). Boundary H atom have drown by light circle.

The geometry of this system also was optimized. It was found that C-C bonds within the limits of hydrogen influence area were elongated also. The energy adsorption (chemisorptions) of second hydrogen has appeared the equal $E_{ch}$ = 3.45 eV, and H-SWNT bond length - R (C-H) = 1.12 Å.

## 3. Multiple adsorption of atomic hydrogen on (6,6) SWNT

Consecutive adsorption of hydrogen atoms of on an external tube surface has been carried out starting from the opened border of SWNT (fig. 4). Such choice of an initial adsorption position is connected that the opened nanotube end, due to two unsaturated bonds on each boundary carbon atom, has higher reactionary ability than all rest tube. Therefore we consider that the binding of atomic hydrogen to SWNT can start from boundary atoms with the greater probability.

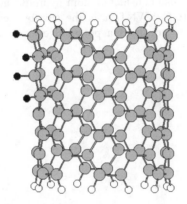

**Fig 4.** The carbon (6,6) nanotube cluster with linear chain of adsorbed hydrogen atom (black circle). Boundary H atom have drown by light circle.

The calculation results of consecutive adsorption energy ($E_{ch}$) and also the H-SWNT bond lengths ($r_{C-H}$) are shown in the table 1. The analysis of research results has found out that the energetically most favorable situation is when the H atom joins C atom placed on the opened nanotube end (site 1 in tab. 1). It can be explained as follows. Because the border is broken the carbon atom has got as a minimum one "free" bond, on which hydrogen adsorbing. Other carbon atoms, which joined H and which settled down by a chain for chosen, have till three σ-bonds (with the neighbors hexagons) and one π-bond, therefore adsorption energy for such positions is much less than at adsorption process on boundary atom C.

In site 1 (when hydrogen atom adsorbing on boundary carbon atom) the C-H bond has appeared equal $r_{c-H}$ =1.13 Å, that is smaller than at adsorption other hydrogen atoms on non-boundary C atoms (tab. 1). It is possible to explain it by formation of σ-bond between H and boundary C and by freedom of a choice of optimum bond angle. For other atoms such freedom are not exist.

This consecutive adsorption also causes SWNT surface deformation. Alternating through one (in a chain) carbon atoms indicate a different behavior: one rise above a surface, others - leave inside the tube (displacement concerning initial situations ~ 0.25 Å). As result the internal and external active center formations take place. On the other hand H atoms can join to both an external tube surface (in our case) and to internal these.

If only external adsorption occurs then, how it follows from the table 1, energetically favorable situation is when the H atoms join the C atoms raised concerning initial surface positions.

## 4. Atomic adsorption on (10,0) SWNT surface

In the work we have also considered atom hydrogen adsorption on (10,0) SWNT (d = 7.9 Å). As well as in (6,6) SWNT case, the step-by-step approach of H atoms to SWNT surface has allowed to construct a potential energy curve of H-SWNT system, which is shown on fig. 5.

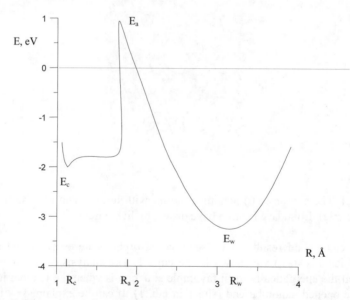

**Fig. 5.** The curve of potential energy of hydrogen atom interaction with (10, 0) SWNT surface as a function of the distance from hydrogen to the tube surface. The sum of reactant energies is determined as 0 eV.

As it is well visible on the curve the mechanism of H binding to (10,0) SWNT qualitatively differs from similar process for a case of (6,6) SWNT. The difference deals with, first, the presence of a energy minimum on the distance R (C-H) = 3.2 Å (in the curve it corresponds to a point $R_W$). The energy minimum can be explained by Wan-der-Waal's interaction of H atom with the tube. In a point $R_W$ carbon atom rises above a tube surface on 0.12 Å. Second, the SWNT surface deformation proceeds completely distinctly from (6, 6) SWNT case. Namely, the attacked surface center rises above a surface on insignificant height in the beginning; the neighbor carbon atoms leave inside the tube insignificantly too. Then in a vicinity of a point $R_a$ there is a sharp jump of carbon atom out from a surface on 0.5 Å. Thus it pulls behind itself the neighbor carbon atoms, and they rise above a tube surface on distance $\delta R \sim 0.1$ Å too. The chemical C-H bond is formed then bond length remains practically constant, and all group C-H comes back in the tube and occupies a situation appropriate to an energy minimum. At minimum the carbon atom output from the tube on $\sim 0.33$ Å. Such behavior is well visible on the energy curve: a flat area on the curve in a vicinity of a point $R > R_c$ (fig. 5).

In order to form the chemical C-H bond it is necessary hydrogen atom to has overcame a potential barrier of height $E_a = 1.0$ eV by over barrier or tunnel way. Using the formula (2) it is easy to calculate a share of H atoms "jumped" an energy barrier $\alpha \sim 10^{-5}$. Reaction rate $V_s$ of classical binding of atomic hydrogen is $\sim 10^{-5}$ c$^{-1}$ cм$^{-2}$. It is obviously that against (6,6) SWNT the adsorption reaction in a considered case can realize more effectively, but the characteristic time of (10,0) SWNT surface saturation by atomic hydrogen will be of the order $10 \div 15$ hours.

For tunnel way the probability of transition w is $\sim 10^{-3}$ c$^{-1}$ and a share of atoms past a barrier w$\beta$ is $\sim 10^{-27}$ c$^{-1}$. However for an efficiency of process it is necessary to create conditions for existence of 1000 moles of atomic hydrogen.

Consecutive adsorption of atomic hydrogen on (10,0) SWNT surface has been carried out also (fig. 6).

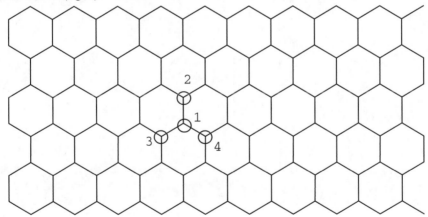

**Fig. 6.** The carbon (10, 0) nanotube cluster with adsorbed hydrogen atom (numbered circle). Boundary H atom have drown by light circle.

At the presence of single H atom, marked 1 in fig. 6, second atom of hydrogen binds, then third H atom was located near to two these and, at last, fourth. Adsorption energy of each bonded H atom, adsorption bond lengths and the distances between the active center and nanotube surface are represented in the table 1 also.

The table results indicate that a connection of two H atoms on the C centers is most energetically favorable. Then at SWNT surface saturation by hydrogen the energy benefit decreases. The bond lengths in each case appear approximately identical. The distances, on which the active center outputs from a surface, in each case have appeared comparable, but the tendency of its increase is observed during a saturation of a surface.

## 5. Atomic hydrogen adsorption inside nanotube

The quantum chemical MNDO calculations within the molecular cluster model of (6, 6) SWNT have been carried out to study the energy and geometry properties of hydrogen atoms adsorbing processes on internal surface. The extended unit cell (EUC) of the cluster consisted of 96 carbon atoms (cluster $C_{96}$) and contained four layers of carbon hexagons in each. The cluster boundary chemical bonds have been saturated by hydrogen atoms. First we have studied the single atom adsorption on the internal site of nanotube surface. The hydrogen attack of (6,6) surface have been simulated by step-by-step approach of H to the carbon atom along a perpendicular to the inside tube surface through carbon atom. Thus the H atom (adatom) has got two degrees of freedom and can freely deviate the perpendicular in two mutually orthogonal directions. While calculating the surface center C has got the degrees of freedom allowing it to output from the tube surface. The surface C atom has been selected in middle of the cluster in order to exclude the influence of the boundary effects.

The dependence of adsorption complex energy as a function of the distance from hydrogen to the tube surface has been constructed and shown in fig. 7.

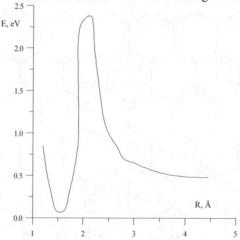

**Fig. 7.** The potential energy curve of the hydrogen adsorption process as a function of the distance from hydrogen to the tube surface. The sum of reactants energy is determined as 0 eV.

The C-H bond length has been found approximately 1.14 Å and corresponds to the energy minimum on the curve (fig. 7). The curve has indicated the adsorption process to be endothermic and stable. The adsorption energy ($E_{ad}$) calculated as the difference between the sum of reactant energy and the energy of the product has appeared equal to 0.15 eV. The activation energy calculated as the energy maximum on the curve is 2.3 eV. Besides that the adsorption has led to the deformation of SWNT surface: the C-C bond increases in the nanotube from 1.44 Å to 1.5 Å after H binding. The surface carbon atom has been displaced inside of the tube up to 0.37 Å. In fig. 7 we have determined the sum of reactant energy as 0.0 eV. Indeed the energy curve represents the potential energy of hydrogen interaction with SWNT.

The energy curve has allowed us to realize the single hydrogenation mechanism of SWNT inside. In the beginning of H attacking when hydrogen is on the tube axis the carbon is displaced outside. The potential energy curve is of quite smooth character. Then C atom starts moving inside of the tube up to 0.5 Å from the surface, the curve increasing to the maximum. At the maximal point the C-H bond of 1.2 Å length is forming. After great overlapping of the atomic orbits the C-H pair moves keeping the bond to the surface up to the point of minimal energy corresponding to the stable state of the system.

At the same time there is a second stable H station inside the tube. It concerns to the H atom location near the tube axis and may be called the physical adsorption of the atom. In fig. 7 it is shown as the energy minimum of about 4.5 Å. The energy of the physical adsorption is 0.48 eV.

We have also investigated the possibility of the multiple adsorption of H atom on the nearest to each other atoms of carbon inside of nanotube. The calculations of various variants of consecutive and simultaneous connections of hydrogen atoms to 2 - 8 carbon atoms have been carried out (table 2). The analysis of the adsorption energies has found out the energy advantage of the consecutive variant. So we consider the mechanism of hydrogen atomic adsorption is as follows: H atom connects to the nanotube surface and deforms it in such a manner that the new active center is being prepared for the successful adsorption of the following hydrogen atom.

**Table 2.** The adsorption energy ($E_{ad}$) per H atom of the simultaneous and consecutive adsorption of hydrogen in the inside (6, 6)-SWNT: 1) the single H atom adsorb, 2) the simultaneous adsorb of 2 H atoms, 3) the consecutive adsorb of 2 H atoms and 4) the consecutive adsorb of 8 H atoms.

| Variants | 1 | 2 | 3 | 4 |
|---|---|---|---|---|
| $E_{ad}$, eV/H | 0.15 | 4.49 | 5.51 | 15.57 |

The investigation of electron structure of adsorption complexes has also shown that in all cases of hydrogen adsorption the electron density transfer has taken place from adatoms to the surface. It has been obtained that the charge perturbation of tube surface induced by the inside adsorption damps slowly than the perturbation induced by the outside one. So, the outside adsorption of H atom leads to the charge redistribution up to the first sphere of the neighboring carbon atoms along the tube perimeter and up to the third sphere along the tube axis. While the inside adsorb process causes the surface charge disturb up to

254

the second sphere along the perimeter and up to the fourth sphere along the axis (table 3). Obviously, it is linked with the character of nanotube surface curvature: inside the tube the adatom is placed closer to the other surface carbon atoms than outside.

**Table 3.** The distribution of charges initiated by adatoms on (6,6)-SWNT surface up to five (1-5) spheres of interaction: a) along the axis, b) on the tube perimeter, I) the outside adsorption [5], II) the inside adsorption.

| Spheres of interaction | | | 1 | 2 | 3 | 4 | 5 |
|---|---|---|---|---|---|---|---|
| Variants | a | I | -0.10 | 0.05 | 0.01 | 0.00 | 0.00 |
| | | II | -0.11 | 0.08 | 0.04 | 0.01 | 0.00 |
| | b | I | -0.10 | 0.01 | 0.00 | 0.00 | 0.00 |
| | | II | -0.09 | 0.01 | 0.00 | 0.00 | 0.00 |

## 6. Regular external and internal hydrogenation

As mentioned above we have considered the outside hydrogenation of (6, 6) SWNT in our early articles [7 - 10]. In those papers the most probable variants of hydrogen atoms bindings to the external tube surface have been determined. The results have led to the version that hydrogen atoms have to connect to both external and internal surfaces simultaneously.

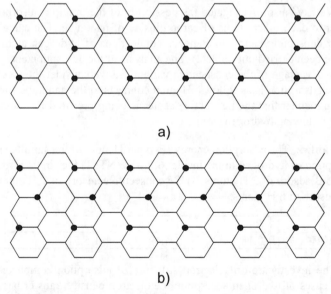

a)

b)

**Fig. 8.** The (6, 6)-SWNT EUC with the sites of hydrogen atoms on the surface: a) H atoms form the rectangular hydrogen superlattice, b) H atoms form the rhombic hydrogen superlattice.

That is why we have studied the case of the internal hydrogenation when the adatoms saturate the inside tube surface periodically. We have considered 2 variants of hydrogen atoms binding to the nanotube surface: 1) H atoms form the rectangular hydrogen superlattice (fig. 8a); 2) H atoms form the rhombic hydrogen superlattice (fig 8b). The IECCC method [26] within the framework of semi-empirical quantum chemical scheme MNDO is used to calculate the energy and geometry properties of (6,6) SWNT hydride. The analysis results in the second variant which is energetically more advantageous: the adsorption energy for 1 variant is 4.3 eV/H but for second – 5.1 eV/H. The optimized C-H bond length has appeared equal to 1.6 Å. This value is longer then that for the case of the single adsorption (1.4 Å). It can be explained by the mutual influence of H adatoms to each other and the surface curvature forces the atoms to stay at greater distances from it.

Besides it has been obtained from the analysis of the IECCC band gap of two SWNT hydrides variants that the studying compounds are keeping the metallic conductivity. It is clear now that in spite of the surface deformation the hydrogen subsystem has given the additional electrons to nanotubes, and to our mind the outside $\pi$-electron system forms the conductive channels due to the overlapping of the orbital wave functions.

At last we have carried out the IECCC calculation of the simultaneous inside and outside hydrogenation of the tube. The rhombic hydrogen superlattice has been used both inside and outside adsorptions as the most energetically advantageous variant. The analysis of the IECCC band gap has shown this material is a narrow gap (about 0.01 eV) semiconductor. The inside and outside hydrogen superlattices are explained to have broken the $\pi$-bonds and the orbital overlapping.

## 7. Conclusions

The atomic hydrogen chemisorptions processes on carbon SWNT surface of two structural types ("arm-chair" and "zig-zag") closed on a diameter (8.1 Å and 7.9 Å accordingly) have been investigated. On the basis of quantum-chemical calculations the curves of potential energy of hydrogen atom interaction with SWNTs have been constructed. The reaction mechanisms of H atom binding to (6,6) and (10,0) SWNT have been investigated. It has appeared that there is a basic difference in mechanisms of these reactions, which consists available Wan-der-Waal's energy minimum for (10,0) - H interaction. That is there is an opportunity of atomic hydrogen physical adsorption on the tube surface. Besides that the height of a potential barrier (energy of activation) appears in 2 times less than for (6,6) - H interaction. A consequence of last fact is more effective quantum yield of hydrogen tunnel carrying reaction.

The basic difference of chemisorptions mechanisms, on our sight, correlates with a distinction of physical properties nanotubes: (6,6) SWNT shows metal conducting properties and (10,0) - semi-conductor. For confirmation of the revealed correlation it is necessary to carry out a lot of calculations of both structural type tubes having various conducting properties, that is a subject of the further research.

Thus, the carried out researches have found out the following. First, external adsorption results in deformation of SWNT initial surface and occurrence, besides the external, internal active center, providing an opportunity of simultaneous surface and volumetric adsorption. And it provides even more active use of SWNT as effective adsorbents both atoms and molecules. Secondly, the adsorption rate of atomic hydrogen appears concerning small that testifies faster to a prevalence of molecular adsorption above atomic process.

Thus the energy and geometry properties of (6,6)-SWNT hydrides have been investigated with the use of the molecular cluster models of solids within the framework of semi-empirical quantum chemical scheme MNDO. The energy curve of the hydrogen adsorption process as a function of the distance from hydrogen to the tube surface has been calculated and analyzed. The curve has shown the possibility of atomic hydrogen stable stations inside the tube. The study of the multiple adsorption of atomic hydrogen has indicated the energy advantage of the consecutive process. The mechanism of the single inside hydrogenation of SWNT has been studied circumstantially.

The attractive results of the quantum chemical calculations of carbon nanotube hydrogenation have been presented in [11 - 20]. There are some differences of the results from ours because the authors of the papers studied either the carbon nanotube molecular hydrogenation or used ab initio method to calculate electronic properties. For instance, ab initio calculations of outside hydrogen adsorption [11] have given the C-H bond energy of about 1.2 eV and the experimental ones about 1.0 eV (see reference in [11]).

The IECCC method has been used to calculate the energy and geometry properties of (6,6) SWNT periodic hydride. Some general properties of the inside and outside hydrides have been found. First, the H adsorption distances in both cases are quite near: 1.4 Å is for the outside case and 1.6 Å for the inside one. Second, (6,6) SWNT hydride prefers to form the rhombic hydrogen superlattice. Third, both the outside and inside hydrides keep the metallic conductivity properties. In the opposite the calculation of the simultaneous inside and outside hydrogenation of (6,6) SWNT has shown the transformation of the band structure of materials: it becomes a narrow gap semiconductor.

## Acknowledgements

We thank Prof. A.O. Litinski (Volgograd State Technical University) for fruitful discussion of the results and important notices and recommendations. This work was supported in part by Russian Basic Research Fund (project 02-03-81008) and INTAS (project 00-237).

## 8. References

1. Dresselhaus M.S., Dresselhaus G., Eklund P.C. Science of Fullerenes and Carbon Nanatubes, Academic Press, 1996, 965 P.
2. Ivanovoskii A.L. Kvantovaja khimia in materialovedenii. Nanotubularnye formy veshestva. Ekaterinburg: UrORAN, 1999, 176 P.
3. Saito R., Dresselhaus M.S., Dresselhaus G. Physical properties of carbon nanotubes, Im perial College Press, 1999, 251 p.
4. Dillon A.C., Jones K.M., Beccedahl T.A., Kiang C.H., Bethune D.S., He-ben M.G. Storage of hydrogen in single-walled carbon nanotubes. Nature 1997; 386(27): 377 -379.
5. Chambers A., Park C., Baker R.T.K., Rodrigues N.M. Hydrogen storage in graphite nanofibers. J. Phys. Chem. B 1998; 102(22): 4253 - 4256.

6. Park C., Anderson P.E., Chambers A., Tan C.D., Hidalgo R., Rodrigues N.M. Further studies of the interaction of hydrogen with graphite nanofibers. J. Phys. Chem. B 1999; 105: 10572 – 10581.
7. Zaporotskova I.V., Litinskii A.O., Chernozatonskii L.A. Adsorbcia atomov H, O, C I Cl na poverhnosti odnosloinih yglerodnih tubulenov. Vestnik VolSU: Seria Mathematika. Physika 1997; 2: 96 – 99.
8. Zaporotskova I.V., Litinskii A.O., Chernozatonskii L.A. Characteristics features of the sorption of light atoms on the surface of a single-layer carbon tubelene. JETP Lett. 1997; 66(12): 841 - 846.
9. Chernozatonsky L.A., Lebedev N.G., Zaporotskova I.V., Litinskii A.O., Gal'pern E.G., Stankevich I.V., Chistyakov A.L. Carbon single-walled nanotubes as adsorbents of light (H, O, C, Cl) and metal (Li, Na) atoms. In book: Adsorption Science and Technology, Brisbane: World Scientific, Australia, 14-18 May 2000, p. 125-129.
10. Zaporotskova I.V., Lebedev N.G., Chernozatonski L.A. Some features of hydrogeniza tion of single-walled carbon nanotubes. In book: Abstracts of invited lectures and con tributed papers "Fullerenes and Atomic Clusters", St.Peterburg: FIZINTEL, Russia, 2-6 July 2001, p. 325.
11. Froudakis G.E. Hydrogen interaction with single-walled carbon nanotubes: a combined quantum-mechanics/ molecular-mecanics study. Nano Lett. 2001; 1(4): 179 - 182.
12. Tada K., Furuya S., Watanabe K. Ab initio study of hydrogen adsorption to single-walled carbon nanotubes. Phys. Rev. B 2001; 63: 155405-7.
13. Yang F.H., Yang R.T. Ab initio molecular orbital study of adsorption of atomic hydro gen on graphite: insight into hydrogen storage in carbon nanotubes. Carbon 2002; 40: 437 – 444.
14. Bauschlicher Jr C.W. Hight coverages of hydrogen on a (10,0) carbon nanotubes. Nano Lett. 2001; 1(5): 223 - 226.
15. Andriotis A.N., Menon M., Srivastava D., Froudakis G. Extreme hydrogen sensitivity of the transport properties of single-wall carbon nanotube capsules. Phys. Rev. B 2001; 64: 193401-4.
16. Yildrim T., Guelseren O., Ciraci S. Exohydrogenated single-wall carbon nanotubes. Phys. Rev. B 2001; 64: 075404-5.
17. Chen P., Wu X., Lin J., Tan K.L. High $H_2$ uptake by alkali-doped carbon nanotubes under ambient pressure and moderate temperature. Science 1999; 285(2): 91-93.
18. Liu C., Fan Y.Y., Liu M., Cong H.T., Cheng H.M., Dresselhause M.G. Hydrogen stor age in single-walled carbon nanotubes at room temperature. Science 1999; 286(5): 1127 - 1129.
19. Ma Y., Xia Y., Zhao M., Mei L. Effective hydrogen storage in single-wall carbon nano tubes. Phys. Rev. B 2001; 63: 115422-6.
20. Chan S. ae al. Chemisorption of hydrogen molecules on carbon nanotubes under high pressure. Phys. Rev. Lett. 2001; 87(20): 205502-4.
21. Lebedev N.G., Zaporotskova I.V., Chernozatonskii L.A. Single and regular hydrogena tion and oxidation of carbon nanotubes: MNDO calculations. International Journal of Quantum Chemistry 2003; 95(3) (accepted).

22. Lebedev N.G., Zaporotskova I.V., Chernozatonskii L.A. Hiral effects of single wall carbon nanotube fluorination and hydrogenation. In book: Abstracts of invited lectures and contributed papers "Fullerenes and Atomic Clusters", St.Peterburg: FIZINTEL, Russia, 30 June - 4 July 2003, p. 251.

23. Evarestov R.A. Klasternoe priblijenie v teorii tochechnih defektov v tverdih telah. Zur nal strukt. Khimii 1983; 24(4): 44 - 61.

24. Dewar M.J.S., Thiel W. Ground states of molecules. 38. The MNDO method. Approximations and Parameters. J. Am. Chem. Soc. 1977; 99: 4899-4906.

25. Dewar M.J.S., Thiel W. A semiempirical model for the two-center repulsion integrals in the NDDO approximation. Theor. Chim. Acta 1977; 46: 89-104.

26. Litinskii A.O., Lebedev N.G., Zaporotskova I.V. A model of an ion-incorporated cova lent cyclic cluster in MNDO calculations of intermolecular interactions in heterogene ous systems. Russian J. Phys. Chem. 1995; 69(1): 175 - 178.

27. Emanuel N.M., Knorre D.G. Kurs himicheskoi kinetiki. Moscow: Visshaya shkola, 1984, 463 p.

28. Stromberg A.G., Semchenko D.P. Fisicheskaya himia. Moscow: Visshaya shkola, 1988, 496 p.

29. Landau L.D., Lifshic E.M. The quantum mechanics. Moscow: Nauka, 1974, 752 p.

# QUANTUM CHEMICAL INVESTIGATIONS OF THE GROWTH MODELS OF SINGLE WALL CARBON NANOTUBES ON POLYHEN RINGS, FULLERENES AND DIAMOND SURFACE

LEBEDEV N.G.[1], ZAPOROTSKOVA I.V.[1], CHERNOZATONSKII L.A.[2]

[1] *Volgograd State University, ul. 2-ya Prodolnaya, 30, Volgograg, 400062, Russia*
*E-mail: nikolay.lebedev@volsu.ru*
[2] *Emanuel Institute of Biochemical Physics, Russian Academy of Sciences, ul. Kosygina 4,*
*Moscow, 117334 Russia, E-mail: cherno@sky.chph.ras.ru*

## Abstract

Quantum-mechanical MNDO calculations were performed for models of the growth of achiral carbon nanotubes from polyene rings of the *cis* and *trans* types, fullerene and diamond (111) surface. Several variants of the growth of nanotubes (by absorption of dimers or trimers or a mixed set of species) were considered. A comparison of the obtained characteristics, in particular, the total energies of sequential absorption, allowed several conclusions to be drawn on the possibility of growth of cyclic nanotubes of the (n, n) and (n, 0) types from polyene rings by absorption of carbon dimers as the most probable process. The orbital-stoihiometric cluster (OSC) model in the framework of quantum chemical MNDO-scheme has been applied to simulate the diamond nanocluster in order to study carbon nanotube generation process on diamond (111) surface. The comparison of the received characteristics of processes, in particular adsorption energy, has shown, that most favourable process is carbon monomer sorption on a pure diamond (111) surface. Processes of nanotube origin on surface quantum dots, which have been simulated by adsorbed Li, Na, K, Be, and H atoms, have shown high efficiency of nanotube growth: all of them proceed without energy barriers.

**Keywords**: A, Carbon nanotubes, Fullerene; D, Absorption, Activation energy, Adsorption properties, Electronic structure.

## 1. Introduction

Starting with 1991, nanotubular carbon forms, that is, closed surface structures with special properties that allow them to be used as interesting peculiar physical and chemical systems, have been the object of extensive theoretical and applied studies in view of prospects for their wide use in many areas as diverse devices and construction elements for nanotechnologies of the nearest future; the list of such areas has constantly been growing [1, 2].

One of the most important problems of the theory that describes the properties of nanotubular systems is the development of correct models of their formation and growth.

259

*T.N. Veziroglu et al. (eds.),*
*Hydrogen Materials Science and Chemistry of Carbon Nanomaterials,* 259-278.

The mechanisms of nucleation and formation of cylindrical nanotubes (NT) are exceedingly interesting and have been discussed in many works (see [3, 4] and reviews [1, 2]). The conclusions drawn in most of these works, however, refer to the kinetics of growth of structures according to the selected procedure and preparation conditions and cannot therefore be considered universal.

A large number of factors influencing the formation of nanotubes and the diversity of the resulting tubular structures has led to the appearance of many models describing the mechanism of growth of NTs. All the suggested models use a certain characteristic of the resulting nanotubular structure. For instance, they can be designed to construct a definite (single- or multilayer) structure, take into calculation a certain nanotube nucleus type (fullerene hemisphere, polyene ring, nanoislet on the surface of a crystallite-base, catalyst particle, etc.), reproduce the selected growth process (at the open nanotube end or by incorporating atoms into a tube wall, growth of an open or closed NT), describe the mechanism of nanotube growth termination, etc. Several models are considered in much detail in reviews [1, 2].

Under above mentioned the nature of generation and mechanism of growing of carbon nanotubes steels by the subject of debates. This work is denoted to quantum-chemical study of the energy features of the processes of generation and further growing of nanotubes on polyene rings of the *cis* and *trans* types, fullerene and (111) diamond surfaces. The similar possibility of NT formation was offered earlier in works [2, 3] and the calculation results were partially submitted in [5 - 7].

For modelling the geometry and electronic structure of (111) diamond surfaces the orbital-stoihiometric cluster (OSC) method [8] modified for calculations of bulk and surface fragments of solids within the framework of the quantum chemical semi-empirical MNDO [9, 10] scheme has been used. The basic features of the method conclude in the following. The solid fragment (cluster) gets out so that atoms forming chemical bonds of located character to be boundary. The boundary atoms bring in cluster orbital basis only located hybrid orbits (LHO) directed inside chosen cluster. Thus orbital, electronic and nuclear structure of the cluster corresponds to the crystal unit formula. OSC cluster assumes the most correct boundary conditions for a diamond crystal, in which the chemical bonds have well expressed $sp^3$-hybrid character.

## 2. The SWNT growth mechanism models on polyhene rings

A mechanism of the growth of single-layer cylindrical carbon nanotubes on polyene rings observed in laser ablation of graphite was suggested in [2]. Ring structures were most characteristic of $C_n$ particles at $10 < n < 40$, and spheroidal structures, at $n > 40$ [1, 2]. At the first stage, in the presence of a catalyst (for instance, cobalt carbide $Co_mC_n$) in plasma experiences a planar carbon ring distortions of the *trans* or *cis* type. At later stages, the addition (edge absorption) of carbon atoms (C), dimers ($C_2$), or trimers ($C_3$) generates diversified cylindrical nanostructures.

It can be suggested that distorted polyene rings twisted in two ways are formed during graphite ablation. The first variant (cylindrical) suggests that the *cis* and *trans* ring forms are fractured along the vertical plane passing through the nearest nodes occupied by carbon atoms. In the second variant (prismatic), the line of fracture passes through the atoms that are involved in bonding between two neighboring carbon hexagons supposed to participate in subsequent growth.

## 2.1. THE GROWTH OF (6, 0) SWNT

We have performed MNDO calculations [5] of the electronic structure and energy characteristics of (6, 0) SWNT growth on polyene rings of the *trans* type twisted as described above (variants 1 and 2). Initially, the carbon ring comprised 12 carbon atoms. The system was hermetically closed around the circumference. All the parameters of the *trans* ring, which was a precursor for the growth of a nanotube, were optimized. Interatomic distances were found to be 1.4 Å.

Several variants of forming the first two layers of hexagons of a (6, 0) nanotube growing by edge absorption were studied:
(1) a sequential absorption of eight carbon trimers ($nC_3$, $n = 8$) on the polyene ring (Fig. 1a);
(2) a sequential absorption of 12 carbon dimers ($nC_2$, $n = 12$) on the polyene ring (Fig. 1b);
(3) a sequential absorption of 24 C atoms on the polyene ring;
(4) a sequential absorption of two times the "trimer + 4 dimers + 1 carbon atom" ($C_3 + 4C_2 + C$) mixed set of species necessary for completing each hexagon layer on the polyene ring (Fig. 1c).

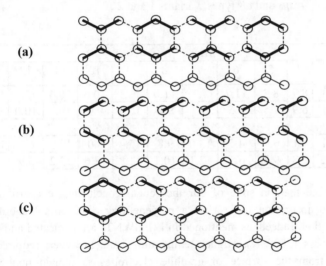

**(a)**

**(b)**

**(c)**

**Fig. 1.** Absorption of (a) 8 $C_3$, (b) 12 $C_2$ and (c) two times $C_3 + 4C_2 + C$ set of species on a *trans* polyene ring.

The energies of absorption calculated sequentially (step-by-step as the nanotube grows) are listed in Tables 1 - 4 for all these variants. A comparison of the total energies for each absorption variant shows that, for the (6, 0) SWNT growth from a *trans* polyene ring, sequential addition of, first, 24 C atoms and, next, 12 $C_2$ dimers is more favorable energetically than the absorption of eight $C_3$ trimers or two times the $C_3 + 4C_2 + C$ set of species. This result is quite realistic, explicable, and expected. Naturally, the formation of atomic carbon or carbon dimers in graphite vaporization is more probable than the formation of more complex structures, and these simpler structures themselves are stabler.

**TABLE 1.** Energies (eV) of sequential absorption of $nC_3$ on polyene rings of the *trans* and *cis* types, variants 1 and 2

| N | 1 | 2 | 3 | 4 | 5 | 6 | 7 | 8 | $E_{ab}^{tot}$ |
|---|---|---|---|---|---|---|---|---|---|
| | | | | | *trans* | | | | |
| 1 | 3.1 | 5.9 | 7.6 | 8.7 | 7.0 | 9.5 | 10.8 | 11.5 | 64.1 |
| 2 | 2.6 | 3.2 | 5.7 | 5.8 | 4.8 | 10.3 | 7.6 | 8.8 | 48.8 |
| | | | | | *cis* | | | | |
| 1 | 5.9 | 6.2 | 7.9 | 6.2 | 10.9 | 6.3 | 5.9 | 5.9 | 55.0 |
| 2 | 8.6 | 9.9 | 5.9 | 5.4 | 7.5 | 12.1 | 5.4 | 13.0 | 67.7 |

**TABLE 2.** Energies (eV) of sequential absorption of $nC_2$ on polyene rings of the *trans* and *cis* types, variants 1 and 2

| n | 1 | 2 | 3 | 4 | 5 | 6 | 7 | 8 | 9 | 10 | 11 | 12 | $E_{ab}^{tot}$ |
|---|---|---|---|---|---|---|---|---|---|---|---|---|---|
| | | | | | | *trans* | | | | | | | |
| 1 | 4.7 | 8.5 | 8.3 | 8.7 | 8.3 | 11.7 | 7.1 | 11.0 | 11.1 | 10.6 | 10.4 | 13.3 | 113.5 |
| 2 | 3.3 | 8.3 | 6.5 | 6.9 | 7.5 | 9.5 | 6.7 | 9.6 | 10.3 | 9.0 | 10.7 | 10.0 | 98.4 |
| | | | | | | *cis* | | | | | | | |
| 1 | 9.0 | 5.8 | 7.4 | 7.1 | 6.9 | 4.5 | 13.0 | 9.8 | 15.4 | 6.0 | 10.2 | 6.7 | 101.7 |
| 2 | 12.0 | 5.6 | 11.2 | 5.8 | 9.7 | 6.1 | 12.1 | 10.1 | 11.8 | 12.6 | 13.2 | 3.9 | 114.1 |

Nevertheless, note that, in spite of our theoretical conclusions concerning the favorableness of atomic absorption (24 C) on a polyene ring, this variant should be excluded from consideration. Indeed, as mentioned in [2], SWNTs are nucleated and grow at fairly high temperatures (3000 K). Higher temperatures are, however, required for carbon that vaporizes from the surface of graphite electrodes to remain nonbonded (atomic). Such temperatures would destroy the yet unstable structure of a growing nanotube. As a result, SWNT would be constantly destroyed under the action of such high temperatures, and the process of such a growth would be fairly improbable.

The next important result of a comparison of the total energies of absorption of carbon compounds on a polyene ring of the *trans* type leads us to conclude that tube building on rings of variant 1 is more favorable energetically. This should result in the formation of cylindrical SWNTs.

**TABLE 3.** Energies (eV) of sequential absorption of nC on polyene rings of the *trans* and *cis* types, variants 1 and 2

| | | | | | | *trans* | | | | | | |
|---|---|---|---|---|---|---|---|---|---|---|---|---|
| n | 1 | 2 | 3 | 4 | 5 | 6 | 7 | 8 | 9 | 10 | 11 | 12 |
| 1 | 5.5 | 5.8 | 7.9 | 7.2 | 8.5 | 6.4 | 8.4 | 6.8 | 8.5 | 6.5 | 8.0 | 10.3 |
| 2 | 4.8 | 4.3 | 9.6 | 6.1 | 6.8 | 6.3 | 8.1 | 5.4 | 8.5 | 5.6 | 7.5 | 8.6 |

| 13 | 14 | 15 | 16 | 17 | 18 | 19 | 20 | 21 | 22 | 23 | 24 | $E_{ab}^{tot}$ |
|---|---|---|---|---|---|---|---|---|---|---|---|---|
| 7.6 | 5.9 | 6.4 | 11.1 | 10.2 | 7.5 | 9.9 | 7.3 | 9.7 | 7.1 | 10.0 | 10.5 | 192.9 |
| 8.0 | 5.5 | 9.5 | 7.8 | 7.3 | 9.3 | 8.7 | 7.5 | 9.8 | 7.1 | 5.9 | 9.7 | 177.7 |

| | | | | | | *cis* | | | | | | |
|---|---|---|---|---|---|---|---|---|---|---|---|---|
| n | 1 | 2 | 3 | 4 | 5 | 6 | 7 | 8 | 9 | 10 | 11 | 12 |
| 1 | 4.7 | 11.9 | 3.9 | 6.1 | 6.0 | 9.6 | 5.2 | 8.0 | 5.7 | 8.5 | 3.7 | 8.7 |
| 2 | 6.5 | 9.4 | 6.0 | 9.5 | 5.9 | 1.9 | 8.9 | 22.6 | 3.1 | 19.8 | 33.5 | 0.5 |

| 13 | 14 | 15 | 16 | 17 | 18 | 19 | 20 | 21 | 22 | 23 | 24 | $E_{ab}^{tot}$ |
|---|---|---|---|---|---|---|---|---|---|---|---|---|
| 9.0 | 5.8 | 7.4 | 7.1 | 6.9 | 4.5 | 13.0 | 9.8 | 15.4 | 6.0 | 10.2 | 6.7 | 153.6 |
| 9.2 | 8.2 | 7.7 | 9.4 | 7.7 | 17.9 | 35.1 | 6.8 | 9.9 | 8.6 | 7.5 | 7.2 | 168.8 |

**TABLE 4.** Energies (eV) of sequential addition of $2(C_3 + 4C_2 + C)$ to polyene rings of the *trans* and *cis* types, variants 1 and 2 (n = 1, the first set, and n = 2, the second set)

| N | 1 | 2 | $E_{ab}^{tot}$ |
|---|---|---|---|
| | | *trans* | |
| Variant –1 | 47.2 | 44.1 | 91.3 |
| Variant – 2 | 39.2 | 40.1 | 79.2 |
| | | *cis* | |
| Variant –1 | 55.1 | 54.3 | 109.5 |
| Variant – 2 | 46.6 | 64.6 | 111.1 |

In analyzing the calculation results, we studied the distribution of charges on the carbon atoms of the base and growing structures. The introduction of "extra" (adsorbed on the ring) carbon atoms was found to somewhat perturb charges. This perturbation influenced the effectiveness of growth of nanotubes and damped as the hexagon layer was completed. It follows from Tables 1--4 that the first added dimer or trimer has the lowest binding energy. The energy of binding subsequent carbon clusters is significantly higher than the energy of the initial addition. This is evidence that the polarization of the polyene ring by the primary cluster increases the effectiveness of sorption.

264

## 2.2. THE GROWTH OF (6, 6) SWNT

The second problem was to study the growth of a single-layer carbon tubulene of the (6, 6) type from a polyene *cis* ring comprising 24 carbon atoms. As with the *trans* ring, the parameters of geometrically closed systems twisted as in variants 1 and 2, which were precursors for the growth of prismatic and cylindrical nanotubes, were optimized. The interatomic distances $r_{C-C}$ remained equal to 1.4 Å.

MNDO calculations of molecular models were performed for several variants of the growth of single-layer armchair-type tubulenes. Sequential absorption of carbon compounds on *cis* rings of types 1 and 2 with the formation of the first two carbon hexagon layers of the (6, 6) nanotube was considered. The following combinations were studied:
(1) a sequential absorption of eight $C_3$ trimers on the polyene *cis* ring (Fig. 2a);
(2) a sequential absorption of 24 $C_2$ dimers on the polyene ring (Fig. 2b);
(3) a sequential absorption of 24 C atoms on the polyene ring;
(4) a sequential absorption of two times the $C_3 + 4C_2 + C$ mixed set of species necessary for the completion of each hexagon layer on the polyene ring (Fig. 2c).

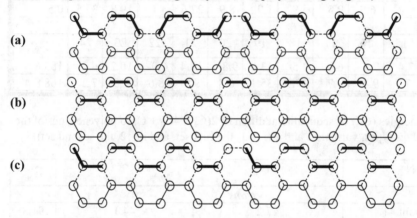

(a)

(b)

(c)

**Fig. 2.** Absorption of (a) $8C_3$, (b) $12C_2$, and (c) two times the $C_3 + 4C_2 + C$ mixed set of species on a *cis* polyene ring.

The results of absorption energy calculations are listed in Tables 1-4. A comparison of the total energies of absorption has led to the conclusions similar to those obtained above. As with (6, 0) nanotubes on a polyene *trans* ring, processes energetically favorable for the growth of (6, 6) tubulenes are the addition of, first, atomic carbon, secondly, $C_2$ dimers, and, thirdly, $C_3$ trimers. The reasoning concerning atomic absorption (see above) remains valid for (6, 6) nanotubes. Variant (3) can therefore be excluded from consideration, and preference should be given to the model with sequential edge absorption of $C_2$ carbon dimers, which are a more probable and stabler configuration, on a polyene *cis* ring.

In addition, an analysis and comparison of the $E_{ab}^{tot}$ energies (Tables 1-4) leads us to conclude that building the tube on a *cis*-distorted ring of variant 2 is more favorable energetically. As a result, a prismatic tubulene should grow. In this respect, (6, 6) tubes differ from (6, 0) ones, the precursor for the growth of which is a *trans*-distorted polyene ring. For (6, 0) tubulenes, the growth of cylindrical structures is more favorable energetically. The difference of the total energies is, however, small, and it can be suggested that nanotubes of both twisting variants should grow. The charges on the carbon atoms of the *cis* ring and the growing (6, 6) tubulene exhibit a behavior similar to that characteristic of the *trans* ring and the growing (6, 0) tubulene.

To summarize, the conclusion can be drawn that sequential absorption of $C_2$ dimers, which are the stablest and simplest carbon structure, is energetically favorable for the growth of both (6, 0) and (6, 6) nanotubes.

## 2.3. THE INFLUENCE OF THE SUBSTRATE HEIGHT

It follows from the preceding reasoning that the absorption of precisely carbon dimers is responsible for axial cylindrical growth of (n, n) and (n, 0) (in this work, n = 6) SWNTs. For this reason, we will use precisely this variant for considering the influence of the base (precursor) height on the energy characteristics of growth of (n, n) and (n, 0) nanotubes.

In conformity with the earlier conclusion that the growth of a prismatic armchair-type SWNT on a *cis*-distorted polyene ring is energetically favorable, we selected precisely this nanotube structure as the object of study. We considered sequential absorption of $C_2$ dimers on a ring containing 24 carbon atoms, which resulted in the formation of the first two hexagon layers (Fig. 2b). The total energy of absorption of 12 dimers was found to be 114.12 eV. Next, a "planar" (6, 0) nanotube containing 96 carbon atoms was selected as a base for growth. This nanotube comprised four hexagon layers (Fig. 2c), and its height was $H_{(6, 6)} \sim 9.7$ Å. Twelve dimers (12 $C_2$) were also absorbed on the edge of this tube, that is, nanotube elongation by two carbon hexagon layers occurred, as with a polyene ring as a base. The total energy of absorption was then found to be 129.08 eV. A comparison of these two results allows us to unambiguously assert that $C_2$ dimers are more actively absorbed on a nanotube than on a ring.

Calculations of the total energies of absorption of 12 $C_2$ dimers on a *trans*-distorted ring of the cylindrical variant comprising 12 carbon atoms and a cylindrical (6, 0) nanotube of finite height $H_{(6, 0)} = 11.2$ Å were performed quite similarly. These calculations gave $E_{ab}^{tot} = 113.52$ eV for absorption on a polyene ring and $E_{ab}^{tot} = 127.34$ eV for absorption on a four-layer SWNT. It follows that sequential addition of dimers to a nanotube of a finite length, which increases the height of the growing nanostructure by two layers of carbon hexagons, is energetically favorable.

To summarize, the molecular MNDO calculations of the models of growth of nanotubes on bases of various heights (polyene rings and four-layer SWNTs) prove that the higher the base on which carbon dimers are absorbed the more effective their addition (we obtained $\Delta E_{ab} \sim 14 - 15$ eV); that is, multilayer structures more strongly attract compound-absorbates than polyene rings.

## 2.4. DISCUSSION AND CONCLUSION

We considered several models of the growth of SWNTs on various precursors of carbon compounds. The energies of absorption for each of the models were calculated by the MNDO method. A comparison of the total energies of absorption of $nC_3$, $nC_2$, and $2(C_3$

+ $4C_2$ + C) for *cis* and *trans* rings (Tables 1-4) allowed us to draw the following conclusions.

(1) Sequential addition of 12 $C_2$ is more energetically favorable for the formation of the first two layers of hexagons of (6, 6) and (6, 0) nanotubes than the addition of eight $C_3$ or the $2(C_3 + 4C_2 + C)$ complex set of species. This is an expected result, because the formation of dimers in graphite vaporization is more probable, and the $C_2$ structure itself is stabler.

(2) Building nanotubes on a *cis*-distorted polyene ring of variant 2 is more favorable energetically. As a result, the so-called prismatic SWNT grows. Conversely, the formation of nanotubes of variant 1 (cylindrical nanotubes) is more favorable for polyene *trans* rings. The energy difference is, however, small, and both twisting variants are therefore possible.

(3) Of all the variants considered in this work, the addition of 24 carbon atoms is most favorable energetically. This follows from theoretical calculations. However in reality, such a nanotube cannot be obtained, because the conservation of carbon involved in absorption in the nonbonded (atomic) state requires maintaining temperatures that would cause the destruction of the yet unstable structure of growing nanotubes.

(4) A comparison of the energies of absorption shows that adding layers of hexagons to a nanotube of a finite height is more favorable energetically than adding them to a polyene ring because such a SWNT more strongly attracts carbon atoms than the polyene ring. Indeed, the ring is a homogeneous system with the predominance of the covalent component of chemical bonds between carbon atoms. Conversely, a tube is an extended object with unsaturated bonds at its end. The ionic component is admixed to covalent edge bonds as a result of interactions with "internal" tube layers, which causes polarization of edge layers and therefore increases the effectiveness of the formation of carbon nanotubes.

To summarize our semiempirical quantum-mechanical MNDO molecular calculations of the mechanisms of growth of carbon nanotubes on polyene rings of the *cis* and *trans* types variants 1 and 2 prove that these structures can play the role of precursors for the formation of achiral (n, n) and (n, 0) nanotubes, whose growth by the addition (absorption) of $C_2$ dimers is the most probable process. It is likely that similar results can be obtained in considering the growth of open achiral SWNTs on a graphene sheet rolled up into a ring or on a fullerene hemisphere, whose circumference is similar to a polyene ring.

## 3. The SWNT growth mechanism models on fullerenes

### 3.1. SWNT GROWTH ON FULLERENE HEMISPHERE

We have carried out MNDO-calculations of an electronic structure and energy characteristics of growth nanotube such as (6, 6) on open border of a fullerene $C_{60}$ hemisphere [6]. All basic structure (cluster), being precursor of growth nanotube, contained 72 atoms of carbon and consist of a hemisphere fullerene and two hexagon carbon layers, last of which included 12 boundary carbon atoms with the unsaturated bonds (fig. 3). The open border, on which adsorption binding of carbon occurred, tested strong enough influence on the part of spherical "cap". The geometrical parameters of basic system have been optimized during calculation. The interatomic distances $r_{C-C}$ of open border of a hemisphere have appeared equal 1.42 Å.

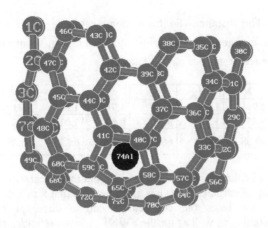

**Fig. 3.** Geometrical model of cluster on the basis of a hemisphere fullerene with the introduced aluminium atom. The figures mark numbers of atoms in molecular cluster model (are used at calculations).

In the work the consecutive absorption on a hemisphere border of various carbon bindings resulting to formation of two carbon hexagon layers of growing nanotube (6, 6) was investigated. The following variants of origin nanotube on fullerene basis were investigated:
1) a consecutive absorption on open cluster border of eight trimers of carbon ($nC_3$, n = 8) (fig. 2a);
2) a consecutive absorption on open border of twelve carbon dimers ($nC_2$, n = 12) (fig. 2b).

By results of quantum-chemical calculation the adsorption energy for the described variants of nanotube growth (tab. 5 and 6) has been determined and the comparison of total absorption energies of 8 trimers and 12 dimers of carbon has been executed. MNDO-calculations have been carried out with complete optimization of cluster geometrical parameters. It was especially important for the second way of nanotube growth because it was quite possible to expect binding of dimers $C_2$ neither parallel to open border of hemisphere fullerene, as it is presented on fig. 2b, but orthogonal, when dimer adsorbs on the neighbor carbon atoms of the border, and then atoms form the bonds among themselves.

**TABLE 5.** Energies (eV) of consecutive absorption of $nC_2$ (n=12), $nC_3$ (n=8) on a fullerene hemisphere for the cases of singlet (S=0) and triplet (S=1) states of initial systems, at presence atom defect Si and its total meanings.

| n | 1 | 2 | 3 | 4 | 5 | 6 | 7 | 8 | 9 | 10 | 11 | 12 | $E_{ab}^{tot}$ |
|---|---|---|---|---|---|---|---|---|---|---|---|---|---|
| | | | | | | $C_2$ | | | | | | | |
| S=0 | 26.2 | 13.4 | 13.3 | 13.3 | 13.3 | 13.8 | 2.4 | 12.6 | 10.9 | 13.9 | 13.4 | 13.6 | 160.1 |
| S=1 | 25.2 | 13.4 | 13.4 | 13.7 | 12.4 | 14.0 | 1.6 | 12.5 | 12.6 | 12.7 | 26.1 | 29.9 | 187.5 |
| Si | 13.8 | 12.4 | 13.1 | 13.3 | 15.4 | 15.5 | 15.9 | 15.5 | 15.9 | 16.7 | 16.4 | 16.9 | 180.8 |
| | | | | | | $C_3$ | | | | | | | |
| S=0 | 8.7 | 9.5 | 10.8 | 9.3 | 12.9 | 13.3 | 14.3 | 14.2 | - | - | - | - | 92.9 |

**TABLE 6.** The distance (R) from atoms-catalysts up to central hexagon of a fullerene hemisphere, charges (q) on catalyst atoms and total absorption energies $E_{ab}^{tot}$ of dimers $C_2$ at the presence of various catalyst atoms.

| Атом | Si | K | Mg | Na | Li | Zn | Ca | Al | Ga |
|---|---|---|---|---|---|---|---|---|---|
| R, Å | 1.83 | 2.09 | 1.46 | 1.95 | 1.76 | 1.63 | 1.71 | 1.87 | 1.91 |
| q | 1.36 | 0.45 | 1.31 | 0.51 | 0.54 | 1.01 | 0.85 | 0.54 | 0.48 |
| $E_{ab}^{tot}$, эВ | 83.49 | 69.68 | 68.31 | 66.70 | 66.26 | 65.73 | 62.25 | 45.31 | 42.55 |

The analysis of calculation results has found out that a real way of adsorption variant 2 is the parallel binding of particles $C_2$. It has appeared also that for (6, 6) nanotube growth on the hemisphere fullerene basis the consecutive carbon dimers (12 $C_2$) adsorption is energetically more favorably as well as for the SWNT growth on polyen rings case (chapter 2).

In order to investigate influence of "radicalness" of attacking particles on energy benefit of processes the comparison of two possible spin states of carbon dimers – singlet and triplet – has been carried out. As it has appeared (tab. 5) the most favourable is triplet spin state of initial system. In this case not coupled electrons are located in the field of single carbon atoms, therefore carrying of electronic density on fullerene cap occurs without preliminary process of unpairing of electrons that, obviously, is energetically more favourable.

Besides the opportunity has appeared to answer a question, what precursor of SWNT growth is more preferable. For this purpose the absorption energy comparisons of 12 dimers $C_2$ binding to fullerene cap (tab. 5) and to polyen rings (tab. 1-4) have been carried out, which result in (6, 6) nanotube growth. It has appeared that energetically more favorable variant is a fullerene hemisphere as precursor of SWNT growth.

## 3.2. SWNT GROWTH ON ENDOHEDRAL FULLERENE HEMISPHERE

It is necessary to note that the synthesis of SWNTs is essentially catalytic process requiring application of various metals (as separate, and their mixes) as catalysts of NT growth [11]. However while it is poorly known about influence of these substances on kinetic of SWNT growth and there are no unequivocal recommendations for the use of those or other metals as effective catalysts.

With the purpose of approach to the decision of the above mentioned problem we have carried out MNDO-calculations of (6, 6) SWNT growth processes on a fullerene hemisphere at the presence of various metal and metal catalysts for finding out of its influence on growth rate of tube. The following elements have been chosen ss catalysts: Li, Na, Mg, Al, Si, K, Ca, Zn and Ga. The atom-catalyst was located inside cluster containing 72 carbon atoms (fig. 3), which is a fragment of endohedral fullerene. The geometry of system was optimized and calculations results are presented in tab. 6. As follows from the table, the less own sizes of catalyst atom, the "more deeply" in a hemisphere it settles down. The atom of defect represents a quantum dot in structure of investigated cluster, which changes a polarization (charging distribution) carbon basis, and, hence, influences processes of origin and further growth of nanotubes.

The quantum-chemical calculations have allowed to calculate the absorption energy of dimers on quantum dot of fullerene hemisphere. The following row of priorities on the greatest total energy of carbon dimers absorption of has been obtained: Si, K, Mg,

Na, Li, Zn, Ca, Al, Ga (tab. 6). As it is shown from the table, the greatest influence on SWNT growth kinetic appears silicon atom. The comparison of total absorption energies has shown that its efficiency is much more than efficiency of other chosen elements, surpassing catalitic ability of Ga almost twice. The influence of atoms K, Mg, Na, Li, Zn, Ca practically is identical, their efficiency also is high enough and below than Si efficiency only on some percents (about 8 %). Aluminum and gallium have appeared the weakest catalysts in the chosen row of elements, them catalytic properties are lower approximately on 40 - 45 % than at other substances.

We also have compared (6, 6) SWNT growth processes at the presence of the catalyst and without it. As the catalyst the silicon (most effective element on the properties) was chosen. SWNT growth processes have been simulated by consecutive absorption of 12 dimers $C_2$, first two hexagon layers have been built. Absorption energies of both variants of growth are given in the table 5.

As it has appeared, the catalyst essentially accelerates nanotube growth that it is reflecting in a difference of total absorption energies of dimers at the presence of the catalyst (181 eV) and without it (160 eV). It has made $\Delta E_{ab} = 20.6$ eV (tab. 5).

### 3.3. CONCLUSIONS

In conclusion of the chapter we shall formulate the basic results received in the job:
1. The total energy of consecutive adsorption of carbon dimers on a fullerene hemisphere has been determined. The carried out comparison of $E_{ab}^{tot}$ for various precursors of nanotube growth – a polyhene ring or a fullerene hemisphere - has found out that last is energetically more preferable.
2. The paramagnetic state of carbon dimers adsorbing on a tube germ increases the efficiency of process of the further growth of carbon nanotube.
3. A row of catalysts influencing on nanotube growth kinetic has been investigated: Si, K, Mg, Na, Li, Zn, Ca, Al, Ga. The priorities in the chosen row of chemical elements have been determined.

### 4. The growth mechanism models of SWNT on diamonds surface

In the work some ways of carbon NT origin and further growth on diamond surface have been investigated: adsorption of atomic carbon, dimers and trimers of carbon [7].

### 4.1. ATOMIC CARBON ADSORPTION ON DIAMOND SURFACE

To study the energy characteristics of carbon NT origin process we have considered the elementary act of atomic carbon adsorption on diamond surface. As diamond surface model the crystal fragment consisting of 3 surface layers of carbon atoms and containing 72 carbon atoms - $C_{36}C_{36}$* cluster, where C* - boundary carbon atoms with one or two $sp^3$-LHOs, has been chosen. If C* atom was boundary surface atom, 2 LHO directed to the cluster neighbour atoms were included in orbital basis from it. If C* atom was closed the cluster border in crystal volume, it brought in to basis only one LHO. C-C bond length is equal 1.54 Å. Diamond surface has not been optimised in the process of calculation, i.e. the rigid lattice approximation has been applied to the surface. The surface fragment is evidently represented in a fig. 4.

270

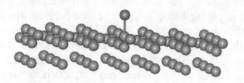

**Fig. 4.** (111) diamond surface cluster with sorbed carbon atom.

The adsorption processes has been simulated by step-by-step approach of C to surface carbon atom along a perpendicular to the surface through carbon atom. Thus the C atom has got two degrees of freedom and can freely deviate a perpendicular in two mutually orthogonal directions. While calculating the surface center C has got the degrees of freedom allowing it to output from the surface. The surface C atom has been selected in middle of the cluster in order to exclude the influence of the boundary effects (fig. 4).

The quantum chemical calculations of interaction energy of atom with diamond surface fragment within the framework of OSC method have allowed us to construct a structure of a potential energy surface of atomic carbon adsorption process on diamond. The potential energy dependence curve as a function of the distance from attacking carbon to the surface has shown in a fig. 5. The energy characteristics of the given process (adsorption energy $E_{ad}$, activation energy $E_a$) are represented in the table 7. The interaction energy on infinite distance between cluster and attacking C atom was accepted as energy zero level.

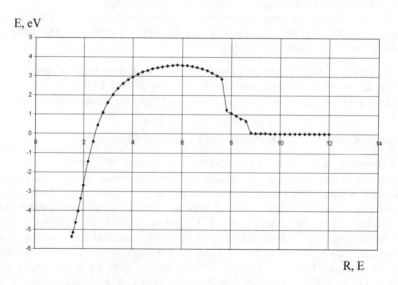

**Fig. 5.** The potential energy curve of carbon atom adsorption on diamond surface as a function of the distance between attacking C atoms and the surface.

**TABLE 7.** The energy characteristics (adsorption energy $E_{ad}$, activation energy $E_a$) of carbon monomer, dimer and trimer adsorption on quantum dots of diamond surface.

| Defect | Li | Na | K | Be | H | - |
|---|---|---|---|---|---|---|
| C | | | | | | |
| $E_a$, eV | - | - | - | - | 0.3 | 3.6 |
| $E_{ad}$, eV | 2.6 | 3.1 | 2.4 | 5.8 | 5.2 | 5.5 |
| $C_2$ | | | | | | |
| $E_a$, eV | - | - | - | - | 3.6 | 1.0 |
| $E_{ad}$, eV | 4.5 | 5.9 | 6.5 | - | 0.4 | 0.7 |
| $C_3$ | | | | | | |
| $E_a$, eV | - | - | - | - | 5.0 | 1.5 |
| $E_{ad}$, eV | 7.3 | 7.1 | 6.8 | - | 0.2 | 0.7 |

The calculation result analysis (fig. 5, tab. 7) has shown that carbon atom has to overcome a potential barrier (activation energy $E_a$) about 3.6 eV in order to connect to the surface centre. The C-C bond formation energy (adsorption energy $E_{ad}$) appears equal 5.5 eV.

Last years investigations [2, 11 - 14] have shown that most effective processes of carbon NT origin proceed on quantum points of a substrate surface as which the various authors considered transition metal atoms adsorbed on a surface. Therefore we have also studied the energy characteristics of atomic carbon adsorption processes in a vicinity of a quantum dot of diamond surface. Atoms of alkaline metals (Li, Na, K), hydrogen and beryllium atoms have been used for simulation of quantum dots. The potential curves of adsorption processes and their energy characteristics are represented in a fig. 6 and in the table 7.

As follows from a fig. 6 practically for all considered quantum dots (except for hydrogen atom) the atomic carbon adsorption processes proceed without barrier. The presence of maximum on potential energy curves does not interfere C atom chemisorption on a diamond substrate, because the initial energy of attacking atom (on the diagram it corresponds 0 eV) considerably surpasses all potential barriers meeting on way of atom. The exception makes of the considered case with hydrogen atom as quantum defect. The given process occurs as a result of overcoming of a small potential barrier about 0.3 eV, which is lower in order than the barrier for a case of a pure surface. Width of the potential barrier makes about 0.3 Å. The given parameters of a potential barrier indicate that C atom chemisorption in a vicinity of this quantum dot is really feasible by a classical and tunnel ways.

### 4.2. (6,0) NANOTUBE FORMATION ON (111) DIAMOND SURFACES

The quantum chemical calculations of consecutive adsorption processes of monocarbon on (111) diamond surfaces have been carried out to study the energy characteristics of "zig-zag" carbon NT formation. First tube layer consisting of 6 carbon atoms formed by the consecutive connection to the surface centres. Second layer (also from 6 atoms) joined to adsorbed atoms of first layer by "bridge" binding. The atoms of third layer bonded to second layer atoms, the atoms of fourth layer closed the formed hexagons. In result the elementary unit (EU) of (6,0) NT containing 6 hexagons on the tube perimeter (fig. 7) has been built. Here we have considered only case when the C atoms in each layer joined to diamond surface near neighbour adsorbed C atom. Adsorbed C atom geometrical parameters have been optimised during quantum-chemical calculations. The calculation results of energy adsorption of (6,0) NT formation are represented on the diagram (fig. 8a).

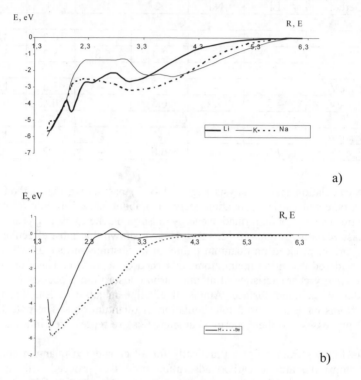

a)

b)

**Fig. 6.** The potential energy curve of carbon atom adsorption on quantum dots of diamond surface: a) Li, K, Na atoms, b) Be and H atoms.

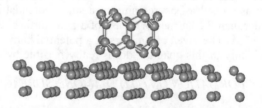

**Fig. 7.** (111) diamond surface cluster with (6,0) NT germ.

The diagram has shown that the process of (6,0) NT formation indicates quasi-periodical energy benefit and the building of various tube layers is not equivalent energetically:

1) first 6 atoms (fig. 8a) are adsorbing with energy benefit about 4-5 eV;
2) following 6 atoms are adsorbing with energy benefit about 10-12 eV;
3) next layer from 6 atoms are adsorbing with energy benefit about 3-5 eV again;

273

4) six atoms, which finish building of NT EU, - with energy about 10 eV.

On our sight similar quasi-periodicity is concerned with that C atoms connected to diamond surface have 3 unsaturated bonds, which increase chemical reactivity of formed NT germ. Second layer atoms have only 2 free bonds therefore their reactivity appears less than previous layer. In whole first layer of sorbed C atoms plays a role of quantum dot of diamond surface, which promotes increasing of adsorption efficiency.

During "first half-period" (fig. 8a), when first layer connects to the surface, the certain property is also observed. In the beginning the energy advantage of process increases up to third atom adsorption. It is possible to explain it again that C atom sorbed on the surface changes the charge distribution of diamond surface and becomes a quantum dot raising adsorption efficiency of other atoms.

The following layers (in opposite first) show oscillating dependence of energy adsorption from number of the joined atoms. It means that at this stage the formation, alongside with σ-bonds, of π-bonds between sorbed C atoms takes place. The large binding energy of second layer atoms (fig. 8a), equal 10 - 12 eV, indicates about it. Each second layer atom forms as a minimum 2 C-C bonds with first layer atoms, appearing in "bridge" site. If to norm the binding energy on each formed bond, the meaning 5 - 6 eV, comparable with binding energy of first layer atoms, to be obtained.

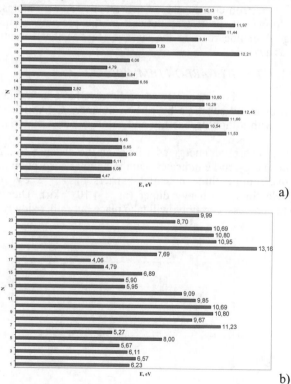

**Fig. 8.** Diagram of consecutive adsorption energy of N atoms C: a) on diamond surface, b) on quantum dot (Li atom) of the surface.

274

Last layer completing the construction of tube germ shows the energy behaviour similar to first layer. It is obvious that the similar tendency will be kept and for all nanotube. One difference, on our sight, will be the decrease of energy benefit of process with the increase of tube length, i.e. the substrate influence will be decreasing. At certain tube length the adsorption energy will leave on saturation and its meaning will be defined by interaction energy with several layers already formed (6,0) NT.

With the purpose to find out, how a quantum dot influences on (6,0) NT germ formation, we have investigated the energy properties of first layer NT chemisorption on (111) diamond surfaces with adsorbed lithium atom on it. The level-by-level binding of monocarbon has been occured by such manner that Li remained at the centre of generating carbon NT. The calculation results of energy adsorption of (6,0) NT formation on a quantum dot of diamond surface are represented on the diagram (fig. 8b).

As it can be seen from a fig. 8b the described above energy property in building of tube layers takes place and at the presence of Li atom on diamond surface. The analysis of the diagrams (fig. 8a and 8b) has shown that Li renders a small influence on energy benefit of NT growth process (adsorption energy of atoms for each site has approximately same value). The essential difference is observed only for first carbon atom layer, for which the binding is more advantageous on 2 eV in average than for a case of pure diamond surface. Such "inertness" of defect is actually apparent, as the chemical C-C bond formation energy is defined mainly by electron-energy state of nearest environment atoms. The quantum dot influence on NT formation, as it was noticed in the previous chapter, is reduced to barrierless binding of monocarbon to the surface (fig. 5) that essentially increases the process efficiency as a whole.

### 4.3. (6,0) NT FORMATION BY CARBON DIMMERS $C_2$

Alongside with carbon monomer adsorption we have investigated a possibility of (6,0) NT origin on diamond substrate by carbon dimmers $C_2$, which are formed at laser ablation of graphite with a high probability and can serve as a material for NT formation [2].

The dimmer adsorption energy as a function the distance to diamond surface has shown in fig. 9. The energy curve indicates that the dimmer adsorption process on diamond surface requires overcoming three potential barriers of 1 - 2 eV height, what is an inconvenient obstacle for quite heavy dimer (m $\approx 4 \cdot 10^{-26}$ kg). Therefore due to dimer energy dispersion the process, certainly, will take place but rather slowly.

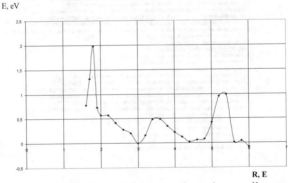

**Fig. 9.** The potential energy curve of carbon dimmer $C_2$ adsorption on diamond surface as a function of the distance between attacking C atoms and the surface.

The energy diagram of NT formation on diamond surface by dimers is shown in fig. 10. As it follows from the diagram the first layer dimer bindings is accompanied by periodic change of energy adsorption. The relative large meanings of energy (~ 10 eV) indicate to the formation of two C-C bonds. One of them is formed between one C atom, belonging dimmer, and surface C, another - between second dimmer C atom and atom C, belonging neighbour adsorbed dimmer.

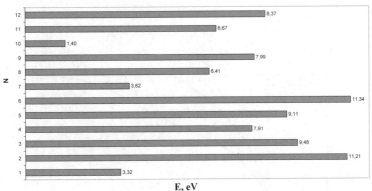

**Fig. 10.** Diagram of consecutive adsorption energy of N dimmers $C_2$ on diamond surface.

The second dimmer layer is adsorbed with smaller energy in comparison with first this, that confirms the tendency, marked in the previous chapter, about saturation of energy with the tube size increase.

The study of dimmer adsorption on quantum dots of diamond surface has led to the results similar monomer adsorption (fig. 11). The figure has shown that at the presence of metal atoms (Li, K, Na) the process is carried out practically without barrier, as well as in a case of atomic carbon adsorption. At the presence of hydrogen atom the potential barrier size is 0.46 eV. Obtained in quantum-chemical calculations the adsorption and activation energies of processes on quantum dots are represented in table 1.

*4.4. (6,0) NT FORMATION BY CARBON TRIMMERS $C_3$*

All calculations for the given model of origin and further growth of carbon NT by trimers are similar to calculations on dimer adsorption. For binding to the surface centre the particle has to overcome a barrier of height ~ 1.6 eV, that is practically inconvenient for heavy trimer but it is also possible due to particle energy dispersion.

The energy diagram of (6,0) NT formation process is analogical to the $C_2$ diagram (fig. 10). Oscillating dependence of binding energy on joined $C_3$ number is well seen. The general tendency to increase the energy benefit of adsorption process within one layer of tube is also reflected on the diagram. The greatest values of the energy are observed at binding of last trimer. In this case the formation of several chemical bonds (on 3.1 eV on everyone) and the steadiest geometrical configuration takes place.

The energy curve of processes adsorption of $C_3$ particle on quantum dots of diamond has analogical behaviour to the $C_2$ (fig. 11). It is seen from the diagram that the

276

tendency reflected in the previous sections is kept here again. The chemical activity of diamond surface to adsorb carbon trimers is increased at the presence of metal atoms (Li, K, Na). On these quantum dots the adsorption process proceeds practically without barrier. At the presence of hydrogen atom the potential barrier size is 0.15 eV. Obtained in quantum-chemical calculations the adsorption and activation energies of processes on quantum dots are represented in table 1 also.

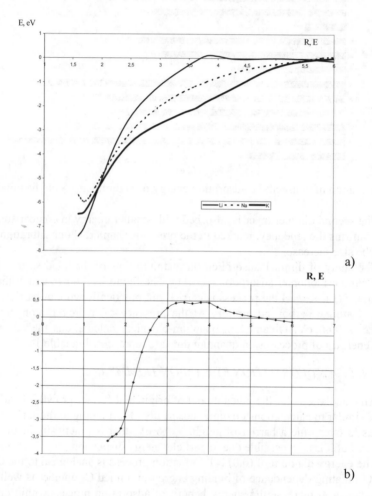

a)

b)

**Fig. 11.** The potential energy curve of carbon dimmer $C_2$ adsorption on quantum dots of diamond surface: a) Li, K, Na atoms, b) H atom.

*4.5. CONCLUSION*

In last chapter we shall resume the basic results and conclusions of the given work:

1. Quantum-chemical semi-empirical calculations have shown that the OSC model simulates adequately the geometrical structure of (111) diamond surfaces and the carbon NT origin and further growth processes by the way of carbon monomer, dimmers and trimmers chemisorption.

2. The basic energy characteristics of the given processes, from which follows that most effective adsorption process takes place on diamond surface quantum dots modeled by adsorbed hydrogen, lithium, sodium, potassium, beryllium atoms, are obtained. For the adsorption of the considered carbon particles in a vicinity of alkaline metal atoms (Li, Na, K) the process occurs without barrier, in a vicinity of hydrogen atom the activation energy is of 0.1 - 0.2 eV that is relatively small value. The received results reflect the general tendency stated above: in quantum dot vicinity the chemisorption processes (still except for H atom) proceed without barrier.

3. From the represented results it is well seen that energetically more favorable is the adsorption process of carbon monomers on a pure diamond surface. However the chemisorption on quantum dots changes a situation with accuracy on the contrary. If to arrange energy advantage of the process on decrease, we shall obtain the following row: trimer – dimer – carbon monomer adsorption. Such property is observed for any metal defect simulating a quantum dot on diamond surface.

4. At last, some words about the possible practical application of the given results. Modern nanotechnologies allow (for example, by scanning tunnel microscope) to create monoatomic defects on solid surfaces. Preparing a diamond surface with quantum dots as sorbed metal atoms, it is possible effectively to synthesise single walled carbon nanotubes of "zig-zag" type with beforehand set diameter. The (111) diamond surface is constructed in such a manner that NT will be practically its continuation.

**5. Acknowledgements**

We thank Prof. A.O. Litinski (Volgograd State Technical University) for fruitful discussion of the results and important notices and recommendations. This work was supported in part by Russian Basic Research Fund (project 02-03-81008) and INTAS (project 00-237).

**6. References**

1. Dresselhaus M.S., Dresselhaus G., Eklund P.C. Science of Fullerenes and Carbon Nanatubes, New York: Academic Press, 1996, 965 p.
2. Ivanovskii A.L., Kvantovaya khimiya v materialovedenii. Nanotubulyarnye formy veshchestva (Quantum Chemistry in Materials Science: Nanotube Forms of Substance), Ekaterinburg: UrORAN, 1999, 176 p.
3. Chernozatonskii L.A. Zarozdenie grafitizirovannih nanotrub na almazopodobnih kristallitah. Himich. Fiz 1997; 16(6): 78-87.

4. Lebedev N.G., Ponomareva I.V., and Litinskii A.O. OSC-model of generation of carbon nanotubes on quantum dots of diamond surface. Abstracts of Papers, 5th Biennial Int. Workshop "Fullerenes and Atomic Clusters", St. Petersburg, 2001, p. P278.
5. Lebedev N.G., Zaporotskova I.V., Chernozatonskii L.A. A quantum-chemical analysis of models of growth of single-layer carbon nanotubes on polyene rings. Russian J. Phys. Chem. 2003; 77(3): 431-438.
6. Zaporotskova, I.V. Lebedev N.G., Chernozatonskii L.A. Simulation of carbon nanotube growth processes on fullerene hemisphere substrate. Zurnal Fizicheskoi Khimii 2003; 77(12): 2254-2257 (accepted).
7. Lebedev N.G., Ponomareva I.V., Chernozatonskii L.A. The orbital-stoichiometric cluster model of carbon nanotube generation on quantum dots of diamond surface. International Journal of Quantum Chemistry 2003; 95(3) (accepted).
8. Litinskii A.O., Lebedev N.G. A model of an ion-incorporated cluster with orbital stoichiometry for calculating interactions between the solid surfaces and gas-phase molecules. Russian J. Phys. Chem. 1995; 69(3): 119- 124.
9. Dewar M.J.S., Thiel W. Ground states of molecules. 38. The MNDO method. Approximations and Parameters. J. Am. Chem. Soc. 1977; 99: 4899-4906.
10. Dewar M.J.S., Thiel W. A semiempirical model for the two-center repulsion integrals in the NDDO approximation. Theor. Chim. Acta 1977; 46: 89-104.
11. Eletskii A.V. Uglerodnie nanotrubki i ih emissionnie svoistva. Uspehi Fiz. Nauk 2002; 172(4): 401 – 438.
12. Saito R., Dresselhaus M.S., Dresselhaus G. Physical properties of carbon nanotubes, Imperial College Press, 1999, 251 p.
13. Dai H. Nanotube growth and characterization. Carbon nanotubes, Topics Appl. Phys. 2001; 80: 29-53.
14. Charlier J.-Ch., Iijima S. Growth mechanisms of carbon nanotubes. Carbon nanotubes, Topics Appl. Phys. 2001; 80: 55-81.

# COVALENT-BINDING CARBON NANOTUBE: SIMULATION OF FORMATION MECHANISMS AND ENERGY CHARACTERISTICS

E.E. MIKHEEVA, L.A. CHERNOZATONSKII, T.Yu. ASTAHOVA

*Institute of Biochemical Physics of RAS, Moscow, 119991 Russia*

The synthesis of carbon nanotube branching structures is an important step in the development of carbon based nano-electronic and micro electro-mechanic devices and because these materials are potentially able to bring in new mechanical and electrical properties. Process of nanotube's polymerization is one of resources of such building. Earlier we considered 3-D structures of polymerized identical nanotubes in the rope [1,2].

There are many types of connecting two closed pares of neighboring nanotube hexagons. In this work we have investigated the most simple and likely formation mechanisms of nanotube bonding through "2+2" cyclo-addition and four common ring connection ("Osawa" connection [3]).

We have considered I-, V-, T-, X- and Y-types of covalent-binding carbon nanotube. We use carbon zigzag (8,0) nanotubes for formation of I-, V-, T- and Y- structures and the (5,5) and (9,0) tubes for X-junctions. The length of caped nanotubes is ~2 nm, and opened tubes have ~5 nm length. Figures 1 (a) and (c) shows optimized geometry scheme of I-type of covalent-binding carbon nanotubes with "2+2" cyclo-addition and "Osawa" connection, respectively. The similar types of tube connections in V-structures are shown in Figures 1 (b) and (d). We have investigated also T- and Y-covalent-binding carbon nanotubes – Figure 2. Optimized structures of four-terminal covalent X-junction are presented on Figure 3. The energy simulation use molecular dynamics method with empirical bond-order potential that was parameterized by Brenner [4].

Cohesive energy values of these structures are in the Table 1, were energies of uncoupled tube fragments ($E_f$), "2+2" and "Osawa" covalent compounds of nanotubes (E), and difference between them ($E-E_f$) describe stable energy characteristics of these structures. It is shown that I-, V- and T-covalent-binding carbon nanotubes formed by "2+2" cyclo-addition have small potential barrier (~ 2 eV), and "Osawa" connections are more stable than free tube elements. The X- and Y-structures have metastable state at ~3.5 eV and ~5 eV higher, respectively, than that of free tube elements – their formation needs pressure and temperature treatments [1] and can be prepared during electron beam heating [5].

Then we have considered in detail the addition reaction with formation "2+2" cyclo-addition X-type connection of the (5,5) and (9,0) nanotubes of diameter 0.7 nm intersecting at a right angle. We has calculated this polymerization process in the temperature range 2100–1100 K for different pressures (1–15 GPa) applied to the intersection region with an area of $10^{-15}$ cm$^2$ by using Brenner potential molecular dynamics [4]. We choose the regime of constant forces applied at right angle to the

*T.N. Veziroglu et al. (eds.),*
*Hydrogen Materials Science and Chemistry of Carbon Nanomaterials,* 279-282.

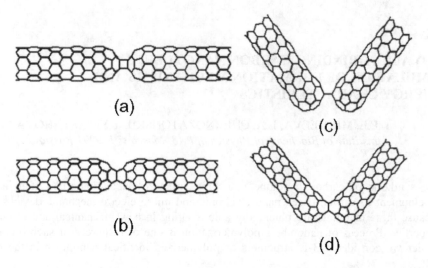

Figure 1. Optimized structures of two-terminal I- and V-junctions with "2+2" cyclo-addition (a, b) and four common ring connection (c,d) of two caped tubes, respectively.

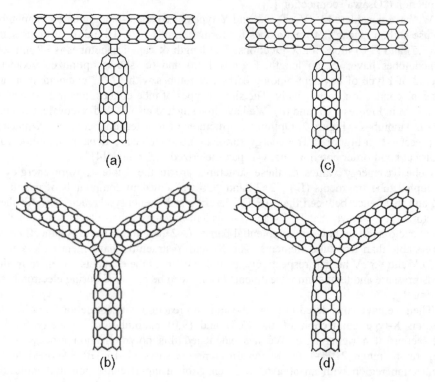

Figure 2. Optimized structures of three-terminal covalent T- and Y-junctions with "2+2" cyclo-addition (a, b) and four common ring connection (c, d) of tubes, respectively.

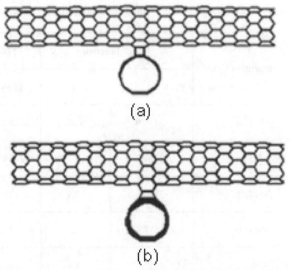

(a)

(b)

Figure 3. Optimized structures of four-terminal covalent X-junction with "2+2" cyclo-addition (a) and four common ring connection (b) of two opened tubes, respectively.

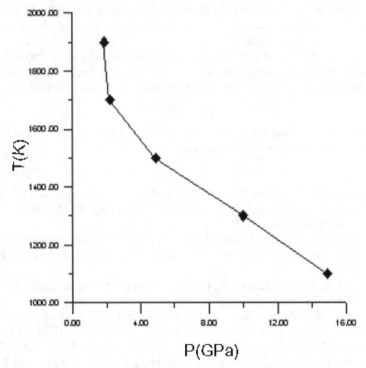

Figure 4. Phase T-P diagram for polymerization of (5,5) and (9,0) nanotubes intersecting at a right angle.

TABLE 1. Energetic characteristics of various types of covalent-binding carbon nanotube.

| Type (atoms) | Free structures $E_f$, eV | "2+2" junction | | "Osawa" junction | |
|---|---|---|---|---|---|
| | | E, eV | $\Delta$, eV | E, eV | $\Delta$, eV |
| "I" – type (316 atoms) | -2238.4 | -2236.9 | -1.5 | -2241.5 | 3.1 |
| "Y" – type (492 atoms) | -2238.4 | -2237.7 | -0.7 | -2238.5 | 0.1 |
| "T" – type (446 atoms) | -3172.5 | -3170.1 | -2.4 | -3173.1 | 0.6 |
| "X" – type (786 atoms) | -5641.0 | -5637.7 | -3.3 | -5637.0 | - 4.0 |
| "Y" – type (492 atoms) | -3487.2 | -3482.8 | -4.4 | -3481.7 | -5.5 |

intersection region to the most upper atomic row of (9,0) nanotube and the most lower one of (5,5) nanotube sections and limited the evolution time by ~1000 of 0.0003 ps steps. The "phase"-separation curve is shown in Figure 4. This diagram allows to predict the necessary conditions for experiment.

The energy simulation has shown that the considered structures are stable and can be used as nano-mechanic elements. The detailed investigation processes of formation covalent-binding carbon nanotube allows to determine the probable conditions of their formation.

The present work is supported by Russian programs: "Topic directions in condensed matter physics" and "Low dimensional quantum structures".

**References**
1. Chernozatonskii, LA., Menon, M., Astakhova, T.Yu. and Vinogradov, G.A. (2001) Carbon Systems of Polymerized Nanotubes: Crystal and Electronic Structures, *JETP Lett.* **74**(9), 467-470.
2. Chernozatonskii, L., Richter, E. and Menon, M. (2002) Carbon Crystals of Covalent Bonded Nanotubes: Energetic and Electronic Structures, *Phys.Rev.B* **65**(24), 241404-241407.
3. Osawa, S., Osawa, E. and Hiroes, Y. (1995) $C_{60}$ dimer structures, *Fullerene Sci. & Technol.* **3**, 565-567.
4. Brenner, D.W., Shenderova, O.A., Harrison, J.A. and Stuart, S.J. (2002) Ni B.& Sinnott SB. A second-generation reactive empirical bond order (REBO) potential energy expression for hydrocarbons, *J. Phys.: Condens. Matter* **14**(4), 783-802.
5. Terrones, M., Banhart, M., Grobert N., Charlier, J.-C., Terrones, H. and Ajayan, P.M. (2002) Molecular Junctions by Joining Single-Walled Carbon Nanotubes, *Phys. Rev. Lett.* **89**(7), 075505.

# TO THE THEORY OF FORMATION IN CAST IRON OF SPHERICAL GRAPHITE

BARANOV A. A., BARANOV D. A.
*Donetsk national technical university*
*ul. Artema, 58, Donetsk, 83000, Ukraine*
*e-mail: BaranovDA@rambler.ru, fmt@fizmet.dgtu.donetsk.ua*

**Abstract.** Is given a ccritical education hypotheses analysis of spherical graphite. Is show determining role non-continuities in origin and graphite growth. Is considered evolution of gas vesiccle during origin and graphite growth.
**Keywords:** spherical graphite, interfacial surface tension, graphitization of ferrous alloys.

## 1. Introduction

The set of hypotheses of formation of spherical graphite at solidification of cast iron are known [1, 2]. One of them, it is preferred, takes into account change of a surface tension on a interfacial area. It is considered, that the modifying by a magnesium or cerium raises interfacial area and it promotes the cutting of graphite particles by basic planes $(0001)_g$ [2]. Discussed hypothesis contains the contradictions arising by comparison of the theory of process of solidification and experiment. According to the thermodynamic theory of phase transformations, the work of formation of a germ of the critical dimension equals to:

$$A = 32 \cdot \left(\frac{M}{\rho}\right)^2 \cdot \frac{\gamma^3 T_0^2}{q} \cdot \frac{1}{(T_0 - T)^2}, \tag{1}$$

where M - molecular weight;
$\rho$ - density;
$\gamma$ - interfacial surface tension;
q - melting heat;
$T_0$ and T - accordingly melting temperature and temperature of supercool-ing of an alloy.

The increase $\gamma$ strongly increases work of formation of germs and should reduce velocity of origin. Actually, the modifying by a magnesium and cerium in thousand time increases number of eutectic colonies, that testifies to heterogeneous character of origin. The groundlessness of a hypothesis, founded on a surface tension is shown also at an explanation of effect over-modification, which, by the way, is Achilles' heel of the majority of hypotheses.

*T.N. Veziroglu et al. (eds.),*
*Hydrogen Materials Science and Chemistry of Carbon Nanomaterials,* 283-290.
© 2004 *Kluwer Academic Publishers. Printed in the Netherlands.*

Insufficient, in our opinion, is the hypothesis, founded on influence of the modifying agent on the parity of velocities of longitudinal and cross growth of graphite. The supporters of this hypothesis, taking into account the experimentally fixed raised (increased) concentrating of a magnesium and cerium in graphite, assume, that the atoms of the modifying agent are adsorbed on graphite in the certain places (e.g., points of increase), and it influence the configuration of graphite. speculativeness of this hypothesis follows from the data on a forming of graphite at external influence. So, as a result of plastic strain the graphite gets look of disks extended along the current of metal. Repeated heatings, in each of which is dissolved more 1%C, practically do not change the extended look of graphite and do not recover the spherical configuration, as it should follow from adsorptive hypothesis. This supervision testifies for the benefit of stated above [3] representations that the graphite configuration in many respects depends on a kind of hollow in a metal ground of cast iron.

The marked deficiencies the hypothesis is deprived, founded on a role of pinhole origin of graphite. "The pinhole" theory is confined to modern concepts about the role of discontinuity flaw in formation and growth of graphite and successfully has passed testing by time. According to them, present in liquid and solid alloys flaws increases Gibss energy, and it is possible to reduce it by covering a free surface by graphite skim. The specified concepts underlie an estimation of influence of pluggings of extrinsic phases on a graphitization of cast iron and steel [3]. It is appeared, that the efficiency of the modifying agent can be connected not with bed role of generating pluggings, then to those discontinuity flaws, which arise on phase boundary. Over-modification, according to considered concepts, is stipulated by formation of large pinholes, which, according to the Stokes equation, are easily deleted from a melt, diverting with them coagulated pinholes.

With formation of pinholes at vaporization of the modifying agent connected occurrence of a free surface with high tensation ($\gamma_s \sim 2$ J·m$^{-2}$). Gibbs energy decreases if the surface of the pinhole will become covered by skim of graphite. The interfacial surface tension makes $\gamma_{g/s} \cong 0,6$ J·m$^{-2}$, and surface-tension of a basal plane of graphite is $\gamma_{g/s} \cong 0,12$ J·m$^{-2}$. As $\gamma_s > \gamma_{g/s} + \gamma_g$, the formation of graphite skim occurs in conditions, at which the melt not even l is not supercooled concerning an eutectic temperature. With formation of an austenitic envelope on hollow graphite the abnormal eutectic solidification begins, at which the growth of graphite is realized due to inflow of carbon through the austenitic envelope.

During solidification of cast iron the pinhole experiences transformations. They are stipulated by change of pressure in the pinhole, graphite precipitation, growth austenite and condensation of gas. In a fig. 1 the scheme illustrates evolution of the pinhole filling graphite is submitted. At the first stage of modification (the segment 0 - 1) is formed a pinhole, which is diminished in the sizes at a temperature drop (segment 1 - 2). In the pinhole

of radius $r_n$ the pressure is created $P_n^0$

$$P_n^0 = P_1 + \frac{2\gamma_s}{r_n},$$  (2)

where $P_1$ - external pressure, including hydrostatic.

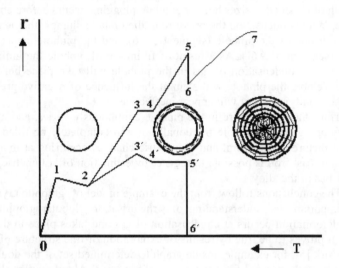

Figure 1. Evolution of gas vesicle and graphite making in him

The matting by graphite reduces a surface-tension almost in 3 times, so capillary pressure component will decrease, and the total pressure in the pinhole will appear other $P_n^1$

$$P_n^1 = P_1 + \frac{2\gamma_{g/s}}{r_g}.$$  (3)

Under influence of a difference of gas pressure $\left(P_n^0 - P_n^1\right)$ volume of pinhole will vary. An internal diameter of hollow graphite spheroid and its outside diameter will increase according to segments 2 - 3 and 2' - 3'. According to calculations:

$$r_g \cong r_n \cdot \left(\frac{\gamma_s}{\gamma_{g/s}}\right)^{\frac{1}{3}}.$$  (4)

Further, at a temperature drop below euclectic, the pinhole is gradually filled by carbon (cut 3 - 4), that promoted by the austenitic envelope, formed on graphite. At some stage, when the thickness of graphite will reach defined one, the growth of graphite by filling-up of the pinhole will be stopped. It can be carried out at the expense of increase of outside radius (curve and, segment 4 - 5). Thus internal radius (the curve B) does not vary. The growth of the outside sizes hollow graphite plugging occurs before condensation of gas in pinhole. With condensation the pressure in the pinhole slumps, that leads to "retraction" of graphite inside of the pinhole (segment 5 - 6), and the pinhole, apparently, completely disappear (segment 5′ - 6′). As a result of filling-up of pinhole the grain dimensions can decrease. If the condensation of gas in the pinhole will take place before occurrence of graphite envelope, the pinhole will slam at the influence of negative pressure in the fluid [4], and the modifying agent will appear ineffective.

Thus, the gas pressure in the pinhole continuously varies, and at condensation of the modifying agent the pinhole fractionally or will completely be filled by graphite. The considered picture of structural and volumetric changes occurring at modification and solidification of cast-iron is possible to register by estimation of volumetric or other physical characteristics of the alloy.

The conclusions follow from the example model of graphite crystallization in gas pinhole, important for understanding of structurisation globular graphite cast iron. The maximal deformation occurs at condensation of gas and takes place in the central parts of plugging. It proves to be true by results of exploration of fine structure of spherical carbon spots. In work [5], for example, inside graphite determined set of the doubles, which number decreases with disposal from the centre of plugging. About similar changes of internal structure of graphite was communicated and in other works, for example [1, 2]. Other important conclusion concerns to necessity of development of caution at the estimation of the information about fine structure of graphite, as the twin and other traces of shears can be formed not at the moment of growth of graphite (dislocation growth), as a result of mechanical deformation and shattering of graphite occurring at filling-up of pinhole.

It is seemed, that the failures pursuing the researchers at attempt to receive spherical graphite by immediate introduction in a melt of hydrocarbons, contradict explained here concepts. These failures, apparently, are connected and to hydrogen deposited to primary pinholes together with the modifying agent. The hydrogen saturates atomic bonding of carbon in the graphite ends surface, forming with strong bridgings. In result the surface of pinholes will not be coated with the graphite slim.

The saving of the spherical configuration of graphite at long-distance stages of solidification of cast iron depends on supercooling temperature of the melt and supersaturation degree of the solid solute by carbon. The influence of the specified factors on longitudinal growth (increase of the area of grids of graphite) and transversal growth (increase of number of grids) is shown in a fig. 2 and is considered in work [6].

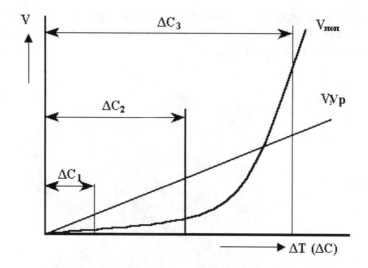

Figure 2. Dominance of supercooling (ΔT) and supersaturation (ΔC) on speed of longitudinal (Vp) and transversal (Vпоп) growth of graphite [6]

Cited in [6] scheme last years has received affirming in work [7]. It is grounded on the analysis of growth of crystals by the normal mechanism or screw dislocations, the authors [7] have received the following dependences of growth rate

$$V = \frac{D}{a_0} \cdot \frac{\Delta S}{k \cdot T},$$ (5)

for normal growth, and

$$V = \frac{D}{a_0} \cdot (\Delta T)^2 \cdot \frac{1}{k_0},$$ (6)

for dislocation mechanism of formation of spirals.

Here: D - diffusion constant;

ΔS - change of an entropy;

ΔT - degree of supercooling.

From comparison of dependence V from ΔT in [7] and scheme in work [6] we can conclude about good conformity of design and experimental data.

According to works [6, 7], at small supercoolings (ΔT₁) or supersaturation (ΔC) the speeds of longitudinal and cross growth are close, so in cast-iron the almost equiaxial crystals of graphite are formed. With increase of supercooling (ΔT₂) velocity of longitudinal growth sold by the normal mechanism association of atoms to prismatic planes of graphite is above than transversal. In such conditions are formed prolated graphite grains. At the very large supercoolings (ΔT₃) velocity of transversal growth accompanying with formation of jogs by the dislocation mechanism, above longitudinal, so the graphite acquires an equiaxial aspect. Sometimes all three stages of growth of graphite are determined on one plugging (fig. 3a). Influence of pinhole on the growth of spherical graphite is shown and after melting in the vacuum [8].

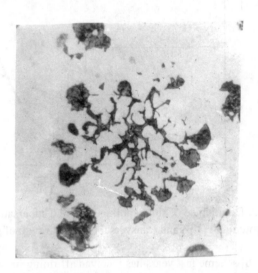

Figure. 3   Structure of cast-iron modified by magnesium,  x1000

The structure of graphite in the large degree is influenced by the mechanism of a release of space in a metal ground of cast iron. At exposition of this mechanism sometimes prefer shift processes - slipping and twinning. Their contribution in growth of spherical graphite is insignificant. It is testified by disposition of oxides and sulphides in graphite. At exploration in a raster microscope the spherical grains are determined in peripheral parts of spherical graphite, that it is impossible to explain by shift processes. The disposal of atoms of iron and silicium is realized by the vacancy mechanism, and foreign particle, as the markers in diffusive pairs, fix a line of phase boundary shifts and gets in graphite.

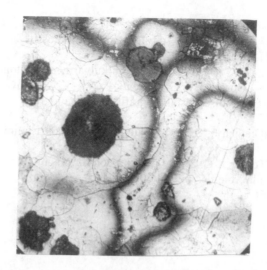

Figure. 3b  Structure of cast-iron modified by magnesium, x1000

In a fig. 3b the structure of high-duty cast iron after complex pickling treatment on an segregation of silicium and structure of a metal ground is shown. As we can see, that besides spherical graphite in a ground of cast iron the shrinkage cavity are determined. The formation of shrinkage cavity is promoted by admixture, which concentrating in a melt were in balance with primary blowholes, should be heightened. Last portions of eucletic fluid enriched with admixtures, are crystallized on boundaries in bounded together austenite and graphite units placed as lines, and contain set of pinholes. Not all of them are coated with graphite, that it is possible to explain by presence at an alloy of admixtures lowering a surface-tension more, than carbon. Such admixtures are oxygen and sulphur, which, by the way, lower a solidifying point of the rests of a melt.

## 2. Conclusions

Thus, example of structural changes occurring at modification of cast iron, formation of pinholes and also their roles in a graphite precipitation, and also influence of super-cooling confirm cited above [3] the mechanism of shaping of spherical graphite in modified cast irons. The considered sample comes out of developments of the theory of a graphitization of ferrous alloys, founded on a dominant role of discontinuity flaws in origin and growth of graphite.

## 3. References

1. Spravochnik po izgotovleniyu otlivok iz vysokoprochnogo chuguna / Pod red. Gorshkova A. A. - M. - K.: Mashgiz, 1961: 290.
2. Taran Yu. N., Chernovol A. V. Chugun s sharovidnym grafitom (50-letniy put' razvitiya teorii i proizvodstva). Metall i lit'e Ukraine, 1996; 6: 4 - 14.
3. Baranov A. A., Bunin K. P. O roli vklyucheniy pri grafitizatsii chuguna i stali //

Liteynoe proizvodstvo, 1964; 7: 26 - 28.

4. Kheyord A. Otritsatel'nye davleniya v zhidkosti,kak ikh zastavit' sluzhit' cheloveku? UFN, 1972; v. 108, №2: 315.

5. Otlivka iz chuguna s sharovidnym i vermikulyarnym grafitom / Zakharchenko E. V., Levchenko Yu. N., Gorenko V. G., Varenik P. A., - K.: Naukova dumka, 1986: 248.

6. Baranov A. A O gabituse kristallov grafita // Kristallografiya, 1964; v. 9, №5: 759-762.

7. Salli I. V., Fal'kevich E. S. Upravlenie formoy rosta kristallov. - K.: Naukova dumka, 1989: 160.

8. Kasperek J., Tellier J. C. Sur le germination du graphite daus des fonts synthetigues élaborées sous vide: role des microbulles gazeuses // Mem. et etud. sci. Rev. met, 1991; v. 88, №1:653-661.

# MODELING OF DEHYDRATION AND DEHYDROGENATION IN PURE AND Ba-, Ca-, Sr- OR Y-MODIFIED ZIRCONIA NANOLAYER

TOKIY N.V., KONSTANTINOVA T.YE., SAVINA D.L., TOKIY V.V.[1]
*Donetsk Physico-Technical Institute NAS Ukraine, 72, R.Luxembourg, 83114 Donetsk, Ukraine,    nat1976@mnogo.ru*
*(1)Donetsk Institute of Social Education, 2, Universitetskaya, 83000 Donetsk, Ukraine,    tokiy@dise.donbass.com*

## 1. Introduction

New developments in the quantum chemistry methodology, aimed at partial incorporation of the periodicity within the cell model, make it possible to obtain new data concerning the electronic structure of surface of nanolayers zirconia and the dopants in it.

The previous our work [1] was dedicated to the cell simulation of the electronic structure of 26 impurity d-elements (Sc, Ti, V, Cr, Mn, Fe, Co, Ni, Cu, Y, Nb, Mo, Tc, Ru, Rh, Pd, Ag, Lu, Hf, Ta, W, Re, Os, Ir, Pt, Au) and activation energies of dehydrogenation of Y-stabilized zirconia nanolayers.

The present paper is dedicated to the simulation of the electronic structure of hydroxyl and terminal water cover of zirconia nanolayer with Ba, Ca, Sr and Y cations, with using the tight-binding theory. Also present work is devoted to definition activation energies of dehydration, dehydroxylation and dehydrogenation in doped nanocrystalline zirconia.

These problems are considered within the framework of the cell model and the band calculations.

During simulation the impurity is placed in the substituting cation site.

## 2. Molecular-orbital treatment

We use the tight-binding theory [2-4]. In the theory of SPD-bonded systems the electronic eigenstates are written in terms of a basis set consisting of a single S state, five D states of each metal atom, three P states on each oxygen atom of zirconia and one S state on each hydrogen atom. In the given work, we use parameters containing values for diagonal terms of zirconia covered by hydrogen  $\varepsilon_s = -5.68$  eV,  $\varepsilon_d = -8.46$  eV,  $\varepsilon_p = -14.13$ eV,  $\varepsilon_{sH} = -13.55$ eV.

On different atoms these orbitals are assumed orthogonal. However atomic orbitals are not eigenfunctions of the considered quantum-mechanical system, as the Hamiltonian matrix elements between orbitals of various atoms are not equal to zero and we use, for account of diagonal elements, the data [2] for energy of s- and d- states of elements.

*T.N. Veziroglu et al. (eds.),*
*Hydrogen Materials Science and Chemistry of Carbon Nanomaterials,* 291-298.

292

The matrix elements of sp- and pd-type were considered on the basis of the pseudo-potential theory for transition metals from [2].

To find the eigenfunctions and eigenvalues of the system energy, it is necessary to diagonalize the symmetric matrix $H_{\mu v}$.

The single-electron orbitals of our cell can be expressed as

$$\Phi_\alpha = \sum_j^{NZr}\left(c_s^j\left|5s^j\right\rangle + \sum_{k=1}^{5}c_{d_k}^j\left|4d_k^j\right\rangle\right) + \sum_j^{NO}\sum_{i=1}^{3}c_{p_i}^j\left|2p_i^j\right\rangle + \sum_j^{NH}c_s^j\left|1s^j\right\rangle,$$

where NZr- is the number of zirconium atoms ;
NO- is the number of oxygen atoms;
NH- is the number of hydrogen atoms;
i - runs along the value of the three coordinate axes;
$c_{v\alpha}$ - solution of the one-electronic equations for a cell:

$$\sum_v^{n}\left(H_{\mu v} - \delta_{\mu v}E_\alpha\right)c_{v\alpha} = 0, \qquad \alpha = 1,2,..n \tag{1}$$

where $E_\alpha$ – the single-electron eigenvalue of the cell energy; $H_{\mu v}$ – the matrix elements between hybridized orbitals.

The band structure of $ZrO_2$ has the forbidden gap. Both bonding orbitals of Zr and O, and orbitals, located only on atoms of oxygen grouped into what can be identified as a valence band. The antibonding orbitals of Zr and O forming,a conduction band, and with a forbidden gap between. Thus electrons from d-levels pass on p- levels of oxygen, providing the ionic character.

The wave function electron, located on impure yttrium, whose factors are the solution of system of the equations (1), is given as a sum composed of 5s- and 4d- orbitals of each zirconium atom, 2p-orbitals of each atom of oxygen, 1s- orbitals of each atom of hydrogen both 5s- and 4d- orbitals of the impurity atom of yttrium.

$$\Phi_\alpha = \sum_j^{NZr}(c_s^j\left|5s^j\right\rangle + \sum_{k=1}^{5}c_{p_k}^j\left|4d_k^j\right\rangle) + \sum_j^{NO}\sum_{i=1}^{3}c_{p_i}^j\left|2p_i^j\right\rangle + c_{3s}^Y\left|5s\right\rangle + \sum_{k=1}^{5}c_{p_k}^Y\left|4d_k\right\rangle + \sum_j^{NH}(c_s^j\left|1s^j\right\rangle$$

where: NZr- is the number of zirconium atoms ;
NO- is the number of oxygen atoms;
NH- is the number of hydrogen atoms;
i - runs along the value of the three coordinate axes;
$c_{v\alpha}$ - solution of the one-electronic equations for a cell.

## 3. Dehydrogenation and dehydroxylation for three-coordinat the OH-groups

As a basis of modeling of dehydrogenation process of transformation of amorphous phase of hydroxide to crystal zirconia we take the model structure of amorphous zirconia proposed in work [5], in which distribution functions obtained by X-ray and neutron diffraction analysis showed, that the atoms were not distributed at random.

In work [5] is submitted a plane model for structure of amorphous zirconia, as a thin plate consisting of a layer of Zr atoms between two layers of oxygen (Fig. 1).

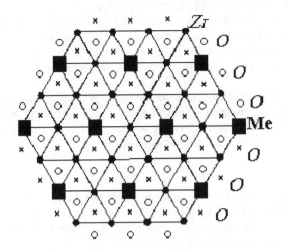

Fig. 1.Model of hydrogen covering amorphous zirconia. ● - Zr· atoms, ■ – Me atoms
(where Me = Zr, Y, Ba, Ca, Sr); O - atoms of oxygen and hydrogen located above Zr of a
plane; x — atoms of oxygen and hydrogen located under Zr plane.

In [1], this model of a plate-like amorphous zirconia is supplemented from above
and from below by planes of hydrogen atoms located above and under the oxygen atoms.

We used a cell of 20 atoms (3 Zr atoms, 1 Me atom, 8 O atoms and 8 H atoms): for
calculation of energy levels of the zirconia plate covered by hydrogen. Our approach is to
superpose periodic boundary conditions on the cell in the (111) plane.

The plates are, of course, hypothetical with respect to both the composition and the
structure. By definition they have the same lattice period and, moreover, show an ordering
arrangement of Me impurity atoms (in reality, these may be distributed within their
sublattices statistically).

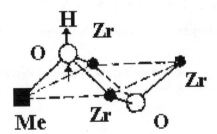

Fig.2. The structure of three-coordinate hydrogen covering of sides of zirconia with
fluorite structure.

The ability of hydroxyl ion to form bridges between ions of metal is well known [6]. According to modern concepts of the structure of hydrogen cover of oxyde [7,8] the reason of emergence of several stretching vibrations of OH - groups in IR-spectra, is that the oxygen of hydroxyl can be in contact to several atoms of metal. The atoms of metal are the nearest neighbours of OH - group, therefore their number should determine the frequency of vibration of OH.

A rough estimate of the bonding energy for the hydrogen and zirconia plate can be obtained by monitoring the total energy of the cell at removing hydrogen. As described in [9] a measure of the total energy of the system is given by the sum of the energies of all the occupied one-electron molecular orbitals. That is,

$$E_{tot} \cong \Sigma \, n_i \in_i,$$

where $\in_i$ is the energy and $n_i$ is the occupancy number of the i-th orbital.

The calculation of total energies of all the occupied one-electron molecular orbitals has been carried out to analyze the electronic structure in Zr4O8H8, Zr3MeO8H7, Zr3MeO8H8, Zr4O8H7, where Me = Ba, Ca, Sr, Y.

We define three-coordinate dehydrogenation energy in undoped zirconia,

$E_{HbZr} = 11.43$ eV.

As calculations show, the activation energy of thermal desorption of hydrogen from a plate of zirconia change at doping.

$E_{HbY} = 11.18$ eV.

$E_{HbBa} = 11.39$ eV.

$E_{HbCa} = 11.39$ eV.

$E_{HbSr} = 11.39$ eV.

A rough estimate of the bonding energy for the hydroxyl and zirconia plate can be obtained by monitoring the total energy of the cell at removing hydroxyl.

The calculation of total energies of all the occupied one-electron molecular orbitals has been carried out to analyze the electronic structure in Zr4O8H8, Zr3MeO7H7, Zr3MeO8H8, Zr4O7H7, where Me = Ba, Ca, Sr, Y.

We define three-coordinate dehydroxylation energy in undoped zirconia:

$E_{HObZr} = 6.05$ eV.

As calculations show, the activation energy of thermal desorption of hydroxyl from a plate of zirconia change at doping.

$E_{HObY} = 5.74$ eV.

$E_{HObBa} = 4.58$ eV.

$E_{HObCa} = 4.61$ eV.

$E_{HObSr} = 4.60$ eV.

## 4. Dehydrogenation and dehydroxylation for terminal the OH-groups

In the present paper is considered a terminal hydroxyl group of type I located at a cation in addition for three-coordinate the OH-group (see fig.3,4)

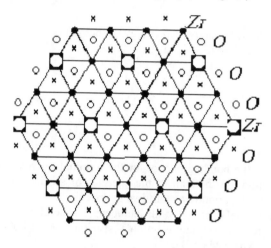

Fig. 3.Model of hydrogen covering amorphous zirconia with terminal OH-groups.
• - Zr· atoms;▫ – Me atoms with oxygen and hydrogen located above Me (where Me = Zr, Y, Ba, Ca, Sr); O - atoms of oxygen and hydrogen located above Zr of a plane; x — atoms of oxygen and hydrogen located under Zr plane.

We used a cell of 22 atoms (3 Zr atoms, 1 Me atom, 9 O atoms and 9 H atoms): for calculation of energy levels of the zirconia plate covered by hydrogen and terminal hydroxyl groups.

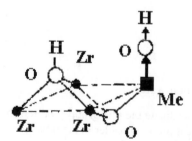

Fig.4. The structure of three-coordinate hydroxyl covering of sides of zirconia with fluorite structure.

A rough estimate of the bonding energy for the hydrogen and zirconia plate can be obtained by monitoring the sum of the energies of all the occupied one-electron molecular orbitals of the cell at removing hydrogen from terminal oxygen.

The calculation of total energies of all the occupied one-electron molecular orbitals has been carried out to analyze the electronic structure in Zr4O9H9, Zr3MeO9H8, Zr3MeO9H9, Zr4O9H8, where Me = Ba, Ca, Sr, Y.

We define terminal dehydrogenation energy in undoped zirconia,

$E_{HtZr}$ = 10.84 eV.

As calculations show, the activation energy of thermal desorption of hydrogen from terminal oxygen of zirconia change at doping.

$E_{HtY}$ = 10.32 eV.

$E_{HtBa}$ = 11.51 eV.

$E_{HtCa}$ = 11.49 eV.

$E_{HtSr}$ = 11.50 eV.

A rough estimate of the bonding energy for the terminal hydroxyl and zirconia plate can be obtained by monitoring the sum of the energies of all the occupied one-electron molecular orbitals of the cell at removing terminal hydroxyl.

The calculation of total energies of all the occupied one-electron molecular orbitals has been carried out to analyze the electronic structure in Zr4O9H9, Zr3MeO8H8, Zr3MeO9H9, Zr4O8H8, where Me = Ba, Ca, Sr, Y.

We define terminal dehydroxylation energy in undoped zirconia,

$E_{HOtZr}$ = 1.30 eV.

As calculations show, the activation energy of thermal desorption of terminal hydroxyl from a plate of zirconia change at doping.

$E_{HOtY}$ = 1.33 eV.

$E_{HOtBa}$ = 0.35 eV.

$E_{HOtCa}$ = 0.37 eV.

$E_{HOtSr}$ = 0.36 eV.

## 5. Dehydration for terminal water

In addition for three-coordinate the OH-groups and terminal hydroxyl group of type I, a terminal water locate at a cation we consider in the present paper (see fig.5,6)

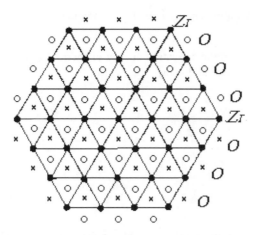

Fig. 5. Model of hydrogen with terminal water covering amorphous zirconia. • - Zr· atoms; ◙ – Me atoms with water located above Me (where Me = Zr, Y, Ba, Ca, Sr); O - atoms of oxygen and hydrogen located above Zr of a plane; x — atoms of oxygen and hydrogen located under Zr plane.

We used a cell of 23 atoms (3 Zr atoms, 1Me atom, 9 O atoms and 10 H atoms): for calculation of energy levels of the zirconia plate covered by hydrogen and terminal water.

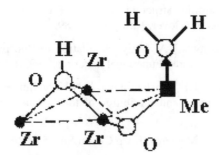

Fig. 6. The structure of hydroxyl and terminal water covering of sides of zirconia with fluorite structure.

A rough estimate of the bonding energy for the terminal water and zirconia plate can be obtained by monitoring the total energy of the cell at removing terminal water.

The calculation of total energies of all the occupied one-electron molecular orbitals has been carried out to analyze the electronic structure in Zr4O9H10, Zr3MeO8H8, Zr3MeO9H10, Zr4O8H8, where Me = Ba, Ca, Sr, Y.

We define terminal dehydration energy in undoped zirconia,

$E_{H2OtZr} = 3.22$ eV.

As calculations show, the activation energy of thermal desorption of terminal water from a plate of zirconia change at doping.

$E_{H2OtY} = 3.02$ eV.

$E_{H2OtBa} = 0.60$ eV.

$E_{H2OtCa} = 0.67$ eV.

$E_{H2OtSr} = 0.63$ eV.

## 6. Conclutions

In present work we predicted influence of every impurity Y, Ba, Ca, Sr the process of thermal desorption of hydrogen, OH- group and terminal water from the surface of zirconia plate nucleus. The results of our calculations are coordinated with experimentally observable disappearance of absorption band of terminal hydroxyl groups in zirconia at doping by Ba, Ca, Sr [8].

## 7. References

1. T.Konstantinova, V.Tokiy, N.Tokiy, D.Savina Computational modeling of electron properties of 26 d-elements in nanolayer Y-doped tetragonal zirconia. CIMTEC'2002

2. W.A.Harrison, Electronic Structure and the Properties of Solids, Freeman, San Francisco, 1980, reprinted by Dover, New York, 1989.

3. N.V.Tokiy,M.V.Grebenyuk,V.V.Tokiy. Simulation of Phosphorus on Diamond surface (111) and its hyperfine structure// Functional materials. 1995,т.2 №2.с.294-295

4. V.V.Tokiy, N.V.Tokiy, T.E.Konstantinova, V.N.Varyuhin, British Ceramic Proceedingc No.60. The Sixth Conference and Exhibition of the European Ceramic Society, 2 (1999) 491

5. J. Livage, K.Doi, and C. Mazieres. J.Am.Ceram.Soc. 51 (1968) 349

6. Kotton F., Uilkinson Dg. Sovremennaya neorganicheskaya khimiya, v.2. M., «Mir», 1969

7. Morterra C., Cerrato G., Ferroni L. // Mater. Chem. Phys. 1994. 37. P. 243

8. A.N. Kharlamov, N.A. Zubareva, E.V. Lunina, V.V. Lunin, V.A. Sadykov, A.S. Ivanova, Gidroksil'niy pocrov i elektronoaktseptornie svoystva poverkhnosti dioksida tsirkoniya, promotirovannogo kationami strontsiya, bariya i kal'tsiya, http://www.chem.msu.su/rus/vmgu/Lunin/welcome.html

9. N.Tokiy, V.N.Varyuhin, V.V.Tokiy // Proceedings of 4-th international symposium on diamond films and related materials. 1999, pp.99-102

# METALLCONTAINING NANOPARTICLES IN CARBOCHAIN POLYMERIC MATRIXES

GUBIN S.P.[1], BUZNIK V.M.[2], YURKOV G.YU.[1], KOROBOV M.S.[1],
KOZINKIN A.V.[3], TSVETNIKOV A.K.[4], DOTSENKO I.P.[1]

[1]N.S. Kurnakov Institute of General and Inorganic Chemistry RAS, Leninskii pr. 31,
Moscow, 119991 Russia.
[2]Boreskov Institute of Catalysis, Siberian Division, RAS, akademika Lavrent'eva pr. 5,
Novosibirsk, 630090, Russia
[3] Research Institute of Physics, Rostov State University, pr. Stachki 194, Rostov-on-Don,
344104, Russia;
[4]Institute of Chemistry, Far East Division, RAS, Stoletiya Vladivostoka pr. 139,
Vladivostok, 690022, Russia.

**Keywords**: nanoparticles, carbochain polymeric matrixes, nanoparticles stabilized of
carbochain polymeric matrixes.

## Abstract

In this paper in order to stabilize nanoparticles, matrices consisting of organic polymers
such as polytetrafluorinethylene (teflon), polyethylene, and polypropylene are used. A
universal method to introduce metal-containing nanoparticles in a teflon matrix has been
developed to allow fabrication of large amounts (kilogram-scale) of nanoparticle polymer
composites. Encapsulation was done by thermal decomposition of the metal-containing
compounds (MRn; M = Fe, Cu, Ni; R = CO, HCOO, $CH_3COO$) in a dispersion system of
polytetrafluorinethylene in mineral oil or in a solution-melt of polyethylene or
polypropylene in mineral oil. For subsequent characterization, we used TEM, EXAFS, X–
ray emission spectroscopy and Mossbauer spectroscopy. The morphology and particle size
distributions were investigated using a high-resolution electron microscope (TEM).

## 1. Introduction

Nanometer size structures are an intermediate form of matter, which fills the gap between
atoms/molecules and bulk materials. Often, these types of structures exhibit exotic physical
and chemical properties different from those observed in bulk three–dimensional materials.
This interdisciplinary field of mesoscopic and nano-scale systems is important to
fundamental physics, as well as for some new technologies. As is well known, some of the
metallic nano–particles are thermodynamically unstable and chemically very reactive. For
many technological applications, materials with nanoparticles embedded in a matrix are
impotant. In this paper for the stabilization of nano-particles the matrices of the organic
polymers as polytetrafluorinethylene (teflon), polyethylene, polypropylene it is used.

299

*T.N. Veziroglu et al. (eds.),*
*Hydrogen Materials Science and Chemistry of Carbon Nanomaterials, 299-306.*

The high reactivity of nanoparticles and their tendency toward spontaneous compaction accompanied by deterioration of basic physical properties make stabilization be a major challenge in fabrication of materials based on metal nanoparticles. The best-developed method of stabilization is embedding nanoparticles in polymer matrices [1]. The known "friability" of the structures of most of partially crystalline carbon-chain polymers forms the basis for the method of introducing metal-containing nanoparticles into solution melts of polymers in hydrocarbon oils [2]. However, this method is inapplicable to "hard" polymer matrices, such as polytetrafluoroethylene.

At the same time, creation of materials that combine opposite properties of its components at the nanolevel is one of the rapidly developing strategies of modern materials science. For example, it is expedient to augment the high thermal and chemical stability of polytetrafluoroethylene with the electrical conductivity on a semiconductor scale, magnetization by means of introducing metal nanoparticles, or optoelectronic properties by introducing so-called quantum dots, nanoparticles of some metal sulfides and selenides. At first glance, it would seem that poor solubility and swelling ability of polytetrafluoroethylene in solvents hinder the introduction of different nanoparticles in this matrix and their uniform distribution over its bulk. In searching for the solution of this problem, we have focused our attention of ultradispersed polytetrafluoroethylene (UPTFE). UPTFE was fabricated by a thermal gas dynamic method, which is suitable for commercial production of UPTFE [3]. The essence of the method is in formation of a finely dispersed UPTFE powder upon the thermodestruction of the block polymer.

## 2. Experimental Details

Metallcontaining nanoparticles on the surface nanogranules polytetrafluoroethylene were prepeared by new methods on based a standart "cluspol" technique [4]. Metallcontaining nanoparticles were prepeared by original "cluspol" technique. This is universal method, which allows the fabrication of large amounts (kilogram-scale) of polymer nanoparticle composites. The encapsulation was done by thermal decomposition of metallcontaining compounds (MRn; M = Fe, Cu, Ni; R = CO, HCOO, $CH_3COO$) in dispersion system of polytetrafluoroethylene in mineral oil; solution-melt polyethylene or polypropylene in mineral oil.

We are discovered that granules ultradispersion polytetrafluoroethylen make fluidized bed on the surfase mineral oil. This effect we are used for nanometallization of nanogranules. An appropriate amount of metallcontaining compounds was added to the high-temperature dispersion system of polytetrafluoroethylene in mineral oil with vigorous stirring. The residual oil was removed by washing with benzene in a Sohxlet apparatus. The resultant powder was dried in vacuum and stored in air. The optimum conditions are developed for the decomposition of MCC in order to introduce highly reactive nanoparticles into the polymeric matrix with concentration of 2-10 wt. %. The particle size of iron oxide was determined by transmission electron microscopy (TEM) on a JEOL JEM-100B high-resolution microscope. The powder was dispersed ultrasonically in ethanol and then spread over a carbon substrate.

Extended Fe $K$ x-ray absorption fine structure (EXAFS) measurements were made with a bench-scale spectrometer [5] equipped with a focusing quartz analyzer. The x-ray source used was a Mo-BSV21 finefocus x-ray tube (20 kV, 30 mA). The sample was thoroughly ground with Apiezon, and the mixture was sandwiched between mylar films. The absorber thickness was such that the ratio of incident to transmitted intensity was about 3. The absorption spectra $\mu,(E)$ were processed by a standard procedure [6]. Fourier analysis of the EXAFS curves, $k^2(\chi)k$, of the samples and standards (δ-Fe and γ-Fe2O3) was carried out in the range $k = 2.7$-$12.5$ Å$^{-1}$.

Mossbauer spectra were recorded on an MC1101E high-speed spectrometer at room temperature (without thermostatic control) and liquid-nitrogen temperature. The source used was rhodium-shielded "Co with an activity of 10 mCi (ZAO Tsiklotron). The isomer shift was measured relative to a-Fe. The spectra were analyzed by a least squares fitting routine using the UNTVEM v.4.50 program.

X-ray diffraction (XRD) measurements were made on powder and pressed samples with a DRON-3 difractometer *(CuKa* radiation, scan speed of 2°/min). Peak positions were determined with an accuracy of ±0.05°. Electron paramagnetic resonance (EPR) spectra were measured with a Varian E-4 X-band spectrometer equipped with a Varian-E257 nitrogen-sweep variable-temperature facility. From the measured spectra, we determined the effective peak-to-peak width $\Delta H_{pp}$, peak-to-peak height $A_{pp}$, and signal intensity $I \cong A_{pp}(\Delta H_{pp})^2$.

## 3. Results and Discussion

### 3.1. STABILIZATION ON THE SURFACE OF UPTFE MATRIX.

Scanning tunneling and transmission electron microscopy [7-8] shows that powder particles several hundreds of nanometers in size have a regular spherical shape and an intricate inner structure composed of tinier particles; they are prone to conglomeration. Inasmuch as UPTFE has a highly developed surface (~1000 m$^2$/g) and the above-mentioned surface functional groups, it is very active and can be stabilized as stable suspensions in some solvents. In this work, metal (Fe, Co, Ni, Cu) formates, acetates, oxalates, ammines, and carbonyls were used as MCCs; thermodestruction of these compounds has been comprehensively studied [2]. Metal content of the resulting composite samples was 5–20 wt %. The size of granules of UPTFE in the range from 100-300 nm are shown in Fig. 1.

Metal-containing nanoparticles are arranged in bunches at the surface of nanoglobules (Fig. 2). The mean size of these nanoparticles is below 6 nm, and the size distribution is rather narrow. The particle shape often differs from the spheroidal shape typical of metal-containing nanoparticles in polyethylene and other polymeric matrices. In some cases, a clearly pronounced polyhedral form of particles is observed.

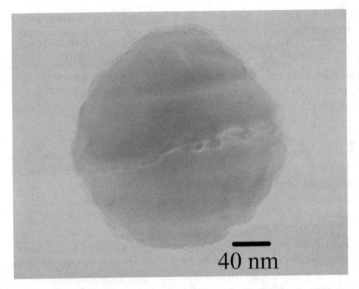

Fig. 1. TEM microphotograph of granule of ultradispersed polytetrafluoroethylene.

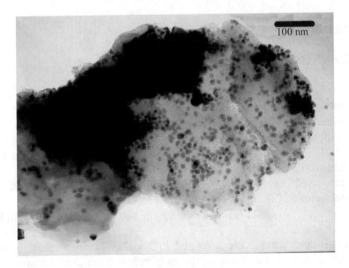

Fig. 2. TEM microphotograph of Ni-containing nanoparticles on the surface of nanogranule UDPTFE. Middle size 18±3 nm.

The XRD pattern of the Fe-containing nanoparticles on the surface of UPTFE shows a strong low-angle peak from polytetrafluoroethylene and weak, broad peaks from an Fe-containing phase, indicative of small particle size (Table 1.).

Table 1. XRD data for Fe-containing compounds (14,03 wt. % Fe) on the surface UPTFE.

| № | D, Å | I/I$_0$, % | d, Å | I/I$_0$, % | d, Å | I/I$_0$, % | d, Å | I/I$_0$, % | d, Å | I/I$_0$, % |
|---|---|---|---|---|---|---|---|---|---|---|
| | This work | | JCPDS PDF data | | | | | | | |
| | | | γ-Fe$_2$O$_3$ (№ 39-1346) | | α-Fe$_2$O$_3$ (№ 33-664) | | β-FeOH (№ 34-1266) | | Fe$_5$O$_7$(OH)·4H$_2$O (№ 29-712) | |
| 1 | 4,900 | 100 | | | | | | | | |
| | | | 3,740 | 5 | 3,684 | 30 | 3,728 | 5 | | |
| | | | 3,411 | 5 | | | 3,333 | 100 | | |
| | | | 2,953 | 35 | | | | | | |
| 2 | 2,667 | 15 | 2,6435 | 2 | 2,700 | 100 | 2,6344 | 25 | | |
| 3 | 2,521 | 20 | 2,5177 | 100 | 2,519 | 70 | 2,5502 | 55 | 2,5 | 100 |
| | | | | | | | 2,3559 | 9 | | |
| | | | 2,2320 | 1 | 2,207 | 20 | 2,2952 | 35 | 2,21 | 80 |
| | | | | | | | 2,1038 | 7 | | |
| | | | | | | | 2,0666 | 7 | | |
| 4 | 2,022 | 20 | 2,0886 | 16 | 2,0779 | 3 | 1,9540 | 20 | 1,96 | 80 |
| | | | | | 1,8406 | 40 | | | | |
| | | | 1,7045 | 10 | 1,6941 | 45 | 1,7557 | 15 | 1,72 | 50 |
| 5 | 1,605 | 15 | | | 1,6033 | 5 | 1,6434 | 35 | | |
| | | | 1,6073 | 24 | 1,5992 | 10 | | | | |
| | | | 1,5248 | 2 | 1,4859 | 30 | 1,5155 | 9 | 1,51 | 70 |
| | | | | | | | 1,5034 | 5 | | |
| 6 | 1,475 | 15 | 1,4758 | 34 | 1,4538 | 30 | 1,4456 | 15 | 1,48 | 80 |
| | | | | | 1,3115 | 10 | | | | |
| | | | | | 1,3064 | 6 | | | | |
| | | | 1,2730 | 5 | 1,2592 | 8 | | | | |

The particle composition is rather complex. Previously [9], we showed that, in the course of formation of Fe nanoparticles in the matrix of the copolymer of ethylene and tetrafluoroethylene (fluoroplastic-40), the matrix is partially defluorinated to produce the $FeF_2$ phase at the surface of the particles. It is not improbable that an analogous process takes place in the matrix under study: as shown by Mössbauer spectra (Fig. 3), the resulting nanoparticles contain, in addition to the metal and metal oxide and carbide phases, iron (II) ions, which are presumably incorporated in iron fluoride. Magnetic measurements give the strongest evidence of the formation of nanoparticles in the material. The fact is that a decrease in the size of a magnetic particle results, under definite conditions, in its single-

304

domain state. In the single-domain state, domain walls are lacking and the external magnetic flux is present, because in a particle whose size is smaller that than some critical value, creation of the magnetostatic energy is more favorable.

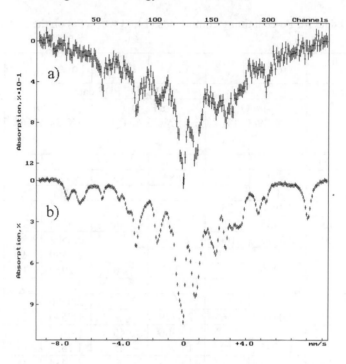

Fig. 3. Mossbauer spectra of sample: (a) containing 4,09 wt. % Fe and (b) containing 14,03 wt. % Fe on the surface of nanogranules UDPTFE.

As judged from the EPR spectral shape, we may assume that the particles in the sample are mainly superparamagnetic at room temperature; this points to their relatively small dimensions and single-domain magnetic structure. However, the internal magnetic energy of each of these particles is quite sufficient for the entire material to be strongly attracted by a constant magnet.

Therefore, we developed the method of nanometalization of ultradispersed polytetrafluoroethylene; the resulting powdery material contains nanoparticles of complex composition, with physical characteristics typical of such particles. The composite powders can be converted by known methods into compact block polytetrafluoroethylene with uniformly distributed metal-containing nanoparticles.

Necessary note, that stabilizing resulting metallcontaining nano-particle for polyethylene and for polytetrafluoroethylene realized of different ways. For in case of polyethylene high pressure matrix stabilizing carry out in all volume of polymer, but in case of ultradispersed polytetrafluoroethylene matrix metallcontaining nanoparticles are bunched at the surface of nanogranules.

## 3.2. STABILIZATION IN VOLUME POLYETHYLENE MATRIX

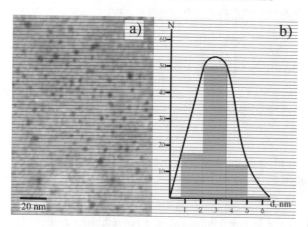

Fig. 4. TEM image of the $Fe_2O_3$ + polyethylene sample (a) and distribution of the size Fe-containing nanoparticles (b).

The formation of nanoparticles was confirmed by high-resolution TEM (Fig. 4). The particle size was determined to be $15 \pm 3$ nm.

On Fig. 5 can see the normalized EXAFS FTM of the $Fe_2O_3$ + polyethylene sample. The FTM of a normalized EXAFS curve passes through maxima at $r_j = R_j - \alpha_j$, where $R_j$ is the mean distance from the absorbing atom to the atoms of the $j$th coordination sphere, and $\alpha_j$ is the phase correction which can be determined from the EXAFS spectra of compounds with known structures or by calculation. As can be seen in Fig. 5, the FTM of the sample studied is similar in shape and peak positions to that of $\gamma$-$Fe_2O_3$ and differs from the FTM of $\alpha$-$Fe_2O_3$ in the position of the second peak and the total number of peaks.

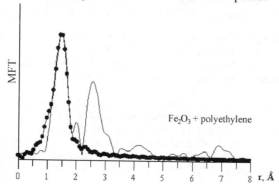

Fig. 5. Normalized Fe $K$ EXAFS FTM for the $Fe_2O_3$ +polyethylene (PE) sample.

As can be seen from Table 2, the model yield different average interatomic distances, equal coordination numbers (within the error of determination), and comparable $R$ factors.

Table 2. Coordination numbers $N_j$, interatomic distances $R_j$, and rms atomic displacements $\sigma^2_j$ in the nearest neighbor oxygen environment of Fe.

| Sample | $N_1$ | $N_2$ | $R_1$, Å | $R_2$, Å | $R_{cp}$, Å | $\sigma^2_1$, Å$^2$ | $\sigma^2_2$, Å$^2$ | R-factor, % |
|---|---|---|---|---|---|---|---|---|
| $Fe_2O_3+\Pi\Theta$ | 2.8 | 2.5 | 1.93 | 2.03 | 1.98 | 0.0025 | 0.0107 | 8 |
| $\gamma$-$Fe_2O_3$ | 2.6 | 2.8 | 1.91 | 2.03 | 1.97 | 0.0019 | 0.0069 | 7 |

Necessary note, that most of the powders, except those containing copper nanoparticles, are amorphous as probed by X-ray diffraction.

Our experimental results demonstrate that we obtained isolated magnetic $Fe_2O_3$ nanoparticles 15 ±3 nm in size embedded in a polyethylene matrix. The nanoparticles are close in structure to bulk $\gamma$-$Fe_2O_3$.

## Acknowledgment

This work was supported by the Russian Foundation for Basic Research (*02-03-32321, , 02-03-32435, 03-03-06386, 03-03-06387, 01-03-32783 и 01-03-32955*), ISTC (project no. 1991), and by program of fundamental research RAS «Fundamental physical and chemical problems of nanosizing systems and nanomaterials».

## 4. References

1. A.D. Pomogailo, A.S. Rozenberg, and I.E. Uflyand, 2000. Nanochastitsy metallov v polimerakh (Metal Nanoparticles in Polymers), Moscow: Khimiya, 2000.
2. S.P. Gubin and I.D. Kosobudskii, 1983. Usp. Khim., 52, 1350-1364.
3. V.M. Buznik, A.K. Tsvetnikov, L.A. Matveenko, 1996. Khimiya v Interesakh Ustoich. Razvitiya, 4, 489-496.
4. I.D. Kosobudskii and S.P. Gubin. 1985. Metallic Clusters in Polymer Matrices: A New Type of Metal-Pilled Polymers, Vysokomol. Soedin., 27(3), 689-695.
5. A.T. Shuvaev, B.Yu. Khel'mer and T.A. Lyubeznova, 1988. Spectrometric Facility for DRON-3 Xray Diffractometers for Investigation of the Short-Range Order in Crystalline and Disordered Systems, Prib. Tekh. Eksp., 3, 234-237.
6. D.L. Kochubei, Yu.A. Babanov, K.I. Zamaraev et al., 1988. Rentgenospektral'nyi metod issledovaniya struktury amorfnykh tel: EXAFS-Spektroskopiya (EXAFS Spectroscopy: A Tool for Probing the Structure of Amorphous Materials), Novosibirsk: Nauka.
7. V.M. Buznik, A.K. Tsvetnikov, B.Yu. Shikunov and V.V. Pol'kin, Perspekt. Mater., 2, 69-72.
8. V.G.Kuryavyi, A.K. Tsvetnikov and V.M. Buznik, 2001. Perspekt. Mater., 3, 57–62.
9. I.D. Kosobudskii, S.P. Gubin, V.P. Piskorskii et al., 1985. Vysokomol. Soedin., 4, 689–695.

# X-RAY STRUCTURAL STUDY OF DEPOSIT FORMED ON ELECTRIC ARC SPUTTERING OF Me₁-Me₂-C COMPOSITES

A. A. ROGOZINSKAYA, D. V. SCHUR, I. I. TIMOFEEVA,
L. A. KLOCHKOV, A. P. SIMANOVSKIY, A. A. ROGOZINSKIY
*Institute for Problems of Materials Science of Ukrainian Academy of
Science, 3, Krzhizhanovsky str., Kiev, 03142 Ukraine*

## 1. Introduction

At present, nanostructural carbon, specifically in the form of nanotubes and fullerenes, is used in different applications. These materials take a special place in electronics due to their physical and chemical properties.

One of the products resulting from electric arc synthesis of fullerenes is deposit - the growth forming on the cathode on the electric arc evaporation of the anode which consists of either graphite or composites containing carbon and different additives. The deposit mainly consists of multi-wall carbon nanotubes and some amount of graphitized carbon mass [1].

As the deposit is formed at high temperatures and under high-power electric field, it is of specific interest as the source for electronic industry.

## 2. Experimental Part and Discussion

Additives introduced into the electrode by intercalation, encapsulation and other methods have a strong effect on physical and chemical, electrical and other properties of the product what allows the development of its application.

Table 1 shows additives introduced in amount of 2-4% and conditions for the deposit production.

TABLE 1. Conditions for deposit production.

| № | Catalyst | Components ratio | $U_B$ | $I_A$ | Speed rate of cathode, mm/min | Deposit: % of the mass of the anode sputtered |
|---|----------|------------------|-------|-------|-------------------------------|-----------------------------------------------|
| 1 | Si | 1 | 32 | 180 | 1,82 | 42,9 |
| 2 | Hf+Ni | 10:1 | 35 | 175 | 1,45 | 48,1 |
| 3 | Hf+Fe | 8:1 | 33 | 180 | 1,68 | 39,7 |
| 4 | Co+Mg | 1:1 | 35 | 175 | 1,77 | 36,9 |
| 5 | Co+Ti | 1:1 | 32 | 180 | 2,0 | 56,8 |
| 6 | Co+W | 1:1 | 37 | 170 | 1,79 | 65,3 |
| 7 | Co+Ni | 3:1 | 40 | 180 | 1,81 | 11,6 |
| 8 | Fe+Ni | 3:1 | 39 | 180 | 1,91 | 27,1 |
| 9 | Fe+Ni+Cr | 1:1:1 | 35 | 180 | 1,88 | 37,9 |

*T.N. Veziroglu et al. (eds.),*
*Hydrogen Materials Science and Chemistry of Carbon Nanomaterials,* 307-312.
© 2004 *Kluwer Academic Publishers. Printed in the Netherlands.*

308

To ascertain the phase composition and structural features of the deposit produced, there exists necessity to perform the X-ray analysis of the deposit with metal catalysts introduced and without them.

In the work represented the X-ray structural study has been performed on the products resulting from electric arc evaporation of graphite and $C+M_1+M_2$ composites (here $M_1$ and $M_2$ - different transition metals-catalysts).

The following composites were used: (Hf+Ni+C); (Hf+Fe+C); (Co+Mg+C); (Co+Ti+C); (Co+W+C); (Co+Ni+C); (Fe+Ni+C); (Fe+Ni+Cr+C); (Si+C).

Preliminary, the deposit was subjected to grinding for 3 h in the planetary mill, in the reactor volume isolated from the external atmosphere, at the grinding rate of 94 rev/min in the $H_2$ medium.

X-ray photography of the ground deposit has been performed on the X-ray apparatus DRON-3M in the filtered CuKa- radiation followed by decoding the diffractograms [2].

The X-ray structural study of the deposit, prepared from the graphite anode without metal catalysts introduced, revealed that the deposit consisted of two phases - hexagonal and rhombohedral graphite (Fig. 1).

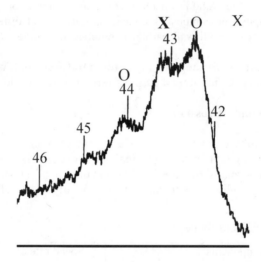

Fig.1. Fragment from the diffractogram of the deposit not containing catalysts. O - hexagonal graphite; X - rhombohedral graphite

When metal catalysts are introduced, the master phase in the deposit remains graphite in its two modifications. According to the intensity of X-ray lines, the hexagonal phase dominates. The fact, that the catalysts introduced increases the interlayer distances for line (002) of graphite, attracts attention (Table 2).

The important parameter is the size of particles in the deposit. The size has been calculated according to the broadening X-ray diffraction line (002) of graphite and

309

assuming that all the broadening is defined by dispersity of particles [3]. The high-temperature method for deposit production is the basis for the calculation above.

As Table 2 shows, the calculated size of particles is in the nano-dimensional range from 62 to 80 nm. Without metal catalysts, the size of carbon particles varies around 90 nm.

TABLE 2. Interplanar distances and the size of graphite particles in deposit.

| № | Catalyst | $d_{002}$, nm | $D_{002}$, nm | T melt., K |
|---|---|---|---|---|
| 1 | C (without cat.) | 0,3412 | 90 | C 4680 |
| 2 | Si | 0,3458 | 69 | Si 1688 |
| 3 | Hf+Ni | 0,3453 | 62 | Hf 2222 |
| 4 | Hf+Fe | 0,3427 | 80 | Fe 1811 |
| 5 | Co+Mg | 0,3461 | 76 | Co 1767 |
| 6 | Co+Ti | 0,3466 | 80 | Ti 1941 |
| 7 | Co+W | 0,3452 | 70 | W 3660 |
| 8 | Co+Ni | 0,3453 | 69 | Ni 1728 |
| 9 | Fe+Ni | 0,3452 | 65 | Fe 1811 |
| 10 | Fe+Ni+Cr | 0,3453 | 63 | Cr 2176 |

The catalysts introduced result in forming new phases [4]. Metal carbides and intermetallic compounds, corresponding to the compositions of compound additions, are formed (Table 3).

When Si is added into the graphite charge, the deposit shows the reflection lines of silicon carbides ß-SiC (cubic modification) and traces of a-SiC (hexagonal modification) (Fig.2)

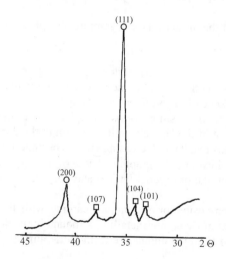

Fig.2. Fragment from the diffractogram of the deposit with Si-catalyst.
o –ß-SiC;  □–a-SiC

TABLE 3. Phase composition of deposit.

| Catalyst | Phase composition of deposit[*] | Lattice constant $a$, Å | Lattice volume $V$, Å$^3$ | Atomic radius $r_{at.}$ Me, Å |
|---|---|---|---|---|
| Si | βSiC<br>α-SiC (little) | 4,358 | 82,768 | Si 1,17 |
| Hf+Ni | HfC stoich.<br>Ni$_3$C (little) | 4,636 | 99,639 | Hf 1,59 |
| Hf+Fe | HfC stoich.<br>Hf(Fe)C | 4,636<br>4,605 | 99,639<br>97,653 | Hf 1,59<br>Fe 1,27 |
| Co+Mg | Mg$_2$C$_3$ (little)<br>Co$_2$C (v. little) | –<br> | –<br> | Mg 1,60<br>Co 1,26 |
| Co+Ti | TiC stoich.<br>CoTi (little) | 4,324 | 80,846 | Ti 1,45<br>Co 1,26 |
| Co+W | WC cubic<br>α-WC hex. (little) | 4,220 | 75,151 | W 1,40 |
| Co+Ni | Ni(Co) solid solut. | 3,529 | 43,950 | Ni 1,24<br>Co 1,26 |
| Fe+Ni | γ-Fe(Ni) solid solut. | 3,595 | 46,462 | Fe 1,27<br>Ni 1,24 |
| Fe+Ni+Cr | γ-Fe(Ni) solid solut.<br>Cr$_3$C$_2$ (little) | 3,598 | 46,578 | Cr 1,28<br>Fe 1,27 |

[*]As in all the deposit the master phase is graphite, this phase is not given in the table.

The (Hf+Ni) composition added into the graphite charge results in forming hafnium carbide HfC with the stoichiometric composition and the lattice constant a = 4.636 A. The traces of nickel carbide Ni$_3$C are seen (Fig.3).

The (Hf+Fe) composition forms HfC phases with the stoichiometric composition and very little amount of the Hf(Fe)C solid solution in the deposit. Substitution of hafnium atoms with iron atoms results in decreasing lattice constant of hafnium carbide (Table 3).

(Co+Mg) added into the graphite charge of the anode results in magnesium carbide Mg$_2$C$_3$ (little amount, the hexagonal lattice) and very little amount of Co$_2$C forming in the cathode deposit.

(Co+Ti) composite forms titanium carbide with the stoichiometric composition and very low amount of CoTi intermetallic compound in the deposit.

Addition of (Co+W) composite forms little amount of both cubic and hexagonal WC in the deposit.

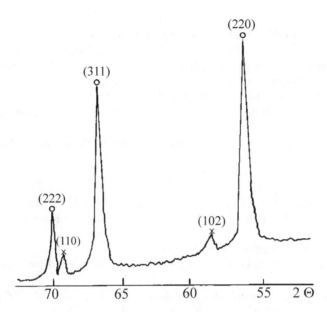

Fig. 3. Fragment from the diffractogram of the deposit with (Hf+Ni) catalyst.
o - HfC; x - Ni$_3$C

When (Co+Ni) is added into the anode charge, the continuous solid solution Ni(Co) is formed in the deposit. The slight increase in the lattice parameter of nickel from 3.524 A to 3.529 A indicates this. This increase is caused by the larger Co atoms present in the nickel lattice.

(Fe-Ni) composition forms the solid solution γ -Fe(Ni) in the deposit. Here, Ni stabilizes the y-Fe lattice. According to literature data, the lattice constant of γ-Fe equals 3.645 A. The observed decrease in the lattice constant of γ -Fe(Ni) is caused by the size of Ni atoms.

After electric arc evaporation of the anode with (Fe-Ni-Cr) additive, low amount of y-Fe(Ni) and 0^02 is observed in the deposit (Table 3).

Therefore, in electric arc evaporation of the composites comprising the mechanical mixture of the alloy and carbon, their constituents interact in the gaseous phase. When added, carbide-forming metals interact with carbon to form corresponding carbides (SiC, HfC, TiC, WC). Moreover, there is possibility of complex carbides forming, for example, Hf(Fe)C.

Detected carbides Mg$_2$C$_3$, Co$_2$C, Ni$_3$C seem to be formed in quick cooling the gaseous phase moving at high velocity under strong electric field. The presence of the cubic tungsten carbide, which is formed and stable only at high temperatures, indicates it. The continuous solid solutions are also formed between the corresponding elements - Fe(Ni), Ni(Co). Formation of intermetallic compounds, for example CoTi, is also possible.

Tendency of metals to interact with carbon in the gaseous phase seems to define the catalytic properties of these metals in this process.

## 3. Conclusions

In the course of the work the following has been established:

a) catalysts added into graphite of the anode results in forming the new phases in the deposit. Transition metals, which are added into the carbon electrodes evaporated in electric arc synthesis of fullerenes, are present in the deposit in the form of simple or complex carbides, and solid metal solutions in these carbides;

b) introduction of catalysts decreases the size of carbon particles in the deposit down to 62-80 nm.

c) properties of the deposit depend on the phases formed by the catalysts introduced.

## 4. References

1. Trefilov V. I., Schur D. V., Tarasov B. P., Shulga Yu. M., Chernogorenko A. V., Pishuk V. K., Zaginaichenko S. Yu. (2001) Fullerenes is future materials basis, Kiev: Izd. ADEF-Ukraine, 148 p.

2. ASTM, (1985) X- ray diffraction date cards.

3. Iveronova V.I., Revkevich G.P. (1972) Teoriya rasseyaniya rentgenovskih luchey, M., 248 p.

4. Samsonov G.V. (1976) Tugoplavkie soedineniya, M., "Metallurgiya", 560 p.

# EFFECT OF HYDROGEN ON DELAYED FRACTURE OF HCP ε - STEELS BASED ON Fe-Mn SOLID SOLUTION

EFROS B.M., BEREZOVSKAYA V.V.(1), GLADKOVSKII S.V.(1), LOLADZE L.V.
*Physics and Engineering Institute, NASc of Ukraine, 72, R.Luxemburg st. 83114 Donetsk, Ukraine*
*(1)Ural State Technical University, 19, Mira St., 620002 Ekaterinburg, Russia*

## 1. Introduction

Among the negative reactions to environmental effects is the degradation of the ultimate strength of materials during the delayed fracture (DF) tests. Resistance to DF depends very much on the phase and structural factors, one being the metastability of materials with respect to deformation, external pressure and temperature [1]. In this article, a high-manganese metastable alloy Fe-20 %Mn-0.05 %C with the 55% content of HCP ε-phase (in initial state) was investigated. The specimens were hardened bay heat treatment and then deformation pre-treatment bay room temperature hydroextrusion (HE) method to a degree $\varepsilon \approx 40$ %. The DF tests were carried out at constant load (according to the scheme of pure bending) at stress $\sigma < \sigma_k$ in distilled water.

## 2. Results and discussion

It is shown that deformation by HE strengthens the studied alloy considerably, yield strength ($\sigma_{0.2}$) are increasing. However, the plasticity characteristics (specific elongation $\delta$ and reduction of area $\psi$) become somewhat lower (Fig.1). The growth of $\psi$ values at $\varepsilon \approx$ 40% can be related to the phenomenon of healing the micro-defects during at HE [2].
Fig.2 illustrates the dependences of crack-propagation specific work (KCT), of chart-term strength ($\sigma_K$), which characterizes the bending strength of alloy, and the threshold stress $\sigma_\Pi$, i.e. the minimal stress for fracture to occur on deformation degree $\varepsilon$ at HE. Past a $\varepsilon \approx 20$-30% bay HE, there is an increase in KCT and a decrease in $\sigma_K$, which may be due to stress relaxation in the process of preliminary deformation or testing. High enough speed of the test on KCT and $\sigma_K$ evaluation difference them from the DF test, and this difference should influence the kinetics of martensite transformations (MT) and, finally, the fracture. That's why; there is no correlation between KCT and $\sigma_K$ and $\sigma_\Pi$ dependences. The results of testing have shown that the HE at $\varepsilon$ to 30% makes the tendency to DF higher, which is evidenced by a decrease in $\sigma_\Pi$.

*T.N. Veziroglu et al. (eds.),*
*Hydrogen Materials Science and Chemistry of Carbon Nanomaterials,* 313-318.
© 2004 *Kluwer Academic Publishers. Printed in the Netherlands.*

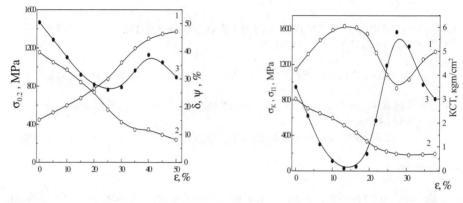

Fig.1. Effect of HE on of yield strength $\sigma_{0.2}$ (1), $\delta$ (2) and $\psi$(3) of studied alloy
Fig.2. Effect of HE on parameters of $\sigma_K$ (1), $\sigma_\Pi$ (2) and KCT (3) of studied alloy

At $\varepsilon > 30$ %, there was an improvement in $\sigma_\Pi$, characteristics and, as a result, the tendency to DF has become lower. The lack of non-monotonic in $\sigma_\Pi$ change in the range of $\varepsilon \approx 20\text{-}30$ % can, in contrast to KCT and $\sigma_\Pi$ change, be due to the difference in kinetics of martensite transformation (MT). Besides, the high-speed tests for KCT and $\sigma_k$ evaluation prevent the corrosive medium effect developing to a considerable extent. In the case of DF test, this factor can be essential, as it affects the process of crack nucleation. It is seen that the investigated alloy 05G20 is most susceptible to corrosion at $\varepsilon \approx 30\%$ bay HE (Fig.3).

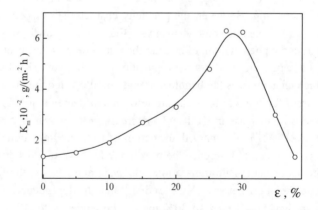

Fig.3. Effect of HE on corrosion rate $K_m$ of studied alloy

The fractography analysis has shown that the character of fracture considerably depends on the degree of preliminary HE and on the type of testing. Thus, in specimen pre-deformed to a $\varepsilon \approx 30$ % there was a dimple-type fracture after the impact tests (Fig.4). The DF tests of similar specimen have resulted in high embrittlement of the studied alloy: the $\sigma_{II}$ level has decreased and the character of fracture has changed.

Investigation of the phase composition in the zone of deformed specimen fracture done before and after the test show that kinetics of $\gamma, \varepsilon \rightarrow \alpha'$-MT is defined by not only the degree $\varepsilon$ of preliminary HE, but by the type of testing and the rate of fracture (Tabl.1). The comparative analysis of fracture in specimens having cracks and subjected to conventional static concentrated-bending and DF test has shown that in the formed case quantity of $\alpha'$-phase in the zone of rupture for the initial (un-deformed) state does not exceed 40 %, while in the latter case it reaches almost 90 %. It could also be noted that when the subcritical crack growth (SGG) is slow, the intensity of MT by HE is lower that in the case of quick supercritical crack propagation (SCP).

TABLE 1. Influence of deformation by HE on phase composition of studied alloy during DF test

| Degree of deformation by HE, % | Phase composition | | | | | | | | | Angle of crack branching, deg |
|---|---|---|---|---|---|---|---|---|---|---|
| | In the initial state | | | In SGG zone | | | In SCP zone | | | |
| | $\varepsilon$ | $\alpha'$ | $\gamma$ | $\varepsilon$ | $\alpha'$ | $\gamma$ | $\varepsilon$ | $\alpha'$ | $\gamma$ | |
| 0 | 55 | - | 45 | 45 | 35 | 20 | 4 | 87 | 9 | 0 |
| 10 | 62 | 6 | 32 | 35 | 50 | 15 | 4 | 88 | 8 | 0 |
| 20 | 67 | 12 | 21 | 30 | 60 | 10 | 6 | 86 | 8 | 128 |
| 30 | 72 | 8 | 20 | 41 | 48 | 11 | 18 | 70 | 12 | 155 |
| 40 | 60 | 5 | 35 | 29 | 62 | 9 | 11 | 68 | 21 | 140 |

It is also shown that with the increase of $\varepsilon$, the content of $\alpha'$-phase is non monotonously growing both in the initial state and after DF in the zone of fracture. None monotonously of the kinetics of $\alpha'$-phase formation is due to stabilization of $\varepsilon \rightarrow \alpha'$-MT (Tabl.2). With allowance for the scheme of loading the specimens, under HE, where the compressive-stress component is predominant, stabilization of the $\varepsilon \rightarrow \alpha'$-MT accompanied by the increase in volume, seems natural. The intensity of $\varepsilon$- and $\alpha'$-martensite formation is shown to depend on method and rate of alloy deformation. Thus, in the case of DF, the $\varepsilon \rightarrow \alpha'$-MT become stabilized at preliminary $\varepsilon \approx 20$ % and retains to degree of deformation $\varepsilon \approx 30$ %, whereas in the case of HE, stabilization is observer after $\varepsilon \approx 30$ %. Stabilization of $\varepsilon \rightarrow \alpha'$-MT may be connected with the formation of deformation twins in HCP $\varepsilon$-martensite along the $\{10\overline{1}2\}_\varepsilon$ - planes [2].

316

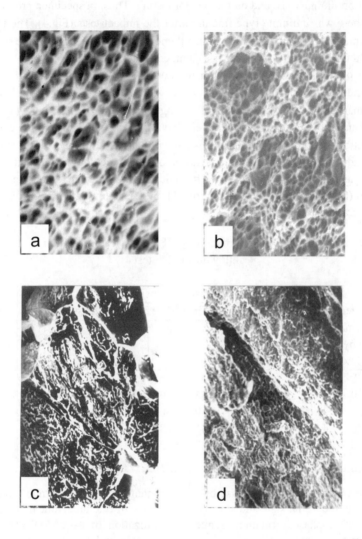

Fig.4. Microfractography of ruptures of studied alloy after different treatment: a – quenching, impact test, x 2000; b – quenching, HE (ε ≈ 30 %), impact test, x 2000; c – quenching, DF water test, x 600; d - quenching, HE (ε ≈ 30 %), after DF water test, x 600

TABLE 2. Influence of deformation by HE and of delayed failure on kinetics of martensite transformations in studied alloy

| Degree of deformation by HE, % | Completeness of martensite transformations, % | Stabilization of transformations | | |
|---|---|---|---|---|
| | | During HE | During DF | |
| | | | IN SGG zone | In SCP zone |
| 0 | - | | | |
| 10 | 13γ→7ε→6α′ | | | |
| 20 | 24γ→12ε→12α′ | | | |
| 30 | 25γ→17ε→8α′ | ε→α′ | | |
| 40 | 10γ→5ε→5α′ | | | |
| 0 | 25γ→10ε→35α′ | | | |
| 10 | 17γ→27ε→44α′ | | | |
| 20 | 11γ→37ε→48α′ | | ε→α′ | |
| 30 | 9γ→31ε→40α′ | | ε→α′ | |
| 40 | 26γ→31ε→57α′ | | | |
| 0 | 36γ→51ε→87α′ | | | |
| 10 | 24γ→58ε→82α′ | | | |
| 20 | 13γ→61ε→74α′ | | | ε→α′ |
| 30 | 8γ→54ε→62α′ | | | ε→α′ |
| 40 | 14γ→49ε→63α′ | | | ε→α′ |

The obtained results show that the necessary, but not the only factor of the DF of HCP ε-FeMn alloys is hydrogen the source of which is a corrosive medium. The accumulation of hydrogen increases the level of residual internal stresses of the alloy, which, in time, attain the values enough for crack nucleation. The processed conditioned by phase transformations are the source of internal stresses. Stabilization of alloys with respect to deformational and thermal effect at HE can essentially decrease the level of residual internal stresses in the HCP ε-FeMn alloy. This is evidenced by the increase in resistance to DF.

318

## 3. Conclusion

he obtained results shown that the HCP $\varepsilon$-FeMn alloy of the Fe-20 %Mn-0.05 %C type deformed by HE undergoes brittle DF along the boundaries of initial austenitic grains. A decreased resistance of alloy extrudates to DF is due to a high-intensive $\gamma \rightarrow \varepsilon$-MT with a certain stabilization of $\varepsilon \rightarrow \alpha'$-MT, especially in the zone of rupture. This results in the lowering of relaxation stability of the alloy. The optimal regime of hardening by HE is deformation to a $\varepsilon \approx 30\text{-}40$ %, which provides the improvement of mechanical properties in complex and the increase in the resistance to DF. It is shown that the induced to a considerable degree by hydrogen degradation of the structural strength of HCP $\varepsilon$-FeMn alloys is a complex problem for material science, chemistry and mechanics of metallic materials.

## 4. References

1. V.I. Sarrak, S.O. Suvorova, E.N. Artemova, Phenomenon of delayed failure in chromium-manganese steel with metastable austenite. Dokl. Acad. Nauk SSSR 1986; 290:1371-1374.
2. S.V. Gladkovsky, B. M. Efros, Hardening the metastable high-manganese steels. FTVD 12 (1980) 87-92.

# SYSTEM COMBINED AUTOMOBILE FEED ON CARBON NANOSTRUCTURES WITH HYDROGEN ADSORBATE APPLICATION

ZAKHAROV A.I. *, KOSTIKOV V.I.[(1)], KOTOSONOV A.S.[(1)],
IVANKOVA T.A., MILOVANOVA O.A.
*Bardin Central Research institute of ferrous metallurgy,
105005, Moscow, 2-ja Baumanskaja, 9/23, Russia
[(1)]State science-research institute of graphite construction materials
111141, Moscow, Electrodnaja, 2
* Fax (095) 777 93 00 E-mail:aizskap@yandex.ru

**Abstract** - Hydrogen providing system for combined automobile feed SKAP includes changeable cylinder filled by carbon nanofibers. SKAP usage reduces fuel consumption and decrease considerably toxicant components content in exist gases of engines. It is given evaluation of ecological effect of SKAP utilization.

Automobile transport gives the most contribution in the environment pollution (80 - 90%). Automobile companies began work out low toxicant content transports and tests for its already long time ago. Having spent great amounts, some firms began serial production hybrid automobiles and electromobiles on hydrogen with «zero exhaust».

In the nearest years some part of automobiles will be changed on expensive ecological pure transport facilities. It will be risen exhaust purity requirements and only new automobiles with neutralizator will be satisfying it. The most part of motor-car transport (~80%) will become lower these requirements. In this case system combined automobile gasoline-hydrogen feed (SKAP) installation will allow provide the required purity of exhausting gases up to required level EVRO-4. SKAP is the most simple and chippest way to drive automobiles until their change on transport with «zero exhaust».

The prospects of hydrogen's usage in motor transport is connected mainly with ecological purity and high motor properties [1].

Usage of hydrogen will allow also to eliminate one of the main automobile engine's lacks fall-off efficiency from 30 up to 10 % on partial loads in conditions of urban exploitation. The hydrogen even at the small addings 1-6 % to gasoline already allows to raise fuel profitability on partial loads by 30-40 %. It is reached due a limited value of ignition for hydrogen more broad, than for gasoline. The limits of ignition in % on volume basis for hydrogen are in an interval 4,7 - 74,2, and for gasoline - 0,59 - 6,0. More demonstrative for ignition limits is counting over volume parts on coefficient of air surplus which has ignition limits for hydrogen from 0,15 to 10 and for gasoline from 0,27 to 1,7 accordingly. Special interest presents lower ignition limit which shows degree of fuel-air mixture impoverishment. Mixture impoverishment limit for hydrogen is several times more than for gasoline. Therefore hydrogen fuel permits regulate engine capacity on broader limits.

*T.N. Veziroglu et al. (eds.),*
*Hydrogen Materials Science and Chemistry of Carbon Nanomaterials,* 319-323.
© 2004 *Kluwer Academic Publishers. Printed in the Netherlands.*

Hydrogen-air mixtures have high speed of combustion. Due to it the work process efficiency is risen.

Flame extinguish distance is width of near wall layer of fuel-air mixture in which oxidation reaction is finished and for hydrogen it is more than for times less. This brings to fuller fuel combustion and reducing toxicant part of carbon-hydrogen composites in exit gases.

Hydrogen on the whole complex of properties: broad concentration limits, high speed of combustion and the high diffusive mobility characterizes itself as the ideal component to fuel for acceleration of a hydrocarbon-air mixture's combustion process. It allows to increase fuel profitability and considerably to lower the contents of toxicant components in exit gases. The estimation of ecological efficiency SKAP at the component only 1 % of hydrogen to gasoline is adduced in work [2]. It is visible that toxicant components are reducing considerably in exit gases at petrol consumption drop.

It were taken approximately 30 million cars in Russia and 600 million on the Earth for evaluation of ecological damage at automobile maintenance. At average charge 15000 km a year an automobile burns 1,5 tons of petrol and throws out about 3000 kg - $CO_2$, 300 kg - CO, 100 kg - $NO_x$ $C_xH_y$ and other admixtures.

Using all the approximate data it was made up the adduced Table, taking into account that SKAP-3 on nanofibers is working only on hydrogen at stoppage and cross-roads and due to it we have efficiency for $CO_2$ - 25%, for CO - 50%, for NO - 5% and $CxH_y$ - 2%.

TABLE 1. Ecological efficiency SKAP-2 (thousand.ton)

| Structure exhaust | Russia | | the Earth | | effãct |
|---|---|---|---|---|---|
| Gàses | $Í_2 = 0\%$ | $Í_2 = 1\%$ | $Í_2 = 0\%$ | $Í_2 = 1\%$ | % |
| $ÑÍ_2$ | 102900 | 77180 | 2058000 | 1543500 | 25 |
| $ÑÍ$ | 9000 | 4500 | 160000 | 80000 | 50 |
| $NÍ_õ$ | 3000 | 2850 | 60000 | 57000 | 5 |
| $Ñ_óÍ_õ$ | 900 | 882 | 18000 | 17640 | 2 |

1 - 6% of hydrogen ($Í_2$) in combustion-mixture of a petrol engine:
■ reduced the toxicant elements of exhaust gases 2-20 times;
■ reduces fuel consumption by 15 - 24 %;
■ increases motor-resource of the engine and its efficiency by 40 % at a mode of an urban cycle.

Automobile consumes energy of liquid carbon - hydrogen fuel (petrol) in order to move. It is kept on the car board. The hydrogen in a gaseous kind has low density, that is a problem, which one restrains broad usage of hydrogen in the transport. The hydrogen on board automobiles can be stored in a gaseous compressed condition, in a liquefied condition and in a chemically bound condition by the way of hydrides. The last time was opened such capability of carbon nanofibers to be adsorbed up to 12% at pressure about 10 MPa.

Usage of compressed hydrogen for engine feed has number of faults. It is spent for compression one/seven part of energy which hydrogen can produce. Cylinders with hydrogen have large weight and volume. Besides high pressure hydrogen cylinders on the board of car in case of crash become explosive device with a big strike strength.

Liquid hydrogen has boiling temperature - 252,4° C and density 0,071 kg/l but it has density only 0,025 kg/l under the pressure of 30 MPa. Consequently, liquid hydrogen more compact than compressed one and present significant interest for transport. In this connection it is necessary to note many examples of liquid hydrogen utilization in cars and buses. However, complication of cryogenic equipment and inevitable fact of hydrogen boiling out at keeping restrain liquid hydrogen usage in transport.

Hydrogen keeping on the board of transport facility in chemical connected condition as hydride present practical interest so as in this case conditions of high safety level are fulfilled at low energy expenditure for hydrogen accumulating in hydrides. It is necessary to mention that hydrogen has higher characteristics on mass and volume in hydrides. It was used intermetallic combination on the basis of Fe-Ti for the keeping hydrogen in the work [3]. This combination is with small additions of Mn and Mm (lanthanum).

Hydrogen absorption in alloys Ti-Fe-Mn is happening in pressure intervals from 0,2 to 4 MPa at 20°C temperature. The maximum hydrogen content at this temperature was about 1,9%. Hydrogen desorption is going in temperature intervals 40-100°C. Main hydrogen part get under desorption at pressure lower than 1 MPa in temperature intervals 20-90°. However, hydrogen content reached less than 2% on mass that is obviously insufficient for feed on hydrogen.

In this work hydrogen keeping is fulfilled by carbon nanofibers at pressure 10 MPa. The mass containment is more than 7% in this case that 3 times higher than in intermetal Fe-Ti.

In these experiments 7 liter cylinder filled by nanofibers is used. Cylinder contains about 5 kg of nanofibers which sorb about 300 gram of hydrogen, having in mind that for the time of the work would be spent 30 kg of petrol. Thus the hydrogen addition to carbon fuel became 1% on mass. Fulfilled test have shown efficiency of hydrogen addition in quantity of 1% is noticeable at exploitation in urban conditions and regime with a big fuel spending and in speeding up uniform addition show small effect. In this connection were elaborated several variants of combined fuel feed system. In this report variant hydrogen consumption proportionally fuel consumption is being considered with possible correction of spending hydrogen quantity on idle move, in urban conditions and at speeding.

Figure 1 shows the scheme of the automobile's power supply system. In a designed system the hydrogen moves passing the carburettor, through laying, directly into a collector of the engine.

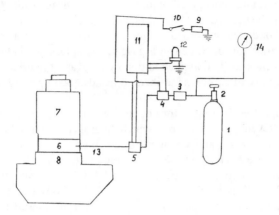

Figure 1. Cylinder, 2. Valve lock, 3. Gas reduction gearbox, 4. Magnet valve, 5. Metering device-injector, 6. Laying, 7. Carburettor, 8. Collector of engine, 9. Accumulator, 10. Ignition key, 11. Control block, 12.Induction coel, 13. Fitting pipe, 14. Pressure gauge.

On the scheme the power supply system by gasoline is not presented, the carburettor 7 with laying 6 are only figured.

The power supply system by hydrogen includes cylinder 1. The hydrogen moves into collector 8 from cylinder through reduction gearbox 3, magnet valve 4, metering device - injector 5, fitting pipe 13. The control block 11 is to open metering device-injector and magnet valve.

Cylinder 1 is allocated in a luggage part of the car. Electromagnet valve 4, a metering device-injector 5 are installed on the wall of the automobile engine part . A pipeline for giving hydrogen from a cylinder through gas reduction gearbox 3 is put in exit of magnet valve. The pipeline with gas is installed on the bottom of the automobile.

Engine with hydrogen addition is working by the next way.

At engine ignition through an ignition lock 10 tension moves on the electronic control block 11, which one opens a magnet valve 4 and the hydrogen through the metering device - injector 5 moves proportionally to rotation numbers of engine into collector 8 of engine . Thus, at actuation of a starter the maiden portions gasoline-hydrogen-air mixture fall in engine , enriched by hydrogen, that makes the engine start-up easy.

On idling mode a minimum shaft revolving frequency is regulated at the opened electric valve 4. As hydrogen-air mixture is combustible in a broad range of concentrations, the consumption of hydrogen can be adjusted close to committed activity for overcoming friction losses of the engine at the idling mode. Thus, the exhaust gases will contain tens times less toxicant components CO, NO and ÑÍ in compare to an activity only on gasoline-air mixture on this mode.

The control block 11 allows to vary by the time interval of the injector's opening, that enables to change the content of hydrogen in a combustion-mixture in depend on a work cycle and quantity of hydrogen in cylinder. Engine can work only with hydrogen on idle move while standing on cross-roads and also in stoppage. A carburettor switches of during that. In this case automobile exhaust is pure in respect of poison components.

Using hydrogen addition we can increase exhaust purity to upper poisons admittances level accepted by European standards. It demands more than 10 years to appear hydrogen electromobiles in exploitation and exchange all the autotransport by it. It is necessary to exploit the cars to their end for the time, besides to do it at higher requirements on exhaust purity.

Thus hydrogen using permit to solve two important tasks: first to exploit automobiles to their end and second to join people to culture of hydrogen application.

## Conclusions

1. Growth of automobile transport was accompanied by arising requirements to exhaust purity level. The most automobile companies come to serial producing of automobiles with «zero exhaust». In this connection hydrogen addition to carbon-hydrogen feed will allow to increase exhaust purity and exploit cars until their change by principal new kind of transport - hydrogen electromobiles.
2. System of combined automobile feed is used for the solving the problem.
   2.1. It is worked out system of combined petrol-hydrogen engine feed [3].
   2.2. It were found methods of carbon-hydrogen nanofibers getting and also contents of hydrogen keepers alloys found. And cheap technology of their production have been worked out [3].
   2.3. Changeable hydrogen accumulator construction have been elaborated.
   2.4. It was elaborated installation project of combined petrol-hydrogen feed system on carburettor automobiles VAZ 23093, GAZ and other types.
3. Changeable hydrogen accumulator (akostat) can easily solve the problem of mastering of hydrogen fuel in automobiles and other transport facilities.

## References

1. Mishenko AI Application of hydrogen for automobile engines. Naukova dumka, Ukraine, Kiev, (1984).
2. J. Haconen, G. Pinhasi, Y. Puterman and E. Sher. Driving cycle simulation of a vehicle motoved by a si engine fueled with $H_2$ - enriched gasoline. Int. J. Hydrogen Energy vol 16 pp 695-702 (1991)
3. Zakharov AI, Ivankov VV and others. Elaboration of combined oil-hydrogen feed for improving of automobile ecological characteristics. International Scientific Journal for Alternative Energy and Ecology № 2, 2002, p.20-24.
4. Zakharov AI, Milovanova OA, Ivankov VV. Alloys - Hydrogen Accumulators as Factor of Reduction in Environmental Stress, Steel, № 4, 2002, p. 84-89

# T-NANOCONSTRUCTIONS ON THE (0001)-SURFACE OF GRAPHITE BASED ON CARBON (6,6)-NANOTUBES

A.P. POPOV, I.V. BAZHIN
*Teachers' Updating Institute, str. Guardeiskaya , 2/51,*
*Rostov-on-Don, 344011 Russia*
*Don State Technical University, Gagarin Sq., 1,*
*Rostov-on-Don, 344010 Russia*

## Abstract

The possibility of existence of T-constructions based on carbon nanotubes, which grow from graphite (0001)-surface and compose the indivisible whole with higher graphite monolayer, is predicted. This construction can be treated also as the limit case of two carbon nanotubes T-junction in the limit when diameter of one of the nanotubes aspires to infinity. The T-construction can be used for creation of some nanoelectronic devises.

*Keywords:* A. Nanotubes; Graphite; C. Molecular simulation; Quantum-chemical
         calculations; Modelling

## 1. Introduction

The creation of some nanoelectronic devices requires rigid fixation of nanotubes orientations in direction perpendicular to crystal surface. One of the possibilities to solve the problem (in the way of synthesis of T-constructions on graphite surface) is discussed in the paper.

## 2. Theoretical and computational

The nanoconstruction of principally new type –open (6,6)-nanotube, which grows from (0001)- surface of graphite and compose the indivisible whole with higher graphite monolayer. The construction in whole is similar to tower; below we use "tower" as term for meaning of construction.

Let's describe the process of "creation" of tower more detailed. One can mark centre of any hexagon on the graphite surface and remove 24 atoms from the first three coordspheres in higher graphite monolayer (this atoms are disposed in foundation of created tower). As result in the higher graphite monolayer the almost round hole appears. In the boundary of the hole 24 carbon atoms from fourth coordsphere are disposed. Namely these atoms participate in formation of firm covalent bonds with 24 carbon atoms disposed at the edge of open carbon (6,6)-nanotube. The construction contains six 7-members cycles near the tower foundation, which are not typical for carbon framework's constructions. But in spite of this, the construction is very stable:

325

*T.N. Veziroglu et al. (eds.),*
*Hydrogen Materials Science and Chemistry of Carbon Nanomaterials,* 325-328.
© 2004 *Kluwer Academic Publishers. Printed in the Netherlands.*

binding energy in 10-20 times exceeds absorption energy of nanotube with graphite surface.

Below we consider geometry and electronic structure not only of the tower, but also of the elements of construction: ring-like graphite monolayer and open (6,6)-nanotube. All the calculations are performed in scheme, which use empirical Tersoff-Brenner method [1,2] for approximate determination of equilibrium geometry with following quantum chemistry calculations in framework of semiempirical PM3 method [3], making also more precise geometrical parameters of equilibrium configuration.

## 2.1. Ring-like sheet of graphite monolayer $C_{192}$

Ring-like sheet of graphite monolayer $C_{192}$ is one of the elements of T-constructon. One can get it from continuous sheet of graphite monolayer $C_{216}$, if throw out 24 carbon atoms disposed near the center of the sheet (fig.1a). All the covalent $sp^2$-like bonds (for exception internal and external edges of ring) are equivalent each other and have length 1.42 Å, what is very close to length of bonds in (0001)-plane of graphite.

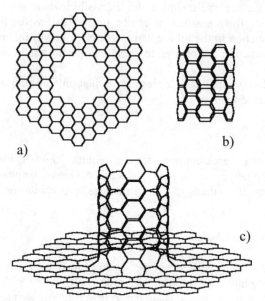

Figure 1. The structures of ring-like graphite monolayer $C_{192}$ (a),
(6,6)-open nanotube $C_{120}$ (b) and T-construction $C_{312}$ (c)

The PM3 method gives the following values for energies of higher occupied and lower unoccupied states $E_{HOMO} = -8.04$ eV, $E_{LUMO} = -3.64$ eV. Therefore, the width of forbidden gap is equal to $E_g = 4.40$ eV. The meaning of heat of formation is $\Delta H = 4172.10$ kcal/mol. The results of DOS calculations is shown in fig. 2.

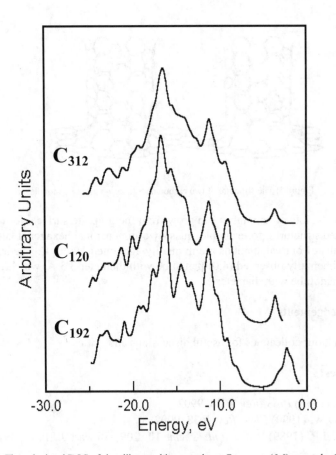

Figure 2. The calculated DOS of ring-like graphite monolayer $C_{192}$, open (6,6) nanotube $C_{120}$ and T-construction $C_{312}$

## 2.2. Open (6,6)-nanotubes $C_{120}$

The structure of open (6,6)-nanotube $C_{120}$ is shown in fig. 1b. The calculated energies of higher occupied and lower unoccupied states are following: $E_{HOMO}$ = -7.92 eV, $E_{LUMO}$ = -3.42 eV. The width of forbidden gap is $E_g$ = 4.51 eV. The heat of formation is $\Delta H$ = 1932.58 kcal/mol. The shape of calculated DOS is given in fig. 2.

## 2.3. T-construction $C_{312}$

The optimized T-construction $C_{312}$ is presentation in fig. 1c. The PM3 method leads the following values for energies of higher occupied and lower unoccupied states $E_{HOMO}$ = -7.79 eV, $E_{LUMO}$= = -3.67 eV. The width of forbidden gap is equal to $E_g$ = 4.11eV. The heat of formation is $\Delta H$ = 4766.43 kcal/mol. The results of DOS calculations is shown in fig. 2.

We consider also construction of two nearly disposed towers (fig. 3). The results of DOS calculations clearly show the essential role of interactions of electron states in nanotubes through graphite monolayer.

328

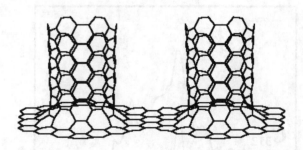

Figure 3. The structure of two carbon nanotubes on graphite monolayer

It's clear the lots of nanotubes can grow from the graphite surface, forming different well-regulated structures, for example, quasi-two-dimensional hexagonal lattice.

The results of our calculations are in quality agreement with [4]. The authors of this paper experimentally observed (4,0)-nanotube with diameter 3.3 Å, which growths from more thick nanotube with diameter 15 Å.

## 3. Acknowledgements

Authors thank our colleagues for useful discussions and critics.

## 4. References

1. Tersoff, J. (1988) *Phys. Rev. B* **38**, 9902.
2. Brenner, D.W. (1990) *Phys. Rev. B* **38**, 9902.
3. Stewart, J.J.P. (1989) *J. Comput. Chem.* **10**, 209; Stewart J.J.P. (1989) *J. Comput. Chem.* **10**, 221.
4. Peng, L.M. et al. (2000) *Phys. Rev. Lett.* **85**, 3249.

# THREE-DIMENSIONAL POLYMERIZED CUBIC PHASE OF FULLERENES C28

A.P. POPOV, I.V. BAZHIN
*Teachers' Updating Institute, str. Guardeiskaya , 2/51,*
*Rostov-on-Don, 344011 Russia*
*Don State Technical University, Gagarin Sq., 1,*
*Rostov-on-Don, 344010 Russia*

## Abstract

Geometrical parameters, heat of formation, energy of higher occupied and lower unoccupied states and density of one-electron states (DOS) for isolated molecule $C_{28}$, dimer $(C_{28})_2$ and cubic cluster $(C_{28})_8$ are determined by using of semi-empirical quantum chemistry PM3-method. The results of calculations allows to assume the existence of polymerized cubic crystal structure of fullerene $C_{28}$.

*Keywords:* A. Fullerenes; Polymerization; Cluster; Crystal structure; B. Cubic symmetry; C. Molecular simulation; Quantum-chemical calculations; Modelling

## 1. Introduction

The first discovered solid phase of fullerenes $C_{60}$ represents typical molecular crystal. Later it was established that high pressure applied to solid $C_{60}$ at high temperature induces polymerization of $C_{60}$.

Now polymerized structures on basis of big and small fullerenes with different dimensionality and symmetry are the subjects a lot of theoretical and experimental investigations.

The aim of our paper is the semiempirical quantum-chemical calculations of equilibrium geometrical configuration and electronic structure of big cuban-like cluster $(C_{28})_8$. This cluster can be considered as a fragment of polymerized crystal phase of $C_{28}$ with simple cubic symmetry.

## 2. Theoretical and computational

The choice of molecule $C_{28}$ as base unit for construction of polymerized structure is not casual. Firstly, $C_{28}$ is typical representative of small fullerenes class, secondly, molecules $C_{28}$ possess Td-symmetry, and besides near each of three reciprocal perpendicular two-fold axes on the opposite sides of molecule six pairs of atoms are disposed, which take part in formation of bridge-like bonds between neighboring molecules in polymerized structure. Bonds of similar kind are observed and well studied in polymerized structure of $C_{60}$. Essential difference from $C_{60}$ is absence in molecules $C_{28}$ the center of inversion and "vertical" planes of symmetry, containing two-fold axes. As sequence, parameter of

329

*T.N. Veziroglu et al. (eds.),*
*Hydrogen Materials Science and Chemistry of Carbon Nanomaterials,* 329-332.

crystal lattice for considered polymerized structure is two-time more than distance between centers of nearest molecules in cluster.

Below we consider consequently isolated molecule $C_{28}$, dimmer $(C_{28})_2$ and only then cluster $(C_{28})_2$. All the calculations are performed in scheme, which uses empirical Tersoff-Brenner method [1,2] for approximate determination of equilibrium geometry with following quantum chemistry calculations in framework of semi-empirical PM3 method [3], making also more precise geometrical parameters of equilibrium configuration.

### 2.1. Isolated molecule $C_{28}$

Among all the small fullerenes $C_{28}$ is the best studied. The isolated molecule $C_{28}$ is the Td-symmetrical cage of 28 carbon atoms, connected by 42 covalent $sp^2$-like bonds (see fig.1). There are three essentially different kinds of bonds in molecule $C_{28}$:

1) 24 bonds with length 1.42 Å, which are the verges of four hexagons;
2) 6 bonds with length 1.52 Å, which connect the vertexes of each pair of hexagons;
3) 12 bonds with length 1.46 Å, which are joined in four vertexes, common for each three neighboring pentagons.

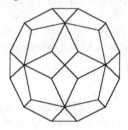

Figure 1. The structure of $C_{28}$

The PM3 calculations leads to the following values for energies of higher occupied and lower unoccupied states $E_{HOMO}$ = -8.485 eV, $E_{LUMO}$ = = -4.550 eV. Therefore, the width of forbidden gap is equal to $E_g$ = 3.935 eV. Calculated meaning of heat of formation is $\Delta H$ = 824.643 kcal/mol. The results of DOS calculations is shown in fig. 2.

### 2.2. Dimer $(C_{28})_2$

Dimer $(C_{28})_2$ represents the simplest polymerized structure, which is formed by two molecules of $C_{28}$. Molecules $C_{28}$ in dimer are connected by two bridge-like bonds, directed along of their common two-fold axis, as is shown in fig. 3. The formation of dimer is accompanied by distortion of geometry of molecules $C_{28}$. The distortion is especially considerable in the nearest neighbor of the intermolecular bridge-like bonds. In this region the lengths of intramolecular bonds are changed and become equal to 1.51 Å instead 1.42 Å and 1.69 Å instead 1.52 Å. The length of intermolecular bonds is equal to 1.52 Å.

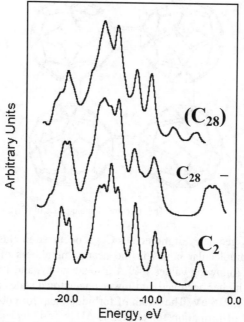

Figure 2. The calculating DOS of $C_{28}$, dimer $(C_{28})_2$ and cuban-like $8C_{28}$

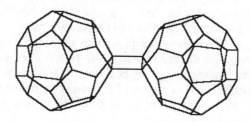

Figure 3. The structure of dimer $C_{28} - C_{28}$

The energies of higher occupied and lower unoccupied states are equal to $E_{HOMO} = -9.086$ eV, $E_{LUMO} = -3.671$ eV. The width of forbidden gap for dimer is $E_g = 5.415$ eV. The heat of formation is $\Delta H = 1494.561$ kcal/mol. The form of calculated DOS is represented in fig. 2.

## 2.3. Cuban-like cluster $(C_{28})_8$

The geometry of cuban-like cluster $(C_{28})_8$ is similar to geometry of cuban molecule: cluster can be got from cuban by replacing each of carbon atoms onto fullerene molecule $C_{28}$. Molecules $C_{28}$ in cluster are connected by 12 pairs of bridge-like bonds directed along three perpendicular two-fold axes (see fig. 4). The length of intermolecular bonds is equal to 1.54 Å.

332

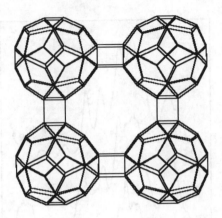

Figure 4. The structure of cubic-phase of $(C_{28})_8$

The distortion of geometry of molecules $C_{28}$ is occurred in cluster also. The length of non-equivalent intramolecular bonds in the nearest neighbors of intermolecular bonds change and take the following values 1.49 Å instead 1.42 Å and 1.695 Å instead 1.52 Å.

The energies of higher occupied and lower unoccupied states are equal to $E_{HOMO} = -7.107$ eV, $E_{LUMO} = -5.216$ eV. The width of forbidden gap for cuban-like cluster is $E_g = 1.891$ eV. The heat of formation for cluster is $\Delta H = 5547.095$ kcal/mol. The shape of DOS is demonstrated in fig. 2.

The comparison of the results of calculation (the energy of HOMO-LUMO states, densities of one-electron states and heat of formation) for isolated molecule $C_{28}$, dimer $(C_{28})_2$ and cuban-like cluster $(C_{28})_8$ allows to suppose the existence of polymerized FCC crystal phase of $C_{28}$. The space symmetry group of lattice coincides with symmetry group of NaCl. Period of the lattice is nearly 10.3-10.5 Å, the width of forbidden gap is about 1.8-2.0 eV. The binding energy can be estimated as 0.96 eV/atom.

Recently the possibility of existence of diamond-like polymerized structure of $C_{28}$ was predicted [4] with binding energy 0.65 eV/atom, but there no experimental confirmation of this prediction.

The results of our calculations are not also in contradiction with earlier investigations [5,6], devoted to theoretical and computational studying of different polymerized phases of classic fullerenes $C_{60}$.

## 3. References

1. Tersoff, J. (1988) *Phys. Rev. B* **38**, 9902.
2. Brenner, D.W. (1990) *Phys. Rev. B* **38**, 9902.
3. Stewart, J.J.P. (1989) *J. Comput. Chem.* **10**, 209; Stewart, J.J.P. (1989) *J. Comput. Chem.* **10**, 221.
4. Guo T. et al. (1992) *Science* **257**, 1661.
5. Iwasa, Y. et al. (1995) *Science* **264**, 1570.
6. Ocada, S. and Saito, S. (1997) *Phys. Rev. B* **55**, 4039.

# NANOSTRUCTURE AND ELECTRONIC SPECTRA OF Cu-C$_{60}$ FILMS

O.P. DMYTRENKO [(1)], M.P. KULISH[(1)], L.V. POPERENKO[(1)],
YU.I. PRYLUTSKYY[(2)],E.M. SHPILEVSKYY[(3)], I.V. YURGELEVICH[(1)]
M. HIETSCHOLD[(4)],F S. SCHULZE[(4)], J. ULANSKI[(5)], P. SCHARFF[(6)]
Kyiv National Shevchenko University, Departments of [(1)]Physics and
[(2)]Biophysics,
Volodymyrska Str., 64, 01033 Kyiv, Ukraine
[(3)]Belarus State University, Skorina pr.,4, 220050 Minsk, Belarus
[(4)]Chemnitz University of Technology, Institute of Physics, Solid Surfaces
Analysis Group, D-09107 Chemnitz, Germany
[(5)]Technical University of Lodz, Faculty of Chemistry, Department of
Molecular Physics, ul. Zeromskiego 116, 90-924 Lodz, Poland
[(6)]Technical University of Ilmenau, Institute of Physics, D-98684 Ilmenau,
Germany

## Abstract
Nanostructure and optical conductivity of single-emulsion Cu-C$_{60}$ films in the wide
temperature range and time of annealing are studied in detail. The essential reconstruction
of electronic spectra and spectral dependence of optical conductivity in comparison with
the pure C$_{60}$ films is revealed.

*Keywords:* Cu-C$_{60}$ films; electronic spectra; optical conductivity; X-ray diffraction;
transmission electron microscopy; ellipsometric spectroscopy

## 1. Introduction

Careful study of the alloying influence of C$_{60}$ fullerites by the metals on the optical, electro-
and photo-transport properties, in essence, is concentrated on the use of alkali metals.
Depending on the type of formed chemical connections their conductivity corresponds to
the metallic and dielectric phases. This behavior of conductivity is explained by the
ionization, observed in the compounds, as the consequence of the hybridization of the states
of C$_{60}$ and metals, which significantly changes the energy-band structure of C$_{60}$ fullerites
[1]. The properties of C$_{60}$ fullerites, alloyed by other metals (M), are studied insufficiently.
The presence of polymorphous heterophase [2-4], which is characterized by the complex
molecular, atomic and imperfect structure, is shown for the film samples, obtained during
the simultaneous spraying of the C$_{60}$ molecules and Cu, Sn and Bi atoms. Naturally to
expect that as a result of the simultaneous condensation of the C$_{60}$ molecules and metal
atoms the intercalation and the presence of different phase states will lead to reconstruction
of electron structure, filling with the electrons of the conductivity band, and also the
polymerization of M-C$_{60}$ structure [5-6]. In connection with this, for investigation of the
alloying influence of metals on the electronic spectra of C$_{60}$ fullerites, is of interest a

*T.N. Veziroglu et al. (eds.),*
*Hydrogen Materials Science and Chemistry of Carbon Nanomaterials,* 333-338.
© 2004 *Kluwer Academic Publishers. Printed in the Netherlands.*

334

comprehensive study of changes in the nanostructure, phase state and optical properties, carried out for one and the same samples.

In this work the single-emulsion Cu-$C_{60}$ films of different compositions, obtained during the vacuum condensation of the thermally activated $C_{60}$ molecules and Cu atoms [2-3] are studied. The thickness of films was ~100 nm. Later on the samples were subjected annealing at temperatures of higher than the room in the region of existence of the fcc phase of $C_{60}$ fullerenes. The phase state and nanostructure were investigated by the methods of X-ray diffraction analysis and transmission electron microscopy (TEM). Optical properties were studied by the method of ellipsometric spectroscopy.

## 2. Results and Discussion

It should be noted that the behavior of the optical conductivity, which is caused by inter-band transitions with the removal from the beginning of optical absorption, essentially differs with the annealing for pure $C_{60}$ fullerites and Cu-$C_{60}$ system even in the case of a comparatively low copper concentration (4 wt.%) (Fig.1 and 2). So, in the case of the pure $C_{60}$ films the position of the absorption bands, connected with the inter-band transitions, with the annealing with different temperatures and holding times practically does not change and corresponds to the known dipole allowed transitions for the isolated $C_{60}$ molecules. The diagram of such transitions is given on the insert to Fig. 1 [7].

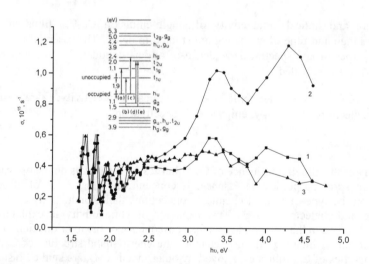

Fig.1. Optical conductivity of the $C_{60}$ films in the initial state after vacuum condensation (1), annealed with 120 °C during 20 min (2) and 200 °C , 300 min (3). Substrate from silicon, the thickness of the film is $2 \cdot 10^3$ nm.

Only annealing at 200 °C during 5 hour leads to the noticeable decrease of optical conductivity and the broadening of absorption bands with the photon energy, which exceed ~3.0 eV, i.e., to reduction in density and delocalization of energy states. X-ray diffraction and electron-microscopic studies testify about the conservation of molecular crystal structure, which corresponds to the fcc phase in this temperature region. At the same time with an increase in the temperature of annealing noticeable is a change in the morphological special features of the $C_{60}$ film, which is characterized by an improvement in their crystallinity. Thus, for the crystalline state of fcc phase the noticeable deviations of the formable bands of electronic states from the narrow molecular orbitals occur with an increase in the crystallinity of $C_{60}$ fullerites. It is possible to assume that the observed delocalization of electronic states is caused also by the polymerization of the fullerite structure due to the appearance together with Van der Waals forces, responsible for the formation of fcc isotropic structure, additional intermolecular interaction. The presence of such interaction is possible as a result of the noticeable oxidation of fullerites with the annealing. It is evident that with the low photon energy the discrete special features of the optical conductivity, which testify about the presence of the donor and acceptor states of oxygen [8] (Fig. 1) occur. With the annealing as an improvement in the crystallinity the position of the energy levels of impurity traps remains constant. At the same time the density of acceptor states decreases due to an increase in the density of the donor levels of oxygen. This growth of optical conductivity because of the quantum light absorption on the donor levels testifies about an increase in the concentration of intercalated oxygen to the detriment of the oxygen concentration, which reacted with the $C_{60}$ molecules [8].

For Cu-$C_{60}$ system with 4 wt.% Cu a difference in the optical conductivity occur both in the visible and ultraviolet ranges of energy of the absorbed photons of light (Fig. 2).

If in the initial state (curve 1) the bands, which correspond to inter-band transitions, remain also sufficiently narrow, then with the annealing even at low temperatures (curves 2, 3 and 4) all bands are strongly washed away near the energy of ~2.1-2.4 eV except the band, caused by $h_u \rightarrow t_{1g}$ ($V_1$-$C_2$, $V_2$-$C_1$) transition.

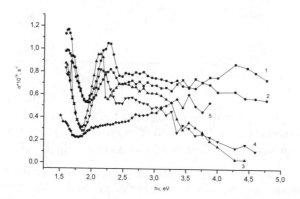

Fig.2. Optical conductivity of the Cu-$C_{60}$ (4 wt.% Cu) films in the initial state after vacuum condensation (1), annealed with 80 °C during 20 min (2), 120 °C, 20 min (3), 180 °C, 60 min (4) and 200 °C, 300 min (5). Substrate from silicon, the thickness of the film is ~100 nm.

The optical conductivity noticeably decreases in the region of high energies. Only annealing at 200 °C during 5 hour (curve 5) leads to an increase in the optical conductivity with these energies. With an increase in the temperature of annealing the absorption band near the energies of ~2.1-2.4 eV displaces to the side of smaller energies, and the value of the optical conductivity in this energy region after an initial increase decreases. By analogy with the chemical compounds of the $C_{60}$ molecules with the alkali metals ($M_6C_{60}$) and in contrast to pure $C_{60}$ fullerites the band, caused by the transition between HOMO-LUMO bands, which are forbidden on the parity in the isolated $C_{60}$ molecules, is observed in the optical conductivity spectrum. The center of this band is fallen to the energy of ~1.7 eV. The value of absorption near this energy with the annealing after initial increase decreases. It is obvious that this behavior of optical conductivity is the result of ionization as the consequence of hybridization between the states of $C_{60}$ and M [1], i.e., it is caused by the charge transfer between the electronic energy bands of $C_{60}$ and M, and also other phase formations, which appear with the vacuum condensation of films and the annealing.

With an increase the content of copper to 20 wt.% (Fig. 3) the conductivity bands, which were observed in $C_{60}$ and Cu-$C_{60}$ (4 wt.% Cu), remain, however, undergo noticeable changes.

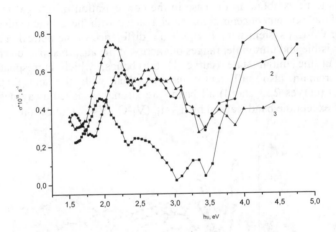

Fig.3. Optical conductivity of the Cu-$C_{60}$ (20 wt.% Cu) films in the initial state after vacuum condensation (1), annealed with 80 °C during 20 min (2) and 200 °C, 300 min (3). Substrate from silicon, the thickness of the film is ~100 nm.

In comparison with Cu-$C_{60}$ system (4 wt.% Cu) the essential displacement of the peaks, which correspond to $V_1 \rightarrow C_1$, $V_1 \rightarrow C_2$ transitions, to the side of high energies occurs. The value of these peaks with the annealing grows. Furthermore, the annealing leads to the appearance of weak additional peaks near the energy of ~1.5 eV. Thus, in the copper – fullerene system, especially with the annealing, the metallic nature of conductivity, which is caused by filling of the conductivity band of fullerenes, begins to be manifested to a greater extent. In turn this filling is determined by the atomic arrangement of copper in the resultant molecular and atomic crystalline phases and their transformations with the annealing. Fig. 4 gives the X-ray of single-emulsion Cu-$C_{60}$ films (4 wt.% Cu) in the initial state of spraying and with the annealing. It is evident the presence of heterophasic structure, which consists of the fcc solid solution of $C_{60}$ fullerite and $Cu_2O$ cuprite. In contrast to pure

$C_{60}$ fullerite, the lattice parameter for which in the initial state of condensation was 0.1414 nm, the parameter of the solid solution of Cu-$C_{60}$ (4 wt.% Cu) decreases to the value of 0.140 nm. The lattice parameter grows with the annealing and it reaches the value of 0.1427 nm (200 °C, 300 min). Furthermore, the additional reflections of type (200), which testify about the disappearance of the rotational degrees of freedom of the $C_{60}$ molecules appear at the diffractogram (Fig. 4).

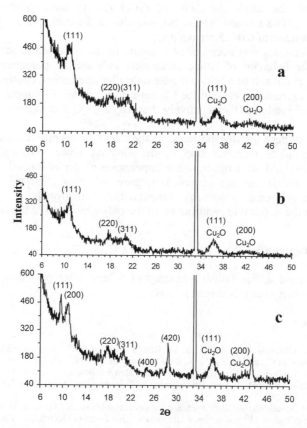

**Fig.4.** X-ray diffraction of the single-emulsion Cu-$C_{60}$ (4 wt.% Cu) films with the annealing: a - the initial state of spraying; b - annealing at 180 °C during 60 min; c - 200 °C, 300 min.

The lattice parameter of $Cu_2O$ cuprite remains with the annealing practically constant and equal to 1.0428 nm. Change with the annealing of the background in the region of small angles testifies about the presence of topological disordering and an improvement in the crystallinity of samples. The presence of structural disordering is confirmed by the high resolution electronic microscopical and electronic diffractional images of Cu-$C_{60}$ crystals.

With an increase in the content of copper the formation of planar defects, in the first place, the packing defects and twins, which are manifested in the appearance of additional diffraction peaks and which can be attributed to the hcp phase noticeable becomes. Usually the origin of planar defects connects with the lamellar arrangement of atoms excess in comparison with the equilibrium solubility, which leads to a change of the fullerite planes piling [2]. So, for the sample with 20 wt.% of Cu at the prolonged annealing (200 °C, 300 min) the co-existence of three nanostructures, which can be attributed to the fcc phase with the lattice parameter of 0.143 nm, to the hcp phase with the lattice parameters of 0.1008 nm and 0.1666 nm, and also to the cubic phase of $Cu_2O$ cuprite with the lattice parameter of 0.0425 nm occurs.

It is obvious that such complex nature of the observed nanostructures in $Cu-C_{60}$ system and the behavior of lattice parameters indicate the polymerization of fullerites, which appear as a result of additional intermolecular interaction, caused either due to direct of the charge exchanges between the $C_{60}$ molecules and the Cu atoms in the internodes or during ionic interaction of the negatively charged $(C_{60} - O_2)^-$ complexes and the positive Cu ions [9].

It is evident that the presence of the above heterophasic nanostructures leads to practically complete reconstraction of the electron levels position, to filling of the conduction band and, as a result, to the appearance of the additional bands of the optical conductivity, which are not observed in pure fullerites. The role of each structural component in this case is obvious, although the total growth of metallic conductivity testifies about the noticeable contribution of the polymerization of the fullerite structure and copper oxide.

## 3. Acknowledgements

O.P.D. is grateful to the Polish Academy of Sciences for providing the Mianowskiego Scholarship to carry out this research work.

## 4. References

1. Wang Y, Holden IM, Rao AM, Lee WL, Bi XX, Ren SL, Lehman GW, Hager GT, Eklund PC. Interband dielectric function of $C_{60}$ and $M_6C_{60}$ (M=K, Rb, Cs). Phys. Rev.B 1992; 45: 14396-14399.
2. Panchokha PA, Toptygin AL. Structural model of the vacuum fullerene-bismuth condensates. ISTFE-12, Diamand Films and Related Material. Proceedings of 12th International Simposium "Thin Films in Electronics", Kharkov 2001; 210-212.
3. Shpilevsky EM. Metal-fullerite films: synthesis, properties, application. ISTFE-15, Diamand Films and Related Material. Proceedings of 15th International Simposium "Thin Films in Electronics", Kharkov 2003; 242-263.
4. Shpilevsky EM, Baran LA, Okatova GP, Gusakova SV. Structure, phase composition and mechanical stress in copper-fullerene films. ISTFE-15, Diamand Films and Related Material. Proceedings of 15th International Simposium "Thin Films in Electronics", Kharkov 2003; 265-269.
5. Rao AM, Eklund PC, Holdeau JL, Margues L, Nunez-Regueiro M. Infrared and Raman studies jf pressure polymerized $C_{60}$. Phys. Rev.B 1997; 55: 4766-4773.
6. Kamaras K, Iwasa Y, Forro L. Infrared spectra of one- and two- dimensional fullerene polimer structures: $RbC_{60}$ and rhombohedral $C_{60}$. Phys. Rev.B 1997; 55: 10999-11002.
7. Kelly MK, Etchegoin P, Fuchs D, Kratschmer W, Fostiropoulos K. Optical transitions of $C_{60}$ films in the visible and ultraviolet from spectroscopic ellipsometry. Phys. Rev.B 1992; 46: 4963-4968.
8. Makarova TL. The electrical and optical properties of monomer and polymerized fullerenes. Fiz.Tekh.Polupr. (in Russian) 2001; 35: 257-293.
9. Katz EA, Faiman D, Tuladhar SM, Shtutina S, Froumin N, Polar M. Electrodiffusion phenomena in thin films. Fiz.Tverd. Tela (in Russian) 2002; 44: 473-475.

# HYDROGENATED AMORPHOUS SILICON CARBIDE FILMS AS PERSPECTIVE TRIBOLOGICAL COATINGS AND SEMICONDUCTOR LAYERS

V.I. IVASHCHENKO, O.K. PORADA, L.A. IVASHCHENKO,
G.V. RUSAKOV, S.M. DUB[*], V.M. POPOV[**]
*Institute of Problems of Material Science, NAS of Ukraine, Krzhyzhanovsky str. 3, 03680 Kyiv,*
[*]*V.N. Bakul Institute of Superhard Materials, NAS of Ukraine, Avtozavodska str. 2, 04074 Kyiv*
[**]*"Microanalytics" Center at State Institute of micro-devices, Pivnicho-Syretska str. 3, 04136 Kyiv*

## Abstract

Amorphous hydrogenated silicon carbide (a-SiC:H) films have been deposited from methyltrichlorosilane (MTCS) in a 40.56 MHz PECVD reactor with an additional substrate bias on crystalline silicon wafers. The films as wear-resistant coatings and semiconductor layers contain about 60-70 and 70-80 at% Si, respectively. Film thicknesses were about 0.2-0.4 μm. Film morphology and scratch test investigations show that the films are homogeneous and have rather good adhesion to the substrates. From nanoindentation tests it follows that a-SiC:H coatings exhibit the hardness up to 10 GPa and elastic modulus of 118 GPa. The annealed at 600 $^0$C samples have higher hardness then the as-deposited films (by about 25%). Ball-on-plane tests have reveled that the abrasive wear resistance of the covered substrates is 1.5-4 times higher than that of the un-covered ones. So, despite the comparatively low hardness, the a-SiC:H films demonstrate good tribological properties. The possible mechanisms describing the tribological properties of a-SiC:H coatings are discussed. a-SiC:H films prepared under special conditions possess good semiconductor properties. Particularly, dark conductivity, photosensitivity and optical band gap were $10^{-9}$-$10^{-11}$ S/cm, $10^3$-$10^4$ and 2.1-2.7 eV, respectively, which enables the films to be used as possible active layers in semiconductor devices.

*Keywords:* Amorphous hydrogenated silicon carbide; PECVD; Methyltrichlorosilane; Wear-resistant coatings; Semiconductor films.

*T.N. Veziroglu et al. (eds.),*
*Hydrogen Materials Science and Chemistry of Carbon Nanomaterials,* 339-346.
© 2004 *Kluwer Academic Publishers. Printed in the Netherlands.*

## 1. Introduction

Many researchers in the field Physical Vapor Deposition (PVD), Chemical Vapor Deposition (CVD) and Plasma Enhanced CVD (PECVD) set as their goal the pursuit of high hardness, believing that this will provide the ultimate level of wear resistance [1-3]. The view [1] is also that wear-resistant coatings (WRC) need to have the ability to absorb deformation, i.e. they have to combine an adequate degree of "elasticity' with adequate hardness to fulfill the needs of the tribological contact conditions. The coatings based on a-SiC:H seem to belong to the latter classification. We have shown earlier [4] that such coatings can serve as protective and wear-resistant layers on cutting tools in processing aluminum or titanium materials. Later, Esteve, Lousa, Martinez *et al* [5] have shown that PVD carbon-rich a-SiC films on silicon wafers have higher wear resistance than the substrates, in spite of the fact that the amorphous films have low hardness (H) and low elastic modulus (E) (H ~6 GPa and E ~ 80 GPa) as compared with those of the crystalline silicon wafers. Hydrogen- free a-$Si_xC_{1-x}$, x~ 0.5, films obtained by laser ablation and triode sputtering deposition [6] exhibit H ~ 30 GPa and E ~240 GPa. As a rule, PECVD hydrogenated films have H and E that are about two times lower than the un-hydrogenated counterparts [6]. In the annealed a-SiC:H films hydrogen was found to undergo high effusion in the temperature range of 400-600 $^0$C [7,8]. So far, the investigations of the effect of annealing on the hardness of a-SiC:H films were not carried out.

a-SiC;H films are also among the most promising materials for the low-cost production for photovoltaic applications [8,9 ]. Usually device-quality films are grown in a PECVD system from silane-mathane mixtures diluted in hydrogen and, a laser degree, from organosilanes [8]. Since the thoroughly investigations of semiconductor a-SiC:H films prepared from methyltrichlorosilane (MTCS) are absent, it is reasonable to evaluate the possibilities of such films as semiconductor materials.

In this work we present the results investigations of Si-rich a-SiC:H films deposited on silicon crystalline wafers. We shortly characterize as-deposited and annealed a-SiC:H films prepared by PECVD from MTCS. It is shown that wear resistance and high hardness cannot correlate in amorphous hydrogenated Si-rich silicon carbide. Although the coatings show low hardness, their good wear resistance and thermal stability make them suitable as WRC in processing soft materials. Besides the tribological investigations, we have studied semiconductor properties of a-SiC:H films. It is shown that a-SiC:H films are also suitable for photovoltaic applications.

## 2. Experimental details

Both the a-SiC:H based coatings and semiconductor films were produced on silicon crystalline wafers by r.f.(40.68 MHz) PECVD by using MTCS as a main precursor. An additional bias was applied to the substrates using a 5.27 MHz generator. The MTCS vapor was delivered into the reaction chamber with hydrogen that passed through a thermostated bubbler with MTCS. The substrate temperature ranged from 200 to 300 $^0$C, the bubbler temperature was 32 C$^0$, the applied bias and $H_2$:MTCS+$H_2$ ratio were –200 V and 5, respectively. The total gas pressure was up to 1 Torr (~ 0.3 Torr in depositing the semiconductor films). The discharge power density was about 0.3 W/cm$^2$. The thin films (0.2-0.4 µm) were deposited during 60-90 min. The coatings were annealed in vacuum (about 10$^{-5}$ Torr) at 450 $^0$C and 600 $^0$C during one hour.

The study of the film morphology and microstructure is carried out with a scanning electron microscope JSM-IC845 (JEOL, Japan). The film structure is examined by electronography on reflection with the help of electron microscope EMP-100 (USSR). The obtained films were tested with a calowear tester using a ball of hard steel rotating against a sample in the presence of diamond paste (0.1µm). The film composition is examined with the help of an Auger spectrometer JAMP-10S (JEOL, Japan) and a secondary ion mass-spectrometer CAMECA IMS-4F (France). The film thickness is investigated by means of a microprofilometer Alpha-Step 200 (Tencor Instruments, USA) and a calowear tester (for preliminary evaluating film thickness). Infrared and optical absorption spectra of the films deposited on glass or silicon crystalline wafers were measured using spectrometers M80 and M400 (Carl Zeiss, Germany), respectively. Electrical characteristics were studied by means of an automatic system for analysis of electrophysical characteristics of semiconductor structures HP 4061A (Hewlett-Packard, USA) and an automatic system for the control of electrical parameters of semiconductor devices HP 4145A (Hewlett-Packard, USA). Hardness and elastic modulus are determined from nanoindentation tests carried out with Nano Indenter-II (MTS Systems Corporation, Oak Ridge, TN, USA). The accuracy of the nanoindentation measurements on loading and indentation depth was ± 75 nN and ± 0.04 nm, respectively. Hardness and elastic modules were determined according to the procedure [10]. The nanoindentaion was carried out under such conditions to minimize the substrate influence. The scratch tests, allowing us to characterize the adhesion of the film to the silicon substrate, were carried out with a modified tester ПМТ-3 (USSR) using the Vickers indenter with a tip radius of 1-3 µm. The load rate was 10-50 mN/s.

## 3. Results and discussion

### 3.1 a-SiC:H FILMS AS WEAR-RESISTANT COATINGS

In order to identify the coating structure, electronography on reflection studies were carried out. The electronograms of as-deposited a-SiC:H coatings obtained at substrate temperatures < 250 $^0$C (not shown) prompts us to conclude that the film structure is amorphous. It is well known [11] that the rise of substrate temperature, especially above 300 $^0$C, promotes the creation of microcrystalline islets in the amorphous matrix. Here, this question specially was not studied.

Chemical composition was determined by secondary ion mass spectroscopy (SIMS). The typical SIMS sputter depth profiles for a-SiC:H films are presented in Fig. 1. Since the SIMS elemental distributions cannot be presented in the absolute scale, we evaluated the silicon content taking into account the etalon β-SiC sample. The silicon concentration was evaluated to vary around 60-67 % depending on deposition parameters. These findings are consistent with the results of the Auger measurements, which show that a-SiC:H films contain about 60-65 at % of Si and up to 25, 6-8 and 0.8 at % of C, O and Cl, respectively. The anomalous behaviour of the elemental distribution is seen to be observed at the coating-surface and coating-substrate regions. In other regions, the elements are distributed more or less uniformly. Since oxygen was not specially supplied into the chamber, its comparatively high content in the films is supposed to be due to the residual oxygen absorbed onto the chamber surface. We also do not exclude that our Auger measurements can overestimate the oxygen content in the film.

The presence of different bond configurations in the films can be revealed from infrared absorption spectra measurements. The results of such measurements are presented in Fig. 2. We have put an accent on the investigation of the infrared absorption in the range of wavenumbers of 400-1200 $cm^{-1}$. In this branch, one can single out prominent features at 480, 670, 800, 1040 and 1100 $cm^{-1}$. The dip centered around 480 $cm^{-1}$ is due to Si atoms bound to other ones in an amorphous states [12]. The structure at 670 $cm^{-1}$ is attributed to Si-C stretching vibrations [13]. The presence of the strong dip at about 800 $cm^{-1}$ is associated with the fundamental transversal optic phonon frequency of Si-C stretching vibrations [14]. The large width of this minimum can be attributed to the different Si-C bond lengths and bond angles in a-SiC:H. The structure around 800 $cm^{-1}$ is mostly revealed in a-SiC:H films obtained from organosilanes [13]. The dip around 1040 $cm^{-1}$ is caused by Si-O stretching, $Si-CH_2$ rocking or twisting and $C-H_n$ rocking or wagging vibrations [11-14]. The absorption band located at 1100 $cm^{-1}$ is associated with the presence silicon oxide (Si-O stretching modes) [11]. Since the oxygen content in the films changes insignificantly (6-8 at %), the transformation of the dip at 1040 $cm^{-1}$ in the films obtained under different deposition conditions can be attributed to changing $C-H_n$ bonding configurations. It is not difficult to see that, on going from the upper curve to the lower one, the dip around 800 $cm^{-1}$ becomes dipper as compared to the one around 1040 $cm^{-1}$, which can be considered as the increase of a number of Si-C bonds and the decrease of a number of $C-H_n$ bonds. Correspondingly, one can expect the enhancement of the strength of the films with increasing substrate temperature in depositing. Strengthening of the dip at 1100 $cm^{-1}$ points to the high oxidation ability of the films deposited at high substrate temperatures. We do not exclude the possibility of the nucleation in the latter films, in which case the creation of crystallites promotes high oxygen adsorption.

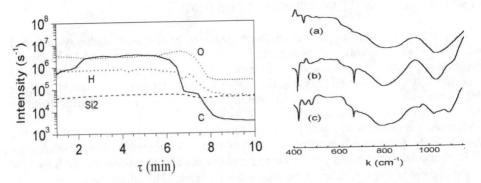

*Figure 1.* Typical SIMS sputter depth profiles. The substrate-coating interface is reached after sputtering during 7 min. Si2 is double ionized silicon ions.

*Figure 2.* Infrared transmission spectra of the coatings deposited at 200 (a), 250 (b) and 300 $^0$C (c).

We have carried out nanoindentation tests of the obtained films. Before testing, two sets of the a-SiC:H samples (M1 and M2, prepared at 200 and 300 $^0$C, respectively) were annealed in   vacuum at 450 $^0$C and 600 $^0$C during 60 minutes. The results of the nanoindentations are summarized in Fig. 3,4. One can see that the films obtained under high temperature conditions have higher hardness than the coatings prepared at low temperatures. The possible cause of such a rise of hardness was discussed above in interpreting the infrared absorption spectra.  From the nanoindentation at 1 mN it follows that the annealed coatings show H and E up to 10 GPa and 118 GPa, respectively, which is higher the analogous characteristics obtained in Ref. [5] (4-6 GPa and 50-80 GPa, respectively) and are very close to those of the stoichiometric PECVD hydrogenated films prepared from a silane-methane mixture (H ~ 15 GPa and E ~ 120 GPa) [6].  While the annealing leads to increasing hardness up to 25 %, in annealing the Young modulus rises only by 3.5 %.

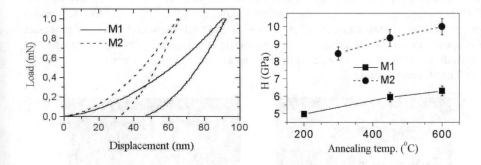

*Figure 3.* Results of the nanoindentation tests of the annealed  (600 $^0$C) M1 and M2 samples.

*Figure 4.* Nanohardness of the M1 and M2 samples as a function of annealing temperature.

We have tested the film behavior under loading effort of 50 mN and not found any film peeling. This preliminary finding is consistent with the results of scratch test investigations under the maximal load of 500 mN. It was found that the critical load parameter was around 300 mN, which points to the comparatively good adhesion of the thin coating (~ 0.2 μm) to the silicon substrate.

The wear coefficient of the coatings was determined by the ball-cratering method (calowear test), using a suspension of 0.1 μm diamond. The diamond particles have the developed surface with a pin-like structure. To determine the wear coefficient (k), the procedure of the measurements [3] was used. It is based on measuring the crater diameters on the clean and covered substrates. The preliminary measurements of film thickness (d), needed for determining wear resistance, were carried out with a microprofilometer. Our films have on average  thicknesses that change inside the region 0.2 < d < 0.4 μm.  Taking into account values d and performing the calowear test with the clean and covered substrates under the same conditions, we have determined the wear coefficients of the substrates and the coatings by evaluating the volume of wear craters. Wear coefficients of

344

the clean and covered by a-SiC:H coatings crystalline silicon substrates turned out to vary on average around $3.0 \times 10^{-13}$ - $7.5 \times 10^{-13}$ m$^2$/N and $1.0 \times 10^{-13}$ -$1.8 \times 10^{-13}$ m$^2$/N, respectively, depending on film preparation conditions. In particular, k decreases as substrate temperature or annealing temperature increases. It follows that the wear resistance of a-SiC:H coatings is 1.5-4 times higher than silicon crystalline wafers. In Fig. 5 we show the wear craters formed as a result of the calowear test on the clean and covered with a-SiC:H coating silicon substrates. One can see that the coating is destroyed to a lesser extent than the substrate under polishing by a steel ball with diamond paste.

Measured nanoindentation hardness and Young's modulus of the coatings show values lower than those of crystalline silicon substrates (H ~ 13.4 GPa and E ~ 168 GPa). However, the wear resistance of the films is unexpected high as compared to that of silicon wafers. We suppose that the main cause of such a phenomenon consists in the low rigidity and the developed amorphous surface of a-SiC:H films. The film surface easy transforms during the sliding process adjusting itself to the friction different surface. The hydrogen effused from the bulk serves as a hard lubrication. Due to a low thermal conductivity of a-SiC:H as compared to c-Si, the friction pieces in the a-SiC:H/c-C contact are heated stronger than the friction surfaces in the c-Si/c-C contact. All these properties and processes are supposed to lead to lowering the friction coefficient, especially, at high speed sliding. As a result, the wear of the amorphous hydrogenated coating will be lower than the one of a crystalline silicon substrate.

*Figure 5.* Wear craters of the un-covered (left panel) and covered by a-SiC:H coating (right panel) silicon substrate. The film thickness is 0.214 μm, the crater diameters are 627 μm (clean substrate) and 607 μm (covered substrate). The wear coefficients of the substrate and coating were evaluated to be $6.366 \times 10^{-13}$ m$^2$/N and $3.325 \times 10^{-13}$ m$^2$/N, respectively.

## 3.2 a-SiC:H FILMS AS SEMICONDUCTOR MATERIALS

a-SiC:H films for photovoltaic application were deposited under the conditions to reach good optoelectronic characteristics. The chamber pressure was reduced to 0.3 Torr. The typical Auger depth profile of the semiconductor films is presented in Fig. 6. It is seen that inside the film the elements are distributed uniformly. The oxygen content reduced to 5 at% as compared with that in the coatings. The silicon content reaches 85 at%. The optical bad gap varies in the range of 2.1-2.7 eV depending on deposition regimes and the level of doping. The band gap ($E_g$) as a function of the ratio $X_C/X_{Si}$, where $X_{C,Si}$ is the carbon and silicon concentrations, is shown in Fig. 7. The value $E_g$ is higher in un-doped films than in boron-doped ones, and increases with increasing the carbon content. The rest characteristics of semiconductor a-SiC:H films were the following. A dark conductivity was $10^{-9} - 10^{-11}$ S/cm (un-doped) and $10^{-5} - 10^{-8}$ S/cm (nitrogen or boron doped). A photosensitivity was $10^3$-$10^4$ (under 100 mW/cm² illumination). So, these films can be recommended for photovoltaic applications [8,9].

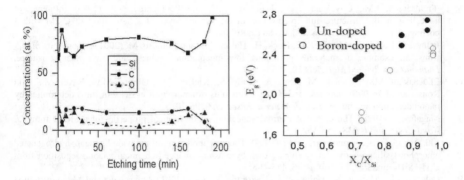

*Figure 6.* Auger depth profile of the semiconductor films.

*Figure 7.* Optical band gap of un-doped and boron-doped a-SiC:H films as a function of composition.

## 4. Conclusions

a-SiC:H films with the predominance of a silicon content were grown on silicon crystalline wafers in a PECVD reactor from methyltrichlorosilane with adding hydrogen at the substrate temperature ranging from 200 °C to 300 °C. The films were prepared for tribological and photovoltaic applications. The coatings were annealed in vacuum at 450 and 600 °C. The comprehensive coating investigation shows that the rise of substrate temperature or annealing temperature leads to increasing nanohardness and elastic modulus. Despite the presence of oxygen (up to 6-8%), micro-voids at the substrate-coating interface and a high silicon content, a-SiC:H coatings exhibit rather good tribological properties. Consequently, they can be recommended for the use as wear-resistant coatings in processing soft materials. We have prepared the series of a-SiC:H films under special conditions to obtained the layers with good optoelecronic characteristics. The optimal growth conditions were singled out to produce the films, suitable in solar cell applications.

## 5. Acknowledgements

This work was supported in part by STCU Contracts No 1591, 1590-C.

346

# 6. References

1. Matthews, A. and Leyland, A. (2002) Developments in Vapour Deposited Ceramic Coatings for Tribological Applications, *Key Engineering Materials* **206-213,** 459-466.
2. Veprek, S. (1997) Electronic and mechanical properties of nanocrystalline composites when approaching molecular size, *Thin Solid Films* **297,** 145-153.
3. Allsopp, D.N. and Hutchings, I.M. (2001) Micro-scale abrasion and scratch response of PVD coatings at elevated temperatures **251,** 1308-1324.
4. Frantsevich, I.M., Gayevska, L.A., Rusakov, G.V. and Ponomarev, S.S. (1985) Influence of annealing on the structure and composition of the surface layer of a hard alloy KTC-2M with a coating, *Reports of AS of UkrSSR* **2,** 81-86 (in Russia).
5. Esteve, J., Lousa, A., Martinez, E., Huck, H., Halac, E.B. and Reinoso, M. (2001) Amorphous $Si_xC_{1-x}$ films: an example of materials presenting low indentation hardness and high wear resistance, *Diamond and Related Materials* **10,** 1053-1057.
6. El Khakani, M.A., Chaker, M., Jean, A., Boily, S., Kieffer, J.C., O'hern, M.E., Ravet, M.F. and Rousseaux, F. (1996) Hardness and Young's modulus of amorphous a-SiC thin films determined by nanoindentation and bulge tests, *J. Material Research* **9,** 96-101.
7. Magafas, L (1998) The effect of thermal annealing on the optical properties of a-SiC:H films, *J. Nono-Cryst. Solids* **238,** 158-162.
8. Liu, Y., Giorgis, F. and Pirri, C.F. (1997) Thermal modification of wide-bandgap hydrogenated amorphous silicon-carbon alloy films grown by plasma-enhanced chemical vapour deposition from $C_2H_2+SiH_4$ mixtures, *Phil. Mag.* **75,** 485-496.
9. Calcagno, L., Martins, R., Hallen A. and Skorupa, W. (ed) (2001) Amorphous and Microcrystalline Silicon Carbides: Materials and Applications, **112,** European Materials Research Society, Amsterdam, pp 640.
10. Oliver, W.C. and Pharr, G.M. (1992) An improved technique for determining hardness and elastic modulus using load and displacement sensing indentation experiments, *J. Mater. Res.* **7,** 1564-1583.
11. Demichelis, F., Grovini, G., Pirri, C.F. and Tresso, E. (1993) Infrared vibrational spectra of hydrogenated amorphous and microcrystalline silicon-carbon alloys. A comparison of their structure, *Phil. Mag.* **68,** 329-340.
12. Ech-chamkin, E., Ameziane, E.L., Bennouna, A., Azizan, M., Nguyen Tan, T.A. and Lopez-Rios, T. (1995) Structural and optical properties of r.f.-sputtered $Si_xC_{1-x}$:O films, *Thin Solid Films* **259,** 18-24.
13. Bullot, J. and Schmidt, M.P. (1987) Physics of Amorphous Silicon-Carbon Alloys, *Physica Status Solidi* (b) **143,** 345-418.
14. Avila, A., Montero, I., Galan, L., Ripalda, J.M. and Levy, R. (2001) Behavior of oxygen doped SiC thin films: An x-ray photoelectron spectroscopy study, *J. Appl. Phys.* **89** (2001) 212.

# VIBRATIONAL SPECTRA AND MOLECULAR STRUCTURE OF THE HYDROFULLERENES $C_{60}H_{18}$, $C_{60}D_{18}$, AND $C_{60}H_{36}$ AS STUDIED BY IR AND RAMAN SPECTROSCOPY AND FIRST-PRINCIPLE CALCULATIONS

A.A. POPOV[a]*, V.V. SENYAVIN[a], A.A. GRANOVSKY[a], A.S. LOBACH[b]

[a]Department of Chemistry, Moscow State University, 119992 Moscow, Russia
[b]Institute of Problems of Chemical Physics RAS, 142432 Chernogolovka, Moscow Region, Russia
*Corresponding author, fax: 7 (095) 932-88-46,
E-mail: popov@phys.chem.msu.ru (A. A. Popov)

## Abstract

IR and Raman spectra of hydrofullerenes $C_{60}H_{18}$, $C_{60}H_{36}$, and deuterofullerene $C_{60}D_{18}$, were thoroughly studied by the conjunction of experimental and computational technique. A perfect correspondence achieved between the first-principle calculations and the observed vibrational spectra for $C_{60}H_{18}$ molecule resulted in a complete assignment of both light and heavy isotopomers spectra. An extended search of the $C_{60}H_{36}$ stable isomeric structures was performed in the frames of PM3 and DFT levels of a theory. Vibrational spectra for 30 most stable isomers were computed and the structure of two dominant $C_{60}H_{36}$ isomers was determined. Interpretation of the experimental spectra as a 3:1 superposition of the $C_1$ and $C_3$ isomers features was fulfilled.

*Keywords*: Fullerene; Computational chemistry, Infrared spectroscopy, Raman spectroscopy; Chemical structure

## 1. Introduction

The first chemical derivative of the fullerene ever synthesized after its discovering was hydrofullerene $C_{60}H_{36}$ [1] obtained via Birch reaction; other methods for production of $C_{60}H_{36}$ were reported since that time [2-5]. As mass-spectrometry has shown [1], the dehydrogenation of the compound occurred above 300 °C with the formation of $C_{60}H_{18}$ hydride, and the same species was detected later as an intermediate product of $C_{60}H_{36}$ thermal decomposition [3-5]. The $C_{60}H_{18}$ compound can also be synthesized individually via hydrogen transfer from 9,10-dihydroanthracene [5] or directly by the reaction of $C_{60}$ with hydrogen at high pressure and temperature [6]. Only the latter procedure led to chemically pure product, while in the hydrogen transfer synthesis unremovable admixtures were always present [5]. Possibly due to this reason the information on properties of $C_{60}H_{18}$ is rather rare. Molecule of $C_{60}H_{18}$ was proposed by Darwish et al. [6] on the base of [1]H NMR spectroscopy to possess $C_{3v}$ symmetry with all the hydrogen atoms located at one hemisphere of the fullerene forming a crown-like belt around 3-fold axis. The same authors

*T.N. Veziroglu et al. (eds.),*
*Hydrogen Materials Science and Chemistry of Carbon Nanomaterials*, 347-356.
© 2004 *Kluwer Academic Publishers. Printed in the Netherlands.*

reported IR spectrum of the compound, but no interpretation of the $C_{60}H_{18}$ vibrational spectra is available up to date. Spectroscopic properties of $C_{60}H_{36}$ samples were studied in more details [2,5-12], however some ambiguity still exists in the molecular structure of $C_{60}H_{36}$ hydrofullerene, the variety of possible structures being proposed on the base of experimental and theoretical studies [3,4,7-15]. Recent NMR data on $^1H$, $^{13}C$ and endohedral $^3He$ nuclei [11] have shown that different isomeric mixtures were obtained by Birch reduction and hydrogen transfer from 9,10-DHA. While various first-principle theoretical approaches [11-16] favor the *T*-symmetry isomer with 4 aromatic rings in the skeleton of the molecule, the product of hydrogen transfer reaction was proved to consist of *C₁* and *C₃* structure in 1:3 ratio. Experimental values of $^3He$ chemical shift in He@$C_{60}H_{36}$ were most consistent with the computed data for isomers, containing three benzene rings in their structure [11], however definite structures of the isomers were not revealed.

Deutero-derivatives of $C_{60}$ were described by a number of authors. $C_{60}D_{24}$ and its IR spectra were reported by Shul'ga et al. [17]. $C_{60}D_{18}$ and $C_{60}D_{36}$ were synthesized by Somenkov et al. [18]. Meletov et al. reported Raman spectra of $C_{60}D_{36}$ samples obtained by high pressure deuteration [10]. Only the brutto composition of the samples derived on the base of elemental analysis and/or X-ray diffraction study was reported in all the papers. No molecular structures were revealed.

In this work we present the results of experimental IR and Raman study of $C_{60}H_{18}$, $C_{60}D_{18}$, and $C_{60}H_{36}$ samples supplemented with first-principle calculations of the structures and vibrational features of the title molecules. Special attention is paid to the elucidation of $C_{60}H_{36}$ molecular structure.

## 2. Experimental details

The sample of $C_{60}H_{36}$ was synthesized via hydrogen transfer from 9,10-DHA to $C_{60}$ as described earlier [5]. In the synthesis of $C_{60}H_{18}$ the hydrogenation methodic was modified so that instead of reported previously 24 h duration of synthesis [3] the mixture of $C_{60}$ and 9,10-DHA was held at 625 K for only 5–7 minutes. The bright-red melt was cooled to 380 K and then heated at 1–2 Pa for 8 hours to remove unreacted 9,10-DHA and formed anthracene. The samples of $C_{60}H_{18}$ and $C_{60}H_{36}$ were further purified from the anthracene impurities by heating at 500 K in high vacuum ($10^{-5}$ Pa) for 5–8 hours. The sample of deuterofullerene with $C_{60}D_{18}$ brutto composition was obtained from the group of Prof. V.A. Somenkov where it was prepared by direct synthesis at 20 MPa and 750 K [18].

Infrared spectra of the samples pelletted in KBr were measured on a Specord M80 and Bruker IFS-113v spectrometers under 0.5–2.5 cm$^{-1}$ resolution. Additional studies were performed on a Nicolet 400 Protege FTIR spectrometer with the pellets formed from pure substances. The FT-Raman spectra were obtained on a Bruker Equinox 55S spectrometer equipped with a FRA 106s module and Nd:YAG 1064 nm excitation laser.

## 3. Computational details

DFT calculations of molecular structures and IR spectra were performed with the PRIRODA package [19] using GGA functional of Perdew, Burke and Ernzerhof (PBE) [20]. To accelerate the evaluation of the Coulomb and exchage-correlation terms the code employed the expansion of the electron density expansion in an auxiliary basis set. TZ2P basis sets {6,1,1,1,1,1/4,1,1/1,1} for carbon and {3,1,1/1} for hydrogen atoms were used throughout the calculations. Harmonic vibrational frequencies and IR intensities were calculated analytically.

Raman intensities were computed numerically at a Hartree-Fock (HF) level within 6-31G** basis set employing PC version [21] of GAMESS (US) quantum chemical package [22]. Semiempirical PM3 calculations of $C_{60}H_{36}$ molecular structures and their relative energies were also performed using PC GAMESS.

Normal coordinate analysis was performed in a complete internal coordinates system using the DISP [23] suite of programs for vibrational calculations.

## 4. Results and Discussion

### 4.1 $C_{60}H_{18}$ and $C_{60}D_{18}$

Fig. 1 presents Schlegel diagram of $C_{60}H_{18}$ viewed along the $C_3$ axis. No experimental data on $C_{60}H_{18}$ molecular geometry are available. However, for its fluorinated analogue, $C_{60}F_{18}$, it was judged from X-ray diffraction [24] that the addition of $X_{18}$ moiety to a fullerene carcass led to the strong "aromaticity" of the hexagon surrounded by the attached atoms (fig.1). This peculiarity was displayed by the planarity of the fluorinated fragment and close lengths of 6/6 and 5/6 edges in the central hexagon.

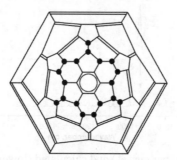

Fig. 1. Schlegel diagram of $C_{60}H_{18}$ molecule.

For $C_{60}H_{18}$ our calculations predict the same effects to take place, though in a lesser extent. The parameters obtained within the geometry optimization at PBE/TZ2P level of theory revealed the reduced bond alteration (only 0.012 Å) in the "benzene" ring and the small values of dihedral angles between the plane of the ring and neighboring bonds (10-11° in $C_{60}H_{18}$ versus 42° in $C_{60}$). Structural parameters of the unfunctionalized hemisphere were close to those of the pristine fullerene molecule optimized within the same PBE/TZ2P approach (1.399/1.453 Å for double/ordinary bonds, respectively).

According to $C_{3v}$ symmetry the vibrations of $C_{60}H_{18}$ molecule span into irreducible representation as:

$$\Gamma_{vib} = 41A_1 + 35A_2 + 76E.$$

117 fundamentals of $A_1$ and E symmetry should be both IR- and Raman-active, thus resulting in rather complicated spectra.

The experimental IR and Raman spectra of $C_{60}H_{18}$ are compared with calculated ones on Fig. 2. Our IR spectrum coincides well with that from ref. [6] whenever they could be compared. Raman spectrum of this hydrofullerene is reported here for the first time. The agreement between experimental and computed spectra is quite satisfactory, the absolute deviation of the predicted wavenumbers from the measured frequencies being less than 8 cm$^{-1}$ for the majority of the bands. The force field and the intensity parameters of $C_{60}H_{18}$ molecule were employed further to simulate the IR and Raman spectra of the deuterofullerene $C_{60}D_{18}$. A perfect correspondence of the calculated data to the experimental spectra of the sample (Fig.3) allows one to conclude that hydrofullerene $C_{60}H_{18}$ and deuterofullerene $C_{60}D_{18}$ (which were obtained by different synthetic pathways) are isostructural. Finally, nearly exact matching of the experimental data for the $C_{60}H_{18}$ and $C_{60}D_{18}$ molecules by the results of calculations enabled us to propose a complete interpretation of the spectra.

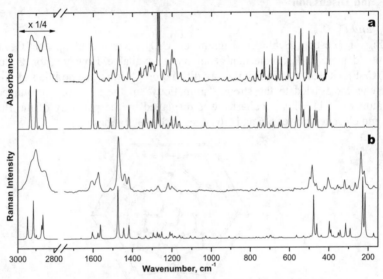

Fig.2. Experimental (upper curves) and calculated (lower curves) IR (a) and Raman (b) spectra of $C_{60}H_{18}$. Calculated frequencies in the C–H stretching region are scaled by the factor of 0.98.

Deformational vibrations of carbon skeleton in $C_{60}H(D)_{18}$ molecule fall into 150 – 800 cm$^{-1}$ region. In Raman spectra they are presented by intense lines at 150–400 cm$^{-1}$. Since the displacements of hydrogen or deuterium atoms for these modes are negligible compared to those of carbon atoms, the spectra of isotopomers are nearly identical in this region.

In IR spectra the vibrations of the carbon skeleton are seen as groups of strong and medium absorptions at 400 – 800 cm$^{-1}$. While for the $C_{60}H_{18}$ molecule the vibrational potential energy is, as before, distributed mostly between $\gamma$(C=C–C), $\gamma$(CCC) and $\gamma$(CC(H)C) internal coordinates, the contributions of $\alpha$(CCD) coordinates becomes apparent in $C_{60}D_{18}$ vibrations above 600 cm$^{-1}$. Hence, the spectra of the carcass vibrations of the isotopomers at higher frequencies – in the vicinity of radial modes – are quite different.

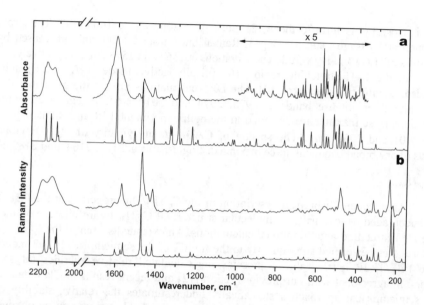

Fig.3. Experimental (upper curves) and calculated (lower curves) IR (a) and Raman (b) spectra of $C_{60}D_{18}$. Calculated frequencies in the C–D stretching region are scaled by the factor of 0.98.

IR and Raman spectra in the 800 – 1100 cm$^{-1}$ are rather poor. Though calculations predict high vibrational density of states at these frequencies due to CC(H)C deformations and C(H)–C(H) stretchings, intensities of these modes are extremely low. Above 1100 cm$^{-1}$ the δ(CCH) deformations arise in the spectra of $C_{60}H_{18}$ molecule, being partially mixed with ν(C–C) vibrations. While Raman intensities of these modes are also quite low, the absorption transition probabilities are rather high and a number of medium bands and a very strong absorption at 1272 cm$^{-1}$ are observed in IR. For the $C_{60}D_{18}$ molecule the CCD deformations are distributed in wider range (600 – 1000 cm$^{-1}$) and possess lower IR intensities. The medium IR bands in the spectra of the heavy isotopomer at 1237, 1294, and 1335 cm$^{-1}$ are caused mainly by the C–C stretching vibrations.

The stretches of a benzenoid ring and the ν(C=C) vibrations in the fullerene-like fragment of the $C_{60}H_{18}$ molecule fall into 1400 – 1700 cm$^{-1}$ region, where the spectra of isotopomers are quite similar again. Noteworthy, that the definite vibrations of an aromatic ring, e.g. the CC stretching modes of a "parent" benzene molecule can be traced in these vibrations of $C_{60}H_{18}$ with only minor changes in their forms. So, a weak Raman line of $C_6H_6$ at 1601 cm$^{-1}$ of $E_{2g}$ symmetry [25] shifts upwards to 1662 cm$^{-1}$ in $C_{60}H_{18}$ molecule (calculated value), but is not observed in experimental spectra due to its low predicted intensity. The very intense IR band of benzene at 1484 cm$^{-1}$ of $E_{1u}$ symmetry [25] transforms to corresponding E-symmetry mode of $C_{60}H_{18}$ and downshifts slightly. With two vibrations of $A_1$ symmetry it composes a very strong IR absorption with a complex shape and a maximum at 1472 cm$^{-1}$.

Other bands in this spectral region belong to the ν(C=C) vibrations. A very strong Raman line at 1472 cm$^{-1}$ is assigned to the mentioned previously $A_1$ mode of hydrofullerene, which is close by its form to the $A_g(2)$ mode of pristine $C_{60}$ at 1468 cm$^{-1}$. The strong IR absorption at 1610 cm$^{-1}$ and a group of Raman lines around 1600 cm$^{-1}$ are caused by the derivatives of $H_u(7)$ and $H_g(8)$ fullerene vibrations at 1567 and 1573 cm$^{-1}$, respectively.

In the C–H stretching region 10 optically active modes ($4A_1$ + $6E$) and two forbidden $A_2$ vibrations are expected. The DFT calculations predict the grouping of these modes into three intense bands at 2925, 2974 and 3010 cm$^{-1}$. In agreement with the calculations three maxima are observed in the both experimental IR and Raman spectra at 2849, 2901 and 2931 cm$^{-1}$. The spectra of $C_{60}D_{18}$ in the vicinity of $\nu$(C–D) vibrations consists of one broad band with three maxima distinguishable at 2127, 2164, and 2180 cm$^{-1}$.

*4.2 $C_{60}H_{36}$*

An adequate theoretical description of the $C_{60}H_{18}/D_{18}$ properties at the PBE/TZ2P level encouraged us to clarify the molecular structure of $C_{60}H_{36}$ hydrofullerene on the base of the DFT structural and vibrational calculations. An extended search of the $C_{60}H_{36}$ stable structures was performed subsequently in the frames of the semiempirical PM3 model and at the PBE/TZ2P level. Table 1 presents the calculated relative energies for selected isomers. In agreement with our previous report [12] and recent results of Clare and Kepert [16], semiempirical approach systematically underestimates the relative stability of the structures containing aromatic cycles. However, more reliable values can be obtained by subtracting the empirically derived quantity of ca 67 kJ/mole per benzenoid ring from the PM3 relative energies, the increment being revealed from the comparison of semiempirical and DFT results. Using this correction we have computed the relative energies of more than 100 $C_{60}H_{36}$ isomers, mostly for those of $C_1$ and $C_3$ symmetry with 3 aromatic rings in their structure.

TABLE 1. Relative energies of selected $C_{60}H_{36}$ isomers computed at PM3 and PBE/TZ2P levels of theory, kJ/mol

| | $\Delta E_{PM3}$ | $\Delta E_{DFT}$ | $\Delta E_{PM3}$ - $\Delta E_{DFT}$ | $N^a$ | $\Delta E_{PM3}$ corr.$^b$ |
|---|---|---|---|---|---|
| $T$(№2)$^c$ | -34.5 | -290.1 | 255.6 | 4 | -302.5 |
| $C_1$ | -70.4 | -272.5 | 202.1 | 3 | -271.4 |
| $C_3$(№64) | -71.4 | -270.9 | 199.5 | 3 | -272.4 |
| $S_6$(№88) | -96.8 | -237.9 | 141.1 | 2 | -230.8 |
| $S_6$(№91) | -34.6 | -171.0 | 136.4 | 2 | -168.6 |
| $D_{3d}$(№1) | -32.7 | -171.0 | 138.3 | 2 | -166.7 |
| $C_3$(№3) | 74.6 | -114.5 | 189.1 | 3 | -126.4 |
| $T_h$ | 0.0 | 0.0 | 0.0 | 0 | 0.0 |

$^a$N – number of aromatic cycles in the structure
$^b$corrected PM3 relative energies
$^c$numeration of the isomers is given after ref. [26]

Then, for 30 isomers with the energy lying within ca 120 kJ/mol above the most stable one, as well as for some less stable structures, which were claimed earlier [2,7,13] to be the major components in $C_{60}H_{36}$ isomeric mixture, the IR spectra were simulated at a PBE/TZ2P level. Detailed data on various isomers stability and spectra will be reported elsewhere; here we discuss the primary results.

While the *T*-symmetry isomer was found to be the most stable form in our calculations and in previous DFT reports [11-13,17], the complexity of experimental vibrational spectra rejected the realization of this structure, at least as the dominant component in the system. Alternatively, the 3:1 superposition of the spectra simulated for the second and the third lowest-energy isomers (namely, those of the $C_1$- and the $C_3$-symmetry, Fig.4b,c) perfectly fitted the experimental IR spectrum (Fig.5a). Raman scattering intensities computed for these two isomers were also found to correspond well with the measured spectrum (Fig.5b). Thus, our vibrational calculations not only confirmed the isomer composition of $C_{60}H_{36}$ suggested on the base of NMR results [11], but also allowed us to determine the definite molecular structures for the major $C_1$ and $C_3$ isomers of the mixture.

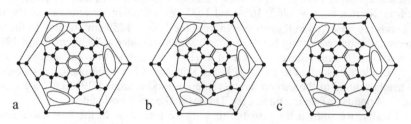

a          b          c

Fig.4. Schlegel diagrams for the most stable isomers of $C_{60}H_{36}$: $T$ (a), $C_1$ (b), and $C_3$ (c).

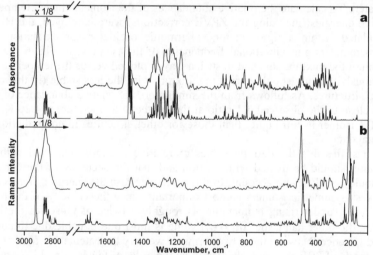

Fig.5. Experimental (upper curves) and calculated (lower curves) IR (a) and Raman (b) spectra of $C_{60}H_{36}$. Theoretical spectra are simulated as 3:1 superposition of calculated ones for $C_1$ and $C_3$ isomers.

Though nearly continuous distribution of the vibrational frequencies in the spectra of the hydrofullerene precludes a peak-to-peak vibrational assignment, tentative interpretation for $C_{60}H_{36}$ can be proposed on the first-principle background. Radial vibrations of the carbon skeleton are observed at 160 – 600 cm$^{-1}$ with medium and high intensities in IR and Raman spectra, respectively, including the strongest Raman lines at 207 and 484 $cm^{-1}$. In the medium range (600 – 1100 $cm^{-1}$) a number of weak bands are detected in Raman scattering and a group of weak and medium peaks in the infrared spectra. Potential energy of these vibrations is distributed mostly between $\gamma$(CCC), q(C–C) and $\alpha$(CCH) internal coordinates. For instance, an intense absorption at 815 $cm^{-1}$ is assigned to the mode with $\gamma$(CCC) and $\alpha$(CCH) contributions being 40% and 30%, respectively. Stretching C–C vibrations become apparent in the 900 – 1100 $cm^{-1}$ interval with quite low intensities. Strong IR and medium Raman bands at 1100 – 1400 $cm^{-1}$ are ascribed to $\delta$(CCH) vibrations with minor contributions from $C(sp^2)$–$C(sp^3)$ elongations. Vibrations of the aromatic cycles fall into 1430 – 1500 and 1630 – 1700 $cm^{-1}$ intervals. They cause the strongest absorption at 1490 $cm^{-1}$ and weak peaks at 1634 and 1672 $cm^{-1}$ in the IR spectrum, and medium Raman lines at 1462, 1654 and 1675 $cm^{-1}$. Stretching vibrations of the isolated double bonds are observed in Raman spectrum in the 1700 – 1750 $cm^{-1}$ region. In the $\nu$(C–H) characteristic region two strong bands at 2850 (with a shoulder at 2830 $cm^{-1}$) and 2910 $cm^{-1}$ are observed in IR and Raman spectra of $C_{60}H_{36}$. According to p.e.d. analysis the low-frequency bands are ascribed to the C–H bonds with two $C(sp^3)$ and one $C(sp^2)$ atoms in the nearest neighborhood of the carbon atom, while the modes around 2910 $cm^{-1}$ are associated with the C–H bonds possessing three $C(sp^3)$ atoms in the neighborhood.

Finally, we address the possible reasons for formation of the less stable structures as the result of the synthesis. In order to clarify this phenomenon the equilibrium composition of the isomeric mixture consisting of the three lowest-energy forms of $C_{60}H_{36}$ was simulated. To calculate the relative abundances of the isomers at each temperature the DFT relative energies (including the ZPVE corrections) were employed, the $H_T$-$H_0$ terms were calculated within a rigid rotator – harmonic oscillator approximation using the geometry parameters and vibrational frequencies of the isomers. In Fig. 6 the fractional composition of the mixture obtained in such a way is plotted versus the temperature as solid lines. Obviously, heating should enrich the mixture with the less stable (in this case – less symmetric) structures. As follows from the simulation, the relative abundance of $C_1$ and $C_3$ isomers is close to the observed 3:1 ratio at the synthetic temperature of 620 K, while the fraction of $T$ isomer is still higher than those for other structures in contradiction with the experimental data.

It testifies that the definite values of relative energies computed within a DFT approximation should be treated with caution and that the accuracy of the computational method should be taken into account. In this particular case the energy difference between the $T$ and less symmetric isomers is most important, since the velocity of the most stable species decrease under rising temperature is roughly a twice of those for the less stable increase. On the other hand, the $\Delta E(T-C_1)$ value is most questionable due to considerable structural differences between these two forms. If we vary the calculated value of 17 kJ/mol within the limits of PBE accuracy claimed to be the order of few kcal/mol [27], and set, e.g. the energy difference of 10 kJ/mol, then we obtain a drastically different picture illustrated on Fig. 6 by dotted lines. Now the $C_1$-isomer becomes the dominant component in the equilibrium mixture already at 300 K, far below the real synthetic conditions. Hence, the formation of a less stable structure might be explained by simple thermodynamic considerations. At the same time, an unknown, strictly speaking, accuracy in the estimation of the isomer relative energy precludes the ascribing of the mixture composition to the thermodynamic factors only, and the kinetic aspects of the synthesis procedure cannot be ruled out.

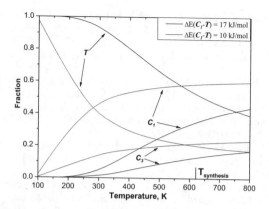

Fig.6. Simulated equilibrium composition of $C_{60}H_{36}$ isomeric mixture, see text for details.

## 5. Conclusions

Vibrational simulations at the PBE/TZ2P level of theory supplemented with Hartree-Fock computations of Raman intensities were shown to be a powerful tool in predictions of IR and Raman spectra for such complicated molecules as hydrofullerenes. Good correspondence between observed and measured spectra enabled us not only to perform a vibrational assignment for $C_{60}H_{18}$ and $C_{60}D_{18}$ molecules, but also to define the molecular structure of dominant $C_1$ and $C_3$ isomers of $C_{60}H_{36}$ hydrofullerene.

## Acknowledgments

This work was supported by RFBR grant No 03-03-32179, a part of computational abilities was provided by Research Computing Center of Moscow State Univeristy. We are thankful to Prof. V.A. Somenkov for supplying us with the sample of deuterofullerene. The assistance of Dr. A.N. Kharlanov in Raman measurements and that of Dr. N.S. Nesterenko in Nicolet spectrometer facilities is gratefully acknowledged.

## 6. References

1. Haufler RE, Conceicao J, Chibante LPF, Chai Y, Byrne NE, Flanagan S, et al. Efficient Production of $C_{60}$ (Buckminsterfullerene), $C_{60}H_{36}$, and the Solvated Buckide Ion. J Phys Chem 1990; 94: 8634–8639.
2. Attalla MI, Vassallo AM, Tattam BN, Hanna JV. Preparation of Hydrofullerenes by Hydrogen Radical Induced Hydrogenation. J Phys Chem 1993; 97: 6329–6331.
3. Rüchardt C, Gerst M, Ebenhoch J, Beckhaus H-D, Campbell EEB, Tellgmann R, et al. Transfer Hydrogenation and Deuteration of Buckminsterfullerene $C_{60}$ by 9,10-Dihydroanthracene and 9,9`,10,10`[D4] Dihydroanthracene. Angew Chem Int Ed 1993; 32(4): 584–586.
4. Darwish AD, Abdul-Sada AK, Langley CJ, Kroto HW, Taylor R, Walton DRM. Polyhydrogenation of [60]- and [70]-fullerenes. J Chem Soc Perkin Trans 2 1995; 2359–2365.
5. Lobach AS, Perov AA, Rebrov AI, Roschupkina OS, Tkacheva VA, Stepanov AN. Preparation and study of hydrides of fullerenes $C_{60}$ and $C_{70}$. Russ Chem Bull 1997; 46(4): 641–648.
6. Darwish AD, Avent AG, Taylor R, Walton DRM. Structural characterization of $C_{60}H_{18}$: a $C_{3v}$ symmetry crown. J Chem Soc Perkin Trans 2 1996; 2051–2054.

356

7. Bini R, Ebenhoch J, Fanti M, Fowler PW, Leach S, Orlandi G, et al. The vibrational spectroscopy of $C_{60}H_{36}$: An experimental and theoretical study. Chem. Phys 1998; 232: 75–94.
8. Okotrub AV, Bulusheva LG, Asanov IP, Lobach AS, Shulga YM. X-ray Spectroscopic and Quantum-Chemical Characterization of Hydrofullerene $C_{60}H_{36}$. J Phys Chem A 1999; 103: 716–720.
9. Bulusheva LG, Okotrub AV, Antich AV, Lobach AS. Ab initio calculation of X-ray emission and IR spectra of the hydrofullerene $C_{60}H_{36}$. J Mol Struct 2001; 562(2-3): 119–127.
10. Meletov K, Assimopoulos S, Tsilika I, Bashkin IO, Kulakov VI, Khasanov SS, Kourouklis GA. Isotopic and isomeric effects in high-pressure hydrogenated fullerenes studied by Raman spectroscopy. Chem Phys 2001; 263: 379–388.
11. Nossal J, Saini RK, Sadana AK, Bettinger HF, Alemany LB, Scuseria GE, et al. Formation, Isolation, Spectroscopic Properties, and Calculated Properties of Some Isomers of $C_{60}H_{36}$. J Am Chem Soc 2001; 123(35): 8482–8495.
12. Popov AA, Senyavin VM, Granovsky AA, Lobach AS. Vibrational Spectra of Hydrofullerenes, $C_{60}H_{18}/C_{60}D_{18}$ and $C_{60}H_{36}$, as Studied by IR Spectroscopy and *ab initio* Calculations. In: Kamat PV, Guldi DM, Kadish KM, editors. Fullerenes – Volume11: Fullerenes For The New Millennium (Proceedings of the International Symposium on Fullerenes, Nanotubes, and Carbon Nanoclusters), New Jersy: The Electrochemistry Society, Inc., 2001, 405–415.
13. Boltalina OV, Buhl M, Khong A, Saunders M, Street JM, Taylor R. The $^3He$ NMR spectra of $C_{60}F_{18}$ and $C_{60}F_{36}$: the parallel between hydrogenation and fluorination. J Chem Soc Perkin Trans 2 1999; 1475–1479.
14. Dunlap BI, Brenner DW, Schriver GW. Symmetric Isomers of Hydrofullerene $C_{60}H_{36}$. J Phys Chem 1994; 98: 1756–1757.
15. Buhl M, Thiel W, Schneider U. Magnetic Properties of $C_{60}H_{36}$ Isomers. J Am Chem Soc 1995; 117: 4623–4627.
16. Clare BW, Kepert DL. Structures, stabilities and isomerism in $C_{60}H_{36}$ and $C_{60}F_{36}$. A comparison of the AM1 Hamiltonian and density functional techniques. J Mol Struct (THEOCHEM) 2002; 589: 195–207.
17. Shul'ga YM, Tarasov BP, Fokin VM, Shul'ga NY, Vasilets VN. Crystalline fullerene deuteride $C_{60}D_{24}$: Spectral investigation. Phys Solid State 1999; 41(8): 1391–1397.
18. Somenkov VA, Glazkov VP, Shil'shtein SS, Zhukov VP, Bezmel'nitsyn VN, Kurbakov AI. Neutron diffraction study of the structure of deuterated and fluorinated fullerene $C_{60}$. Met Sci Heat Treat 2000: 42: 319–325.
19. Laikov DN. Fast Evaluation of Density Functional Exchange-Correlation Terms Using the Expansion of the Electron Density on Auxiliary Basis Sets. Chem Phys Lett 1997; 281: 151–156.
20. Perdew JP, Burke K, Ernzerhof M. Generalized gradient approximation made simple. Phys Rev Lett 1996; 77(18): 3865–3868.
21. Granovsky AA. PC GAMESS URL: http://classic.chem.msu.su/gran/gamess/index.html
22. Schmidt MW, Baldridge KK, Boatz JA, Elbert ST, Gordon MS, Jensen JH, et al. General Atomic and Molecular Electronic-Structure System. J Comput Chem 1993; 14(11): 1347–1363.
23. Yagola AG, Kochikov IV, Kuramshina GM, Pentin YA. Inverse Problems of Vibrational Spectroscopy. Zeist: VSP. 1999.
24. Neretin IS, Lyssenko KA, Antipin MY, Slovokhotov YL, Boltalina OV, Troshin PA, Lukonin AY, Sidorov LN, Taylor R. $C_{60}F_{18}$, a flattened fullerene: Alias a hexa-substituted benzene. Angew Chem Int Ed 2000; 39(18): 3273–3276.
25. Goodman L, Ozkabak AG, Thakur SN. A Benchmark Vibrational Potential Surface – Ground-State Benzene. J Phys Chem 1991; 95(23): 9044–9058.
26. Clare BW, Kepert DL. The structures of $C_{60}F_{36}$ and new possible structures for $C_{60}H_{36}$. J Mol Struct (THEOCHEM) 1999; 466: 177–186.
27. Ernzerhof M, Scuseria GE. Assessment of the Perdew-Burke-Ernzerhof exchange-correlation functional. J Chem Phys 1999; 110 (11): 5029–5036.

# HYDROGEN SEGREGATION IN THE RESIDUAL STRESSES FIELD

N.M.VLASOV, I.I.FEDIK[1]
*Scientific Research Institute Scientific Industrial Association "Luch",*
*Podolsk, Moscow region, Russia*

## Abstract

The process of hydrogen absorption in a cylindrical cladding with residual stresses has been considered. The latter ones occur in the material of products when carrying out different technological operations. The level and character of the residual stresses distribution change kinetics of diffusion migration of hydrogen atoms. The process of hydrogen absorption is described by a non-stationary diffusion equation in the field of forces under corresponding initial and boundary conditions. The analytical relations for the field of hydrogen atoms concentration have been given. Possibility of decreasing the near-surface barrier for diffusion migration of the hydrogen atoms into the volume of the metal are discussed. The analytical analysis results are attracted for explaining physical mechanism of the hydrogen atoms absorption with the residual stresses.

**Keywords:** Hydrogen atoms, Residual stresses, Hydrogen absorption kinetics.

## Nomenclature

| | |
|---|---|
| C | Hydrogen concentration |
| $C_1$ | Hydrogen concentration with absorption from the inner surface |
| $C_2$ | Hydrogen concentration with absorption from the outer surface |
| D | Hydrogen diffusion coefficient |
| k | Boltzmann's constant |
| T | Absolute temperature |
| $C_0$ | Average concentration of hydrogen atoms |
| $r_0$ | Inner cladding radius |
| R | Outer cladding radius |
| $\mu$ | Shear modulus |
| $\nu$ | Poison's ratio |
| $\omega$ | Angle of rotation of the cladding cut edges |
| $\sigma_{ll}$ | First invariant of the residual stresses tensor |
| $\delta\upsilon$ | Change of metal volume when placing hydrogen atoms |
| $C^1p$ and $C^2p$ | Equilibrium concentrations of hydrogen atoms on the inner and outer cladding surfaces |

---

[1] Corresponding author. Fax: +7-095-239-1749, E-mail address: iifedik@podolsk.ru (I.I. Fedik).

357

*T.N. Veziroglu et al. (eds.),*
*Hydrogen Materials Science and Chemistry of Carbon Nanomaterials,* 357-362.
© 2004 *Kluwer Academic Publishers. Printed in the Netherlands.*

## 1. Introduction

The residual stresses occur in the material of the products when carrying out different technological operations. The level and character of the residual stresses distribution have an essential effect on the diffusion processes kinetics. Thus, for example, substitution impurities of a large atomic radius migrate into the area of tensile stresses, and corresponding impurities of a small atomic radius migrate into the field of compression stresses In this way decomposition of the solid solution in the field of residual stresses occurs [1]. The purpose of this paper is analyzing the hydrogen absorption kinetics in the cylindrical cladding with the residual stresses. Choosing such a model system is caused by the following reasons. Firstly, it is possible to obtain residual stresses of different signs within the cylindrical cladding by cutting and adding (excluding) part of the material with the following connecting the edges of the cut. Secondly, logarithmic dependence on a radial coordinate of the first invariant of the residual stresses tensor provides obtaining the exact analytical solution of the diffusion equation in the field of forces. And at last, proposed model analysis has a direct analytical application. Hydrogen absorption is considered to be one of the main tasks of hydrogen engineering.

For obtaining the residual stresses in the cylindrical cladding, do as advised below. The edges of the cladding cut are moved apart through angle $\dot{\omega}$ and some deficit material is put there. The surrounding of the inner cladding surface is in the tensile state, and the area of the outer surface is in the compression state. The hydrogen atoms interact with the residual stresses of the opposite signs. The interaction potential for dimensional effect is defined by the known relation [2]

$$V = -\frac{\sigma_{\ell\ell}}{3}\delta\upsilon \tag{1}$$

where $\sigma_{ll}$ - first invariant of the residual stresses tensor, $\delta\upsilon$ _ change of volume of the metal when placing the hydrogen atoms. It is interesting to be noted that magnitude of $\delta\upsilon$ is approximately equal for all metals and it is of $3 \cdot 10^{-30}$ m$^3$ [3]. For a hydrogen atom $\delta\upsilon > 0$ and therefore at $\sigma_{ll} > 0$ (tensile stress), potential V takes a negative value. It corresponds to attraction of the hydrogen atoms to the area with the tensile stresses. In other words, the hydrogen absorption process depends on the character of the residual stresses distribution. The first invariant of the residual stresses tensor takes a form (flat deformation) [4]

$$\sigma_{ll} = \frac{\mu\omega(1+v)}{2\pi(1-v)}\left\{1 + 2\ln\frac{r}{R} + \frac{2\left(\frac{r_0}{R}\right)^2}{1-\left(\frac{r_0}{R}\right)^2}\ln\frac{r_0}{R}\right\}, \tag{2}$$

where $\mu$ - shear modulus, $v$ - Poison's ratio, $\omega$ -turning angle of the cladding cut edges, $r_0$ and R - inner and outer cladding radius. Under other equal conditions the sign of $\sigma_{ll}$ depends on the angle of turning the cladding cut edges. We take by convention $\omega < 0$, if $\sigma_{ll} > 0$ onto the inner cladding surface.

## 2. Hydrogen absorption kinetics

The hydrogen atoms absorption within the accepted model system is described by a non-stationary diffusion equation in the field of potential V under corresponding initial and boundary conditions [5]

$$\frac{1}{D}\frac{\partial C}{\partial t} = \Delta C + \frac{\nabla(C\nabla V)}{kT}, \qquad r_0 < r < R, \qquad (3)$$

$$C(r,0)=0, \qquad C(r_0,t)=C_p^1, \qquad C(R,t)=C_p^2,$$

where D – coefficient of the hydrogen atoms diffusion, $C_p^1$ and $C_p^2$ - equilibrium concentrations of hydrogen atoms on the inner and outer cladding surfaces, , k- Boltzmann's constant, T- absolute temperature. Physical sense of the initial and boundary task conditions (3) is quite obvious. At the initial time moment the hydrogen atoms concentration within the cladding is equal to 0. The boundary conditions at $r = r_0$ and $r=R$ mean that the equilibrium concentration is kept on the boundaries of the area in accordance with the interaction potential. The process of the hydrogen absorption runs from the side of the inner and outer surfaces of the cylindrical cladding. Without loss a generality, let's consider the following character of the residual stresses distribution: extension on the inner surface and compression on the outer one. Such distribution of the residual stresses "extracts" the hydrogen atoms from the side of the outer surface and "retards" the hydrogen absorption from the inner surface.

Diffusion migration of the hydrogen atoms depends on the gradient of potential V. Thus, the constant relations (2) do not influence the diffusion process. In this case $\Delta V=0$ as V is a harmonic function. Taking into account (1) and (2), task (3) is mathematically formulated as follows:

$$\frac{1}{D}\frac{\partial C}{\partial t} = \frac{\partial^2 C}{\partial r^2} + \frac{1-\alpha}{r}\frac{\partial C}{\partial r}, \quad r_0 < r < R, \qquad (4)$$

$$C(r,0)=0, \qquad C(r_0,t)=C_p^1, \qquad C(R,t)=C_p^2.$$

Non-dimensional parameter $\alpha$ specifies the relation between the connection energy of a hydrogen atom with the residual stresses to the energy of heat motion

$$|\alpha| = \frac{\mu|\omega|(1+v)\delta\upsilon}{3\pi(1-v)kT}$$

In case of $|\alpha| \ll 1$, the residual stresses field is considered to be a small disturbance of a diffusion flow of the hydrogen atoms at the expense of the concentration gradient. For $|\alpha| \gg 1$ the residual stresses field gives main contribution into the diffusion process. When $|\alpha| \approx 1$, the diffusion flows of the hydrogen atoms are compared at the expense of concentration gradients and potential V. Let's evaluate magnitude $|\alpha|$ for the system Zr-H. Taking kT$= 10^{-20}$ J; $\mu=4\cdot10^{10}$ Pa; $v=0.3$; $|\omega| =0.3$ rad; $\delta\upsilon=3\cdot10^{-30}$ m$^3$ we obtain $|\alpha| \approx 1$. Thus, take the non-dimensional parameter $\alpha$ being equal 1 in absolute value. This parameter sign depends on the character of the residual stresses distribution in the cylindrical cladding. In case of $\omega<0$ (tensile stresses on the inner surface), $|\alpha| = -1$. When changing the sign of the residual stresses, parameter $\alpha$ takes a positive value, i.e. $\alpha=1$. The hydrogen absorption from the inner and outer surfaces of the cladding obeys different laws. For qualitative analyzing the process kinetics, principle of superposition of the two tasks solution should be used. The first task describes the hydrogen absorption from the outer side of the surface, and the second one does it from the inner side. The principle of superposition follows from the diffusion equations linearity.

For $\alpha = -1$, the atoms concentration field is found from the task solution

$$\frac{1}{D}\frac{\partial C_1}{\partial t} = \frac{\partial^2 C_1}{\partial r^2} + \frac{2}{r}\frac{\partial C_1}{\partial r}, \qquad r_0 < r < R, \qquad (5)$$

$$C_1(r,0)=0, \quad C_1(r_0,t)=C_p^1, \quad C_1(R,t)= 0.$$

The residual stresses change symmetry of the diffusion equation: the diffusion process within the cylindrical cladding obeys the law of spherical symmetry. Conversion of the coordinate dependence decreases a rate of forming a concentrating profile from the hydrogen atoms. The physical essence of retarding the hydrogen absorption kinetics is caused by reducing the tensile stresses in the radial direction up to their change with the compression stresses. The hydrogen absorption from the outer cladding surface is described by the equation (non-dimensional parameter changes the sign)

$$\frac{1}{D}\frac{\partial C_2}{\partial t} = \frac{\partial^2 C_2}{\partial r^2}, \qquad r_0 < r < R, \qquad (6)$$

$$C_2(r,0)=0, \quad C_2(r_0,t)=0, \quad C_2(R,t)= C_p^2.$$

And again the residual stresses change the symmetry of the task: the diffusion process in the cylindrical cladding obeys the law of flat symmetry. The hydrogen absorption rate increases from the side of the outer boundary. Acceleration of the process kinetics is caused by reducing the compression stresses in the radial direction, which are changed with the tensile stresses. The solution of tasks (5) and (6) is well known. Thus, it is not difficult to obtain the solution of task (4). This solution describes the hydrogen absorption kinetics from the two surfaces of the cylindrical cladding at given character of the residual stresses distribution

$$C=C_1+C_2 = \frac{C_p^1 r_0(R-r)+C_p^2 r(r-r_0)}{r(R-r_0)} +$$

$$+\frac{2}{\pi}\sum_{n=1}^{\infty}\frac{(-1)^n r C_p^2 - r_0 C_p^1}{r}\cdot\frac{\sin\frac{\pi n(r-r_0)}{R-r_0}}{n}\exp\left(-\frac{\pi^2 n^2 D t}{(R-r_0)^2}\right). \qquad (7)$$

The hydrogen absorption kinetics obeys the exponential law and depends on the equilibrium concentrations on the area boundaries. The latter ones is defined by the level of the residual stresses on the corresponding boundaries

$$C_p^1 = C_0 \exp\left(\frac{\sigma_{\ell\ell}\delta\upsilon}{3kT}\right)_{r=r_0}, \quad C_p^2 = C_0 \exp\left(\frac{\sigma_{\ell\ell}\delta\upsilon}{3kT}\right)_{r=R} \qquad (8)$$

where $C_0$ - concentration of hydrogen atoms on the cladding surfaces after dissociation of molecules. On the inner surface (tensile stresses) $C_p^1 > C_0$, and on the outer one (compression stresses) $C_p^2 < C_0$.

## 3. Conclusions

Therefore, the hydrogen absorption is defined by two parameters: equilibrium concentration on the area boundaries and the symmetry of the diffusion equation. We will make it clear in details. The equilibrium concentration of the hydrogen atoms on the inner cladding surface ($r=r_0$) exceeds a corresponding value on the outer surface ($r=R$) at the expense of the tensile stresses. However, the spherical symmetry of the diffusion equation retards the process kinetics in comparison with the cylindrical symmetry. The equilibrium concentration of the hydrogen atoms for the outer cladding surface is reduced in relation to the inner surface. On the other hand, the flat symmetry of the diffusion equation accelerates the process of the hydrogen absorption in comparison with the cylindrical symmetry. Let's estimate the values of the equilibrium concentrations in accordance with relations (8). In case of

$\dfrac{r_0}{R} = 0.9$; $\mu = 4 \cdot 10^{10}$ Pa; $\nu = 0.3$; $|\omega| = 0.3$ rad; $kT = 10^{-20}$ J; $\delta \upsilon = 3 \cdot 10^{-30}$ m$^3$ we will obtain

$C_p^1 = 1{,}04 C_0$ and $C_p^2 = 0{,}97 C_0$. The equilibrium concentrations of the hydrogen atoms on the boundaries are negligibly differ from the average value of $C_0$.

Thus, the hydrogen absorption changing is exclusively caused by different symmetry of the diffusion equation in the near-surface layers of the cladding. Retarding the absorption process kinetics takes place for the spherical symmetry (inner surface), and accelerating the hydrogen atoms diffusion is characteristic for the flat symmetry (outer surface). Stationary distribution of the hydrogen atoms concentration is settled in correspondence with the interaction potential. The character of the residual stresses distributions changes the hydrogen absorption kinetics only. In this case a possibility of controlling a rate of the hydrogen atoms migration into the volume of the metal appears. It is sufficiently to make

such a stressed state, which leads to increasing in the volume of the metal $\dfrac{\partial C}{\partial t}$. This is

reached by creating on the surface the compression stresses converting into the tensile stresses. In this case the hydrogen atoms are "displaced" out of the near-surface area into the volume, i.e. acceleration of the absorption process takes place.

Given character of the stresses distribution also decreases a magnitude of the near-surface barrier for the hydrogen atoms migration. The physical essence of such a process is in the following: hydrogen absorption within metals includes step-by-step physical adsorption, chemical adsorption, storage of the hydrogen atoms in the near-surface layer and their diffusion migration into the volume of the metal. The stage of passing the hydrogen atoms from the near-surface area into the volume remains less understandable. This passing is connected with overcoming the energy barrier of the surface. The latter one occurs at the expense of the image force and near-surface dilatation. Therefore, energy preferability of a hydrogen atom placing in the near-surface layer occurs (some interatomic distances). In case of the near-surface area is subjected to the tensile stresses, such preferability decreases in part. Accelerating the hydrogen absorption kinetics in metals is experimentally observed when alloying by the substitution impurities of a small atomic radius or applying thin surface films made of the material with a small atomic radius as well [6, 7].

Such processes are accompanied by changing the character of the stressed state. In the scope of the proposed model the following qualitative picture can be observed. The substitution impurities of a small atomic radius reduce the parameter of the crystal lattice in correspondence with Vegard's rule. In this case, reducing of the lattice parameter in the volume of the metal exceeds the corresponding value near the surface. Therefore, the tensile stresses occur in the volume of the metal, and the compression stresses occur near the surface. In this way the near-surface layer for the hydrogen absorption reduces. The situation reminds forming the self-balanced system of the thermal stresses for a negative value of a linear expansion coefficient. The areas having an elevated temperature are in the tensile state, and the areas having a reduced temperature are in the compression state. When applying the thin surface films made of the material with a small atomic radius, the similar stressed state is realized: tension – in the volume; compression – near the surface. The tensile stresses are one of the features for the thinnest film. While increasing the film thickness, the epitaxial dislocations are formed on the boundary, and the stressed state of the metal in the near-surface area is restored.

Thus, the level and character of the residual stresses distribution within the cylindrical cladding makes an effect on the hydrogen absorption kinetics. For accelerating the absorption process the compression stresses are required to be created on the corresponding surface, then they are changed with the tensile stresses in the volume of the metal. Such a character of the distribution reduces the energy barrier of the surface as well. This promotes to equalizing the chemical potential of the hydrogen atom within all the volume of the metal including the near-surface layers. Under the other equal conditions the hydrogen absorption rate increases.

## 4. References

1.  Vlasov N.M., Fedik I.I. Decomposition of Solid Solution in Residual Stresses Field. Dokl.Acad. Nauk (Physics) 2002; 382 (2): 186-189.
2.  Teodosiu C. Elastic models of crystal defects. Moscow: Mir, 1985.
3.  Kolachev B.A. Hydrogen brittleness of metals. Moscow: Metallurgy, 1985.
4.  Lurie A.I. Theory of Elasticity. Moscow: Nauka, 1970.
5.  Vlasov N.M. Fedik I.I. Diffusion of Interstitial Impurities through Cylindrical Cladding with Residual Stresses. Dokl. Acad. Nauk (Physics) 2002; 384 (3):324-327.
6.  Libowitz G.G., Maeland A.J. Hydride formation by BCC solid solution alloys. Materials Science Forum. 1988; 31: 176-196.
7.  Pick M.A., Greene M.G., Strongin Myron. Uptake rates for hydrogen by Nb and Ta: effect of thin metallic overlayers. J. of the Less-Common Metals. 1980; 73, 89-95.

# NANOPARTICLES OF METALS AND OXIDES ON SURFACE OF CARBON FIBERS ARE EFFECTIVE CATALYSTS OF CHEMICAL TRANSFORMATIONS OF EPOXY OLIGOMERS

V.I. DUBKOVA

*Institute of General and Inorganic Chemistry of National Academy of Sciences of Belarus, 9 Surganov str., Minsk, 220072, Belarus*

## Abstract

Influence of the carbon fibers containing into theirs structure nanoparticles of metals or theirs oxides (copper, zinc, titanium, zirconium, lead, molybdenum, iron) on the thermochemical transformations of the epoxy oligomers of various chemical structures has been studied. On the one hand, a catalytic effect of the Me-CF$_s$ on decomposition of the epoxy oligomers has been revealed. On the other hand, an intensive formation of an insoluble gel-fraction is observed. At the same time a new exothermic effects in the 200-300 °C temperature range are taken place. A conclusion about concurring reactions of thermal decomposition of the epoxy oligomer and formation of three-dimensional polymer occurring on the Me-CF$_s$ – oligomer interface has been made on the basis of the thermal, chemical analysis and also the electron paramagnetic resonance and infrared spectroscopy data. It is revealed that nanoparticles of metals or theirs oxides uniformly distributed on the carbon fibers surfaces play the dominating role in these processes. It is made the conclusion that the Me-CFs are perspective for creation of the carbon fibers nanocomposites with the special properties.

*Key words:* A. Carbon fibers, B. mixing, B. heat treatment, C. thermal analysis (DTA and TGA), C. electron paramagnetic resonance, D. chemical structure

## 1. Introduction

All increasing interest to the carbon nanostructure systems is caused by the broad capabilities of their effective utilization in the different areas of science, engineering and nanotechnologies. The activities of a technological direction in nanochemistry are oriented on the analysis of the chemical transformations into nanosystems at creation and using theirs in production, and also at developing of the nanotubulenes and nanocomposites, possessing by the unique properties. A subject of the intensive researches last years is also nanodimension particles of the metals, which have the properties different significantly from the ones usual to the volumetric metals. It is caused that they are began to employ at catalysis and at creating of the different technical devices more broad. In this connection it is necessary to consider element-containing carbon fibers in particular metal-containing (Me-CF$_s$) under a new angle of view. Me-CF$_s$ were developed in the Institute of General and Inorganic Chemistry of

*T.N. Veziroglu et al. (eds.),*
*Hydrogen Materials Science and Chemistry of Carbon Nanomaterials,* 363-374.

National Academy of Sciences of Belarus still in the seventieth years of the past century [1]. It was established that the ultra dispersed particles of some metals (nickel, cobalt, copper, platinum) including in a structure of a carbon fiber despite of their rather small contents show a high catalytic activity in the reactions of hydrocarbon transformations [2, 3]. The activated carbon fibers containing of the metals or the oxides of cobalt, copper, nickel, chromium or manganese as an active component were recommended for using as the effective conversion catalysts of a carbon monoxide of [4].

As it is known that, on the one hand, the nanodimension particles of the metals have so a high supply of energy that they can form the chemical combinations with the unusual properties [5]. and, on the other hand, the carbon fibers are one of the perspective fillers of the composite materials [6]. It was of interest to consider the processes which are taking part on the Me-CF$_s$ – polymer binder interface. In the given work the generalized and selected outcomes of the researches of an influence of the metal - containing inclusions of nanodimension values uniformly distributed on a surface of the carbon fibers on the thermochemical transformations of the separate epoxy oligomers, as of the matrix most frequently used in the carbon fiber composites are presented.

## 2. Experimental

We have studied the metal-containing carbon fibers (Cu-, Zn-, Ti-, Zr-, Sn-, Pb-, Mo-, Fe- -CF$_s$ ) obtained on the basis of hydrated cellulose fibers and a cotton fabric. Production of the fibers including of the different elements of the D. I. Mendeleyev Periodic Table is described in a monograph [1]. The content of the metals in the carbon fibers is 0,3 - 34,5 weight %. The structure of the Me-CF$_s$ was studied by the X-ray powder diffraction method and by infra-red spectroscopy. The X-ray diffractometer, DRON-3, was used. The accuracy of the Bragg angular determination was $0.02°$. Cu K$_\alpha$ radiation (Ni- filter ) were applied.

The metals in the CFs are as the separate elements (Fig. 1, curve 5) or their oxides (Fig. 1, curve 2-4 ) or in a mixture of element with its oxide (Fig. 1, curve 6 ) . Dimensions of the metal-containing particles uniformly distributed on the CF$_s$ surfaces are in limit from 8-10 up to 100 and more nanometers. Unmodified carbon fibers (CFs), that is not containing of metal (Fig 1, curve 1) , obtained on the basis of the same initial raw material and under the same carbonization conditions, were used for comparison. As an oligomeric matrix we have used several commercially available epoxy oligomers of various chemical structures: ED-20 epoxydian and DEG-1 low-molecular weight aliphatic resins; alicyclic diepoxides of acetal and ester type (UP-612 and UP-632 respectively), UP-650D alicyclic oligomer with two homogeneous glycide groups and UP-650T alicyclic-aliphatic triepoxide. Also laboratory -made element - containing epoxy oligomer were investigated, which provide some special features to the final product.

It is well known that liquid-solid transition of the epoxy oligomers is accompanied by intensive heat release. That is why differential thermal analysis (DTA) was used in estimation of the effect the Me-CF$_s$ on the structural transformations of the epoxy oligomers. The samples were prepared according to the method described in an earlier paper [7]. We have used a Paulik-Paulik-Erdey derivatograph (MOM, Hungary, Budapest) with the 20-500°C temperature range, in an air stream (10 ml/min) and in the

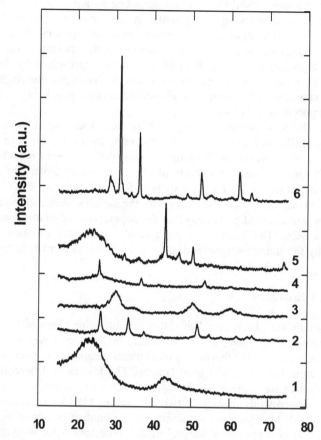

Fig 1. X-ray diffraction patterns of $CF_s$ unmodified (1) and $Sn$-$CF_s$ (2), $Zr$-$CF_s$ (3), $Mo$-$CF_s$ (4), $Cu$-$CF_s$ (5), $Pb$-$CF_s$ (6).

furnace atmosphere, i.e. under the conditions of reduced oxygen supply. These conditions are similar to those in the manufacturing in closed molds. The heating rate was 5 deg/min. Aluminium oxide was the reference material.

The electron paramagnetic resonance (EPR) spectrums were recorded on a ERS-230 radiospectrometer (Germany), $\lambda = 3$ cm. Samples of the oligomers and Me-CFs previously were warmed up at temperature 70 °C and 110°C during:0.5 and 4 h accordingly. Mixing of the Me-$CF_s$ with an oligomer realized directly before of measuring. Heating of the composition was realized immediately in the resonator of a radiospectrometer from 20 °C up to 260 °C. Recording of a spectrum was made after warming up of a system at prescribed temperature during 10 minutes. For obtaining of an accumulation curve of the radicals at constant temperature the samples after exposition during the fixed time is refrigerated quick in fluid nitrogen and a EPR spectrum recorded at temperature -196 °C. The identical experiments were made for all individual components.

Titration was used to determine the concentration of the epoxy groups in the oligomers and in their composites with Me-$CF_s$ before and after thermal processing. We have followed the well established method described by Kline [8] based on the reaction of hydrochloric acid with epoxy groups in an organic solvent and producing chlorohydrin. For calculating the conversion of epoxy groups during heating, the contents of epoxy groups in the initial oligomer not treated thermally was assumed to be 100 %. The degree of hardening was assumed to be represented by the gel-fraction, which is by the ratio of the weight of the insoluble binder to the original amount of binder. The extraction of the samples with boiling acetone was being carried out in the Soxhtlet apparatus during 24 hours.

Several methods of preparation of samples were used for IR spectroscopy. The method of liquid film was used for the oligomers and their mixtures with fibers [9]. The infrared spectra of absorption in the range of 4000-400 cm$^{-1}$ were recorded on a UR-20 ("Karl Tseys", UENA, Germany) spectrophotometer. The oligomer or its mixture with Me-CF$_s$ was deposited on a KBr plate so as to form a thin layer of 10-15 μm and was covered by the other salt plate. The plates with the samples were then placed in a heated cell. It had been ascertained previously that the components of mixture were not affected by the plates used. The Me-CF$_s$ with grafted oligomers on theirs surface were investigated by the immersion method, that is, by pressing them with the KBr powder [10].

## 3. Results and discussion

The investigations have been established that the following phenomena take place. The metal containing carbon fibers in overwhelming majority of cases render an essential influence on behaviour of the thermochemical transformations of the epoxy oligomers of a different chemical nature under joint heating. The results of differential thermal and chemical analysises are evidence of that. As it is shown from Fig. 2 the curves contours of DTA are changed while heating of two-component Me-CF$_s$ + alicyclic diepoxide of acetal type (UP-612) or chlorine-containing epoxy oligomer compositions. The shifts of the maximums of heat releases relatively of the same curves for individual (not filled) oligomers and their compositions with the unmodified fibers are observed.

Catalytic action of the metal containing carbon fibers on the process of the thermal destruction of the epoxy oligomers shows clearly. If for the epoxy oligomers unreinforced the thermal decomposition begins at temperatures above 300 °C, that is conformed with literary data [11], a temperature of a beginning of an oligomer thermal destruction is considerably diminished at presence of the Me-CF$_s$, and that is in the greater extent than is more content of a metal in the CF$_s$ (Tab. 1). At the same time the registration of an intensive signal in the EPR spectrums begins, that speaks for the benefit of an initiation influencing of the Me-CF$_s$ on the process of thermal degradation of an epoxy oligomer.

As it is visible from Fig. 2 the exothermic effects also appear at lower temperatures and in series of cases a nature of heat releases changes also (VI- VIII). So, if for chlorine-containing epoxy oligomer not filled and for its compositions with CFs unmodified two stages of the exothermic effect in the 200-370 °C range of the temperatures can be clearly seen on the DTA curves and the second stage is characterized by more intensive heat releases and more appreciable loss of mass by an oligomer on a second stage: 3,7-3,71 weight % and 34, 3-35,7 weight % accordingly then for a composition of the oligomer with the Mo-CF$_s$, for example, only one

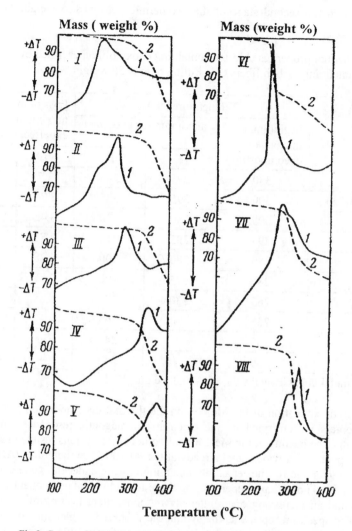

Fig 2. Curves of DTA (1) and TG (2) of UP-612 diepoxide (I–V), chlorine-containing epoxy oligomer (VI–VIII), filled by CF$_s$ unmodified (V, VIII) and by Zn-CF$_s$ (I), Fe-CF$_s$ (II), Cu-CF$_s$ (III), Sn-CF$_s$ (IV), Mo-CF$_s$ (VI), Pb-CF$_s$ (VII)

exothermic peak on the DTA curves is observed (intensive and narrow at 245° C) which is corresponded to a maximum of decomposition rate by the oligomer. The total loss of mass is 25,7 weight % moreover.

For the compositions of the chlorine-containing epoxy oligomer with the Pb-CF$_s$ the redistribution of the intensity of the exothermic peaks is observed on the DTA curves. For all this in spite of more appreciable heat release on a first stage of exothermic effect, the loss of mass in the given range of the temperatures is small (4, 28 weight %). Decrease of a mass loss is also observed and in a range of the temperatures corresponding to the second stage of the exothermic peak (18, 5 weight %) (see the Tab.1).

TABLE 1. Thermal properties chlorine containing epoxy oligomer, filled by metal containing carbon fibers $CF_s$. Filling ratio is 50 weight % %.

| Me-CF$_s$ | Temperature (°C) | | | Loss of mass for maximum of exoeffect (weight %) | |
|---|---|---|---|---|---|
| | beginning of intensive decomposition | maximum for exoeffect | | | |
| | | I | II | I | II |
| – | 330 | 295 | 345 | 3,7 | 34,3 |
| CF$_s$ unmodified | 300 | 295 | 325 | 3,71 | 35,7 |
| Zr-CF$_s$ | 305 | 280 | 320 | 3,42 | 28,5 |
| Sn-CF$_s$ | 295 | 295 | 315 | 2,85 | 22,8 |
| Pb-CF$_s$ | 280 | 280 | 300 | 4,28 | 18,5 |
| Zn-CF$_s$ | 280 | 275 | 290 | 2,85 | 12,86 |
| Cu-CF$_s$ | 265 | 230 | 280 | 0,85 | 25,7 |
| Fe-CF$_s$ * | 255 | – | 265 | – | 15,0 |
| Fe-CF$_s$ ** | 235 | – | 245 | – | 29,3 |
| Mo-CF$_s$ | 230 | – | 245 | – | 25,7 |

* Percent of metal is 7,97 weight %; ** –20,0 weight %.

The catalytic activation of the Me-CF$_s$ on the thermal decomposition process of the epoxy oligomers is confirmed by the electron paramagnetic resonance studies. It is established that the heating of the Me-CF$_s$ – oligomer compositions results in appearance in the EPR spectrums a wide signal already at the 80-100°C temperatures ($\Delta H$=9.5-17.0 mT), not characteristic for the separately taken ingredients under the same conditions of heat treatment and not being a superposition those for individual components. Narrowing of a signal and an increasing of its intensity by consequent heating of the system are observed. The spectrum represents the symmetric unsolved singlet similar one desribed in literature for the epoxydian oligomer in the present of a juvenile surface of the metals [12]. It is explained such spectrum by formation of the radicals owing to destructive catalytic processes in organic matrix on a quick-formation surface [13, 14].

The signals observed on the EPR spectrums for the compositions of the carbon fibers with the epoxy oligomers are similar written in literature [8], forming at pyrolysis of the epoxy compounds. This is confirmation of the initiation influence of an active surface of the fibers on the thermal transformations of the epoxy oligomers. The time lag of the Me-CFs–epoxide two-phase system at constant temperature results to extreme dependence of concentration of the radicals on time. So, for example, for the Pb-CFs – ED-20 system the maximal quantity of the paramagnetic centers at the 220 °C is observed after treatment of the composition during 20 minutes. The Neyman [15] have been offered a free-radical mechanism of destruction of the epoxies. For all this the decomposition reactions of macromolecules and macroradicals can be tracked by their recombination in which the sequence of involvement of the different radicals depends on the conditions of reaction passing and on a composition of a mixture.

Into consideration the fact that in our case the particles of a metal or its oxide are in a highly dispersive state on a surface of a carbon fiber carrier it is possible to believe that the EPR signals in the examined systems are stipulated but the free radicals which are generated under catalytic thermal destruction of the epoxy oligomers. The signal observed in the EPR spectrum belongs to radicals not primary, probably, the end –ones, extremely reactive therefore short-lived and not recorded in the spectrums, but to the steady macroradicals generated as a result of the radical reactions which one can be already fixed by the EPR method.

Decrease of the beginning temperature of the paramagnetic centers registration into the Me-CFs –epoxy oligomer two-component system by 80-130 °C as compared with written in the literature [12] one may explain by the presence in a system of a carbon surface which as it is known can promote the process of formation of the free radicals [16]. The curves of a change of a quantity of paramagnetic centers in system under heating up of the Pb-CF$_s$ – epoxy oligomer heterogeneous composition are shown in Fig. 3,b. It is taken notice of a complex course of the curves for the Me-CF$_s$–epoxide compositions (curves 3, 4). At the outset a linear decrease of the EPR quantity is observed, then their increase and following repeated decrease is taken place unlike the compositions of the oligomers with CF$_s$ unmodified (curves 1, 2).

It was possible to believe that the Me-CF$_s$, activating of the catalytic decomposition of the epoxy oligomers, promote simultaneously their further structurization. It was revealed that the beginning of an increase of quantity of the paramagnetic centers in the system correlates with a start of an exotermic reaction generated in a system (see Fig. 2). As it is noted above in the interval of the temperatures of 200-300 °C, that is in the conditions preceding of the epoxy oligomer homopolymerization, the new exothermic effects not characteristic for the individual epoxy oligomrers and for the mixtures of the oligomers with CF$_s$ unmodified are appeared. The mass losses at this temperature range are highly insignificant. Therefore, if the first branch of the kinetic curves of a paramagnetic centers accumulation (Fig. 3, b) it is possible to refer to deactivation of the initially generated radicals caused by the lightly volatility products, keeping in the initial oligomers, the subsequent increase of quantity of the paramagnetic centers is conditioned primarily already by thermal transformation of the epoxy oligomers under influence of the Me-CF$_s$. Symbatum dependence is observed and for the EPR spectrums line width at heating up of the epoxy oligomers with the Me-CF$_s$. As it follows from Fig 3, a, the primary sharp narrowing of a signal with increase of a temperature is stopped and a width of line becomes constant approximately. The change of width of a line in the EPR spectrums under heating of the compositions confirms the assumption above-mentioned about presence in system of two independent processes, namely, removal of the lightly volatility products at a primary stage of a heating and a formation of the new radicals owing to processes of the thermochemical transformations of the epoxy oligomers.

It is revealed also that the following increase of heat release in the Me-CF$_s$ – oligomer system under its heating is accompanied by simultaneous decrease of concentration of the paramagnetic centers formed after a stage of their increase (Fig 3, b, curves 3, 4} and by forming of three-dimensional structure of a polymer (see Table 2). It testifies already about suppression of the destructive processes developing in system and reversal of the proceeding reactions in the side of the prevailing process of formation of

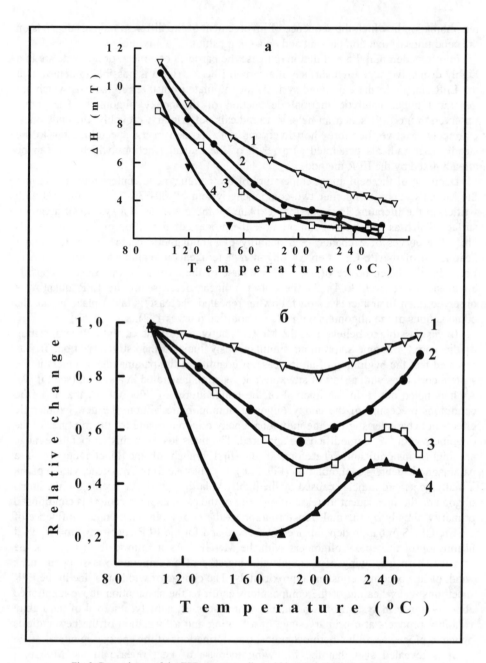

Fig 3. Dependence of the EPR spectrums line width, (*a*), and relative change of a quantity of paramagnetic centers, (*б*), on temperature of treatment of the ED-20 epoxy oligomer (1,2) and UP-612 diepoxide (3,4), filled by CFs unmodified (1,3) and by Pb-CFs (2, 4) .

the epoxy polymer. For the compositions of the epoxides with CF$_s$ unmodified in a temperature range of 200-300°C the processes of accumulation of the free radicals at an increase of temperature are continuing (Fig. 3, b, curves 1, 2) and a formation of a three-dimensional product is not observed.

Comparing the data obtained with the results of the infra-red spectroscopy and the chemical analysis pointing out by a conversion of the epoxy groups and a formation of an unsolvable gel-fraction at joint heat treatment of the oligomers with the Me-CF$_s$ (Table 2), it is possible to consider that the new exothermic effects are conditioned by inside- and intermolecular transformations of the epoxy oligomers under influence of the Me-CF$_s$ surfaces. The determining role in these processes is played of the metal containing inclusions relating to a phase of a metal or its oxide finely divided on the carbon monofibers surfaces and shown clearly on the electron - microscopically pictures. Catalyzing of the destruction process of the epoxy oligomers on the initial stage of a heat treatment nanoparticles of the metals or their oxides on surface of carbon  fibers interact in further with ones and the products of their transformation with formation of a new phase. It is confirmed by the IR-spectroscopy studies (there are the  new absorption bands at 420 cm$^{-1}$, 880 cm$^{-1}$ and 1720 cm$^{-1}$) and also by a dependence of an amplitude of signal appearing in the electron paramagnetic resonance' spectra (symmetrical unsolved singlet, g = 2,028) on the concentration of a metal in a system. The further transformations result in a formation of an unsolvable three-dimensional polymer (see Table 3) with the included atoms of a metal in the molecular chain and of the radicals stabilized at the three-dimensional structure. This is agreed with the data of the work [17], in which it is shown an opportunity of an existence of the steady free radicals in the carbon structures with a high concentration of cross-links, and also in the glass polymers [18]. A conformation of an active role of the metal-containing inclusion into the CF$_s$ on a transformation of the examined epoxy compounds in interphase zone is the dependence of an yield of a unsolvable product on percentage of a metal in the CFs and accordingly on a degree of filling of an oligomer by the fibers (Tab. 3).

The data obtained have allowed to propose the radical mechanism of the epoxy oligomer destruction on the metal containing surface of the CF$_s$ at simultaneous interaction of the active superficial centers of the CF$_s$ with the epoxy oligomer by analogy with the reactings of transformations of the epoxides on the colloidal particles of metals written by Natanson etc. [11]. The results of the IR-spectroscopy and of the kinetic studies of chemical interaction of the Me-CF$_s$ with oligomers testify for the favour of pronounced. Sigmoid form of the conversions curves by the epoxy groups with the induction periods and acceleration ones (Fig. 4, a), sharp dependence of a transformation rate of an epoxy oligomer on temperature (Figure 4, b), high values of an effective energy of an activation of the process of chemical interaction nanoparticles of a metal with the epoxides (183 kgJ/mol for a composition of the Pb-CF$_s$ with ED-20, for example) are observed. The formation of the chemical bonds between the nanoparticles of a metal or its oxide and the epoxide results in a grafting of the oligomer to a carbon fiber surface.

The epoxy oligomer grafted to surface of CF$_s$ is arranged uniformly on all external surface of the mono-fiber , forming some kind of " a shirt" of a polymer and insulating it from the other mono-fibers. In a process of an increase of an insoluble fraction an interfibre volume is stuffed and the system converts into monolith. The studies of the cured compositions by the method of an electron microscopy were shown a formation on

TABLE 2. Amount of cured oligomer after heat treatment of its composition with metal-containing CFs (filling – 50 weight %, temperature – 200 °C, time –3 hour)

| CF$_s$ | Yield of gel-fraction (weight %) | | | | | |
|---|---|---|---|---|---|---|
| | ED-20 | DEG-1 | UP-612 | UP-632 | UP-650D | UP-650T |
| – | 0,0 | 1,1 | 3,1 | 1,3 | 17,4 | 6,9 |
| CF$_s$ unmodified | 0,0 | 4,8 | 4,9 | 1,2 | 11,3 | 7,6 |
| Sn-CF$_s$ | 20,3 | 70,3 | 57,7 | 31,2 | 50,3 | 75,6 |
| Fe-CF$_s$ | 48,5 | 22,6 | 46,4 | 31,5 | 34,1 | 42,7 |

TABLE 3. Gain in weight of cured fractions of ED-20 epoxydian oligomer non-extractable from a surface of the Fe-CF$_s$ in dependence on the content of metal in fiber

| The content of iron in the CF$_s$ (weight %) | 0,3 | 2,7 | 5,7 | 9,0 |
|---|---|---|---|---|
| Gain in weight of polymer on the Fe-CF$_s$ (weight %) | 26,4 | 48,9 | 86,2 | 122,4 |

a surface of the fibers and for some compositions in an interfibers space of the crystalline structures distinguishing on the form in dependence on a state of a metal in the CF$_s$, conditioned, probably, particulate diffusion of the metal containing inclusions in a volume of a formed polymer and a following growth of the crystals on a free surface. The researches indicate also relation of a nature of the thermal transformations to chemical structure of the oligomers studied under influencing of the same kind of the Me-CF$_s$. The selectivity is watched concerning preferential structuring of the oligomers under influence of the Me-CF$_s$.

## 4. Conclusions

Process of thermochemical transformations of the epoxy oligomers of various chemical structures in the presence of the carbon fibers containing into theirs structure nanoparticles of metals or theirs oxides has been studied. A conclusion about concurring reactions of thermal decomposition of the epoxy oligomer and formation of three-dimensional polymer occurring on the Me-CF$_s$ – oligomer interface under their joint heating has been made on the basis of the thermal and chemical analysis, EPR and IR spectroscopy data. Nanoparticles of metals or theirs oxides play the leading role in these processes.

Catalytic effect of the Me-CF$_s$ on decomposition of the epoxy oligomers has been revealed. The onset temperature of the oligomer decomposition considerably decreases

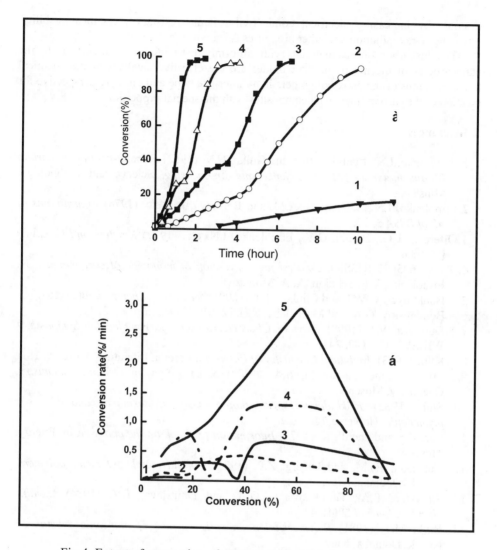

Fig 4. Extent of conversion of epoxy groups (a) and conversion rate of epoxy groups (б) while heating at 160 °C (1), 170 °C (2), 180 °C (3), 190 °C (4), 200 °C (5) of ED-20 epoxy oligomer filled by Pb-CFs. Ratio of Pb-CFs - oligomer is 1:1.

in the presence of the Me-CF$_s$, and the decomposition proceeds with higher rate. A new exothermic effect is revealed in the 200-300°C temperature range, intensive formation of an insoluble gel-fraction is observed, which is not characteristic of the individual epoxide compounds and mixtures of them with CF$_s$ non-modified. Taking into consideration of the factors detected the conditions of the directional regulation by the reactions of the epoxy oligomers transformations on the active surfaces of the Me-CF$_s$ were determined. The capability of developing of the composite material on a base of the Me-CF$_s$ and the studied oligomers possessing with the improved characteristics , for example by increasing electrical conductivity, flame resistance, chemical stability, screening X-ray radiation and other properties is shown.

Thus the carbon fibrous materials with uniformly distributed on their surfaces by the nanoparticles of the metals or their oxides are the effective catalysts of the chemical transformations of the epoxy oligomers in an interface zone and ones are perspective for creation of the carbon fibers nanocomposites with the special properties.

## 5. References

1. Ermolenko, I.N., Lyubliner, I.P. and Gulko, N.V. (1982) *Element-containing carbon fibrous materials. Minsk: Science and Engineering,* Science and Technology, Minsk.
2. Ermolenko, I.N., Saphonova, A.M. and Belskaya, R.I. etc. (1976) *Informations of AS of BSSA* **5**, 17.
3. Olfereva, T.G., Bragin, O.V., Ermolenko, I.N. etc. (1977) *Kinetics and Catalysis* (2), 933.
4. Patent 695695 (USSR). *Catalyst for conversion of monoxide of carboneum* I. N. Ermolenko, V.P. Schukin, A. A. Morozova
5. Boukhtiyarov, V. I. and Clinko, M. G. (2001) *Success of Chemical* **70**(2), 167.
6. Boudnitskiy, G.A. and Machalaba, N.H. (2001) *Chemical fibers* **2**, 4.
7. Dubkova, V.I. (1999) *Polymer Characterization. Macromolecular Symposia.* - WILEY-VCH **148**, 241.
8. Kline (1963) *Analytical Chemistry of Polymers*, Foreign Literature, Moscow, **1**.
9. Mironov, and Yankovski, S.A. (1985) *Spectroscopy in Organic Chemistry*, Chemistry, Moscow.
10. Smit (1982) *Prikladnaya spektroskopia. Osnovy: tekhnika, analiticheskoe primenenie*, Universe, Moscow.
11. Lee, Kh. and Nevill, K. (1973) *Information textbook on the epoxy resin.* Energy, Moscow.
12. Natanson, En.M. and Ulberg, Z.P. (1971) *Colloid metal and metal polymers.* Naook. Doomka, Kiev.
13. Sirovatka, L.A., Gorokhovskiy, G.A. and Dmitrieva, T.V. (1984) *Ukraine chemical journal* **50**(4), 430.
14. Bryk, M.T. (1981) *Polymerization on solid surface of inorganic substances.* Naouk. Doomka, Kiev.
15. Neyman, M.B. (1964) *Ageing and stabilization of polymers*, Science, Moscow.
16. Francisco Rodrígues-Reinoso (1998) *Carbon* **36**, 3, 159.
17. Quinn, G.A., Seymour, R.C. (1978) *Investigation of carbonized polymers by electron paramagnetic resonance*, Moscow.
18. Van Krevelen, D.V. (1976) *Properties and chemical structure of polymers*, Chemistry, Moscow.

# THEORY OF TRANSPORT PHENOMENA ON PLASMA – METAL HYDRYDE INTERFACE

V.N. BORISKO, S.V.BORISKO, D.V. ZYNOV'EV, Ye.V. KLOCHKO
*V.N.Karazyn Kharkov National University, 31 Kurchatov Av.,*
*61108, Kharkov, Ukraine*
*A.N. Podgorny Institute of Mechanical Engineering Problems of*
*NAS Ukraine, 2/10 Pozharsky Str.,61046, Kharkov, Ukraine*
*E-mail: Borisko@htuni.kharkov.ua*

## Abstract

In this paper the theory of hydrogen vibrational excited molecule emission stimulated by plasma ion bombardment of metal hydride cathode surface in condition of gas discharge is discussed. The conception of direct momentum transfer due to collisions of both high-energy plasma ions and atoms with hydrogen atoms, dissolved in metal matrix, is assumed the theory as a basis. It is shown, that the efficiency of this process can be enough high to provide the hydrogen vibrational excited molecule emission out of metal hydride. It is shown, that the intensity of desorbed hydrogen flow linearly depends on flux of bombarding particles. The calculations, based on this mechanism, qualitatively agree with experimental data.

## 1. Introduction

In case of use of hydrogen isotopes as a plasma-forming medium, the hydride forming getter materials on a basis of Zr-V alloys are perspective. The metal hydride electrodes for gas-discharge devices manufactured of these alloys, allow to combine in one device of functions of compact storage, purification from gas impurities and programmed leak-in of hydrogen isotopes into useful volume of the gas-discharge chamber, as well as pumping of gas excess. The temperature stability of operating mode of metal hydride system for gas supply of vacuum plasma is complicated problem. Distinctive feature of autostabilized on pressure mode of gas discharge using metal hydride cathode is single-value functional dependence of hydrogen pressure in gas discharge chamber on discharge current [1]. In case of Penning gas discharge cell this dependence has non-monotonous character. In [2] we attempted of theoretical explanation of such dependence in case a glow discharge with metal hydride cathode. The theory [2] was based on thermal effect of both high energy hydrogen ions and atoms impinge upon the metal hydride cathode. However considerable quantitative discrepancies between results of calculations [2] and experimental data [1] indicate that this model does not take into account some essential aspects of interaction between the high energy hydrogen particles and metal hydride cathode.

*T.N. Veziroglu et al. (eds.),*
*Hydrogen Materials Science and Chemistry of Carbon Nanomaterials,* 375-382.

In particular, in [3] it was shown, that irradiation of metal hydride samples by both the electron and ion beams in high vacuum condition stimulates emission of vibrational - excited hydrogen molecules from the sample surface. In experiments [1, 3] the polymetallic compositions on the basis of Zr-V getter alloys were used. These alloys are capable to reversible hydrogen absorption of low-pressure hydrogen at room temperature as well as to desorbtion by heating up to moderate temperatures (150-4000 C) [4,5]. Those materials cosist of mixture of individual zirconium phases and cubic C15 or hexagonal C14 Laves phases [6]. Besides of high hydrogen sorptive capacity Laves phase in alloys on the basis of zirconium are activated in "weak" conditions and have high rate of absorption and desorption of hydrogen [7]. The pressure of dissociation of number hydrides of Zr-V intermetallic compounds at a room temperature is much less atmospheric ones (usually no more than 10–8 Pa) [5, 7]. In case of $ZrV_2H_x$ about 50 % of the accumulated hydrogen is desorbed in temperature interval from 400 K up to 500 K. In [8, 9] it is shown that concentration of vibrational - excited hydrogen molecules desorbed into vacuum from metal hydride surface on the basis of Zr-V alloys considerably exceeds the equilibrium ones. The authors [9] explain this effect by difficult energy exchange between desorbed hydrogen molecule and metal matrix.

Aim of the paper is a theoretical investigation of autostabilized on pressure mode of a glow discharge using metal hydride cathode.

## 2. Model

Mentioned above characteristic properties of both the desorption and absorption of hydrogen can be explanted from the common point of view if to assume, that in this case limiting stage is the recombination of hydrogen atoms on a surface of the material, as well as during absorption - dissociative chemosorption of the hydrogen molecules. Accepting this assumption it is possible to show [10, 11], that both the desorbed and absorbed hydrogen flow by metal hydride is proportional to a difference of values of pressure above the metal hydride and pressure of metal hydride dissociation at present temperature.

The particle balance on the metal hydride surface is:

$$
\frac{d\theta}{dt} = \frac{2\alpha P(1-\theta)^2}{N_S \sqrt{2\pi MkT}} - v_{ds}\theta^2 \exp\left(-\frac{2E_{ds}}{kT}\right) +
$$
$$
+ \frac{N_0}{N_m}(1-\theta)v_{out}\exp\left(-\frac{E_{out}}{kT}\right) - \theta\left(1-\frac{N_0}{N_m}\right)v_{in}\exp\left(-\frac{E_{in}}{kT}\right)
$$

(1)

where $\theta$ – the degree of covering, $\alpha$ – the attachment probability of the hydrogen molecule that imping on the material surface, P – the hydrogen pressure, $N_S$ – the adsorption site density on material surface, where hydrogen atoms can is adsorbed, M – mass of hydrogen molecule, k – Boltzmann constant, T – the temperature of the metal hydride, $E_{ds}$ – the activation energy of hydrogen desorption, $v_{ds}$ – the preexponential factor for hydrogen desorption rate, $N_m$ – the bulk density of hydrogen atoms in subsurface layer of the metal hydride, $N_m$ – the largest bulk density of hydrogen atoms in metal hydride, $E_{out}$ – the activation energy of hydrogen transition from surface to bulk of metal hydride, $v_{in}$ – the preexponential factor for rate of hydrogen transition from surface to bulk of metal hydride.

The first term in equation (1) is a frequency of hydrogen molecule dissociative chemosorption by unit area of the material surface. The second term in this equation is a frequency of hydrogen desorption into gas phase from unit area of the material surface. Next to last and last items in right hand of equation (1) are transition rates of hydrogen atoms out of the metal bulk to its surface and reverse direction, respectively. Their algebraic sum is diffusion flux of hydrogen atoms through the subsurface layer of the metal hydride:

$$N_S\theta\left(1-\frac{N_0}{N_m}\right)v_{in}\exp\left(-\frac{E_{in}}{kT}\right)-N_S\frac{N_0}{N_m}(1-\theta)v_{out}\exp\left(-\frac{E_{out}}{kT}\right)=-D\lim_{z\to0}\mathrm{grad}(N), \quad (2)$$

where $D$ is the constant of hydrogen atom diffusion in metal hydrogen bulk, $N$ is the hydrogen atom density in metal hydride bulk.

From expressions (1), (2) follows:

$$\frac{d\theta}{dt}=\frac{2\alpha P(1-\theta)^2}{N_S\sqrt{2\pi MkT}}-v_{ds}\theta^2\exp\left(-\frac{2E_{ds}}{kT}\right)+\frac{D}{N_S}\lim_{z\to0}\mathrm{grad}(N). \quad (3)$$

On the assumption of steady state the equation becomes

$$\frac{2\alpha P(1-\theta)^2}{N_S\sqrt{2\pi MkT}}-v_{ds}\theta^2\exp\left(-\frac{2E_{ds}}{kT}\right)=-\frac{D}{N_S}\lim_{z\to0}\mathrm{grad}(N). \quad (4)$$

Following model of operation [10], we assume, that the chemosorption is a limiting stage of hydrogen desorption. In this case the rates of all other stages of the process so much are great, that the hydrogen density gradient in metal hydride can be neglected.

Then, taking into account, that: $\mathrm{grad}(N)\equiv0$, $P=P_{eq}$ from equation (3) we obtain:

$$\frac{\theta^2}{P_{eq}(1-\theta)^2}=\frac{2\alpha}{N_Sv_{ds}\sqrt{2\pi MkT}}\exp\left(\frac{2E_{ds}}{kT}\right). \quad (5)$$

At small derivation from equilibrium, i.e. when $\dfrac{|P-P_{eq}|}{P_{eq}}\ll1$, it is possible to consider that the expression for flow of absorbed hydrogen molecules becomes:

$$\Gamma=\frac{2\alpha(1-\theta)^2}{\sqrt{2\pi MkT}}(P-P_{eq}) \quad (6)$$

where $\theta$ degree of covering is determined by equation (5).

In condition of intensive ion bombardment of metal surface the degree of covering tends to zero. From this it follows that:

$$\Gamma=\frac{2\alpha}{\sqrt{2\pi MkT}}(P-P_{eq}) \quad (7)$$

The particle balance on the vacuum-metal hydride interface (Fig.1) is:

$$\frac{2\alpha}{\sqrt{2\pi MkT}}(P-P_{eq})=Y_{sp}J_b-\Lambda_{imp}J_b. \quad (8)$$

where: $Y_{sp}$ – the factor of ion stimulated desorption by direct transfer of momentum to hydrogen subsystem of metal hydride, $\Lambda$ – factor that determines implantation of high energy hydrogen ions and atoms into metal hydride, $J_b$ – density flux of high energy hydrogen ions and atoms impinge the metal hydride surface. The first term in right hand of this equation is flow of hydrogen into gas out off metal hydride by ion stimulated desorption. The second term is flow of hydrogen atoms and ions implanted into metal hydride sample.

As the flow of impinging high-energy particles is low, so it can be neglect to temperature effect on hydrogen desorption rate. In this case $T$ temperature of sample and corresponding to it $P_{eq}$ equilibrium pressure are constant. The carried out comparative estimations of the contributions both of ion stimulated desorption and implantation flows show, that in condition of our experiment the energy of bombarding ions is less 10 keV. So, effect of implantation is negligibly small.

In this case equation (8) is:

$$\frac{2\alpha}{\sqrt{2\pi MkT}}\left(P-P_{eq}\right)= Y_{sp}J_b .$$  (9)

Radiating from this, the following model of interaction of ion flow with a metal hydride sample is discussed. During collision of high energy hydrogen particle beam with hydrogen subsystem of metal hydride the most of the energy can be transferred. If these particles are scattered to sharp angle on atoms of crystalline lattice, the probability of their exit into vacuum will be considerable.

The atoms of hydrogen sublattice, to which the energy was transferred can donate a part of this energy to other hydrogen atoms. This process causes an avalanche of secondary particles. The avalanche advances so long as the primary particle will be broken up to energy approximately to equal T temperature of metal hydride lattice.

The mechanism of reflective ion stimulated desorption is realized as follows.

The high-energy particles bombarding metal hydride, are scattered to sharp angle from point of lattice of metal matrix (Fig.2), with some probability can leave out of metal hydride or transmit a considerable part of the energy to atoms of the hydrogen subsystem. Thus, these particles cause an avalanche of secondary particles in hydrogen subsystem.

Let's enter the following labels: $\Delta S$ – cross-section of a scattering on an acute angle of a primary particle on a knot of a crystalline lattice; $\Delta_0 S$ – complete cross-section of a scattering of a primary particle on a knot of a crystalline lattice $\delta S_H$ – cross-section of a scattering of a primary particle on atom of a hydrogen sublattice of metal-hydride; $S_H$ – the section area of an interstice of a lattice of metal-hydride, which is occupied by atom of hydrogen; $S = a2$ – sectional area of a unit cell of a crystalline lattice of metal-hydride, where $a$ – is a constant of a crystalline lattice of metal-hydride (see Fig. 1);

Transmitted flow of the high-energy hydrogen atoms through the first crystallographic plane $j_1'$ (on Fig. 2 is figured by black dotted arrow) is equal:

$$j_1'=\left(1-\frac{\Delta_0 S}{S}\right)J_b$$  (10)

Further, the flow of high energy particles $j_1'$, hit in the first layer of metal-hydride (limited between first and second crystallographic planes), in which $x$ characterizes

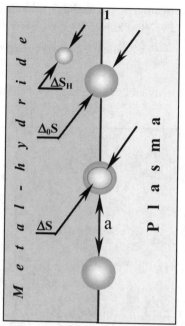

Fig.1.    Plasma – metal hydride
interface

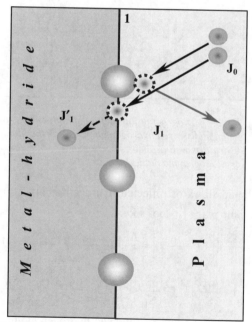

Fig. 2.    Reflection and penetration of high-energy
hydrogen atoms through first crystallographic plate of
metal hydride.

quantity of the dissolved hydrogen, and endures seldom collisions with the dissolved atoms (see Fig. 3). The probability of such collisions is determined by product of transmitted flow $j_1'$ on $\dfrac{\delta S_H}{S_H} x$, where $x = \dfrac{N}{N_m}$. Expression for flow of primary

hydrogen atoms, which has passed through the first layer looks like: $j_1' \left( 1 + \dfrac{\delta S_H}{S_H} x \right)$.

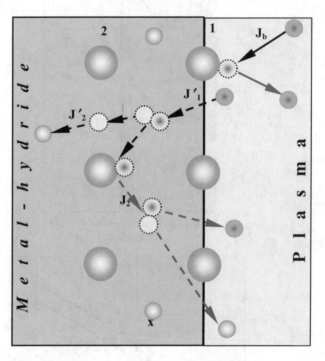

Fig.3.    Diagram of interaction of high-energy hydrogen atoms with the dissolved hydrogen atoms inside the first layer of metal hydride as well as their reflection and penetration into second crystallographic plane.

Similarly we found flows of reflected hydrogen atoms $j_2$ and passed through the second crystallographic plane $j_2'$ look like:

$$j_2 = \frac{\Delta S}{S} j_1' \left( 1 + \frac{\delta S_H}{S_H} x \right) = \frac{\Delta S}{S} \left( 1 - \frac{\Delta_0 S}{S} \right) \left( 1 + \frac{\delta S_H}{S_H} x \right) J_b \,,$$

$$j_2' = \left( 1 - \frac{\Delta_0 S}{S} \right) j_1' \left( 1 + \frac{\delta S_H}{S_H} x \right) = \left( 1 - \frac{\Delta_0 S}{S} \right)^2 \left( 1 + \frac{\delta S_H}{S_H} x \right) J_b .$$

Recursion expression for flows of reflected hydrogen atoms $j_k$ and passed through k-crystallographic plane $j_k'$ look like:

$$j_k = \frac{\Delta S}{S}\left(1 - \frac{\Delta_0 S}{S}\right)^{k-1}\left(1 + \frac{\delta S_H}{S_H}x\right)^{k-1} J_b, \tag{11}$$

$$j_k' = \left(1 - \frac{\Delta_0 S}{S}\right)^{k}\left(1 + \frac{\delta S_H}{S_H}x\right)^{k-1} J_b. \tag{12}$$

The part of flow is reflected from points of k-crystallographic plane and get out into gas phase $j_{k_0}$ looks like (see Fig. 4):

$$j_{k_0} = j_k\left(1 - \frac{\Delta_0 S}{S}\right)^{k-1}\left(1 + \frac{\delta S_H}{S_H}x\right)^{k-1}. \tag{13}$$

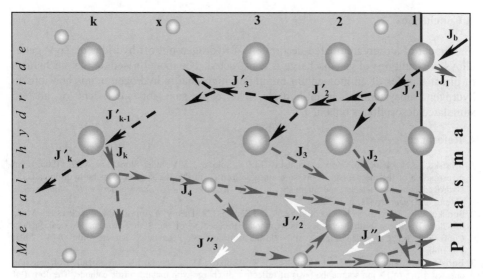

Fig.4. *Propagation dynamics of high-energy hydrogen atoms in Metal hydride body with uniform concentration of dissolved hydrogen.*
$j_i''$ *flows of penetrated particles are marked as black dashed arrows.*
$j_i''$ *flows of reflected particles are marked as red dashed arrows.*

Having substituted to expression (13) values $j_k$ from expression (11), finally we shall receive:

$$j_{k_0} = \frac{\Delta S}{S}\left(1 - \frac{\Delta_0 S}{S}\right)^{2k-2}\left(1 + \frac{\delta S_H}{S_H}x\right)^{2(k-1)} J_b \tag{14}.$$

Within of given model the coefficient of an ion stimulated desorption looks like:

$$Y_{sp} = \frac{\Delta S}{S}\left(1 - \frac{\Delta_0 S}{S}\right)^{2k-2}\left(1 + \frac{\delta S_H}{S_H}x\right)^{2(k-1)} \tag{15}$$

Dependence of hydrogen pressure in the chamber on flow of particles impinges the metal-hydride, takes account (9) and (15) is:

$$P = \frac{Y_{sp}}{K(T)} J_b + P_{eq},$$ (16)

where $K(T) = \dfrac{2\alpha}{\sqrt{2\pi MkT}}$.

For following parameters:

$S = 8.15 \ 10^{-16} \text{cm}^2$, $\Delta 0S = 6 \ 10^{-17} \text{cm}^2$, $\Delta S = 3.8 \ 10^{-17} \text{cm}^2$, $\delta SH = 3 \ 10^{-17} \text{cm}^2$, $SH = 4.41 \ 10^{-16} \text{cm}^2$, $P_{eq} = 10^{-8} \text{Pa}$, $\alpha \approx 1$, $x = 0.4$, $M = 3.32 \ 10^{-27} \text{kg}$, $T = 300 \text{K}$, $J_b = 6.25 \ 10^{18} \text{m}^{-2} \text{c}^{-1}$ the pamper hydrogen in chamber is: $P = 3{,}8 \ 10\text{-}5 \text{Pa}$. In this case the coefficient of ion-stimulated desorption is: $Y_{sp} = 0{,}43$.

## 3. Conclusions

The model of an ion-stimulated desorption of hydrogen out off hydrides of Zr-V getter alloys in conditions of gas discharge is constructed. The model based on the mechanism of direct transfer of an momentum impinging high energy hydrogen atoms and ions to hydrogen subsystem of metal-hydride. The expression for coefficient of an ion stimulated desorption is obtained.

## 4. References

1. Borisko, V.N., Klochko, Ye.V., Lototsky, M.V., Solovey, V.V. and Shmal`ko, Yu.F. (2002) An investigation of the mechanism of pressure autostabilization in the discharge with metal-hydride cathode. Hydrogen materials science and chemistry of metal hydrides, *NATO Science Series. II. Mathematics, physics and chemistry* **71**, 217-235.
2. Borisko, V.N., Borisko, S.V., Klochko, Ye.V. and Sereda, I.N. Theory of Autostabilized on Pressure Mode of Glow Discharge With Metal Hydride Cathode. Proceeding of the 30th European Physical Society Conference on Controlled Fusion and Plasma Physics. St Petersburg, Russia, July 7-11, 2003, P-4.104.
3. Borisko, V.N., Borisko, S.V., Klochko, Ye.V., Ryabchikov, D.L., Sereda, I.N. Investigation of emission characteristics of H- ion source the basis of reflective discharge with metal hydride cathode, The Jornal of kharkiv National University. Physical series "Nuclei, Particles, Fields". 2002, #. 569, Issue 3/19, 91-93 p.
4. Yartys, V.A., Zavaliy, I.Yu. and Lototsky, M.Yu. The absorptions low-pressure hydrogen on the basis of modified Cr-V and Cr-V-Fe alloys by oxidation additions. // Coordination chemistry, 1992, v. 18, # 4, p. 409-423. (in Russian).
5. Podurets, L.N., Sokolova, Ye.I., Kost, I.Ye., Kuznetsov, N.T. (1989) The interaction of hydrogen with Cr-V alloys, *Journ. of inorganic chemistry* **34**(2), 311-315. (in Russian).
6. Tretyachenko, L.A., Prima, S.B., Suhaya, S.A. (1975) Levis`s phases and phases diagrams of the third system which created by transient metals with Zr, In book Physical chemistry of condensed phases of superhard materials and their interfaces, Kyiv: Naukova Dumka, 167-175. (in Russian).
7. Semenenko, K.N., V.N. Verbetskiy, Mitrohin, S.V., V.V. Burnashova (1980) The research of interaction with hydrogen of Zr intermetallic compounds crystallizable in structural types of Levis`s phases, *Journ. of inorganic chemistry* **25**(7), 1731-1736. (in Russian).
8. Shmalko, Yu.F., Lototsky, M.V., Klochko, Ye.V. and Solovey, V.V. (1995) The formation of excited H species using metalhydrides, *J. Alloys and Compounds* **231**, 856-859.
9. Shmalko, Yu.F., Borisko, V.N., Klochko, Ye.V., Solovey, V.V. (2000) About vibrational excitation of hydride molecules, which desorbed out off metal hydride, Repots of National Science Academy of Ukraine 11, p. 91-95. (in Ukraine).
10. Martin, M., Gommel, C., Borkhart, C., Fromm, E. (1996) Absorption and desorption kinetics of hydrogen storage alloys, J. Alloys and Compounds **238**, 193-201.
11. Bloch, J. (2000) The kinetics of a moving metalhydride layer, J. Alloys and Compounds **312**, 135-153.

# THE ULTRADISPERSE FORMATIONS OF FREE CARBON IN ALLOYS OF IRON

BARANOV D.A., BARANOV A.A., LEIRICH I.V.
*Donetsk national technical university*
*e-mail: BaranovDA@rambler.ru; fmt@fizmet.dgtu.donetsk.ua.*
*ul. Artema, 58, Donetsk, 83000, Ukraine*

**Abstract.**
The report is devoted to ways of forming of a disperse structure and properties of the iron alloys containing up to 15 % (vol.) free carbon as spherical particles with a diameter ~ 1 microns. Receptions of disperse structures of free carbon in surface sections of products is for practical purposes especially important, and the technology surveyed here allows implementing it in many cases.

**Keywords:** carbon, plazma-treatment, cycling treatment, high-strength cast iron.

## 1. Introduction

The experimental development of methods and their approbation was preceded with the thermodynamic analysis of processes of formation and dissolution of free carbon in iron alloys. According to its results, origin of stressing probably only in pores, which matting of a surface leads to decrease of a surface energy of an alloy. According to the analysis, the radius ($r_\kappa$) a flat nucleus of the critical size on a free surface is equal:

$$r_K = \frac{t * \gamma_\perp}{\left[\gamma_S - \left(\gamma_{SI_g} + \gamma_{II}\right)\right] + t \cdot \Delta f} \quad , \tag{1}$$

where    t - thickness of a film of free carbon;
$\gamma_\parallel$ and $\gamma_\perp$ - accordingly the surface tension along and across a carbon film;
$\gamma_s$ and $\gamma_{s/g}$ - the surface tension of a solute and a phase-to-phase surface accordingly;
$\Delta f$ - change of energy of Gibbs.

From the thermodynamic analysis it is follows, that increasing of a surface energy of a solute and diminution of face surfaces of carbon films promotes their formation. For origin are especially effective fine pores in supersaturated enough carbon solution. According to calculation, at dissolving cement carbide in Fe-C alloys the diameter of a nucleus of the critical size makes ~ 10 nanometers [1]. A film of the free carbon, formed in pores the greater size, grow as a result of delivery of carbon from solution. The destiny hollow formations of free carbon by the size less than 10 nanometers are less determined. The chemical potential of carbon in them is higher, than in an environmental solution, and films growth becomes possible only at origination in immediate proximity new energy and coordination fluctuations. At the same time a hollow film of free carbon less than the critical size can long-term to be maintained in an alloy as with dissolution of them power-intensive free surfaces are exposed, that thermodynamically it is not justified. Dissolution of shell films conjugates to development of face surfaces of free carbon and demands simultaneity of acts of a leaving of atoms of carbon and arrival of atoms of iron.

*T.N. Veziroglu et al. (eds.),*
*Hydrogen Materials Science and Chemistry of Carbon Nanomaterials, 383-386.*

Thus, the problem of reforming of the ultra disperse, including nancrystalline, formations of free carbon is reduced to making in an iron alloy of fine pores set with a diameter = 10 nanometers. At presence of sufficient their number in an iron-carbon alloy of the eutectic makeup are formed hollow formations of free carbon, the distance between which is close to their diameter. For making similar condition in steel and pig iron, electromachining, a flowage, and a temperature cycling and formation gas bubble up to a hardening of a melt were used.

Thermocycling treatment with the martensitic transformation gave the formation in steel up to $10^9$ cm$^{-3}$ of particles of free carbon. At the martensitic transformation numerous microcracks in which of a supersaturated solute free carbon is oozed are formed. At variation of the top temperature of a cycle and a cooling medium it is possible to change the sizes of microcracks and their number, so also number of formations of free carbon. The magnification of number of particles appreciably after three cycles and the further temperature cycling affected poorly. Efficiency of a temperature cycling can be increased systematicly from a cycle to a cycle by dipping of the top temperature of a cycle due to what redistribution of pores is loosened at the moment of dissolution of carbon formations and the degree of their dissolution is restricted.

## 2. Technique of experiment

The number of presipitation of free carbon increases, if a high-strength cast iron to deform at temperatures above 1000°C. So, warp by a upsetting on 75 % of pig-iron, the composition and procedure of testing are given in works [2, 3], gave the formation of numerous plugging free carbon (« a graphitic rash »), disposed separately from the deformed particles, who contains in an initial state. Results of exploration of formation of "rash" testify, that the nature of it is interlinked to origination in pig iron at the moment of warp of microseparations which surface is coated with a film of free carbon. Carbon film provided maintenance and development pores because of decrease of a surface energy. Fine particles of "rash" had the spherical shape at the considerable forming of large particles so the strain resistance of pig iron under influence of stressing of free carbon varies.

## 3. Results and discussion

The surface activity of carbon manifested in a carbon coating of cracks, promotes development of shattering. It is especially appreciable at high-temperature warp of a high-strength cast iron. Above 700°C the strain resistance of graphite is higher, than austenite. At the same time in rolled and forged pig-iron at graphite "tails" and the "whiskers" testifying the preferred deformation of stronger phase – graphite, are formed [3]. In the mechanism of origination of carbon appendixes at large particles of graphite the major role the polycrystalline of spherical particles is played due to which in the graphite, having only two systems of slipping, at mechanical action form numerous microcracks. It promotes replacement of carbon in generatored pores and formation of carbon coating on their surface. Orientation of "tails" and "whiskers" testify about a volume, that in a cracks formation the tangential and normal stresses are effective also.

The application of plastic deformation can change the form of graphite from spherous to disk, bar and fibrous at treatment of preliminary graphitized iron and steel. Deformation by rolling and forging with drawing out of 10 – 15 allows to decrease the cross-section area of graphite inclusions in some times and imparts to rolled forged iron goods the structure and property of composition material. Bars and disks of graphite contact with surface of metallic base of deformed alloy mainly by basis pranes $(0001)_g$. Due to tecsture of graphite particles by deformed high-carbon alloys they impart the anisotropy of physic-chemical and mechanical properties, which provide sharply rising of quality of goods. The corrosion resistance in concentrate solutions of acid under influence of deformation changes in interval 2.0 – 2.5 times, but wear-strength in conditions of dry friction of sliding in dependence of direction of contrbody rotation relative to deformed iron changes on order. Anisotropy of deformed high-strength iron show same at decarburization and high-temperature oxidation. So, change of form of graphite particles and their dispergation essential influence on properties of high-carbon alloys and imparts to them anisotropic character.

Intensive fine crushing of presipitation of free carbon was observed at electromachining. Depending on energy parameters of machining, in surfaces of hot deformed pig iron the band with the disperse cementite-austenitic eutectic at which post heating the ultra disperse presipitation free carbon (fig. 1) were formed.

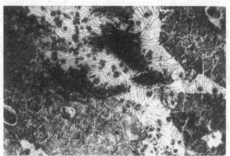

Fig. 1. Globular rules in ledeburite zones of thermal dominanc, x200

After heating cementite quickly receive properties of graphite. Diameter of formation of free carbon and distance between them are 1 mcm (fig. 2a).

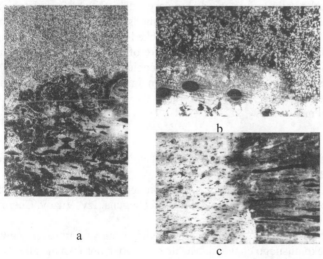

Fig. 2. Kohesion zone structure after self-control of deformed high-strength cast-iron attached to 850°C - 5 min (a, x500), 900°C - 5 min (b, x250) and 1000°C –10 min (c, x100)

386

Immediately after plazma-treatment microstrength of zone reaches to 20 GPa. Strength drops at heating, but exposure of 5 min at 850°C sufficient to total decay of cementite. With increasing of temperature of annealing the amount of graphite particles decreases (fig. 2b and c). Super fast dissolving is caused by sharp magnification of number of carbon formations up to $10^{11}$ cm$^{-3}$. The nature of sharp increment of number of presipitation of free carbon is interlinked to formation of gas bubble in a melt, shrinkable micropores and a congestion of carbon clusters on phase-to-phase surfaces cementite eutectic during its crystallization.

What near one graphite inclusion, which existed in alloy and deformed at forging, form at plasmic heating to 1150 - 1200°C. Dozens of fine graphite spheroids are evidence about saving ability of deformed high-strength iron to unormous eutectic crystallization after kohesion. Detected heredity has explanation, if take into account, that near spherous graphite in base of high-strength iron the increased content of magnesium is detected ("galo"). According this melting product, which arise on contact surface of graphite with base, enriched by dissolved magnesium, which in time of cooling of melting product exudes in form of bubbles and assists to it's unormous eutectic crystallization. Sharply increasing of number of graphite particles in hearth of kohesion is evidence about small sizes of bubbles, which assist the graphite origin, but significant cooling of melting product after plasmic melting facilitates the formation and growth of graphite in spherous form. Let's mark, at prasmic treatment of grey iron with plane graphite in kohesion zone at cooling the colonial structure, which is result of normal eutectic crystallization. It co-ordinates with literature data, which received at investigation of structure of weld connection of iron [4].

## 4. Conclusions

The given data testify about a volume, that, using methods of thermomechanical processing iron-carbon alloys, it is possible to vary a dispersity of formations of free carbon in a wide range. Besides overlapping existing methods of a dispersion of formations of free carbon and a variation key parameters of the checked modes of treatments, engaging other views of the action providing forming of structure of high-carbon alloys of iron, a nickel and a cobalt with high density of submicroscopic volumetric defects is of interest. Receptions of disperse structures of free carbon in surface sections of products is for practical purposes especially important, and the technology surveyed here allows implementing it in many cases.

## 5. Reference

1. Baranov A. A. Fazovye prevrascheniya i termotsiklirovanie metallov. K.: Naukova dumka. 1974: 232.
2. Baranov A. A., Baranov D. A. Perspectivy tekhnologiy, osnovannykh na sovmeschenii goryachey deformatsii i termicheskoy obrabotki chuguna. Izv. Vuzov. Chernaya metallurgiya. 2002; 7: 34 – 40.
3. Baranov D. A., Nesnov D. V. Kompyuterne modelyuvannya formozminy grafitu pri deformatsii vysokomitsnogo chavunu. Metaloznavstvo ta obrobka metaliv. 2002; 4: 13 – 17.
4. Metallografiya svarnykh soedineniy chuguna / Pod. red. Gretskogo Yu. Ya. K: Naukova dumka. 1987: 192.

# SIMULATION OF FULLERENE IRRADIATION AND FRAGMENTATION BY PARTICLE BEAMS

N.V. MAKARETS[1], V.V. MOSKALENKO[1], YU.I. PRYLUTSKYY[2], O.V. ZALOYILO[2]
*Departments of Physics[1] and Biophysics[2], Kiev National Shevchenko University,*
*Vladimirskaya Str., 64, 01033 Kiev, Ukraine. e-mail: mmv@mail.ups.kiev.ua*

## Abstract

The simple computer model for calculation of a fragmentation of free fullerenes by electrons and singly charged ions has been proposed. It gives qualitatively correct mass spectrums of fragments and after fitting of its parameters to the experimental data it can be used for calculations in case of the supported fullerenes and a fullerite.

**Keywords**: Fullerene; Particle irradiation; Modeling, Molecular simulation; Fragmentation.

## 1. Introduction

Study of fullerenes fragmentation mechanisms [1-6] is an actual problem because its solution will allow to modify their properties. Present models of the free fullerenes fragmentation are differed by role of projectile energy losses in elastic and inelastic collisions, and also by contribution of electronic and vibrational interactions in energy interchange in a fullerene after its collision with a particle. One of difficulties at the experiment results interpretation is absence of precise models for calculation of elastic and inelastic energy losses in collision of an ion especially multi-charged ion with a nanocluster. Quantum mechanical "*ab-initio*" calculations [7] have shown, that they essentially differ from predictions within the concept of the "stopping power" in solids. Apparently, the cause of it is an intermediate scale of these objects, which are great in comparison with atom, and are very small to be a crystal.

## 2. Model and methods

The purpose of this work is to study a role of elastic and inelastic energy losses in a fullerene fragmentation processes by using a simplified computer simulation of those situations where only one kind of losses dominates. Such conditions can be implemented on an experiment, but at computer simulation it is possible to trace separate contributions in many details. For this a computer simulation of gaseous $C_{60}$ molecules irradiation by electron and one-charged ion beams has been carried out. The most investigated interval of projectile velocities lies from 0.1 up to 2 a.u. (1 a.u.= $C/137$ = $2.188 \cdot 10^4$ A/ps) in the experiments [1-5] for fullerenes fragmentation by charged particles. In our calculations the

*T.N. Veziroglu et al. (eds.),*
*Hydrogen Materials Science and Chemistry of Carbon Nanomaterials,* 387-390.
© 2004 *Kluwer Academic Publishers. Printed in the Netherlands.*

electron energies were changed from 0.5 to 2 кeV, and velocities of the H, Li, C, Ar ions – in the above energy range. Calculations of the particle energy losses have been carried out within the concept of the "stopping power" for graphite whose electronic orbital are similar to those in a fullerene. Several analytical and empirical models of electronic stopping and nuclear scattering have been used. The obtained specific elastic and inelastic energy losses of ions and electrons then have been enumerated for an average energy, which can be transferred to the target atoms. The results obtained by us within the framework of various models differ among themselves up to 30 %. It is necessary to note that those semi-empirical models which yield good results for stopping in a solid can not yield good results for a nanostructure [7]. Thus we had received estimations for the averaged kinetic and electronic excitation energies of fullerene's atoms. Typical values of obtained quantities: a) the enumerated elastic losses of ions H, Li, C and Ar with velocities from 0.1 up to 2 a.u. are not higher 2, 18, 40 and 180 eV/atom, respectively; b) the enumerated inelastic losses of electrons with energies from 0.5 up up to 2 keV decrease from 10 up to 5 eV/atom. The electronic excitation of the fullerene atoms permanently increases with increase of a projectile's energy, and the its kinetic energy decreases. Velocities of ions for which elastic and inelastic losses are equal each other do not exceed 0.3 a.u.

Since the averaged transferred energies are small in comparison with initial energy of particles hence its trajectory can be presented by a straight line transited a fullerene in a random fashion. The time of flying a particle through a fullerene is less than 1fs that is much less than typical period of $C_{60}$ oscillations (more then 10 fs) and it allows to view the molecule's atoms as immovable during the collision. We chose random evenly distributed parameters of a particle trajectory and found its neighbour atoms. A random kinetic energy or/and energy of electronic excitation were given to each of them. We distributed these energies proportionally to differential cross section of scattering on the BZ potential [8] for ions and on the cut-off Coulomb potential for electrons. Thus we found the energy contribution of a projectile into the electronic and vibration systems of a fullerene after its collision.

Further we proportioned the electronic losses of projectile between the valence electrons of the excited atom by random way. We took into account only the ground, excited (with HOMO-LUMO transition energy) and ionization states. Using the obtained electronic states of atoms, the interaction potentials of atoms in the fullerene has been composed. Interaction of atoms in the ground state has been described by the Brenner potential [9] and for excited and ionized atoms we took into account the corresponding multipoles members. For the two ionized atoms it is the Coulomb repulsion of charges, for the two excited atoms it is a dipole-dipole interaction, and between the ionization or excited and neutral atoms it is a charge or a dipole and induced dipole interaction. We used a dielectric susceptibility of carbon atom in the ground state ($\alpha = 7.6$ a.u.) and the dipole moment of an excited atom of carbon ($p=1.385$ D).

The elastic losses were enumerated into the initial velocities of atoms, which are directed in correspondence to an impact parameter of the projectile-atom collision.

Finally we calculated an evolution of the impacted and excited fullerene by the molecular dynamics method. Evolution of the electronic subsystem has not been considered. The behaviour of a molecule was simulated on the interval time up to 300 fs.

# 3. Results and discussion

It was set, that efficiency of fullerene fragmentation (ratio of number of the fragmented fullerenes to the number of the irradiated ones) by an electron is lower on several orders than for a light ion, and this efficiency is decreased with energy faster than the total energy losses of electron. It can be explain by the fact that electrons of the above energies transmit energy as a rule in the inelastic channel and it carries out only to excitation and ionization of a fullerene. Since the HOMO-LUMO transition in the $C_{60}$ molecule is equal about 7 eV the efficiency of excitation and ionization and switch on of the corresponding strong Coulomb interaction is suppressed in the our model. We shall mark, that the averaged values of electron energy losses (about 10 eV/atom) are obtained after average of the all possible energy transfers on the sharp forward-directed differential cross section. Therefore the energy transfers with values in several tens and even hundreds eV can occur relatively frequently. They give the noticeable contribution to fullerene fragmentation. It was set also, that electronic excitation and ionization leads to that kind of fragmentation where all ionized and excited atoms are involved as well as their nearest neighbors. Thus the ionized atoms left the excited fullerene alone more often than into a composition of some fragment.

For the ions irradiation we had investigated the particle energy dependence of the following quantities: amount of fullerenes fragments, amount of the broken bonds, mass of the fragments, charge-state of fragments, the kinetic energy and direction of outgoing fragments. Some of these quantities can be measured in an experiment. Two spectrum of fragments mass distribution are presented on the Fig. 1.

They has been calculated for an irradiation of the free fullerene molecule by $Ar^+$ ions with energies 10 keV and 420 keV. They are obtained after calculation of fragments amount with all masses for about 1000 statistical trials for each ions energy. Trials differed by the trajectories and by the energy transferred to atoms on each of them. The obtained data transmit main features of these two situations for which the relation of elastic and inelastic losses are equal 3.3 and 0.33 correspondingly. From them it follows, that the basic yield of fragmentation under preference of the elastic energy losses is the dimers $C_2$ and the fullerene fragments $C_{60-2n}$ which remain after fragmentation. The fullerene destruction on the many lonely atoms dominates at fragmentation owing to the electronic excitation. It is stimulated by the Coulomb repulsion. Qualitatively these distributions are like to the data of [1-5], however there is also discrepancy which is essential in the interval of medial fragment mass. Their amount is obviously underestimated especially in case of electronic excitation. We think that it is linked with the fixing of electron excitation on the several separate atoms in our model. As a result the atoms leave a molecule very fast while the excitation reallocation on some group of atoms could delete larger atoms amount with smaller velocity.

390

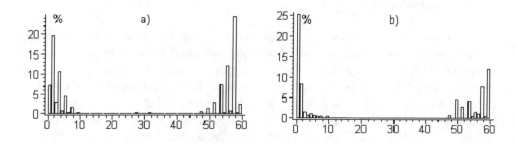

Fig. 1. Spectrum of fragments mass calculated for an irradiation of the free fullerene $C_{60}$ by $Ar^+$ ions with energies: a) E =10 keV (0.1 a.u.); b) E =420 keV (0.65 a.u.).

## 4. Conclusions

The proposed model gives qualitatively correct mapping of a fullerene fragmentation by particles. After improvement and fitting of some parameters under the experiment data it can be used for rather prompt calculations of fragmentation both the free fullerenes, and supported fullerenes and fullerite films, where *"ab-initio"* quantum mechanical calculations are meanwhile unavailable.

### Acknowledgement

This work was supported by INTAS grant № 2136.

## References

1. Nakai Y, Itoh A, Kambara T, Bitoh Y, Awaya Y. The fragment ion distribution of $C_{60}$ in close collision with fast carbon ions. J Phys B: At Mol Opt Phys 1997; 30:3049-58.
2. Opitz J, Lebius S, Tomota S, Huber BA, Moretto-Capelle P, et al. Electronic excitation in $H^+$-$C_{60}$ collisions: Evaporation and ionization. Phys Rev A 2000; 62:022705.
3. Bordenave-Montesquieu D, Moretto-Capelle P, Bordenave-Montesquieu A, Rentenier A. Scaling of $C_{60}$ ionization and fragmentation with the energy deposited in collisions with $H^+$, $H^+_2$, $H^+_3$, and $He^+$ ions (2-130 keV). J Phys B: At Mol Opt Phys 2001; 34: L137-46.
4. de Vries J, Hoekstra R, Morgenstern R, Schlathölter TJ. Cq+-induced excitation and fragmentation of uracil: effects of projectile electronic structure. J Phys B: At Mol Opt Phys 2002; 35:4373-81.
5. Reinköster A, Siegmann B, Werner U, Huber BA, Lutz HO. Multi-fragmentation of $C_{60}$ after collision with $Ar^{z+}$ ions. J Phys B At Mol Opt Phys 2002; 35:4989-97.
6. Campbell EEB, Rohmund F. Fullerene reactions. Rep Prog Phys 2000; 63:1061-109.
7. Kunert t, Schmidt R. Excitation and fragmentation mechanisms in ion-fullerene collisions. Phys Rev Lett 2000; 86(23):5258-61.
8. Ziegler JE, Biersack JP, Littmark J. The Stopping and Range of Ions in Solids, vol 1, New York: Pergamon Press. 1999.
9. Brenner DW. Empirical potential for hydrocarbons for use in simulating the chemical vapor depositions of diamond films. Phys Rev B 1990; 42(15):9458-71.

# LITHIUM IN NANOPOROUS CARBON MATERIALS PRODUCED FROM SiC

I.M.KOTINA [(1)], V.M.LEBEDEV, A.G.ILVES, G.V.PATSEKINA,
L.M.TUHKONEN, A.M.DANISHEVSKII[(1)], S.K.GORDEEV[(2)],
M.A.YAGOVKINA[(3)]
*Petersburg Nuclear Physics Institute, Gatchina 188350, Leningrad district,
Russia*
[(1)] *Physico-Technical Institute, St.Petersburg 192037, Russia*
[(2)] *Central Science Research Institute for Materials, St.Petersburg 191014,
Russia*
*(3) "Mechanobr-Analyt" Co, St.Petersburg 199026, Russia*
*E-mail address:kotina@pnpi.spb..ru (I.M.Kotina)*

## Abstract

Bulk nanoporous carbon materials (NPCMs) obtained from SiC by chlorination have a
high total porosity up to 70 and nanoporosity about 50%. These materials have been
intercalated with lithium at the temperature range 30-200 °C by vacuum deposition and
diffusion. The major analytical tools for study included neclear reaction method and X-ray
diffraction. The diffusion coefficient of lithium was found to be a function of the diffusion
duration. The effect of the interaction of lithium with pore walls and intercalation process
on the value of $D_{Li}$ is discussed. It was found that the phase composition lithiated samples
depends on the ratio between the deposition and diffusion duration. The condition for
producing the samples uniformly impregnated by Li have been determined. Time stability
obtained NPC:Li composites have been studied.

**Key words**: bulk nanoporous carbon, lithium, insertion, diffusion coefficient, phase
composition, intercalation, cluster.

## 1. Introduction

Carbon materials with high–developed surface look promising for the use in lithium
rechargeable batteries [ 1-2 ]. Nanoporous carbon materials (NPCM) produced from
carbides have high open nanoporosity [ 3 ] that makes possible a sufficiently high lithium
diffusion coefficient. Therefore investigation of the lithium intercalation process in such
materials is of particular interest. Of fundamental importance is investigation of lithium
intercalated nanoporous carbon materials for an understanding of their structural
characteristics.

In this report, the results of study of intercalated bulk nanoporous carbon materials
(NPCMs) with lithium are presented. The following questions have been studied: the
lithium diffusion mechanism, relationship between the structure of the initial samples and
the phase composition of the NPCM:Li composites, their stability.

*T.N. Veziroglu et al. (eds.),*
*Hydrogen Materials Science and Chemistry of Carbon Nanomaterials, 391-398.*
© 2004 Kluwer Academic Publishers. Printed in the Netherlands.

## 2. Experimental

Bulk NPC samples used in this study were produced from an intermediate product on the base SiC powders by a high temperature chlorinating process [ 4 ]. We used two type NPC samples: C/SiC/ B and C/SiC/A. All have high total porosity - up to 70% but are distinguished by nanoporosity and pyrocarbon (PC) content. So, the samples of C/SiC/ B type contain ~10 wt.% pyrocarbon and have nanoporosity ~30%.; the samples of (C/SiC/A) type do not contain PC and have nanoporosity ~40%. It should be noted that nanoporosity was determined by impregnating with benzene.

First to minimize the effect of adsorbed water the samples being studied were vacuum annealed at 300 °C for 70 h. Lithium insertion was carried out via vacuum evaporation and subsequent diffusion at temperatures ranging from 30 to 200 °C. After cooling the lithiated samples were kept in an atmosphere of dry nitrogen.

The total quantity of lithium evaporated on NPC samples was monitored with a Si sample. It is known that in this temperature interval the diffusion of lithium into silicon does not essentially take place [5]. In such a situation the total quantity of evaporated Li is calculated by integration of the Li concentration profile near surface. The nuclear reaction $7Li(p,a)^4He$ was used to measure the concentration profiles of lithium [ 6]. From the experimental diffusion profiles one can calculate the diffusion coefficient D by using the solution of the Fick's diffusion equations [ 7 ]. In the one-dimensional case for a semiinfinite crystal with a constant surface concentration of diffusing atoms under the assumption that the diffusing atoms do not interact with each other, the solution to these equations is:

$$N(x) = N_0 \left[ 1 - erf\left( \frac{x}{2\sqrt{Dt}} \right) \right] \tag{1}$$

where No and Nx are concentrations of Li atoms on the surface and at distance x from the surface, D is diffusion coefficient, t –time.

It was not infrequently impossible to give our experimental curves of lithium concentration profile by the equation (1) with a single D value. Because of this, we estimated DLi , using Eq. (1) in 10 points of the experimental curve N(x) and then calculated the average value of the of the diffusion constant <DLi>. Experimental results show that the value of <DLi> determined this way might be used as a characteristic parameter for the lithium diffusion process in the different NPCMs being studied at the same experimental conditions.

## 3. Results

It turned out that the value of <DLi> in the C/SiC/A samples (type A) is less than the one in the C/SiC/ B samples (type B) in the temperature range under study (Table1) .

In addition, it has been found that the variation in the value of the lithium diffusion coefficient as a function of diffusion duration depends on NPC type (Table2).

TABLE 1. Values of diffusion coefficients in the samples of NPC. Diffusion time 5 min

| Sample | Number | T, 0C | $<D_{Li}>*10^9$ cm$^2$/sec |
|--------|--------|-------|----------------------------|
| C<SiC>B | 2 | 30 | 6.6 |
| C<SiC>A | 24 | 30 | 3,5 |
| C<SiC>B | 2 | 80 | 33 |
| C<SiC>A | 24 | 80 | 2.7 |
| C<SiC>B | 21 | 100 | 21 |
| C<SiC>A | 14 | 100 | 7.7 |

TABLE 2. Diffusion coefficients at different durations of diffusion process. Temperature of diffusion 80$^0$C.

| Sample | Number | $<D_{Li}>*10^9$cm$^2$/s $t_{dif}$ =5' | $<D_{Li}>*10^9$cm$^2$/s $t_{dif}$ =10' | $<D_{Li}>*10^9$ cm$^2$/s $t_{dif}$ =20' |
|--------|--------|------|------|------|
| C<SiC>B | 11 | 33.4 | 23 | 4.1 |
| C<SiC>A | 24 | 2.7 | | |
| | 13 | | 5.4 | 2.4 |

From Table 2 we can see that an increase of the diffusion duration from 10 to 20 min. results in a decrease of <DLi> in the C/SiC/ B sample by a factor of approximatly 6. At the same time in the C/SiC/ A sample the <DLi> value varies less than by a factor of 2 in this time interval.

Structure of NPCM:Li composites was studied by X-ray diffraction method using radiation with wavelength 1.789 Å. Lithium was introduced into NPC samples being studied at a temperature of 80°C with diffusion duration of 40 minute.

Several reflexes conditioned by phases of $Li_2C_2$ , $Li_2CO_3$ and by the Li-graphite intercalation compounds were observed on the X-ray diffraction patterns. To estimate the graphite interlayer distance and the lithium intercalate sandwich thicknesses we used the expansion of complex peaks on the pattern in the angle interval $2\theta$ =16 – 35 ° . It should be noted that peaks resulting from this expansion were broad. The relative content of the intercalation and graphite phases were determined by the integrating of the corresponding peaks. Therewith, the total content of these phases in the sample under study was taken as 1. Obtained data are shown in Tabl.3.

TABLE 3. Phase composition lithiated samples. Temperature of diffusion $80^0C$, diffusion duration 40 min.

| Literature data | | | C/SiC/B:Li | | C/SiC/A:Li | |
|---|---|---|---|---|---|---|
| X in LixC6 | Stage index | d (0002), Å | d (0002), Å | Relative intensity | d (0002), Å | Relative intensity |
| 0 | 0 (graphite) | 3.355 | 3.38 | 0.1 | 3.42 | 0.75 |
| 0.25 | 3 | 3.46 | 3.48 | 0.4 | | |
| 0.5 | 2 | 3.53 | | | | |
| 1.0 | 1 | 3.706 | 3.62 | 0.5 | 3.71 | 0.25 |

One can see from the table that much of the graphitized clusters are intercalated in the C/SiC/B sample. As this takes place, the intercalation phases of different stage index (1 and 3) coexist in the C/SiC/B:Li composite. Let us note first of all that the graphite interlayer distance as well as the third-stage interlayer distance in the lithiated C/SiC/B sample is more than one in graphite, as the first-stage interlayer distance is less. In the lithiated C<SiC>A there is a small share of graphitized clusters, which are intercalated as LiC6. The attention has been given to the fact that in this case the first- stage interlayer distance is more than one in the intercalated C/SiC/B. As it seen from Table3, the total relative content of intercalation phases in the C/SiC/B sample is considerably more than one in the C/SiC/A sample. Afterwards we used the C/SiC/B material to create NPC-Li composites.

For the purpose to obtain NPC-Li composites with Li uniform distribution we investigated the effect of the Li deposition rate and the diffusion duration on the Li concentration profile. We have also examined the phase composition of the obtained composites. It was stated that ratio between deposition and diffusion rates defines the lithium concentration profile as well as the phase composition of the lithiated samples. It should be also noted that Li insertion in the samples did not result in a pure single-stage compound as determined by XRD patterns. Coexistence of several phases was always observed. Some of the results are shown in Fig.1. In this case the sample was 1mm in thickness, Li have been evaporated during 20 min and the duration of lithium diffusion process was 30 h. Lithium concentration profiles and XRD patterns were measured on both sides of the sample under study.

As we see in Fig.1, from the front side of the sample (pattern a) intercalation phases $LiC_6$, $LiC_{12}$, $LiC_{24}$ and large volume of carbonate $Li_2CO_3$ and carbide $Li_2C_2$ phases are registered. At the same time only small quantity of $Li_2CO_3$ and $Li_2C_2$ phases is seen from the backside of the sample (b). It turned out that Li concentration gradient is available in the sample at this condition of the Li impregnation. No evidence of $Li_2CO_3$ and $Li_2C_2$ phases existence was obtained when the sample was polished from the front side at the depth of 0,4 mm. However, the intercalation phases $LiC_6$, $LiC_{12}$. $LiC_{24}$ are observed in this case. It is beleived that the formation of $Li_2CO_3$ and $Li_2C_2$ phases are connected with the lithium accumulation in pores followed by the formation of Li clusters. The kinetic of this process is apparently determined by the relation between the Li evaporation time and the diffusion duration . Therefore we used pulse evaporation process with a duration of the evaporation pulse much less than the interval between pulses in order to avoid of Li accumulation in pores. This provided an opportunity to produce the samples uniformly impregnated by lithium and containing mainly intercalation phases.

In using NPC:Li intercalation compounds, the important property of them is the time stability. We investigated the stability of obtained NPC:Li composites. The change in relative content of $Li_2CO_3$ phase in time was taken as a measure of the composite's stability. It was found that the stability of obtained composites is determined both by preparation and storage conditions. Some of obtained data are tabulated in Table4.

We can see that as the sample is stored in the execator with absorbent, the average rate of the formation of $Li_2CO_3$ phase is less than one when it is stored at air. In other words, being stored in the execator the NPC:Li composite keeps its substance for longer time. It should be noted that the stability of the NPC:Li composite depends also on the concentration of lithium inserted in the NPC sample. As we see from Table4, the average rate of $Li_2CO_3$ phase formation in the time of the storage at air is more in the factor of 4 at front side where the Li concentration is $2.7*10^{21}$ cm-3 than one at the back side where Li concentration is $8.5*10^{20}$ cm-3 .

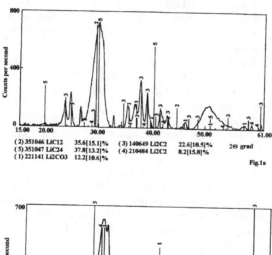

Figure 1. X-ray diffractional patterns of lithiated C/SiC/B sample for front side (a) and back side (b). Diffusion time 30 hours.

## 4. Discussion

We have produced the insertion of lithium in the different bulk NPCMs via vacuum deposition and diffusion. The results presented in Table 1 show that in the NPCM of type A the value of <DLi> is less than one in the NPCM of type B at the diffusion duration of 5 min..

In the NPCMs diffusion of lithium can occur along pores and within the Van der Waals space between every pair carbon sheets in the quasi-graphite clusters. On the diffusion of lithium along pores a resistance to the flux of the diffusing atoms arises due to the interaction between the Li atoms and pore walls. The volume of nanopores in the C/SiC/A samples is larger than one in the C/SiC/B samples and nanopore size on the average is smaller [3]. As a result, in the A type samples the total wall surface is larger and hence such resistance is larger. This factor can define the less value of <DLi> in the C/SiC/A samples. Besides, relative volume of ordered and enough big clusters in the samples of A type is less than one in B type samples. Thus the Li diffusion within the Van der Waals space also provides a less value of <DLi> in the C/SiC/A samples. However consideration must be given to the attractive interaction between lithium atoms located in the intercalation sites inside of Van der Waals spaces followed by the formation of staged phases. It will be inhibit the diffusional motion of lithium and thereby will cause the diffusion coefficient to a decrease. With consideration the foregoing it is expected that considered effect will be more pronounced in the C/SiC/B samples. This assumption is supported by experimental results. As it seen from Table 2, the <DLi> value decreases with an increase in the diffussion duration in the both samples. But an increase of the diffusion duration from 10 to 20 min. results in a decrease of <DLi> in the C/SiC/B sample by a factor of approximatly 6, and at the same time in the C/SiC/ A sample the <DLi> value varies less than by a factor of 2 in this time interval. The attention has been given to the fact that at an increase of the diffusion duration from 5 to 10 min the value of diffusion coefficient in the C/SiC/B sample decreases only slightly. Apparently the value of <DLi> is determined by the interaction diffusing atoms with pore walls, when a diffusion duration is too short that the intercalation process manifested itself. One can think that such situation prevails in the case of 5 min. diffusion duration. Thus one or another mechanism of diffusion effects on the value of <DL>i according to the diffusion duration.

As shown in Table3, the relative content of intercalation phase is more in the C/SiC/B sample than one in the C/SiC/A sample (75% and 25 % respectivly). This is consistent with a large volume of graphite-like clusters in the C/SiC/B sample compared to C/SiC/A one. As this takes place, obtained values of the interlayer graphite distances indicate that graphite-like clusters is more ordered in the C/SiC/A sample. But in both samples clusters are less ordered than ones in graphite. It is of interest that the interlayer distance of first stage compound in the C/SiC/B sample is considerably less than one in graphite. It is likely that this is caused by the fact that the formation of this stage does not whilst still complete.

In some cases we registered $Li_2CO_3$ and $Li_2C_2$ phases. The measurments of the Li concentration profiles and phase composition in the obtained NPC-Li composites showed that the formation of $Li_2CO_3$ phase is strongly dependent on concenration of lithium in the investigated samples (Fig.1). We suppose that this fact is associated with the interaction lithium atoms with each other in pores. Such interaction gives rise to the accumulation lithium inside the pores followed by the formation of metal clusters. Then avalaible Li clusters react with atmospheric $CO_2$ that is the cause of $Li_2CO_3$ phase formation. As for $Li_2C_2$ phase, it is believed that the case of its formation is a direct interaction between impregnated Li atoms and host atoms of NPC. This assumption is based on our observation that the formation of $Li_2C_2$ phase takes place at the larger, as distinct from the formation of $Li_2CO_3$ phase, Li concentration in NPC hosts.

We conducted lithium pulse evaporation with the pulse duration rather less than the interval between the pulses to restrict the formation of Li metallic cluster in the pores from which afterwards $Li_2CO_3$ phase arises. This enable us to obtain lithiated samples of NPC with only slightly content of $Li_2CO_3$ and $Li_2C_2$ phases.

As we see in Table 4, a time stability of the lithiated NPCMs depends on the concentration of the inserted lithium and storage conditions. The larger the Li concentration, the less the stability of obtained composites. It is believed that rate of the formation of Li clusters during storage controls the stability. The larger Li concentration, the larger rate of the formation of the Li clusters and therefore the less the stability of lithiated samples. Storage in the execator with absorbent increases stability because in this case the content of water vapour decreases.

TABLE 4. Relative content of $Li_2CO_3$ phase in the sample

| Side of the sample | After three weeks in the execator with absorbent | After three weeks in the execator and 2 weeks at air | The average rate of the formation $Li_2CO_3$ in the execator | The average rate of the formation at air | Li concentration cm-3 |
|---|---|---|---|---|---|
| front | 7% | 15% | 2.3% in week | 4.0% in week | $2.7*10^{21}$ |
| back | 4% | 6% | 1.3% in week | 1.0% in week | $8.5*10^{20}$ |

## 5. Conclusions

1. Introduction of lithium into NPCMs by evaporation and subsequent diffusion makes possible obtaining of stable lithium-carbon composites which contain intercalation phases in the main. This encourages NPCM to be used as Li containers.

2. Given method of Li introduction with the use of XRD phase analysis makes it possible to conduct a rapid meaningful comparison between the various NPCMs, to reveal if there are ordered graphitized clasters in them and thereby to evaluate suitablitity of NPCM samples for Li containers and for rechargable Li batteries.

3. Lithium diffusion in NPCM occurs along pore walls and within the Van der Waals space between every pair carbon sheets in the quasi-graphite clusters. One or another mechanism of diffusion effects on the value of $<DL>i$ according to the diffusion duration. The value of $<DLi>$ is big enough.

4. Different intercalation phases appear depending on NPCM carbon skeleton structure, duration of diffusion process and concentration of introduced Li.

398

The work was supported by Intas grant: 00-761 and Russian program "Controllable synthesis of fullerenes and other atomic clusters".

## 6. References

1. Buiel E., Dahn J.R., J.Electrochem. Soc. 1998, 145, 1977.
2. Buiel E., Dahn J.R., Electrochimica Acta 1999, 45, 121.
3. Kyutt R.N,. Smorgonskaya E.A, Danishevskii A.M, Gordeev S.K, Grechinskaya A.V, Phys.Solid State 1999, 41, 1484 (in Russian).
4.Gordeev S.K,.Vartanova A.V, Zh. Prikl. Khim. 1994, 66 , 1080 (in Russian).
5. Semiconductor Detectors, ed.by G.Bertolini and A.Coche, (1968), North-Holland Publishing Company-Amsterdam, (1968) 33.
6. I.M.Kotina, V.M.Lebedev, G.V.Patsekina, Proceedings of International Conference on Nuclear Physics "Clustering phenomena in nuclear physics" (50 Meeting on Nuclear Spectroscopy and Nuclear Structure, June 14-17, 2000, St-Petersburg, Russia), St-Petersburg, (2000), 396.
7. B.I.Boltaks, Diffusion in semiconductors, edit. H.J.Goldsmit, Acsdemic Pres, NY, 1963 p.93-128

# SCIENTIFIC-TECHNICAL PREREQUISITES IN UKRAINE FOR DEVELOPMENT OF THE WIND-HYDROGEN PLANTS

V.A. GLAZKOV, A.S. KIRICHENKO, B.I. KUSHNIR, V.V. SOLOVEY,
A.S. ZHIROV, D.V. SCHUR
*State Design Office "YUZHNOYE", 3 Kryvoriska St.,*
*Dnepropetrovsk, 49008, Ukraine*
*Institute of Mechanical Engineering Problem of the National Academy of*
*Science of Ukraine, 2/10 Pozharsky St., Kharkov, 61046, Ukraine*
*Institute for Problems of Materials Science of NAS of Ukraine,*
*3 Krzhizhanovsky St., Kiev, 03142, Ukraine*

## 1. Introduction

Analysis of world-wide trends to windpower engineering development makes it possible to conclude that this ecologically clean and renewable type of power engineering will become one of the key sources for satisfaction of energy needs of the world community in the immediate future. In spite of all windpower engineering intriguing capabilities one cannot but emphasize a drawback peculiar to it in connection with irregularity of power produced and so makes it necessary to search for rational technologies providing power production in periods of time when wind load is not available. To solve this problem, it is necessary to build up a system, which provides energy storage with following consumption.

The wind-hydrogen plant (WHP) is one of possible variants of such system.

The concept of WHP based on two basic directions of use of hydrogen as energy carrier in power engineering:
1. Energy accumulation in independent uninterrupted power supply systems.
2. Increase of economic efficiency of power stations by accumulation of superfluous energy in hydrogen and use of this energy at o'clock "peak" and emergencies.

According to it within the framework of the STCU project Uzb-23j we develop:
1. Energy-technological complex (ETC) of moderate power (200-250 kW) independent use for the areas removed from industrial networks, providing uninterrupted power supply of consumers in areas with sufficient wind potential.
2. ETC of high power (500-600 kW) universal use for accumulation of superfluous energy in hydrogen and use of this energy according to schedules of consumption.

Let us note that similar complexes are developing by the Norwegian experts for power supply of the manned islands removed from centralized networks. Creation of such complex on island Utsira, that will provide with the electric power only due to use of no conventional renewed sources of primary energy is offered.

The scientific and technical preconditions created by State Design Office "YUZHNOYE" and Institute of Mechanical Engineering Problem of the National Academy of Science of Ukraine at performance of conversion works in area of wind-hydrogen power engineering are put in a basis of our development.

*T.N. Veziroglu et al. (eds.),*
*Hydrogen Materials Science and Chemistry of Carbon Nanomaterials,* 399-404.
© 2004 *Kluwer Academic Publishers. Printed in the Netherlands.*

400

Works under project Uzb-23j are not completed. Preliminary results on problems of creation the autonomous wind-hydrogen plant (AWHP) representing the greatest practical interest now are offered your attention.

Basic scheme of AWHP submitted in Figure 1.

AWHP provides use of distillate in the convertible closed cycle.

Creation of such complex demands the decision of some scientific and technical problems, both under AWHP, and under its basic parts.

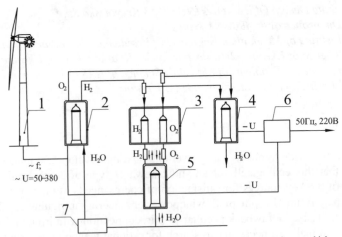

1 - Wind turbine; 2 - Electrolyzer; 3 - Products storage system (hydrogen and oxygen high pressure balloons); 4 - Electrochemical generator (fuel cell); 5 - Convertible electrolyzer; 6 - Invertor; 7 - Water supply system;

Fig. 1a - Autonomous wind - hydrogen plant( (AWHP) principle scheme (hydrogen - oxigen version)

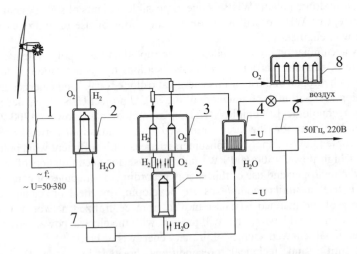

1 - Wind turbine; 2 - Electrolyzer; 3 - Products storage system (hydrogen and oxygen high pressure balloons); 4 - Electrochemical generator (fuel cell); 5 - Convertible electrolyzer; 6 - Invertor; 7 - Water supply system; 8 - commercial product

Fig. 1b - Autonomous wind - hydrogen plant( (AWHP) principle scheme (hydrogen - air version)

## 2. Results and discussion

At present, Institute for Problems in Machinery of NAS (Ukraine) and "Yuzhnoe" SDO have developed the project wind-hydrogen plant (WHP) with power of 200–600kW and hydrogen storage (accumulator). The wind-hydrogen plant is intended for conversion of wind energy (at a wind speed $\geq$ 3 m/s.) into AC electric power (voltage – 220/380V, frequency- 50Hz) and production of ecologically clean energy carrier, i.e. hydrogen, as well as gaseous oxygen as a commercial product.

## 3. AWHP Conception

The fuel cells using hydrogen as fuel put in a basis of autonomous wind-hydrogen plant. Advantages of the plants using the fuel sells are:
- the high efficiency reaching 70 - 75 %;
- absolute ecological cleanliness;
- noiselessness;
- high power consumption.

This advantages are confirmed with experience of the fuel cells use in spacecrafts ("Gemini", "Apollo", «Space Shuttle» - the USA; "Buran" - the USSR), submarines (a German submarine of type 212), overland vehicles.

Direct connection of wind turbine to electrolyzer that does not need «qualitative» electric power puts in the concept of AWHP. It opens the big opportunities for creation of AWHP simple on a design and operation.

A technical opportunity of creation high-pressure electrolyzers, capable to work in a mode of fuel cells and convertible modes allows choosing the following optimum AWHP scheme:
- three electrolyzers of 35 kW capacity working in a mode of the hydrogen generator;
- three electrolyzers working in a mode of the electrochemical generator (fuel cell);
- three electrolyzers working in convertible mode.

This scheme makes possible because of the analysis of situation to connect convertible electrolyzers for accumulation or use of hydrogen.

## 4. Wind turbine conception

In a basis of the wind turbine (WT) concept - application of the synchronous generators providing independent (without a network) work. It is offered WT with two generators, each of which has individual capacity smaller in comparison with rated WT power. It allows to increase WT energy output on small winds. With the same purpose, WT operating mode at the fixed pitch angle of blades with variable frequency of rotor rotation is chosen.

At work only on electrolyzer practically all electric power, developed by wind turbine on small winds, is used for production of hydrogen. It is advantage of WT with hydrogen accumulation of energy in comparison with traditional work of WT in a network, at which the electric power produced is not used up to achievement of synchronous frequency of rotor rotation.

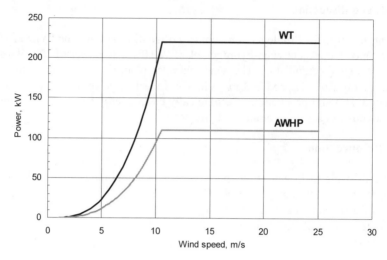

Fig. 2. Power vs. wind speed

## 5. Electrolyzer conception

The proposed design of electrolyzer uses new method of separation of gases (hydrogen and oxygen) in time i.e. the process of work of electrolytic system becomes cyclic and consists of alternative cycles of hydrogen and oxygen release.

Separated in time the processes of gas release are possible in the case of accumulation of one of the water electrolysis products in electric-chemically active compound occurring in electric-chemical cell in liquid or solid phase (active electrode). It does not result in considerable volume change of this compound and allows to get the second component at the passive electrode in a form of gas without any dividing membranes. Then with a change of polarity the cycle of the accumulated component release takes place on the electrodes. With this the gas pressure can be limited only by strength of constructional elements and a threshold of gas solubility in electrolyte. In practice the real level of pressure does not exceed 70.0 MPa. The proposed variant of electrolyzer would provide production of 5 nm$^3$ hydrogen per hour under pressure of 120 atm without compressor.

Stable current of 2.2 kA m$^3$ per hour provides powering of cell 7. In this case voltage on electrodes depends on the degree of oxidation (restoration) of active electrode mass and it is defined by both inner resistance of electrodes and inner EMF. The electrolyzer fuel cell working temperature fluctuates in limits of 353-373 K and does not demand forced cooling. Control for electrolyzer work parameters is realizing by specially developed electronic control block.

The advantage of hydrogen accumulator in use is that it can store hydrogen at a high pressure and generate electric power, working as a fuel cell in the absence of wind. Hydrogen produced can find the following use:
- heating green-houses by catalytic burning of hydrogen;
- provision of working medium applied to hydrogen-oxygen torches for soldering, welding, cutting and heat treatment of metal products;
- heating of lodging spaces with catalytic heaters;

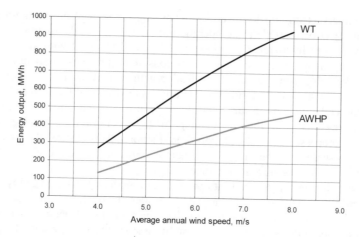

Fig. 3. Energy output vs. average annual wind speed

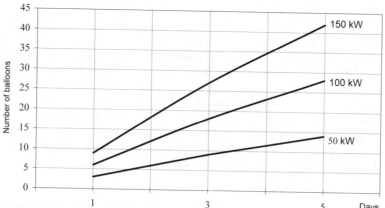

Fig. 4. Number of balloons with hydrogen for uninterrupted power supply under calm conditions

- as an engine fuel;
- as a filling gas for sounding balloons;
- as fuel for electrochemical generation (fuel cell).

This work describes the method, which allows to keep activity of the electrode materials at a sufficiently high level and to extend the list of the materials suitable for design of fuel cells. This method is connected with practical implementation of keeping the activity of the electrode materials at the given level. This is achieved by setting the operation regimes providing self-regeneration capability of the electro-chemical systems.

As a porous electrode, this system uses the spongy metal deposited on a perforated or a mesh substrate in order to keep the shape of the porous electrode as needed. The merit of the porous deposits is their higher activity as compared to the solid-phase dispersed powder to be in common use for electrode design.

The specific feature of the proposed technology is that the fuel cell uses the electro-chemical regeneration system making it possible to move away (dissolve or transfer it onto the auxiliary electrode) followed by depositing the sponge on the main electrode including the electrolyte used during operation of the fuel cell in the routine regime.

The oxygen and hydrogen electrodes are fabricated of metal with activity meeting both fuel components. The electrode design considered presents mesh tubular frames made of nickel. Between the active electrodes with spongy coating are arranged the passive electrodes fabricated of the same material as the frame of the active electrodes.

When the cell functions in the electrolyzer regime, the gas emission on the active electrode is prevented from the metal sponge of the electrode frame owing to running the reversible processes viz. «dissolution – production». This ensures the basic concept of the high-pressure hydrogen generator characterizing in that the cycles of hydrogen and oxygen liberation are separated in time.

To transfer the electrolysis cell into the fuel cell mode the thicker spongy coating is electrochemically produced on the active electrodes. In operation of the system this coating needs lesser current densities. In the course of metal spongy deposition the passive mesh electrode is used as an insoluble anode. On completion of the sponge coating process the switching of oxygen and hydrogen electrodes is conducted. The passive mesh electrode is cut off from electrical lines and not used when system functions in the fuel cell mode.

At regular intervals as the activity of the spongy electrodes reduces, the system is transferred to the steady state mode of operation. The process of complete dissolution of the spongy mass and deposition of the new one takes place. The metal mass used for preparation of the spongy mass is reduced to a minimum providing thereby saving of time and energy consumption for electrode regeneration.

The system functions under high pressure (15 MPa) and is equipped with a special-purpose control system ensuring the reliable and safe operation of the plant.

The proposed plant compares favorably to the known wind-driven power plants of the similar capacity offered in the world market as to its engineering standard, simplicity of assembly and attendance, reliability and safety.

## 6. Resume

Use of the wind-driven power plant WPP-200-600 kW will provide annual saving of up to 330-970 ton of coal or 160-500 thousand $m^3$ of natural gas. The proposed plant will reduce environmental pollution with harmful smoke emissions formed during burning of organic fuel.

The plant may be connected to the centralized power-supply system or to operate in self-sustained regime.

## 7. References

1.  Solovey, V., Zhirov, A., Shmal'ko, Yu., Lototsky, M. and Prognimak, A. (2001) A reversible electrolyzer-fuel cell for autonomous power plant using renewable energy sources. HYPOTHESIS IV. Hydrogen Power – Theoretical and Engineering Solution International Symposium. Stralsund-Germany, 354-357.
2.  Zhirov, A.S., Solovey, V.V. and Makarov, A.A. Ustrojstvo dlya polucheniya vodoroda. Patent Ukrainy N 29852A 1998.
3.  Zhirov, A.S., Solovey, V.V., Plichko, V.S. and Makarov, A.A. Ustrojstvo dlya polucheniya vodoroda i kisloroda vysokogo davleniya. Patent Ukrainy N 29853□ 1998.

# SYNTHESIS AND STRUCTURAL PECULIARITIES OF THE EXFOLIATED GRAPHITE MODIFIED BY CARBON NANOSTRUCTURES

SEMENTSOV Yu.I.[(1)], PRIKHODKO G.P.[(1)], REVO S.L.[(2)],
MELEZHYK A.V.[(3)*], PYATKOVSKIY M.L.[(3)], YANCHENKO V.V.[(3)]
[(1)]Institute of Surface Chemistry of NAS of Ukraine, 17, Generala Naumova Str., Kyiv, 03164, Ukraine
[(2)] Kyiv National University by Taras Schevchenko, 2, Glushkova prospect, build.1, Kyiv, 03127, Ukraine. E-mail: revo@phys.univ.kiev.ua
[(3)] TMSpetsmash Ltd., 2A, Konstantinovskaya Str., Kyiv, 04071, Ukraine.
* e-mail: olex-ndr@nbi.com.ua

## Abstract

Carbon-carbon composite materials consisting of exfoliated graphite (EG) support and carbon nanotubes grown on its surface have been synthesized by catalytic chemical deposition method. Acetylene-hydrogen-argon gas mixture at 600°C was used as source of carbon. Nickel and iron compounds were investigated as catalysts in this process. The most catalytically active EG support was obtained by impregnation of intercalated graphite with aqueous ferric chloride solution with subsequent drying and thermal expansion of graphite matrix at 900-1000°C. It was detected by XRD method formation of iron crystallites under reductive conditions at 600-900°C. The growing of carbon nanotubes results fist in substantial increase of surface area, from 30-35 $m^2/g$ for initial EG support to 110 $m^2/g$ at 35% mass content of the grown carbon. However sample with 85% content of deposited carbon has surface area only 42 $m^2/g$. After purifying this sample from amorphous carbon surface area increases to 233 $m^2/g$. If effective catalyst is used the EG support is convenient for carbon nanotubes growing due to very big pore volume.

## Keywords:

Exfoliated graphite; Catalytically grown carbon; Carbon/carbon composites; Chemical vapor deposition; Scanning electronic microscopy; X-ray diffraction.

## 1. Introduction

Carbon-carbon composite materials (CCCM) are considered to be indispensable for use in advanced technologies due to the combination of low density with the unique thermal, mechanical, biomedical properties. In spite of the fact that these material are developed long ago and are used in different areas of technology, this field of science and technology continues to develop dynamically. New possibilities in designing carbon structures and materials have appeared due to discovery of fullerenes, carbon nanotubes,

T.N. Veziroglu et al. (eds.),
Hydrogen Materials Science and Chemistry of Carbon Nanomaterials, 405-414.
© 2004 Kluwer Academic Publishers. Printed in the Netherlands.

and different forms of nanoporous carbon. Adjusting of growing conditions and catalyst composition together with modifying of surface with different functional groups makes it possible to obtain composite carbon materials with necessary physical and structural parameters such as pore geometry, conductivity, different nature of surface, texture, macrostructure [1-5]. On the basis of these materials it is possible to create the supercapacitors, electrochemical energy storage devices, nanoelectronic devices, selective adsorbents, materials for hydrogen storage, articles with high mechanical parameters.

Our study is devoted to synthesis and study of CCCM based on thermally exfoliated (exfoliated) graphite and carbon nanostructures.

CCCM based on thermally exfoliated graphite (EG) are known for a long time [6]. When pressed into compact body EG exhibits unique elastic-plastic characteristics. Introduction of carbon fibers into these materials enlarges their strength parameters in 3-5 times [7].

Surface area of EG is relatively small (20-50 $m^2/g$, usually near 30 $m^2/g$), but due to very large specific pore volume EG is an oleophilic adsorbent with extremely high capacity [8]. Because of use of EG as adsorbent in medicine and ecology it would be very important to increase its surface area. One possible way could be syntheses of carbon nanostructures on the surface of EG particles. So, in [9-11] new carbon material was proposed, representing itself a mixture of EG and carbon nanocrystals (nanotubes), as efficient adsorbent for removing oil-products contamination from water, as well as removing toxic substances from blood, and curing of skin diseases. This material is formed at explosive decomposition of graphite intercalated with perchloric acid and similar highly energetic compounds [9]. The idea of creation CCCM combining high pore volume of EG with high surface area of carbon nanotubes (CNT) appears to be very promising. However, the preparation method proposed in [9] possesses a little opportunity for varying the structure of these materials. More broad possibilities for designing carbon nanostructures on EG surface could be realized by growing of carbon nanostructures on EG surface, for instance by catalytic hydrocarbon decomposition method.

The purpose of our work is to study the possibility of EG surface modifying with carbon nanodimensional structures (nanofibers, nanotubes, amorphous carbon) as a result of chemical vapor deposition under different regime and creation of carbon nanodimensional structures (nanofibers, nanotubes, amorphous carbon).

## 2. Experimental

Crystal structure, morphology and surface area of EG were studied by methods of XRD, scanning electron microscopy, transmission electron microscopy. BET surface area of samples was determined by the argon desorption method. The EG modifying was done by pyrolysis of carbon containing substances (acetylene, benzene, methylene chloride) in argon or argon-hydrogen mixture. Iron, nickel, and cobalt chlorides, acetylacetonates, formate-acetates were studied as catalysts precursors for nanodimensional carbon growth. The compounds named above were embedded into electrochemically intercalated graphite, with subsequent thermal expansion of treated graphite (thermal shock) at 900-1000°C. In another method, EG was impregnated with solutions of above-mentioned compounds with subsequent thermal treatment at moderate temperature (500-600°C) sufficient for conversion of the catalyst precursors to metals and/or metal oxides. Introducing of alumina and zinc oxide into EG was performed by impregnation of electrochemically intercalated graphite with aluminum and zinc nitrates with subsequent drying and thermal expansion at 900-1000°C. The phase composition of the catalysts introduced was checked by XRD method.

## 3. Results and discussion

The EG flake electron microscope image is shown in Fig 1.

Fig. 1. EG flake (x100000).

Elongated (cylindrical-like) formations 20-30 nm in diameter are easily seen. Apparently, such image appears when one or several graphite planes are corrugated or rolled up. Obviously that presence of such cavities could be a reason of the restriction of pressed EG density which does not exceed 2.0 g/cm$^3$ (X-ray density of graphite is 2.24 g/cm$^3$) and provides the springy features of such material up to maximum possible density.

However, probably content of carbon present in these "semi-tubular" nanostructures is too small for manifestation of adsorption characteristics typical for CNT.

In connection with aforesaid it is interesting to note also highly dispersed carbon which forms at exothermal decomposition of dry graphite oxide. This carbon is characterized by surface area 500-600 m$^2$/g and its structure is composed of strongly twisted and considerably disordered carbon layers. Possibly this carbon contains nanocrystalline fragments similar to ones observed in carbon material obtained by explosive decomposition of intercalated graphite as invented in [9-11].

Taking into account the data available the method of catalytic gas-phase carbonization of hydrocarbons and other carbon containing compounds (for instance CO, mixture of $CO_2$ and $H_2$) can be regarded as most effective for deposition of nanodimensional carbon on different supports including graphite [12, 13]. At that the nature and structure of catalyst particles which is determined by synthesis method in its order critically determine efficiency of the whole process and structure of nanodimensional carbon obtained.

When using graphite, including EG, as a substrate for CNT growing, it should be taken into account that substrate itself (probably, defects of its structure) can initiate the growing of carbon. It is very probable that carbon so formed will reiterate the structure of substrate (template). So, in Fig. 2 there are shown XRD patterns of EG samples, on which carbon was deposited (45% with respect to EG mass) by decompositions of vapors of methylene chloride at 530°C (the gas-carrier - argon).

Since after deposition of substantial amount of carbon no additional peaks or broad halo appear in XRD one may suppose that carbon formed from methylene chloride decomposition in these conditions repeats the structure of the graphite substrate. Herewith interlayer distance $d_{002}$ for sample with deposited carbon practically repeats the parameter of EG lattice ($d_{002}EG = 0.3368$ nm, $d_{002}EG$+pyrolytic carbon=0.3367 nm, $d_{002}EG$+ catalyst + pyrolytic carbon = 0.3361 nm). However increase of the peak half-width is indicative of increase of microtension and/or reduction of areas of coherent dispersion. Herewith surface area of the material decreases from 35 $m^2/g$ (starting EG) to 7.2-9.0 $m^2/g$ because of obliteration of micropores.

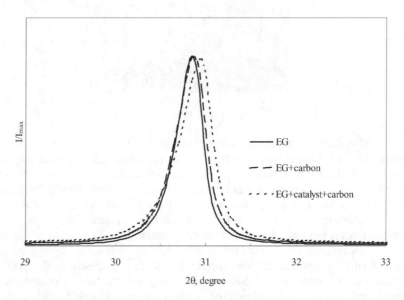

Fig.2. XRD reflection (002) of EG and EG modified with pyrolitic carbon deposited from methylene chloride vapor without catalyst and with catalyst (4% $Al_2O_3$+1%ZnO). Radiation Co $K_\alpha$.

Thereby, the EG substrate, both pure and modified with alumina and zinc oxide, is capable to initiate the growing of graphite layers when using of methylene chloride as source of carbon.

One should take into account that highly dispersed metallic catalysts at high temperature are capable to react with carbon substrate with formation of carbides. This can also change expected direction and velocity of the process. It is probable also that metal atoms or clusters formed from metal oxide precursors are localized predominantly on defects of graphite lattice and surrounding graphite structure acts as a structure promotor for graphite layer growing.

Probably, because of reasons noted above the catalysts of iron-nickel-cobalt group behave themselves differently on graphite support compared to metaloxide supports. So, authors of the work [12] could not reach good yield of CNT from acetylene (600°C) on crystalline graphite promoted with iron compounds (carbon deposit was only 3.4% with respect to mass of the graphite support) in spite of the fact that iron both pure and dispersed on metal oxide supports is an effective catalyst for CNT growing from gas mixtures containing acetylene.

Experiments carried out by us have shown that nickel compounds deposited on EG exhibit moderate catalytic activity in the system considered. So, nickel-containing EG was prepared by impregnation of intercalated graphite by solution of nickel chloride with addition of monoethanolamine, with following drying and thermal expansion at 900°C. EG so obtained was magnetic that was indicative of metallic nickel particles formation. When treated in gas mixture containing argon and 3% vol. acetylene (3 h, 700°C) deposit of nanodimensional fibrous carbon was obtained on EG surface. The deposit mass was 110% with respect to EG support mass. Only partial covering of EG surface with fibrous carbon was observed (Fig.3). Surface area of the sample obtained was 75 m²/g.

Fig. 3. EG particle covered with carbon nanofibers. Catalyst - nickel

Much more high yield of deposited carbon (up to 570% with respect to mass of EG substrate ) was achieved with use iron compounds as catalysts of acetylene carbonization at 600°C. The reaction duration was 1-2 h, gas mixture contained 2-10% vol. acetylene, 10% vol. hydrogen, and argon - rest. However, catalytic activity of iron-containing EG critically depends on preparation method. Most active samples of promoted EG were obtained by impregnation of intercalated graphite with saturated ferric chloride solution (calculated 5% of elemental iron to mass of graphite), with following drying at 130°C in air and thermal expansion (thermal shock) at 900°C in quartz reactor at limited contact with air. The iron-containing EG was noticeably magnetic. Analysis gives the content of iron (in all possible forms) in the EG sample to be 86% with respect to starting amount of iron. Probably, 14% of iron was lost at thermal expansion as volatile ferric chloride. It is noteworthy that natural crystalline graphite treated with concentrated aqueous

410

solution of ferric chloride expands at thermal shock. This indicates that intercalation of graphite takes place, possibly with complex ferric chloride anions like $[Fe_x(OH)_yCl_z]^{3x-y-z}$.

It is known that iron interacts with carbon at high temperature with formation of carbide, $Fe_3C$. For clarification of whether this compound was formed in conditions of our experiments EG was treated with concentrated aqueous solution of ferric chloride (calculated 50% of elemental iron with respect to mass of graphite), dried 5 h at 120°C in air, then heated to 500°C in air, and finally reduced in hydrogen 2 h at 900°C. XRD pattern of this sample (Fig. 4) shows peaks of graphite matrix, fairly intensive peaks corresponding to metallic iron, and does not reveal any peak which could be attributed to iron carbide.

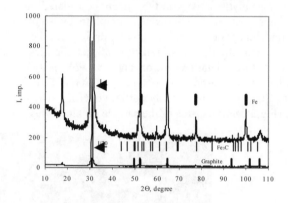

Fig. 4. XRD pattern of EG promoted with iron (EG treated with ferric chloride solution with following drying, heating to 500°C and reducing with hydrogen 2 h at 900°C)

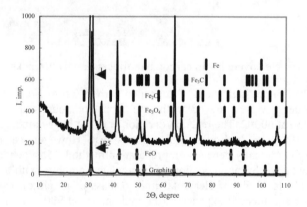

Fig. 5. XRD pattwrn of EG promoted with iron (treating with ferric formate-acetate solution, then heating to 500°C in argon without reducing)

In Fig. 5 there is shown XRD pattern of sample obtained by impregnation of EG with ferric formate-acetate solution (calculated 50% of elemental iron with respect to mass of graphite) with following drying at 120°C in air and heating to 500°C in argon flow. The sample was strongly magnetic. In this instance additional peaks in XRD possibly refer to magnetite, $Fe_3O_4$.

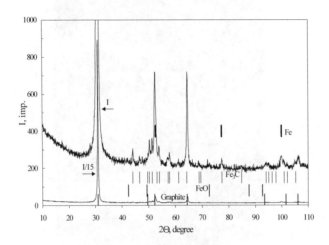

Fig. 6. XRD pattern of EG promoted with iron (treating with ferric formate-acetate solution, then heating in argon to 500°C and reduction in hydrogen and carbon monoxide at 600°C)

In Fig. 6 there is shown XRD pattern of the previous sample which was additionally reduced at 600°C 0,5 h in hydrogen and 2 h in carbon monoxide. In this instance peaks are present which refer to metallic iron. However, in spite of larger mass content of iron compounds these peaks are much less intensive than for sample reduced in hydrogen at 900°C (Fig. 4). Possibly at 600°C reduction of iron oxides with CO is not complete.

In Fig.7 there are presented images of EG particles covered with deposited nanodimensional carbon obtained by acetylene pyrolysis (600°C). EG was promoted with ferric chloride as described above. The "hair" presents itself carbon filaments about 100 to 1000 nm in diameter .

412

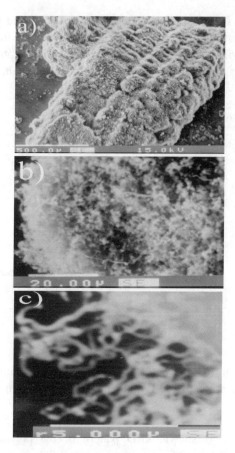

Fig.7. EG particle covered with carbon nanofibers (a - x70, b - x1000, c - x5000). Catalyst - iron

It was revealed that when depositing of carbon on EG surface from acetylene-hydrogen-argon mixture in conditions of our experiments surface area of samples increases in the beginning, but then drops. So, at mass content of deposited carbon 35% (EG 65%) surface area of the sample was 110 m$^2$/g (for starting EG, promoted with iron, 31 m$^2$/g). If one consider the obtained material as mechanical mixture of EG with deposited carbon, than surface area of pure deposited carbon can be calculated to be 226 m$^2$/g. This value is ordinary for CNT (usually 100-400 m$^2$/g) When the deposition process was continued to mass content of deposited carbon 85% (EG 15%), surface area of material obtained was only 42 m$^2$/g. Possibly low surface area is caused by obliteration of pores with amorphous carbon, relative rate of formation of which increases in time. To remove amorphous carbon the sample was treated 5 h in boiling 59% nitric acid, with following washing by water and drying at 100$^\circ$C. Accordingly to [2] purifying of CNT in this conditions results in removing of amorphous carbon, opening of carbon nanotubes and attaching of carboxyl and hydroxyl groups at ends of tubes and at structure defects. The sample processed in such way was strongly hydrophilic. When stirred with water the nanofibers deposited had tendency to separate from EG support. After heating at 600°C in argon hydroxyl groups were removed ant the material had not tendency to separate. Surface area of the material purified as described above was 233 m$^2$/g, and its mass was 82% with respect to mass of starting material.

Thereby, one may suppose that after too prolonged deposition of carbon access of acetylene molecules to catalytically active iron-containing centers becomes hampered, or they become deactivated, and relative rate of amorphous carbon growth increases. This results in decrease of measured surface area.

Probably this negative phenomena can be avoided by selecting the more efficient catalyst or it is necessary to purify the product.

We have also carried out experiments with depositing of carbon on EG promoted with nickel from benzene vapor (gas-carrier argon) at 800-1000° C. In conditions of our experiments probably the process of amorphous carbon growing dominated. Addition of hydrogen into gas mixture reduced deposition of amorphous carbon, however, we could not obtain samples with sufficiently high surface area. The system with benzene appears to be less acceptable for growing of nanodimensional carbon on EG.

## 4. Conclusions

Thereby, thermally exfoliated graphite as nanodimensional cluster-assembled system promoted by compounds of VIII Group metals appears to be a suitable substrate for growing nanodimensional carbon fibrous material and for creation of nanodimensional CCCM. Due to small appearing density EG and its very large pore volume the mass of deposited nanodimensional carbon can many times exceed the mass of the substrate. Depending on mode of after-treatment both hydrophobic and hydrophilic samples containing surface polar groups can be obtained.

414

**References**

1. 1-st International Conference "Carbon: fundamental problems of science, materials and technology". – Thesis. Moscow, 2002.
2. Shpak A.P., Kunitskiy Yu.A., Karbovskiy V.L. Cluster and nanostructured materials. Kiev, "Academkniga", 2001, 588 p.
3. Eletskiy A.V. Carbon nanotubes and their emission properties. Uspekhi fizicheskih nauk (Russia), 2002, vol. 172, No 4, p. 401-438.
4. Dhakate S.R., Wathur R.B., Dhami T.L. Mechanical Properties of Unidirectional Carbon-carbon Composites as a Function of Fiber Volume Content. Carbon Science 2002;3(3):127-132.
5. Jain P.K. Mahajan Y.R., Sundararajan G., Okotrub A.V., Yudanov N.F., Romanenko A.I. Development of Carbon Nanotubes and Polymer Composites Therefrom. Carbon Science 2002;3(3):142-145
6. Chung D.D.L. Review. Exfoliation Graphite, J Mater. Sci. 1987;22(12):4190-4199.
7. Pyatkovskiy M.L., Sementsov Yu.I., Chernysh I.G. Synthesis and investigation of properties of carbon-graphite composite materials. Chemical Industry of Ukraine, 1996, 3:42-45.
8. Inagaki M., Kawahara A., Konno H. Sorption and recovery of heavy oils using carbonized fir fibers and recycling. Carbon 2002;40:105-111.
9. Petrik V. I. Method for removing oil, petroleum products and/or chemical, pollutants from liquid and/or gas and/or surface United States Patent Application 20030024884 A1, February 6, 2003.
10. Petrik V.I. Method for curing of skin illnesses characterized by discharge and device for embodying the method. Pat. of Russia 2199350 C1, 27.02.2003.
11. Petrik V.I. Method for purifying blood plasma from uric acid and creatinine. Pat. of Russia 2199351 C1, 27.02.2003.
12. Hernadi K., Fonseca A., Nagy J.B., Bernaerts D., Lucas A.A. Fe-catalyzed carbon nanotube formation. Carbon, 1996, vol. 34, No 10, pp. 1249-1257.
13. Sheem K.-Y., Yoon S.-Y. Negative active material for lithium secondary battery and manufacturing method of same. United States Patent 6440610, August 27, 2002.

# MODELLING OF TDS-SPECTRA OF DEHYDRATING

Yu.V. ZAIKA, I. A. CHERNOV
*Institute of Applied Mathematical Research of*
*Karelian Research Centre of RAS,*
*185610, Pushkinskaya str. 11, Petrozavodsk, Russia*

## Abstract

In the paper several mathematical models of dehydration kinetics of metals for the experimental method of thermal desorption spectrometry (TDS) are presented. Only high-temperature peaks of TDS-spectra are considered. This allows (assuming diffusion is fast) consider only ordinary differential equations. In this case inverse problem of parametric identification of the models on experimental data becomes much simpler.

*Key words:* Thermal desorption spectrometry, dehydration kinetics of metals, mathematical modelling

## 1. Introduction

Power engineering and energy management problems are now among the most important scientific subjects. Due to growing ecological restrictions special attention is paid to possibility of using hydrogen as energy carrier. Materials which are suitable for its storing and transporting are being intensively searched for. Experimental research work demands numerical methods for estimating different kinetic parameters and for modelling hydrogen interaction with solids with new physical and chemical ideas taken into account. In this paper we discuss some models applied to the experimental method of thermal desorption spectrometry (TDS). This method is widely used in experimental work. A sample of studied material previously saturated with hydrogen is placed into an evacuated vessel with further constant evacuation and relatively slow heating. The desorption flux of hydrogen from the sample (measured by a mass spectrometer) gives information about hydrogen interaction with the studied material. Equilibrium regularities are sufficiently well studied (PTC diagrams). Growing interest is paid to kinetics of hydrating/dehydrating of metals. Among others we would like to mention the works [1–6] that contain further bibliography. The work [4] became for the authors a starting point of the results presented here. We consider the material from this work as the modelled object. This material is convenient because it allows identificating models both for low-temperature phase (when diffusion is significant) and for high-temperature phase. Mathematical justification of the boundary-value problem for the TDS method is given in [7,8]. In this paper we consider only the high-temperature phase and thus suppose that diffusion is relatively fast.

*T.N. Veziroglu et al. (eds.),*
*Hydrogen Materials Science and Chemistry of Carbon Nanomaterials,* 415-426.

The aim of the paper is to model the cases when dehydrating dynamics may be described using ordinary differential equations. Inverse problems of parameter identification for distributed systems (with diffusion equation in both phases) are difficult to solve and are not mentioned in this paper. Parameters of simplified models will serve as initial approximations. In the sequel material is not important, so physical and chemical terminology will be used marginally.

## 2. Basic assumptions

The material is a powder, we model a single particle. Consider a sphere of radius $L$ with a hydride core of radius $\rho$ inside ($\beta$ phase). A spherical layer with width $L - \rho$ is metal with dissolved hydrogen ($\alpha$ phase). The following conjectures are usually used:

1.  hydrogen atoms come to the surface (it is especially important) from the bulk and desorb forming molecules;
2.  due to porosity of the material molecules are formed and desorb directly from the bulk (near-to-surface area).

Both variants will be considered in the sequel. Surface is important not only because of its physical role. Due to presence of admixtures and oxides the particle is usually covered by different films. The fact that experimental data depend on hydride pounding is also an argument for the significant role of the surface. Let us introduce the following notation. Suppose $c(t)$ is the concentration of hydrogen dissolved in $\alpha$ phase at time $t$, $Q$ is the concentration of hydrogen in hydride ($\beta$ phase), $q(t)$ is the surface concentration. Usually the linear heating is used: $T(t) = \upsilon t + T_0$. Here we assume that $c(t, r) = c(t)$ because of spherical symmetry, quick diffusion, and the fact that $L \ll 1$, $T \gg 1$. Diffusion in the hydride is relatively slow and has no time to change the concentration. This means that in the shrinking core of radius $\rho(t)$ $Q = const$ during dehydrating with monotonic heating (peaks of TDS spectrum).

We model the desorption flux density with a square dependence on the surface concentration (or on the concentration in the near-to-surface area): $J(t) = b(T)q^2(t)$ ($J(t) = b(T)c^2(t, L)$). The desorption coefficient is denoted $b$ in both cases, but it has different sense and different units. We assume also that all parameters depend on temperature in an Arrhenius way, in particular $b(T) = b_0 \exp\{-E_b /[RT]\}$, $b(t) \equiv b(T(t))$.

## 3. The simplest model

Suppose that hydride decomposition is fast enough to maintain the equilibrium concentration in the solution. Desorption outflow is compensated by hydride decomposition. Desorption is bulk. Then $c(t) = \bar{c}$, $\bar{c} < Q$, until $\rho(t) > 0$. This means that the concentration remains unchanged until there is some hydride left. After hydride vanishes $\rho(t) = 0$ and $c(t)$ will decrease. Let us consider these two stages. The balance equation for decomposing hydride is

$$QV(L) = QV(\rho(t)) + c(t)(V(L) - V(\rho(t))) + S(L)\int_0^t b(\tau)c^2(\tau)d\tau \cdot$$

Here $V(r)$ is for the volume of a sphere of radius $r$, $S(r)$ is for its surface. Differentiating on $t$ we obtain

$$0 = (Q - \bar{c})\dot{\rho} + b(t)c^2(t), \rho(0) = L \Rightarrow \rho^3 = L^3 - \frac{3\bar{c}^2 L^2}{Q - \bar{c}} \int_0^t b(\tau)d\tau \qquad (1)$$

From this explicit expression for the hydride core radius one can (numerically) obtain the time $t_s$ when hydride vanishes ($\rho(t_s) = 0$).

Now let us consider hydrogen balance on the second stage ($t \geq t_s$, $\rho \equiv 0$):

$$[c(t + \Delta t) - c(t)]V(L) = -b(t)c^2(t)S(L)\Delta t + o(\Delta t).$$

Dividing this on $\Delta t$ and considering a limit we obtain

$$L\dot{c} = -3bc^2, \quad c(t_s) = \bar{c} \Rightarrow c(t) = \bar{c}\left(1 + \frac{3\bar{c}}{L}\int_{t_s}^t b(\tau)d\tau\right). \qquad (2)$$

In spite of simplicity of this model (only two parameters: $b_0$ and $E_b$), it gives rather good results if the particle size distribution is taken into account. Low number of unknown parameters allows using this model as a source of approximations for more complex inverse problems of parameter identification on experimental data.

On Fig. 1 one can see a typical desorption flux density $J(T)$ (a solid line) together with scaled curves of concentration $c(T)$ (dashed) and radius $\rho(T)$. Parameters are: $b_0 = 6 \cdot 10^{-21}$ cm$^4$/s, $E_b = 100$ KJ, $\bar{c} = \eta Q$, $\eta = 0.15$, $Q = 1.5 \cdot 10^{22}$ cm$^{-3}$, $\dot{T} = 0.3$ K/s, $T(0) = 670$ K, $L = 0.005$ cm.

Note that in the one-particle approximation existence of two stages leads to a spike on the curve. Such spikes are typical for quick-diffusion models because at the point when hydride vanishes the flux from hydride has discontinuity. There is no diffusion which could smoothen it. The curve $\rho(T)$ is similar for all models considered in the paper, therefore in the sequel it is not shown; a point when hydride vanishes is noted by a tiny circle.

Let the density of particle radii distribution be $N(L)$. Formally $0 < L < \infty$ but for $L > L_{max}$ there is hardly any particle (the density is almost zero). The mean flux density from a powder with density $N$ is calculated as follows:

$$\bar{J}(t) = \frac{\int_0^\infty J(t, L)S(L)N(L)dL}{\int_0^\infty S(L)N(L)dL}. \qquad (3)$$

Numerically it is sufficient to consider $20 - 50$ radii to obtain rather smooth mean curve. On Fig. 2 there is a mean desorption flux density curve from a powder with normal distribution of particle radii. The mean radius is $\bar{L} = 0.005$ cm and standard deviation is $\sigma = 0.001$ cm. Parameters are the same as above. The tiny circle shows the time of hydride disappearance in the particle of the mean radius.

418

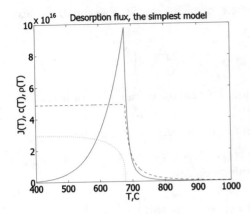

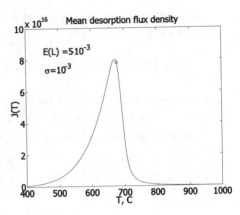

Figure 1. The simplest model, desorption flux density, a single particle

Figure 2. The simplest model, desorption flux density, powder

## 4. A switching model

Desorption is bulk: $J(t) = b(t)c^2(t)$. We suppose that for the certain time desorption is low enough, so the equilibrium concentration is maintained in the solution. Hydride decomposes with the rate necessary to compensate the desorption outflow. When the desorption flux becomes high enough to overcome the maximum possible hydride decomposition flux, $c(t)$ will decrease.

Denote this time by $t_s$. We suppose the highest possible density of flux from the hydride phase is $I(t) = k(t)Q$. Parameter $k(t) = k(T(t))$ is Arrhenius with respect to the temperature. The switch condition follows from the fact that the fluxes must be equal (remember that $c \equiv \bar{c}$ if $t < t_s$):

$$kQS(\rho) = b\bar{c}^2 S(L) \Rightarrow kQ\rho^2 = b\bar{c}^2 L^2, \quad t = t_s. \tag{4}$$

Consider time segment $[0, t_s]$. We have $c(t) \equiv \bar{c}$, $\dot{c} = 0$. The radius $\rho(t)$ satisfies (1). Substituting (1) in (4), we get an equation for $t_s$:

$$1 - \frac{3}{L}\frac{\bar{c}^2}{Q - \bar{c}^2}\int_0^{t_s} b(\tau)d\tau - \left(\frac{b(t_s)\bar{c}^2}{k(t_s)Q}\right)^{\frac{3}{2}} = 0.$$

It can be effectively numerically solved, for instance, by Newton's method. Now consider the time segment $t \geq t_s$. Hydride decomposes with the highest possible rate. Concentration change is determined by difference of the desorption flux and the flux from the hydride phase. The total balance is:

$$QV(L) = QV(\rho(t)) + c(t)(V(L) - V(\rho(t))) + S(L)\int_{t_s}^t b(\tau)c^2(\tau)d\tau$$

Differentiating on $t$, we get

$$-b(t)c^2(t)S(L) = Q\dot{V}(\rho) + \dot{c}(t)[V(L) - V(\rho)] - c(t)\dot{V}(\rho) \Rightarrow$$
$$\Rightarrow -b(t)c^2L^2 = \rho^2\dot{\rho}[Q - c(t)] + \dot{c}(t)[L^3 - \rho^3]/3. \tag{5}$$

Now from the balance near the phase bound:

$$[Q - c(t)]\dot{\rho}(t) = I(t), \qquad I(t) = k(t)Q, \rho(t_s) = \rho_{t_s}. \tag{6}$$

Initial data $\rho_{t_s}$ are defined by the previous stage $(t < t_s)$. Substituting (6) in (5), we get a differential equation for $c(t)$:

$$(L^3 - \rho^3)\dot{c} = 3[I(t)\rho^2 - b(t)c^2L^2] \quad I(t) = k(t)Q, \quad c(t_s) = \bar{c}. \tag{7}$$

Here are the model equations together:

$$0 < t < t_s : \qquad c \equiv \bar{c}, \qquad \dot{\rho} = \frac{b\bar{c}^2L^2}{(Q - \bar{c})\rho^2}, \quad \rho(0) = L,$$

$$t > t_s : \qquad \dot{\rho} = -\frac{kQ}{Q - c}, \qquad \dot{c} = 3\frac{kQ\rho^2 - bc^2L^2}{L^3 - \rho^3}, \qquad c(t_s) = \bar{c}.$$

Desorption flux density curves typically have the spike similar to that of the simplest model. Size distribution results in significant smoothing.

## 5. A model with initial cover

We suppose here that the flux balance from the very beginning determine the dynamics of the dissolved hydrogen. Desorption is bulk. To avoid singularity in the obtained differential equations at $t = 0$ we should consider an "initial cover": a thin layer of metal with dissolved hydrogen around the hydride core. Initial cover appears at the preliminary stage of the TDS experiment before the heating starts.

We model the density of the hydrogen flux from the hydride phase as $I(t) = k(t)Q[1 - c(t)/\bar{c}]$. The factor $1 - c(t)/\bar{c}$ 1 describes how concentration of dissolved hydrogen influences on the flux from the hydride phase: the lower the concentration, the greater the flux. If the concentration is near to equilibrium, the flux will be significantly lower than the highest possible.

Considering balance equations we obtain the equations for $c(t)$ and $\rho(t)$ in the same way as (6) and (7):

$$(L^3 - \rho^3)\dot{c} = 3[I(t)\rho^2 - b(t)c^2L^2] \quad I(t) = k(t)Q, \quad c(t_s) = \bar{c}. \tag{8}$$

$$[Q - c(t)]\dot{\rho}(t) = I(t), \qquad I(t) = k(t)Q, \rho(t_s) = \rho_{t_s} \tag{9}$$

Note that $I(0) = 0$. Hence, $\dot{c}(0) > 0$. The spike on the desorption flux density curves (determined by hydride flux discontinuity) is present only if the potential hydride decomposition rate is much higher than the rate of the desorption outflow. Below are some pictures. Parameters are as follows: $b_0 = 6 \cdot 10^{-21}$ cm$^4$/s, $E_b = 100$ KJ, $\bar{c} = \eta Q$, $\eta = 0.15$, $Q = 1.5 \cdot 10^{22}$ cm$^{-3}$, $\dot{T} = 0.3$ K/s, $T(0) = 670$ K, $L = 0.005$ cm, $k_0 = 5 \cdot 10^{-5}$ cm/s, $E_k = 0$ KJ, $\rho_0 = 0.96L$.

Together with the desorption flux density curves J here are the scaled concentration curves $0.5c / max(c) \cdot max(J)$. There is no spike (the curve is smooth), because J

420

changes its sign continuously. It is possible that the nominator in (8) changes sign, so that $\dot{c}(t) > 0$ for some $t$. This will happen if hydride decomposition parameter k grows quickly and so the hydride decomposition flux I surpasses the desorption flux J. For the Fig. 4 the following parameters change: $E_b = 90$, $k_0 = 5$, $E_k = 100$. One can see the local minimum of J.

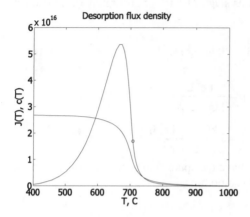

Figure 3. The model with initial cover, desorption flux density, a single particle

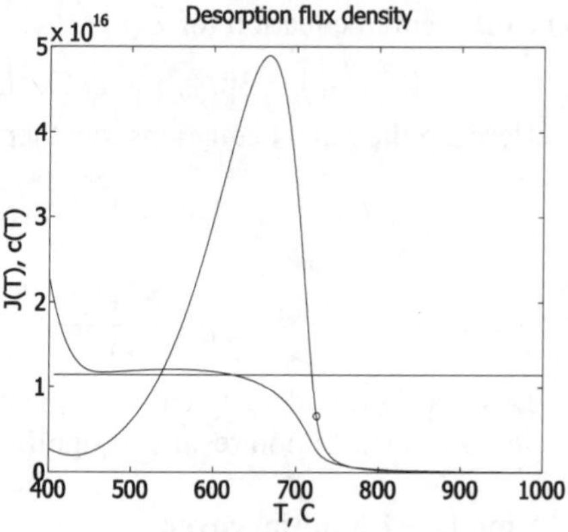

Figure 4. The model with initial cover, another example

## 6. A model with relatively fast diffusion

In this section we cancel the supposition that diffusion is mathematically momentary (so $c(t,r) = c(t)$ ). Instead we suppose that diffusion in the solution is quick enough compared to processes on the phase bound and on the surface, but not quick enough to make $c(t,r)$ constant with respect to $r$.

Let the function $c(t,r)$ satisfies the diffusion equation in spherical co-ordinates

$$c_t = D(t)[c_{rr} + 2c_r / r], \qquad \rho(t) < r < L, \qquad D(t) = D_0 \, exp\{-E_D /[RT]\}.$$

Suppose processes on the bounds (r = ½(t); L) are relatively slow (so the concentration profile changes slowly, $c_t \approx 0$ ) and the diffusion coefficient D is large enough; then we can consider $D(t)[c_{rr} + 2c_r / r] \approx 0$. Integrating these second order equation with respect to r we get the "quasi-stationary" concentration profile: $c(t,r) = A(t) + B(t) / r$. Thus concentration as a function of r is always stationary (a hyperbola); parameters A and B are determined by the flux balance. As the fluxes change, the stationary distribution changes, but slowly compared to diffusion. The sense of the prefix "quasi" is that $A = A(t)$, $B = B(t)$ and $\dot{A} \approx 0$, $\dot{B} \approx 0$ relatively.

Desorption is still bulk. Let us consider a balance equation (transforming a volume integral):

$$QV(\rho) + 4\pi \int_{\rho}^{L} r^2 c(t,r) dr + 4\pi L^2 \int_{0}^{t} b(\tau) c^2(\tau, L) d\tau = \text{const}.$$

Differentiating on t (the lower limit $\rho$ depends on t ) we get:

$$Q\rho^2\dot\rho(t)-c(t,\rho)\rho^2\dot\rho(t)+\int_\rho^L r^2\dot c(t,r)dr+L^2b(t)c^2(t,L)=0 .$$

Denote by $I(t)$ density of the flux from the hydride phase. It will be specified later. The equation

$$[Q-c(t)]\dot\rho(t)=I(t) \tag{10}$$

serves as a condition on the phase bound (called Stefan condition in general theory of problems with moving bound). Indeed, the bound shifts during the time $dt>0$ on $d\rho=\dot\rho dt$ . In the spherical layer of volume $4/3\cdot\pi[\rho^3-(\rho-d\rho)^3]$ concentration decreases from $Q$ to $c(t,\rho)+o(dt)$ by means of the flux on the phase bound. The equation (10) follows from

$$4\pi\rho^2[Q-c(t)]d\rho=-I(t)4\pi\rho^2 dt .$$

Finally, (using $c(t,r)=A(t)+B(t)/r$ ) we get

$$-I(t)\rho^2(t)+\int_\rho^L r^2\dot c(t,r)dr+L^2b(t)c^2(t,L)=0,$$

$$I(t)\rho^2(t)-L^2b\left[A+\frac{B}{L}\right]^2=\dot A\frac{L^3-\rho^3}{3}+\dot B\frac{L^2-\rho^2}{2} . \tag{11}$$

This equation is connected with (8) in the following sense: if $B\equiv0$ (hyperbola transforms to line) then this equation coincides with (8).

To determine $A(t)$ and $B(t)$ we need one more equation. Hydride decomposition flux $I$ must be equal to diffusion flux (continuity of the flux), so

$$I=-Dc_r(t,\rho)=DB\rho^{-2}, \quad [Q-A-B/\rho]\dot\rho=-DB\rho^{-2} .$$

1. Let $I(t)=k(t)Q$ . Then $B=kQ\rho^2D^{-1}$ and $\dot B=f(\rho,A,B;k,Q,D)$ . Substituting $\dot B$ in (11), we obtain a system of differential equations:

$$\dot\rho=f_1(\rho,A,B), \quad \dot A=f_2(\rho,A,B), \quad B=kQ\rho^2D^{-1} .$$

Initial data are determined from the condition that the initial concentration on the phase bound is equilibrium:

$$\rho(0)=\rho_0<L, \quad B(0)=k(0)Q\rho_0^2D^{-1}(0),$$

$$c(0,\rho_0)=A(0)+B(0)\rho_0^{-1} \Rightarrow A(0)=\bar c-k(0)Q\rho_0D^{-1}(0) .$$

2. Now let us consider a more complicated case. Suppose hydride decomposition flux density satisfies $I(t)=k(t)Q[1-c(t,\rho)/\bar c]$ ; then

$$DB\rho^{-2}=kQ\left[1-\frac{A+B/\rho}{\bar c}\right] \Rightarrow B=\frac{kQ[\bar c-A]\rho^2}{D\bar c+kQ\rho} . \tag{12}$$

Differentiating on t , substituting in (11) and solving with respect to $\dot A$ we obtain a differential equation for $A$ . Initial data are determined in a similar way as above: from $c(0,\rho_0)=\bar c$ it follows that $I(0)=0$ , thus $B(0)=0$ , $A(0)=\bar c$ . Note that singular at $\rho\to L-0$ factors in (11) make an initial cover necessary ( $\rho(0)=\rho_0<L$ ). It appears at

the preliminary stage of the TDS experiment. However, it is possible to consider a singular mathematical problem (in case $\rho(0) < L$) if necessary.

Here are some pictures. On Fig. 5 one can see a plot of desorption flux density for the first of the considered models ($I = kQ$). Parameters are: $b_0 = 6 \cdot 10^{-21}$ cm$^4$/s, $E_b = 120$ KJ, $\bar{c} = \eta Q$, $\eta = 0.15$, $Q = 1.5 \cdot 10^{22}$ cm$^{-3}$, $\dot{T} = 0.3$ K/s, $T(0) = 670$ K, $L = 0.005$ cm, $k_0 = 5 \cdot 10^{-4}$ cm/s, $E_k = 70$ KJ, $\rho_0 = 0.96L$, $D_0 = 8 \cdot 10^{-3}$ cm$^2$/s, $E_D = 50$ KJ.

One can notice a special "footstep". If something of the sort is present on the experimental curves, then it is an argument for choosing this model for a certain material. On Fig. 6 there is a plot of the desorption flux density J for the second case ($I = kQ[1 - c(t,\rho)/\bar{c}]$). Parameters are: $b_0 = 6 \cdot 10^{-21}$ cm$^4$/s, $E_b = 110$ KJ, $\bar{c} = \eta Q$, $\eta = 0.15$, $Q = 1.5 \cdot 10^{22}$ cm$^{-3}$, $\dot{T} = 0.3$ K/s, $T(0) = 670$ K, $L = 0.005$ cm, $k_0 = 5 \cdot 10^{-5}$ cm/s, $E_k = 0$ KJ, $\rho_0 = 0.96L$, $D_0 = 8 \cdot 10^{-3}$ cm$^2$/s, $E_D = 50$ KJ. Note that both curves are smooth, with no spike despite a single particle is considered.

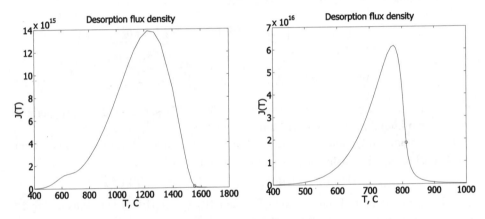

Figure 5. The model with relatively fast diffusion, I = kQ

Figure 6. The model with relatively fast diffusion, $I = kQ[1 - c(t,\rho)/\bar{c}]$

## 7. Models with surface

1. Let us consider a model that explicitly takes the surface processes into account. Remember that $q(t)$ denotes surface concentration. We assume that the surface and the bulk concentrations are connected by a condition of quick solution: $c(t) = g(t)q(t)$, $g(t) = g(T(t)) \approx$ const in the range of TDS peak, $T \in [T^-, T^+]$. It can be obtained from $Dc_r(t,L) = k^-(T)q(t) - k^+(T)c(t,L)$ (difference of the solution into bulk flux and the coming back to surface flux is taken away by diffusion) if both terms at the right part are significantly greater than the left part (when $T \gg 1$ the surface is active) and the activation energies of $k^+$, $k^-$ are almost equal (the surface is isotropic): $g = k^- / k^+$.

The balance equation is

$$QV(\rho)+c(t)[V(L)-V(\rho)]+q(t)S(L)+S(L)\int_0^t b(\tau)q^2(\tau)d\tau = \text{const}.$$

Differentiating on $t$ and using (9) we get:

$$[Q-c(t)]\dot{V}(\rho)+\dot{c}(t)[V(L)-V(\rho)]+\dot{q}(t)S(L)+S(L)b(t)q^2(t)=0,$$

$$\dot{q}(t)=-b(t)q^2(t)+I(t)\rho^2(t)L^{-2}-\dot{c}(t)[L^3-\rho^3(t)]/(3L^2).$$

Substituting the connection between the surface and the bulk concentrations and denoting $a(t)=1+g(t)[L^3-\rho^3(t)]/(3L^2)$, we obtain a model

$$a\dot{q}=-bq^2-q\dot{g}[L^3-\rho^3]/(3L^2)+I(t)\rho^2/L^2, \tag{13}$$

$$\dot{\rho}=-I/(Q-c(t)), \ \rho(0)=L, \ I=kQ[1-c(t)/\overline{c}], \tag{14}$$

$$c(t)=g(t)q(t), \ c(0)=\overline{c}, \ q(0)=\overline{c}/g(0). \tag{15}$$

If the surface is isotropic, i. e. $g(t)=g(T(t))=\text{const}$, $T\in[T^-,T^+]$, then (13) becomes simpler: $a\dot{q}=-bq^2+I(t)\rho^2/L^2$. On Fig. 7 there are model and experimental (see sec. 8) curves J together with a scaled $c(t)$. Parameters are:

$b_0=8\cdot10^{-10}$ cm$^2$/s, $E_b=202$ KJ, $g_0=2.25\cdot10^4$ cm$^{-1}$, $E_g=0$ KJ, $\overline{c}=\eta Q$, $\eta=0.15$, $Q=7.76\cdot10^{23}$ cm$^{-3}$, $\dot{T}=0.3$ K/s, $T(0)=670$ K, $L=7.5\cdot10^{-5}$ cm, $k_0=0.15$ cm/s, $E_k=100$ KJ.

2. Here we will consider a more complex condition of solution on the surface. Assume that surface concentration is driven by the flux balance:

$$\dot{q}(t)=-b(t)q^2(t)+I_s(t).$$

Here $I_s(t)$ is for density of the flux from the bulk to the surface. It will be defined later. Differentiating the previous balance equation on t and substituting the expressions for $\dot{q}$, $\dot{\rho}$ $(I_h=I)$ we get

$$b(t)q^2(t)-I_s(t)+\rho^2(t)I_h(t)/L^2-\dot{c}(t)[L^3-\rho^3(t)]/(3L^2)-b(t)q^2(t)=0.$$

Finally we have a system of differential equations

$$\dot{q}(t)=-b(t)q^2(t)+I_s(t), \ q(0)=q_0, \tag{16}$$

$$(Q-c(t))\dot{\rho}(t)=-I_h(t), \quad \rho(0)=\rho_0<L, \tag{17}$$

$$(L^3-\rho^3(t))\dot{c}(t)=-3(I_s(t)L^2-\rho^2(t)I_h(t)), \ c(0)=\overline{c}. \tag{18}$$

Flux densities $I_h(t)$, $I_s(t)$ can be modelled by the formulae

$$I_h(t)=k(t)Q\left[1-\frac{c(t)}{\overline{c}}\right], \ I_s(t)=g(t)c(t)\left[1-\frac{q(t)}{\overline{q}}\right].$$

Note that all equations follow from the conservation laws. The expressions for the flux densities may be varied according to the physical and chemical ideas.

## 8. Some notes and comparing with the experiment

We assumed that an evacuating system is powerful enough and therefore the amount of desorbed hydrogen that returns back to the surface is negligible. If this is not true we should substitute $bc^2 - \mu sp$ $\left(bq^2 - \mu sp\right)$ for $bc^2$ $\left(bq^2\right)$ in all the equations. Here $\mu$ is a kinetic constant ($\mu \approx 1.46 \cdot 10^{21}$ mol cm$^{-2}$ s Torr), $s(t) = s(T(t))$ is a material-specific parameter, $p(t)$ is the pressure of molecular hydrogen in the vessel. However one should keep in mind that additional *a priori* unknown parameters significantly complicate the inverse problem of parameter identification. Even the auxiliary problem of determination the desorption flux $J = bc^2 \left(bq^2\right)$ on measured (noisy) pressure $p(t)$ from the equations

$$p(t) = \theta_1 \int_0^t exp\{(\tau - t)/\theta_0\} J(\tau) d\tau, \quad J = [\dot{p} + p/\theta_0]/\theta_1, \quad \theta_i = const,$$

is numerically ill-posed and demands regularization procedures.

Besides we assumed that the concentration in the hydride phase is constant $(Q)$. It is possible to take hydrogen solution into hydride phase into account. In this case diffusion in the hydride is also considered quick. Then $Q$ is the minimal possible concentration of hydrogen in hydride, $c_* \geq Qc$, is the total concentration of hydrogen in the hydride phase. One additional stage should be added to any model: the phase bound is still (this means that hydride doesn't decompose), dissolved in hydride hydrogen comes to the solution and desorbs.

From the balance equations one easily obtains $V(\rho_0) \dot{c}_*(t) = b(t)c^2(t)S(L)$. The stage ends when $c_*(t) = Q$. After that any model which was considered above may be used. Finally we present some results of processing an experimental curve kindly given to authors by Prof. I. E. Gabis. We don't describe the experiment and the material (erbium hydride [4]) here, because the presented pictures are just illustrative. On the plots we compare the experimental desorption flux density curve with those obtained from the models described above. Curves on Fig. 8, 9 are for model of sec. 5 (a single particle and a powder with normal radii distribution $\left(\overline{L} = 0.005 cm, \sigma = 0.0025 cm\right)$. $c(t)$.

Parameters are: $b_0 = 6 \cdot 10^{-18}$ cm$^2$/s, $E_b = 157(155)$ KJ, $\overline{c} = \eta Q$, $\eta = 0.15$, $Q = 1.7 \cdot 10^{22} \left(1.3 \cdot 10^{22}\right)$ cm$^{-3}$, $\dot{T} = 0.3$ K/s, $T(0) = 670$ K, $L = 0.005$ cm, $k_0 = 5 \cdot 10^{-5}$ cm/s, $E_k = 35$ KJ, $\rho_0 = 0.96L$. One can see that it is difficult to choose a model. Therefore a single experiment with the linear heating doesn't provide enough information. A series of experiments with different heating rates and even different heating laws is necessary to specify the dehydration mechanism. Changing $\upsilon = \dot{T}$ shows that even different heating rates can fail to distinguish the models. Thus, more complicated cases should be considered. Fig. 11, 12 are examples of $T(t) = 1.2T_0 - 0.2T_0 cos(2\pi t / 500) T(t)$, $T_0 = 670$ K, and a "sawtooth" heating (period is 500 sec., $T_0 = T_{min} = 670$ K, $T_{max} = 938$ K, heating and cooling rates are equal, rates are equal). Together with the desorption flux density there are the scaled $c(t)$ curve (dashed line) and the radius curve $\rho(t)$ (dotted).

At the decreasing stage the model curves J are slightly below the experimental. This is because the models don't take defects of the material (traps) into account. Degassing of these traps results in a little increase of the desorption.

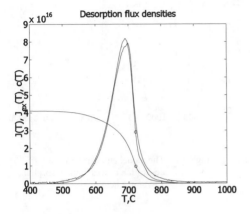

Figure 7. The model and experimental desorption curves, surface desorption

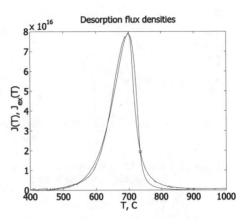

Figure 8. The model and experimental desorption curves, bulk desorption from a single particle

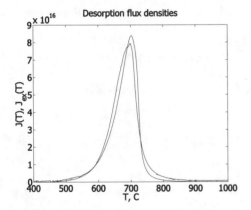

Figure 9. The model and experimental desorption curves, bulk desorption from powder

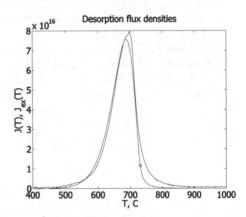

Figure 10. The model and experimental desorption curves, bulk desorption, another radius

426

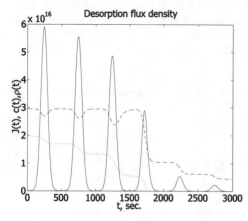

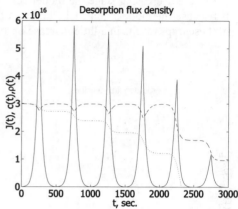

Figure 11. The model and experimental desorption curves, bulk desorption, sinusoidal heating

Figure 12. The model and experimental desorption curves, bulk desorption, "sawtooth" heating

## 9. References

1. Lufrano, J., Sofronis, P. and Birnbaum, H.K. (1998) *J. Mech. Phys. Solids* **46**, 1497.
2. Varias, A.G. and Massih, A.R. (2002) *J. Mech. Phys. Solids* **50**, 1469.
3. Castro, F.J. and Meyer, G. (2002) *J. Alloys Comp.* **330332**, 59.
4. Gabis, I., Evard, E., Voit, A., Chernov, I. and Zaika, Yu. (2003) *J. Alloys Comp.* **356357**, 353.
5. Remhof, A. (2003) *J. Alloys Comp.* **356357**, 300.
6. Fernandez, J.F. and Sanchez, C.R. (2003) *J. Alloys Comp.* **356357**, 348.
7. Zaika, Yu.V. (1996) *Comp. Maths Math. Phys.* **36**, 1731.
8. Zaika, Yu.V. and Chernov, I.A. (2003) *Int. J. Maths Math. Sci.* **23**, 1448.

# SOME SCHEMATICS OF USE OF HYDRIDE DEVICES IN THE AUTOMOBILE

SHANIN YU.I. [*]
*FSUE SRI SIA "Luch"*
*Zheleznodorozhnaya 24, Moscow area, Podolsk, Russia, 142100*

## Abstract

Steady tendency to use of hydrogen in the automobile can promote distribution of being available hydride process engineerings in motor industry.
In activity the problems connected to creation of the hydride accumulator of hydrogen and the heat pump are analyzed. The accumulator can supply by hydrogen fuel elements or engine of internal combustion. The heat pump working in a mode of a refrigerator, can be used as the automobile conditioner or refrigerator on the basis of a lorry (auto refrigerator).

## Nomenclature

| | |
|---|---|
| $T$ | Temperature (°C, K) |
| $\Delta H$ | Enthalpy of hydride formation [J kg$^{-1}$] |
| $\Delta X$ | Quantity of transferable hydrogen [kg] |
| $C$ | Heat capacity [J kg$^{-1}$ K$^{-1}$] |
| HS | Hydrogen storage (accumulator) |
| HTH (A), LTH (B) | High-temperature and low-temperature hydride |
| HHP | Hydride heat pump |
| CE | Engine of internal combustion |
| EC | Efficiency coefficient |
| COP | Efficiency a HHP (relation of energy of an obtained cold to energy, loiter on heating) |

*Indexes*

| | |
|---|---|
| h, m, l | Concerns to levels of temperatures of activity a HHP: high, mean, low |

*Key words:* hydride, hydrogen, automobile, accumulator of hydrogen, hydride heat pump.

## 1. Introduction

Hydrogen - most perspective fuel of the future. The hydrogen ecological is pure, has good motor properties and it the reserves at ocean practically are unlimited. A main direction of

---

[*] Corresponding author e-mail: syi@luch.podolsk.ru; fax: 007(096)634582

*T.N. Veziroglu et al. (eds.),*
*Hydrogen Materials Science and Chemistry of Carbon Nanomaterials,* 427-435.
© 2004 *Kluwer Academic Publishers. Printed in the Netherlands.*

428

activities on introduction of hydrogenous process engineerings for a truck - creation of automobiles with fuel elements (FC) because of rigid polymeric electrolytes (PEM FC) and electric drive, automobiles with a CE and automobiles with a hybrid propulsion system /1/. The hydrogen or can be reserved onboard the automobile, or is worked out during motion (for example, by conversion) of usual fuel (petrol, methanol, diesel fuel, natural gas etc.) in hydrogen in several catalytic reactors. The progress and problems connected to storage of hydrogen by various ways (compression, liquefied hydrogen converted hydrides, black, fullerens etc.), explicitly are covered in many activities, partially in activity of the author /2/.

Because of hydrogenous process engineerings tens demonstration are created is exemplar of automobiles of various models. The demonstration auto run of hydrogenous automobiles are calculated in hundreds thousands of kilometers /1/.

The metal hydride technologies already today successfully enter into a market. In main it of NiMeH$_y$ of the battery and accumulators. Now in many countries the compact safe systems of storage of hydrogen because of hydrides /1/ (including for automobiles) issue, the new alloys storages of hydrogen are created.

The steady tendency to use of hydrogen in a truck and the large thermal losses of automobile propulsion systems can promote broad distribution of hydride technologies in motor industry.

The problems arising at designing of hydride devices will below be considered: the accumulator of hydrogen (HS) and hydride thermal pump (HHP). It is supposed, that, basic elements of technical devices and some solutions in general is known to the reader the given subjects at a realization of similar devices /3,4/.

## 2. Hydride accumulators of hydrogen

To storage of hydrogen all growing requirements are presented. For example, the American program on promoting hydrogenous technologies /5, June 2003 / puts a problem to demonstrate onboard systems of storage of hydrogen with a capacity 6 weights. % in 2010 and 9 weights. % in 2015. Not pressing in advantages and shortages of all methods of storage of hydrogen we shall tell only, that the accumulation of hydrogen in converted metal hydrides for today is most technically realizabled direction. The problems connected to storage of hydrogen in converted hydrides will below be considered only.

The high sorption capacity is a main requirement, which satisfaction will make the application of hydrides is competitive capable. The modern metal hydride materials allow to reserve 1.5-2.5 weights. % of hydrogen (alloys FeTi and magnesium of a system Mg$_2$Cu). It is obvious unsufficiently the accumulator of hydrogen. It is necessary to have a sorption capacity more than 3 weights. %. As the candidates for such hydrides the alloys based on V (VTiCrMn), alloys TiCrMo (up to 3 %), composite structures of the nano-sizes of an alloy CaMg2 (up to 5.4 weights can be considered. % at the temperature of desorptions in 350°C) /6/. The capacity on hydrogen tightly is connected to such physical sizes, as a heat of formation of a hydride, temperature and pressure of a sorption (absorption/desorption), ease of activation of an alloy. The value of these parameters is significant at a technical realization of the hydride accumulator of hydrogen. Is expedient to have a low heat of formation of a hydride, since at charge (discharge) of hydrogen in the accumulator it is possible to use the heat exchanger with smaller output for a output (input) of excessive heat. The pressure at a desorption of hydrogen should be sufficient for a further feed of a propulsion system of the automobile (fuel cells, carburettors a CE). The lower pressure of a sorption of hydrogen promotes a drop of a structure weight and does the accumulator more safe. Technically acceptable limit sees pressure no more than 25 atm (2.5 MPas).

Value of desorption temperature - major parameter for a technical realization of the accumulator. As in the automobile there is a source of high temperature at a level 100°C and high thermal capacity (fluid cooling the engine), it is expedient to use this resource for a desorption of hydrogen. In the CE is one more heat source of a high potential. The exhaust gases on exit from an outlet collector have temperature at a level 700-1100°C (in an exhaust muffler 150-450°C). Small quantity of gas, low it thermal capacity and high temperature does their application as a heat source for a desorption of hydrogen problematic.

Use the HS on means of transport limits it the weight -dimensional characteristics. Is expedient to have high density of an initial hydride for minimization of volume a HS. The high requirements to cleanness of charged hydrogen or to impossibility of a metal hydride are presented also to be poisoned with gas impurity contained in hydrogen.

The hydrogenation of an initial alloy can pass outside of the hydride device and then the conditions of a hydrogenation can be specially selected and are sustained. But if the hydrogenation of an alloy happens immediately in the hydride device, the physical conditions of a hydrogenation (pressure, temperature) should not hardly differ from the working conditions of the device. In turn device should be designed with allowance for realizations of processes of a hydrogenation in it.

The important property of a metal hydride is the durability to a disproportionation in cycles absorption/desorption. For example, WE-NET Project (Japan) /6/ requires, that the metal hydride should save 90 % of the initial capacity on hydrogen after 5000 cycles charging - discharging. A completely necessary condition for a potential hydride is it the low fire danger and explosion safety.

But even, if the hydride will be discovered which will satisfy to majority of the indicated physical requirements, further it the application will be determined by fabricability and cost. Therefore problem of a selection of a hydride for a HS is optimization and at it the solution quite reasonablly to apply computer technologies /2,7,8,9/. There can be constructed various models of selection and ranking of hydrides, but a necessary condition of correct application of computer programs is the availability of base of valid data on properties of hydrides.

Other technical problem is the creation in a HS rather highly of transcalent hydride beds. At charging the HS (if, certainly, is not substituted completely whole HS) it is necessary, that the duration of processes of an absorption of hydrogen did not exceed reasonable limits. Therefore effective heat conduction of hydride beds in a HS should be at a level of 5-10 W/(m·K). The heat conduction of hydride powders does not exceed 0.1-1.0 W/(m·K). Proceeding from it, in hydride beds the well structured highly porous matrix from high of thermal conductivity materials (copper, aluminum), filled powdered hydride by an alloy should be organized. The porosity of such matrix owes there is at a level 90-95 %, and thickness of walls of voids at a level of 0.1 mm. Application of foam-metals, goffered and punched foils here is perspective. From a point of view of heat exchange in hydride beds there is a problem of transfer of a heat from (to) to hydride beds to (from) the heat-carrier. Also it is necessary to supply a good thermal contact of a metal matrix to shell details with the purpose of a drop of contact resistance by transfer of heat flows. In too time the hydride beds should supply a fast output of liberated hydrogen from a bed at insignificant hydraulic resistance.

At use of external charge the HS by hydrogen is necessary alongside with association of hydrogenous hoses to provide connection of hoses for cooling fluid (for removal of excessive heat selected at a response of a hydrogenation). Or to provide in the means of transport of the device (refrigerating type), which could cool fluid below temperatures of an environment (or up to it of a level) and to use it for cooling of a hydride at charge it by hydrogen.

The heat source is necessary for desorption of hydrogen from a hydride. For automobiles completely working on hydrogen (fuel cells, the CE), except the basic heater is necessary to provide the installation on a HS of the additional heater (for example, electric heater with a feed from the electrical accumulator). The application of the additional heater is expedient until then, when during a desorption of hydrogen the excessive heat release of a propulsion system in the basic heater can be used.

Thus, the accumulator of hydrogen should from itself represent rather original heat and mass transfer the device, in which the processes external (from fluid to a hydride bed) and internal (inside hydride beds) heat transfer should be matched.

The hydride compressor (HC) can be used for recompression of hydrogen up to the necessary level with the purpose of its further application as fuel. Then in addition it is possible to select hydrides, which have low pressure of a dissociation. The HC works completely on the other hydride, than the HS, is additional to a HS by the device, in which the quantity of a hydride at 10-20 of time is less, than in a HS. For the drive a HC the sources with two various temperatures (for example, fluid at the temperature of air and fluid, cooling the engine,) are required. The capability of control a level of pressure with the help of HC will allow to increase equilibrium pressure of hydrogen and by that to expand range of hydrides accepted to use in the hydride accumulator of hydrogen. The HC on a cold propulsion system both coordination of activity a HS and HC is necessary also to decide problems of initial start-up.

## 3. Hydride heat pumps

In any propulsion system of the automobile (existing or perspective) the thermal processes were and will be far from efficiency. Maximum EC the modern automobile CE in a mode of the greatest moment does not exceed 30%. At lower turnovers of the engine and at urban maintenance EC is reduced till 10-12 of % /10/. In the long term application of fuel cells promises to increase EC of the installation till 60-85 of % (at salvaging of heat), but now EC of such installations is at a level 35-46 % (on manufacture of an electricity). The remaining energy in the various forms of a heat (exhaust gases cooling fluids, radiation and convection) is dispersed in an environment.

Excessive, overflow heat of a propulsion system can serve for generation of a cold with the help of hydride heat pump (HHP). Two levels of temperature - high ($T_h = 100\text{-}200°C$) and mean ($T_m$ = 20-40°C are necessary for operation a HHP which corresponds to temperature of an environment in summer period). At such levels of temperatures of heat sources the hydrides can be selected, which with the help of on a low-temperature hydride is possible to achieve negative temperatures ($T_l$ - adiabatic temperature on a hydride) in range from -5 up to -50°C /9/. Such level of temperatures enough that it was possible to construct the conditioner with temperature of an air on an output 10-20°C or refrigerating chamber with temperature -5...-15°C. Hydride heat pump does not require for the drive of mechanical moment, works, as the periodic heat-transport device, and uses for the drive only flows of heat-carriers. On obtaining of a cold the useful power of the engine is not expended, therefore it can be considered as a charge-free spurious yield and, therefore, thus the efficiency of use of fuel is increased.

The HHP concerns to a class of sorption heat pumps. It has obvious advantages connected to a peculiar working skew field - by a hydride and hydrogen. The shortages a HHP are connected that the responses of a sorption of hydrogen are chemical and are characterized by a certain kinetics and limited number of converted cycles without a degradation of a working skew field.

A HHP, as well as all thermal machines, are characterized by efficiency or EC. It is possible precisely to designate problems connected to designing a HHP, analyzing it thermal efficiency. The efficiency of thermal machines is calculated, normalizing useful thermal power on expended power. For a refrigerating cycle efficiency the HHP (with allowance for influences of a design and operational mode) is calculated under the formula:

$$COP_r = \frac{K_m\left[\Delta H_B \Delta X_B - (C_{Ph})_B(1+\varphi_{CB})(1-\lambda_B)K_{KB}(T_m - T_l)\right]}{\left[\Delta H_A \Delta X_A + (C_{Ph})_A(1+\varphi_{CA})(1-\lambda_A)K_{KA}(T_h - T_m)\right]},$$

where $\Delta H$ - enthalpy (or heat) formation of a hydride; $\Delta X$ - maximum quantity of hydrogen participating in a cycle; $C_{Ph}$, $C_{pк}$ - thermal capacity of a hydride and material of a design accordingly; $\varphi_c = C_{pк}/C_{Ph}$; $K_m$ - ratio between weight of a hydride A (high-temperature) and B (low-temperature), ensuring carry of maximum quantity of hydrogen; $K_K$ - factor of a design, i.e. relation of a structure weight, in which is made a hydride, to weight of a hydride; $T_h$, $T_m$, $T_l$ - accordingly temperature of high, mean and low temperature levels a HHP; $\lambda$ - degree of regeneration of a heat in a cycle.

It is natural for effective activity the HHP is necessary to maximize size $COP_r$. The actual efficiency differs from a physical maximum of efficiency definable by the relation of enthalpies of hydrides B and A additional addends which are included in a numerator and a denominator of fraction. These addends also determine influence of thermal capacity of a design to efficiency a HHP considerably reducing it. It is connected to necessity of cyclical passing of preparatory stages in a cycle (pre-heating and pre-cooling) and with losses called by it. The efficiency hardly depends on size of double-side deviations from mean temperature ($T_m$-$T_l$) and ($T_h$-$T_m$), decreasing at their increase. For increase of efficiency it is necessary to be aimed:

1) Maximum to increase quantity transferable in a response of hydrogen $\Delta X$;
2) By selection of a hydride for a refrigerator as opposed to the hydrogenous accumulator it is necessary to be aimed that the enthalpy of formation of a hydride was as much as possible. Therefore it is necessary to increase a heat of formation of a hydride and in

particular of low-temperature hydride **B** (for example, application of hydrides on base of vanadium);

3) Maximum to reduce total thermal capacity of a design, i.e. to reduce a factor $K_K$.

There is one more effective way of decreasing of losses of a heat in a cycle which opens the whole direction in creation a HHP and schemes of operation a HHP. This way - regeneration of a heat. The degree of regeneration of a heat in a cycle $\lambda$ is quantity of a heat under the relation to all loiter which was it useful the HHP used for a pre-heating of other sorber. The duration of stage of regeneration is insignificant on a comparison with duration of a cycle (for example, 1 min at a level 15 min), but the efficiency it is great since thus the maximum initial temperature potential $(T_h-T_m)$ is used. Usually $\lambda$=0.25-0.4. It is necessary to remember that the realization a HHP with regeneration becomes complicated because of occurrence of new elements (heat-transport outline between identical sorbers) and stages of a cycle.

Problem of hydride beds for sorbers a HHP same as well as at the accumulator of hydrogen. The main problems of a realization a HHP are connected to heat exchange of hydride beds with the heat-carrier. In HHP two kinds of hydrides and each from them should periodically interact with heat-carriers of different temperature. Because of a recurrence of process of absorption/desorption of hydrogen by hydrides it is necessary periodically to heat up and to cool sorption devices. The activity of the device at two levels of temperature results in a difficult control system of flows of the heat-carrier. Therefore to heat exchangers the contradictory requirements are presented. On the one hand it is necessary to organize sufficient heat transfer from hydride beds to external heat-carriers, on the other hand it is necessary to minimize quantity of the heat-carrier located in a contact to hydride beds.

Especially problems, connected to heat exchange, carefully should be decided at use of gas heat-carriers (yields of combustion a CE, air). If in case of application of fluids the external heat exchange can be organized with sufficient efficiency for the gas heat-carrier it is necessary to develop an external heat-transport surface of the hydride device. Thus weight of a design is increased and the efficiency a HHP drops. Besides it will be necessary to decide a problem of a drop of contact resistance in places of interface of fins and hydride sorber.

The operation a HHP is possible only at application as a minimum of two sorbers, which are connected among themselves by hydrogenous line. In a line the additional elements such can be installed which clear hydrogen (filters) and control it current (stop or reverse clackes). Lock elements can and be not. Then the process of exchange by hydrogen passes at approximately equal pressure in various sorbers and the main process managing mass transfer is the heat exchange between a hydride filling and heating up environment. Intensity of heat exchange in low-temperature and high-temperature sorbers should be matched so that certain quantity of hydrogen selected in one from sorbers without delay was absorbed in the other sorber. The version a HHP without a valve especially easily is realizable for a modular principle of a construction of sorbers (when the necessary power the HHP is typed by adding completely of autonomous modules). As the exchange processes in a HHP have non-stationary (exponential) character there is a heavily initial part of process and more flat, but extended stage of process described by small gradients of temperatures and efficiency. For want of a lock valve in a hydrogenous line the common duration of a cycle because of more languidly current processes is increased. The availability of a stop valve allows to force auxiliary operations of a cycle (pre-heating and pre-cooling) for the score of more heavily heat exchange. Besides at a by-pass of hydrogen (at open of a valve) the responses of a hydrogenation go faster, since the kinetics of a chemical response is proportional to a difference of pressure of hydrogen in a system and equilibrium pressure of hydrogen above a hydride at given temperature. The main part of a reaction passes for much smaller time than at version without a valve. Therefore installation of a valve is desirable for obtaining the best characteristics on power a HHP.

There is a positive experience of development of experimental models a HHP which are aimed at application in the automobile /11/. Two types a HHP in the specially created and tested refrigerating experimental models of the installations were realized:

1. With use of heat from combustion of liquid fuel with temperature 180-200°C (imitating operation of an exhaust system a CE) for cooling of an air up to a minus 3°C at periodic operational mode with one couple of hydride sorbers. As a high-temperature hydride (HTH) was used $LaNi_{4.6}Al_{0.4}H_x$, as a low-temperature hydride (LTH) $MmNi_{4.15}Fe_{0.85}H_y$. The operation of the installation has confirmed a reality of creation of the automobile or ship refrigerator. Because of this experimental model the design concept of the refrigerator installation of a continuous operation (drum-type type) with gas heat-carriers was offered. The hydrides are previously selected and the technical parameters of the refrigerator with output of 2 kW are appreciated.

2. With use as a power source of a heated water with temperature 80-90°C. On a cold part a HHP (at application of a water as the carrier of a cold) the temperatures 1-4°C were reached, the use of antifreeze (i.e. cooling fluid of the automobile) allows to receive temperature up to a minus 5°C. In quality the HTH was used $LaNi_5H_x$, in quality a LTH - $MmNi_{4.15}Fe_{0.85}H_y$. The mean cooling power of the installation has made 550-660 W. The obtained results were simulated to the device of the conditioner on the automobile where as a source of a heat it is supposed to use cooling a CE fluid. The level of its temperature in 100-110°C allows to calculate on obtaining of low temperature for a level -15°C. Thus the cooling of the conditioner can achieve 1-3 kW and temperature of a cold air 0-10-20°C.

Improving of the characteristics the HHP is possible to achieve applying a two-stage cycle /5/. Thus the number of hydrides is increased up to three, and temperature $T_1$ is reduced in relation to a single-stage cycle on 10-15°C. In case of multistage cycles availability of the stop hydrogenous equipment in a channel of hydrogen it is necessary.

The correct (optimum) operation control a HHP is the independent problem solved with the help of mathematical simulation and experimental operational development realized HHP. Let's mark that there are optimum times of changing of heat flows in a system at specific time of a cycle for obtaining maximum refrigerating effect per unit of bringed heat from a high-temperature source.

Preliminary selection of hydrides for a HHP and mathematical simulation of their activity (one-dimensional non-stationary case) can be made with the help of computer programs /12,13/. Thus the registration of a hysteresis, drop of a plateau, propulsion of pressure for two hydrides is possible. The model set of equations involves non-stationary one-

434

dimensional equations of thermal balance in hydride sorbers with allowance for of thermal effects at an absorption/allocation of hydrogen. In computer calculations the experimental data on equilibrium isotherms in systems a metal alloy hydrogen are used. The heat conduction of hydride beds was calculated and correlated with heat conduction determined experimentally. In calculations the time cyclogramme of operation a HHP is determined: distribution of temperature of design elements on a radius of a cylindrical hydride bed, contents of hydrogen in sorbers, pressure of hydrogen, energy and thermal power. The research of influence of main parameters of a mode of the heat pump (pressure of charge of hydrogen, temperatures and rates of the heat-carrier, time of a cycle) on it the power characteristics can be carried out. The comparison of computational and experimental results has shown /13 / that the developed one-dimensional model is even rather far from the registration of all actual design features a HHP. It can use for qualitative study of influence of various parameters and approximate quantitative assessments (maximum differences from an experimental data 40-60 % aside of overestimate of results). And nevertheless realized mathematical simulation allows operatively to carry out calculations of various designs a HHP. With the help of simulation for short time and at minimum costs (without manufacturing of devices in "metal") are possible are to revealed the best approaches to development by a HHP and them practically accessible parameters.

### 4. Hydride launching device /2,9/

In a cold season for start-up a CE the energy selected at an absorption of hydrogen by a hydride can be used. To have two bottles with a hydride of a different chemical structure jointed pipeline through the valve enough. One from hydrides is saturated by hydrogen at high pressure, other has not hydrogen and to be at low pressure. At open of the valve the hydrogen has directed in a bottle with a hydride at low concentration of hydrogen. Thus temperature of this bottle is increased and selected heat with the help of heat exchanger can be used for a heating a CE. It is enough of selected heat to increase temperature of motor oil or antifreeze.
Creation of a hydride launching device now is possible which can be made as the independent aggregate. For connection with the CE technically is necessary to execute insertion of the heat exchanger of the device in a cooling engine channel. Thus it is necessary to provide the installation of a small electric pump for supply of antifreeze at the switched off engine. After start-up and warm-up of the engine the launching hydride device with the help of waste heat can be to initialize before the following use.

### 5. Conclusions

The implementation problems of hydrides in a automobile transportation are connected to shortages of hydrides and first of all with their unsufficient capacity on hydrogen.
The problems of a technical realization of hydride devices are solvable and require further development.

# 6. References

1.  Malyshenko SP, Pekhota FN. Today and tomorrow of hydrogenous power engineering. Energia 2003; 1: 2-8 (in Russian).
2.  Shanin YuI. Use of hydride devices in a vehicle. In: Veziroglu TN, Zaginaichenko SY, Schur DV, Trefilov VI editors. Hydrogen Materials Science and Chemistry of Metal Hydrides. NATO science series. Series II: Mathematics, Physics and Chemistry. Kluwer Academic Publishers, 2002. 82. p.313-320.
3.  Ron A. A hydrogen heat pump as a bus air conditioner. J. Less-Common Metals 1984; 104: 259-278.
4.  Izhvanov LA, Solovey AI, Frolov VP, Shanin YuI. Metal hydride heat pump - new type of heat converter. Int. J. Hydrogen Energy 1996; 11/12: 1033-1038.
5.  Hydrogen, Fuel Cells and Infrastructure Technologies Program. Multi-Year Research, Development and Demonstration Plan. U.S. Department of Energy, Energy Efficiency and Renewable Energy. Draft (June 3, 2003).
6.  Materials site World Energy NETwork (http://www.enaa.or.jp/WE-NET/contents_e.html).
7.  Shanin YuI. Choice of hydrides for two-stage metal hydride chemical heat pumps. In: Abstracts book by 5-th intern. confer. "Hydrogen materials science and chemistry of metals hydrides", ICHMS`97. Ukraine, Yalta, September 02-08 1997. p.256. (in Russian).
8.  Shanin YuI. Preselection of hydrides for hydride heat pumps. In: Abstracts book by 6-th intern. confer. "Hydrogen materials science and chemistry of metals hydrides". Ukraine, Yalta, (02-08) .09.1999. p.304-305. (in Russian).
9.  Shanin YuI. Selection of hydrides for automobile hydride devices. Int. Scientific J. for Alternative Energy and Ecology (ISJAEE) 2002; 3: 50-53. (in Russian).
10. Mishchenko AI. Application of hydrogen for automobile engines. Kiev: Naukova dumka, 1984. (in Russian).
11. Astakhov BA, Afanasev VA, Bokalo SYu, Izhvanov LA, Solovey AI, Frolov VP, Shanin YuI. Development of installation based on metal hydride heat pump for heat and cold generation. In: Abstracts book by 6-th intern. confer. "Hydrogen materials science and chemistry of metals hydrides". Ukraine, Yalta, (02-08).09.1999. p.360-361.
12. Fedorov EM, Shanin YuI, Izhvanov LA. Simulation of hydride heat pump operation. Int. J. of Hydrogen Energy 1999; 24: 1027-1032.
13. Shanin YI. Simulation of hydride heat pump operation with reference to vehicle refrigerating devices. In: Veziroglu TN, Zaginaichenko SY, Schur DV, Trefilov VI editors. Hydrogen Materials Science and Chemistry of Metal Hydrides. NATO science series. Series II: Mathematics, Physics and Chemistry. Kluwer Academic Publishers, 2002. 82. p.97-106.

# ELECTRONIC-MICROSCOPIC INVESTIGATION OF NANOSCALE PRODUCTS OF CATALYTIC PYROLYSIS OF TOLUENE

P.M. SYLENKO[1], A.M. SHLAPAK, S.N. KAVERINA,
D.V. SCHUR, S.O. FIRSTOV, V.V. SKOROKHOD

*Institute for Problems of Materials Science of NAS of Ukraine,
3 Krzhizhanovsky St., Kiev, 03142, Ukraine*

## Abstract

Carbon nanotubes, polyhedral nanoparticles and films were synthesized by pyrolysis of toluene using iron powder as the catalytic agent. Transmission electron microscope and scanning electron microscope were utilized to study the geometrical shape and size of nanoparticles and nanotubes. The optimum temperature range to produce nanotubes and nanoparticles by pyrolysis of toluene is 850-900 °C. The produced carbon films consist of nanotubes.

## 1. Introduction

Last years problems on manufacture and examination of properties of nanomaterials are payed the special attention due to of their unique properties [1-2]. One of most promising nanomaterials is carbon nanotubes (CNTs). After their discovery in 1991 by Iijima [3] the numerous publications have appeared, in which the methods of their manufacture and their properties are considered. The results of CNTs investigations into properties have allowed forecasting the huge effect of their application.

So, at present it is considered that the most promising areas of application of CNTs are electron technics (diodes, transistors, integrated plans, flat displays etc.), energetic (secondary generators of gases, chemical cells), analytical technique (sensors, probes for microscopes), structural materials (reinforcing fillers of composites) [4-5] etc.

Among methods of CNTs synthesis the most spread and the most explored is the method of nanotubes synthesis by sputtering of a graphitized electrode in the helium plasma during the electric arc discharge [5-6]. For the first time CNTs were produced by this method. It allows to produce both single-walled and multi-walled nanotubes, using different catalytic agents. At the same time, this method has such essential restrictions, as necessity of using high electrical power equipment, complexity in controlling parameters of the technological process and impossibility of its use for continuous synthesis of nanotubes.

---

[1] Corresponding author, e-mail: dep69@ipms.kiev.ua

*T.N. Veziroglu et al. (eds.),*
*Hydrogen Materials Science and Chemistry of Carbon Nanomaterials,* 437-446.
© 2004 *Kluwer Academic Publishers. Printed in the Netherlands.*

Attractive and one of the most promising methods of CNTs manufacture is also the hydrocarbons pyrolysis method. The method is simple enough and at the same time it uncloses broad opportunities for producing carbon nanotubes both in laboratory and in industrial conditions.

The intensive researches performed by many scientist last years show, that in some cases, for example in manufacturing field emitters by the single-stage method, only hydrocarbons pyrolysis method is effective.

Carbon nanotubes are mainly produced by the hydrocarbons pyrolysis method of acetylene [7]. In some papers also is informed about CNTs producing from methane [5], ethane [6], acetone [4], ethanol [8] and other chemical compounds, which contain carbon. Only papers [9-10] mention about opportunity of producing nanotubes from toluene, but there no data about parameters of the process and properties of CNTs. Authors have investigated the process of CNTs synthesis of by pyrolysis of toluene.

## 2. Experimental

Investigation of carbon nanotubes synthesis by pyrolysis of toluene was carried out on the installation which key diagram is shown in a Fig. 1.

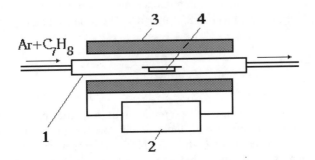

Figure 1. The diagram of the installation for pyrolysis of toluene: 1-quartz reactor, 2-temperature controller, 3-electrical furnace, 4-substrate with the catalytic agent.

The vapor-gaseous mixture of argon and toluene was injected to quartz reactor 1, in which the substrate 4 with the catalytic agent was placed. Iron powder with 1-3 microns particles has been used as the catalytic agent. In the synthesis zone the temperature has been kept constant with the help of the resistance furnace 3 and temperature controller 2. The toluene/argon ratio was 1:20.Study of CNTs synthesis conducted in a temperature range 600-1100 °C.

## 3. Results and discussion

Investigation of samples of the synthesized material has been conducted on the transmission electron microscope JCM 100CX. The analysis of the results obtained about the temperature effect on synthesis of carbon nanoscale materials (Fig. 2-5) shows that at the temperature 700 °C (Fig. 2) the produced materials represents carbon particles of

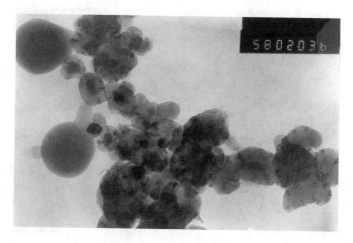

Figure 2. Carbon nanoparticles obtained at temperature 700 °C, (X 58 000).

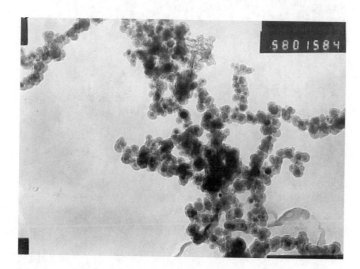

Figure 3. Carbon nanoparticles obtained at temperature 800 °C, (X 58 000).

440

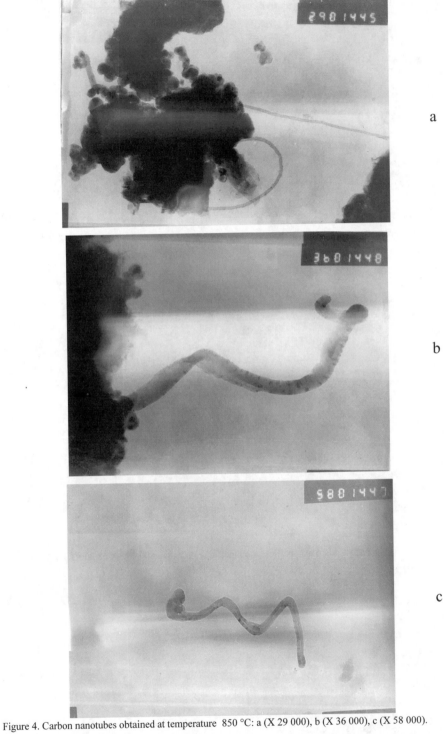

Figure 4. Carbon nanotubes obtained at temperature 850 °C: a (X 29 000), b (X 36 000), c (X 58 000).

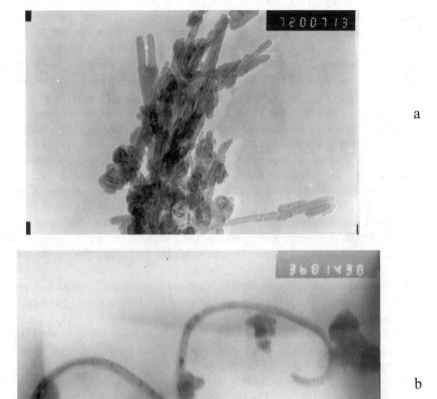

a

b

Figure 5. Carbon nanotubes obtained at temperature 900 °C: a (X 72 000), b (X 36 000).

442

different sizes and mostly the rounded shape. Inside some rounded particles have a dark kern, which has higher electronic density and may be the catalytic agent. Nanotubes and nanofibers in the synthesis products are practically absent. At the temperature 800 °C (Fig. 3) among the carbon particles, integrated in aggregates, the single multi-walled nanotubes are seen. Here, the size of carbon particles is much smaller in comparison with that of particles, synthesized at 700 °C. As a matter of fact, every particle of the catalytic agent is covered with the uniform layer of carbon. With rising temperature up to 850-900 °C (Fig. 4-5) the mixture of carbon particles and nanotubes is observed. In one cases nanotubes joint with the massive concentration of mixed carbon particles, in others – they have the varied geometrical shape, namely: direct and arched, which join arrays of carbon particles, single nanotubes of the spiral shape, bent half-axes, which, the most probably also consist of nanotubes. In some samples of deposit we observed splice nanotubes, and their cross-section has both circle and hexagon (Fig. 5a). For the first time presence of the nanotubes having the polyhedral shape has been discovered by Gogotsi with the colleagues [11] With rising temperature up to 1000 °C we observed both cylindrical nanotubes, and nanofibers, which ends have the cone-shaped form (Fig.6a). At this temperature of synthesis the nanotubes of the Y- similar shape were also produced (Fig. 6b). Manufacture and application of nanotubes having such a shape is described in detail in the review of Yeletsky [5 ].

In the temperature range 800-1000 °C, besides nanotubes, the carbon polyhedrons are also obtained. Their typical view is shown in fig. 7. Presence of polyhedral particles in products of carbon nanofibers synthesis is also observed in paper [12].

Besides the mixture of carbon particles and nanotubes the carbon films was also investigated. Films deposited on the reactor walls during synthesis of nanotubes. The scanning electron microscopy (Superprobe 733) revealed rather long ropes of tubes on the surface of such films which thickness is 20-30 microns. In particular intervals there was a knot on a rope (Fig. 8). The film was supposed to consist of nanotubes. Paper [13] indicated possibility of manufacturing the carbon films, which consist of nanotubes.. To check this supposition the films has been shattered in the agate mortar and its fragments have been investigated on the transmission electron microscope. The structure of the carbon film proved not to differ from the deposited products of pyrolysis on the catalytic agent (Fig.9). Fragments of nanotubes are seen in the photograph as well as in Fig.5. So, at catalytic pyrolysis of toluene, it is necessary to take into account processes occurring both on the surface of the Fe-catalytic agent, and inside the reactor for synthesis. The similar conclusions were also made in paper [14], in which the authors studied pyrolysis of acetylene.

Besides pyrolysis products synthesized on the catalytic agent and carbon films, deposited on the reactor walls, were also investigated carbon structures formed in gas phase of the reactor (Fig. 10). These structures were caught by the fluid gate, located at the exit of the reactor (It is not shown in Fig. 1). As can be seen in figure, nanotubes are transparent and their ends have hemispherical vertex.

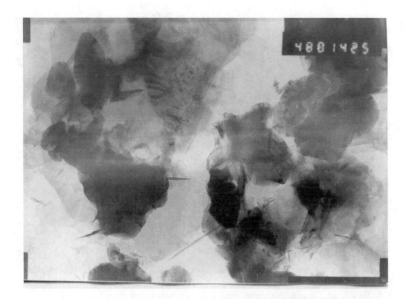

a

b

Figure 6. Products of pyrolysis of toluene at temperature 1100 °C, a (X 48 000), b (X48 000).

444

Figure 7. Carbon particle of the polyhedral shape, (X 100 000).

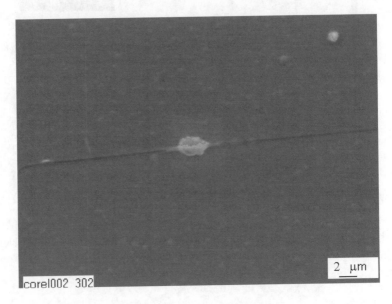

Figure.8. Nanotube rope on a carbon film surface.

Figure 9. Structure of a carbon film, (X 58 000).

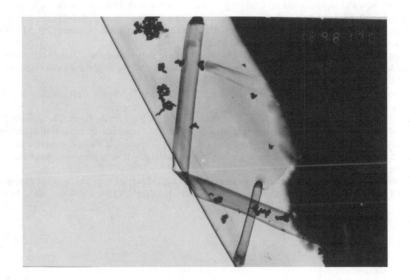

Figure 10. Structure of pyrolysis products, formed in gas phase of the reactor, (X 48000).

## 4. Conclusions

-  Multi-walled carbon nanotubes of the varied geometrical shape: cylindrical and hexahedral, rectilinear and curvilinear, and also Y-similar were produced by catalytic pyrolysis of toluene.

- The optimum temperature range for nanotubes synthesis  from toluene using Fe catalytic agent is 850-900 °C.

- Carbon films, which were deposited on the reactor walls during pyrolysis of  toluene, consist of nanotubes.

## 5. References

1. Skorokhod, V.V. etc. (1988) Nanocrystaline materials, Kyiv.
2. Trefilov, V.I, Schur, D.V., Tarasov B.P.etc. (2001) Fullerenes - basis of materials  of the future, Kyiv.
3. Iijima S. (1991)  Helical microtubules of graphitic carbon, *Nature*,.**354**, 56-58.
4. Rakov, E.G. (2001) Chemistry and application of carbon nanotubes, *Successes of  chemistry*,**10**, 934-973 (in Russian).
5. Eletsky,  A.V. (2002) Carbon nanotubes and their emissive properties, *Successes  of  physical sciences,* **4**, 401-438 (in Russian).
6. Ebbesenn,T.W. and Ajayan, P.M.  (1992) Large-scale synthesis of carbon nanotubes, *Nature*,.**358**, 220-222.
7. Resasco, D. E., Kitiyanan, B., Harwell, J. H., Alvarez, W.. Method of producing carbon nanotubes,Pat. 6333016 USA, Int. Cl. D01F 009/127.  Filed.03.09.1999, Pub.25.12.2001.
8. Redkin, A.N. and Melerevich, L.V. (2003) Obtaining of carbon nanofibers and  nanotubes by a method of a super fast heating of vapors of ethanol, *Inorganic      materials*, **4**, 433-437.
9. Tennent, H. G., Barber, J. J., Hoch, R,  Carbon fibrils and method for   producing same, Pat. 5165909 USA, Int. Cl. D01F 009/127, Filed.   01/10/1990, Pub. 24.12.1992.
10. Zhou, W.,  Ooi,  Y. H.,  Russo, R. at al. (2001) Structural characterization and diameter-dependent oxidative stability of single wall carbon nanotubes  synthesized by  the catalytic decomposition of CO,*Chem. Phys. Letters,* **350**, 6-10.
11. Gogotsi,Yu., Libera,  J.A., Kalashnikov, N. at al. (2000) Graphite Polyhedral  Crystals, *Science*, **290**, 317-320.
12. Forry, L., and Schonenberger, C. (2001) Carbon nanotubes, materials for the future, *Europhysics News,* **3**,
13. Obedkov, A.M., Zaitsev, A.A., Domrachev, G.A. at al. (2003) Examination of  multi-walled carbon nanotubes obtained by thermal decomposition of mixtures of ferrocene or diethylferrocene and benzol.- Abstracts of 2 International conference "Carbon": fundamental problems of a science, materials, technology,      Moscow (in Russian).
14. Schur, D.V., Motysina, Z.A. and Zaginaichenko, S.Yu. (2002) Study of Physico-Chemical Processes on Catalyst if the Course of the Synthesis of Carbon Nanomaterials, *NATO Science Series*,  **82**, .235-243.

# APPLICATION OF THE METAL HYDRIDE ACTIVATION EFFECT OF HYDROGEN ISOTOPES FOR PLASMA CHEMICAL TECHNOLOGIES

Yu.F. SHMAL'KO, Ye.V. KLOCHKO
*A.N. Podgorny Institute for Mechanical Engineering Problems of National Academy of Sciences of the Ukraine,*
*2/10 Pozharsky Str., Kharkov-46, 61046 Ukraine*

## Abstract

Earlier authors were offered a new method of increase of rate and selectivity of plasma chemical reactions with participation of hydrogen isotopes. This effect based on sorption activation of hydrogen by metal hydrides. Is shown that the sorption activation result in thermo emission of atomic hydrogen, ionized hydrogen atoms and molecules and excited hydrogen molecules from metal hydride surface. Ways of control various plasma chemical reactions by use of the sorption activation effect are discussed.

*Keywords:* Hydrogen isotopes, plasma chemical reactions, thermo sorption activation, metal hydrides.

## 1. Introduction

One of the major ways of the chemical reactions speed and selectivity intensify is creation of superequilibrium concentration of the vibration-excited molecules in the reacting gas medium. Now for this purpose two ways are used - the gas discharge and a laser beam irradiation of a reacting gas mixture [1] . In works [2, 3] we was shown, that alongside with the mentioned above ways there is the new method based on features of interaction of some alloys and compounds with chemically active gases.

Interesting peculiarity of convertible interaction hydrides forming metals and alloys with gaseous hydrogen is the thermodesorbig activation effect. This phenomenon proves to be true direct and indirect results of numerous experiments and is that processes of a metal hydrides formation and decomposition are accompanying by significant deviations of a near surface gas phase layers condition from thermodynamic balance. Thus, emissions of the excited molecules [2], hydrogen atoms [4], nuclear and molecular ions [5, 6] from a metal hydrides surface in a gas phase are observing. Mass - spectrometer researching of hydrogen, which desorbed from a metal hydrides surface , have shown, that such hydrogen has the sections of ionization by electronic impact of molecules $H_2$ increased on 30-50 % and lowered on 0,3-0,5 eV potentials of occurrence of molecular ions $H_2^+$, in comparison with equilibrium hydrogen . Last value is close to energy of the first oscillatory level of a molecule of hydrogen in basic electronic condition $X^1\Sigma_g^+$. Based on these results, and significant lifetime of the observed excited molecules of desorebed hydrogen the assumption was stating, that, the given excitation has oscillatory

*T.N. Veziroglu et al. (eds.),*
*Hydrogen Materials Science and Chemistry of Carbon Nanomaterials,* 447-455.
© 2004 *Kluwer Academic Publishers. Printed in the Netherlands.*

character and takes place in a course of an atoms $H$ recombination on a surface of hydride after hydrogen desorption [2].

In the present paper results carried out by authors of the experimental researches directed on control of plasma chemical reactions by using of vibration-excited molecules of hydrogen, which desorbed from metal hydrides are systematizing.

## 2. Features of chemical displays of a hydrogen activation effect by metal hydrides

Well-known catalytic activity of an intermetallic compounds hydrides in the reactions connected to carry of hydrogen, as a rule, spoke occurrence of atomic hydrogen near to a surface of hydride [7, 8]. However now there are no convincing experimental proofs of this assumption. High recombination factors for hydrogen atoms on metal catalyst surfaces [9] also raise the doubts that the determining role in the kinetic reactions improvement with participation of hydrogen is connecting with atoms H.

In papers [3,10] results of mass - spectrometer measurements of structure neutral components of the magnetron discharge hydrogen plasma were resulted. This research has shown that using metal hydride cathode stimulates various plasma chemical reactions with participation of hydrogen. Typical mass - spectra neutral components of the magnetron discharge hydrogen plasma (at a current of discharge $I_p$=100 mA, hydrogen filling from a cylinder in absence of metal hydride sample (1) and in the medium of the hydrogen contacting with a metal hydride sample (2)) are given on fig. 1. The analysis of the received data shows, that values intensity peaks of hydrogen ions $(H^+, H_2^+, H_3^+)$ in both cases remain relatives, but in structure of other components of the gas medium significant changes are observed. So, amplitude of peaks with $m/e$ = 12-17 and $m/e$ = 26-29 at presence a metal hydride sample are increased in some times. Simultaneously intensity of $O_2$ peak ($m/e$ = 32) considerably decreases, and for water ($m/e$ = 18) changes of intensity insignificant.

On fig. 2 characteristic dependencies on a discharge current ($I_p$) relations partial pressure of atomic oxygen $O$ ($m/e$=16), radical $OH$ ($m/e$=17), molecules of water $H_2O$ ($m/e$=18) and molecular oxygen $O_2$ ($m/e$ = 32).

In plasma of the discharge with the metal hydride cathode ($P_i$) to similar values partial pressure ($P_i^o$) for the control discharge are submitting. In control experiments, the discharge with the same geometry in the medium of the hydrogen filled from a cylinder, with the cathode from stainless steel, was using. In case of the gas discharge with the metal hydride cathode, essential decrease of molecular oxygen concentration is observing in comparison with similar value in absence of metal hydride. Together with it appreciable growth of concentration of radicals $OH$ ($m/e$ = 17) and molecules $H_2O$ ($m/e$ = 18) is observed. On the other hand, in process of growth of a discharging current the intensity of peaks of the same components at presence and in absence metal hydride tends to rapprochement. It is necessary to note, that in conditions of realization of experiments [3, 10] full pressure in a discharge cell was proportional to a current, which is increase of a current caused increase of pressure. Hence, observably deviations of componental plasma structure, caused by using the metal hydride cathode, decrease with growth of pressure. However, increase of gas pressure and the connected to it growth of collisions frequency results in increase of speed of a $VT$-relaxation of the vibration-excited molecules. Thus, the influence of metal hydride on the chemical reactions kinetic proceeding in plasma of the discharge may be explaining if to assume issue of vibration-

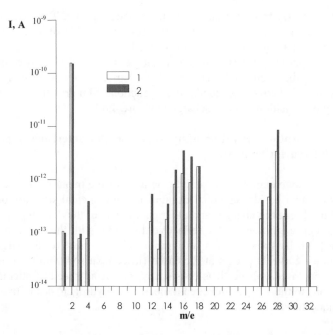

Fig. 1. Mass - spectra neutral components of magnetron discharge hydrogen plasma (a current of discharge $I_p$=100 mA), removed at leak-in hydrogen from a cylinder in absence metal hydride sample (1) and in the medium of the hydrogen contacting with metal hydride (2)

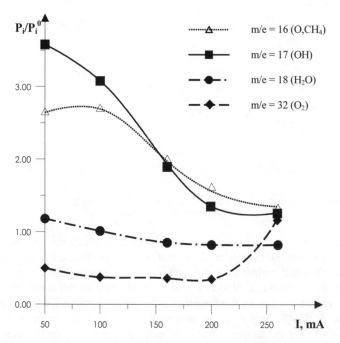

Fig. 2. Dependencies of relative amplitudes of peaks of mass - spectra of plasma on a discharge current

450

excited molecules $H_2$ from a metal hydride cathode surface. The offered explanation coordinated to results of mass - spectrometer researches [11] of a gas phase reactions kinetic with participation of the vibration-excited hydrogen.

Besides in paper [12] it is showing, that using a metal hydride source of hydrogen in technologies of a diamond synthesis in glow discharge results in substantial increase of quality of synthesized coverings, increase of a diamond film growth rate, and to decrease of probability of transition of the discharge in an arc mode.

## 3. Influence of sorption activation of hydrogen on characteristics of gas discharges with the metal hydride cathodes

The mentioned above explanation catalytic activity of the metal hydride cathode results in a conclusion, that besides essential changes in componental structure of plasma, this procedure should result in changes and other parameters of the discharge.

On fig. 3, curve ignitions of the reflective discharge with the metal hydride cathode are showing [13]. It is visible, that an ignition of the reflective discharge at presence metal hydride (a curve 1) occurs at increased (in comparison with control - a curve 2) values of the enclosed voltage. It is causing by the raised speed of sticking of the low energy electrons to the vibration-excited molecules of hydrogen, desorpting from a metal hydride cathode surface.

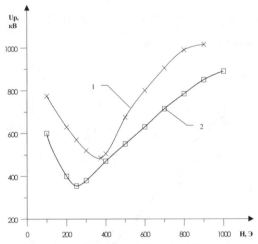

Fig. 3. Curve ignitions Penning discharge in the Hydrogen medium at pressure $P=3 \cdot 10^{-2}$ Pa (cathode material: 1 - metal hydride; 2 – steel).

In result, the near cathode area is impoverished on electrons and development of electronic avalanches is complicated. Therefore, for performance of a condition of existence of the self-maintained discharge it is necessary to increase speed of development of electronic avalanches that is achieving by increase of the enclosed voltage. The raised voltage of the magnetron discharge burning with the metal hydride cathode [13] is for the same reason observed. On fig. 4 are submitted volt-ampere characteristic of the magnetron discharge with the metal hydride cathode and the stainless steel cathode. The researches carried out by us [2, 10] have shown, that use of

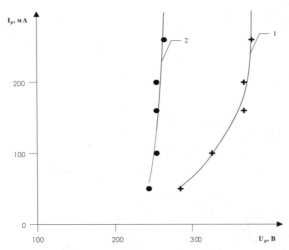

Fig.4. Volt - ampere characteristic of the magnetron discharge burning in the hydrogen medium:
1 - metal hydride cathode; 2 - stainless steel cathode

the metal hydride cathode results in significant changes of spatial distribution of a hydrogen plasma density in a discharge cell. On fig. 5, distributions of plasma in the magnetron discharge with the metal hydride cathode and the stainless steel cathode are showing. In process of distance from the cathode, spatial distribution of plasma density becomes homogeneous. Distribution of plasma density in volume of the vacuum chamber is defining by dynamics of hydride decomposition. Presence metal hydride does not result in appreciable changes of an electron temperature in plasma. The received results gave the basis to assume, that in discharge with the metal hydride cathode the increased output of hydrogen negative ions is possible. For check of this hypothesis, we carried out special researches [14-16] processes of formation of negative ions in the reflective and magnetron discharges. They have convincingly shown that the vibration-excited hydrogen molecules desorption from a metal hydride cathode surface unequivocally results in essential increase of concentration of negative ions $H^-$ in near-surface areas. As the cross section of process dissociate sticking slow electrons to the vibration-excited molecules of hydrogen quickly grows with increase of oscillatory quantum number $v$ [17], concentration of a hydrogen negative ions in near-cathode area is rather sensitive to a power condition of molecules of hydrogen desorpting from a metal hydride surface [13]. Use of the metal hydride cathode in a plasma source of ions results in increase of an output of ions $H^-$ approximately on the order of size (fig.6) [14-17].

## 4. Mass - spectrometer researches of a hydrogen molecules power conditions, desorpting from a metal hydride surface

Experiments [18-20] were operating by a method of removal of efficiency ionization curves (EIC) of a protium and deuterium molecules by electronic impact. Receiving EIC represented dependencies of a registered current of ions $H_2^+$ $(D_2^+)$ from an ionizing electrons energy. By mathematical processing of these dependencies [19], influence of thermal disorder of ionizing electrons on form EIC was eliminating and sets of values of

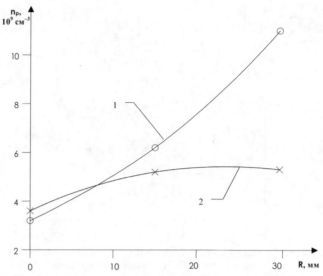

Fig. 5. Distribution of a plasma density on radius of magnetron discharge at height of 15 mm above a surface of the cathode at a discharge current of $I_p$=100 mA (1-metal hydride cathode; 2- stainless steel cathode)

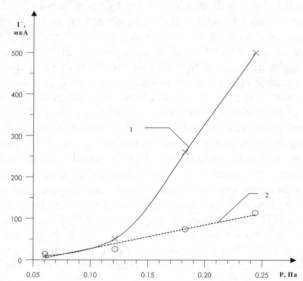

Fig. 6. Dependence of current I⁻ of negative ions H⁻ taken from Penning-discharge on pressure of hydrogen P (1-metal hydride cathode; 2- stainless steel cathode)

vertical potentials of ionization for desorpting hydrogen molecules are determined. According to curves of potential energy $H_2$ and $H_2^+$ [1, 21], the vertical ionization potential of an initial molecule of hydrogen with oscillatory quantum number $v$, forming a molecular ion with oscillatory quantum number $v'$, calculate on a ratio:

$$PI\ (v \to v') = PIA - E_v + E_{v'}, \qquad (1)$$

Where **PIA** - adiabatic ionization potential of a hydrogen molecule; $E_v$, $E_{v'}$ – vibration energy of a neutral hydrogen molecule at a level $v$ and a molecular ion at a level $v'$, accordingly.

Thus, the received set of vertical potentials of ionization depends on the vibration - excited conditions of hydrogen molecules desorbing from a metal hydride surface. The example of result of processing experimental EIC for protium is submitting in Table 1. The ionization potentials appropriate to vibration-excited conditions of initial molecules of the desorbed hydrogen are allocating fat with a font. At similar processing EIC for equilibrium, molecular hydrogen of such potentials it was revealed not. Thus, the appreciable part hydrogen molecules desorbing from a metal hydride is in the vibration-excited condition.

TABLE 1.Vertical potentials of hydrogen ionization (eV), desorbed from a sample $(Zr_{56}V_{35}Fe_8)H_x$

| Oscillatory quantum number | | Theoretical potential of ionization | Result of the experimental EIC processing |
|---|---|---|---|
| $H_2$ (v) | $H_2^+$ (v') | | |
| 1 | 0 | **14,910** | **14,90** |
| 1 | 1 | **15,174** | **15,17** |
| 0 | 0 | 15,427 | 15,43 |
| 1 | 3 | **15,667** | **15,66** |
| 0 | 2 | 15,943 | 15,94 |
| 0 | 3 | 16,182 | 16,18 |
| 0 | 4 | 16,405 | 16,40 |

For specification of vibration levels and a finding of a force field of molecules comparative research of vibration spectra of their isotope versions is widely used. Within the framework of the Born-Opengeimer approximation it is supposing, that at replacement of isotopes, distribution of electronic density, equilibrium internuclear distances, function of potential energy, and force constants remain constant. Distinctions in weights of nucleus results only in change of kinetic energy of varying nucleus that causes distinctions of frequencies of vibration of isotope versions of molecules. From the received experimental EIC for hydrogen [18] and deuterium [20] was calculated potential of occurrence of ions. Results of processing are submitting in Tabl. 2.

TABLE 2. Potentials of the hydrogen isotopes ionization

| Isotope of hydrogen | Adiabatic ionization potential, eV | Measured ionization potential, eV | Measured displacement of ionization potential, eV | Excitation Energy for first oscillatory level, eV |
|---|---|---|---|---|
| $H_2$ | 15,426 | 14,91 | 0,52 | 0,546 |
| $D_2$ | 15,467 | 15,08 | 0,39 | 0,386 |

For two-nuclear molecules, in harmonious approximation, the relation of frequencies of fluctuations of isotope versions of the molecules, which are taking place at the first oscillatory level, depends only on weights of atoms making molecules. In case of hydrogen isotopes, frequency of fluctuations of a protium molecule $H_2$ in $\sqrt{2}$ times more frequency of fluctuations of a deuterium molecule $D_2$. The relation of the experimentally measured displacement of ionization potential for desorbed protium and deuterium makes 1, 35 (Tabl. 2). The given size, within the framework of an allowable error of measurements, is close to the above-stated isotopic effect size. The received result is direct experimental acknowledgement of the hypothesis stated earlier by authors [18] about vibration excitation of molecules of hydrogen during them desorbing from a metal hydrides surface.

## 5. Conclusions

In the present paper results of the experimental researches carried out by authors directed on management of reactions kinetic with use of vibration-excited molecules of hydrogen, desorbed from a metal hydrides surface are stated. It is showing, that using a metal hydrides composite material - convertible sorbents of hydrogen of low pressure as electrodes of discharge devices of technological purpose effectively allows influencing on different parameters of the plasma-chemical reactions with participation of hydrogen.

## 6. References

1. Capitelly, M., editor. (1986) Non Equilibrium Vibration Kinetics. Topics in Current Physics: Springer Verlag, Berlin.
2. Shmal'ko, Yu.F., Klochko, Ye.V., Solovey, V.V., a.a. (1995) The formation of excited H species using metal hydrides, *Journ. Alloys and Compounds*, 231.
3. Klochko, Ye.V., Popov, V.V., Shmal'ko, Yu.F., a.a. (1999) Investigation of plasma interaction with metal hydride, *Int. Journ. of Hydrogen Energy*, 24.
4. Savvin, N.N., Mjasnikov, I.A., Lobashina, N.E. (1985) Non-stationary emission of hydrogen atoms from a surface as-grown layer metal on glass at the presence of molecular hydrogen, *Journ. of physical chemistry* **59**, 10 (in rus.).
5. Bachman, C.H., Silbergt, P.A. (1958) Thermionic ions from hydrogen palladium, *Journ. Apl. Phys.* **29**, 8.
6. Shapovalov V.I. (1978) About the form of existence of hydrogen in iron and an opportunity of proton issue from metals, Proceedings of High schools. *Ferrous metallurgy*, 8 (in rus.).
7. Lobashina, N.E., Savvin, N.N., Mjasnikov, I.A. (1984) Research of the spill-over mechanism $H_2$ on the put metal catalysts. III. A nature of migrating particles. *Kinetic and Catalyst* **25**, 2 (in rus.).
8. Lobashina, N.E., Savvin, N.N., Mjasnikov, I.A. (1983) Formation and transport of hydrogen atoms from the metal - activator on a surface of the carrier (spill-over effect) and in a gas phase, *Reports of Academy of Sciences of USSR* **268**, 6 (in rus.).
9. Lavrenko, V.A. (1973) Recombination of hydrogen atoms on a solid state surface, *Nauka*, Kiev. (in rus.).
10. Klochko, Ye.V., Popov, V.V., Shmal'ko, Yu.F. a.a. (1997) Sorption and electro transfer characteristics of hydrogen-gettering material in contact with a hydrogen plasma, *Journ. Alloys and Compounds*, 261.
11. Dodonov, A.F. (1985) Mass-spectrometer researching the elementary gas-phase reactions proceeding in system $H+O_2+O_3$. Reactions $H+OH$ ($v \geq 6$) $\rightarrow H_2$ ($v=5$)$+O$, $H+H_2$ ($v=5$)$\rightarrow H_2$ ($v<4$)$+H$, *Chemical Physics* **4**, 10 (in rus.).
12. Pashnev, V.K., Strel'nitskij, V.E., Shmal'ko, Yu.F. a.a. Highly Effective Setup for Diamond Coating Deposition. In: Proceedings of the Sixth Applied Diamond

Conference/Second Frontier Carbon Technology Joint Conference (ADC/FCT 2001), Y. Tseng, K. Miyoshi, M. Yoshikawa a.a. editors. Auburn, Alabama (USA): 2001.

13. Shmal'ko, Yu.F., Borisko, V.N., Klochko, Ye.V., a.a. (2000) About vibration excitation of the hydrogen molecules desorbed from metal hydrides, *Proceedings of National Academy of Sciences of the Ukraine*, 11 (in ukr).

14. Shmal'ko, Yu.F., Solovey, V.V., Strokach, A.P., a.a. (1994) Application of Metal Hydrides in Ion Sources, *Z. Physik. Chem.*, 183.

15. Borisko, V.N., Klochko, Ye.V., Shmal'ko, Yu.F., a.a. (1998) A technological plasma source of negative ions. Problems of nuclear sciences and engineering. Physics of radiating damages and radiating material sciences, **3**(69), **4**(70) (in rus).

16. Bitnaja, I.A., Borisko, V.N., Klochko, Ye.V., a.a. (1998) A source of negative ions on the basis of planar magnetron, *Proceedings of Kharkiv University. Nucleus, particles, fields*, 421 (in rus).

17. Allan, M., Wong, S.F. (1978) Effect of Vibrational Exitation on Dissociative Attechment in Hydrogen, *Phys. Rev. Letters* **41**, 26.

18. Valujskaja, S.B., Skripal', L.P., Shmal'ko, Yu.F., a.a. (1989) Research of a hydrogen activation process by metal hydrides. II. Mass-spectrometer definition of potential and section of a hydrogen ionization. Problems of nuclear sciences and engineering, *Nucl. engineering and technology*, 1 (in rus.).

19. Shmal'ko Yu.F., Solovey V.V., Klochko Ye.V. a.a. (1995) Mass-spectrometry determination of vibration excited states of hydrogen molecules desorbed from the metal hydrides surface, *Int. Journ. of Hydrogen Energy* **20**, 5.

20. Shmal'ko, Yu.F., Klochko, Ye.V., Lototsky, M.V. (1996) Influence of isotopic effect on the shift on the ionization potentials of hydrogen desorbed from metal hydride surface, *Int. Journ. of Hydrogen Energy* **21**, 11/12.

21. Gaydon, A.G. (1947) Dissociation Energies and Spectra of Diatomic Molecules, London.

# VIBRATIONAL SPECTRA OF C$_{60}$ POLYMERS: EXPERIMENT AND FIRST-PRINCIPLE ASSIGNMENT

V. M. SENYAVIN*, A. A. POPOV, A. A. GRANOVSKY

*Chemistry Department, Moscow State University, 119892 Moscow, Russia*
*Corresponding author: fax 7-095-9328846, E-mail:*
*senyavin@phys.chem.msu.ru (V. M Senyavin)*

**Abstract**

The infrared and Raman spectra of orthorhombic and tetragonal polymers of C$_{60}$ are thoroughly studied and interpreted on the base of DFT vibrational calculations of their finite fragments. Complete assignment of the spectra serves to establish the genetic relationships of the polymer vibrations to the parent fullerene modes and to follow the vibrations associated with some particular intramolecular motions.

## 1. Introduction

Infrared and Raman spectroscopy was demonstrated to be a convenient and reliable tool to follow the derivatization of an extraordinarily high-symmetric C$_{60}$ fullerene molecule, particularly in the processes of 2+2 cycloaddition reactions [1] leading to the dimer and orthorhombic (O), tetragonal (T) and rhombohedral (R) polymers of C$_{60}$ [2,3]. The exhaustive symmetry considerations dealing with the predictions of the IR and Raman spectra of the C$_{60}$ polymers on the base of parent modes splitting and activation were reported [2,4] and several tentative assignments of the spectra were proposed [3,4].

At the same time, the plausible assignment of the spectra should be based on adequate calculations. Vibrational spectra of dimer, 1D chain and medium size oligomers were computed within various hybrids of LDA and tight-binding approaches (DF-TB [5] and QMD [6]), but these data could hardly serve as a fundamental background for vibrational studies due to insufficient accuracy of the methods. Recently [7,8] more reliable results were obtained for O, T, and R phases with the use of improved wavefunction cut-off parameter in the framework of QMD and assignment of the far-IR spectra of the polymers was proposed.

Nevertheless, the complete interpretation of the vibration spectra of the C$_{60}$ polymers at an adequate quantum chemical level is still unavailable being a highly demanding problem. We have started our computational study of the fullerenic systems by an ab initio and DFT calculations of the (C$_{60}$)$_2$ dimer molecule [9]. Present work is aimed to spread such approaches to the polymers of C$_{60}$ by combining the superiority of high-level quantum chemical calculations and the reliable models to simulate the spectra of polymeric species.

## 2. Experiment and calculations

The samples of the investigated O and T polymeric phases were described previously [3]. Briefly, they were selected from some hundreds of the p,T-treated C$_{60}$

*T.N. Veziroglu et al. (eds.),*
*Hydrogen Materials Science and Chemistry of Carbon Nanomaterials,* 457-466.
© 2004 *Kluwer Academic Publishers. Printed in the Netherlands.*

samples synthesized by Dr.V.A.Davydov group in the vicinity of the conditions stated in [3] to provide more than 90 mol per cent purity [2]

Infrared spectra of the compounds/KBr pellets were measured on a Bruker-113v FT-spectrometer in a 400-4000 cm$^{-1}$ range under 0.5 cm$^{-1}$ resolution. Raman spectra of the O phase were measured on a Bruker Equinox 55S spectrometer equipped with a FRA 106s module and Nd:YAG excitation laser operated at 1064 nm line. Experimental Raman spectrum of the T phase is that published previously [2,3]. It was obtained on a Dilor XY set under 562.8 Kr$^+$ line excitation and was supplied to us by Dr. V.N. Agafonov.

DFT calculations were performed with the PRIRODA package [10] using GGA functional of Perdew, Burke and Ernzerhof (PBE) [11] and TZ2P {6,1,1,1,1,1/4,1,1/1,1} basis sets. To accelerate the evaluation of the Coulomb and exchange-correlation terms the code employed an expansion of the electron density in an auxiliary basis set. Polarizability tensor derivatives were computed numerically at HF/6-31G level with PC version [12] of GAMESS (US) quantum chemical package [13].

## 3. Results and discussion
### 3.1 Linear trimer (C$_{60}$)$_3$

The simplest molecule containing a real fragment of polymerized fullerene is a linear trimer, (C$_{60}$)$_3$, reported [14] to be HPLC isolated as an individual substance (solvated by o-dichlorobenzene) and characterized by UV and IR spectroscopy. Besides an apparent self interest this molecule can serve as a model to estimate the abilities of modern computational approaches and to test the schemes for the polymeric structures and spectra treatment.

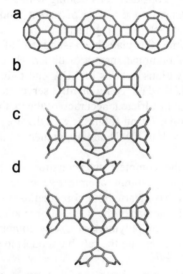

Figure 1. The structures of the trimer (C$_{60}$)$_3$ molecule (a), of the substructure tested for the force field locality (b) and of the model molecules used for simulations of the O (c) and T (d) polymers spectra.

The first-principle vibrational calculations of such a giant molecule (Fig.1a) meet virtually the limits of common computational equipment. Indeed, the calculations in TZ2P basis set (which was shown to be sufficient for fullerenic systems [15]) includes 4500 basis functions and demands for a Gbyte-volume operative memory and some dozens of Gbytes disc space during several hundreds of hours to operate at more than 1 Gbyte processor frequency. As the result of these calculations we have obtained the spectrum of $(C_{60})_3$ molecule (Fig 2a) satisfactorily matching the experiment [14] and closely resembling calculated [9] and observed [2,16] spectra of the dimer $(C_{60})_2$. The geometric parameters were also close to those calculated earlier [9] for $(C_{60})_2$ molecule.

Searching for the model to simulate the spectra of polymers we have used the results obtained for $(C_{60})_3$ molecule and tested extensively the approach of Cui and Kertesz [17], proposed initially for the quite small links. Within this method the force field of a polymer is approximated by that of its repetitive link corrected for the interactions with some neighboring links, i.e.

$$\mathbf{F}'(\mathbf{k}) = \mathbf{F}(0,0) + \sum_{q=\pm 1}^{\infty} \mathbf{F}(0, q)\exp(ikqa)$$

where matrices $\mathbf{F}(0,q)$ correspond to the interactions of the central link with $q$-th link, $a$ is the lattice parameter and $k$ is the reciprocal lattice vector. As it was shown in [17], the series converge rapidly and frequencies obtained on the base of rather small oligomeric molecules perfectly reproduce those computed within the periodic boundary conditions.

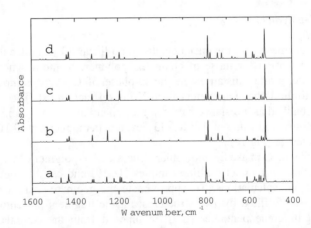

Figure 2. (a) IR spectrum of $(C_{60})_3$ molecule; (b)-(d) IR spectra simulated for O polymer: using $(C_{60})_3$ molecule, using fragment (b) on Fig.1, and using the model molecule (c) on

At the first stage using the complete force field and the intensity parameters of a trimer molecule the spectrum of a linear $(C_{60})_n$ polymer was simulated (Fig.2b), which satisfactorily reproduced the experimental data for the orthorhombic polymerized phase of $C_{60}$ (see below). At the second stage, the localizability of inter-link interactions was studied, which was of prime importance for the purposes of a present study. To elucidate the extent of significant interactions between the carbon atoms in neighboring $C_{60}$ spheres the spectra of infinite chain were simulated using the force constant matrices corresponded to several hypothetical substructures of a trimer molecule. The regarded fragments combined one $C_{60}$ moiety and several carbon atoms in each neighboring sphere, the number of atoms was increased consequently until the deviations from the complete solution became negligible. Simulations of the chain polymer IR spectrum with the force field of the $C_{60}(C_{14})_2$ structure (Fig.1 b) illustrates that the account of 14 carbon atoms nearest to the central link is sufficient to yield the vibrational properties of a polymer with the accuracy of complete solution (Fig. 2 b, c). Hence, the methodic of the subsequent polymers spectra calculations consists in the high-level first-principle computing of the force fields and intensity parameters of their representative fragments embracing the central bucky ball surrounded by the three belts of carbon atoms from each neighboring $C_{60}$ sphere. For the convenience, the unsaturated valences of the boundary carbon atoms were artificially filled with the nitrogen (triple-coordinated) or oxygen (double-coordinated) atoms which contributions were not accounted for in the spectra simulations. An example of such structure used to model the spectra of O polymer is shown on Fig.1c, the results of simulation – at the top of Fig.2. To characterize the quality of approximation it can be noted that the root mean square deviation in the frequencies obtained for this model molecule from those calculated with the complete force field of a trimer constitutes 1.8 cm$^{-1}$, while the intensity distribution of the spectral bands is nearly identical. The structure of a fragment used to obtain the spectra of T-polymer is given on Fig.1d.

### 3.2 O- and T-polymers of $C_{60}$
### 3.2.1 Geometry

The geometry parameters calculated for a linear $(C_{60})_3$ molecule are believed to be a good approximation for those of chain $C_{60}$ polymer. In the absence of experimentally determined inter-atomic distances in the O-phase of $C_{60}$, the computed values can be compared with the lattice parameters from X-ray diffraction [3]. The calculated parameter of 9.119 Å (obtained as a distance between the centers of the four-member rings) coincides well with the observed value of 9.098 Å [3], even better agreement (9.106 Å) was found for the model molecule (Fig.1c).

Unlike the O phase the molecular structure of T-polymer of $C_{60}$ was determined by X-ray diffraction on monocrystalline samples [18,19] and can be compared directly with the results of our calculations. To approximate the geometry parameters of T-phase the optimization was fulfilled for two structures: for the fragment presented in Fig.1d and for the branched fullerene pentamer $(C_{60})_5$, composed from the central unit with four $C_{60}$ moieties attached after the tetragonal motif (for the atoms in the outer hemispheres of the latter the dz basis set was used). The parameters obtained for these two structures agreed with each other within 0.001 Å for the intra-ball bonds and 0.003 Å for the inter-ball bonds. Therefore, in table 1 only the parameters computed for the first structure (for which the vibrational calculations were done) are compared with experiment; the table contains also the values calculated for O polymer modeled by the trimer molecule.

Table 1. Experimental (T phase [18,19]) and calculated for O and T polymers bondlengths, Å [a]

| | Calculated | | X-ray | |
| | O | T | T [18] | T [19] |
|---|---|---|---|---|
| Double bonds | | | | |
| **j** | 1.385 | 1.388 | 1.367(6) | 1.365(4) |
| **k** | 1.403 | 1.390 | 1.377(6) | 1.382(4) |
| **l** | 1.403 | 1.405 | 1.388(6) | 1.390(4) |
| **m** | 1.398 | 1.396 | 1.370(10) | 1.367(8) |
| **n** | 1.398 | | | |
| Ordinary bonds | | | | |
| **a** | 1.522 | 1.519 | 1.521(5) | 1.515(4) |
| **b** | 1.437 | 1.433 | 1.432(5) | 1.436(4) |
| **c** | 1.453 | 1.475 | 1.460(8) | 1.455(6) |
| **d** | 1.452 | 1.450 | 1.457(6) | 1.450(4) |
| **e** | 1.446 | 1.440 | 1.423(6) | 1.434(4) |
| **f** | 1.481 | 1.476 | 1.476(7) | 1.472(6) |
| **g** | 1.445 | 1.426 | 1.400(6) | 1.416(4) |
| **h** | 1.453 | 1.521 | 1.529(5) | 1.529(4) |
| **i** | 1.450 | 1.454 | 1.461(7) | 1.456(6) |
| Four-member rings | | | | |
| **o** | 1.601 | 1.603 | 1.620(10) | 1.585(8) |
| **o-o'** | 1.597 | 1.593 | 1.600(10) | 1.605(8) |
| **n** | | 1.598 | 1.580(10) | 1.562(9) |
| **n-n'** | | 1.593 | 1.560(10) | 1.575(8) |

[a] see Fig. 3 for bonds notation

Bonds notation is given on the Shlegel diagram in Fig.3.

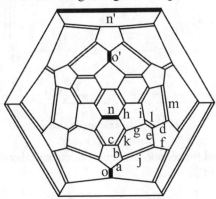

Figure 3. Shlegel diagram for the monomeric fragments of O and T polymers with bonds notation

As can be seen, calculations overestimate slightly the length of the bonds, especially for "double" bonds. As for the geometry of 4-member rings, neither ours, nor literature [8] calculations do not reproduce the prominent experimental differences in the bondlengths, which are parallel and normal to the layer plane. The same is true for the inter-ball distances: if the experimental values with the necessity of $P_4$ space group dictates the identical distances "along" and "across" the layer, our calculations showed 0.1 Å difference between them.

### 3.2.2 Vibrational spectra

The monomeric links of both O and T polymers possess the $D_{2h}$ point symmetry, therefore, the expected number of the bands in the spectra of polymers and the manner of the parent modes splitting and activation upon polymerization is the same for these two phases [2]. Hence, from the symmetry point of view they are equivalent and only an adequate calculation could predict the values of splitting and the extent of activation of silent $C_{60}$ modes.

Calculated spectra are compared with experimental data on Figs.4-5.

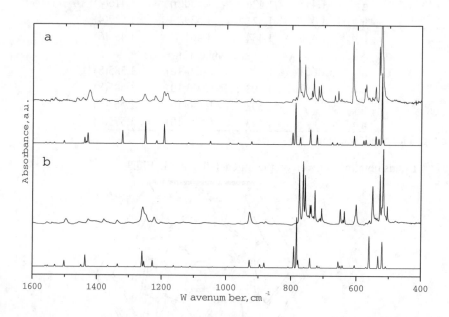

Figure 4. Experimental (upper curves) versus calculated (bottom curves) IR spectra of O (a) and T (b) polymers of $C_{60}$.

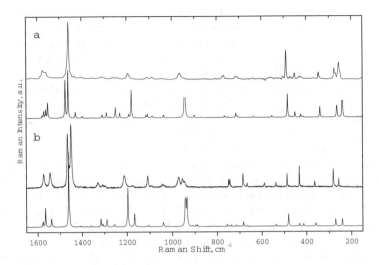

Figure 5. Experimental (upper curves) versus calculated (bottom curves) Raman spectra of O (a) and T (b) polymers of $C_{60}$.

As can be seen, the correspondence between them is quite satisfactory. Computed IR spectra perfectly match the observed ones in the region above 800 cm$^{-1}$ for both polymers, while some discrepancy is found for the T polymer spectrum in the intensity distribution within the 700-800 and 500-600 cm$^{-1}$ regions. Computed Raman spectrum for O polymer agrees well with experiment in the whole spectral range with the only exception for the line at 967 cm$^{-1}$, which position is downshifted by the DFT calculations with the considerable overestimation of its intensity. The origin of this phenomenon lies in the fact that this vibration is localized mostly on the saturated 4-member fragment of the molecule, which is hardly be described correctly with the PBE functional suited well for the conjugated systems [15]. The same is true for the some mentioned above IR-active vibrations in the vicinity of 800 cm$^{-1}$. Experimental Raman spectrum of the T-phase possess a distinctive pre-resonant character [2] and only the positions of the lines could be compared with the results of calculations.

In spite of a few discrepancies marked the calculations performed allowed us to fulfill the complete interpretation of the $C_{60}$ polymers vibrational spectra and to ascribe all the observed bands to the definite molecular motions. The assignment of the spectra was done both within the normal coordinate analysis and in terms of the parent $C_{60}$ modes. The conjunction of these two approaches allowed us as to follow the parent modes transformation upon polymerization as to reveal the vibrations caused by the structural fragments formed as a result of 2+2 cycloaddition.

The lowest-frequency $H_g(1)$ parent mode splits considerably upon polymerization to 251/256/271/277/346 cm$^{-1}$ and 257/279/282/363 cm$^{-1}$ multiplets in Raman spectra of O and T phases, respectively, in close accordance with the symmetry predictions; $H_g(2)$ mode causes the lines at 416, 420, 426, 431 and 452 cm$^{-1}$ in O, and 415/430 cm$^{-1}$ doublet in T polymer. $A_g(1)$ line shifts slightly from its position in $C_{60}$ molecule to derive the 491 and

485 cm$^{-1}$ lines in spectra of O and T polymers. Low-intensity Raman lines of O and T polymers in a 500-600 cm$^{-1}$ region are assigned to $G_g(2)$ mode derivatives. $H_g(3)$ mode causes the broad peak at 714 cm$^{-1}$ in O and splits significantly in the spectra of T polymer to form the 667, 701 and 710 cm$^{-1}$ features. One of its components is mixed with that of $H_g(4)$ mode to cause the 685 cm$^{-1}$ line in the spectrum of T phase.

For some vibrations in a 900-1200 cm$^{-1}$ interval the large contribution from the C-sp$^3$ atoms motion was found. In Raman spectra the stretching vibrations within the 4-member cycle are observed at 967, 1036, and 1196 cm$^{-1}$ for O polymer and at 971, 1034, and 1174 cm$^{-1}$ for the T phase. More complex stretching-bending modes of the cycle are displayed as a shoulder at 950 cm$^{-1}$ in the spectrum of O polymer and as a 945/955 cm$^{-1}$ doublet in T polymer spectra. The torsions of the cycle as a whole around the axes connecting the $C_{60}$ moieties cause the lines at 1047(O) and 1044(T) cm$^{-1}$.

The appearance of $H_g(5)$ mode of $C_{60}$ as the doublets at 1089/1111 and 1092/1109 cm$^{-1}$ is quite similar in the spectra of O and T polymers, while the $H_g(6)$ components are seen in the O-phase spectrum only as the lines at 1244, 1262 and a shoulder at 1300 cm$^{-1}$. Within the 1400-1600 cm$^{-1}$ Raman shifts the intense lines at 1457 (O) and 1465 (T) cm$^{-1}$ have their origin in the $A_g(2)$ parent vibration, those at 1398/1434 (O) and 1406/1449 (T) cm$^{-1}$ are the derivatives of $H_g(7)$ mode, and the group of the peaks in a 1560-1580 cm$^{-1}$ range are caused by the $H_g(8)$ $C_{60}$ mode.

In infrared spectra the most intense are the groups of the bands in the 500-600 and 700-800 cm$^{-1}$ range. In the lower-frequency interval the derivatives of IR-active $F_{1u}(1)$ and $F_{1u}(2)$modes of $C_{60}$ are observed. The former causes the most intense and medium absorptions at 524/531 (O) and 522/555 (T) cm$^{-1}$ and contribute with the $H_u(2)$ mode components to the forms of 520 (O) and 510 (T) cm$^{-1}$ vibrations. The $F_{1u}(2)$ mode splits to the medium and weak bands within 542-572 and 532-565 cm$^{-1}$ intervals in the spectra of O and T polymers, respectively.

In a 600-700 cm$^{-1}$ region the most remarkable IR feature, characteristic for all the polymers and the dimer of $C_{60}$ is observed, namely the strong band near 610 cm$^{-1}$, which appear as the first manifestation of polymerization process [3] and retains its position throughout all the phases within 606-612 cm$^{-1}$ interval. As our calculations showed this vibration is connected with a specific movement of the fullerene pentagons attached to the 4-member ring. Along with this band the triplets of $H_u(3)$ derivative within the 640-670 cm$^{-1}$ interval are observed for both polymers.

The 700-800 cm$^{-1}$ spectral range embraces the derivatives of $F_{2u}(2)$ parent mode (the 710/717 and 711/718 doublets in O and T spectra) and those of $H_u(4)$ and $G_u(3)$ modes (as the strong bands near 760 and 770 cm$^{-1}$ for both polymers). In accordance with experiment the high IR intensity was calculated for the bands in the 780-800 cm$^{-1}$ interval, though their frequencies were overestimated by calculations within 10-15 cm$^{-1}$. As mentioned above these bands are caused by the translational movements of the 4-membered rings. Other IR-active vibrations of the saturated cycles possess the higher frequencies and are of less intensity. Their very characteristic position [2] was perfectly matched by the calculations.

## 4. Conclusions

The model to simulate the vibrational spectra of fullerene polymers using the quantities of their finite fragments was realized and thoroughly tested. Within this approach the results of the high-level quantum chemical force fields and intensity parameters calculations were transferred to the polymeric species without deterioration. Complete interpretation of the infrared and Raman spectra of O and T $C_{60}$ polymers on the base of DFT calculations was used to establish the genetic relatioships of the polymer vibrations to the parent fullerene modes and to follow the vibrations associated with some particular intramolecular motions. The combined experimental and theoretical study of the fullerene polymers vibrational spectra have confirmed the purity of the phases studied and revealed the frames and limits of the computational approaches.

### Acknowledgements
This work was supported by RFBR grant No 03-03-32179 and in part by grant No 03-07-96842-r2003jugra_v. We are thankful to Dr. V. A. Davydov for supplying us with the samples of O and T polymers and to Dr. V. N. Agafonov for the Raman spectrum of T phase. The assistance of Dr. A. N. Kharkanov in Raman measurements is gratefully acknowledged.

### References
1. Iwasa Y, Arima T, Fleming RM, Siegrist T, Zhou O, Haddon RC, et al. New Phases of $C_{60}$ Synthesized at High Pressure. Science 1994; 264: 1570–1572.
2. Senyavin VM, Davydov VA, Kashevarova LS, Rakhmanina AV, Agafonov V, et al. Spectroscopic properties of individual pressure-polymerized phases of C60. Chem Phys Lett 1999; 313: 421–425.
3. Davydov VA, Kashevarova LS, Rakhmanina AV, Senyavin VM, Ceolin R, Szwarc H, et al. Spectroscopic study of pressure-polymerized phases of $C_{60}$. Phys Rev B 2000; 61(18): 11936–11945.
4. Kamaras K, Iwasa Y, Forro L. Infrared spectra of one- and two-dimensional fullerene polymer structures: $RbC_{60}$ and rhombohedral $C_{60}$. Phys Rev B 1997; 55(17): 10999 - 11002.
5. Porezag D, Pederson MR, Fraunheim Th, Kohler Th. Structure, stability, and vibrational properties of polymerized $C_{60}$. Phys Rev B 1995; 52(20): 14963 – 14970.
6. Adams GB, Page JB, Sankey OF, O'Keeffe M. Polymerized $C_{60}$ studied by first-principles molecular dynamics. Phys Rev B 1994; 50(23): 17471 – 17479.
7. Zhu ZT, Musfeldt JL, Kamaras K, Adams GB, Page JB, Davydov VA, et al. Far-infrared vibrational properties of linear $C_{60}$ polymers: A comparison between neutral and charged materials. Phys Rev B 2003; 67:045409.
8. Zhu ZT, Musfeldt JL, Kamaras K, Adams GB, Page JB, Davydov VA, et al. Far-infrared vibrational properties of tetragonal $C_{60}$ polymer, Phys Rev B 2002; 65: 85413.
9. Senyavin VM, Popov AA, Granovsky AA. Ab initio and DFT study of cohesive energy and vibrational spectra of a fullerene dimer, $C_{120}$. Book of abstracts, 201st Meeting of The Electrochemical Society. Philadelphia (Pennsylvania, USA): Electrochemical Society, 2002; Abs. No 1014.
10. Laikov DN. Fast Evaluation of Density Functional Exchange-Correlation Terms Using the Expansion of the Electron Density on Auxiliary Basis Sets. Chem Phys Lett 1997; 281: 151–156.
11. Perdew JP, Burke K, Ernzerhof M. Generalized gradient approximation made simple. Phys Rev Lett 1996; 77(18): 3865–3868.

12. Granovsky AA. PC GAMESS URL: http://classic.chem.msu.su/gran/gamess/index.html.
13. Schmidt MW, Baldridge KK, Boatz JA, Elbert ST, Gordon MS, Jensen JH, et al. J Comput Chem 1993; 14(11): 1347–1363.
14. Komatsu K, Fujiwara K, Tanaka T, Murata Y. The fullerene dimer $C_{120}$ and related carbon allotropes. Carbon 2000; 38: 1529–1534.
15. Senyavin VM, Popov AA, Granovsky AA, Davydov VA, Agafonov VN. Ab initio and DFT-bazed assignment of the vibrational spectra of polymerized fullerenes. Full Nanotub Carbon Nanostruct, in press.
16. Komatsu K, Wang G-W, Murata Y, Tanaka T, Fujiwara K, Yamamoto K, Saunders M. Mechanochemical synthesis and characterization of the fullerene dimmer $C_{120}$. J Org Chem 1998; 63: 9358.
17. Cui CX, Kertesz M. Quantum-mechanical Oligomer Approach for Calculation of Vibrational Spectra of Polymers. J Chem Phys 1990; 93(7): 5257–5266.
18. Narymbetov B, Agafonov V, Davydov VA, Kashevarova LS, Rakhmanina AV, Dzyabchenko A.V, et al. The crystal structure of the 2D polymerized tetragonal phase of $C_{60}$. Chem Phys Lett 2003; 367: 157–162.
19. Chen X, Yamanaka S. Single-crystal X-ray structural refinement of the 'tetragonal' $C_{60}$ polymer. Chem Phys Lett 2002; 360: 501–508.

# FEASIBILITY OF HYDROGEN ENERGY PRODUCTION THROUGH NATURAL GAS STEAM REFORMATION PROCESS IN THE UAE

A. KAZIM
*United Arab Emirates University*
*Department of Mechanical Engineering*
*P.O. Box 17555 Al-Ain, U.A.E.*
*Fax: +971-3-7623-158*
*E-mail: akazim@uaeu.ac.ae*

## Abstract

This paper presents a feasibility study of producing environmental friendly hydrogen energy through natural gas steam-reformation process in the UAE, which has estimated natural gas reserves of $224.9 \times 10^9$ GJ. The current findings demonstrated that the cost of hydrogen energy produced by steam-reformation is approximately \$8.2/GJ, which is 10%, 45% and 57% less than hydrogen produced through biomass gasification, solar-hydrogen and wind-hydrogen, respectively. Moreover, The total amount and cost of carbon emitted by natural gas steam-reformation are calculated to be 48 Kg-$CO_2$/GJ and \$6.7/GJ, respectively. These values were considered to be more favorable in comparison with other fossil fuels and hydrogen producing methods.

*Keywords:* Hydrogen energy, PEM fuel cell, natural gas steam-reformation, carbon emission

## 1. Introduction

United Arab Emirates (UAE), which is considered to be one of the leading countries in the world in natural gas production, currently produces approximately 3 billion cubic feet per day (bcf/d), and it is expected to reach 4 bcf/d by the year 2005. Energy Information Administration EIA, estimated that UAE's reserves of the natural gas are about $224.9 \times 10^9$ GJ, which is considered world's fourth-largest natural gas reserves after Russia, Iran and Qatar. Over 90% of UAE's reserves of natural gas are in the emirate of Abu Dhabi, which has been supplying Japan most of its liquefied natural gas (LNG) since 1977 [1]. In order to fulfill the rapid demand of natural gas in the energy market, UAE has embarked major NG-intensive projects, specially the multi-billion \$US Dollars Dolphin project, which interconnects pipelines between Qatar, UAE and Oman and with possible links to Indian subcontinent [2].

In recent years, the environmental concerns of the developed countries and their determination of implementing the Kyoto-commitment has led them to seek a suitable environmentally-friendly energy alternative that could be relied on for future energy needs. Many scientists, energy economists, and energy policy makers believe that hydrogen energy, which possesses significant characteristics such as being

*T.N. Veziroglu et al. (eds.),*
*Hydrogen Materials Science and Chemistry of Carbon Nanomaterials,* 467-472.

environmentally clean, storable, transportable and inexhaustible, could play a key role in fulfilling the global energy demand without any environmental sacrifices. Presently, utilization of hydrogen energy to run fuel cells for a clean power generation has gained worldwide attention. Fuel cells especially proton exchange membrane fuel cells (PEMFC) have been playing a significant role in industrial, utilities and transportation sectors of many developed countries such as Japan, USA and Germany. Thus, UAE could play a key role in supplying the developed countries with hydrogen energy produced from natural gas reformation in order to fulfill their Kyoto commitment.

The objective of the current study is to conduct a feasibility study of hydrogen energy production by natural gas steam-reformation process in the UAE. The analysis is to be performed at various scenarios (%) of natural gas used in hydrogen production. Moreover, the total cost of hydrogen produced by 10% of NG as well as the total amount and cost of carbon emitted by natural gas steam-reformation are to be compared against other fossil fuels and hydrogen producing methods.

## 2. Technical, economical and environmental considerations

Production of hydrogen can be achieved through either utilizing conventional resources such as oil and natural gas steam-reformation, coal gasification or by renewable resources such as biomass gasification, solar-hydrogen and wind-hydrogen and hydropower electrolysis processes. Globally, around 400 billion cubic meters of hydrogen are produced annually with natural gas steam-reformation process, is considered to be the most common and cost effective hydrogen producing method [3]. Approximately, 48% of worldwide hydrogen energy is produced from this process; even experts have recommended mass production of small-scale natural gas reformers, supplying direct hydrogen for stationary fuel cells and at vehicle refueling stations.

Natural gas steam-reformation is a two-step process in which four parts hydrogen are produced from one part methane and two parts water at high operating temperatures and pressures and in the presence of a catalyst. This process is relatively efficient and inexpensive and it can be made still more efficient through cogeneration, which utilizes waste heat. Hence, making steam-methane particularly attractive for local use.

Hydrogen production is based on large-scale plants operating at an average capacity factor of 90%, capable of producing 36,000 GJ of $H_2$ per day, which is sufficient to fuel 500,000 proton exchange fuel cell vehicles (PEMFCV). On the other hand, small-scale plants would be capable of producing 180 GJ of $H_2$ per day, which is sufficient to fuel approximately 2,200 PEMFCV's [9]. However, hydrogen production through large-scale plants is more favorable for high-hydrogen demanding applications such as transportations and electric power generation.

Practically, there are two major drawbacks associated with natural gas steam-reformation process. Firstly, steam-reformation method is not the best method for hydrogen production due to low level of operating efficiency. Secondly, the emission associated with this type of process emits substantial amount of carbon dioxide, which could have a major impact on the environment.

The efficiency of the steam-reforming process, which is defined as the ratio of the heating value of the produced hydrogen over the energy input of the raw material in terms of fuel and electricity, is about 65-75% [4]. On the other hand, efficiency of renewable electrolysis technology, which refers to the process of splitting water powered by an electric current to produce hydrogen, is about 80-85%. Since hydrogen production through steam reformation of natural gas is highly exothermic process, the plant generates more steam than it consumes. Thus, energy efficiency of the steam-reformation plant could be drastically increased by at least 10% if the excess steam generated by the plant is used for another source [5], and it could compete with renewable electrolysis technology.

Experimentally, natural gas is considered to be the least polluting fossil fuel as opposed to oil and coal. It contains mixture of various components such as methane ($CH_4$), ethane ($C_2H_6$), propane ($C_3H_8$) and other types of hydrocarbons, carbon dioxide ($CO_2$), etc., which results difficulty in predicting the performance of the steam reformation plant [6]. Of all the above-mentioned components, $CO_2$ could be the major emitted component that should be dealt with extreme caution due to its greenhouse effect. By using technologies to capture $CO_2$ for either safe, long-term storage or use as a commercial commodity or carbon sequestration technologies, that will lower carbon emission to the minimum. Thus, hydrogen production through renewable resources would be more environmentally attractive as compared with natural gas reformation. Research from the US National Renewable Energy Laboratory has found that hydrogen produced from wind electrolysis results in greenhouse gas emissions that are one-twelfth those of a large natural gas reformer [7].

Generally, cost of hydrogen produced from steam-reformation process, fluctuates with the fluctuating price of natural gas. However, it is proved to be more favorable in the near-term than hydrogen produced from renewable methods (approximately $0.65/kg if it is consumed at site and $1-$2/kg for hydrogen produced by electrolysis) [8]. On the other hand, cost of hydrogen produced from natural gas is predicted to increase by 70% in the long-term due to increase in the cost of natural gas as a result of its massive utilizations and its imminent scarcity in the near future [9].

In order to reduce the carbon emission to 1990 level set by the Kyoto protocol, a suggested carbon emission tax of $140/tonne-$CO_2$ was set in accordance with the current study presented by L. Horwitz [10]. He concluded that a $100 per tonne of $CO_2$ would not be sufficient to bring emissions back to 1990 levels and a tax of $200 per tonne of $CO_2$ would be more than enough to eliminate growth in emissions. Department of Energy, suggested a carbon tax of $14, which seems to be unrealistic and it will not be sufficient to meet the level set in Kyoto-protocol.

## 3. Results and discussion

The current analysis was performed based on the UAE's current production of natural gas, which is about $1.1 \times 10^9$ GJ/year as opposed to the average production in the next 100 years period of $2.11 \times 10^9$ GJ/year to fulfill the demand in the energy market [11]. Depending on the pace at which the government is welling to shift toward a cleaner production of energy, the analysis was conducted at various scenarios (%) of natural gas to be used in hydrogen production through steam reforming process. In Figure 1, a cost breakdown was conducted on hydrogen production at NG utilization rate of 10%, 50% and 100% of the country's present NG production. Clearly, the higher the utilization rate of natural gas (%), the more cost effective hydrogen is produced and stored. For instance, at 100% utilization of NG, the capital cost of NG steam reforming plant would be approximately $1.4/GJ, which is 50% less than 10% of utilized NG. Similarly, the

470

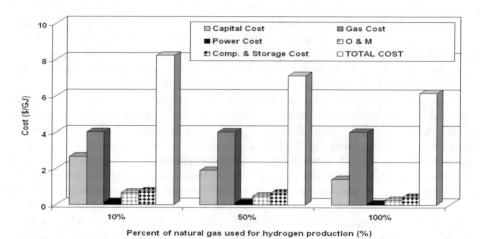

Figure 1: Cost breakdown of hydrogen production at various percentages of current production of natural gas.

operation and maintenance cost (O&M) and compression and storage cost of hydrogen at 100% NG utilization would be $0.265/GJ and $0.45/GJ, respectively. These values are estimated to be 60% and 45% less than hydrogen produced at 10% NG utilization. In general, the total cost of hydrogen produced at 100% NG utilization would be approximately $6.15/GJ, which is 25% less than at 10% NG utilization if an average cost of natural gas is set to be $4/GJ [11].

Cost of hydrogen produced by 10% of NG steam reforming is considered to be more attractive than hydrogen produced by biomass gasification or by solar-hydrogen or wind-hydrogen methods as depicted in Figure 2. However, the capital cost of NG steam-reforming plant is $1.225/GJ more than biomass gasification plant, which is $1.45/GJ. Nevertheless, the capital cost of NG steam-reforming plant is 3 times less than the capital cost of solar-hydrogen or wind-hydrogen plants. Furthermore, utilization of renewable resources (solar and wind) requires electrolysers, which could add more to the total cost of hydrogen. An average cost of solar-PV assisted electrolyser would be about $3.9/GJ of hydrogen produced. And, the average cost of wind-power assisted electrolyser would be about $5.5/GJ of hydrogen produced. The total cost of hydrogen produced by 10% of NG is $8.2/GJ, which is 10% less than hydrogen produced through biomass gasification. Moreover, the total cost of hydrogen produced by 10% of NG is 45% and 57% less than solar-hydrogen and wind-hydrogen energy systems, respectively.

The total cost of solar-hydrogen and wind-hydrogen are considered to be far more than other hydrogen energy producing methods, although they do not contribute any environmental damages such as carbon emission as depicted in Figure 3. The total amount and cost of carbon emitted by natural gas steam-reformation are respectively estimated to be 48 kg-$CO_2$/GJ and $6.7/GJ, which are regarded to be the least among other fossil fuels and hydrogen production methods. Conversely, the total amount and cost of carbon emitted by coal gasification to produce synthetic fuels are considered to be the highest among other sources and methods, and they are estimated to be 145 kg-$CO_2$/GJ and $20/GJ, respectively. The emission cost presented in the study is based on $140/tonne-$CO_2$, which could be set to reach the limit assigned in Kyoto. This value is more reasonable than the value suggested by the EPA of $14/tonne-$CO_2$ [10].

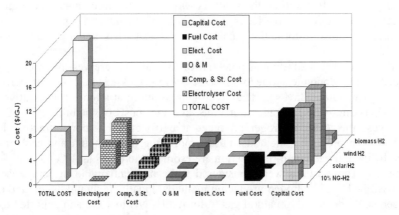

Figure 2: Cost comparison between hydrogen production by 10 % NG at current production and other types of hydrogen producing methods.

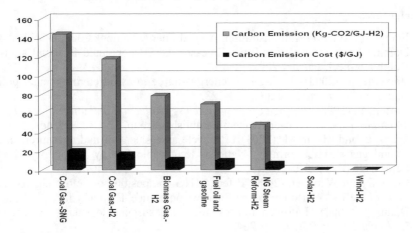

Figure 3: Amount and cost of carbon emission from various types of energy producing methods.

At a worst-case scenario, hydrogen could be produced at a 10% of current NG production, which is plausible for the UAE and increase the quota gradually as the demand increases. In addition, hydrogen could be added to the natural gas and exported overseas as a mixture, which could have some advantages when compared with natural gas alone. For instance, a mixture of hydrogen and natural gas can be used in internal combustion engines and fuel burners, emitting less pollutant, such as $CO_2$, $SO_x$, $NO_x$, HCs and CO to the atmosphere [12, 13]. Also, the amount of hydrogen in the $H_2$/NG mixture could be increased in hydrogen content with the increase in the percentage of NG used in the reformation process.

## 4. Conclusions

Feasibility study on hydrogen energy production in the United Arab Emirates through natural gas steam reformation process was carried out. The analysis was performed at various scenarios (%) of natural gas used in hydrogen production through steam reforming process. In addition, the analysis was conducted to compare the total cost of hydrogen produced by 10% of NG as well as the total amount and cost of carbon emitted by natural gas steam-reformation against other fossil fuels and hydrogen producing methods.

With the country's substantial potential of natural gas reserves, hydrogen could be produced at an economical rate of at least $6.15/GJ based on 100% of current NG utilization. This rate is 25% less than hydrogen produced at a more conservative rate of 10% NG taking into consideration the average cost of natural gas of $4/GJ. Nevertheless, the cost of hydrogen produced at a 10% of NG is more attractive than any other hydrogen producing methods. Furthermore, the total amount and cost of carbon emitted by natural gas steam-reformation are estimated to be 48 kg-$CO_2$/GJ and $6.7/GJ, respectively. These values were considered to be more favorable in comparison with other fossil fuels and hydrogen producing methods. Finally, it should be pointed out that hydrogen could be added to natural gas to form a $H_2$/NG mixture and exported overseas taking into consideration, a gradual increase of hydrogen content in the mixture in the event of increase in the percentage of NG used in the steam-reformation process.

## 5. References

1. Energy Information Administration, Annual Energy Outlook 2002, DOE/EIA-0383.
2. Dolphin Press Information: Press Releases, CFSB appointed financial advisor for Dolphin project, January 2002.
3. Valverde, S. (2002) Hydrogen as energy source to avoid environmental pollution, *Geofisica Internacional* **41**(3), 223-228.
4. Yurum, Y. (1995) *Hydrogen energy systems- production and utilization of hydrogen and future aspects*, Kluwer Academic Publisher group, The Netherlands.
5. Spath, P. and Mann, M. (2001) Life cycle assessment of hydrogen production via natural gas steam reforming, National Renewable Energy Laboratory, Technical Report, NREL/TP-570-27637.
6. Chan, S. and Wang, H. (2000) Effect of NG composition on reforming $H_2$ and CO, Proc. of the 13th World Hydrogen Energy Conference, Beijing, China, 200-206.
7. Dunn, S. (2002) Hydrogen futures: toward a sustainable energy system, *Int. J. Hydrogen Energy* **27**(3), 235-264.
8. Johansson, T., Kelly, H., Reddy, A. and Williams, R. (1993) Renewable energy - sources for fuels and electricity, *Island Press*.
9. Kazim, A. (2002) Production of hydrogen energy from the biomass resources of the United Arab Emirates, Proc. of the First Ukrainian International Conference on Biomass Energy, Kiev, Ukraine, Sept. 23-27, 2002, O50.
10. Horwitz, L.M. (1996) The Impact of carbon taxes on consumer living standards. In an economic perspective on climate change policies, American Council for Capital Formation Center for Policy Research, Washington, D.C., February, 119-157.
11. Kazim, A. and Veziroglu, T.N. (2001) Utilization of solar-hydrogen energy in the UAE to maintain its share in the world energy market for the 21st century, *Renewable Energy* **24**(2), 259-274.
12. Bunger, U. and Zittel, W. (1994) Hydrogen in the public gas grid- a feasibility study about its application and limitations for the admixture within a demonstration project for city of Munich, Proc. of 10th World Hydrogen Conf. in Miami, 127-138.
13. De Souza, S. and Da Silva, E.P. (1998) The integration of secondary energy and natural gas by hydrogen, Proc. of 12th World Hydrogen energy Conference, Buenos Aires, Argentina, 23-32.

# ISOTOPE EFFECTS IN THE QUASIELASTIC MÖSSBAUER ABSORPTION OF $^{57}$Fe IN NbH$_{0.78}$ AND NbD$_{0.76}$

R. WORDEL R* AND F.E.WAGNER
*Physik-Department E 15, Technische Universität München,*
*D-85747 Garching, Germany*
*Corresponding author. E-mail address: Rainer_Wordel@macnews.de (R. Wordel).*

## Abstract

The quasielastic Müssbauer absorption in NbH$_{0.78}$ and NbD$_{0.76}$ has been studied with dilute, substitutional $^{57}$Fe probes. The quasielastic line can be observed within the region of the β-phases between about 260 K and 340 K. The spectra were evaluated on the basis of a model in which only nearest hydrogen neighbours are assumed to contribute to the displacements and hyperfine interactions of the Müssbauer atoms. This yields values for the displacements $u_1$ caused by a single nearest interstitial and for the mean times of residence $\tau_{Fe}$ of the Müssbauer probes between diffusion-induced cage jumps. The temperature dependence of the latter follows Arrhenius' law with activation energies of $U$= 0.30 meV for the hydride and $U$ = 0.35 meV for the deuteride, while the displacements are $u_1$ = 0.019 nm in the hydride and $u_1$ = 0.016 nm in the deuteride.

*Keywords:* Niobium hydride; Müssbauer spectroscopy; Isotope effects

## 1. Introduction

Diffusing interstitials in metal-hydrogen systems give rise to fluctuating lattice strains. The resulting motion of the metal atoms resembles a diffusion within a cage, with a jump rate that reflects the jump rate of the neighbouring hydrogen atoms. When a Müssbauer atom executes such a cage motion, the coherence of the emission or absorption process of γ-rays is partly destroyed. This leads to the appearance of a quasielastic line in the Müssbauer spectrum [1]. The width of this line increases with the jump rate, while the fraction of the Müssbauer spectrum that is subject to the quasielastic broadening increases with the magnitude of the displacements of the Müssbauer atom. When the diffusion is very slow at low temperatures, the broadening vanishes and the whole spectrum becomes elastic. At high temperatures, in the limit of fast diffusion, the quasielastic line becomes so broad that it can no longer be distinguished from the nonresonant background. What then remains is the truly elastic fraction of the Müssbauer spectrum, which remains narrow even in the case of fast diffusion. The existence of a truly elastic fraction is typical for diffusion of the Müssbauer atom within a limited region of space. When the Müssbauer atom diffuses freely through the lattice, the whole spectrum broadens and eventually disappears [2].

Quasielastic Müssbauer absorption has mainly been observed in proteins (see, e. g., ref. [3]) and in metal-hydrogen systems [4-14]. In most of the experiments on hydrides, the existence of the quasielastic line has been deduced indirectly from the drop of intensity that

*T.N. Veziroglu et al. (eds.),*
*Hydrogen Materials Science and Chemistry of Carbon Nanomaterials,* 473-480.
© 2004 *Kluwer Academic Publishers. Printed in the Netherlands.*

ensues when the quasielastic line merges into the background. Only for $^{57}$Fe in NbH$_{0.78}$ [9,13] and in VD$_{0.84}$ [10] has the broad line so far been observed directly by measuring spectra in a sufficiently wide velocity range and with the statistical accuracy needed when the intensity of the quasielastic line is smeared out over a wide velocity region. We now report on the direct observation of the quasielastic Müssbauer absorption line of $^{57}$Fe in NbD$_{0.76}$. These experiments were undertaken in a search for isotope effects in the diffusion rates and displacements of the iron solutes in the Nb-H and Nb-D systems.

## 2. Experimental Details

The absorber was prepared as described previously [9,13] from a 26 mg·cm$^{-2}$ thick foil of Fe$_{0.0032}$Nb$_{0.9968}$. The deuterium content was determined gravimetrically and by outgassing after the Müssbauer experiments. The spectra were recorded with a sinusoidal velocity waveform at a maximum velocity of ±12 mm·s$^{-1}$ into 4096 channels of a multichannel analyzer, the large number of channels ensuring that the structure of the narrow, central part of the spectrum could be resolved. While the absorber temperature was varied in a gas-flow cryostat, the source of 50 mCi $^{57}$Co in Rh was always kept at room temperature. The γ-rays were detected with a Kr/CH$_4$ proportional counter, with single channel windows set on both the 14.4 keV photopeak and the 1.8 keV escape peak. The setting of the single channel windows and the intensity of the nonresonant background were monitored at regular intervals and found to be constant during the whole time of the experiments. The areas under the individual spectra can therefore be compared with each other without corrections for a changing nonresonant background.

## 3. Results and Discussion

The Müssbauer spectra of $^{57}$Fe in NbH$_{0.78}$ obtained at temperatures between 223 and 350 K, the region where the transition from the slow to the fast diffusion regime takes place, are shown in Fig. 1(a), while Fig. 1(b) shows the Müssbauer spectra of $^{57}$Fe in NbD$_{0.76}$ obtained at temperatures between 200 and 360 K. The plots covering the whole velocity range show the presence of the quasielastic component (shown in red) which broadens when the temperature increases, while at the same time its depth diminishes. The data points are the sum of 8 adjacent points of the original spectrum, since good resolution is of less importance in the wings of the pattern than good statistical accuracy. The separate plots of the central parts of the spectra in Fig. 1 reveal the quadrupole doublet structure with the good resolution resulting from the large number of data points. They also show the decrease of the intensity of the central part of the spectra when the quasielastic line escapes observation.

The transition from the static to the fast diffusion limit takes place within the β-phase, where the deuterium occupies a superlattice that is expected to be the same as that in β-NbH$_x$ [15]. The spectra could therefore be fitted with the same theoretical approach as those of $^{57}$Fe in NbH$_{0.78}$ [9,13]. This model assumes that only *nearest* deuterium neighbours contribute noticeably to the displacements of the Müssbauer atoms [16-18] and that only

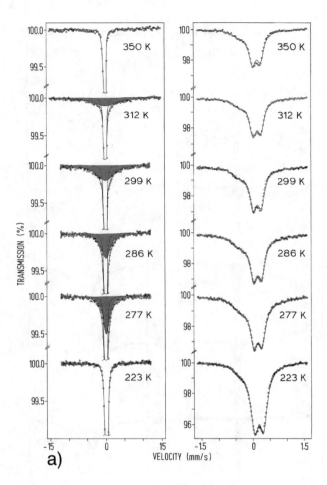

Fig. 1.

476

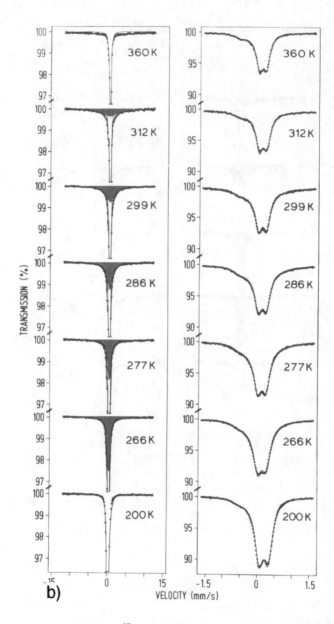

Fig. 1. Müssbauer spectra of $^{57}$Fe in NbH$_{0.78}$ (a) and NbD$_{0.76}$ (b) taken at various temperatures with a maximum velocity of $\pm 12$ mm·s$^{-1}$. The plots on the left show the whole velocity range, but only the part of the spectra near the baseline. Those on the right show the central parts of the same spectra on an expanded velocity scale.

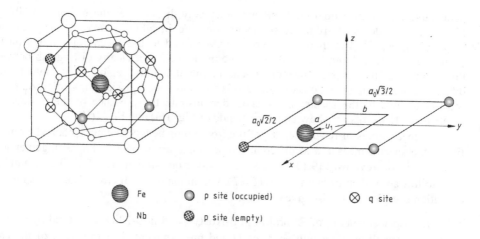

Fig. 2. Environment of a substitutional iron atom in β-NbH, (left). Normally only the p sites are occupied by hydrogen. On the right the rectangle formed by these sites and the coordinate system used in the model description are shown. When one of the four p sites is empty and this vacancy jumps around the Fe atom, the latter moves between the corners of the similar, smaller rectangle with the half diagonal $u_1$

four nearest interstitial sites forming a rectangle around a metal atom (Fig. 2) may be occupied in the β-phase. At a deuterium-to-metal ratio of $x = 0.75$ three of these sites should be occupied on the average, unless the environment of the iron impurities is modified by a repulsive or attractive interaction with the interstitials. Such interactions have been found to be small in the hydrides and deuterides of niobium [6,9,11-13]. The model simplifies the situation somewhat by assuming that three of these four sites are *always* occupied, and that the deuterium diffusion can be described as a hopping of the vacancy on the fourth corner of the rectangle around the Fe probe. The displacements of the iron atoms can in this case be imagined as a shift towards the vacant corner by a length $u_1$, which is equal to the displacement caused by a single nearest neighbour, since the influence of the two interstitials at opposite corners of the rectangle cancels. The cage diffusion of an iron atom can thus be depicted as a hopping between neighbouring corners of a rectangle [9,13] with the diagonal $2u_1$. The model description includes the fluctuating electric field gradient caused by the deuterium neighbours within the framework of a point charge model. The isomer shift does not fluctuate, since the number of nearest deuterium neighbours does not change. The formalism allowing the lineshape to be calculated exactly within the framework of this model has been described previously [9,13]. The relevant free parameters that have to be adjusted are the displacement $u_1$, which determines the fraction of the Müssbauer spectrum that broadens into the quasielastic line and is assumed to be independent of temperature within the rather narrow range of interest, and the hopping rate $\tau_{Fe}^{-1}$ defined as the probability per unit time for an iron probe to hop from a given corner of the displacement rectangle to any of the two neighbouring ones. The electric field gradient causing the electric quadrupole splitting of the central peak can be described by an effective charge on the neighbouring interstitials [9,13], which is also treated as an adjustable parameter. The electric field gradients are practically the same as those found in $NbH_{0.78}$ [9,13], as was deduced from measurements on $^{57}Fe$ in $NbD_{0.79}$ covering only a small velocity range [12]. The theoretical model curves fitted to the spectra are shown on the right side of Fig. 1(a) and Fig. 1(b). Since the model does not clearly distinguish between the elastic and the quasielastic component in the transition region from slow to fast diffusion, simple superpositions of narrow (elastic) and wide (quasielastic) Lorentzian lines were also fitted to the spectra in order to demonstrate the presence of the quasielastic component. These fits are shown on the left side of Fig. 1(a) and Fig. 1(b).

The mean times of residence, $\tau_{Fe}$, of the iron atoms between jumps obtained from the fits of the spectra measured at different temperatures are shown in Fig. 3, together with those obtained previously [9,13] in the same way from similar data for $^{57}Fe$ in $NbH_{0.78}$. Both follow an Arrhenius law $\tau_{Fe,0} \cdot \exp(U/kT)$, within the limits of error. The corresponding activation energies $U$ are also given in Table 1.

Table 1. Displacements $u_1$ of Fe probes in $\beta$-$NbH_{0.78}$ and in $\beta$-$NbD_{0.76}$, caused by a single intersitial, as well as activation energies $U$ and pre-exponential factors $\tau_{Fe,0}$, of the cage diffusion jumps of the iron impurities.

|  | $u_1$ (nm) | $U$ (meV) | $\log(\tau_{Fe,0})$* |
|---|---|---|---|
| $NbH_{0.78}$, ref. [9] | 0.019±0.001 | 0.30±0.02 | −1.4±0.6 |
| $NbD_{0.76}$, this work | 0.016±0.001 | 0.35±0.02 | −1.8±0.6 |

\* $\tau_{Fe,0}$ in units of $10^{-12}$ s.

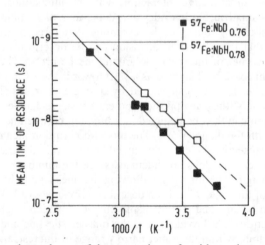

Fig. 3. Temperature dependence of the mean time of residence between cage jumps of the iron probes in $NbH_{0.78}$ (open symbols) and in $NbD_{0.76}$ (full symbols). The Arrhenius lines fitted to the data correspond to the parameters listed in Table 1.

The values of the pre-exponential factors, $\tau_{Fe,0}$, have rather large uncertainties because of the limited temperature range in which Müssbauer spectroscopy can provide values for the mean residence times. The results compiled in Table 1 reveal an isotope dependence of both the activation energies $U$ and the displacements $u_1$. That the mean square displacements of the $^{57}$Fe are larger in $\beta$-NbH$_x$ than in $\beta$-NbD$_x$ has been deduced before from measurements within a small velocity range only [12]. In the $\alpha$-phases, this isotope dependence is absent or at least much weaker [12]. The straightforward explanation of these observations is that the $u_1$ exhibit a substantial isotope dependence in the $\beta$-phase, but a much smaller one in the $\alpha$-phase. An alternative interpretation could be that the jump geometry in the $\beta$-phase is different in the hydride and in the deuteride, while such differences are small or absent in the $\alpha$-phase. Model calculations for different jump geometries that may be able to shed some light on this question are in progress. At present, no definite interpretation of the isotope effects in the intensity of the quasielastic line can be given. Fig. 3 shows that near 300 K the mean times of residence in the hydrides are about a factor two shorter than in the deuterides. This difference does not depend much on the model used to fit the data; it is, for instance, also found when the quasielastic component is simply fitted by a Lorentzian line. A normal isotope effect of about the same magnitude also exists in *dilute* interstitial solutions of H and D in Nb near 300 K [19]. One should note, however, that the absolute values of the residence times in the dilute systems are shorter by about three orders of magnitude than those observed for the cage jumps of the iron impurities. Although the latter are not identical to the mean residence times of the interstitials, they differ from them only by a factor that depends on the jump geometry and is expected to be close to unity [9,13]. The activation energies derived from the $\tau_{Fe}$ should thus be the same as those for the interstifials. The isotope dependence of the activation energies $U$ obtained from the present data (Table 1) is small and may also reflect differences in the jump geometries of hydrogen and deuterium.

## 4. Conclusions

Müssbauer spectroscopy is a useful tool to study dynamic processes in metal hydrides, although in experiments with substitutional probes one does not study the unperturbed hydride lattice. The use of different models for the interpretation of data like those presented here may help to improve the understanding of the jump processes and blocking conditions in the Nb-H and Nb-D systems.

## Acknowledgement

We thank the Deutsche Forschungsgemeinschaft for financial support of this work.

## References

1.      Krivoglaz MA, Respetskii SP. Fiz Tverd Tela 1966; 8(10):2908–2918 (Engl. Transl.: Theory of diffusional broadening of the Müssbauer line in nonideal crystals. Sov Phys - Solid State 1967;8:2325–2332).

480

2. Dattagupta S, Schroeder K. Müssbauer spectrum for diffusing atoms including fluctuating hyperfine interactions. Phys Rev B 1987;35:1525–1546.

3. Parak F, Heidemeier J, Nienhaus GU. Protein structural dynamics as determined by Müssbauer spectroscopy. Hyperfine Interactions 1988;40:147–158.

4. Blдsius A, Preston RS, Gonser U. Müssbauer study of the diffusion of stored hydrogen in $^{57}$Fe-doped titanium. Z Phys Chem Neue Folge 1979;115:187–199.

5. Prübst F, Wagner FE, Karger M. Diffusion induced reduction of the $^{57}$Fe Müssbauer fraction in β-PdH$_x$. J Phys F: Met Phys 1980;10:2081–2091.

6. Wagner FE, Wordel R, Zelger M. Influence of interstitial diffusion on the Müssbauer spectra of iron solutes in VD$_{0.78}$ and NbH$_{0.84}$. J Phys F: Met Phys 1984;14:535–547.

7. Wordel R, Wagner FE. Müssbauer study of transition metal impurities in niobium hydride. J Less-Common Met 1984;101:427–436.

8. Wagner FE, Prübst F, Wordel R, Zelger M, Litterst FJ. $^{57}$Fe Müssbauer study of hydrogen diffusion in the hydrides an deuterides of vanadium, niobium and palladium. J Less-Common Met 1984;103:135–144.

9. Wordel R, Litterst FJ, Wagner FE. Quasielastic Müssbauer absorption of $^{57}$Fe in niobium hydride. J Phys F: Met Phys 1985;15:2525–2545.

10. Berneis M, Trager J, Wordel R, Zelger M, Wagner FE, Butz T. Müssbauer and perturbed angular correlation studies of the distribution and diffusion of hydrogen near impurities in metal hydrogen systems. Z Phys Chem Neue Folge 1985;145:129–140.

11. Zelger M, Wordel R, Wagner FE, Litterst FJ. Quasielastic resonance absorption of $^{57}$Fe in the hydrides and deuterides of vanadium and niobium. Hyperfine Interactions 1986;28:1009–1012.

12. Wordel R, Wagner FE. Müssbauer Study of diffusion and lattice distortions in the hydrides and deuterides of niobium. J Less-Common Met 1987;129:271–278.

13. Wordel R. Untersuchung von Niobhydrid und Niobdeuterid mittels der Müssbauerspektroskopie und der gestürten Gamma-Gamma Winkelkorrelation. Thesis. Technical University of Munich, 1988.

14. Zelger M, Monitto DC, Rinser HJ, Wagner FE. Müssbauer study of $^{57}$Fe in the hydrides and deuterides of tantalum. Z Phys Chem Neue Folge 1989;164:779–784.

15. Schober T, Wenzl H. The systems NbH(D), TaH(D), VH(D): structures, phase diagrams, morphologies, methodes of preparation. In: Alefeld G, Vülkl J, editors. Hydrogen in Metals II, Topics in Applied Physics, Vol. 29. Berlin: Springer-Verlag, 1978. p. 11–71.

16. Metzger H, Behr H, Peisl J. Atomic displacements close to point defects from static Debye-Waller factor measurements. Solid State Commun. 1981;40:789-791; The static Debye-Waller factor of defect-induced lattice displacements. Z Physik B 1982;46:295-299.

17. Metzger H, Behr H, Peisl J. The Debye-Waller factor due to static displacement around hydrogen in niobium. J Less-Common Metals 1982;88:159-164.

18. Behr H, Keppler HM, Steyrer G, Metzger H, Peisl J. Atomic displacements close to hydrogen in Nb, Ta and V. J Phys F: Met Phys 1983;13:L29-L31.

19. Vülkl J, Alefeld G. Diffusion of hydrogen in metals. In: Alefeld G, Vülkl J, editors. Hydrogen in Metals I, Topics in Applied Physics, Vol. 28. Berlin: Springer-Verlag, 1978. p. 321–348.

# ELECTRICAL RESISTANCE OF BINARY ORDERED ALLOYS WITH HCP STRUCTURE IN THE PRESENCE OF IMPURITY ATOMS OR THERMAL VACANCIES

Z.A. MATYSINA, S.Yu. ZAGINAICHENKO, D.V. SCHUR,
A.Yu. VLASENKO
*Dnepropetrovsk National University,*
*72, Gagarin ave., Dnepropetrovsk, 49000 Ukraine*
*Institute for Problems of Materials Science of NAS of Ukraine, lab. # 67,*
*3, Krzhizhanovsky str., Kiev, 03142 Ukraine*

The electrical resistance of alloy is determined by the conduction electron scattering in the crystal lattice when the potential field of the crystal shows a broken periodicity [1-5]. This scattering is characterized by probability of the electron transition between different equilibrium states.

The calculation of residual resistance has been performed in present paper for the cases when in disordered state the positions of lattice sites form the Bravais lattice and each lattice site is the centre of crystal symmetry. But the phenomenon of ordering is observed also in a series of alloys with crystal lattices of more low symmetry, for example alloys with hexagonal close-packed lattice. Moreover the impurity atoms and thermal vacancies have effect on electrical resistance. So the consideration of these cases is of physical interest.

Below is given in the context of zone theory the theoretical investigation and the computer calculation of residual resistance at first for binary ordered alloys with h.c.p. lattice of AB and $AB_3$ composition [6] with comparison of theoretical results with experimental data for $MgCd_3$ and $Mg_3Cd$ alloys [7] and then the resistance of these alloys containing impurity atoms or thermal vacancies [7, 8].

The calculations have been performed in some simplifying assumptions. The correlation between substitution of lattice sites by atoms of different sorts is not taken into consideration. The crystal lattice is assumed to be geometrically ideal, i.e. it is free from any geometric distortions in overall concentration range of alloy. The differences in energies for interaction between conduction electrons and lattice ions of A, B, C kind are considered to be small values.

## 1. Binary alloy A-B

The hexagonal close-packed lattice, as known, consists of two simple hexagonal sublattices, inserted one into another. In this case h.c.p. crystal lattice of stoichiometric composition AB comprises simple sublattices, one of each consists of sites of first type only (legal for the A atoms) and second consists of sites of second type (legal for the B atoms). In lattice of stoichiometric composition $AB_3$ the sites of first and second type form the both sublattices. The figure 1 shows the investigated structures $B8_1$ and $DO_{19}$ of AB and $AB_3$ alloys.

481

*T.N. Veziroglu et al. (eds.),*
*Hydrogen Materials Science and Chemistry of Carbon Nanomaterials,* 481-488.
© 2004 *Kluwer Academic Publishers. Printed in the Netherlands.*

482

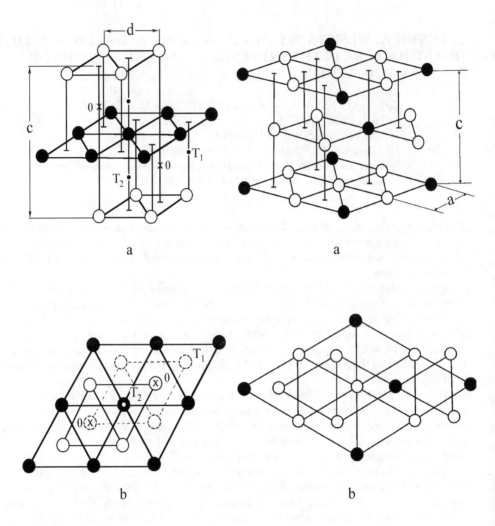

Figure 1. The superstructures B8$_1$ and DO$_{19}$ of AB and AB$_3$ alloys in three-dimensional presentation (a) and in projection on the plane perpendicular to $c$ axis.

●○ are the sites of first and second type legal for A and B atoms.

The consideration of residual resistance reduces to the calculation of transition probability between states of conductivity electron (one state is characterized by wave vector $\vec{\kappa}$, another state with vector $\vec{\kappa}'$) or to the calculation of the square of matrix element modulus $V'_{\vec{\kappa}\vec{\kappa}'}$ of disturbing energy, constructed on the wave functions of zero-order approximation of different states with the $\vec{\kappa}$ and $\vec{\kappa}'$ vectors. As an zero approximation we take the quite ordered alloy from effective ions, each ion creates the potential equal to average potential of alloy ions. The wave function of electrons in zero-order approximation is in the form of Blokh function

$$\Psi_{\vec{\kappa}}(\vec{r}) = e^{i(\vec{\kappa}, \vec{r})} u_{\vec{\kappa}}(\vec{r}), \tag{1}$$

where $u_{\vec{\kappa}}(\vec{r})$ is the periodical function with the lattice constants.

The disturbing energy is deviation of electron potential energy from the average energy

$$V'(\vec{r}) = \sum_{s=1}^{N} \sum_{\chi=1}^{\mu} \left[ V^{\chi s}\left(\vec{r} - \vec{R}_s - \vec{h}_\chi\right) - \overline{V}^{\chi}\left(\vec{r} - \vec{R}_s - \vec{R}_\chi\right) \right], \tag{2}$$

where we make the summation with respect to all N elementary cells and to $\mu$ sites in the cell, $\vec{h}_\chi$ is the vector, drawn from the zero site of cell to $\chi$ site, the $\chi$ site can be of two kind $\chi_1$ and $\chi_2$ (legal respectively for A and B atoms),

$$\vec{R}_s = s_1\vec{a}_1 + s_2\vec{a}_2 + s_3\vec{a}_3 \tag{3}$$

is the vector of lattice sites ($\vec{a}_1, \vec{a}_2, \vec{a}_3$ are the basis vectors, $s_1$, $s_2$, $s_3$ are integers), $V^{\chi s}$ and $\overline{V}^\chi$ are potential energies of electron correspondingly in the field of ion, situated in the site $s\chi$, and in the field of effective ion.

As the result of calculation the square of matrix element of disturbing energy takes the form

$$\left| V'_{\vec{\kappa}\vec{\kappa}'} \right|^2 = N \sum_{j=1}^{2} \sum_{\chi_j=1}^{\lambda_j} \left( \left| V_{\vec{\kappa}\vec{\kappa}'}^{\chi_j} \right|^2 - \left| \overline{V}_{\vec{\kappa}\vec{\kappa}'}^{\chi_j} \right|^2 \right), \tag{4}$$

where $\lambda_j$ is the number of sites in the cell of j type (j = 1, 2).

By definition of the average

$$\overline{V}_{\vec{\kappa}\vec{\kappa}'}^{\chi_j} = P_A^{(j)} V_{A\,\vec{\kappa}\vec{\kappa}'}^{\chi_j} + P_B^{(j)} V_{B\,\vec{\kappa}\vec{\kappa}'}^{\chi_j},$$
$$\overline{\left| V_{\vec{\kappa}\vec{\kappa}'}^{\chi_j} \right|^2} = P_A^{(j)} \left| V_{A\,\vec{\kappa}\vec{\kappa}'}^{\chi_j} \right|^2 + P_B^{(j)} \left| V_{B\,\vec{\kappa}\vec{\kappa}'}^{\chi_j} \right|^2, \tag{5}$$

where

$$V_{\alpha\,\vec{\kappa}\vec{\kappa}'}^{\chi_j} = \int_\Omega \Psi_{\vec{\kappa}'}^* V_\alpha\left(\vec{r} - \vec{h}_{\chi_j}\right) \Psi_{\vec{\kappa}} d\Omega, \qquad \alpha = A, B. \tag{6}$$

The integrals in (6) are taken over the base region of crystal. $P_\alpha^{(j)}$ are a priori probabilities for substitution of the site of j type by atoms of $\alpha$ sort. The last-named are determined by the $c_A$, $c_B$ concentrations of A, B components and degree $\eta$ of long range order

$$P_A^{(1)} = c_A + (1-v)\eta, \qquad P_B^{(1)} = c_B - (1-v)\eta,$$
$$P_2^{(1)} = c_A - v\eta, \qquad P_B^{(2)} = c_B + v\eta, \tag{7}$$

where $v$ is the relative concentration of first type sites ($v = 1/2$ for AB alloys and $v = 1/4$ for $AB_3$ alloys).

Substitution of formulae (5) into (4) gives the expression for the square of matrix element of disturbing energy

484

$$\left|V'_{\vec{\kappa}\vec{\kappa}'}\right|^2 - N\left( P_A^{(1)}P_B^{(1)}\sum_{\chi_1=1}^{\lambda_1}\Delta_{\chi_1} + P_A^{(2)}P_B^{(2)}\sum_{\chi_2=1}^{\lambda_2}\Delta_{\chi_2} \right). \tag{8}$$

The values $\Delta_{\chi_1}$ and $\Delta_{\chi_2}$ are defined by formulae

$$\Delta_{\chi_j} = \left| \int_\Omega e^{i(\vec{\kappa}-\vec{\kappa}',\vec{r})}\left[V_A\left(\vec{r}-\vec{h}_{\chi_j}\right)- V_B\left(\vec{r}-\vec{h}_{\chi_j}\right)\right]u_{\vec{\kappa}}(\vec{r})u_{\vec{\kappa}'}^*(\vec{r})d\Omega \right|^2 \tag{9}$$

($j = 1, 2$), where $\Omega$ is the volume of base region of crystal. The $\vec{r}' = \vec{r} - \vec{h}_{\chi_j}$ variable is entered into (9) instead of $\vec{r}$ and it is indicated again by $\vec{r}$. Then we have

$$\Delta_{\chi_j} = \left| \int e^{i(\vec{\kappa}-\vec{\kappa}',\vec{r})}\left[V_A(\vec{r})- V_B(\vec{r})\right]u_{\vec{\kappa}}(\vec{r}+\vec{h}_{\chi_j})u_{\vec{\kappa}'}^*(\vec{r}+\vec{h}_{\chi_j})d\Omega \right|^2. \tag{10}$$

Furthermore, we restrict our consideration to the approximation of near free electrons. In this case the $u_{\vec{\kappa}}$ and $u_{\vec{\kappa}'}$ values can be considered as independent of coordinates, the $\Delta_{\chi_j}$ values are found to be equal

$$\Delta_{\chi_1} = \Delta_{\chi_2} = \Delta. \tag{11}$$

As this takes place, the formulae for $\left|V'_{\vec{\kappa}\vec{\kappa}'}\right|^2$ and for residual resistance derived in [3] for alloys with more higher degree of symmetry are true.

The residual resistance proportional to the square of matrix element of disturbing energy (8) in view of relations (7) and (11) is determined by formula

$$\rho_0 = A\left[c_A(1-c_A)-v(1-v)\eta^2\right], \tag{12}$$

where A is the value independent of $c_A$ and $\eta$.

We shall now look at the special cases.

**Alloy AB.** Putting in this case $v = 1/2$ in formula (12), we obtain the following formula for resistance:

$$\rho_0 = A\left[c_A(1-c_A)-\frac{1}{4}\eta^2\right]. \tag{13}$$

**Alloy AB₃.** Here, setting $v = 1/4$, the dependence of electric resistance on alloy composition and long range degree becomes [6]

$$\rho_0 = A\left[c_A(1-c_A)-\frac{3}{16}\eta^2\right]. \tag{14}$$

The last formula is comparable with literature experimental data [9] on measurements of electric resistance made for MgCd₃ and Mg₃Cd alloys of stoichiometric composition.

According to ordering theory [10, 11] the order-disorder transition for AB₃ alloys is the first-kind phase transition, in this case abrupt change of long range order $\eta_0$ at transition temperature is found to be equal $\eta_0 = 0,46$ for alloys of stoichiometric composition ($c_A = 1/4$).

Figure 2 illustrates theoretical curves and experimental points that give the dependence of full resistivity $\rho$ of MgCd₃ and Mg₃Cd alloys on annealing temperature. The approximate formula is used for the construction of theoretical plot

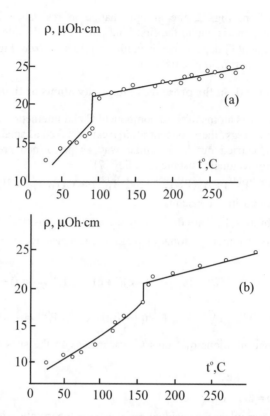

Figure 2. The temperature dependence of electrical resistance for MgCd$_3$ (a) and Mg$_3$Cd (b) alloys of the stoichiometric composition. Curves correspond to the theory, points correspond to the experimental data.

$$\rho = \rho_0 + BT = \frac{3}{16} A(1 - \eta^2) + BT. \tag{15}$$

In doing so it is assumed that part BT of electric resistance, associated with thermal oscillations, does not vary with change in long range order. The B constant is defined from the slope of experimental curves (that coincide practically with straight lines) for disordered states of alloy, when $T > T_0$, where $T_0$ is the ordering temperature. This temperature is equal to $T_0 = 87°C$ for MgCd$_3$ alloy and $T_0 = 155°C$ for Mg$_3$Cd alloy. The A constant is determined by one of experimental points lying below the ordering temperature. The temperature dependence of $\eta$ is taken from statistic theory of ordering. The last dependence is determined by the relation

$$\ln \frac{\left(c_A + \frac{3}{4}\eta\right)\left(c_B + \frac{1}{4}\eta\right)}{\left(c_A - \frac{1}{4}\eta\right)\left(c_B - \frac{3}{4}\eta\right)} = 2\frac{\omega + \omega'}{kT}\eta, \tag{16}$$

where $\omega + \omega'$ is ordering energy of alloy ($\omega$ and $\omega'$ correspond to the different interatomic distances between the nearest neighbouring pairs of atoms at $c \neq 1{,}63a$).

As is obvious from Fig. 2, the abrupt change in resistance is observed at $T_0$ temperature and this jump is due to the first-kind phase order-disorder transition. Below the $T_0$ temperature the $\rho(T)$ dependence is nonlinear. As seen from Fig. 2, the agreement between theory and experiment is satisfactory.

## 2. The ordered alloy AB$_3$ in the presence of impurity atoms or thermal vacancies C

Considering vacancies as atoms of third component C and interaction energy with nearest atoms as zero-point energy, their concentration $c$ can be taken as small: $c \ll 1$.

The calculation, carried out in a similar way as performed previously, gives the following formula for residual resistance of alloy [7]

$$\rho = A_1(P_A^{(1)}P_B^{(1)} + 3P_A^{(2)}P_B^{(2)}) + A_2(P_A^{(1)}P_C^{(1)} + 3P_A^{(2)}P_C^{(2)}) + A_3(P_B^{(1)}P_C^{(1)} + 3P_B^{(2)}P_C^{(2)}), \quad (17)$$

where $A_1$, $A_2$, $A_3$ are positive constants.

A priory probabilities $P_\alpha^{(j)}$ are defined by the alloy composition, the order parameter and concentration $c$ of impurity atoms or thermal vacancies according to the following formulae [12]

$$P_A^{(1)} = (c_A + \frac{3}{4}\eta)(1+c)^{-1}, \quad P_B^{(1)} = (c_B - \frac{3}{4}\eta - \delta)(1+c)^{-1}, \quad P_C^{(1)} = 4c_1(1+c)^{-1},$$
$$\quad (18)$$
$$P_A^{(2)} = (c_A - \frac{1}{4}\eta)(1+c)^{-1}, \quad P_B^{(2)} = (c_B + \frac{1}{4}\eta + \frac{1}{3}\delta)(1-c)^{-1}, \quad P_C^{(2)} = \frac{4}{3}c_2(1-c)^{-1},$$

where $c_1$, $c_2$ are concentrations of atoms C (vacancies) on the sites of first and second type, in this case

$$c = c_1 + c_2, \quad c \ll 1,$$
$$\delta = 3c_1 - c_2, \quad \delta \ll 1. \quad (19)$$

In view of last formulae, the residual resistance of binary alloy is expressed by the formula (correct to linear relatively small values $c$ and $\delta$):

$$\rho = A_1(c_A c_B - \frac{3}{16}\eta^2) + c[A_2 c_A + A_3 c_B - 2A_1(c_A c_B - \frac{3}{16}\eta^2)] - \frac{1}{4}(A_1 - A_2 + A_3)\delta\eta. \quad (20)$$

The first component of this formula corresponds to residual resistance if binary alloy without impurity C.

The availability of vacancies in alloy can change in one or another form both the abrupt change in resistance at $T=T_0$ and the slope of $\rho(T)$ plot at $T>T_0$ depending on the values of $A_1$, $A_2$, $A_3$ constants. For alloys of stoichiometric composition at $2(A_2+3A_3)<3A_1$ the discontinuous change in resistance in the point $T=T_0$ is decreased, and the slope of the curve for $\rho(T)$ at $T>T_0$ becomes negative due to impurity or vacancies C.

It should be noted that at rather high annealing temperatures of alloy the quadratic component and exponential component can appear in the formula for the temperature dependence of resistance $\rho(T)$ besides the linear term. The quadratic component is due to anharmonicity of oscillations and the component $\exp\dfrac{Q}{kT}$ is proportional to the concentration of thermal vacancies

$$\rho(T) = \rho_1 + B_1 T + B_2 T^2 + B_3 \exp\frac{Q}{kT}, \quad (21)$$

(for $AB_3$ alloys these components do not play a significant part in the range of ordering temperatures $T \geq T_0$), Q is the energy of vacancies formation, $B_1$, $B_2$, $B_3$ are positive constants ($B_2 \ll B_1, B_3$).

Experimental investigations of $\rho(T)$ dependence (21) at different annealing temperatures can allow the evaluation of Q energy of vacancies formation [3].

The concentration dependence of residual resistance $\rho(c)$ can also essentially be modified in the presence of impurity or vacancies in the alloy. The character of this change is determined by the relation between $A_1$, $A_2$, $A_3$ constants and by values of energies for interatomic interaction.

Fig. 3 demonstrates the plots of resistance versus concentration $\rho(c_A)$ for completely disordered and maximum ordered alloys. The $\rho(c_A)$ plots for binary alloys without impurities and vacancies are shown by continuous curves, these dependences for alloys with impurities and vacancies are plot as dotted curves. By virtue of the fact that the total tendency for a decrease of vacancies concentration manifests itself for composition close to stoichiometric (especially in ordered alloys), the effect of vacancies on electrical resistance is stronger at concentrations $c_A$, $c_B$ that differ from $c_A = 1/4$, $c_B = 3/4$.

For disordered alloys in the absence of vacancies the $\rho(c_A)$ dependence is a parabola with maximum at concentration $c_A = 0,5$. Such dependence is in agreement with experimental data for many alloys. Data for gold-silver and gold-copper alloys show especially good agreement with theory [3]. At the expense of impurity or vacancies presence there appear the asymmetry of $\rho(c_A)$ curve, the shift of maximum to one or another side from the point $c_A = 0,5$ and also the shift up and down from this point for binary alloy without impurity and vacancies. The existence of such changes of $\rho(c_A)$ curve makes it possible to judge about the presence of impurity atoms or vacancies in alloy and to evaluate both the $A_1$, $A_2$, $A_3$ constants and energetic parameters of interatomic interaction.

The electrical resistance of the maximum ordered alloy that does not contain impurities and vacancies is determined by the formula:

$$\rho_1 = A_1 \cdot \begin{cases} c_A(c_B - 3c_A) & \text{at} \quad c_A \leq \dfrac{1}{4}, \\[2mm] c_B(c_A - \dfrac{1}{3}c_B) & \text{at} \quad c_A \geq \dfrac{1}{4}. \end{cases} \tag{22}$$

This dependence $\rho_1(c_A)$ (solid curve in Fig.3(b)) was observed for gold-copper alloys [3]. Impurities or vacancies can cause an interesting effect: residual resistance of pure A,B components, annealed at the same temperature as the alloy, can appear much higher than that of the ordered alloy of stoichiometric composition. This is determined by the slight solubility of impurity or practical absence of vacancies in the ordered alloy of stoichiometric composition. Therefore, the appearance of such special feature in the $\rho(c_A)$ curve will point to the fact of the presence of impurity atoms or vacancies in the alloy.

Hence, the performance of experiments on determination of temperature and concentration dependences of residual resistance of alloys can permit us to judge on the presence of impurities or vacancies in the alloy, to evaluate the ordering level of alloy, its ordering temperature, ordering energy, concentration of impurity and vacancies and energy of vacancies formation.

488

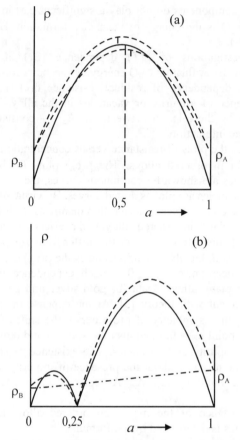

Figure 3. The concentration dependence of residual resistance for completely disordered (a) and maximum ordered (b) alloys with the *hcp* lattice. Solid curves correspond to the absence of vacancies, the dotted ones correspond to the presence of vacancies.

## 3. References

1. Payerl'e, R. (1956) The quantum theory of solids, Moscow: I.L. (in Russian).
2. Krivoglaz, M.A. and Smirnov, A.A. (1958) Theory of ordering alloys, M.: GIFML, 388 p. (in Russian).
3. Smirnov, A.A. (1960) Theory of electric resistance of alloys, Kiev: Izd. AN USSR, 148 p. (in Russian).
4. Kittel, Ch. (1967) The quantum theory of solids, M.: Nauka, 492 p. (in Russian).
5. Zaiman, Dj. (1974) Principles of solids theory, M.: Mir, 472 p. (in Russian).
6. Matysina, Z.A., Nosar, A.I. and Smirnov, A.A. (1963) Ordering in alloys with hexagonal lattice and its effect on electric resistance, *Ukr. Phys. J.* **8**, 339-343. (in Russian).
7. Matysina, Z.A. Doctor's thesis.
8. Matysina, Z.A. and Matysina, E.A. (1971) The effect of vacancies on residual resistance of ordered alloys, *Izv. vuzov. Physika*, No 10, 84-88. (in Russian).
9. Yonemitsu, K., Sato (1958) *Journal Phys. Soc. Japan* **13**, 15-19.
10. Matysina, Z.A. and Smirnov, A.A. (1959) The ordering theory of alloys with hexagonal close-packed lattice, *Fiz. tv. tela*, 286-290. (in Russian).
11. Matysina, Z.A. and Smirnov, A.A. (1960) Up to the ordering theory of alloys with h.c.p. lattice, *Ukr. fiz. zhurn.* **56**, 458-471. (in Russian).
12. Babaev, Z.M., Matysina, Z.A. and Mekhrabov, A.O. Phase diagrams and atomic order of ternary substitutional alloys Dep. N4411-83. – M.: VINITI. – 1983. – 16 p. Izv. vuzov. Fizika. – 1984. – N1. – P. 128. (in Russian).

# HYDRIDING PROPERTIES OF MAGNESIUM-SALT MECHANICAL ALLOYS

E.Yu. IVANOV, I.G. KONSTANCHUK, V.V. BOLDYREV
*Tosoh SMD, Inc., 36000 Gantz Road, Grove City, Ohio, USA*
*Institute of Solid State Chemistry, Siberian Branch of R.A.S.*
*Kutateladze 18, 630128 Novosibirsk, Russia*
*Fax: 7 (3832) 321 550; E-mail: irina@solid.nsc.ru*

## Abstract

Mechanical alloying of magnesium with inorganic salt addition has been demonstrated to promote metal powdering and modify the metal surface. This leads to an acceleration of the hydriding and dehydriding reactions of obtained materials and to quite high hydrogen capacities (about 5.5-6 wt.%). The investigated salts (NaF, NaCl, $MgF_2$ or $CrCl_3$) had a different influence on the metal surface properties, which was reflected in reaction kinetics, in particular at first hydriding. $MgF_2$ and NaCl peel off of the oxide layer from magnesium surface facilitating hydride nucleation. $CrCl_3$ may additionally act as catalysis for hydrogen chemisorption. The formation of $NaMgF_3$ ternary fluoride has been observed at the initial stages of hydriding of Mg-NaF mechanical alloys. This ternary fluoride was found to play an active role in the hydriding process.

***Keywords:*** hydrogen absorbing materials, mechanical alloying, gas-solid reactions

## 1. Introduction

The high theoretical hydrogen capacity of $MgH_2$ (7.6 wt.%) attracts attention of many researches to magnesium and magnesium-based alloys as perspective materials for hydrogen storage. The main goal of the investigations is improving of hydrogen storage characteristics of these materials because, from a practical application point of view, kinetics of hydrogen absorption by magnesium and decomposition of $MgH_2$ are not sufficiently fast even at temperatures as high as 573-623K. Moreover, the theoretical hydrogen capacity practically is never achieved, usually being at a level of about 5 wt.%.

Some peculiarities have been shown to be inherent in the interaction of magnesium with hydrogen. The kinetics of hydriding at the first and subsequent cycles differ very much from each other. The overall kinetics of first hydriding of magnesium is determined by the statistical cracking of an oxide layer usually covering the magnesium particles and hydride nucleation on the metal sites formed [1]. This leads to a long induction period and sigmoid shape of the kinetic curve.

The rate of magnesium hydriding at the initial stages of second and subsequent cycles and the rate of decomposition of $MgH_2$ are limited by the dissociative adsorption (desorption) of hydrogen on the metal surface. The nuclei of magnesium hydride are formed on the metal surface, with an interface propagating along the metal surface in the

489

*T.N. Veziroglu et al. (eds.),*
*Hydrogen Materials Science and Chemistry of Carbon Nanomaterials,* 489-502.
© 2004 *Kluwer Academic Publishers. Printed in the Netherlands.*

layer of adsorbed hydrogen [2]. After the nuclei are overlapping, the rate of the reaction is controlled by the diffusion of hydrogen through the product layer. As a rule, a total transformation of magnesium into hydride is not achieved because the hydride layer becomes impermeable before the reaction is finished.

Bearing all this in mind, one can conclude that improving of hydrogen storage characteristics of magnesium and magnesium-based alloys may be achieved by decreasing alloy particle size and modification of their surface with catalytic additives.

Mechanical alloying (MA) is regarded as one of the most effective methods for obtaining fine materials with various microstructure, composition and components content. It was successfully used in preparing various composite materials for hydrogen storage purposes. Mechanical alloying of magnesium with other metals, non-metals and oxides usually improved its hydriding properties (see [3-10], for example). Special surface-active additions capable of impeding aggregation processes are desired for obtaining fine magnesium powder. Some organic compounds [11,12] and graphite [13,14] have been used for this purpose.

The least attention has been paid to the inorganic salts as an addition not only to magnesium-based materials but also to other metal-hydrogen systems. It may be because of the difficulty to expect of catalytic activity of inorganic salts in hydrogen chemisorption processes. Only recently information appeared about the very important role of titanium and iron chlorides in reversible hydrogenation of alkali alanates [15,16] and improving of hydrogen storage properties of Mg-Ni composite by ball milling of Mg and Ni under hydrogen pressure in the presence of $CrCl_3$ [17]. Although it should be noted that Suda and his co-workers as early as in 1990 proposed and now are intensively developing the method of surface modification of hydrogen absorbing alloys by treatment before hydriding of metallic powders in aqueous solution containing $F^-$ anions [18-22].

In this study, we demonstrate the use of inorganic salts (e.g. NaF, NaCl, $MgF_2$ or $CrCl_3$) as an additive to magnesium or magnesium alloys in the course of mechanical alloying and its effect on hydriding properties of materials obtained in the result of this treatment.

## 2. Experimental

Magnesium powder (Alfa; particle size, 50-100 μm) and NaF, $MgF_2$, $CrCl_3$ or NaCl salts ("pure" grade) were used in the experiments.

Mechanical alloying was carried out with a high-energy planetary centrifugal mill of the AGO-2 type (acceleration about 400 m·s$^{-2}$) with water-cooled vials. Stainless steel vials and balls of 5 mm in diameter were used. Ball mass to sample mass ratio was 100:1. To avoid oxidation, after loading the samples, the vials were evacuated and filled with argon.

Specific surface of the samples was measured by the BET method with accuracy ± 10%.

The iron content in the samples was controlled by atomic absorptive analysis.

The surface chemistry was determined by Energy Dispersive Spectroscopy (EDS).

X-ray powder diffraction data were obtained with Cu K$_\alpha$ radiation.

Electron microscopic studies were carried out using a Jeol JEM 2000 FX II electron microscope at an accelerating voltage of 200 kV.

All preparatory work was carried out in air without any precautions.

The interaction with hydrogen was investigated using a constant-volume reactor made of stainless steel. The amount of absorbed or evolved hydrogen was determined by following the changes of pressure in the reactor. The pressure was measured by means of a piezoresistive transducer with overall precision better than 0.2%. Before hydriding, the reactor was evacuated and heated in dynamic vacuum about 5 Pa to the required temperature of the experiment. Temperature was maintained at an accuracy of $\pm$ 1 K.

## 3. Results and discussion

Mechanical alloying of magnesium with addition of NaF, $MgF_2$, NaCl or $CrCl_3$ inorganic salts leads to comminution of magnesium particles and to increase of specific surface of the sample. The specific surface increases with the increasing of salt content and time of mechanical alloying and depends on the nature of the salt used. For example, the specific surface of Mg-5 wt.% NaF sample mechanically alloyed for 5 min. was equal to 0.15 $m^2/g$ whereas Mg-5 wt.% NaCl sample prepared under the same conditions had a specific surface five times higher (0.75 $m^2/g$). The powdering of magnesium during mechanical alloying in the presence of salt addition may be explained by salt action as surface-active substance toward magnesium.

The change of magnesium particles size cannot explain sufficient improving of hydrogen absorption kinetics in the first cycle of hydriding (Fig.1). Magnesium sample prepared by cathode spraying (particles size of 20 μm, estimated surface area 0.17 $m^2/g$) exhibit low sigmoid kinetics with a long induction period [1]. Almost all investigated Mg-salt mechanical alloys start to absorb hydrogen at the highest rate. Only Mg-NaCl mechanical alloy exhibited a sigmoid kinetic curve at first hydriding in our experiments, but without an induction period.

The absence of the induction period at first hydriding and the beginning of the reaction at maximal rate are evidence for the presence on the sample surfaces of places where hydrogen can easily be adsorbed and dissociated. It may be Fe clusters which appeared in the samples as impurities during mechanical alloying. Indeed, according to the results of chemical analysis the iron content increased from 0.03 wt.% (original Mg) to a value that is approximately equal for all samples (0.2-0.3 wt.%) after mechanical alloying. But the iron content is quite low. Moreover, it cannot explain the different shape of the kinetic curves of the first hydriding of the investigated mechanical alloys, particularly at the beginning of the reaction (Fig.1b). Modification of the magnesium surface by salt additives takes place during MA, with distinctive features inhering in each salt.

The Mg-NaF particles had quite a flat surface, the surface of the Mg-NaCl mechanical alloy particles looked like a cobblestone pavement and the slightly erosive surface is observed in the case of MA Mg-$CrCl_3$ (Fig. 2), the particles shape being in the form of flakes for all MA.

$MgF_2$ was less effective than other salts. Both the impurity of Fe and the peeling off of the oxides and hydroxides from the magnesium surface during MA may explain the start of the hydriding reaction at the maximum rate in this case, but the hydrogen capacity is low.

The highest initial reaction rate at first hydriding was found in samples prepared with the addition of $CrCl_3$. In this case we suspect not only the appearance of a magnesium

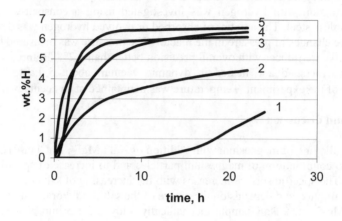

**a**

**b**

Figure 1. First hydriding at T=623 K, $P_{H2}$=1.5 MPa
a – long-time experiments; b – initial part of Fig.1a.
1 – Mg, prepared by cathode spraying, fraction with particles size 20 μm [1];
2÷5 – mechanical alloys prepared for 5 min.
2 - MA Mg-5wt.% $MgF_2$; 3 - MA Mg-5wt.% $CrCl_3$; 4 - MA Mg-5wt.% NaF; 5 - MA Mg-5wt.% NaCl;

metallic surface due to the peeling off effect but also a catalytic action of $CrCl_3$ (the same assumption has been made in [17]) or, more probably, reduction of $CrCl_3$ by magnesium during MA with the formation of Cr clusters which may be additional sites for hydrogen chemisorption.

There is no known catalytic activity of NaF or NaCl in the hydrogen chemisorption reaction. These salts can affect the reaction rate at first hydriding by the destruction of the compact oxide layer on the magnesium surface. The different shape of the kinetic curves of first hydriding of Mg-5%NaF and Mg-5%NaCl mechanical alloys (Fig.1b) allowed us to suppose a different mechanism of the NaCl and NaF action (the Fe content was the same in these samples – 0.23 wt.%).

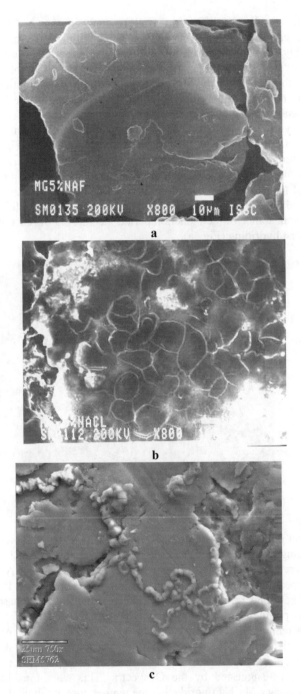

Figure 2. SEM image of particles of Mg-5%NaF (a), Mg-5%NaCl (b) and Mg-5%CrCl$_3$ (c) after mechanical alloying for 5 min.

According to X-ray diffraction (XRD) analysis the NaCl reflections became broadened after MA and then remained unchanged after hydriding. It means that NaCl, more probably, acts mechanically on the magnesium surface during MA and destroys magnesium oxides, the oxide layer being thin or defective but not removed completely. Due to this, the nucleation of $MgH_2$ can proceed only on definite locations having a broken oxide layer. Like pure magnesium [1] the kinetics of the first hydriding of the Mg-NaCl mechanical alloy is determined by the processes of formation of a metallic surface and hydride nucleation on it. The separate nuclei of the magnesium hydride are observed on definite and defect locations of the MA surface (Fig. 3). This is quite a common picture for reactions determined by nucleation and nuclei growth. The oxide layer seems not to be so compact on MA Mg-NaCl as on pure magnesium. The nucleation of $MgH_2$ is easier and an induction period is absent. A more compact oxide layer appeared on the MA Mg-NaCl surface after long storing of the samples in air. This leads to the appearance of the induction period, decreasing of reaction rate and hydrogen capacity at first hydriding (Fig.4a).

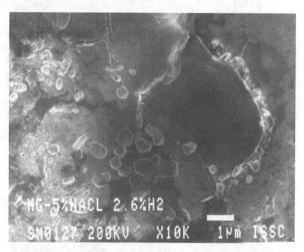

Figure 3. SEM images of MA Mg-5%NaCl hydrided up to 2.6 wt.% H

A different picture has been observed in the case of Mg-NaF mechanical alloys both in the hydriding reaction kinetics and the XRD data. Like for NaCl, broad reflections of NaF are observed in the X-ray diffraction pattern (Fig. 5) after MA. But the reflections of NaF disappear and new reflections appear in the XRD pattern at the very beginning of hydriding. These new reflections may be related to the $NaMgF_3$ phase.

The X-ray photoelectron spectroscopy (XPS) analysis of Mg-NaF performed at room and elevated temperature revealed the change in chemical state of magnesium under heating to 623 K and higher (Fig. 6). All of the Mg spectra show a shift to lower binding energy that might be indicative of an increase in the amount of Mg metal or the presence of Mg oxide, as suggested by the O spectra. This cannot exclude or confirm the formation of ternary Mg-Na fluoride, but the temperature at which the changes in spectra are observed coincides with that of our experimental conditions.

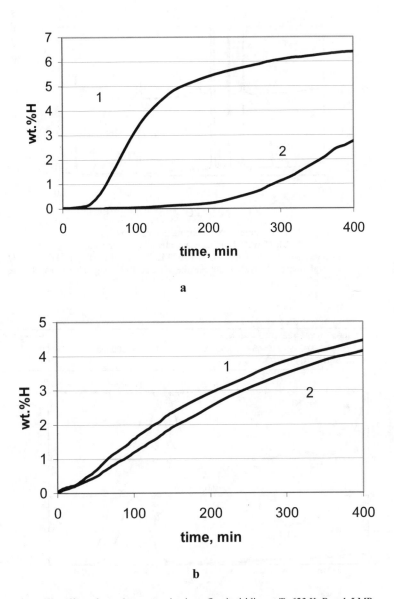

Fig. 4. The effect of samples storage in air on first hydriding at T=623 K, $P_{H2}$=1.5 MPa
a – Mg-5wt.% NaCl, 5 min of MA ; b- Mg-5wt.% NaF, 5 min of MA
1 - stored in air for 0.5 h before hydriding; 2 - stored in air for 2 months before hydriding

NaF spreads along the magnesium surface and forms on it a fluorine-contained layer during mechanical alloying. The EDS analysis performed on different parts of the sample confirmed that the fluorine layer is rather uniform (concentration of F varied from 1.96 to 6.36 at.%). In contrast to NaCl, the fluorine-contained layer formed on the

496

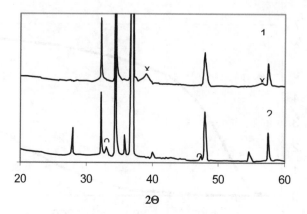

Figure 5.  X-ray diffraction patterns
1 - Mg+5 wt.% NaF, mechanically alloyed for 5 min
2 - Same sample after hydriding at T=623 K, $P_{H2}$=1.5  MPa ( up to 0.69 wt.%H).
x – NaF phase; o -   $NaMgF_3$ phase; the other peaks belong to Mg or $MgH_2$ phases

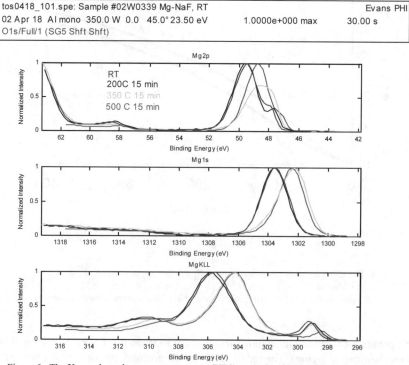

Figure 6.  The X-ray photoelectron spectroscopy (XPS) analysis of Mg-5wt.%NaF mechanical alloy

surface protects the MA Mg-NaF particles from the action of oxygen, water and other surface pollutants. The protective properties of NaF may be confirmed by almost unchanged kinetics of first hydriding of MA Mg-5%NaF after long storage of the sample in air (Fig. 4b).

The surface oxide layer is destroyed during heating of the sample to the experimental temperature due to the reaction of NaF with MgO or Mg(OH)$_2$ and formation of NaMgF$_3$. Separate experiments revealed that peaks of the NaMgF$_3$ phase are observed in the XRD pattern of the Mg-NaF sample heated at 623 K in argon atmosphere or vacuum (without hydrogen) for 15-30 min.

Then, after the reactor is filled with hydrogen the formation of hydridofluoride is possible. A hydridofluoride of the NaMgH$_2$F composition has been obtained in [23,24] under experimental conditions close to that used in this work. NaMgH$_2$F is isostructural to NaMgF$_3$ with very close parameters of the orthorhombic unit cell [24,25]. These two phases practically are not distinguishable from each other in the XRD pattern at low concentrations in particular.

It is quite probable that diffusion of hydrogen proceeds easily in the NaMgH$_2$F phase because "H$^-$ and F$^-$ anions are equally distributed over the two anionic sites in a disordered way" [24]. Like a sponge, the NaMgH$_2$F layer absorbs hydrogen and supplies it to the subsurface of the magnesium, where MgH$_2$ begins to be formed at a definite concentration of hydrogen. However, an impermeable layer is not formed because the recrystallization processes begin on the surface shortly after.

The recrystallization processes seem to be complex and include spatial reconstruction of both the hydridofluoride and the magnesium hydride phases. The EDS analysis of the samples hydrided up to 2.4 wt,%H revealed areas containing sodium atoms along with areas without sodium admixture.

Separate crystallites of MgH$_2$ grow from magnesium hydride layer (Fig. 7) exhausting the latter and making it thinner. The thickness of the hydride layer seems to increase but be maintained at a low level for some period of reaction due to two opposite processes: growth as a result of hydrogen absorption and dissolution at the expense of recrystallization. The diffusion retardation does not arise up to degree of transformation $\alpha \approx 0.54$ ($\approx 4$ wt.%H). The overall reaction course is determined by interface moving into the particle bulk at this period.

The recrystallization processes manifest themselves at final stages of hydriding too. The hydride layer is less compact on Mg-NaF surface in comparison with that on Mg-NaCl at the same degree of transformation.

The second and subsequent hydriding-dehydriding cycles of all the investigated Mg-salt mechanical alloys exhibited a similar and relatively fast reaction rate at hydriding (Fig. 8a) and (with the exception of MA Mg-MgF$_2$) a hydrogen capacity at a level of 5.5-6 wt.%.

The apparent energy of activation (E$_a$) for the hydriding processes has been estimated from Arrhenius plots of $\ln\{v/(P_0-P_{eq})\}$ versus $1/T$, where $v$ is the initial rate of hydriding, $P_0$ is the hydrogen pressure in the reactor and $P_{eq}$ is the equilibrium hydrogen pressure for MgH$_2$. A comparison of the obtained data with E$_a$ for hydriding of pure magnesium (ranging between 89 and 96 kJ/mol [26,27,28]) may provide additional confirmation of our suggestion as to the catalytic activity of CrCl$_3$. E$_a$ for the hydriding of MA Mg-5wt.%CrCl$_3$ was measured to be 70±2 kJ/mol. E$_a$ for the hydriding of MA Mg-5wt.%NaF was found to be higher (79±2 kJ/mol). E$_a$ = 90±4 kJ/mol for the hydriding of

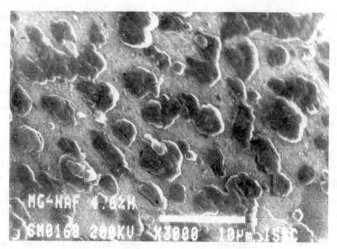

Figure 7. SEM images of MA Mg-5%NaFl hydrided up to 4.6 wt.%H

MA Mg-5%MgF$_2$ is almost the same as for pure Mg. It is evident that MgF$_2$ does not exhibit catalytic activity.

The decomposition of the hydrides formed as a result of hydriding of MA Mg-salts also proceeded faster than pure MgH$_2$ (Fig. 8b), being comparable with that for Mg-5 wt.%Fe mechanical alloy [3] in the case of CrCl$_3$.

The preparation of more complex mechanical alloys, for example, containing salt and metal-catalyst, may lead to a further improvement of hydrogen storage characteristics of the obtained material. Fig. 9 illustrates the action of NaF additives on the hydriding Mg-1%Ni mechanical alloy. The initial reaction rate for MA Mg-1%Ni-5%NaF is the same as for MA Mg-1%Ni, but the first mechanical alloy is hydrided to a higher degree of transformation and achieves a hydrogen capacity of 6 wt.% whereas the hydrogen capacity of Mg-1%Ni mechanical alloy is not larger than 2 wt.% at first hydriding [4]. Both the reaction rate and hydrogen capacity is higher in the case of MA Mg-1%Ni-5%NaF at second and subsequent cycles of hydriding.

## 4. Conclusions

The mechanical alloying of magnesium with some inorganic salts addition leads to powdering of metal and simultaneous modification of metal particle surface, which improve the hydrogen storage characteristics of the material obtained.

The action of salts is different and depends on their nature.

The search for optimal compositions of the mechanical alloys with salt additions may greatly improve the hydrogen storage characteristics of magnesium-based alloys and other probable alloys on the basis of hydrogen absorbing metals.

The huge variety of salts promises a multiplicity of properties of materials produced in such a way. This is supported by the fact that even the actions of rather similar salts (halides investigated in this work) are very different in the course of both mechanical alloying and hydriding.

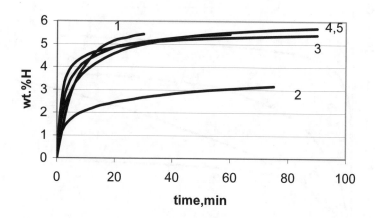

**a**

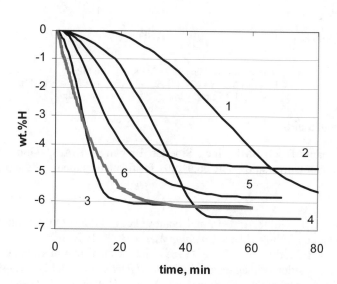

**b**

Figure 8. Hydriding kinetics (at T=623 K, $P_{H2}$=1.5 MPa) - (a) and dehydriding kinetics (at T=623 K and $P_{H2}$=0.1 MPa) – (b) of activated mechanical alloys
1 – Mg; 2- MA Mg-5wt.% $MgF_2$; 3 - MA Mg-5wt.% $CrCl_3$; 4 - MA Mg-5wt.% NaF;
5 - MA Mg-5wt.% NaCl; 6- MA Mg-5wt.%Fe [1].

500

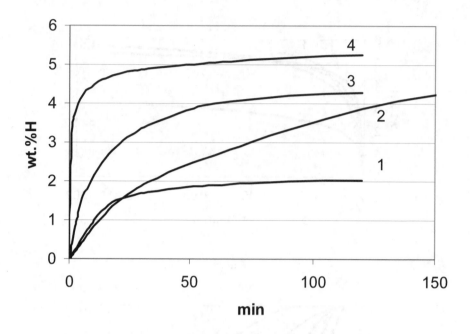

Figure 9.  Hydriding (at T=583 K, $P_{H2}$=1.6 MPa)
1,3 - MA Mg-1wt.%Ni [4] ; 2,4 - MA Mg-1wt.%Ni-5wt.%NaF
1,2 – first hydriding; 3,4 – second hydriding

## 5. Acknowledgements

The authors thank Tosoh SMD, Inc., Siberian Branch of R.A.S. (Grant No. 31) and Program "Russian Universities" UR 06.01.001 for financial support of the present study.

## 6. References

1. Gerasimov, K.B., Goldberg, E.L. and Ivanov, E.Yu. (1987) On the kinetic model of magnesium hydriding, *J Less-Common Met.* **131**, 99-107.
2. Gerasimov, K.B. and Ivanov, E.Yu. (1985) The mechanism and kinetics of formation and decomposition of magnesium hydride, *Materials Lett.* **3**(12), 497-499.
3. Ivanov, E., Konstanchuk, I., Stepanov, A. and Boldyrev, V. (1987) Magnesium mechanical alloys for hydrogen storage, *J. Less-Common Met.* **131**, 25-29.
4. Stepanov, A., Ivanov, E., Konstanchuk, I. and Boldyrev, V. (1987) Hydriding properties of mechanical alloys Mg-Ni, *J. Less-Common Met.* **131**, 89-97.
5. Konstanchuk, I., Ivanov, E., Darriet, B., Pezat, M., Boldyrev, V. and Hagenmüller, P. (1987) The hydriding properties of a mechanical alloy with composition Mg-25 wt.% Fe, *J. Less-Common Met.* **131**, 181-189.
6. Khrussanova, M., Terzieva, M., Peshev, P., Konstanchuk, I. and Ivanov, E. (1989) Hydriding kinetics of mixtures containing some 3d-transition metal oxides and magnesium, *Zeitschrift für Phys. Chem. N.F.* **Bd. 164**, 1261-1266.

7. Khrussanova, M., Terzieva, M., Peshev, P., Konstanchuk, I. and Ivanov, E. (1991) Hydriding of mechanically alloyed mixtures of magnesium with MnO₂, Fe₂O₃ and NiO, *Materials Research Bulletin* **26**, 561-567.

8. Liang, G., Boily, S., Huot, J., Van Neste, A. and Schulz, R. (1998) Hydrogen absorption properties of a mechanically milled Mg-50 wt.% LaNi₅ composite, *J. Alloys Comp.* **268**, 302–307.

9. Oelerich, W., Klassen, T. and Bormann, R. (2001) Metal oxides as catalysts for improved hydrogen sorption in nanocrystalline Mg-based materials, *J. Alloys Comp.* **315**, 237-242.

10. Khrussanova, M., Grigorova, E., Mitov, I., Radev, D. and Peshev, P. (2001) Hydrogen sorption properties of an Mg-Ti-V-Fe nanocomposite obtained by mechanical alloying, *J. Alloys Comp.* **327**, 230-234.

11. Imamura, H., Sakasai, N. and Fujinaga, T. (1997) Characterisation and hydriding properties of Mg-graphite composites prepared by mechanical alloying as new hydrogen storage materials, *J. Alloys Comp.* **253-254**, 34-37.

12. Imamura, H., Takesue, Y., Akimoto, T. and Tabata, S. (1999) Hydrogen-absorbing magnesium composites prepared by mechanical grinding with graphite: effects of additives on composite structures and hydriding properties, *J. Alloys Comp.* **295**, 564-568.

13. Imamura, H., Tabata, S., Takesue, Y., Sakata, Y. and Kamazaki, S. (2000) Hydriding-dehydriding behavior of magnesium composites obtained by mechanical grinding with graphite carbon, *Int. J Hydrogen Energy* **25**, 837-843.

14. Bouaricha, S., Dodelet, J.P., Guay, D., Huot, J. and Schulz, R. (2001) Activation characteristics of graphite modified hydrogen absorbing materials, *J. Alloys Comp.* **325**, 245-251.

15. Bogdanovic, B. and Schwickardi, M. (1997) Ti-doped alkali metal aluminium hydrides as potential novel reversible hydrogen storage materials, *J. Alloys Comp.* **253-254**, 1-9.

16. Sandrock, G., Gross, K., Thomas, G., Jensen, C., Meeker, D. and Takara S. (2002) Engineering considerations in the use of catalyzed sodium alanates for hydrogen storage, *J. Alloys Comp.* **330-332**, 696-701.

17. Yu, Z., Liu, Z. and Wang, E. (2002) Hydrogen storage properties of the Mg-Ni-CrCl₃ nanocomposite, *J. Alloys Comp.* **333**, 207-214.

18. Wang, X.L. and Suda, S. (1995) Surface characteristics of fluorinated hydriding alloys, *J. Alloys Comp.* **231**, 380-386.

19. Wang, X.L., Haraikawa, N. and Suda, S. (1995) A study of the surface composition and structure of fluorinated Mg-based alloys, *J. Alloys Comp.* **231**, 397-402.

20. Liu, F.G. and Suda, S. (1996) Hydriding behavior of F-treated Mg2Ni at moderate conditions, *J. Alloys Comp.* **232**, 212-217.

21. Sun, Y.M., Gao, X.P., Araya, N., Higuchi, E. and Suda, S. (1999) A duplicated fluorination technique for hydrogen storage alloys, *J. Alloys Comp.* **295**, 364-368.

22. Higuchi, E., Li Zh., P., Suda, S., Nohara, Sh., Inoue, H. and Iwakura, Ch. (2002) Structural and electrochemical characterization of fluorinated AB2-type Laves phase alloys obtained by different pulverization methods, *J. Alloys Comp.* **335**, 241-245.

23. Bouamrane, A., de Brauer, C., Soulié, J.-P., Létoffé, J.M. and Bastide, J.P. (1999) Standard enthalpies of formation of sodium-magnesium hydride and

502

hydridofluorides NaMgH$_3$, NaMgH$_2$F and NaMgF$_2$H, *Thermochimica Acta* **326**, 37-41.

24. Bouamrane, A., Laval, J.P., Soulié, J.-P. and Bastide, J.P. (2000) Structural characterization of NaMgH$_2$F and NaMgH$_3$, *Mater. Res. Bull.* **35**, 545-549.

25. Ronnebro, E., Noreus, D., Kadir, K., Reiser, A. and Bogdanovic, B. (2000) Investigation of the perovskite related structures of NaMgH$_3$, NaMgF, and Na$_3$AlH$_6$, *J. Alloys Comp.* **299**, 101-106.

26. Stander, C.M. (1977) Kinetics of formation of magnesium hydride from magnesium and hydrogen, *Z. Phys. Chem. N.-F.* **104**, 229.

27. Gerasimov, K.B. (1986) Reversible interaction of magnesium with hydrogen, *Ph.D thesis*, Novosibirsk, (in Russ.).

28. Fernándes, J.F. and Sánchez, C.R. (2002) Rate determining step in the absorption and desorption of hydrogen by magnesium, *J. Alloys Comp.* **340**, 189-198.

# HYDROGEN SORPTION AND ELECTROCHEMICAL PROPERTIES OF INTERMETALLIC COMPOUNDS La$_2$Ni$_7$ AND La$_2$Ni$_6$Co

E. LEVIN, P. DONSKOY, S. LUSHNIKOV, V. VERBETSKY,
T. SAFONOVA, O. PETRII
*Moscow State University*
*Chemistry Department*
*Leninskie Gory 3, Moscow, 119992 Russia*
*E-mail: levin@elch.chem.msu.ru*

## Abstract

Intermetallic compounds (IMCs) La$_2$Ni$_7$ and La$_2$Ni$_6$Co were obtained by arc melting technique followed by annealing. Samples were characterized by means of X-ray powder diffraction (XRD). Crystal structure of α-La$_2$Ni$_7$ was refined using single crystal X-ray data. Hydrogen desorption isotherms were obtained for La$_2$Ni$_7$ and La$_2$Ni$_6$Co at T=273K and T=295K. Electrochemical discharge properties of La$_2$Ni$_7$ and La$_2$Ni$_6$Co were investigated. The introduction of Co in alloy led to decreasing of hydride equilibrium pressure.

*Keywords:* nickel-metal hydride battery, hydrogen storage alloy, discharge capacity, single crystal X-ray diffraction, crystal structure

## 1. Introduction

At present, in line with the increasing demand for portable batteries, active studies are carried out in a search for new compounds for optimization of electrode materials. One type of such batteries is nickel metal hydride (MH) batteries. These batteries are widely applied today because of their high energy density, recycling possibility and environmental safety. This paper deals with materials for negative electrodes for Ni-MH batteries. For this purposes wide range of compounds are in use [1-4]. For our studies we have chosen two Ce$_2$Ni$_7$-type [5] compounds: La$_2$Ni$_7$ and La$_2$Ni$_6$Co which had never been tested for their electrochemical properties. Crystal structure of La$_2$Ni$_7$ was refined using single crystal X-ray data. Hydrogen sorption and electrochemical properties were investigated.

## 2. Experimental

Alloys were prepared using arc-melting technique from the mixtures of initial metals under argon atmosphere. As known [6], La-based compounds of this structure type are bimorphic, hence all compounds were annealed for 240 h at 1153K to reach homogeneity and for low temperature phase stabilisation. The phase composition of alloys was determined using X-ray powder diffraction data. X-Ray powder diffraction (XRD) patterns were taken at room temperature by using a DRON 3M diffractometer

503

*T.N. Veziroglu et al. (eds.),*
*Hydrogen Materials Science and Chemistry of Carbon Nanomaterials,* 503-510.

(CuK$_\alpha$ radiation, graphite monochromator) in the range of 2Θ=20°-80° with a step of 0.05°. Single crystal diffraction data were collected at room temperature on Enraf-Nonius CAD-4 diffractometer. Structure was refined with JANA-2000 program package [7]. The results of data collection and structure refinement are summarized in Table 1.

Pressure-composition isotherms were obtained using Siewert's type apparatus at a pressure below 25 atm. The samples were hydrogenated by step-by-step pressure increasing. Absorption isotherms were not recorded.

The electrodes for electrochemical experiments were prepared by mixing alloy powder with copper powder in weight ratio 1:4. The mixture was cold-pressed to a 7-mm pellet at a pressure 150 atm/cm$^2$. All electrochemical experiments were carried out using a three-electrode quartz cell with Hg/HgO as the reference electrode, Pt wire as the counter electrode, and MH as the working electrode. The electrolyte was 6M KOH. For measurements, a P-5827M potentiostat was used. The data were collected using X-Y recorder. The cut-off potential was –0.6V.

TABLE 1. Summary of crystallographic information for La$_2$Ni$_7$

| Estimated chemical formula | La$_2$Ni$_7$ |
|---|---|
| Crystal system | Hexagonal |
| Space group | P6$_3$/$mmc$ |
| Cell constants $a$, $c$ (Å) | 5.0577(4)   24.7336(22) |
| Volume (Å$^3$) | 547.926 |
| Z | 4 |
| D$_{calc}$, g·cm$^{-3}$ | 8.345 |
| Radiation/ Wavelenghth (Å) | MoK$_\alpha$/ 0.71073 |
| θ range (°) for cell determination | 9.47 – 22.39 |
| Linear absorption coefficient (cm$^{-1}$) | 38.54 |
| Temperature (K) | 293 |
| Crystal shape | Arbitrary |
| Colour | Metallic |
| Diffractometer | CAD4, Graphite monochromator |
| Data collection method | ω- θ |
| Absorption correction | Rocking curve |
| No. of measured reflections | 3653 |
| No. of observed independent reflections | 388 |
| Criteria for observed reflections | I > 3 σ (I) |
| Data collection θ limits (°) | 2 – 35 |

| Range of $h, k, l$ | $0\rightarrow h\rightarrow 8$, $-8\rightarrow k\rightarrow 8$, $-39\rightarrow l\rightarrow 39$ |
|---|---|
| $R_{int}$ | 0.025 |
| Refinement on | F |
| Extinction | Anisotropic, Becker-Coppens, Type I, Gaussian distribution |
| $R/R_w$    (I > 3σ(I)) | 0.042/0.063 |
| Goodness of fit | 1.01 |
| No. of reflections used in refinement | 386 |
| No. of refined parameters | 26 |
| Weighting scheme | $w = (\sigma^2(F) + (0.045F)^2)^{-1}$ |
| $(\Delta/\sigma)_{max}$ | 0.004 |
| $\Delta r_{max}$ (e/Å$^{-3}$) +/- | +3.47/-4.18 |

## 3. Results and discussion

XRD patterns of alloy powders are given in Figure 1. Comparison of experimental and calculated patterns revealed absence of impurities in samples.

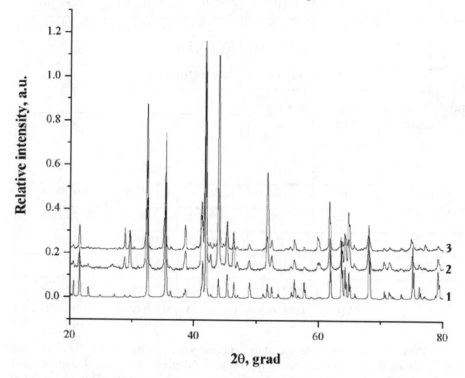

Figure 1. XRD patterns of obtained samples: 1 - calculated pattern for La$_2$Ni$_7$, 2 – experimental pattern for La$_2$Ni$_7$, 3 – experimental pattern for La$_2$Ni$_6$Co.

Structure refinement was carried out using the model from [5]. Coordinates and atomic displacement parameters (ADP) are given in Table 2. Refinement with isotropic temperature factors converged at R=0.042, wR=0.063 and in case of anisotropic temperature factors refinement resulted in R=0.035, wR=0.049. Unfortunately, rocking curve method does not produce perfect absorption correction; and, thus, the refinement produced slightly negative atomic displacement tensors for two Ni atoms, so they are not listed. During refinement we detected no significant differences from [5] in atomic site parameters and no structure peculiarities. All temperature factors were at least two times lower as compared with those given in [5]. Refined atomic coordinates and ADPs are summarised in Table 3.

TABLE 2. Atomic coordinates and ADPs from Cromer, D. [5].

| Atomic Site | Site simmetry | x | Y | z | $B_{iso}$, Å$^2$ * |
|---|---|---|---|---|---|
| Ce1 | 4f | 1/3 | 2/3 | 0.0302(2) | 0.62(7) |
| Ce2 | 4f | 1/3 | 2/3 | 0.1747(2) | 0.61(6) |
| Ni1 | 2a | 0 | 0 | 0 | 0.50(21) |
| Ni2 | 4e | 0 | 0 | 0.1670(4) | 0.64(16) |
| Ni3 | 4f | 1/3 | 2/3 | 0.8334(4) | 0.60(14) |
| Ni4 | 6h | 0.8351(18) | 0.6702(18) | 1/4 | 0.41(8) |
| Ni5 | 12k | 0.8338(14) | 0.6676(14) | 0.0854(2) | 0.67(7) |

TABLE 3. Refined atomic coordinates and ADPs.

| Atomic site | Site symmetry | x | y | z | $B_{iso}$, Å$^2$ |
|---|---|---|---|---|---|
| La1 | 4f | 1/3 | 2/3 | 0.03013(3) | 0.19(2) |
| La2 | 4f | 1/3 | 2/3 | 0.17422(3) | 0.20(2) |
| Ni1 | 2a | 0 | 0 | 0 | 0.22(4) |
| Ni2 | 4e | 0 | 0 | 0.16898(7) | 0.30(3) |
| Ni3 | 4f | 1/3 | 2/3 | 0.83616(8) | 0.29(3) |
| Ni4 | 6h | 0.8341(3) | 0.6682(1) | 1/4 | 0.11(2) |
| Ni5 | 12k | 0.8350(2) | 0.6700(1) | 0.08806(4) | 0.16(2) |

Pressure-composition-temperature curves were measured for T=273K and T=295K. From these data, pressure-composition isotherms were plotted. Figures 2, 3 show the isotherms for La$_2$Ni$_7$ and for La$_2$Ni$_6$Co at equal temperatures. Starting point of desorption isotherms was determined from values of hydrogen sorbed. As seen from these figures, equilibrium hydride pressure decreases with an introduction of Co into alloy, but even for La$_2$Ni$_6$Co the equilibrium pressure is higher than 1 atm. From these isotherms we can see, that at T=273K the plateaus have an increased slope as compared with isotherms taken at room temperature (295 K). Also hydrogen does not desorb totally. These facts can be explained by the slow kinetics of reaction of IMCs with hydrogen at lowered temperatures, which is revealed in the increasing slope that indicating a nonequilibrium process. For clarifying these phenomenon the hydrogen desorption rate should be slower. But for primary estimation of hydrogen-sorption properties we can use already obtained data.

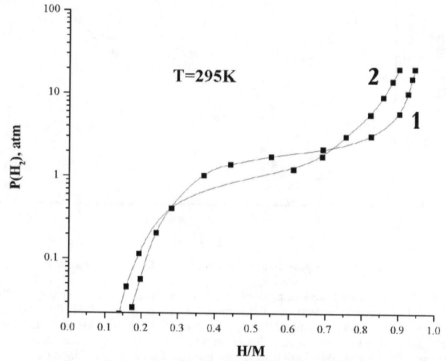

Figure 2. Hydrogen desorption isotherms at T=295 K. 1 – La$_2$Ni$_7$, 2 – La$_2$Ni$_6$Co.

As far as we know, there is only one article dealing with La-based hydrides of such a structure type [8]. Although the values of hydrogen absorbed well agree with these in literature (H/M=0.89 from [8] and H/M=0.94 from our absorption data) there is no previously reported data concerning pressure values of absorption or desorption at all. H/M=0.94 value obtained for La$_2$Ni$_7$ is close to that well known for LaNi$_5$ H/M=1. Although our compound contains more rare-earth metal than LaNi$_5$-type alloys it would be very interesting to investigate Ce-based compounds of such type. There is also only one article dealing with Ce$_2$Ni$_7$ and its Co derivative [9]. These data indicate that the equilibrium pressure of Ce-based compounds is very low (~0.01 atm). On the other hand, introducing Ce into AB$_5$-type alloys leads to catastrophic increasing [10] of equilibrium pressure. Inasmuch as all commercial rare-earth containing alloys are based on mischmetal, compounds of the Ce$_2$Ni$_7$-type can be a cheaper alternative to LaNi$_5$-based compounds. But this question needs more attention and further investigation.

508

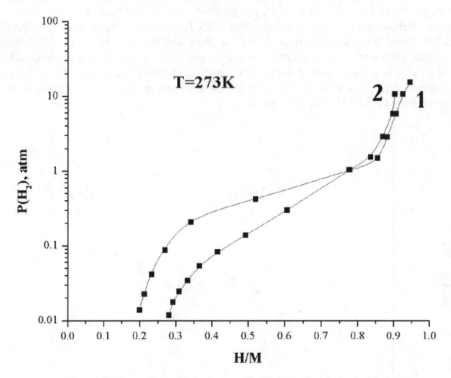

Figure 3. Hydrogen desorption isotherms at T=273 K. 1 – La$_2$Ni$_7$, 2 – La$_2$Ni$_6$Co.

Figure 4 shows the electrochemical discharge curves at current density j=100 mA/g and corresponding sections of desorption isotherms at T=295 K. Pressure for discharge curves was calculated using the Nernst equation. As we can see, electrochemical curves are much shorter in terms of H/M ratio than desorption isotherms. This is caused by high dissociation pressure for La$_2$Ni$_7$ and La$_2$Ni$_6$Co which results in incomplete adsorption-desorption in electrochemical conditions. Moreover, in the discharge curve for La$_2$Ni$_7$, two regions with different slopes exist. The upper (in term of equilibrium pressure) plateau corresponds to α↔β hydride phase transition. The presence of the second plateau can be explained by the contribution of hydrogen ionization. We can suggest that no quantitive measurements can be made on such samples without altering chemical composition in order to low equilibrium pressure.

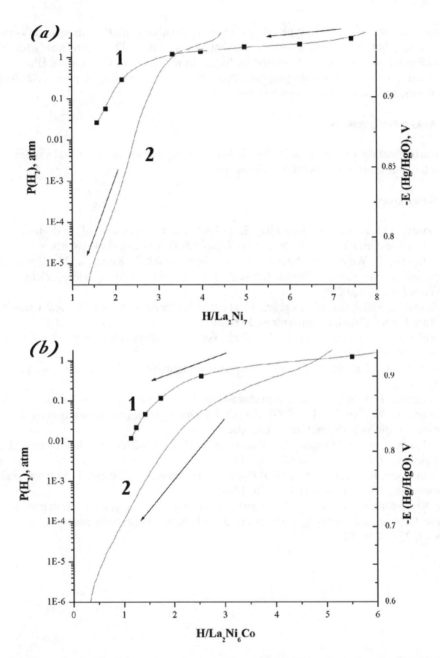

Figure 4. Electrochemical discharge curves and corresponding sections of desorption isotherms; a ($La_2Ni_7$): 1 – desorption isotherm, 2 – discharge curve; b ($La_2Ni_6Co$): 1 – desorption isotherm, 2 – discharge curve.

510

## 4. Conclusions

Thus, the studied $Ce_2Ni_7$-type alloys showed relatively high hydrogen adsorption capacities, but low electrochemical discharge capacities. This correlates with high equilibrium pressure of the La-based hydrides. Introduction of Co into the alloy led to decreasing of the equilibrium pressure. Probably, the introduction of pressure-reducing elements, such as Ce, Mn, V can be promising.

## 5. Acknowledgements

We are grateful to Mironov A.V. for his help in carrying out single crystal experiments and for censorious remarks during discussion.

## 6. References

1. Petrii, O, Vasina, S, Korobov, I. (1996) Electrochemistry of hydride-forming intermetallic compounds and alloys, *Usp. Khim.* **65**(3), 195-210 (in Russian).
2. Kleperis, J., Wojcik, G., Czerwinski, A., Skowronski, J., Kopczyk, M., Beltowska-Brzezinska, M. (2001) Electrochemical behavior of metal hydrides, *Solid State Electrochem.* **5**, 229-249.
3. Sakai, T., Matsuoka, M., Iwakura, C. (1995) Handbook of the Physics and Chemistry of Rare Earths, Elsevier, Amsterdam, 138-180.
4. Fuller, T., Newman, J. (1995) Modern Aspects of Electrochemistry, Plenum Press, New York, 359-382.
5. Cromer, D., Larson, A. (1959) The crystal structure of $Ce_2Ni_7$, *Acta Cryst.* **12**, 855-858.
6. Okamoto, H. (2002) La-Ni (Lanthanum-Nickel), *J. Phase Equilibria*, **23(3)**, 287-288.
7. Petricek, V., Dusek, M. (2000). *Jana2000. The crystallographic computing system.* Institute of Physics, Praha, Czech Republic.
8. Maeland, A., Andressen A., Videm, K. (1976) Hydrides of lanthanium-nickel compounds, *J. Less-Common Met.*, **45**, 347-350.
9. Van Essen, R., Buscow, K. (1980) Hydrogen sorption of Ce-3d and Y-3d intermetallic compounds, *J. Less-Common Met.*, **70**, 189-198.
10. Klyamkin, S., Verbetsky, V., Karih, A. (1995) Thermodynamic particularities of some $CeNi_5$-based metal hydride systems with high dissociation pressure, *J. Alloys Comp.*, **231**, 479-482.

# HYDRIDE PHASES IN $Sm_2Fe_{17}$ –$NH_3$ SYSTEM

V.N. FOKIN, YU.M. SHUL'GA, B.P. TARASOV, E.E. FOKINA,
I.I. KOROBOV, A.G. BURLAKOVA, S.P. SHILKIN
*Institute of Problems of Chemical Physics of Russian Academy of Sciences,*
*Chernogolovka 142432, Moscow Region, Russia*
*E-mail: btarasov@icp.ac.ru*

## Abstract

The interaction of the intermetallic compound $Sm_2Fe_{17}$ with ammonia was investigated at an initial pressure of ammonia of 0.6–0.8 MPa in an interval of the temperatures of 150–500°C in the presence of 10 mass% (from the quantity of intermetallide entered in reaction) of $NH_4Cl$ as a process activator. It was established that depending on temperature of interaction in $Sm_2Fe_{17}$ – $NH_3$ system there is both the process of hydrogenation with a formation of hydride phase of initial intermetallic compound, and the disproportionation process with an isolation of hydride phase of new intermetallide $SmFe_3$. The reaction products are the high-dispersed powders. The magnetic properties of products of an interaction of $Sm_2Fe_{17}$ with ammonia were investigated.

*Keywords:* Intermetallic compound; Ammonia; Hydride; Nitride

## 1. Introduction

The intermetallic compound $Sm_2Fe_{17}$ finds the wide application in technology of the constant magnets in the condition of the nitride and carbide derivatives of compositions $Sm_2Fe_{17}N_x$ and $Sm_2Fe_{17}C_x$ which have significant magnetic energy, saturation magnetization, single-axis magnetic anisotropy at 20°C, the high Curie temperature (750 K) and are corrosion-resistant [1–5].

   The compound $Sm_2Fe_{17}$ forming on peritectic reaction at 1280°C crystallizes in a rhombohedral syngony ($Th_2Zn_{17}$-type structure) with parameters of an elementary cell $a = 8.57$, $c = 12.44$ Å [6,7]. The interaction of $Sm_2Fe_{17}$ with hydrogen under pressure up to 3 MPa at the temperatures of 150–250°C for ~1 h results in a formation of the hydride phases of compositions $Sm_2Fe_{17}H_{5.2-5.5}$ [5]. On data of the X-ray structural analysis, the hydride of composition $Sm_2Fe_{17}H_{5.2}$ has the following parameters of a lattice: $a = 8.745$, $c = 12.68$ Å. In hydrogen atmosphere at 420°C the hydride $Sm_2Fe_{17}H_{5.0}$ is exposed to hydrogenolysis reaction with complete decomposition of intermetallide hydride and formation of the samarium dihydride $SmH_2$ and metallic $\alpha$-Fe. The authors of work [8] communicate to the higher temperature of decomposition (750°C) of hydride phase on a basis of $Sm_2Fe_{17}$ received as a result of first hydrogenation of initial intermetallide.

*T.N. Veziroglu et al. (eds.),*
*Hydrogen Materials Science and Chemistry of Carbon Nanomaterials,* 511-518.
© 2004 *Kluwer Academic Publishers. Printed in the Netherlands.*

The nitriding of $Sm_2Fe_{17}$ prepared by a method of hydride dispergation was carried out by two ways [4]: by ammonothermal synthesis in ammoniacal medium under pressure up to 100 MPa at the temperature of 400–450°C or by an action of molecular nitrogen on intermetallide under pressure up to 8 MPa at 450°C with simultaneous grinding of a solid phase. The mentioned treatments result in an introduction of 3.0–3.5 atoms of nitrogen on an elementary cell of a crystalline lattice of intermetallic compound, that increases the interatomic distances Fe–Fe and by that raises the characteristics of exchange interaction of "magnetic" atoms of iron. Such process of introduction allows to receive a material with the improved magnetic properties.

In this connection in the present work the experimental examination of a preparation of high-dispersed powders of $Sm_2Fe_{17}$, containing in a crystal lattice some of hydrogen and nitrogen, was carried out by interaction of intermetallide with ammonia, generating nitrogen and hydrogen, in the presence of ammonium chloride as a reaction promoter at various temperatures.

## 2. Experimental

The initial samples of $Sm_2Fe_{17}$ were prepared by alloying the charge from samarium (99.7%) and iron (99.9%) in the vacuum induction furnace in an argon atmosphere. In the charge the superfluous quantity (4%) of samarium was entered for a compensation of metal waste at melting. The bake-out of an alloy was carried out in the vacuum furnace for 50 h at the temperature of 1200°C.

$Sm_2Fe_{17}$ powders with a particle size of 100 mcm were prepared by crushing the alloy bead in a metallic mortar with subsequent treatment of a material in the vibromill at 20°C in inert atmosphere for 15 min and with screening the necessary fraction. The specific surface of such powder is $0.06$ $m^2/g$.

For a measurement of magnetization a powder of $Sm_2Fe_{17}H_{5.1}$ was prepared by a method of hydride dispergation. After realization of 10 cycles of hydrogenation–dehydrogenation a sample was hydrogenated and treated by carbon monoxide under pressure of 0.5 MPa for 5 h. The prepared powder had a specific surface of $0.1$ $m^2/g$.

Ammonium chloride was dried by an evacuation for 9 h at 150°C. The ammonia, desiccated by metallic sodium, had a purity of 99.99%.

The X-ray graphical investigations were performed on an automatic assembly consisting of a diffractometer ADP-1 (Cu $K_\alpha$- and Cr $K_\alpha$-emissions) and a computer IBM PC/AT. The determination error of the interflat distances did not exceed 0.005 Å.

A specific surface of the samples was determined on an amount of low-temperature adsorption of krypton after the removal of volatile products from a solid phase under a vacuum of $1.3 \times 10^{-3}$ Pa at a temperature of 300°C for 15 h and was calculated by the Brunauer–Emmet–Teller method. The determination error was ±10%.

The contents of hydrogen in the reaction products were defined by a standard method of burning the samples in a stream of oxygen. The contents of nitrogen were determinated by the Kjeldal method. The contents of chlorine were analyzed by a method of turbidimetry.

The curves of magnetization of the investigated samples were measured with the help of vibrating magnetometer PARC M-4500.

The process of intermetallide hydronitriding was investigated at an initial pressure of ammonia of 0.6–0.8 MPa by using ammonium chloride, as activator, added in the quantity of 10 mass% from the quantity of $Sm_2Fe_{17}$ entered in the reaction.

The mixture of $Sm_2Fe_{17}$ and $NH_4Cl$ powders was subjected to vibration milling at room temperature for 30 min. The interaction of the prepared mixture with ammonia was conducted in the laboratory plant of high pressure with a capacity of 60 cm$^3$. The given quantity of prepared mixture (1.0–1.5 g), placed in a stainless-steel container, was loaded into a reactor-autoclave; traces of oxygen and moisture were removed under a vacuum of ~1 Pa at room temperature for 30 min; the ammonia was brought up to a pressure of 0.6–0.8 MPa and the reactor-autoclave was kept at the same temperature for 30 min. Further, the reactor was heated up to various temperatures for 3 h, was cooled up to ~20°C and was again heated up. In the reaction considered there is an increase of pressure in the system at the expense of an evolution of hydrogen and nitrogen, the process termination is determined by a cessation of the pressure change.

After the realization of a given quantity of heating–cooling cycles, ammonia was discharged into a buffer vessel. The reaction products were unloaded in an inert atmosphere and analyzed.

The wash of hydronitriding products for removal of $NH_4Cl$ and other possible chlorine-containing impurity was carried out on following technique developed by us. The weighed of a sample (~0.5 g) was filled in 10 ml of absolute ethanol and the mechanical mix was carried out for 3 h. The procedure, as a rule, was repeated twice. If the contents of chlorine in a sample, by results of the chemical analysis, were 1–2 mass%, the procedure of washing was repeated, replacing ethanol by distilled water. Sometimes the boiling of a mixture was applied. The traces of water from a sample were deleted by treatment by absolute ether at hashing for 10 min. The products were analyzed on the contents of hydrogen, nitrogen, and chlorine.

## 3. Results and discussion

On the data of the X-ray phase analysis, the annealed sample $Sm_2Fe_{17}$ is single-phase. On the diffractogram of this sample there are only reflections corresponding to the rhombohedral $Th_2Zn_{17}$-type structure with the constants of a crystal lattice $a = 8.556$, $c = 12.43$ Å that well agrees with the literary data [2]. The reflections of $\alpha$-Fe are not revealed on diffractograms.

The results of a treatment of $Sm_2Fe_{17}$ by ammonia in the presence of $NH_4Cl$ are given in Table 1.

It is established that at the temperatures up to 300°C the intermetallide $Sm_2Fe_{17}$ is relatively proof in ammoniacal medium: in these conditions the metal matrix of the intermetallic compound is kept, but there is a gradual accumulation in its of some of nitrogen and hydrogen, that results in a formation of new phases of an introduction.

As follows from the data of Table 1, the intermetallide interaction with ammonia at 150°C for 26 h (sample 1) results in a formation of crystalline hydride phase of composition $Sm_2Fe_{17}H_x$ with lattice constants $a = 8.640$, $c = 12.60$ Å. After the wash of reaction product from ammonium chloride by ethanol (twice) the nitrogen contents have decreased with 2.7 mass% up to 0.5 mass%, the product contains 0.4 mass% of hydrogen.

TABLE 1. Conditions and investigation results of the interaction of $Sm_2Fe_{17}$ with ammonia

| Sample no. | Conditions of synthesis | | | Phase composition of interaction products | Lattice constants of interaction products | | Specific surface of interaction products |
|---|---|---|---|---|---|---|---|
| | T (°C) | Time (h) | $P_{NH3}$ (MPa) | | $a$ (Å) | $c$ (Å) | $(m^2/g)$ |
| 1 | 150 | 26 | 0.72 | $Sm_2Fe_{17}H_x$ | 8.640 | 12.60 | 1.9 |
| 2 | 200 | 24 | 0.70 | $Sm_2Fe_{17}H_xN_y$ | 8.768 | 12.78 | 7.8 |
| 3 | 250 | 29 | 0.68 | $Sm_2Fe_{17}H_xN_y$ | 8.767 | 12.78 | 4.3 |
| 4 | 300 | 26 | 0.68 | $Sm_2Fe_{17}H_xN_y$ + $SmFe_3H_x$ traces | 8.760 | 12.77 | 2.8 |
| 5 | 350 | 24 | 0.80 | $Sm_2Fe_{17}H_xN_y$ + $SmFe_3H_x$ + $\alpha$-Fe + $\gamma$-Fe$_4$N + | 5.188 2.8669 3.8013 | 24.83 – – | 2.9 |
| 6 | 400 | 26 | 0.72 | $Sm_2Fe_{17}H_xN_y$ traces + $SmFe_3H_x$ + $\alpha$-Fe + $\gamma$-Fe$_4$N + | 5.181 2.8660 3.7968 | 25.21 – – | 2.4 |
| 7 | 450 | 26 | 0.76 | $\alpha$-Fe + $\gamma$-Fe$_4$N + SmN | 2.8652 3.7988 5.0511 | – – – | 2.2 |
| 8 | 500 | 24 | 0.62 | $\alpha$-Fe + SmN + $\gamma$-Fe$_4$N traces | 2.8671 5.0415 | – – | 1.7 |

The specific surface of obtained hydride phase makes 1.9 $m^2/g$. For comparison it is necessary to note that the hydrogenation of $Sm_2Fe_{17}$ at the temperature of 200°C by hydrogen selected from the accumulator on a basis of $LaNi_5$ results in considerably lower value of a specific surface of a sample (0.1 $m^2/g$) even after realization of 10 cycles of hydrogenation–dehydrogenation.

The rise of interaction temperature up to 200°C and the realization of reaction for 24 h (Table 1, sample 2) give rise to the conditions for formation of hydride-nitride phase on a basis of $Sm_2Fe_{17}$ with the extended parameters of an elementary cell ($a = 8.768$, $c = 12.78$ Å). The specific surface of nitriding products makes 7.8 $m^2/g$, that on the order exceeds value of a specific surface of products prepared by a method of hydride dispergation (see above). This fact is explained to that ammonia as the carrier of the chemically connected hydrogen and nitrogen promotes joint introduction of atoms of these elements into a crystalline lattice of the intermetallide. The estimation of the average size of particles calculated in an approximation of their ball form has confirmed a formation of high-dispersed powder: the average size of particles makes 105 nm.

After washing a product by ethanol (twice) the contents of chlorine have decreased, on the data of the chemical analysis, with 2.8 mass% up to 0.5 mass%, and further boiling in water (10 min) has lowered a quantity of chlorine up to 0.3 mass%. It is necessary to note that the washing process besides increases a specific surface of a material with 7.8 till 9.4 $m^2/g$. The indicated value of a specific surface (9.4 $m^2/g$) is caused by increase of the contents of nitrogen in samples: after realization of all operations of washing the contents of nitrogen in a sample make 1.1 mass%, the contents of hydrogen are 0.3 mass%.

The heating of a mixture of $Sm_2Fe_{17}$ with $NH_4Cl$ in ammonia at 250°C during 29 h (Table 1, sample 3) does not change neither direction of hydronitriding process, nor its final result: the hydride-nitride high-dispersed phase is formed with parameters of a lattice $a = 8.767$, $c = 12.78$ Å. The specific surface of a sample makes 4.3 $m^2/g$, that corresponds to the size of the particles of ~0.2 mcm. The wash of a hydronitriding product from chlorine-containing compounds by ethanol (twice) has resulted in reduction of the contents of chlorine up to 0.9 mass% (against 2.3 mass% before operation of washing), and after boiling in water (10 min) the chlorine contents have reduced up to 0.6 mass%. The nitrogen contents in a sample at the end have decreased with 3.1 up to 2.5 mass%, and the contents of hydrogen – with 0.9 up to 0.3 mass%.

On the data of the X-ray phase analysis, at realization of interaction reaction of $Sm_2Fe_{17}$ with ammonia at 300°C (Table 1, sample 4) the decomposition of hydride-nitride phase begins with a formation of a hydride phase on the basis of new intermetallide of composition $SmFe_3H_x$ and of α-phase of metallic iron. The occurrence of a new phase of intermetallide testifies that under the indicated conditions alongside with nitriding process the reaction of disproportionation takes place. The specific surface of products of an interaction at 300°C is 2.8 $m^2/g$ that testifies about high dispersity of a material (0.3 mcm). The washing of nitriding products by ethanol reduces the contents of chlorine up to 0.8 mass%, the subsequent washing by hot water does not change this quantity of chlorine in a sample, at this the contents of nitrogen make 3.6 mass%.

At the temperatures of 350–400°C at an interaction of $Sm_2Fe_{17}$ with ammonia (Table 1, samples 5 and 6) there is a further decomposition of a hydride-nitride phase $Sm_2Fe_{17}H_xN_y$. But if at 350°C the relation of the phases on a basis $Sm_2Fe_{17}$ and $SmFe_3$ carries the equimolar character, then at 400°C in a mixture the hydride phase $SmFe_3H_x$ prevails with the following parameters of an elementary cell: $a = 5.181$, $c = 25.21$ Å. The process of disproportionation is accompanied by occurrence of metallic iron ($a = 2.8669$ Å) in the nitriding products. On the diffractograms of nitriding products at 350°C the strips of reflection of a cubic phase of nitride iron of composition $\gamma$-$Fe_4N$ ($a = 3.8013$ Å) are developed. The appearance of this phase is caused by interaction of iron isolated at disproportionation with ammonia. The specific surface of nitriding products at 350°C makes 2.9 $m^2/g$, at 400°C – 2.4 $m^2/g$.

The interaction of the initial reagents at 450°C (Table 1, sample 7) is characterized by the finish of the processes of disproportionation and decomposition of new hydride phase on a basis of $SmFe_3$: the phases of the intermetallides disappear, and α-Fe and the nitrides of samarium and iron of compositions SmN and $\gamma$-$Fe_4N$ are the basic products of nitriding at this temperature. The obtained phases have the following constants of a lattice: $a = 2.8652$ Å for α-Fe, $a = 5.0511$ Å for SmN, $a = 3.7988$ Å for $\gamma$-$Fe_4N$. The high-dispersed homogeneous mixture has enough high specific surface equal 2.2 $m^2/g$.

The hydronitriding results of $Sm_2Fe_{17}$ at 500°C (Table 1, sample 8) are similar to the data received for process, carried out at 450°C, except for a phase of iron nitride which is unstable under the given conditions.

Thus, the interaction of intermetallic compound $Sm_2Fe_{17}$ with ammonia in the presence of $NH_4Cl$ can be described by the reactions, the schemes (1–4) of which are shown below:

$$Sm_2Fe_{17} + NH_3 \xrightarrow[NH_4Cl]{150-300°C} Sm_2Fe_{17}H_xN_y \tag{1}$$

$$Sm_2Fe_{17} + NH_3 \xrightarrow[NH_4Cl]{350°C} Sm_2Fe_{17}H_xN_y + SmFe_3H_x + \alpha\text{-}Fe \tag{2}$$

$$Sm_2Fe_{17} + NH_3 \xrightarrow[NH_4Cl]{400°C} SmFe_3H_x + \alpha\text{-}Fe + \gamma\text{-}Fe_4N \tag{3}$$

$$Sm_2Fe_{17} + NH_3 \xrightarrow[NH_4Cl]{450-500°C} SmN + \gamma\text{-}Fe_4N + \alpha\text{-}Fe + H_2 \tag{4}$$

The presented schemes allow precisely enough to look after the transition process on the steps from the initial $Sm_2Fe_{17}$ to the products of its complete decomposition (SmN, $\gamma$-$Fe_4N$, and $\alpha$-Fe) at an increase of the temperature and to note the special role of ammonia in these processes which conclude in decomposition of a metallic matrix of $Sm_2Fe_{17}$. Certainly, the given schemes do not describe all processes occurring at treatment of the intermetallide by ammonia in the promoter presence. So, since 400°C, on the diffractograms of the interaction products there are the peaks of samarium chlorides which pass in a solution at operations of washing. The study of a promoter role will be continued in following our works.

The certain attention deserves the fact of increase of a specific surface of hydronitriding products with a rise of the interaction temperature (from 1.9 $m^2/g$ at 150°C up to 7.8 $m^2/g$ at 200°C) that testifies to significant increase of powder dispersity. This process, probably, is also promoted by accumulation of some quantity of nitrogen in a lattice of the intermetallide.

The hydride-nitride of composition $Sm_2Fe_{17}H_xN_y$ which forms in process of $Sm_2Fe_{17}$ hydronitriding and exists at 200–300°C brings in the significant contribution to change of the magnetic characteristics of the initial intermetallide.

According to the literary data [9], the correct measurements of coercivity $H_c$ and saturation magnetization $\sigma_s$ for samples $Sm_2Fe_{17}$ can carried out only in high enough fields. The measurements of hysteresis loop carried out in present work at $-1T \leq H \leq 1T$ were called once again to emphasize this feature of intermetallide $Sm_2Fe_{17}$ and to show character of the changes of its magnetic properties at a treatment by ammonia.

In Figure 1 the curves of magnetization for hydride phase $Sm_2Fe_{17}H_{5.1}$ received by a method of hydride dispergation (curve 1) and for hydronitriding product of $Sm_2Fe_{17}$ at 200°C (curve 2; Table 1, sample 2) are submitted. These curves appreciably differ among themselves. From the curves of a hysteresis measured at $-1T \leq H \leq 1T$ the values of coercivity $H_c'$ (T) and saturation magnetization $\sigma_s'$ (emu/g) (Table 2) are estimated for hydride phase $Sm_2Fe_{17}H_{5.1}$ and hydronitriding products of $Sm_2Fe_{17}$ received at 150, 200 and 250°C (Tables 1 and 2, samples 1–3).

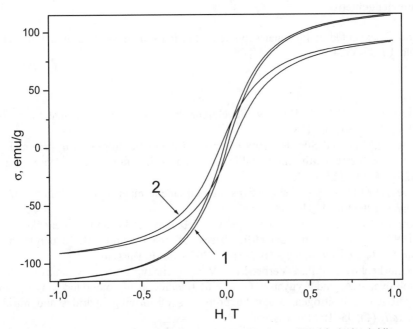

Figure 1. Curves of magnetization for hydride phase $Sm_2Fe_{17}H_{5.1}$ (curve 1) and for hydronitriding product of $Sm_2Fe_{17}$ by ammonia at 200°C (curve 2).

TABLE 2. Values of $H_c'$ and $\sigma_s'$ certain from hysteresis curve

| Sample no. | $H_c'$ (T) | $\sigma_s'$ (emu/g) |
|---|---|---|
| $Sm_2Fe_{17}H_{5.1}$ | $95 \times 10^{-4}$ | 114 |
| 1 | $160 \times 10^{-4}$ | 95 |
| 2 | $323 \times 10^{-4}$ | 90 |
| 3 | $291 \times 10^{-4}$ | 94 |

It is shown that the hydrogenation does not influence on coercivity which remains practically equal to zero for samples $Sm_2Fe_{17}H_x$ (see for example, [10]). The small value of $H_c'$ at investigated hydride phase $Sm_2Fe_{17}H_{5.1}$ is connected, apparently, to a presence at it the trace quantities of $\alpha$-Fe, not fixed by the X-ray phase analysis. The high enough value of $H_c'$ for nitriding products of $Sm_2Fe_{17}$ (Tables 1 and 2, samples 1–3) means that into a metallic lattice of intermetallide not only atoms of hydrogen, but also nitrogen introduce.

It is slightly unusually that the dependence of $H_c'$ from the nitriding temperature of the intermetallide by ammonia passes through a maximum at 200°C. Certainly, $H_c'$ is a function of many parameters, and while it is impossible unequivocally to explain the observable behavior of this dependence. In all probability, at low temperatures the introduced atoms of nitrogen and hydrogen are distributed in the crystallites of intermetallide by a not homogeneous image, that causes an inhibition at turn of spins in an external field.

## 4. Acknowledgements

This work is executed at financial support of the Russian Basic Research Foundation (grant No. 01-03-33241).

## 5. References

1. Buschow, K.H.J. (1991) New developments in hard magnetic materials, *Rep. Prog. Phys.* **54**(9), 1123–1213.
2. Coey, J.M.D. and Sun, H. (1990) Improved magnetic properties by treatment of iron-based rare earth intermetallic compounds in ammonia, *J. Magn. Magn. Mater.* **87**(3), L251–L2.
3. Coey, J.M.D. (1995) Perspectives in permanent magnetism, *J. Magn. Magn. Mater.* **140–144**, 1041–1044.
4. Kravchenko, O.V., Burnasheva, V.V., Terechov, A.D. and Semenenko, K.N. (1998) Synthesis and scientific bases of manufacture of "supernitromag" $Sm_2Fe_{17}N_3$, *Zh. Obshchei Khimii* **68**(5), 705–708 (in Russian).
5. Menushenkov, V.P., Verbetsky, V.N., Lileev, A.S., Salamova, A.A., Bobrova, A.A. and Ayuyan, AG. (1996) Interaction of compound $Sm_2Fe_{17}$ with hydrogen and nitrogen. Magnetic properties of forming hydrides and nitrides, *Metally* (1), 95–100 (in Russian).
6. Buschow, K.H.J. (1971) The samarium-iron system, *J. Less-Common Met.* **25**(2), 131–134.
7. Lyakishev, N.P., editor. (1997) Diagrams of a condition of double metallic systems: Reference book, Vol. 2. Moscow: Mashinostroenie.
8. Kwon, H.W. and Park, S.U. (2000) Study on the effect of repeated HDDR on the coercivity of $Sm_2Fe_{17}N_x$ material, *J. Alloys Compounds* **302**, 261–265.
9. Coey, J.M.D., Sun, H. and Hurley, D.P.F. (1991) Intrinsic magnetic properties of new rare-earth iron intermetallic series, *J. Magn. Magn. Mater.* **101**(1–3), 310–316.
10. Nikitin, S.A., Ovchenkov, E.A., Tereshina, I.S., Salamova, A.A. and Verbetsky, V.N. (1998) Synthesis of threefold hydrides and nitrides of $R_2Fe_{17}$ (R – Y, Tb, Dy, Ho, Er) and influence of elements of introduction ($H_2$, $N_2$) on magnetic anisotropy and magnetostriction, *Metally* (2), 111–116.

# OPTICAL INVESTIGATION OF HYDROGEN INTERCALATION-DEINTERCALATION PROCESSES IN LAYERED SEMICONDUCTOR γ-InSe CRYSTALS

Yu.I. ZHIRKO, Z.D. KOVALYUK [(1)], M.M. PYRLJA [(1)],
V.B. BOLEDZYUK [(1)]
*Institute of Physics, National Academy of Science of Ukraine,*
*46, Prospekt Nauki, Kyiv, 03037 Ukraine, UA,*
*Tel. 38 (044) 265-14-64, E-mail: zhirko@nas.gov.ua*
[(1)] *Chernivtsy Branch of Frantsevych Institute for Problems of Material*
*Science, National Academy of Science of Ukraine,*
*Iriny Vilde str., 5, Chernivtsy 1, 58001*

## Abstract

Processes of hydrogen electrochemical intercalation into layered InSe crystals were investigated. It was ascertained that for concentrations $x<2$ hydrogen forms the state of "quasi-liquid monolayer" from $H_2$ molecules in the van-der-Waals gap, which results in the increasing interlayer lattice parameter, while for $x>2$ atomic hydrogen being built into interstices due to quantum-size effects for the gap and $H_2$ molecules. It was found that the observed at $T=80K$ non-monotonic shift of the exciton absorption peak $n=1$ with growing $x$ stems from the increasing dielectric permeability of the crystal $\varepsilon_0$ due to presence of hydrogen in the van-der-Waals gap. A linear growth of $\varepsilon_0(x)$ at $x<0.5$ results in 30% decrease of the exciton binding energy. At $x>0.5$ 2D localization of exciton motion in the crystal layer plane takes place, which causes reduction and then at $x>1$ stabilization of sizes both for the exciton and quantum well. As to $H_x$InSe samples, their deintercalation degree increases linearly from 60% at $x\to0$ up to 80% at $x\to2$.

*Keywords:* hydrogen, intercalation, layered crystals, InSe, exciton localization, quantum wells.

## 1. Introduction

$A^3B^6$ semiconductor compounds including layered InSe crystals attract investigator interest because of heterostructures based on them possess good photosensitivity [1, 2] and find their application in solar cells [3, 4].

By convention, the great attention of researchers is paid to intercalation, that is to insertion of atoms and molecules into interlayer space of a layered crystal (the so-called «van-der-Waals gap») where weak molecular bonds act between crystal layers. Appearance of great impurity amounts in the gap can essentially change the intrinsic electrical resistance of the crystal from semiconductor to metallic values in the direction perpendicular to the layer, can cause a growth of the crystal dielectric permeability [5],

*T.N. Veziroglu et al. (eds.),*
*Hydrogen Materials Science and Chemistry of Carbon Nanomaterials,* 519-530.
© 2004 *Kluwer Academic Publishers. Printed in the Netherlands.*

appearance of supplementary (local) interlayer vibrations in its phonon spectrum [6] and even crystal exfoliation when intercalation reaches a sufficient level [7].

At the same time, as shown in [8-10], layered crystals are rather promising for hydrogen energetic as operating elements in systems for hydrogen accumulation. The concentration of hydrogen in them can reach values $x = 3...5$, where $x$ is the amount of introduced atoms per one formula unit of the crystal bulk.

This work consist of a two experimental parts, discussion and conclusions. In the first part, we report about our investigations of hydrogen intercalation-deintercalation processes in layered InSe crystals. In the second one, the influence of intercalated hydrogen on optical properties of these crystals is reported.

## 2. Experimental part

Bulk n-type InSe single crystals were grown using the Bridgman method from the non-stechiometric $In_{1.05}Se_{0.95}$ melt. The free carrier concentration was $10^{15}$ cm$^{-3}$ in these crystals. InSe single crystals had γ-modification (as shown in Fig. 1, conventional hexagonal cell of them consists of three $In_2Se_2$ molecules belonging to three crystal layers, the spatial group $C^5_{3v}$). Lattice parameters of γ-InSe are as follows: $C_0 = 25.32$Å, $a_0 = 4.001$ Å; intralayer nearest-neighbor distances in layers are $C_{In-In} = 2.79$ Å, $C_{In-Se} = 2.65$ Å, respectively; $\angle\varphi = 119.3^0$, the layer thickness $C_1 = 5.36$ Å, the distance between layers $C_i = 3.08$ Å and interlayer distance $C_{Se-Se} = 3.80$ Å [11,12].

2.1. Intercalation

Hydrogen intercalation of single crystal γ-InSe samples with the thickness 10 to 20µm was carried out using the electrochemical method from 0.1normal solution of hydrochloric acid with sweeping electric field in the galvanostatic regime. A special choice of the optimal current density enabled to obtain homogeneous in their composition intercalated with hydrogen $H_xInSe$ samples in the range of concentrations $0 < x \leq 2$. The hydrogen concentration in $H_xInSe$ was determined via the quantity of electrical charge transported through the sample placed into the special cell. In more detail, the intercalation method has been described in [10].

Recently [8,9], NMR investigations of ε-GaSe crystals (with the elementary cell consisting of two crystal layers and spatial group $D^1_{3h}$) intercalated with hydrogen showed that hydrogen enters into the interlayer space at $x < 2$, while at $x>2$ it penetrates into interstices of GaSe layer package; $x$ can reach values up to $x=5$. Authors of [8,9] reckon that atomic hydrogen being in the adjacent gaps of ε-GaSe creates hydrogen pairs, which results in decreasing the gap owing to interaction between hydrogen atoms. Contrary to their supposition, we assume that atomic hydrogen entering into the van-der-Waals gap creates $H_2$ molecules that occupy positions schematically shown in Fig. 1.

Starting from this concept, let us estimate the concentration and pressure of $H_2$ molecules in the van-der-Waals gap of $H_xInSe$ at $x=2$. As the volume of γ-InSe elementary cell is equal $V = 351$ Å$^3$, the concentration of elementary cells is $N_0 = 1/V = 2.85 \cdot 10^{21}$см$^{-3}$. At $x=2$, one $H_2$ molecule corresponds to one $In_2Se_2$ molecule. Hence, $H_2$ molecular concentration is equal $N = 3N_0 = 8.55 \cdot 10^{21}$ cm$^{-3}$. Using Clapeyron's equation for the ideal gas pressure

$$P = Nk_BT \ , \tag{1}$$

where $k_B$ is the Boltzmann constant and $T$ is the absolute temperature, we can deduce that the pressure caused by $H_2$ molecules in InSe van-der-Waals gap is equal to 9.4 MPa at $T= 80K$ and 35.4 MPa at $T= 300K$. This pressure, as we believe contrary to [8,9], is sufficient to cause increase of the interlayer distance.

The dependence of $\gamma$-InSe crystal parameters on pressure is studied rather well. In accord with [12], the hydrostatic pressure increase of $10^5$ to $8 \times 10^9$ Pa results in essential drop of the van-der-Waals gap due to the decrease of the distance $C_{Se-Se}$ from 3.8 Å down to 3.3 Å (13%). In this case, the distance $C_{In-In}$ decreases from 2.79 Å down to 2.69 Å (3.58%), $C_{In-Se}$ from 2.65 Å down to 2.59 Å (2.26%). However, as a consequence of the angle $\angle\varphi$ growth from $119.3^0$ up to $121.3^0$ (1.5%), the width of the crystal layer $C_l$ decreases only by 0.26% from 5.36 Å down to 5.345 Å.

Performing an extrapolation of results [12] in our case, when $H_2$ molecules do not compress but flare layer packages, we can estimate that under the pressure $P=35.4$ MPa ($x=2$, $T=300K$) the distance $C_{Se-Se}$ should increase from 3.8 Å up to 3.808 Å, and the constant $C_0$ of $\gamma$-InSe crystal should grow, respectively, by $0.024 \pm 0.001$ Å; at $P=9.4$ MPa ($x=2$, $T=80K$) $C_0$ grows by 0.009 Å.

Note that $C_0$ parameter increase by 0.02 Å is really observed by us when performing X-ray diffraction study of $H_2$InSe crystals at $T=300K$. The same data about growing $C_0$ parameter were observed also in [5] when intercalating GaSe crystals with molecular fluorine.

Before the following discussion of absorption spectra, let us note several important points that are worth to be considered.

*First.* For $H_2$ molecule, the critical temperature at which freezing-out of rotational modes begins is equal to $90K$, in accordance with the classical expression $T_r = h^2/8\pi^2 \cdot J \cdot k_B$, where $J = mr^2$ is the rotational moment of inertia for this molecule, $m = 3.34 \ 10^{-27}$ kg is $H_2$ molecule mass, $r = 0.74 \cdot 10^{-8}$ cm means $H_2$ molecule radius, $h$ and $k_B$ are Planck's and Boltzmann's constants, respectively. When $T < T_r$, the quantity of rotating $H_2$ molecules begins to reduce rapidly, however, even at lower temperatures it remains above zero as a consequence of the Maxwell velocity distribution for molecules.

*Second.* As seen from Fig. 1, the van-der-Waals gap width is modulated periodically: in positions of In atoms it is larger than in positions of Se atoms. This gap can be described as a layer of closely packed parallelepipeds, at the ends of which pyramids are placed. The volume of this body (cavity) is equal to 50.5 Å$^3$, and for an ideal crystal, when defects of the dislocation type are absent the relative gap volume comprises 43% of the crystal bulk. It is obviously larger than the lower estimate 37% obtained using the ratio $C_i/(C_i+C_l)$.

*Third.* Geometrical sizes of $H_2$ molecule and van-der-Waals gap width of InSe crystal have such values that causes quantum-size effects. Using the well-known expression [13] for localization of a particle in the quantum well with infinite rectangular walls

$$E = \frac{\pi^2 \hbar^2}{2d^2 M} \ , \tag{2}$$

and taking $d$ as the gap width $C_i = 3.08$ Å as well as $M$ being the $H_2$ molecule mass, we can deduce the value of the localization energy $E = 1.1$ meV. This dimensional quantizing implies that in one state (cavity), according to the Pauli principle, more than

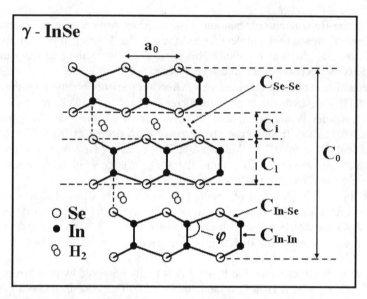

Figure 1. 2D sketch illustrate positions of a native atoms in γ-InSe crystal layers and molecular $H_2$ intercalated into the van-der-Waals gap.

two (Fermi particles) $H_2$ molecules with opposite directions of their spins (similar to para- and orthohydrogen) cannot be placed. However, $H_2$ molecule and cavity sizes nave such values that more than one molecule cannot be arise in this cavity. In the opposite case, $H_2$ molecule outer shells should be taken up with each other till distances approximately 1.5 - 2 Å and fixed in this position. But it is hardly possible in view of their strong repulsion. Therefore at $x > 2$, the only one $H_2$ molecule remains, while abundant atomic hydrogen should penetrate into interstices of crystal layers, which is confirmed by NMR investigations [8,9] with the clear evidence of hydrogen entering into interstices of GaSe at $x > 2$.

**Fourth.** In accord with [14], under conditions: $P = 9.4$ MPa and $T = 80K$, the quantity of hydrogen being in a condensed state is approximately 18%, while considering effect of cooling $H_2$ molecules due to their localization in the gap plane, one can expect that the quantity of liquid hydrogen in gap can reach up to 25%.

In addition to abovementioned properties of molecular hydrogen in the van-der-Waals InSe gap at $T=80K$, we can say that for low concentrations of $H_2$ molecules in the gap, in view of their practically 2D motion, their specific heat at temperatures lower than $90K$ will be $C_v \cong RT$, where $R$ is the universal gas constant. Unfortunately, the discussion of $H_2$ molecule specific heat behavior is beyond the scope of this work and will be considered some later using supplementary experimental data.

This preliminary discussion enabled us to draw the following conclusions:

- Hydrogen in atomic state entering into InSe van-der-Waals gap forms $H_2$ molecules that have a tendency to occupy translation ordered states with the growth of molecular concentration.

- When $x=2$, $H_2$ molecules fill all the translation ordered positions with the period in the layer $a_0 = 4.001$ Å and between layers $C_{H2-H2} = 8.57$ Å in the InSe crystal van-der-Waals gap, which is shown in Fig. 1. At $T=80K$, when rotation of $H_2$ molecules is practically absent, and there is a clear tendency to ordering, this state of molecular hydrogen in InSe van-der-Waals gap can be titled as the state of a «quasi-liquid monolayer» that causes a pressure resulting in growing $C_0$ parameter. With increasing temperature up to the room values, $H_2$ pressure grows up to P=35.4 MPa and $C_0$ parameter grows, too. At T = $110^0C$, molecular hydrogen escapes from the crystal.

- At $x > 2$, except the abovementioned process of $H_2$ molecule creation in the van-der-Waals gap, atomic hydrogen begins to incorporate into interstices of the crystal lattice due to quantum-size effects in the gap and that of $H_2$ molecules repulsion.

## 2.2. Absorption

Investigations of InSe and $H_xInSe$ absorption spectra were carried out at $T=80K$ using the modified spectrometer IKS-31 with 600 mm$^{-1}$ diffraction grating and resolution not worse than 0.7 meV. As a light source, we used incandesent filament lamp KGM-24 supplied from the stabilized direct current unit B5-21.

InSe crystals are characterized with a weak interaction of 3D Wannier excitons with homopolar optical $A^/$-phonons. Therefore, when calculating the exciton absorption spectra, we took into consideration effects of broadening the exciton states using the standard convolution procedure [15] for theoretical values of the absorption coefficient $\alpha(\hbar\omega)$ in the Elliott's model with the Lorentzian function $f(\hbar\omega) = \Gamma/[\pi(E^2+\Gamma^2)]$ in the Toyozawa's model, where $\Gamma$ is the half-width at half-maximum which is usually associated with the lifetime $\hbar/2\Gamma$.

The half-width at a half-maximum of a ground ($n=1$) and excited ($n>1$) exciton absorption bands at fixed temperatures in general form were obtained in [16]

$$\Gamma_i^2(T) = \Gamma_i^{/2}(T) + \Gamma_{inh}^2 = \Gamma_i^{/2}(0) \cdot [\Theta_i + \beta_i' \cdot n^*(T)]^2 + \Gamma_{inh}^2, \qquad (3)$$

where index $i = 1, 2, 3 \dots n \dots c$ attributed with ground ($n=1$), excited ($n>1$) and continual ($c=\infty$) exciton states, and parameters

$$\beta_i' = \begin{cases} \sqrt{2}[(\beta_i^2-1)/\beta_i] & at \ n>1 \\ \beta_i & at \ n=1 \end{cases}, \quad \Theta_{n=1} = \begin{cases} \sqrt{2} & at \ R_0 > \hbar\Omega \\ 1 & at \ R_0 < \hbar\Omega \end{cases}, \quad \Theta_{n>1} \equiv 1.$$

In Eq.(3) $\Gamma_{inh}$ is the temperature independent additive inhomogeneos component of the exciton absorption band half-width. It does not depend on temperature and is due to exciton scattering on the crystal lattice defects; $\Gamma_i'(T)$ is the homogeneous component. It increases with crystal temperature and is due to exciton scattering by phonons. $\Gamma_i'(0) = g^2[\hbar\Omega(R_0/i^2 - \hbar\Omega)]^{1/2}$ , where $g$ is the exciton-phonon coupling constant. $n^*(T) = [\exp(\hbar\Omega/k_BT) - 1]^{-1}$; $k_B$ is the Boltzmann constant; $\beta_i = 1 + [(R_0/i^2 + \hbar\Omega)/(R_0/i^2 - \hbar\Omega)]^{1/2}$; $R_0$ is the exciton binding energy; $\hbar\Omega$ is the energy of effective phonon at which exciton is being scattered. As well as for InSe $R_0 = 14.4$ meV and $\hbar\Omega =13.8$ meV ($A^/$ - homopolar phonon) then in Eq.(3) $\Theta_{n=1} = \sqrt{2}$ .

For evaluation of correct values for homogeneous half-width of ground $\Gamma'_1(0)$, excited $\Gamma'_n(0)$ and continual $\Gamma'_c(0)$ exciton states at T=0 the dependence $\Gamma^+_n(0)=\Gamma^+_c(0)-[\Gamma^+_c(0)-\Gamma^+_1(0)]/n^2$ obtained in [16] where taken into account. Note that this dependence obtained from a semiclassical principles is a common case for experimental dependence $\Gamma^+_n=\Gamma^+_c-(\Gamma^+_c-\Gamma^+_1)/n^2$ obtained by Le Toulec et al in [17]. In more detail fitting procedure of our calculation spectra to experimental exciton absorption of InSe crystals has been described in [15].

Open circles in Fig. 2a show the absorption spectrum of $H_{0.07}InSe$ crystal at $T=80K$, partial reflection of the incident onto the crystal light being taken into consideration ($n_0=[n_0^{\perp 2}\cdot n_0^{\parallel}]^{1/3} = 3.5$ [15]). There also represented are the following absorption bands: $n=1$ (curve 1) and $n=2$ (curve 2) for exciton states, band-to-band transition (curve 3) and the shallow defect level situated by 45 meV lower than the conduction band (curve 4). The curve 5 was drawn taking into consideration the contribution of exciton states for $n=1$ to $n=4$, band-to-band transition, shallow defect level (45 meV) and deep trap level situated by 130 meV higher than the valence band top. It describes reasonably the experimental absorption spectrum of $H_{0.07}InSe$ crystal.

The experimental value $K_1 = 40 eVcm^{-1}$ corresponding to the integral intensity of $n=1$ exciton absorption band in $H_{0.07}InSe$ crystals at $T= 80K$ and $\Gamma_1 = 4.4$ meV is practically equal to $K^0$ - the classical value for InSe crystal ($K_1 = 44$ $eVcm^{-1}$), which is in full accordance with the analytical dependence of $K_1$ on half-width $\Gamma_1$ obtained in [15] for InSe crystals:

$$K_1(\Gamma_1) = K^0 \cdot \sqrt{\Gamma_1^2 - 2\Gamma_1'^2(0)}\Big/\Gamma_1 . \qquad (4)$$

Therefore, we shall assume that polariton effects associated with $K_1$ growth do not take place, and $K_1 \cong K^0$.

At the same time, when introducing hydrogen into interlayer space, one can observe the following peculiarities (see Fig. 2b and 2c):

1.  Even at $x = 0.07$, the exciton absorption band $n = 1$ is shifted to short-wave side by 0.8 meV relatively to that of pure InSe crystal. Note that the energy position $E_1$ of the exciton absorption peak with $n = 1$ for pure InSe crystal at $T=80K$ is in full accordance with the analytical dependence:

$$E_1(T) \text{ [meV]} = 1336.7 - 65/[\exp(162/T)-1] \qquad (5)$$

obtained in [18] for the shift of the forbidden gap with temperature in the range $T=4.2 - 300K$.

Further increase of the hydrogen concentration results in a non-monotonic shift of the band in the whole and its peak in particular. In Fig. 2c, depicted with open circles is the experimental dependence of $E_1$ on $x$ for $H_xInSe$ crystals. For instance, in the range $0<x\leq0.5$, the peak energy position shifts upward by $\Delta E_1 = 4.5$ meV from $E_1=1327.5$ meV up to $E_1=1332.0$ meV; at $0.5<x\leq1$ one can observe its decrease down to the energy $E_1=1329.5$ meV; the further increase of hydrogen amount in interlayer space does not result in changing $E_1(x)$ values.

2.  Along with $E_1$ shift, we observed the change of the half-width $\Gamma_1$ of the exciton absorption band $n=1$ with the growing hydrogen concentration. In Fig. 2d solid triangles is $\Gamma_1(x)$ dependence. It can be seen that $\Gamma_1$ increases with $x$ in the range $0<x\leq1$ and at $x>1$ becomes practically constant.

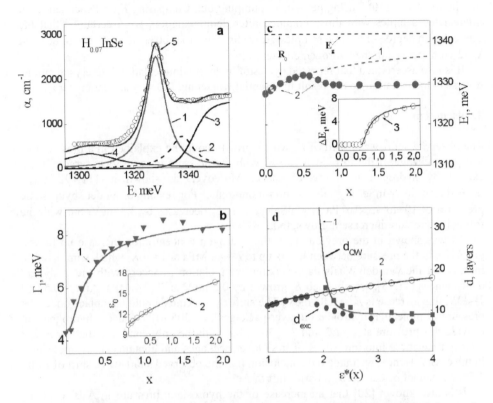

Figure 2. a – open circles experimental and curve 5 calculated exciton absorption spectra of $H_{0.07}$InSe crystal at
$T = 80K$. Curves 1 – 4 are the bands of $n$=1,2 exciton states, of band-to-band transition and of a
shallow donor states.

b – solid triangles experimental $\Gamma_1$ dependence on $x$. Curve 1 - plotted in accordance with Eq.(15). On
insert – open circles experimental $\varepsilon_0$ dependence on $x$. Curve2 - plotted according to Eq. (9).

c – solid circles experimental shift of $n$=1 exciton absorption band maximum $E_1$ on $x$. Curve denoted as
$E_g$ show the stable energy position of $E_g$ on $x$ increase. Curve 1 - plotted in accordance with
Eq.(6), curve 2 with account of exciton localization in layer plane. $R_0$ – show decrease of exciton
binding energy with $x$, when exciton localization ignored. On insert - open circles experimental
dependencies of $\Delta E_1$ on $x$. Curve 3 plotted in accordance with Eq. (13).

d – solid squares experimental values for quantum well thickness $d_{QW}$ on $\varepsilon^*(x)$ parameter. Circles and
solid circles - experimental dependencies for exciton diameter $d_{exc}$ on $\varepsilon^*(x)$ parameter without
and with the account of exciton localization. Curve 1 and 2 plotted according to Eqs. (2) and
(10).

Note that investigations of the hydrogen concentration influence on optical properties
of InSe were carried out using two groups of samples. In the former one, we performed
intercalation till some definite concentration, and then the optical study was carried out.
In the latter one, intercalation was realized in the step-by-step manner using the
following scheme: intercalation of InSe → optical investigations of $H_x$InSe → further
intercalation of $H_x$InSe → optical investigations ..., $x$ being increased step-by-step from
0 up to 2. In both cases, concentration dependences $E_1(x)$ coincided (see Fig. 2c).

Deintercalation (cutting out) of hydrogen from $H_xInSe$ crystals was carried out for 3 to 9 hours at $T = 110^0C$ using permanent pumping out. Comparing $E_1(x)$ dependences in intercalated samples with those obtained after deintercalation, we ascertained that the intercalation degree of $H_xInSe$ samples grows linearly from 60% at $x \rightarrow 0$ up to 80% at $x \rightarrow 2$ with increasing the hydrogen concentration.

It is also ascertained that repeated cycles of hydrogen intercalation into crystal and its reverse deintercalation do not result in essential worsening InSe crystal parameters.

## 3. Discussion

The observed short-wave shift of $E_1$ with $x$ growth could be explained in a most simple way by a change of the forbidden gap $E_g$ with the pressure growth, this pressure being caused by hydrogen in interlayer space. Moreover, as shown in [11,19], $E_g(P)$ dependence in γ-InSe crystals is non-monotonic. For example, under hydrostatic pressures of $10^5$ to $4.2 \times 10^8$ Pa in γ-InSe crystals, $E_g$ decreases by 3.3 meV, but with the following pressure increase it grows, too.

As was shown in the previous part, the hydrogen concentration increase up to $x=2$ gives rise to the pressure between layers up to $P=9.4$ MPa at $T=80K$, which stimulates an increase of the van-der-Waals gap. Making extrapolation in accord with the analytical dependences [11,18], we obtained $E_g$ growth by 0.2 meV at $P=9.4$ MPa. At $x = 0.5$ and $T=80K$, this increase is $E_g = 0.05$ meV. The shift $E_g = 4.5$ meV could take place under the pressure of molecular hydrogen on layer packages $P = 205$ MPa. Indeed, hydrogen can provide this pressure at $x = 12$ and $T = 300K$ but with the only condition that for $x>2$ it does not penetrate into interstices of InSe layer packages and contains six $H_2$ molecules in the cavity. Hence, we made the conclusion that the observed short-wave shift of $E_1$ by 4.5 meV cannot be caused by a baric shift of $E_g$.

It is also known [20] that an increase of the hydrostatic pressure in $A^3B^6$ crystals, contrary to $A^2B^6$ и $A^3B^5$ ones, results in increasing $\varepsilon_0^{\parallel}$ component of the crystal dielectric permeability. As it was shown by direct measurements of the capacitance [20], $\varepsilon_0^{\parallel}$ of InSe in the range $10^5$ to $10^9$ Pa grows almost linearly by 15%. Thereof, in accord with the mean $\varepsilon_0$ value that in anisotropy layered crystals is defined as $\varepsilon_0 = [\varepsilon_0^{\parallel} \cdot \varepsilon_0^{\perp 2}]^{1/3}$ (where, accordingly to [15], $\varepsilon_0^{\perp} = 10.9$, $\varepsilon_0^{\parallel} = 9.9$, and indexes $\parallel$, $\perp$ correspond to parallel and perpendicular orientation relatively to the optical axis C), $\varepsilon_0$ grows by 5%. This $\varepsilon_0$ increase should result in decreasing the exciton binding energy $R_0$. Indeed, the decrease of $R_0$ from 14.2 meV down to 12.9±0.4 meV at $P=1$ GPa was reported in [11]. The obtained in [11] $R_0$ value and observed in [20] $\varepsilon_0$ increase by 5% are in a good agreement with the following dependence written in the simplified form:

$$R_0 \text{ [meV]} = 13605 \cdot \mu / \varepsilon_0^2 \qquad (6)$$

combining $R_0$ with $\varepsilon_0$ and $\mu$ - reduced exciton mass (in InSe $\mu = [m_e^{-1} + m_h^{-1}]^{-1} = 0.12m_0$ [15], where $m_0$ is the mass of the free electron).

Extrapolating data of [11, 20] onto our case of the van-der-Waals gap growth, one can say that at $x=2$ ($P= 9.4$ MPa) $\varepsilon_0$ should decrease by 0.7% and $R_0$ increase by 0.2 meV, which at $x=0.5$ and $\Delta\varepsilon_0(P) = 0.2\%$ results in $\Delta R_0(P) = 0.06$ meV and the respective long-wave shift of $E_1$.

The performed analysis of obtained by various authors data upon the pressure influence on the *energy* structure of InSe crystal enables one to draw the conclusion that the short-wave shift of $E_1$ by 4.5 meV cannot be reasonably explained by $R_0(P)$, $\varepsilon_0(P)$ and $E_g(P)$ dependences, as $R_0(P)$ and $E_g(P)$ contributions into this shift are opposite at $x=0.5$ ($P=2.4$ MPa) and their total contribution is only 0.01 meV, which is considerably lower than the observed one.

At the same time, this behavior of exciton parameters $E_1(x)$ and $\Gamma_1(x)$ gives grounds to deem that the short-wave shift of the exciton absorption band is caused by the $\varepsilon_0$ increase stemming from hydrogen presence in interlayer space, which in accord with Eq. (6) results in $R_0$ decrease.

To make our discussion convenient, we shall describe the $\varepsilon_0$ increase caused by molecular hydrogen in interlayer space introducing an additive to $\varepsilon_0^\parallel$ parameter $\varepsilon^*(x)$ increasing with $x$ and characterizing the degree of dielectric permeability anisotropy for the medium comprised by the exciton in the following form:

$$\varepsilon_0(x) = [\varepsilon_0^{\perp 2} \cdot \varepsilon_0^\parallel \cdot \varepsilon^*(x)]^{1/3} \equiv \varepsilon_0 \cdot \varepsilon^*(x)^{1/3},\qquad(7)$$

where parameter

$$\varepsilon^*(x) = \prod_{j=1}^{w} \varepsilon_i(x) \equiv \varepsilon_i(x)^w.\qquad(8)$$

$w$ is the degree index characterizing the quantity of interlayer space comprised by the exciton diameter $d_{exc}$, and $\varepsilon_i(x)$ is the depended on $x$ dielectric permeability of the van-der-Waals gap, which is equal to that of vacuum $\varepsilon_i(0) = \varepsilon_v = 1$ in the case of pure InSe.

To simplify our estimation procedure, we shall consider that hydrogen introduction into interlayer space changes neither $E_g$ nor $\mu$. Then, the short-wave shift of $E_1(x)$ in the range $0<x=0.5$ (see Fig. 2c) is caused by $R_0$ decrease from 14.5 down to 10.0 meV, which results in practically linear growth of the obtained experimental $\varepsilon_0(x)$ values from 10.5 to 12.6 in accord with Eq.(6) (see insert of Fig. 2b) and is described rather well by the following dependence:

$$\varepsilon_0(x) = 10.5 \cdot (1 + 1.5 \cdot x)^{1/3}\qquad(9)$$

shown with the curve 2. The curve 2 in Fig. 2c represents the dependence $R_0(x)$ for $0<x=2$ obtained according to Eq.(6) in the case when $\varepsilon_0(x)$ corresponds to Eq.(9). As it can be seen, the curve 2 well describes the experimental dependences $R_0(x)$ in the range $0<x=0.5$. In accord with Eq.(7), $\varepsilon_0^\parallel(x)$ increases from 9.9 up to 16.9 due to the growth of parameter $\varepsilon^*(x)$ from $\varepsilon_v = 1$ up to 1.7. In accordance with the expression

$$a_{exc}\,[\text{nm}] = 0.053 \cdot \varepsilon_0(x)/\mu\qquad(10)$$

the exciton radius grows also linearly with changing $\varepsilon_0(x)$ from 4.7 up to 5.7 nm, which is equivalent to the increase of the exciton diameter from $d_{exc} = 11.3$ up to 13.5 crystal layers (layer thickness of InSe $d_{layer} = 8.44$Å). Thereof, for $w = 13$ in agreement with Eq.(8) at $x=0.5$ ($P= 2.3$ MPa) and $T = 80K$, one can deduce $\varepsilon_i = 1.04$.

Considering Eq.(9) defining the linear dependence of $\varepsilon_0$ on $x$, which is associated with hydrogen introduction into interlayer space, we may perform a linear extrapolation of $\varepsilon_0(x)$ to $x=1$. Obtained data will be as follows: $\varepsilon_0 = 14.25$ and parameter $\varepsilon^*(x) = 2.5$. $R_0$ should decrease down to 6.45 meV, while the exciton diameter should increase to $d_{exc} = 15.3$ crystal layers. For $w = 15$ at $x=1$ ($P = 4.7$ MPa) and $T=80K$, one can determine $\varepsilon_i = 1.062$.

In the similar way, for $x=2$ we can obtain $\varepsilon_0 = 16.67$ and $\varepsilon^*(x) = 4.0$, $R_0 = 5.73$ meV, $d_{exc} = 17.9$ crystal layers. For $w = 18$ at $x=2$ ($P = 9.4$ MPa) and $T=80K$, one can determine $\varepsilon_i = 1.081$. It is clear seen that $\varepsilon_i$ value also grows with $x$ and begins to

528

approach to experimental data inherent to the dielectric permeability of condensed hydrogen [14], which is 1.253 and 1.230 at the temperature 14 and 20.5 K, respectively.

The obtained numeric values for dependences of parameter $\varepsilon^*$ and $\varepsilon_i$ on $x$ are well described by the following analytical expressions:

$$\varepsilon^*(x) = \varepsilon_v + 1.5 \cdot x \tag{11}$$

and

$$\varepsilon_i(x) = \varepsilon_v + 0.06 \cdot x^{1/2}, \tag{12}$$

where $\varepsilon_v = 1$. It is seen that $\varepsilon_i$ growth causes increasing the crystal dielectric permeability as a whole, which leads to the experimentally observed $R_0$ decrease in full accordance with Eqs.(6) and (9), as shown in Fig. 2c.

However, at $0.5 < x < 1$, $E_1(x)$ shift changes its sign by the opposite one, and at $1 < x < 2$ $E_1$ position is stabilized. This behavior of $E_1(x)$ allows us to posit that $\varepsilon_i(x)$ increase causes a growth of $a_{exc}$. As a consequence, parameter $\varepsilon^*(x)$ characterizing the degree of anisotropy inherent to medium comprised by the exciton grows so much that at $\varepsilon^* = 1.7$ ($x = 0.5$) the exciton motion is localized in the plane of layers. Further increase of $\varepsilon_i(x)$ increase causes reducing the exciton radius and quantum well width $d_{QW}$. This concept is confirmed by the experimental data for $\Delta E_1(x)$ that are shown with open circles in the insert of Fig. 2c. These experimental data are rather well approximated by the following dependence:

$$\Delta E_1(x) \text{ [meV]} = 13/\exp[x/2 \cdot (x-0.5)] \tag{13}$$

represented by the curve 3 in this figure. The total $E_1(x)$ dependence taking into account the increase of the medium dielectric permeability and exciton localization in the crystal layer plane is depicted by the curve 4.

In Fig. 2d shown with squares is the dependence of the localized exciton quantum well width on the parameter $\varepsilon^*(x)$ obtained in accordance with Eq.(2) and experimental $\Delta E_1(x)$ data. For convenience, we chose the thickness of the crystal layer as a unit of the $d_{QW}$. In the insert, open circles show the dependence of the exciton diameter $d_{exc}$ on parameter $\varepsilon^*(x)$ without taking into account exciton localization, while solid circles do that considering the localization in the QW.

As seen from this figure, in the range of parameter values $1 < \varepsilon^*(x) < 1.7$ the QW does not appear ($d_{QW} = \infty$), and we deal with the ordinary growth of $d_{exc}$ (open circles) bound with $\varepsilon_i(x)$ increase. At $1.7 < \varepsilon^*(x) < 3.0$, strong localization of the exciton takes place in the layer plane, and the QW narrows from 107 down to 12.8 layers. At the same time, $d_{exc}$ decreases from 13.5 down to 8.8 layers.

Further $x$ increase, in accord with Eq.(11), causes $\varepsilon^*(x)$ parameter growth from 2.5 up to 4.0 and decrease the QW and exciton diameter down to $d_{QW} = 12.8$ layers and $d_{exc} = 10.5$ layers. $R_0$ and $d_{exc}$ are stabilized the final level, and the exciton is localized in the crystal layer plane. Experimental data of $d_{QW}(\varepsilon^*)$, where [see Eq.(11)] $\varepsilon^* = f(x)$, is well approximated by the following expression:

$$d_{QW}(\varepsilon^*) \text{ [layers]} = 9 + 0.4/(\varepsilon^*(x) - 1.7)^3 \tag{14}$$

represented by the curve 1 in Fig. 2d.

Note that numeric $d_{exc}$ values at $\varepsilon^*(x) > 1.7$ were obtained in approximation that increasing the parameter up to $\varepsilon^*(x) > 1.7$ should be followed by the condition $\varepsilon_i(x)^w \leq 1.7$. As $\varepsilon_i(x)$ value grows with $x$, the quantity of layers comprised by the exciton $w$ should be decreased.

This statement is confirmed by the experimental dependence of the half-width of $n$=1 exciton absorption band $\Gamma_l$ on $x$ depicted in Fig. 2b. Indeed, the experimentally observed $d_{exc}$ growth in the range $x<0.5$ ($\varepsilon^*<1.7$) results in increasing probability of exciton scattering by point defects and increasing the initial value of homogeneous broadening the ground as well as excited exciton states $\Gamma_i^+(0) = g^2[\hbar\Omega(R_0/i^2 - \hbar\Omega)]^{1/2}$. Despite the more smooth $\Gamma_i^2(T)$ dependence, the total exciton scattering rises at low temperatures ($T$=80$K$), which results in experimentally observed $\Gamma_l$ growth. In the range $x>0.5$, the exciton localization causes $\varepsilon_i(x)$ growth and $d_{exc}$ decrease, which results in stabilization of exciton and quantum well sizes and reaching a stationary value of the probability for excitons to be scattered by defects. The experimental $\Gamma_l(x)$ dependence can be represented with the following function:

$$\Gamma_l(x) \text{ [meV]} = \Gamma_l \cdot \{1+ 1/\exp[1/(\Gamma_l \cdot x)]\} \tag{15}$$

depicted by the curve 1. Here $\Gamma_l = 4.4$ meV.

One can say that the obtained by us experimental results upon 2D exciton localization (taking place due to the growth of the crystal dielectric permeability anisotropy parameter) with $\varepsilon_0^\|$ are very close to [21] where the behavior of polaron excitons in parabolic quantum dots were considered and shown that the dot size decrease results in increasing the exciton binding energy.

## 4. Conclusions

Processes of hydrogen electrochemical intercalation of layered InSe crystals were investigated. It was ascertained that hydrogen in atomic state enter into van-der-Waals gap and forms $H_2$ molecules which have a tendency to occupy translation ordered states with the growth of hydrogen concentrations $x$.

It is shown that for $x$=2 molecular $H_2$ occupy in van-der-Waals gap all translation ordered positions with the period in the layer $a_0 = 4.001$ Å and between layers $C_{H2-H2} = 8.57$ Å (which is shown in Fig. 1). At $T$=80$K$, when rotation of $H_2$ molecules is practically absent, this state of molecular $H_2$ in the gap can be titled as the state of a «quasi-liquid monolayer» of $H_2$ molecules that causes a pressure resulting in growing interlayer lattice parameter $C_0$. At $x>2$, except the abovementioned process of $H_2$ molecule creation in the van-der-Waals gap, atomic hydrogen begins to incorporate into interstices of the crystal lattice due to quantum-size effects in the gap and that of $H_2$ molecules.

It is shown that at T = $110^0C$ molecular $H_2$ escapes from $H_xInSe$ and degree of its deintercalation increases linearly from 60% at $x\rightarrow0$ up to 80% at $x\rightarrow2$. Repeated cycles of hydrogen intercalation-deintercalation do not result in essential worsening of InSe crystal parameters, which open the possibility to use them as hydrogen accumulators possessing the high sorption capability.

It was found that the observed at $T$=80$K$ non-monotonic shift of the exciton absorption peak $n$=1with growing $x$ stems from the increasing dielectric permeability of the crystal $\varepsilon_0$ due to presence of hydrogen in the van-der-Waals gap. A linear growth of $\varepsilon_0(x)$ at $x<0.5$ results in 30% decrease of the exciton binding energy. At larger concentrations (when the anisotropy parameter $\varepsilon^*(x)$ grows), 2D localization of exciton motion in the crystal layer plane takes place, which at $x>0.5$ causes reduction and then at $x>1$ stabilization of sizes both for the exciton and quantum well.

530

## 5. Acknowledgements

This work was partially supported by the Basic Research Fund of Ukraine (Project No F7/310-2001).

## 6. References

1. Lebedev, A.A, Rud', V.Yu. and Rud', Yu.V. (1998) Photosensitivity of geterostructures porous silicon-layered $A^{III}B^{VI}$ semiconductors, *Fiz. Tekch. Polupr.* **32**(3), 353-355.
2. Katerinchuk, V.N., Kovalyuk, Z.D. and Kaminskii, V.M. (2000) Geterotransitions p-GaSe–n-recrystallized InSe, *Pisma Zhurn.Tehn. Fiz.* **26**(2), 19-22.
3. Martines-Pastor, J., Segura, A. and Valdes, J.L. (1987) Electrical and photovoltaic properties of indium-tin-oxide/p-InSe/Au solar cells, *J. Appl. Phys.* **62**(4), 1477-1483.
4. Shigetomi, S. and Ikari, T. (2000) Electrical and photovoltaic properties of Cu-doped *p*-GaSe/*n*-InSe heterojunction, *J. Appl. Phys.* **88**(3), 1520-1524.
5. Grigorchak, I.I., Netjaga, V.V., Gavrilyuk, S.V. and Drapak, S.I. (2002) Structure and physical properties of InSe and GaSe intercalated by molecular fluorine, *Zhurnal Fizychnyh Doslidzhen* **6**(2), 193-196 (in Ukrainian).
6. Zhirko, Yu.I., Zharkov, I.P., Kovalyuk, Z.D. and Shapovalova, I.P. (2002) Phonon spectra of doped and intercalated ε-GaSe crystals, *Fizychnyj sbirnyk NTSh.* **5**(4), 237-245 (in Ukrainian).
7. Anderson, S.H. and Chung, D.D.L. (1984) Exfoliation of intercalated graphite, *Carbon* **22**(3), 253-263.
8. Kovalyuk, Z.D., Prokipchuk, T.P., Seredjuk, A.I. and Tovstyuk, K.D. (1987) Investigation of impurity state in hydrogen intercalates of gallium selenium by NMR method, *Fiz. Tverd. Tela* **29**(7), 2191-2193.
9. Kovalyuk, Z.D., Prokipchuk, T.P., Seredjuk, A.I., Tovstyuk, K.D., Golub, S.Ja. and Vitkovskaja, V.I. (1988) NMR in hydrogen subsystem of $H_xGaSe$ introduction junctions, *Fiz. Tverd. Tela* **30**(8), 2510-2511.
10. Kozmik I.D., Kovalyuk, Z.D., Grigorchak, I.I. and Bakchmatyuk, B.P. (1987) Preparation and properties oh hydrogen intercalated gallium and indium monoselenides, *Isv. AN SSSR Inorganic materials* **23**(5), 754-757.
11. Goni, A., Cantarero, A., Schwarz, U., Syassen, K. and Chevy, A. (1992) Low-temperature exciton absorption in InSe under pressure, *Phys. Rew. B.* **45**(8), 4221-4226.
12. Olguin, D., Cantarero, A., Ulrich, C. and Suassen, K. (2003) Effect of pressure on structural properties and energy band gaps of γ-InSe, *Phys. Stat. Sol. (b)* **235**(2), 456-463.
13. Landau, L.D. and Lifshits, E.M. (1974) Quantum Mechanics, Moscow: *Izd. Nauka.* (in Russian).
14. Malkov, M.P. (1985) Hand-book on physical-chemical fundamentals of cryogenics, Moscow, *Energoatomizdat.* (in Russian).
15. Zhirko, Yu.I. (1999) Investigation of the light absorption mechanisms near exciton resonance in layered crystals. N=1 state exciton absorption in InSe, *Phys. Stat. Sol. (b)* **213**(1), 93-106.
16. Zhirko, Yu.I. and Zharkov, I.P. (2003) Investigation of some mechanisms for formation of exciton absorption bands in layered semiconductor *n*-InSe and *p*-GaSe crystals, *Semiconductor Physics Quantum Electronics and Optoelectronics* **6**(2), 134-140.
17. Le Noulec, R., Piccioli, N. and Chervin, J.C. (1980) Optical properties of the band-edge exciton in GaSe crystals at 10K, *Phys. Rev. B.* **22**, 6162-6170.
18. Camassel, J., Merle, P., Mathieu, H. and Chevy, A. (1978) Excitonic absorption edge of indium selenide, *Phys. Rev. B.* **17**(12), 4718-4725.
19. Ulrich, C., Olguin, D., Cantarero, A., Goni, A.R., Syassen, K. and Chevy, A. (2000) Effect of pressure on direct optical transitions of γ-InSe, *Phys. Stat. Sol.(b)* **221**(2), 777-787.
20. Errandonea, D., Segura, A., Munoz, V. and Chevy, A. (1999) Pressure Dependence of the Low- Frequency Dielectric Constant in III-VI Semiconductors, *Phys. Stat. Sol. (b)* **211**(1), 201-206.
21. Senger, R.T. and Bajaj, K.K. (2003) Polaronic exciton in a parabolic quantum dot, *Phys. Stat. Sol.(b)* **236**(1), 82-89.

# QUANTUM TOPOLOGY AND COMPUTER SIMULATION OF CONFINED HYDROGEN ATOM INSIDE SPHERICAL-FORM GAP

S.A. BEZNOSYUK*[a], D.A. MEZENTSEV[a], M.S. ZHUKOVSKY[a],
T.M. ZHUKOVSKY[b]
[a] *Department of Physical Chemistry, Altai State University,
66, Lenin St., 656049, Barnaul, Russian Federation
E-mail: beznosyuk@chemwood.dcn-asu.ru*
[b] *Department of General Physics, Altai State Technical University,
46, Lenin St., 656099, Barnaul, Russian Federation*

## Abstract

Nanostructural behavior of hydrogen atom is described by using some specific effects of quantum-sized confinement. Computer simulation of self-assembling confinement of single electron in compact subspace of H-atom illustrates theory. It is shown that finite wave function of the confined electron in the ground state includes two parts: an electrodynamics kink-mode and an electrostatic cusp-mode. By way of minimizing functional of ground state energy some curves of energy with respect to radius of spherical gap of confinement were obtained. It is found out that there are two equilibrium states of hydrogen atom. One of them corresponds to infinite H-atom. Another state describes finite H-atom. Difference between them lies in that infinite wave function includes only electrostatic cusp-mode. It is an atomic 1s - orbital. Finite wave function, besides the 1s-cusp-mode, includes small addend of the kink-mode. The addition seriously changes features of H-atom. Equilibrium radius decreases from infinity to 3.1 $a_0$. The "collapse" of H-atom is accompanied by appearance of specific changing in chemical features, which could be using for understanding of hydrogen behavior in nanostructures.

*Keywords:* Nanostructure; Hydrogen storage; Computational chemistry; Modeling

## 1. Introduction

Up to now no evidence has been found for hydrogen condensation in nanotubes [1]. There is no clearness in the question about specific conditions and mechanisms to obtain higher density of hydrogen in and on carbon nanostructures compared with liquid hydrogen. The theoretical calculation of the stability of metal hydrides is very complex, if not possible. Since when the hydrogen exothermically dissolved in the solid-solution at a small hydrogen to metal ratio the metal lattice expends proportional to the hydrogen concentration by approximately 2-3 $10^{-3}$ nm$^3$ per hydrogen atom. But at greater hydrogen concentrations in many cases the crystal structure of the metal hydrides change upon the phase transition.

*T.N. Veziroglu et al. (eds.),*
*Hydrogen Materials Science and Chemistry of Carbon Nanomaterials,* 531-538.
© 2004 *Kluwer Academic Publishers. Printed in the Netherlands.*

The well-known molecular spin-orbital scheme of calculation of "electron-deficient" bonding in metal tetraborides $LiBH_4$ has been discussed. In this case the hydrogen atom bringing lithium and boron atoms possesses a single electron, that is fewer than apparently required to fill the bonding orbital, based on the criterion that normal bonding orbital involving two atoms contains two electrons.

The interaction of hydrogen with the electronic structure of host material is very complicated. To reach improvements and new discoveries in nanotechnology of hydrogen production, storage and conversation, researchers have to turn to *ab initio* approaches. According to theory of material composition the hierarchy of nano structures is based on set of postulates of the thermo field dynamics (TFD) [2], and the quantum field chemistry (QFC) [3]. Early [4-6] we suggested some exact mathematical statements and decision of the problem of self-assembling effects in nanosystems in the frame of the quantum topological approach. Preliminary computer simulation of self-assembling confinement of a single electron and H atom inside cubic form gap was published (refs. 4 and 5, respectively). Below we show some exact results of computer simulation of H-atom confined in a spherical gap of low-dimensional nanosystem.

## 2. Basic approaches

According to QFC theory generating of the condensed state hydrogen atom caused by electron confining as is shown by Fig. 1. A common kink-cusp wave function $\varphi_{kc}$ of self-assembling H-atom electron is a result of energetic instability of homogeneous free electron infinite ground state, which is described with zero-valued wave function $\varphi_0$ in arbitrary compact volume $\Omega$.

As a result of spontaneous broken of local electrodynamics symmetry a wave function $\phi_{kc}$ is represented with two orthogonal components [3,5]:

$$\phi_{kc} = \phi_k + \phi_c ; \tag{1}$$

where $\phi_k$ represents a homogeneous electrodynamics mode, and $\phi_c$ gives an inhomogeneous electrostatics mode of single electron wave function $\phi_{kc}$ in Coulomb potential of nuclear P. Correspondingly, $\phi_k$ describes kink-shaped kinematics density wave (KDW) state of nonlocal "extension" electron, and $\phi_c$ describes a cusp-shaped probability density wave (PDW) state of local "point" electron.

The ground state energy includes two parts, as follows:

$$E_{kc}[\phi_{kc}] = E_k[\phi_k] + E_c[\phi_c] \tag{2}$$

First term on the right site of Eq.2 is energy of the extension electron in KDW-state [4]. In atomic units ($e = m = \hbar = 1$) one gets next expression:

$$E_k[\phi_k] = \sum_i \int_\Omega \frac{|\phi_k|^2}{4 < \Delta r_i^2 >} + \int_\Omega \frac{l(l+1)|\phi_k|^2}{\sum_i < \Delta r_i^2 >} - \int_\Omega \frac{2|\phi_k|^2}{r}, \tag{3}$$

where $\{i = x;y;z\}$ is a reference axis set of Cartesian coordinates and by definition:

$$< \Delta r_i^2 > = \frac{\int r_i^2 |\phi_k|^2}{\int |\phi_k|^2}. \tag{4}$$

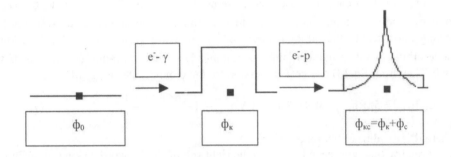

Figure 1. Scheme of spontaneous broken of dynamical symmetries of confined single electron in H-atom:
(e⁻-γ) – spontaneous broken of local electrodynamics symmetry;
(e⁻-p) – spontaneous broken of local electrostatic symmetry

First term in Eq.3 represents kinetic energy of $\phi_k$ as an electrodynamics mode of $\phi_{kc}$ in subspace $\Omega$ caused by the existence of quantum (e⁻-γ)-electrodynamics fluctuations inside spherical form gap. Second term in Eq.3 represents kinetic energy of $\phi_k$ in rotational KWD-state with quantum number $1 = 0,1,2...$of the extension electron inside spherical form gap. Third term describes an electrostatic energy of the extension electron in KWD state.

In Eq.2 the functional $E_c[\phi_c]$ represents a quantum mechanical energy of a point single electron in PDW-state $\phi_c$ as an electrostatic mode of $\phi_{kc}$:

$$E_c[\phi_c] = \int_\Omega |\nabla\phi_c|^2 - \int_\Omega \frac{2|\phi_c|^2}{r}. \tag{5}$$

Here the first term is a kinetic energy, and second term is an electrostatic potential energy of the point electron in PDW state.

Exact ground state of compact H-atom could be found by minimizing of functional (2) in respect to two modes of the single electron wave function $\phi_{kc}$ under constrains of the unite norm: $\|\phi_{kc}\| = 1$, and the orthogonality of modes: $<\phi_k|\phi_c> = 0$.

In a limit case, assigning that $\phi_{kc} = \phi_c$, we obtain standard energy functional $E_{kc} = E_c$, which is equal to well-known functional of infinite H-atom. In another limit case, assigning that $\phi_{kc} = \phi_k$, we obtain an energy functional $E_{kc} = E_k$ for $1 = 0$ that describes finite H-atom with homogeneous in volume $\Omega$ kink boundary distribution of the extension electron density $n_k$. The kink density and a nonlocal gauge phase $\lambda_k$ define the electrodynamics kink-mode $\phi_k$, as follows [3]:

$$\phi_k|_\Omega = n_k^{1/2} \exp(i\lambda_k) \tag{6}$$

Properties of such extension electron state were previously discussed [4]. Set of such kind extension electron states describes transmission of a strong correlated electron swarm through interatomic regions of channels and interfaces in low-dimensional nanosystem [5].

534

As regards to a common case, that there are both modes. The cusp-mode $\phi_c$ has a form of modified atomic 1s-orbital, which is finite in the subspace $\Omega$ (see Fig.1). Taking into account the orthogonality: $< \phi_k | \phi_c >= 0$ we must assign value 1 to kink mode orbital quantum number, since the atomic 1s-orbital (cusp-mode) corresponds to $l = 0$. Then from Eq. (3) one gets expression for the ground state energy of $\phi_{kc}$:

$$E_{kc}[\phi_{kc}] = 5\pi Rn_k + (40/9)\pi Rn_k + \int_\Omega |\nabla\phi_c|^2 - \int_\Omega \frac{2(n_k + |\phi_c|^2)}{r}, \qquad (7)$$

where R is a radius of the spherical gap $\Omega$.

Formally, first and second terms on the right site of Eq.7 give some kind of quantum electrodynamics surface energy of confined hydrogen atom caused by spontaneous broken of local electrodynamics symmetry of the nonlocal "extension" electron. In the limit of infinite volume of spherical gap this surface energy vanishes together with kink density $n_k$. As a result one gets the functional $E_c$ of an infinite H-atom described by PDW-state of a "point" electron.

### 3. Computational scheme of variation procedure

Using the variation principal, ground state wave functions of a single electron confined in H-atom for kink-shaped density, for cusp-shaped density, and for kink-cusp- shaped density of electron distribution may be obtained by minimizing functionals that are given Eqs. (3), (5), and (7), respectively. We have to find localized on compact carrier electron density distribution satisfying the minimum of system energy requirements in external electrostatic field and that is constrained the only kinematics restrictions, as follows:

$$\int_\Omega (n_k + n_c) = 1; n_k = |\phi_k|^2; n_c = |\phi_c|^2; < \phi_k | \phi_c >= 0. \qquad (8)$$

There are many possibilities for the representing the cusp-mode $\phi_c$. But, for the numerical simulation of the system localized on compact carriers the finite-elements function is natural choice. The B-splines are relevant to such function's class. Introducing of B-splines with degenerated nodes is natural method allowing us to describe electron density distribution near atomic nuclei - cusp points. Denoting ``basis" functions as $b_\alpha(\vec{r})$, we expand single-particle wave functions $\phi_c$ through:

$$\varphi_c = \sum_\alpha b_\alpha(\vec{r})C_\alpha; \qquad (9)$$

$$O_{\alpha,\beta} = \int b_\alpha^*(\vec{r})b_\beta(\vec{r})d\vec{r}; \qquad (10)$$

$$G_{\alpha,\beta} = \int \nabla b_\alpha^*(\vec{r})\nabla b_\beta(\vec{r})d\vec{r}. \qquad (11)$$

Using (11) the kinetic energy expression can be transformed into following form:

$$T_c = \int |\nabla\varphi|^2 = (C^*GC). \qquad (12)$$

Using matrix elements of the electrostatic energy in field of nuclear ($V_{nucl}(\vec{r}) = -2/r$)

$$V_{\alpha,\beta} = \int b_{\alpha}^{*}(\vec{r})V_{nucl}(\vec{r})b_{\beta}(\vec{r})d\vec{r}, \qquad (13)$$

the expression of potential electrons energy in the nuclear field can be transformed in the following form:

$$U_c = \int V_{nucl}|\varphi_c|^2 = (C^*VC) \qquad (14)$$

Summing all the contribution, we can obtain the following expression for the total energy of electron in state $\varphi_c$ :

$$E_c = (C^*GC) + (C^*VC) \qquad (15)$$

A general class of trial functions of electron density distributions $n_{kc} = n_k + n_c$ has cusp-singularity on a nucleus and kink-boundary form on external borders of a sphere-confining atom of hydrogen. In this case the variation procedure includes self-consistent approximation of a minimum energy in the ground state with step-by-step minimizing of functional (7) with respect to value of $n_k$, and matrix column (C) describing $n_c$.

## 4. Results and discussion

Fig.2 shows computed ground state hydrogen energy curves for kink-, cusp- and kink-cusp-density distributions of single electron inside spherical form gap. The same, as in previous computer experiments [5] for confinement of H-atom inside cubic form gap, on the set of kink-cusp-functions $\phi_{kc}$ the numerical variation of the ground state energy $E_{kc}$ of confined spherical form hydrogen atom has shown presence of two stable ground states.

First of them is realized at infinite value of the sphere-support radius R. It corresponds to the ground state of the infinite free atom of hydrogen with energy -13.6 eV (-1.31 kJ/mol). Another ground state corresponds to compact spherical atom of hydrogen that is stable inside a sphere of radius $R = 3.1a_0$ (0.16 nm). This ground state describes the confined atom of hydrogen inside the condensed state spherical gap. Energy of finite spherical hydrogen atom is 0.27 eV (26 J/mol) below the energy of the infinite free H-atom. This value of confinement bond energy of H-atom is equal to observable an average hydrogen bond energy in supramolecules and in hydrogen "electron-deficient" metal tetraborides such as $LiBH_4$ It means that single electron mechanism of such kind single hydrogen bonds is probably caused by effect of hydrogen atom confinement.

Let us analyze some electronic properties of confined H-atom. According to the potential curves on Fig.2 the ground state of H-atom corresponds to a confinement volume 17 $10^{-3}$ $nm^3$. In metal hydrides a hydrogen atom are compressed at interstitial position down to real atomic volume nearly by 2-3 $10^{-3}$ $nm^3$. It is meaning that equilibrium confinement volume of H-atom provides efficient electronic overlap integrals between hydrogen and atoms of metal only on neighboring coordination spheres. It becomes apparent as short-range force of adhesion.

536

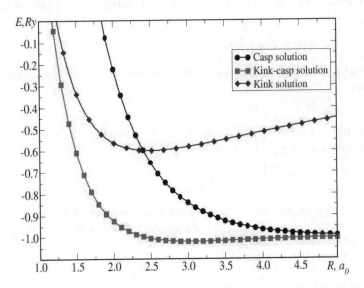

Figure 2. Curves of single electron energy in confined hydrogen atom for different ground states and models of electron density distribution

From Fig. 2 one can see that around equilibrium the energy of kink-cusp-density distribution is essentially lower than the energy of cusp-density one. It is resulted from appearance of $\phi_k$ -mode describing kink-shaped KDW- state of nonlocal extension electron. A contribution of kink-shaped density $n_k$ is comparatively small, but it appreciably decreases a rate of cusp-shaped density rise in neighborhood of nuclear. A superposition of kink- and cusp- modes (according to Eq.1) exerts nonlinear influence on energy of confined hydrogen atom. It is caused by kinematics constrains of Eq. (8) and principal difference between states of "extension" electron and "point" electron. A distinctive feature of the extension electron is a nonlocality of its energy functional $E_k$ . (Eq. 3) and the appearance of a quantum gyroscopic effect caused by p-orbital number ($1 = 1$) of KDW-state.

On Fig.3 we show schematic spectrum of single electron energy in different ground states and exiting states of hydrogen atom for equilibrium radius (R = 0.16 nm) and for nonequilibrium extended radius (R = 0.50 nm) of confinement. First characteristic feature of this spectrum is an energy band E (kink) of extension electron KDW states. This energy band borders upon zero energy level. Second characteristic property is that infinite atomic stationary single electron states, which eigenvalues lying inside of the energy band E (kink), are destroyed in confined H-atom as there are not their wave function nodes inside given compact confinement volume $\Omega$. In case of radius R = 0.16 nm there is only E(cusp|1s) level. In limit of R→ ∞ width of the energy band E (kink) tends to zero, therefore number of infinite atomic stationary single electron states tends to infinity. For example, for R = 0.50 nm there are three eigenvalues of such states: E(cusp|1s), E(2sp), E(3spd).

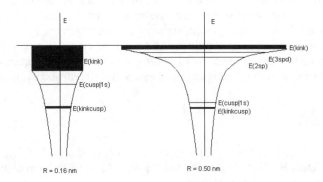

Figure 3. Spectrum of single electron energy of ground states and exiting states of hydrogen atom for
equilibrium radius (R = 0.16 nm) and nonequilibrium radius (R = 0.50 nm) of confinement

As is shown on Fig. 3 the exact ground state of confined H-atom has line spreading
of energy level E(kinkcusp). It is caused by existence in extension electron KDW-state
some quantum ($e^-$-$\gamma$)-electrodynamics fluctuations localized inside spherical form gap.
Instead of destroyed atomic stationary PDW-states new nonsteady KDW-states become
real inside compact gap of confinement. Such non-steady process of hydrogen atom
confinement underlies some synergetic effects of self-assembling in hydrogen materials
[6].

From properties of electronic structure of confined hydrogen atom it is followed that
using of well-known LCAO MO method for calculation of metal hydrides and hydrogen
"electron-deficient" metal tetraborides such as $LiBH_4$ is not correct, in principle. There is
full analogy with well-known problem of tight-binding scheme calculation of electron
spectrum in the condensed state and molecules, when energy eigenvalues lie in energetic
gap between tops of self-consistent electron potential energy and zero level of energy
[7]. Namely, all infinite electron atomic states lying inside this energy gap are really
destroyed. They vanish. Therefore any linear combination of such destroyed infinite
atomic orbitals is insignificant.

Taking into consideration strong interatomic electron exchange-correlation effects it
is clear that changing finite wave function self-assembling confinement of H-atom
essentially influences on overlap integrals between adjoining atoms in material [6]. That
is why theoretical calculation of the stability of metal hydrides is very complex, and self-
assembling phenomena of hydrogen in nanostructures and sypramolecules are such
complicated for theoretical analyzing.

Obtained curves of ground state energy of confined hydrogen atom let us understand
very complicated interaction of hydrogen with the electronic structure of host material.
The existence of two equilibrium stable ground states is very important for new sight on
explanation of quantum mechanical mechanism of transportation of hydrogen atoms
inside crystal grain, intercrystalline boundaries, amorphous substances and low-
dimensional nanostructures. In all cases the finite forms of interparticle potentials are
some results from modifications of electron density distribution inside confined block of
hydrogen atom, which is strictly positioning and regular moving under influence of

definite conditions of the confinement. On the other hand, we come to the important conclusion about the special role of atoms of hydrogen in realization of the phenomena of self-organizing in different nanosystems. Namely, the spontaneous transitions between equilibrium stable ground states of H-atom differing by the form and volumes of the compact support should in themselves cause spontaneous nanostructural transformations. Therefore atoms of hydrogen have major importance for processes of self-organizing in water, biological substance and different accumulators of hydrogen.

## 5. References

1.  Zuttel, A. (2003) Materials for hydrogen storage, *Materials Today* **9**, 24-33.
2.  Umezawa, H., Matsumoto, H. and Tachiki, M. (1982) Thermo field dynamics and condensed state, Amsterdam.
3.  Beznosjuk, S.A., Minaev, B.F., Dajanov, R.D. and Muldachmetov, Z.M. (1990) Approximating quasiparticle density functional calculations of small active clusters: strong electron correlation effects, *Int. J. Quant. Chem.* **38**(6), 779-797.
4.  Beznosjuk, S.A., Beznosjuk, M.S. and Mezentsev, D.A. (1999) Electron swarming in nanosructures, *Computational Material Science* **14**, 209-214.
5.  Beznosyuk, S.A. (2002) Modern quantum theory and computer simulation in nanotechnologies: quantum topology approaches to kinematic and dynamic structures of self-assembling processes, *Material Science & Engineering C.* **19**(1-2), 369-372.
6.  Beznosyuk, S.A., Kolesnikov, A.V., Mezentsev, D.A., Zhukovsky, M.S. and Zhukovsky, T.M. (2002) Dissipative processes of information dynamics in nanosystems, *Materials Science & Engineering C.* **19**(1-2), 91-94.
7.  Ziman, I.M. (1971) The calculation of Bloch functions. Solid State Physics, Vol.26, Academic Press, New York.

# CALORIMETRIC INVESTIGATION OF THE HYDROGEN INTERACTION WITH $Ti_{0.9}Zr_{0.1}Mn_{1.3}V_{0.5}$

E.Yu. ANIKINA, V.N. VERBETSKY
*Chemistry Department, Lomonosov Moscow State University,*
*119899 Moscow, Russia; E-mail: verbetsky@hydride.chem.msu.ru*

## Abstract

The interaction of hydrogen with nonstoichiometric $Ti_{0.9}Zr_{0.1}Mn_{1.3}V_{0.5}$ Laves phase compound at pressure up to 60 atm and temperature range from 336 to 413 K has been studied by means of calorimetric and P-X isotherm methods. The obtained results allow to propose the existence of two hydride phases in the $Ti_{0.9}Zr_{0.1}Mn_{1.3}V_{0.5}$ - $H_2$ system under this conditions.

*Keywords:* Intermetallic Compounds (IMC); Hydrides; Calorimetry;
$Ti_{0.9}Zr_{0.1}Mn_{1.3}V_{0.5}$-$H_2$ system

## 1. Introduction

A lot of works are devoted to investigations of multicomponent alloys on the base Zr-Ti-Mn with Laves phase structure $AB_2$. These alloys are perspective materials for using them in different stationary and mobile hydrogen accumulators, as they posses fast kinetics, a large hydrogen storage capacities and resistance to degradation during cycling. However, in the most works devoted to the hydrogen interaction with these alloys, the measurements were carried out by means of plotting of pressure – composition isotherms and thermodynamic characteristics for studied system were calculated using Van't – Hoff equation. The works, in which thermodynamic properties of intermetallic compound (IMC) – hydrogen system was investigated by calorimetric method, are significantly less.

In the present work as subject under study the alloy with nonstoichiometric composition $Ti_{0.9}Zr_{0.1}Mn_{1.3}V_{0.5}$ was chosen. As it is known, in the alloy with nonstoichiometric composition $AB_{2-x}$ the atoms of A-component occupy in structure of alloy crystallochemical position of B-component. In this alloy A position are occupied by atoms of Zr and Ti, and Mn, V and over-stoichiometric Ti atoms are distributed in the B positions. As a matter of fact, the compound $Ti_{0.9}Zr_{0.1}Mn_{1.3}V_{0.5}$ is correctly described as a $AB_2$, which can be written $(Ti_{0.89}Zr_{0.11})(Mn_{1.39}V_{0.54}Ti_{0.07})$. Earlier the hydrogen interaction with this alloy has been investigated by means of P(pressure) – X (concentration) measurements in a conventional Sievert's type apparatus [1]. It has been found that in the region two phases coexistence $P_{plat.}$ is not constant, but increases with % conversation of hydride phase, that is characteristic of multicomponent alloys. For this reason it is especial interest to measure the thermodynamic properties of this system in the plateau region by calorimetric method.

539

*T.N. Veziroglu et al. (eds.),*
*Hydrogen Materials Science and Chemistry of Carbon Nanomaterials,* 539-546.

## 2.Experimental

The $Ti_{0.9}Zr_{0.1}Mn_{1.3}V_{0.5}$ sample was obtained by arc melting of the pure components in the furnace with tungsten nonconsumable electrode on a copper water cooled hearth under purified argon pressure 1.5 atm. The starting materials had purity: titanium – 99.99%, zirconium – 99.99%, manganese – 99.9%, vanadium – 99.99%. Since manganese is more volatile than the other elements the mixture was prepared with 4% excess of Mn compared to the desired final stoichiometry. The sample was turned over and remelted four times and then annealed at 1100 – 1150 K for 240 h in quartz ampoule under residual argon pressure 0.01 atm to ensure homogeneity. Then the sample was studied by X-ray, electron microscopy, electron probe analysis. X-ray analysis of initial alloy was performed on DRON–2 diffractometer (Cu Kα radiation, Ni filter). Electron-probe analysis was done in JXA–733 microanalyses complex with LINK–2 microanalytical system. From the X-ray analysis the crystal structure of the sample was characterized as C14–type Laves phase and no secondary phase was found. The cell parameters and data of quantitative analysis of the alloy composition are listed in Table 1.

TABLE 1. Chemical and phase composition of the $Ti_{0.9}Zr_{0.1}Mn_{1.3}V_{0.5}$ alloy

| Alloy composition | | | | Phase composition | Cell parameters, A | | Composition of IMC, normalazed to $AB_2$ |
|---|---|---|---|---|---|---|---|
| Ti | Zr | V | Mn | | a | c | $(Ti_{0.89}Zr_{0.11})(Mn_{1.39}V_{0.54}Ti_{0.07})$ |
| 32 | 4 | 18 | 46 | C 14 | 4.92 | 8.07 | |

TABLE 2. Temperature dependence of reaction enthalpy for the $Ti_{0.9}Zr_{0.1}Mn_{1.3}V_{0.5}$ – $H_2$ system

| Temperature , (K) | Range | $|\Delta Hdes.|\pm\delta$, (kJ/mol $H_2$) |
|---|---|---|
| 336 | $0.80 \leq x \leq 1.55$ | 31.47±0.42 |
| 354 | $0.60 \leq x \leq 1.10$ | 30.22±0.36 |
| | $1.20 \leq x \leq 2.00$ | 36.22±0.43 |
| 373 | $0.70 \leq x \leq 1.20$ | 30.13±0.50 |
| | $1.30 \leq x \leq 1.90$ | 35.21±0.37 |
| 393 | $0.60 \leq x \leq 1.00$ | 29.35±0.31 |
| | $1.10 \leq x \leq 1.70$ | 33.07±0.58 |
| 413 | $0.30 \leq x \leq 0.90$ | 28.02±0.33 |
| | $1.00 \leq x \leq 1.70$ | 33.04±0.42 |

Twin-cell differential heat-conducting calorimeter of Tian–Calvet type connected to the apparatus for gas dose feeding was applied to measure the dependences of differential molar enthalpy of desorption (ΔHdes.) and equilibrium hydrogen pressure on hydrogen concentration in $Ti_{0.9}Zr_{0.1}Mn_{1.3}V_{0.5}$ and reaction temperature. Hydrogen was obtained by desorption from $LaNi_5$-hydride. The apparatus scheme was described elsewhere [2]. The hydrogen concentration in the sample was calculated using the Van – der – Waals equation of state for pressure below 20 atm and using the modified Van – der – Waals equation for pressure above 20 atm [3]. The accuracy of the measurements was taken as square of average value $\delta = \sqrt{\sum \Delta^2 [n(n-1)]^{-1}}$ ($\Delta$, deviation from average value; n, number of measurements).

Desorption relative molar enthalpy $\Delta$Hdes was determined from the heat effect of the reaction:

$$Ti_{0.9}Zr_{0.1}Mn_{1.3}V_{0.5} H_X + y/2 H_2 \leftrightarrow Ti_{0.9}Zr_{0.1}Mn_{1.3}V_{0.5} H_{X+Y} \qquad (1)$$

Since $Ti_{0.9}Zr_{0.1}Mn_{1.3}V_{0.5}$ reversibly reacts with hydrogen the same sample ($12639.4 \cdot 10^{-6}$ mole $Ti_{0.9}Zr_{0.1}Mn_{1.3}V_{0.5}$) was used in all experiments.

## 3. Results and discussion

The calorimetric method applied in this work permits to obtained calorimetric data simultaneously with P–X isotherms. The $Ti_{0.9}Zr_{0.1}Mn_{1.3}V_{0.5} - H_2$ system was studied in the temperature range 336 – 413 K and hydrogen pressure from 0 to 60 atm. As the duration of this experiment was about 1 year it was necessary to control of sample characteristics. In this purpose desorption of hydrogen at 336 K was repeated from time to time. Despite the multicomponent nature of the alloy the P–X isotherms and calorimetric data for studied alloy were very reproducible, namely, in the limits of sensitivity of the definition all parameters were fully reproducible.

Fig. 1 shows the P–X dependencies (X = [H]/ [AB$_2$]; P, equilibrium pressure), obtained in the temperature range 298 – 413 K. For all temperatures the desorption isotherms are presented and for 336 K also the absorption isotherm. As one can see from fig. 1 the $Ti_{0.9}Zr_{0.1}Mn_{1.3}V_{0.5} - H_2$ system has a wide α–region at ambient temperature. The length of α–region significantly decreases with increasing of the experimental temperature and plateau slopping on P–X diagrams increase. At experimental temperature above 336 K there is discontinuity in the plateau region when $dP_{H2}/dx$ reaches a maximum value at x~1. The similar observations have been made on the other alloys in previous works [4,5]. With rising of the temperature the location of this point didn't change significantly. It should be marked one peculiarity of obtained P–X diagrams. As it has been mentioned above for 336 K the absorption and desorption isotherms were measured. From fig. 1 it can be seen that at 336 K the plateau region for the desorption isotherm is 0.5 < x < 2.2, and for the absorption one is 0.50<x<1.90. There is no practically hysteresis up to value x=1.6. But at x>1.6 pressure hysteresis appears and P–X reversibility appears only at x>2.2, though the sharp change of the slope of the P–X curve is observed at x=1.7. On can suggest that in the range 1.7<x<2.2 there also exists a two-phase equilibrium. The similar phenomenon for AB$_2$ – H$_2$ has been described in [5, 6] for $Zr(Fe_{0.75}Cr_{0.25})_2$ and ZrCrFe – H$_2$ systems and was accounted for the formation of two hydrides.

Fig. 2 represents the plot of the enthalpy of desorption as a function of X at 336 K. There is the initial range 0.20<x<0.60 where the function |$\Delta$Hdes. |=f(x) has linear dependence which can be described by the first order equation

$$y = -13.72 x + 42.97 \qquad (2)$$

Further, at x>0.60 the character of the |$\Delta$Hdes.|=f(x) dependence changes sharply, namely, in the range 0.60<x<0,75 the values of |$\Delta$Hdes.| decrease from ~35 kJ/ mol H$_2$ to ~25 kJ/ mole H$_2$ and then in the hydrogen concentration range 0.75<x<0.80 this one increases up to the plateau values, corresponding α↔β transition. In the range

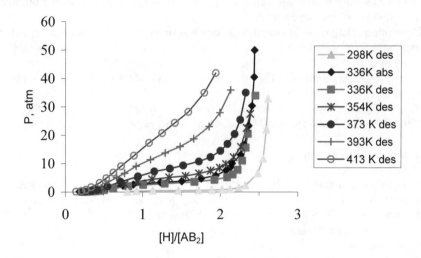

Figure 1. Absorption and desorption isotherms for the $Ti_{0.9}Zr_{0.1}Mn_{1.3}V_{0.5} - H_2$ system

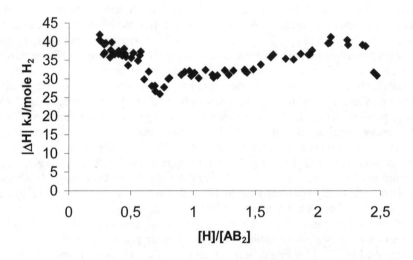

Figure 2. Desorption enthalpy vs. composition at 336 K

$0.80 < x < 1.55$. values of $\Delta Hdes$. remain constant and after that $(x>1.6)$ the function of $|\Delta Hdes.|=f(x)$ changes in accordance with the linear law

$$y = 10.89x + 16.53 \tag{3}$$

On the P–X isotherm (see fig. 2) this phenomenon is corresponded with the pressure hysteresis in the range $1.70 < x < 2.2$. I.e. compared the data, obtained from P–X and $|\Delta Hdes. |=f(x)$ dependences at 336 K, one can suggest that in this hydrogen concentration region $\beta \leftrightarrow \gamma$ transition occurs which is accompanied by additional thermodynamic processes. This phenomenon, characteristic for multicomponent alloys, was described in the work [7].

With increasing experimental temperature the position of minimum values of $|\Delta Hdes.|$, which took place in the $\alpha$-region at 336 K, shifts towards the range of the lower hydrogen concentrations (to compare x=0.75 at 336 K and x=0.50 at 393 K), and the length of $\alpha$ – phase decreases (see fig. 2–4). On the plot of $|\Delta Hdes.|$-X dependence at 353 K one can assign two region with constant values of $|\Delta Hdes.|$: $0.60<x<1.10$ $|\Delta Hdes.|=30.22 \pm 0.71$ kJ/ mol $H_2$ and $1.20<x<2.00$ $|\Delta Hdes.|=36.15 \pm 1.00$ kJ/ mol $H_2$.

Rising the experimental temperature from 353 to 393 K does not influence the shape of $|\Delta Hdes.|$-X thought the region boundaries $\alpha \leftrightarrow \beta$ and $\beta \leftrightarrow \gamma$ shift to the lower hydrogen concentrations and the values of $|\Delta Hdes.|$ decrease slightly.

At 413 K the plot of $|\Delta Hdes.|$-X drastic changes its shape (see fig. 5). There is no minimum value of $|\Delta Hdes.|$, which could be seen on the plots of $|\Delta Hdes.|$-X isotherms at lower temperatures. Three regions can be selected on the obtained $|\Delta Hdes.|$-X diagram with constant values of $\Delta Hdes$. Initial range $0<x<0.3$ corresponds to the formation of $\alpha$–solution hydrogen in IMC ($|\Delta Hdes.|=41.79 \pm 0.92$ kJ/ mol $H_2$). At x=0.30 the values $|\Delta Hdes.|$ sharply decrease and in the range $0.30<x<0.90$ $|\Delta Hdes.|=28.02 \pm 0.68$ kJ/ mol $H_2$). And at last there is the third region of constant values of $|\Delta Hdes.|$ in the range $1.00<x<1.70$ $|\Delta Hdes.|=33.04 \pm 0.97$ kJ/ mol $H_2$.

The existence of $\beta$–and $\gamma$–hydride phases of $Ti_{0.9}Zr_{0.1}Mn_{1.3}V_{0.5}$–$H_2$ system revealed in the work may be caused by two factors. Firstly, it could be supposed that the formation of $\beta$-hydride under rising temperature is accompanied by the transformation of metal sublattice. However, the existence of two hydride phases could be connected with peculiarity of the structure of this IMC. As mentioned above investigated IMC has nonstoichiometric composition. It is known, that in the hexagonal $AB_2$ Laves phase hydrogen atoms mainly occupy two kinds of interstitial sites: $[A_2B_2]$ and $[AB_3]$. Using the data of deuteride $(Ti_{0.89}Zr_{0.11})(Mn_{1.47}V_{0.31}Ti_{0.22})D_{2.8}$ –structure research based on compositionally familiar IMC studied in this work it has been found [8], that in $(Ti_{0.89}Zr_{0.11})(Mn_{1.47}V_{0.31}Ti_{0.22})D_{2.8}$ deuterium atoms occupy three sites 24(l), 12(k)$_1$ and 6(h)$_1$ with $[A_2B_2]$. But since V and Ti, which have strong affinity to hydrogen, as well as Mn occupy some B sites there is a possibility to regard an interstitial site $[A_2B_2]$ as formed by active (Ti, Zr, V), which conditionally signed as A*, and inactive (Mn) to hydrogen components. On the base carried out calculations the authors [8] drew the conclusion that 24(l) transforms to $[A^*_4]$ and $[A^*_3B]$ in higher extent than 12(k)$_1$ and 6(h)$_1$ and therefore 24(l) positions are occupied by hydrogen at hydride formation in the first order. The authors have concluded that $(Ti_{0.89}Zr_{0.11})(Mn_{1.47}V_{0.31}Ti_{0.22})D_{2.8}$ with consideration of distribution deuterium atoms among interstitial sites may be written as $(Ti_{0.89}Zr_{0.11})(Mn_{1.47}V_{0.31}Ti_{0.22})D^1_{1.86}D^2_{0.69}D^3_{0.25}$, $D^1$, $D^2$, $D^3$ – deuterium atoms, arranged in 24(l), 12(k)$_1$ and 6(h)$_1$ respectively.

To make analogy it could be supposed that in our case the formation of $\beta$-hydride corresponds the occupation by hydrogen atoms positions 24(l), and the formation of $\gamma$ – hydride is connected with the occupation by hydrogen atoms the positions 12(k)$_1$ and 6(h)$_1$. As well as it is possible that namely the transformation the part of $[A_2B_2]$ into $[A^*_4]$ and $[A^*_3B]$ could be the reason so prolonged $\alpha$-region. In the work [9] this phenomenon in explained the same reasons.

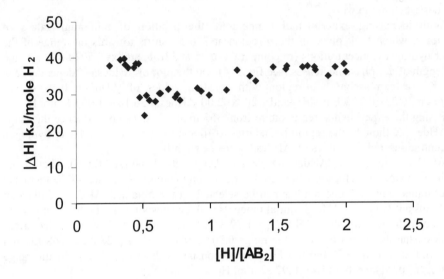

Figure 3. Desorption enthalpy vs. composition at 353 K

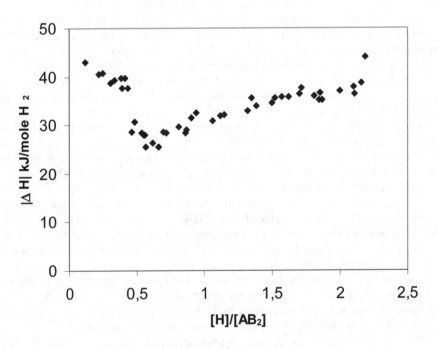

Figure 4. Desorption enthalpy vs. composition at 373 K

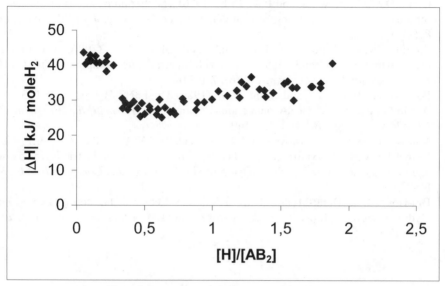

Figure 5. Desorption enthalpy vs. composition at 413 K

## 4. Conclusions

On the base of obtained data it can be concluded that during hydrogen interaction with $Ti_{0.9}Zr_{0.1}Mn_{1.3}V_{0.5}$ at the temperature range 336-413 K two hydride phases forms: β- and γ-hydrides. It has been found that the boundaries of these phases and the values of ΔHdes. depend on temperature.

## 5. Acknowledgements

The work has been partially supported by the RFBR Grant № 03-03-33023 and 02-02-17545.

## 6. References

1. Mitrokhin, S.V., Bezuglaya, T.N. and Verbetsky, V.N. (2002) Structure and hydrogen sorption properties of (Ti,Zr)-Mn-v alloys, *J. of Alloys and Compounds* **330-332**, 146-151.
2. Anikina, E.Yu. and Verbetsky, V.N. (2002) Calorimetric investigation of the hydrogen interaction with $Ti_{0.9}Zr_{0.1}Mn_{1.1}V_{0.1}$, *J. Alloys and Compounds* **330-332**, 45-47.
3. Hemmes, H., Driessen, A. and Griessen, R. (1986) Thermodynamic properties of hydrogen at pressure up to 1 Mba and temperature between 100 and 1000 K, *J. Phys. C: Solid State Phys* **19**, 3571-3585.
4. Flanagan, T.B., Luo, W. and Clewley, J.D. (1993) The characterization of multi-component metal hydrides using reaction calorimetry, *Z. Phys. Chem.*, 35-44.

5.  Grant, D.M., Murray, J.J. and Post, M.L. (1990) The thermodynamics of the system for $Zr(Fe_{0.75}Cr_{0.25})_2$ –$H_2$ using of heat conduction calorimetry, *J. Solid State Chem.* **87**(2), 415-422.

6.  Sirotina, R.A., Umerenko, E.A. and Verbetsky, V.N. (1996) Calorimetric investigation of hydrogen interaction with intermetallic compound ZrCrFe–$H_2$, *Russ. Neorganicheskie materially* **32**(6), 710-714.

7.  Post, M.L., Murray, J.J. and Grant, D.M. (1989) The $LaNi_5$-$H_2$ system at T=358 K: An investigation by Heat-Conduction Calotimetry, *Zeitshrift fur Physikalische Chemie Neue Folge. Bd.* **163**, 135-140.

8.  Mitrokhin, S.V., Smirnova, T.N., Somenkov, V.A., Glazkov, V.P. and Verbetsky, V.N. (2003) Structure of (Ti,Zr) –Mn-V nonstoichiometric Laves phases and $(Ti_{0.9}Zr_{0.1})(Mn0_{.75}V_{0.15} Ti_{0.11})D_{2.8}$ deuteride, *J. Alloys and Compounds* **356-357**, 80-83.

9.  Bin-Hong Lin, Dong-Myung Kim, Ki-Young Lee and Jai-Young Lee (1996) Hydrogen storage properties of $TiMn_2$-based alloys, *J. Alloys and Compounds* **240**, 214-218.

# INTERACTION IN NbVCo-H$_2$ AND NbVFe-H$_2$ SYSTEMS UNDER HYDROGEN PRESSURE UP TO 2000 ATM

S.A. LUSHNIKOV, V.N. VERBETSKY
*Moscow Lomonosow M.V. State University, Chemistry Department,
119992 Leninsky Hills, Moscow, Russia;
E-mail address: lushnikov@hydride.chem.msu.ru*

## Abstract

Interaction IMC (intermetallic compounds) NbVCo and NbVFe with hydrogen in wide temperatures and pressures intervals have been investigated with PCT (pressure-composition) method. X-ray analyses has been shown that hydrogen introduction in lattice was accompanied with volume increasing without rearrangement of metallic atoms.

*Keywords:* Intermetallic Compounds (IMC); Hydrides; Hydrides; High pressure of
hydrogen

## 1. Introduction

For the first time ternary Laves phase in Nb-V-Co and Nb-V-Fe systems have been described in [1]. Intermetallic compounds AB$_2$ with Laves phases structure are perspective material for hydrogen storage. Under usual pressure (up to 100 atm.) hydrides, formed in AB$_2$-H$_2$ system, are able to absorb about two weigh percent of hydrogen. In such systems A-component is usually titanium, zirconium and earth-rare elements. At present work interaction IMC, containing niobium as A-component with hydrogen has been investigated.

## 2. Experimental

Samples of NbVCo and NbVFe alloys have been obtained by electrical furnace in inert atmosphere. For homogenization obtained samples were quenched at 850$^0$C during 240 hours in evacuated quartz ampoules. Investigation of interaction IMC with hydrogen was carried out on apparatus of high pressure (up to 2000 atm.), described in [2]. For calculation of hydrogen amount in obtained hydride phases in situ the equation of high pressed hydrogen have been used [3]. Fixing of the composition obtained hydrides of high pressure has been conducted by transferring them into passive state. For this sample of hydride obtained under high pressure first was cooled at temperature of liquid nitrogen and then has been left at air about forty minutes at this temperature.

Amount of hydrogen in the passive hydrides has been checked with thermal desorption. X-ray analyses of alloys and hydrides have been conducted on DRON-2

547

*T.N. Veziroglu et al. (eds.),*
*Hydrogen Materials Science and Chemistry of Carbon Nanomaterials,* 547-552.
© 2004 *Kluwer Academic Publishers. Printed in the Netherlands.*

(Co,CuK$_\alpha$-emission, Ni filter). Obtained X-ray data were refined with the Rietveld method.

## 3. Results and discussions

X-ray analyses of IMC NbVCo and NbVFe have shown that obtained samples are single-phase and have MgZn$_2$ structure type. Obtained X-ray data are presented in table1 and on figures 1(a, b).

TABLE 1. Results of X-ray analyses of IMC

| Composition of IMC | Cell parameters, Å | V,Å$^3$ | Atoms | Position | Rw-factor,% |
|---|---|---|---|---|---|
| NbVCo | a=4,932±0,002; c=8,03±0,01 | 169 | Nb $V_{0,75}Co_{0,25}$ $V_{0,42}Co_{0,58}$ | 4(f) 2(a) 6(h) | 3,62 |
| NbVFe | a=4,932±0,002; c=8,06±0,01 | 170 | Nb $V_{0,98}Fe_{0,02}$ $V_{0,34}Fe_{0,66}$ | 4(f) 2(a) 6(h) | 5,19 |

Data from table 1 have been shown that in the NbVCo and NbVFe structure Nb atoms occupied 4(f) position. In NbVCo 2(a) position preferentially occupied V and Co atoms, and in NbVFe 2(a) position practically fully occupied of V atoms.

Position 6(h) atoms of B-component occupied approximately equally, in correlation ($V_{0,42}$ +$Co_{0,58}$) for NbVCo and ($V_{0,34}$ +$Fe_{0,66}$) for NbVFe.

Remarkable interaction hydrogen with NbVCo starts under increasing of pressure to 355 atm. and room temperature. Under such conditions hydrogen absorption has been occurred to composition $NbVCoH_{0,3}$. Plateau on absorption isotherm with pressures interval from 350 atm. to 750 atm.  is corresponded to further absorption of hydrogen and forming  hydride with composition $NbVCoH_{2,2}$ (fig.2(a)). Subsequently pressure increasing in this system is accompanied with hydrogen absorption and forming  hydride phase $NbVCoH_{2,9}$ under pressure 1800 atm.

However on desorption isotherm plateau at this temperature hasn't been observed. Under decreasing of pressure gradual hydrogen desorption is occurred until composition $NbVCoH_{1,2}$ under pressure 10 atm.

Unlike from inactive IMC, the $NbVCoH_{1,2}$ rapidly interacts with hydrogen even under low temperatures. Decreasing temperature to (-50$^0$C) in this system leads to additional hydrogen absorption and forming hydride $NbVCoH_{3,8}$ under pressure 1800 atm. At this hydrogen absorption occurs monotonously, without a plateau on the isotherm.

In NbVFe-H$_2$ system interaction of NbVFe with hydrogen starts under 300 atm., as in case with NbVCo. After exceeding  the pressure more 1200 atm., remarkable hydrogen absorption has been observed. This is corresponded of the plateau on the absorption isotherm (fig. 2(b)) with pressure interval from 1200 atm. to 1280 atm. Under this phase composition has been changed from $NbVFeH_{1,1}$ until $NbVFeH_{2,2}$.

Subsequently pressure increasing to 1600 atm. leads to forming hydride phase $NbVFeH_{3,0}$. Hydrogen desorption under pressure decreasing has been occurred without plateau on isotherm. Under this phase composition was equally $NbVFeH_{1,2}$ under 10 atm.

Intensity

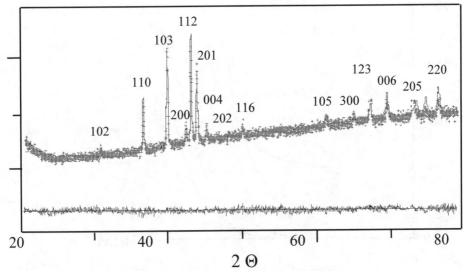

Figure 1 (a). Diffractogram of NbVCo

Intensity

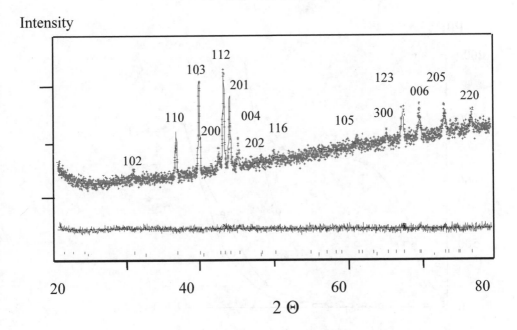

Figure 1 (b). Diffractogram of NbVFe

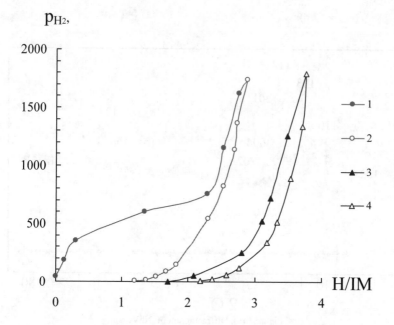

Figure 2 (a). Hydrogen absorption (1,3) and desorption (2,4) isotherms in NbVCo-H₂ system at temperatures 20⁰C (1,2) and -50⁰C (3,4)

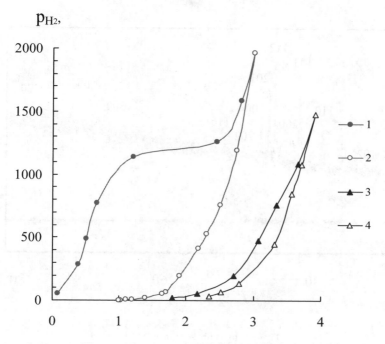

Figure 2 (b). Hydrogen absorption (1,3) and desorption (2,4) isotherms in NbVFe-H₂ system at temperatures 20⁰C (1,2) and -50⁰C (3,4)

After decreasing temperature to (-50$^0$C) in this system hydrogen absorption was occurred monotonously by formed hydride phase until composition NbVFeH$_{3,9}$ under 1500 atm.

Inactive NbVFe interacts with hydrogen at increasing temperature to 350$^0$C and pressure about 50 atm. Obtained hydride phase has composition equally NbVFeH$_{0,8}$ at 35 atm. and room temperature.

After the sample was evacuated at temperature 350$^0$C and hydrogen was fully extracted, active NbVFe interacted with hydrogen under lower pressure – about 20 atm. at room temperature. Obtained hydride also has composition NbVFeH$_{0,8}$.

Obtained samples of hydrides under high pressure are unstable. Passivation these hydrides, carried out before X-ray analyses has shown their less composition comparatively of composition calculated in situ.

Results of X-ray analyses of synthesized hydrides are presented in table 2 and diffractograms are on fig.3 (a,b).

TABLE 2. X-ray data of hydrides IMC NbVCo and NbVFe

| Composition | Str. type | a, Å | $\Delta$ a/a % | c, Å | $\Delta$c/c % | V,Å$^3$ | $\Delta$V/V% | Rw,% |
|---|---|---|---|---|---|---|---|---|
| NbVCoH$_{2,5}$ | MgZn$_2$ | 5,140±0,002 | 4,2 | 8,39±0,03 | 4,5 | 191 | 13,0 | 9,7 |
| NbVFeH$_{2,2}$ | MgZn$_2$ | 5,145±0,001 | 4,3 | 8,43±0,02 | 4,6 | 193 | 13,5 | 8,7 |

Carried out X-ray analyses have shown that under forming NbVCo and NbVFe hydrides, structure type of initial IMC didn't change. Data presented in table 2 have shown, that forming hydrides in NbVCo-H$_2$ and NbVFe-H$_2$ systems is accompany equable expansion of their lattice and increasing of «a» and «c» parameters. Cell increasing volume of hydrides formed under high pressure NbVCoH$_{2,5}$ and NbVFeH$_{2,2}$ are equally 13,0% and 13,5%. From X-ray data, refined by Rietveld method have been obtained, that introduction of hydrogen atoms in lattice didn't lead to relocation of metallic atoms.

## 4. Conclusions

High pressure of hydrogen allowed making synthesis of hydride phases based on IMC NbVCo and NbVFe with Laves structure. After first absorption cycle active samples interact with hydrogen even low temperature.

X-ray analyses have shown that forming of hydride phases is accompanied with equable volume expansion of lattice of initial IMC with introduction in lattice hydrogen atoms. Relocation of metallic atoms in lattice hasn't occurred.

This work was supported with Russian Foundation for Basic Research, grant 03-03-33023.

552

Intensity

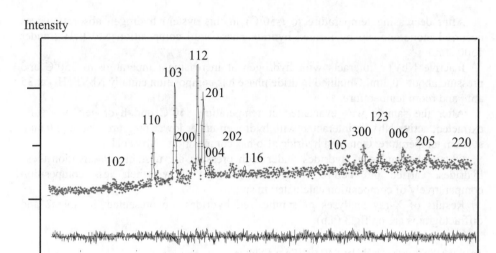

Figure 3 (a). Diffractogram of NbVCoH$_{2.5}$

Intensity

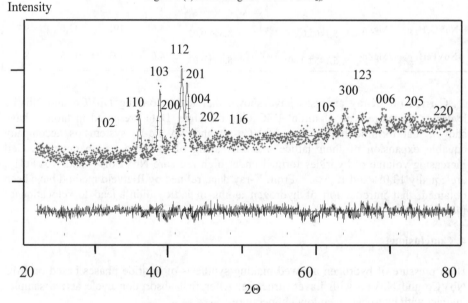

Figure 3 (b). Diffractogram of NbVFeH$_{2.2}$

## 5. References

1.  Teslyk, M.Yu. (1969) Metallic compounds with structure of Laves phase, *M/Nauka*, 1-196.
2.  Verbetsky, V.N., Klyamkin, S.N., Kovriga, A.Yu. and Bespalov, A.P. (1996) Hydrogen interaction with RNi$_3$ (R-rare-earth) Intermetallic compounds at high gaseous pressures, *Int. J. Hydrogen Energy* 11/12, 997-1000.
3.  Hemmes, H., Driessen, A. and Griessen, R. (1986) Thermodynamic properties of hydrogen at pressure up to 1Mba and temperatures between 100 and 1000K, *J.Phys.C: Solid State Phys.* **19**, 3571-3585.

# EFFECT OF HYDROGENATION ON SPIN-REORIENTATION PHASE TRANSITIONS IN $R_2Fe_{14}BH_X$ (R = Y, Ho, Er) COMPOUNDS

I.S. TERESHINA, G.S. BURKHANOV, O.D. CHISTYAKOV,
N.B. KOL'CHUGINA, S.A. NIKITIN [1)], H. DRULIS [2)]
*A.A. Bajkov Institute of Metallurgy and Material Science RAS,*
*Leninski pr. 49, Moscow, 119991, Russia; E-mail: teresh@ultra.imet.ac.ru*
[1)] *Department of Physics, M.V. Lomonosov Moscow State University,*
*Leninskie Gory, Moscow,119992, Russia*
[2)] *Institute of low temperatures and structure research, PAS,*
*ul. Okolna 2, 50-950 Wroclaw, Poland*

**Abstract**

$R_2Fe_{14}B$ compounds with R = Y, Ho, Er were hydrogenated to the composition $R_2Fe_{14}BH_x$ where x = 2.3 – 2.5. Magnetic characteristics of the present compounds were investigated over the temperature range 4.2 to 800 K and at applied fields up to 1 kOe. A systematic study of the influence hydrogen on Curie temperature and the spin reorientation transitions (SRT) in $R_2Fe_{14}BH_x$ was reported. The SRT temperatures were detected by measuring temperature dependences of the magnetization and ac magnetic susceptibility. It was found that the temperatures of Curie and spin-reorientation transition increases upon hydrogenation.

*Key words:* Magnetizm of metal hydrides; Rare earth – transition metal systems;
Spin reorientation

## 1. Introduction

The $R_2Fe_{14}B$ compounds absorb readily hydrogen at room temperature and form stable hydrides. The hydrogen atoms incorporated in the $R_2Fe_{14}B$ (R - rare-earth metal) intermetallic compound's crystalline lattice cause changes in their magnetic properties [1-5]. The changes in magnetic characteristics of $R_2Fe_{14}B$ compounds are of great fundamental and technological interest. Hydrogen expands the lattice and modifies the magnitude of the magnetic moment of the 3d - sublattices, the 3d-3d, 4f-3d exchange interactions and also the magnetocrystalline anisotropy. The technological interest in hydrogen absorption by $R_2Fe_{14}B$ compounds is twofold: 1) as a means of raising the Curie temperatures; 2) in hydrogen treatment such as HDDR (hydrogenation, disproportionation, desorption and recombination process) to produce high-coercive magnets. The aim of this work is to study the effect of hydrogenation of the magnetic phase transitions in the $R_2Fe_{14}B$ (R = Ho and Er) compounds.

## 2. Experimental

Samples of $Y_2Fe_{14}B$, $Ho_2Fe_{14}B$ and $Er_2Fe_{14}B$ were prepared by arc melting under purified argon gas from starting raw materials of at least 99.99 % wt purity. After

*T.N. Veziroglu et al. (eds.),*
*Hydrogen Materials Science and Chemistry of Carbon Nanomaterials,* 553-556.
© 2004 *Kluwer Academic Publishers. Printed in the Netherlands.*

554

melting the as-cast samples were wrapped in a tantalum foil, sealed in quartz tubes and annealed in an argon atmosphere for 14 days at $900^0$ C. X-ray diffraction and metallographic methods have been used to check the purity of the samples.

The hydrides were formed by exposing the host metals at room temperature to high purity hydrogen at about $2\cdot10^5$ Pa. We prepared the following samples: $Y_2Fe_{14}BH_{2.5}$, $Ho_2Fe_{14}BH_{2.3}$ and $Er_2Fe_{14}BH_{2.5}$. The resultant powder was checked to be single-phase by the standard X-ray powder diffraction. It was found that annealed host samples and their hydrides are single phase with the expected tetragonal structure.

Thermomagnetic analysis (TMA) was used to measure the Curie temperature in a field of 1 kOe and in the temperature range 4.2 - 800 K with samples in the form of rough chunks. The Curie temperatures $T_C$ were obtained by plotting $M^2$ vs T and extrapolating to $M^2 = 0$. The spin-reorientation temperatures $T_{SR}$ were determined from the cusp-shaped part of the magnetization M vs T curves for $Er_2Fe_{14}BH_x$ (x = 0; 2.5) and from measurements of the ac susceptibility for $Ho_2Fe_{14}BH_x$ (x = 0; 2.3) compounds.

## 3. Results and discussion

No changes in the structure type of the $R_2Fe_{14}B$ componds were found to be observed upon hydrogenation; however, the change of unit cell volume is equal to 1 % per H atom (see Table 1).

TABLE 1. Crystallographic parameters of $R_2Fe_{14}B$ and their hydrides

| Compounds | a(Å) | C(Å) | V(Å$^3$) | ΔV/V(%) |
|---|---|---|---|---|
| $Y_2Fe_{14}B$ | 8.75 | 12.04 | 921.81 | - |
| $Y_2Fe_{14}BH_{2.5}$ | 8.85 | 12.10 | 947.70 | 2.81 |
| $Ho_2Fe_{14}B$ | 8.74 | 11.97 | 914.36 | - |
| $Ho_2Fe_{14}BH_{2.3}$ | 8.82 | 12.05 | 937.40 | 2.52 |
| $Er_2Fe_{14}B$ | 8.73 | 11.94 | 909.98 | - |
| $Er_2Fe_{14}BH_{2.5}$ | 8.82 | 12.02 | 935.06 | 2.75 |

The effect of hydrogenation of the Curie temperature of these compounds has been studied, as it was mentioned above. Examples of measurements of Curie temperature are shown in fig. 1 and 2. Fig. 1 shows a TMA of $Y_2Fe_{14}B$ compounds and its hydride. We have observed the magnetic moment change in process of thermal cycle (heating – cooling), which indicated the outgasing of hydrogen during the thermomagnetic measurements. It was found that the hydrogenation leads the increase in the temperature of magnetic ordering ($T_C$) that, on the average, is ~ 25 - 30 K per H atom. A similar trend is observed for $Er_2Fe_{14}B$ compounds and its hydride (see Fig. 2). It is known that $T_C$ is very sensitive to the Fe-Fe distances, therefore the $T_C$ increase after hydrogenation is attributed to increase Fe-Fe exchange interaction in the $R_2Fe_{14}BH_x$ compounds (or reduction of the negative exchange interactions associated with the Fe-Fe configurations).

It was studied the effect of hydrogenation on the spin-reorientation temperature. Figs. 2 and 3 show temperature dependences of the magnetization and magnetic susceptibility polycrystalline $Er_2Fe_{14}BH_x$ (x = 0 and 2.5) and $Ho_2Fe_{14}BH_x$ (x = 0 and 2.3) compounds, respectively. It is seen that the hydrogenation causes the shift of experimental dependences of the $R_2Fe_{14}BH_x$ compounds to the high temperatures.

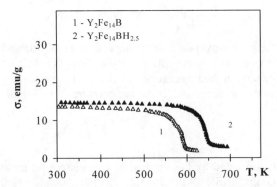

Figure 1. Temperature dependences of the magnetization of polycrystalline $Y_2Fe_{14}BH_X$ compounds (x = 0 and 2.5) measured in a magnetic field of 1 kOe

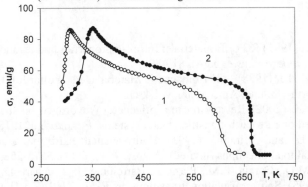

Figure 2. Temperature dependences of the magnetization of polycrystalline $Er_2Fe_{14}BH_X$ compounds (x = 0 and 2.5) measured in a magnetic field of 1 kOe

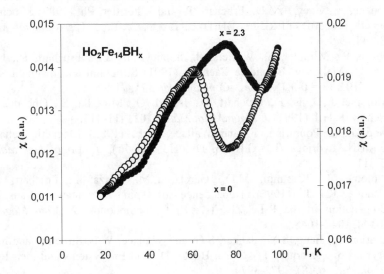

Figure. 3 Temperature dependences of the magnetic susceptibility of polycrystalline $Ho_2Fe_{14}BH_X$ compounds measured in a magnetic field of 10 Oe.

## 4. Conclusions

We present a set of experimenyal data on the dependence of Curie and spin reorientation transition temperatures on hydrogen content in $R_2Fe_{14}BH_X$ (x = 0; 2.3-2.5) compounds.

It was found that upon hydrogenation, 1) the Fe-Fe exchange interaction in the $R_2Fe_{14}BH_X$ compounds increases and 2) the temperature of spin-reorientation transition also increases.

## 5. Acknowledgments

The work has been supported by Department of Chemistry and materials science RAS program № 8, Grant NSH -205.2003.2, RFBR Grant № 02-02-16523 and № 04-03-32194.

## 6. References

1. Herbst, J.F. (1991) $R_2Fe_{14}B$ materials: Intrinsic properties and technological aspects, *Review of Modern Physics* **63**(4), 819-898.
2. Buschow K.H.J. (1988) In Ferromagnetic Materials, ed. Wolfarth E.P. and Buschow K.H.J., North-Holland, Amsterdam, vol.4, ch.1.
3. Pourarian, F. (2002) Review on the influence of hydrogen on the magnetism of alloys based on rare earth-transition metal systems, *Physica B* **321**, 18-28.
4. Fruchart, D. and Miraglia (1991) Hydrogenated hard magnetic alloys from fundamental to applications (invited), *J. Appl. Phys.* **69**(8), 5578-5583.
5. Kuz'min, M.D., Garcia, L.M., Plaza, I., Bartolome, J., Fruchart, D. and Buschow, K.H.J. (1995) Spin reorintation transitions in $R_2Fe_{14}ZH_X$ (Z = B, C) compounds, *J. Magn.Magn.Mater.* **146**, 77-83.
6. Dalmas De Reotier, P., Fruchart, D., Potonnier, L., Vaillant, F., Wolfers, P., Yaouanc, A., Coey, J.M.D., Fruchart, R. and L'heritier, Ph. (1987) Structural and magnetic properties of $RE_2Fe_{14}BH(D)_x$; RE = Y, Ce, Er, *J. Less-Common Met.* **129**, 133-144.
7. Obbade, S., Miraglia, S., Wolfers, P., Soubeyroux, J.L., Fruchart, D., Lera, F., Rillo, C., Malaman, B. and le Caer, G. (1991) Structural and magnetic study of $Ho_2Fe_{14}BH_x$ (x = 0-3.1), *J. Less-Common Met.* **171**, 71-82.
8. Bartolome, J., Luis, F., Fruchart, D., Isnard, O., Miraglia, S., Obbade, S., and Buschow, K.H.J. (1991) *J. Magn.Magn.Mater.* **101**, 411-413.
9. Zhang, L.Y., Pourarian, F. and Wallace, W.E. (1988) Magnetic behavior of $R_2Fe_{14}BH_x$ hydrides (R = Gd, Tb, Dy, Ho and Er), *J. Magn.Magn.Mater.* **71**, 203-211.
10. Bartolome, J., Kuz'min, M.D., Garcia, L.M., Plaza, I., Fruchart, D. and Buschow, K.H.J. (1995) Dependence of spin reorientation transition on hydrogenation in the $R_2Fe_{14}ZH_x$ (Z = B, C) compounds, *J. Magn.Magn.Mater.* **140-144**, 1047-1048.
11. Pourarian, F., Jiang, S.Y., Sankar, S.G. and Wallace, W.E. Structure and magnetic properties of quaternary $Pr_{2-x}R_xCo_{14}B$ (R = Dy and Er) systems and their hydrides, *J. Appl. Phys.* **63**(8), 3972-3974.

# THE CLUSTER GROWTH MECHANISM OF NANOSTRUCTURED DIAMOND

V. MELNIKOVA
*Technical centre, National Academy of Science, Kiev 04070, Ukraine*
vmelnikova@ukr.net

## Abstract

In this paper were investigated the structure of globular and cauliflower-like diamond particles obtained by HFCVD from methane-hydrogen gas mixture. It is established sphaerocrystalline structure. For understanding of principles of diamond growth were analysed various models of the nucleation. On the reason of the surface structure was suggested the mechanism of growth CVD diamond which is conditioned by the formation clusters in vapor phase, following deposition on the surface of diamond particles and coalescence by them. It is characterised the dispersion of ballas-type and cauliflower-like diamond films.

**Keywords:** crystal growth, CVD diamond, cluster, cauliflower-like and ballas-type structure

## 1. Introduction

Diamond synthesis at low pressure (CVD) is wide applied technology in materials science for obtaining of diamond crystals, films and coatings. There are many physical and chemical methods of the crystallisation of epitaxial, textural and polycrystalline diamond films [1-4]. As initial substances are used: gases (all homologous methane series), liquids (methanol, acetone, alcohol etc.) and solids (coal). The peculiarity of CVD technology is the large concentration advantage of hydrogen to carbon gas for the effective etching of graphite. The CVD process is extremely sensitive to change of technological parameters owing to that the wonderful globular and polyhedral morphologies occur. The degree of dispersion is great also from nanostructures up to macrocrystals. The diamond morphology depends on elementary processes of deposition of carbon atoms from a gas phase. The discussion about the mechanism of diamond growth is conducted [5-9] but the full clearness in consideration of elementary acts of a condensation is not. Films with ballas-type and cauliflower-like structure are not enough studied. The aim of this investigation is the structural research into globular forms and analysis of possible models of their growth from the vapor.

## 2. Experimental methods

Diamond films and isolated crystals have been obtained in quartz reactor by hot filament CVD method on regimes described in [10-12]. Total pressure of methane-hydrogen gas mixture change from 5 up to 60 Torr, concentration $CH_4/H_2 = 0.5\text{-}3$ vol.%, W-filament temperature $(T_f)$ - 1900-2200°C. Diamond films were grown on the various mirror-polished substrates W, Si, $Al_2O_3$, $Al_2O_3$-TiC, $Si_3N_4$, $B_4C$, $SiO_2$, SiC at the temperature $(T_s)$ 500-1000°C. The latter was measured by means of thermocouple mounted on the backside of substrate

*T.N. Veziroglu et al. (eds.),*
*Hydrogen Materials Science and Chemistry of Carbon Nanomaterials,* 557-562.

holder. The growth rate in these conditions change from 1 to 10 μm/h, the maximum thickness was obtained 100 μm.

The crystal morphology and fine microstructure was studied by X-ray diffraction, scanning and transmission electron microscopy via preparation of samples by means of replicas and ionic thinning.

## 3. Results

### 3.1 *Morphology of films*

Chemical vapor deposition of diamond films goes through four stages on time: nucleation, growth of nucleuses, coalescence (coagulation) and formation continuous layers. The CVD kinetics is determined by substrate temperature mainly. Other parameters (pressure and methane concentration in hydrogen) exercise influence on distribution of impurities and lattice defects most of all. The depended on a supersaturating degree (actually on $T_s$ ) nucleation density is small: $10^7 cm^{-2}$ at 600°C and $10^5 cm^{-2}$ at 800°C - in comparison with ones by deposition of metallic films that equal $10^{11} cm^{-2}$. Diamond particles on the first stage of growth have nearly ideal octahedral, cubooctahedral, pseudopentagonal shape at $T_s > 750°C$ or globular ones at lower temperature. Continuous films are formed after coagulation (Fig.1). The pressure and concentration $CH_4$ is smaller the diamond crystallinity is better. The spherical form is distorted by a growing of triple joints in continuous film without porosity. The profile of diamond crystals is shown on Fig. 2.

All substrates must be distributed in two types: reacting and nonreacting to carbon. The gas mixture at high temperature can be react to substrate having great chemical affinity to carbon. From materials which were used as substrates only W and Si intensive interact to carbon. Carbides $WC+W_2C$ on W and β-SiC on Si substrates are synthesised on the first stage of deposition. Their thickness is nearly 1 μm. The size of WC, SiC grains in carbide films is less than 0.5 μm. The diamond nucleuses begins after formation of continuous carbide layer. The deposition on other substrates has not an incubation period of carbonisation preceding diamond growth therefore the deposition rate increases. The high density distribution of diamond particles and the maximum rate of growth 10 μm/h was obtained on substrate $Al_2O_3$-20%TiC which do not interact to carbon. The basic macroscopic element of structure is the globule and microscopic - the grain in globules.

The macro- and microlevel of dispersion are defined according to sizes of spherical particles and including grains. Besides the variety of globular diamond is characterised by the ratio of particles and grains sizes. The ballas-type diamond particles consist of nanograins of size about 10-50 nm. Cauliflower-like films consist of grains larger 100 nm. Two level of dispersion determine macro- and microscopic roughness too. The fine surface cauliflower-like structure which illustrate of this size effect is good to seen on SEM images (Fig. 1 b). Microroughness of ballas-type films can look at TEM patterns under a method of replicas best of all (Fig. 1 d). It is marked that a roughness of diamond films with polyhedral structure are formed by pyramidal peaks of facets {111} usually.

### 3.2 *Microstructure of diamond films*

It was established the monophase condition of the globular CVD diamond. By microdiffraction and darkfield technique of TEM was determined the polycrystalline spherulite structure. The each particle consist of numerous spicular or wedges-shaped grains (Fig. 3). A high degree of dispersion and thermostresses bring about the complex absorption and diffraction contrast on the TEM pictures

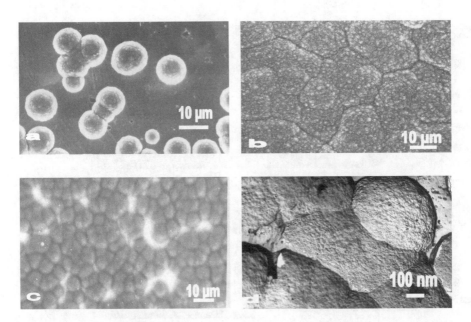

Fig. 1. SEM (a-c) and TEM (replica) (d) images of diamond crystals and films deposited on W substrates: $CH_4/H_2 = 3$ vol.%, p = 60 Torr, $T_f = 2100\ °C$, t = 10 h, $T_s = 750$ (a), 700 (b), 650 $°C$ (c,d).

Fig. 2. Cross-sectional SEM patterns of diamond crystals deposited on W substrates.

. The sectorial constitution shows that the growth begins from one centre and goes at all directions with the equal rate therefore globules are a hemispherical shape. The nucleation stage is connected to homogeneous formation of clusters of the critical size on cold substrate. The density distribution of growth centres on a substrate is $10^5 - 10^7\ cm^{-2}$. The low density of critical nucleus of diamond modification is evidence of low probability of association of carbon atoms with $sp^3$ - hybridization.

### 3.3  Growth mechanism

There are two alternate models of a crystallisation from a gas phase. One of them is conditioned by a layering deposition of carbon atoms on steps of crystal, other by formation clusters direct in activated methane-hydrogen gas mixture and deposition of clusters as united whole on the surface particle with following coalescence.

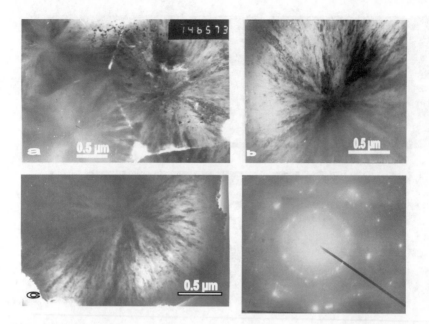

Fig. 3. TEM images (a-c) and microdiffraction pattern (d) of CVD globular diamond particles deposited on Si substrates by  $CH_4/H_2$ =3 vol.%, p = 24 Torr, $T_f$ = 2100 °C, t = 5 h, $T_s$ = 650 °C.

All possible elementary processes were analysed for the CVD growth model. The search of structural elements or building blocks responded for both mechanism crystallisation were carried out. There were investigated the growth surface of the films with different degree of continuous. The interesting results were obtained at the research of surface layers of globular and cauliflower-like particles. The structure of surface layers differ from ones of the volume. The surface is coated with clusters of the size 5-20 nm (Fig. 4). And just this fact can serve as evidence of the cluster growth mechanism.

The necessary supersaturation at substrate temperature lower 750°C is reached not only on a surface but also in near-substrate space therefore the homogeneous nucleation of clusters of diamond modification from carbon atoms can happen in a gas phase in adjoining areas. Clusters but atoms are deposited on a surface film the more so that this surface has segments with negative curvature which are potential wells for their adhesion.

The cluster mechanism of growth exercise influence on internal and superficial structure of films. There are processes of a coalescence and coagulation in a system of deposited clusters in the porous shell around particle. The diamond has extremely low diffusivity therefore collective recrystallization goes slowly. The clusters join by coagulation mainly thanks to that the film is characterised by the nanocrystalline structure (Fig. 5). Great amount of clusters which determine the low limit of grain sizes look as an accumulation of impurities or as incoherent additives. The upper grain size reach up to 100 nm by collective recrystallization.

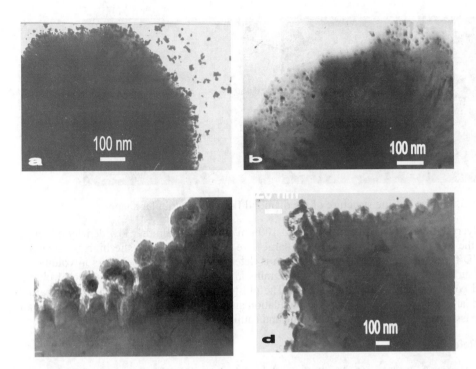

Fig. 4. The surface structure of globular (a, b) and cauliflower-like (c, d) diamond particles.

So diamond sphaerocrystals grow through absorption, reorientation and recrystallization of clusters. Two type of crystals with spherulitic and cauliflower-like structures arise at the substrate temperature 650°C and 750°C accordingly which exercise influence on intensity of the recrystallization. The structure is no perfection and far from the monodispersion.

The microroughness of globular particles is due to clusters on the surface films (Fig. 4 a). Ones can be discovered via replicas (Fig. 1 d). Higher substrate temperature about 750°C is greatly conducive to the collective recrystallization of clusters by formation of cauliflower-like diamond (Fig. 4 c). The result was that surface is coated higher grains which can see via SEM (Fig 1 b). Wonderful morphology occur through two-level dispersion. Thin films consist of diamond particles of size 20-50 μm. The surface of each particle including grains of sizes larger than 100 nm is view like cauliflower.

## 4. Conclusions

The TEM and SEM methods were used for investigation of fine structure of the HFCVD diamond films consist of globular and cauliflower-like crystals. The cluster structure of surface made it possible to build the model of growth from vapor phase. As against the polyhedral shapes of diamond, when for growth it is necessary to create continuous steps for adsorption of carbon atom, in this model the clusters occur in a gas phase then are deposited on surface and coalescence with diamond particles. The offered model can be explained in conformity with colloidal theory of nucleation.

562

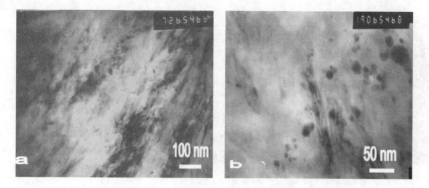

Fig. 5. Intracrystalline structure of the diamond films with globular shape.

Polycrystalline globules can grow up to 100 µm owing to the rather low density of centres of the homogeneous nucleation. The large size of particles cause the high roughness of CVD films. The microlevel of dispersion is determined by sizes of clusters in volume and on surface of particles and by sizes of grains after coagulation or coalescence of clusters and collective recrystallization partly. The cluster mechanism of growth of each particle and languidly current processes of coagulation and coalescence of grains allow to form and to passivate nanocrystalline sphaerolitic and cauliflower-like structure.

## 5. References

1.    Haubner R., Lindlbauer A., Lux B. Diamond deposition on Cr, Co, Ni substrates by MW plasma CVD. Diamond Rel. Mat. 1993, 12(2): 1505-1515.
2.    Singh J. Novel techniques for selective diamond growth on various substrates. J. Mater. Eng. 1994, 3(3): 378-385.
3.    Venter A., Neethling. The effect of filament temperature on the growth of diamond using hot filament chemical vapor deposition. Diamond Rel. Mat. 1993, 3(1): 168-172.
4.    Jang X., Klages C.-P. Heteroepitacsial diamond growth on (100) silicon. Diamond Rel. Mat. 1993, 12(2): 1112-1113.
5.    Singh J., Vellaikal M. Microstructural evolution of diamond growth on iron silicide/silicon substrates by hot filament CVD. Surface coatings technology. 1994; 64(3): 131-137.
6.    Stockel R., Graupner R., Janischowsky K. Initial stages in the growth of polycrystalline diamond on silicon. Diamond Rel. Mat. 1993, 12(2): 1467-1472.
7.    Euckervort W.J.P., Janssen G., Vollenberg W. CVD diamond growth mechanisms as identified by surface topography. Diamond Rel. Mat. 1993, 2(5-7): 997-1003.
8.    Norgard C., Matthews A. Two step diamond growth for reduced surface roughness. Diamond and Rel. Mat. 1996, 5:332-337.
9.    Ece M., Oral B., Patscheider J. Nucleation and growth of diamond films on Mo and Cu substrates. Diamond and Rel. Mat. 1996,5:211-216.
10.  Melnikova V.A., Kolesnichenko G.A., Naidich Yu.V. Sphaerolitic crystalization of diamond films from the vapour phase. Dopovidi NAN of Ukraine, 1996, 9: 99-104.
11.  Melnikova V.A., Kolesnichenko G.A., Naidich Yu.V. Investigation by electron microscopy of graphite distribution in CVD-diamond films. Proceedings 8 International Symposium "Thin Films in Electronics" (Charkov, Russia), 1997, 3; 106-108.
12.  Melnikova V.A., Kolesnichenko G.A., Naidich Yu.V. Effect of carbonization of substrates on growth CVD-diamond films. Adhesion of melts & soldering of materials, Kiev, Naukova dumka, 1997, 33; 60-64.

# SPECIFIC FEATURES IN THERMAL EXPANSION OF $YFe_{11-x}Co_xTiH$ SINGLE CRYSTALS

E. TERESHINA [a], K. SKOKOV[b], L. FOLCIK[c], S. NIKITIN [a], H. DRULIS [c]

[a] Department of Physics, M.V. Lomonosov Moscow State University,
    Leninskie Gory, Moscow,119992, Russia; E-mail: jane@rem.phys.msu.su
[b] Department of Physics, Tver State University,
    Geljabova str.33, Tver, 170002, Russia
[c] Polish Academy of Science, Trzebiatowski Institute of Low Temperature
    and Structure Research, 59-950 Wroclaw 2, P.O. Box 1410, Poland

## Abstract

Thermal expansion and its anomalies on single crystals of the system $YFe_{11-x}Co_xTiH_y$ ($0 \leq x \leq 2$, $y = 0$; 1) are investigated by the tensometric method in the temperature range 77 – 300 K. The temperature dependences of the linear thermal expansion coefficient $\alpha(T)$ are obtained. It is found that the host compounds with uniaxial magnetic anisotropy exhibit two anomalies in the $\alpha(T)$ at $T_1 \approx 170$ - 180 K and $T_2 \approx 260$ - 270 K. The hydrogenation shifts the observed anomalies toward lower temperatures.

*Keywords*: Rare-earth compound, single crystal, hydride, thermal expansion coefficient

## 1. Introduction

The rare-earth (R) - transition metal (M) intermetallics have always attracted a great interest due to their potential for application on permanent magnets [1]. Among them, the $RFe_{11-x}Co_xTi$ with the $ThMn_{12}$-type structure are in the focus of interest [2-4]. Ti plays the role of the stabilizing element. These compounds interact actively with hydrogen gas and form stable hydrides [5,6]. Upon hydrogenation $RFe_{11-x}Co_xTi$ intermetallics exhibit substantial changes in their magnetic properties. Magnetoelastic properties have not been studied extensively up to now [7,8]. The aim of the present work was to investigate the effect of hydrogen on thermal expansion in $YFe_{11-x}Co_xTi$ ($0 \leq x \leq 2$) compounds.

## 2. Sample preparation and experimental technique

Single-crystal samples of $YFe_{11-x}Co_xTi$ alloys in the form of plates 7–10x5–8x1–1.5 mm in size were used in the measurements. The alloys were prepared according to the technique described earlier [9]. The composition and the distribution of elements were checked by x-ray fluorescence probe microanalysis. The final testing of the surfaces of single-crystal samples was performed at the Trzebiatowski Institute of Low Temperature and Structure Research (Wroclaw, Poland). The distribution of

563

*T.N. Veziroglu et al. (eds.),*
*Hydrogen Materials Science and Chemistry of Carbon Nanomaterials, 563-568.*
© *2004 Kluwer Academic Publishers. Printed in the Netherlands.*

elements and the composition were determined from the ratio between the intensities of the Fe and Ti $K_\alpha$– lines and the $L_\alpha$–lines of rare-earth elements. The composition was derived by the standard method with ZAF corrections.

In the course of single-crystal plates preparation, we determined the crystal lattice type and the unit cell parameters and checked their correspondence to the space group by x-ray diffraction analysis. The parameters obtained are in good agreement with the available data [10]. The samples were cut from an ingot according to the orientation of the crystallographic axes with respect to the surface. The plates prepared were investigated by x-ray topography.

The plates chosen for our investigations contained aggregates of large-sized single-crystal blocks (three or four blocks) with a misorientation of no more than 2°. According to the estimates the surface of single crystals was of good quality. As a rule the two opposite surfaces of the plate had a virtually identical block character. Strain gauges were cemented to the surface of the single-crystal sample along the [001] and [110] crystallographic directions.

The procedures of preparation the hydrides of these compounds also were described in Ref. [11]. We have prepared the $RFe_{11}TiH_x$ ($x \approx 1$) hydrides without decrepitation of the single-crystal samples. The thermal expansion was measured by the tensometric technique. The strain gauges used in the measurements were fabricated from a strain-sensitive wire without a noticeable galvanomagnetic effect. The nominal length of the strain gauges was 5 mm, and their resistance was equal to about 100 $\Omega$. The strain gauge factor $S$ was equal to 2.15 over the entire temperature range. During the measurements, one of the strain gauges was cemented to the sample and the other (temperature compensating) gauge was cemented to a thin quartz plate that was kept against the sample. Both gauges were connected across the opposite arms of a Wheatstone bridge. The resistances of the operating and temperature- compensating strain gauges differed by no more than 1%. The circuit was calibrated by measuring of the out-of-balance bridge signal when a standard resistor of 0.1 $\Omega$ was connected in the circuit. Polycrystalline nickel was used as a reference sample.

Since the main purpose of the present work is to investigate the transformation of the magnetic structure in the vicinity of spin-reorientation transitions in the studied compounds, and since the reorientation mechanism is associated primarily with the deviation of magnetic moments from the [001] crystallographic direction, this paper presents the results of measurements of the thermal expansion only along the direction in the sample plane. The temperature dependencies measured along the [110] direction do not provide radically new information and will not be discussed in this work. The samples to be measured preliminarily cooled down to 80 K. Then samples were heated at the rate of less than 1 K/min. Measurements were performed in the temperature range 80 – 400 K. The sample was mounted to the electromagnet gap in the cryostat which allowed us to measure thermostriction in the temperature range 80 – 400 K.

The error in measurements of the elongation length did not exceed 3%.

The temperature dependence of the linear thermal expansion coefficient $\alpha(T)$ was obtained by numerical differentiation of the temperature dependence of thermal expansion $\Delta l/l(T)$, that is:

$$\alpha(T) = \frac{1}{l} \cdot \frac{d}{dT} l(T).$$ (1)

According to our estimates, the error in determination of the thermal expansion coefficient did not exceed 10%.

## 3. Results and discussion

In this work particular emphasis was placed on the $YFe_{11}Ti$ and its hydride compounds. Yttrium ions have no localized magnetic moments. This makes it possible to investigate the specific features of thermal expansion of iron sublattice in $RFe_{11}Ti$ compounds. $YFe_{11}Ti$ possesses uniaxial magnetic anisotropy of easy magnetization axis (EMA) type over the entire temperature range of magnetic ordering. Fig. 1 shows the temperature dependence of the linear thermal expansion coefficient $\alpha(T)$ for the $YFe_{11}Ti$ and its hydride single crystals. This dependence exhibits two clearly pronounced features at $T_1 = 180$ K and $T_2 = 265$ K.

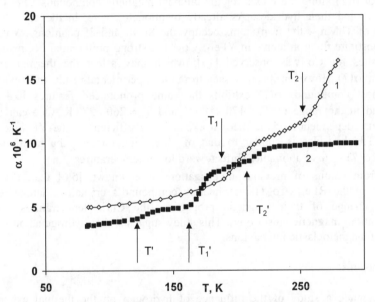

Figure 1. The temperature dependencies of the linear thermal expansion coefficient along the **c** axis for the $YFe_{11}Ti$ (1) and its hydride (2)

The complex behavior of the $\alpha(T)$ dependence can be explained by the fact that iron ions in the lattice of $YFe_{11}Ti$ compounds occupy three crystallographically nonequivalent positions ($8i$, $8j$, and $8f$) and can be treated as ions forming three sublattices. Note that the magnetic moments localized on iron atoms occupying different crystallographic positions in $YFe_{11}Ti$ considerably differ from each other in magnitude and are equal to $1.92\mu_B$ ($8i$), $2.28\mu_B$ ($8j$), and $1.8\mu_B$ ($8f$), where $\mu_B$ is Bor's magneton [12]. It is known that iron atoms can interact ferromagnetically or antiferromagnetically when the distance between them is larger or smaller than 2.4 Å, respectively. If the Fe–Fe distance in particular pairs is less than the critical value they possess negative exchange interactions. The temperature dependencies of the magnetization for particular iron sublattices can have a different shape, which leads to the features observed in the thermal expansion of the $YFe_{11}Ti$ single crystal. As noticed early [13] that the hydrogen concentration: 1 atom per $RFe_{11}Ti$ formula unit corresponds to full occupancy of the interstitial 2b site according to the neutron diffraction experiment. This interstitial site can be seen as a pseudooctahedron with two rare-earth ions and four iron atoms 8j at the corners. It is observed, that the hydrogenation influences the thermal expansion of the 3d sublattice in $RFe_{11}Ti$. The positions of the observed $\alpha(T)$ – curve anomalies shifts toward lower temperatures ($T_1' \approx 160$ K, $T_2' \approx 220$ K). The $YFe_{11}TiH$ hydride exhibits additional $\alpha(T)$ – curve anomaly in the temperature range $T \approx 120$ K (see Fig. 1). It is possible that this anomaly have nonmagnetic origin and are caused by processes of the hydrogen atoms ordering in the $ThMn_{12}$-type crystal lattice. The ordering of hydrogen atoms, occurring at temperature decreasing, can cause significant increasing of the lattice parameters (how it was observed in the case of $ErFe_{11}TiH$ [14]).

Fig. 2 and Fig. 3 show the temperature dependencies of the linear thermal expansion coefficient $\alpha(T)$ along the c axis for the uniaxial magnetic compounds $YFe_{10}CoTi$ and $YFe_9Co_2Ti$ and their hydrides. As already mentioned above, in the crystallographic structure of $ThMn_{12}$, the iron atoms occupy the 8i, 8j and 8f positions. When cobalt atoms substitute for iron atoms in $YFe_{11-x}Co_xTi$, a strong preferential occupation on 8f and 8j sites for Co was observed [15], which may affect the thermal expansion coefficient $\alpha(T)$. However, it can be seen from the experimental results that upon heating of the host compounds $\alpha(T)$ exhibits the same pronounced features like $YFe_{11}Ti$ compound in the range of $T_1 \approx 170 - 180$ K and $T_2 \approx 260 - 270$ K. One can find that temperature dependence of the thermal expansion coefficient of the $YFe_{10}CoTiH$ and $YFe_9Co_2TiH$ hydrides is similar to that of $YFe_{11}TiH$: namely, the positions of the observed $\alpha(T)$ – curve anomalies shifts toward lower temperatures.

So, from results of previous investigations it is known [5,6] that, insertion of hydrogen in the $RFe_{11-x}Co_xTi$ intermetallic compound's crystalline lattice results in apparent change of their magnetic properties (the Curie temperatures, saturation magnetization, magnetic anisotropy). This study suggests, that hydrogenation leads to a change of magnetoelastic interactions.

## 4. Conclusions

We have made a study of the influence of hydrogen on the thermal expansion of $YFe_{11-x}Co_xTi$ ($0 \leq x \leq 2$) compounds. It was found that these compounds with uniaxial magnetic anisotropy exhibit pronounced anomalies in the temperature dependences of the thermal expansion coefficient $\alpha(T)$ at $T_1 \approx 170 - 180$ K and $T_2 \approx 260 - 270$ K. It is observed, that the hydrogenation influences the thermal expansion of the $YFe_{11-x}Co_xTi$ shifting the anomalies toward lower temperatures.

## 5. Acknowledgements

We are very grateful to Irina Tereshina for useful discussion. The work has been supported by the Federal Program on Support of Leading Scientific School Grant NSH -205.2003.2 and RFBR Grants № 02-02-16523, № 03-02-06435.

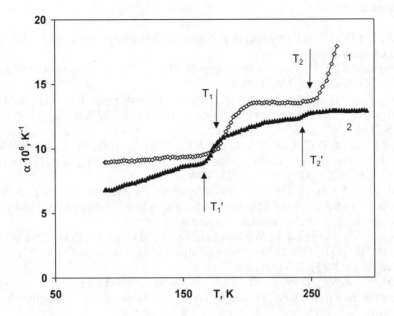

Figure 2. The temperature dependencies of the linear thermal expansion coefficient along the **c** axis for the YFe$_{10}$CoTi (1) and its hydride (2)

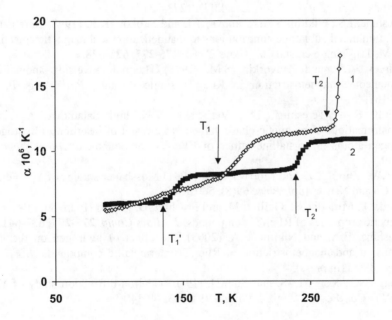

Figure 3. The temperature dependencies of the linear thermal expansion coefficient along the **c** axis for the YFe$_9$Co$_2$Ti (1) and its hydride (2)

## 6. References

1. Coey, J.M.D. (1996) Rare-earth Iron Permanent Magnets, Coey J. M. D., Ed. Oxford, UK: Clarendon, p. 36.
2. Jurczyk, M. (1990) Magnetic behaviour of $YFe_{10.8-x}Co_xT_{1.2}$ systems (T=W and Re), *J. Less. Common Met.* **166**, 335-341.
3. Liang, J.K., Huang, Q., Santoro, A., Wang, J.L. and Yang, F.M. (1999) Magnetic structure and site occupancies in $YFe_{11-x}Co_xTi$ (x=1,3,7,9), *J. Appl. Phys.* **86**(4), 2155-2160.
4. Gu, Z.F., Zeng, D.C., Liu, Z.Y., Liang, S.Z., Klaasse, J.C.P., Bruck, E., de Boer F.R. and Buschow, K.H.J. (2001) Spin reorientations in $RFe_{11-x}Co_xTi$ compounds (R = Tb, Er, Y), *J. Alloys Comp.* **321**, 40-45.
5. Fujii, H. and Sun, H. (1995) Interstitially modified intermetallics of rare-earth 3d-elements in Handbook of Magnetic Materials, series Ferromagnetic Materials, Ed. Buschow K.H.J., vol. 9, Elsevier, Amsterdam, p. 304-404.
6. Skourski, Y., Tereshina, I., Wirth, S., Drulis, H., Mattern, N., Eckert, D., Nikitin, S. and Muller, K.-H. (2002) Magnetocrystalline anisotropy of $SmFe_{11-x}Co_xTiH$, *IEEE Trans. Magn.* **38**(5), 2931-2933.
7. Andreev, A.V. (1998) Thermal expansion anomalies and spontaneous magnetostriction in $RFe_{11}Ti$ single crystals, *Philosophical Magazine B* **77**(1), 147-161.
8. Zubenko, V.V., Tereshina, I.S., Telegina, I.V., Tereshina, E.A., Luchev, D.O. and Pankratov, N.Yu. (2001) Specific Features in Thermal Expansion of $RFe_{11}Ti$ Single Crystals, *Physics of the Solid State* **43**(7), 1273-1277.
9. Tereshina, I.S., Nikitin, S.A., Ivanova, T.I. and Skokov, K.P. (1998) Rare-earth and transition metal sublattice contributions to magnetization and magnetic anisotropy of $R(TM, Ti)_{12}$ single crystals, *J. Alloys Comp.* **275-277**, 625-628.
10. Andreev, A.V. and Zadvorkin, S.M. (1998) Thermal expansion anomalies and spontaneous magnetostriction in $RFe_{11}Ti$ single crystals, *Phil. Mag. B.* **77**(1), 147-161.
11. Nikitin, S.A., Tereshina, I.S., Verbetsky, V.N. and Salamova, A.A. (2001) Transformations of magnetic phase diagram as a result of insertion of hydrogen and nitrogen atoms in crystalline lattice of $RFe_{11}Ti$ compounds, *J. Alloys Comp.* **316**, 46-50.
12. Li, Z.W., Zhou, X.Z. and Morrish, A.H. (1991) Mossbauer studies of $YTi(Fe_{1-x}M_x)_{11}$ (M=Co and Ni), *J. Appl. Phys.* **69**(8), 5602-5604.
13. Isnard, O., Miraglia, S., Guillot, M. and Fruchart, D. (1998) Hydrogen effects on the magnetic properties of $RFe_{11}Ti$ compounds, *J. Alloys Comp.* **275-277**, 637-641.
14. Tereshina, E.A. and Nikitin, S.A. (2003) The effect of hydrogen on the thermal expansion and magnetostriction of $RFe_{11}Ti$ intermetallic compounds, *IEEE Trans. Magn.* **39**(5) (in press).
15. Yang, Y.-C., Kong, L.-S. and Sun, H. (1990) Neutron-diffraction study of $Yco_{11}Ti$ and $YTi(Co_{0.5}Fe_{0.5})_{11}$, *J. Appl. Phys.* **67**(9), 4632-4634.

# STUDY OF STRUCTURE, HYDROGEN ABSORPTION AND ELECTROCHEMICAL PROPERTIES OF $Ti_{0.5}Zr_{0.5}Ni_yV_{0.5}Mn_x$ SUBSTOICHIOMETRIC LAVES PHASE ALLOYS

T.A. ZOTOV, V.N. VERBETSKY, O.A. PETRII, T.Y. SAFONOVA
*Moscow State University, Department of Chemistry,*
*Leninskie Gory 3, Moscow, 119992 Russia*
*E-mail: Verbetsky@hydride.chem.msu.ru*

## Abstract

We continue our previous study [14, 15] of hydrogen absorption and electrochemical properties of $Ti_{0.5}Zr_{0.5}Ni_yV_{0.5}Mn_x$ nonstoichiometric Laves phase alloys. Two series of alloys $Ti_{0.5}Zr_{0.5}Ni_{1.0}V_{0.5}Mn_x$ (x = 0.1- 0.4, $AB_{1.6}$ – $AB_{1.9}$) and $Ti_{0.5}Zr_{0.5}Ni_{1.15}V_{0.5}Mn_x$ (x = 0.1-0.55, $AB_{1.75}$ – $AB_{2.2}$) were prepared and investigated. Their electrochemical discharge characteristics were compared with results of $Ti_{0.45}Zr_{0.55}Ni_{0.7}V_{0.45}Mn_x$ (x = 0.1-1.5, $AB_{1.25}$ – $AB_{2.6}$) and $Ti_{0.45}Zr_{0.55}Ni_{0.85}V_{0.45}Mn_x$ (x = 0.1 1.35, $AB_{1.4}$ – $AB_{2.65}$) alloys corresponding to our previous study. It was shown, that the rate capability increases as the nickel content in alloy series increases. However the discharge capacity of $Ti_{0.5}Zr_{0.5}Ni_{1.0}V_{0.5}Mn_x$ alloys doesn't significantly increase as compared with $Ti_{0.45}Zr_{0.55}Ni_{0.85}V_{0.45}Mn_x$ alloys. Moreover, the discharge capacity decreases with the further increase in the nickel content in $Ti_{0.5}Zr_{0.5}Ni_{1.15}V_{0.5}Mn_x$ alloys. The values of discharge capacities at discharge current densities 100 mA/g are in a range of 310 – 390 mAh/g.

*Keywords:* MH electrodes, intermetallic compounds (IMC), crystal structure, hydrogen absorption, discharge capacity

## 1. Introduction

Nickel metal hydride (Ni-MH) batteries are widely adopted today due to their significant discharge capacity, high rate discharge ability (rate capability) and environmental safety [1, 2]. Numerous investigations were concerned with the application of the Laves phase intermetallic compounds (IMC) in Ni-MH technology in the recent time [1-9]. Ti-Zr-Ni-V-Mn Laves phase alloys demonstrate rather good hydrogen absorption and electrochemical properties [1-9]. However, at present, the influence of nickel and manganese on hydrogen absorption and electrochemical characteristics is not quite studied and systematized.

Two maximums of electrochemical discharge capacity for $Ti_{0.45}Zr_{0.55}Ni_{0.7}V_{0.45}Mn_x$ (x = 0.1-1.5, $AB_{1.25}$ – $AB_{2.6}$) and $Ti_{0.45}Zr_{0.55}Ni_{0.85}V_{0.45}Mn_x$ (x = 0.1-1.3; $AB_{1.4}$ – $AB_{2.6}$) systems were found to correspond to the compositions about $AB_{1.5}$ and $AB_2$ [14, 15]. $Ti_{0.45}Zr_{0.55}Ni_{0.85}V_{0.45}Mn_x$ alloys showed better electrochemical characteristics [14-15].

*T.N. Veziroglu et al. (eds.),*
*Hydrogen Materials Science and Chemistry of Carbon Nanomaterials,* 569-577.
© 2004 *Kluwer Academic Publishers. Printed in the Netherlands.*

Thus, it is of interest to investigate this system in the substoichiometric region ($AB_{<2}$) with higher nickel content. The two series of alloys $Ti_{0.5}Zr_{0.5}Ni_{1.0}V_{0.5}Mn_x$ (x = 0.1- 0.4, $AB_{1.6} - AB_{1.9}$) and $Ti_{0.5}Zr_{0.5}Ni_{1.15}V_{0.5}Mn_x$ (x = 0.1-0.55, $AB_{1.75} - AB_{2.2}$) were prepared for this purpose.

## 2. Experimental

Alloys were prepared by arc-melting the mixtures of pure initial metals under argon atmosphere. With the purpose of homogenization, the alloys were annealed for 240 h at 850°C in quartz ampoule under vacuum with subsequent quenching in cold water. The phase composition of alloys was examined on polished cross sections of the ingots by a scanning electron microscope (SEM) with an energy dispersive X-ray analyzer (EDXA JXA-733 with LINK-2 microcomputer system). The crystal structure and lattice parameters were determined by means of powder X-ray diffraction (URD-6 diffractometer with $CuK_\square$). Conditions were: scan range 25-85 ($2\theta$) with a step of $0.02^0$ ($2\theta$) and a sampling time of 1 second per step. The refinement of diffraction profiles was performed using the Rietveld method [13].

The hydrogen absorption properties were studied by measuring PCT isotherms using a Siewert's type apparatus at a hydrogen pressure below 50 atm.

The electrochemical discharge properties were studied using an automatic galvanostat connected with a personal computer. The electrochemical experiments were carried out in a three-electrode electrochemical glass cell with Hg/HgO electrode as the reference. The electrolyte was 6 M KOH solution. The MH electrodes were prepared by cold-pressing of the mixture of IMC hydride powder (20%) with copper powder (80%) in a pellet. The preliminary activation of MH electrodes involved boiling in 6 M KOH at about 115°C for 1.5 h before the first polarization. The activation conditions were chosen accordingly [12] study. It was found that after preliminary activation the discharge capacity of MH electrodes studied reaches its maximum in 2—4 cycles. Samples were charged at a current density of 100 mA/g for 4 - 6 h. The discharge cut-off potential was – 650 mV against the Hg/HgO electrode. The discharge capacity was checked at current densities of 100, 200, 400, and 600 mA/g. The rate capability of the samples was estimated as ratio of the discharge capacities at 400 and 100 mA/g current densities.

## 3. Results and discussion

The results obtained by using electron microscopy, electron probe microanalysis, and powder X-ray diffraction analysis indicate that samples were single phase pseudo binary IMC with Laves phase C14 structure. The small amount (1 – 2%) of nonstoichiometric zirconium oxide with cubic lattice (Fm3m, 225) is present in all alloys. The lattice parameters of the C14 Laves phase were found to decrease with an increase in the manganese content in the alloys series and the coefficient "n" in $AB_n$ (Tab. 1).

The element distribution in the Laves phase structure in $Ti_{0.5}Zr_{0.5}Ni_{1.0-1.15}V_{0.5}Mn_x$ series is similar to our previous $Ti_{0.45}Zr_{0.55}Ni_{0.7-0.85}V_{0.45}Mn_x$ series [14, 15]. It is known [4, 10, 11] that a part of titanium atoms that occupy the positions of A-atoms in the $AB_2$—Laves structure can also occupy the positions of B-atoms in substoichiometric IMC. In superstoichiometric IMC, a part of manganese atoms can pass from B positions

TABLE 1. Structural characteristics of IMC

| Alloy composition | n in $AB_n$. | $R_{wp}$,[1] | $S^2$ | $RF^3$ | a, E | c, E | V, $E^3$ |
|---|---|---|---|---|---|---|---|
| $Ti_{0.5}Zr_{0.5}Ni_{1.0}V_{0.5}Mn_{0.1}$ | $AB_{1.6}$ | 8.46 | 1.044 | 4.03 | 4.977 | 8.119 | 174.19 |
| $Ti_{0.5}Zr_{0.5}Ni_{1.0}V_{0.5}Mn_{0.2}$ | $AB_{1.7}$ | 7.55 | 1.058 | 3.45 | 4.965 | 8.093 | 172.76 |
| $Ti_{0.5}Zr_{0.5}Ni_{1.0}V_{0.5}Mn_{0.3}$ | $AB_{1.8}$ | 5.51 | 1.074 | 1.95 | 4.953 | 8.075 | 171.53 |
| $Ti_{0.5}Zr_{0.5}Ni_{1.0}V_{0.5}Mn_{0.4}$ | $AB_{1.9}$ | 5.25 | 1.032 | 1.73 | 4.934 | 8.044 | 169.59 |
| $Ti_{0.5}Zr_{0.5}Ni_{1.15}V_{0.5}Mn_{0.1}$ | $AB_{1.75}$ | 7.01 | 1.323 | 2.24 | 4.954 | 8.083 | 172.02 |
| $Ti_{0.5}Zr_{0.5}Ni_{1.15}V_{0.5}Mn_{0.2}$ | $AB_{1.85}$ | 6.41 | 1.175 | 2.15 | 4.949 | 8.065 | 171.10 |
| $Ti_{0.5}Zr_{0.5}Ni_{1.15}V_{0.5}Mn_{0.3}$ | $AB_{1.95}$ | 5.41 | 1.070 | 1.67 | 4.937 | 8.044 | 169.78 |
| $Ti_{0.5}Zr_{0.5}Ni_{1.15}V_{0.5}Mn_{0.4}$ | $AB_{2.05}$ | 7.30 | 1.053 | 3.50 | 4.932 | 8.036 | 169.28 |
| $Ti_{0.5}Zr_{0.5}Ni_{1.15}V_{0.5}Mn_{0.5}$ | $AB_{2.15}$ | 6.62 | 1.330 | 2.24 | 4.925 | 8.021 | 168.48 |
| $Ti_{0.5}Zr_{0.5}Ni_{1.15}V_{0.5}Mn_{0.55}$ | $AB_{2.2}$ | 6.23 | 1.032 | 3.60 | 4.919 | 8.009 | 167.82 |

to A positions [4, 10, 11]. The occupation of metallic sites in the C14 structure was refined by using the Rietveld method. The minimization of R-factors showed that the best fit is achieved for the following model (Tab. 2). In the sub-stoichiometric ($AB_{<2}$) region (Ti, Zr)(Ti, Ni, V, Mn)$_2$ with a low manganese content, it was found that vanadium atoms occupy totally the 2(a) and partially 6(h) sites of the C14 structure; titanium, nickel and manganese atoms occupy the 6(h) sites. The increase of manganese content leads to displacement vanadium atoms from 6(h) sites. The further increase of manganese content result in titanium displacement from 6(h) sites. And lastly, part of manganese atoms pass to 4(f) positions (Tab. 2).

TABLE 2. The C14 structure sites occupancy for $Ti_{0.5}Zr_{0.5}Ni_yV_{0.5}Mn_x$ alloys

| Alloy Composition | Position in C14 structure | | |
|---|---|---|---|
| | (4f) | (2a) | (6h) |
| Substoichiometric $AB_{2-x}$ | Ti+Zr | V | Ti+Ni+V+Mn |
| | Ti+Zr | V+Mn | Ti+Ni+Mn |
| Stoichiometric $AB_2$ | Ti+Zr | V+Mn | Ni+Mn |
| Superstoichiometric $AB_{2+x}$ | Ti+Zr+Mn | V+Mn | Ni+Mn |

There are 7 types of tetrahedral interstities in C14 Laves phase structure where hydrogen atoms can localize.

However, definite set of interstities preferable for hydrogen occupation exists for Laves phase hydrides with any composition. Thus, most available interstities are 24(l), 12($k_1$), 6($h_1$) and 6($h_2$) with [$A_2B_2$] tetrahedral type [10, 11]. In our case, we can see that the affinity for hydrogen decreases in tetrahedral interstities, because the content of

---

[1] $R_{WP} = [(\Sigma_{wi}[y_i(exp)-y_i(calc)]^2)/(\Sigma_{wi}[y_i(exp)]^2)]^{1/2}$.

[2] $S = [(\Sigma_{wi}[y_i(exp)-y_i(calc)]^2)/(N-P)]^{1/2}$, (N – number of experimental points, P – number of refined parameters.

[3] $R_F = (\Sigma|[I(exp)]^{1/2}-[I(calc)]^{1/2}|)/(\Sigma[I(exp)]^{1/2})$.

hydrogen forming elements (Zr, Ti and V) in 4(f), 2(a) and 6(h) C14 structure positions decreases. We can suppose, that hydrogen capacity and hydride stability will decrease with an increase in the manganese content.

The equilibrium hydrogen pressure increases with an increase in the manganese content in each series (Fig. 1, Tab. 3). $Ti_{0.5}Zr_{0.5}Ni_{1.15}V_{0.5}Mn_{0.55}$ alloy failed to absorb hydrogen at pressure below 50 atm. The slightly increasing hydrogen equilibrium pressure with an increase in the nickel content was observed when the $Ti_{0.5}Zr_{0.5}Ni_{1.0}V_{0.5}Mn_x$ and $Ti_{0.5}Zr_{0.5}Ni_{1.15}V_{0.5}Mn_x$ samples with similar $AB_n$ composition were compared. However, the influence of manganese content on hydrogen equilibrium pressure is stronger than that of nickel content. The addition of nickel slightly decreased the hydrogen capacity as compared with previous series [14, 15]. The total hydrogen capacity of samples is in a range of 1.8 — 1.9 mass. % and apparently independent on the alloy composition. However, the hydrogen capacity of samples at hydrogen pressures of 1-5 atm., corresponding to the charged MH-electrode, is about 1.5 – 1.7 mass. %. The hydrogen weight content decreases as the manganese content increases in these conditions. The comparison of $Ti_{0.5}Zr_{0.5}Ni_{1.0}V_{0.5}Mn_x$ and $Ti_{0.5}Zr_{0.5}Ni_{1.15}V_{0.5}Mn_x$ series shows that an increase in the nickel content results in a hydrogen content decrease.

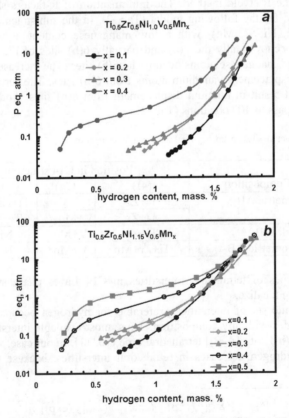

Figure 1. PCT desorption curves for $Ti_{0.5}Zr_{0.5}Ni_{1.0}V_{0.5}Mn_x$ (a) and $Ti_{0.5}Zr_{0.5}Ni_{1.15}V_{0.5}Mn_x$ (b) alloys at room temperature

TABLE 3. Hydrogen absorption and electrochemical properties of $Ti_{0.5}Zr_{0.5}Ni_yV_{0.5}Mn_x$ alloys

| Alloy | n in $AB_n$. | $C_{100}$,[4] mAh/g | Rate capability[5] | $H_2$ mass. % at 40 atm | $P_{eq}$,[6] atm. | $-\Delta H$, kJ/mol |
|---|---|---|---|---|---|---|
| $Ti_{0.5}Zr_{0.5}Ni_{1.0}V_{0.5}Mn_{0.1}$ | $AB_{1.60}$ | 389 | 0.86 | 1.88 | <0.01 | 41.7 |
| $Ti_{0.5}Zr_{0.5}Ni_{1.0}V_{0.5}Mn_{0.2}$ | $AB_{1.70}$ | 323 | 0.91 | 1.86 | 0.04 | 39.0 |
| $Ti_{0.5}Zr_{0.5}Ni_{1.0}V_{0.5}Mn_{0.3}$ | $AB_{1.80}$ | 339 | 0.94 | 1.95 | 0.09 | 36.9 |
| $Ti_{0.5}Zr_{0.5}Ni_{1.0}V_{0.5}Mn_{0.4}$ | $AB_{1.90}$ | 355 | 0.91 | 1.78 | 0.32 | 36.5 |
| $Ti_{0.5}Zr_{0.5}Ni_{1.15}V_{0.5}Mn_{0.1}$ | $AB_{1.75}$ | 350 | 0.73 | 1.80 | 0.05 | 45.7 |
| $Ti_{0.5}Zr_{0.5}Ni_{1.15}V_{0.5}Mn_{0.2}$ | $AB_{1.85}$ | 335 | 0.90 | 1.80 | 0.14 | 41.3 |
| $Ti_{0.5}Zr_{0.5}Ni_{1.15}V_{0.5}Mn_{0.3}$ | $AB_{1.95}$ | 318 | 0.97 | 1.85 | 0.15 | 40.6 |
| $Ti_{0.5}Zr_{0.5}Ni_{1.15}V_{0.5}Mn_{0.4}$ | $AB_{2.05}$ | 327 | 0.98 | 1.83 | 0.64 | 34.5 |
| $Ti_{0.5}Zr_{0.5}Ni_{1.15}V_{0.5}Mn_{0.5}$ | $AB_{2.15}$ | 307 | 0.95 | 1.81 | 1.58 | 32.1 |

From equilibrium pressures at different temperatures (23, 50, 80, 90°C), the change of enthalpy was estimated according to the Vant-Hoff equation. Table 3 shows that the thermodynamic stability of IMC hydrides decreases with the manganese content and n in $AB_n$ relation increases in each series. The thermodynamic stability of IMC hydrides increases with the nickel content increases.

The discharge curves at current density 100 mA/g are shown in Fig. 2 (except $Ti_{0.5}Zr_{0.5}Ni_{1.15}V_{0.5}Mn_{0.55}$ sample). The change of eqilibrium potential corresponds to an increase in hydride desorption equilibrium pressures in PCT curves. The values of discharge capacities at discharge current densities 100 mA/g are in a range of 310 – 390 mAh/g.

The increase of nickel content involves a decrease in the hydride forming elements (Zr, Ti and V) in alloys. So, the hydrogen capacity of $Ti_{0.5}Zr_{0.5}Ni_{1.0-1.15}V_{0.5}Mn_x$ alloys is lower than for $Ti_{0.45}Zr_{0.55}Ni_{0.7-0.85}V_{0.45}Mn_x$ alloys. However, we expected that electrochemical activity would increase with an increase in the nickel content, because in dynamic regime of electrochemical experiment the hydrogen capacity of MH-electrode is only partly realized.

Dependences of discharge capacity on manganese content and B / A ratio are shown in Fig. 3. The shape of the curves for $Ti_{0.5}Zr_{0.5}Ni_{1.0-1.15}V_{0.5}Mn_x$ samples is similar to that of previous $Ti_{0.45}Zr_{0.55}Ni_{0.7-0.85}V_{0.45}Mn_x$ series (Fig. 4). In all cases, we have two maximums of discharge capacity at compositions close to $AB_{1.5}$ and $AB_2$.

---

[4] Discharge capacity at current density 100mA/g
[5] Rate capability is $C_{i=400}/C_{i=100}$
[6] Desorption equilibrium pressure

574

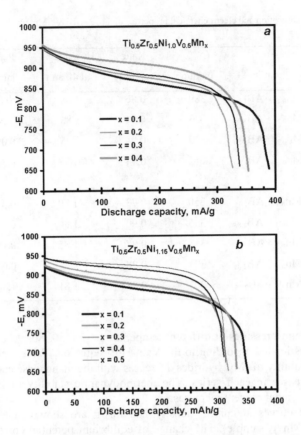

Figure 2. Discharge curves for $Ti_{0.5}Zr_{0.5}Ni_{1.0}V_{0.5}Mn_x$ (a) and $Ti_{0.5}Zr_{0.5}Ni_{1.15}V_{0.5}Mn_x$ (b) alloys at room temperature and current density 100 mA/g

The rate capability of MH-electrodes is characterized by the discharge capacity ratio at current densities of 400 and 100 mA/g. Table 3 shows that the rate capability increases in each series with the manganese content. The rate capability of $Ti_{0.5}Zr_{0.5}Ni_{1.15}V_{0.5}Mn_x$ samples is the highest for the whole $Ti_{0.5}Zr_{0.5}Ni_yV_{0.5}Mn_y$ series. But discharge capacities of $Ti_{0.5}Zr_{0.5}Ni_{1.15}V_{0.5}Mn_x$ samples are lower as compared with $Ti_{0.5}Zr_{0.5}Ni_{1.0}V_{0.5}Mn_x$ samples.

The existing maximums of discharge capacity in alloys series can be explained by the effect of two opposite factors. The first factor decreases the hydrogen capacity with an increase in the manganese content. In contrast, the decrease in thermodynamic stability with an increase in the manganese content leads to an increase in the hydrogen desorption rate and makes electrochemical desorption more complete. This results in the appearance of the maximums, minimums, or "kink" ($Ti_{0.45}Zr_{0.55}Ni_{0.7}V_{0.45}Mn_x$, x = 0.35) of discharge capacities in $Ti_{0.5}Zr_{0.5}Ni_yV_{0.5}Mn_y$ series.

The $Ti_{0.5}Zr_{0.5}Ni_{1.15}V_{0.5}Mn_{0.1}$ sample has a maximum discharge capacity in $Ti_{0.5}Zr_{0.5}Ni_{1.15}V_{0.5}Mn_x$ series at the current density 100 mA/g, but its discharge capacity at 600 mA/g is minimum ($C_{400}/C_{100}$ = 0.73). The decrease of rate capability was observed for $Ti_{0.5}Zr_{0.5}Ni_{1.0}V_{0.5}Mn_{0.1}$ (0.86), $Ti_{0.45}Zr_{0.55}Ni_{0.85}V_{0.45}Mn_{0.25}$ (0.80) and

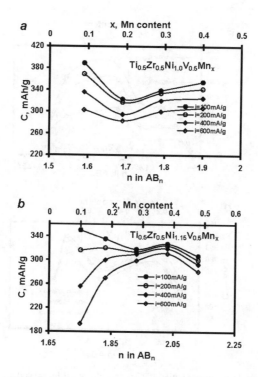

Figure 3. Dependence of discharge capacities of $Ti_{0.5}Zr_{0.5}Ni_{1.0}V_{0.5}Mn_x$ (a) and $Ti_{0.5}Zr_{0.5}Ni_{1.15}V_{0.5}Mn_x$ (b) alloys on manganese content and $AB_n$ relation

$Ti_{0.45}Zr_{0.55}Ni_{0.7}V_{0.45}Mn_{0.35}$ (0.33) alloys, which have maximum ("kink" for $Ti_{0.45}Zr_{0.55}Ni_{0.7}V_{0.45}Mn_{0.35}$) discharge capacity in substoichiometric region of each series (Fig. 3, 4, Tab. 3).

We have $_{shown}$ (Tab. 3), that only two alloys have rather good electrochemical characteristics ($C_{100} > 350mAh/h$; $C_{400}/C_{100} > 0.85$), due to their significant discharge capacity and high rate discharge ability (rate capability). All MH-electrodes were tested in equal experimental conditions (see the experimental part) without studies on the improvement of discharge characteristics. We can suppose, that application of special technique of performance and activating of MH-electrodes can lead to significant improvement of discharge properties for some alloys.

## 4. Conclusions

The structure, hydrogen absorption and electrochemical properties of $Ti_{0.5}Zr_{0.5}Ni_{1.0-1.15}V_{0.5}Mn_x$ alloys were studied in this paper which completes the investigation of $Ti_{0.5}Zr_{0.5}Ni_yV_{0.5}Mn_x$ system with Laves phases. Maximums of hydrogen absorption and electrochemical discharge capacities were found for the $Ti_{0.45}Zr_{0.55}Ni_yV_{0.45}Mn_x$ system in stoichiometric composition of about $AB_{1.5}$ and $AB_2$. The possible explanation to the appearance of the discharge capacity maximum in substoichiometric region was given. The maximum discharge capacity in that region was about 390 mAh/g.

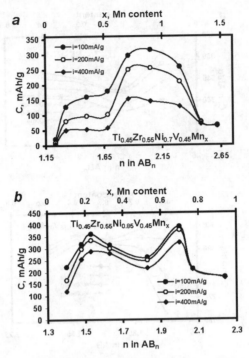

Figure 4. Dependence of discharge capacities of $Ti_{0.45}Zr_{0.55}Ni_{0.7}V_{0.45}Mn_x$ (a) and $Ti_{0.45}Zr_{0.55}Ni_{0.85}V_{0.45}Mn_x$ (b) alloys on manganese content and $AB_n$ relation

## 5. References

1.  Kleperis, J., Wyjcik, G., Czerwinski, A., Skowronski, J., Kopczyk, M. and Beltowska-Brzezinska, M. (2001) Electrochemical behavior of metal hydrides, *J. Solid State Electrochem.* **5**, 229-249.
2.  Petrii, O.A., Vasina, S.Ya. and Korobov, I.I. (1996) The electrochemistry of the hydride forming IMC and alloys, *Usp. Khim.* **65**(3), 195-210.
3.  Kim, D.-M., Jang, K.-J. and Lee, J.-Y. (1999) A review on the development of $AB_2$-type Zr-based Laves phase hydrogen storage alloys for Ni-MH rechargeable batteries in the Korea Advanced Institute of Science and Technology, *J. All. Comp.* **293-295**, 583-592.
4.  Yoshida, M. and Akiba, E. (1995) Hydrogen absorbing-desorbing properties and crystal structure of the Zr-Ti-Ni-Mn-V $AB_2$ Laves phase alloys, *J. All. Comp.* **224**, 121-126.
5.  Lee, H.-H., Lee K.-Y. and Lee, J.-Y. (1997) The hydrogenation characteristics of Ti-Zr-V-Mn-Ni C14 type Laves phase alloys for metal hydride electrodes, *J. All. Comp.* **253-254**, 601-604.
6.  Jung, J.-H., Lee, H.-H., Kim, D.-M., Jang, K.-J. and Lee, J.-Y. (1998) Degradation behavior of Cu-coated Ti-Zr-V-Mn-Ni metal hydride electrodes, *J. All. Comp.* **266**, 266-270.

7. Kim, J.-S., Paik, C.H., Cho, W.I., Cho, B.W., Yun, K.S. and Kim, S.J. (1998) Corrosion behavior of $Zr_{1-x}Ti_xV_{0.6}Ni_{1.2}M_{0.2}$ (M = Ni, Cr, Mn) $AB_2$-type metal hydride alloys in alkaline solution, *J. Power Sources* **75**, 1-8.

8. Song, X., Zhang, Z., Zhang, X., Lei, Y. and Wang, Q. (1999) Effect of Ti substitution on the microstructure and properties of Zr-Mn-V-Ni $AB_2$ type hydride electrode alloys, *J. Mater. Res.* **14**(4), 1279-1285.

9. Kim, D.-M., Lee, H., Cho, K. and Lee, J.-Y. (1999) Effect of Cu powder as an additive material on the inner pressure of a sealed-type Ni-MH rechargeable battery using a Zr-based alloy as an anode, *J. All. Comp.* **282**, 261-267.

10. Westlake, D.G. (1983) Hydrides of intermetallic compounds: a revue of stabilities, stoichiometries and preferred hydrogen sites, *J. Less-Common Met.* **91**, 1-20.

11. Yartys, V.A., Burnasheva, V.V., Semenenko, K.N., Fadeeva, N.V. and Solov'ev, S.P. (1982) Crystal chemistry of $RT_5H(D)_x$, $RT_2H(D)_x$ and $RT_3H(D)_x$ hydrides based on intermetallic compounds of $CaCu_5$, $MgCu_2$, $MgZn_2$ and $PuNi_3$ structure types, *J. Hydrogen Energy* **7**(12), 957-965.

12. Verbetsky, V.N., Petrii, O.A., Vasina, S.Ya. and Bespalov, A.P. (1999) Electrode materials based on hydrogen-sorbing alloys of $AB_2$ composition (A = Ti, Zr; B = V, Ni, Cr), *J. Hydrogen Energy* **24**, 247-249.

13. Izumi, F.A. (1997) Rietveld-refinement program RIETAN-94 for angle-dispersive X-ray and neutron powder diffraction, National Institute for Research in Inorganic Materials, Japan, 35 P.

14. Zotov, T.A., Verbetsky, V.N. and Petrii, O.A. (2002) The Investigation of Hydrogen Absorption and Electrochemical Properties of Alloys Ti-Zr-Ni-V-Mn with Laves Phases in Nonstoichiometric Region, *Hydrogen Materials Science and Chemistry of Metal Hydrides* **82**, 229-234.

15. Petrii, O.A., Mitrokhin, S.V., Verbetsky, V.N. and Zotov, T.A. (2002) Hydrogen absorption and electrochemical properties Ti-Zr-Ni-V-Mn alloys, *Proc. Int. Symp. on Metal Hydrogen Systems*, p. 130.

# EFFECT OF STRESS ON ACCUMULATION OF HYDROGEN AND MICROSTRUCTURE OF SILICON CO-IMPLANTED WITH HYDROGEN AND HELIUM

A. MISIUK[a,*], J. RATAJCZAK[a], A. BARCZ[a], J. BAK-MISIUK[b],
A. SHALIMOV[b], B. SURMA[a,c], A. WNUK[c], J. JAGIELSKI[c],
I.V. ANTONOVA[d]

[a] Institute of Electron Technology,
  Al. Lotnikow 46, PL-02-668, Warsaw, Poland
  E-mail address: misiuk@ite.waw.pl
[b] Institute of Physics, PAS, Al. Lotnikow 32/46, PL-02-668 Warsaw, Poland
[c] Institute of Electronic Materials Technology,
  Wolczynska 133, PL-01-919 Warsaw, Poland
[d] Institute of Semiconductor Physics, RAS,
  Lavrentieva 13, 630090 Novosibirsk, Russia

## Abstract

Effect of stress exerted by hydrostatic pressure (HP) of Ar, up to 1.1 GPa (HT–HP treatment) during annealing at up to 1070 K, on hydrogen storage and microstructure of Czochralski grown silicon (Cz-Si) co–implanted with hydrogen and helium (Cz-Si:H,He, with different sequences of the H- and He-enriched areas) and of reference Cz-Si:H was investigated.

The HT-HP treatment of Cz-Si:(H,He) at $\leq$ 920 K results in a formation of $H_2$ / He filled cavities and bubbles; hydrogen out–diffuses completely at $\geq$ 1070 K. Contrary to the case of Cz-Si:H, hydrogen out-diffusion from Cz-Si:(H,He) is more pronounced under HP than at $10^5$ Pa. Oxygen accumulation in the damaged areas at 720 K is strongly suppressed under HP. Effect of HP on decreased out-diffusion of He atoms, affecting creation of the H–Si bonds and the hydrogen diffusivity, seems to be responsible in part for the observed phenomena.

*Keywords:* Cz-Si, hydrogen, helium, implantation, diffusion, high pressure, treatment

## Nomenclature

| | |
|---|---|
| BE(TO) | transverse optical phonon replica of boron bound exciton |
| $c_o$ | concentration of oxygen in interstitial positions |
| $c_H$ | concentration of hydrogen |
| Cz-Si | Czochralski grown single crystalline silicon |
| D | dose of implanted ions |
| E | energy of implanted ions |
| EHD(TO) | electron – hole droplet |

*T.N. Veziroglu et al. (eds.),*
*Hydrogen Materials Science and Chemistry of Carbon Nanomaterials,* 579-592.
© 2004 *Kluwer Academic Publishers. Printed in the Netherlands.*

| FE(TO) | transverse optical phonon replica of free exciton |
| HP | enhanced (high) hydrostatic pressure |
| HT | enhanced (high) temperature |
| HT-HP | enhanced temperature - enhanced hydrostatic pressure |
| PL | photoluminescence |
| $R_p$ | projected range of implanted ions |
| Si:H | Cz-Si implanted with hydrogen |
| Si:H,He | Cz-Si implanted with hydrogen and helium, with $R_{pH} < R_{pHe}$ |
| Si:He,H | Cz-Si implanted with helium and hydrogen, with $R_{pH} > R_{pHe}$ |
| SIMS | secondary ions mass spectrometry |
| XRRSM | X-Ray reciprocal space map (-ping) |
| XTEM | cross-sectional transmission electron microscopy |

*Greek letters*

$\lambda$          wavelength

## 1. Introduction

Enhanced hydrostatic pressure (HP) at annealing (HT-HP treatment) of hydrogen-implanted oxygen-containing Czochralski silicon (Cz-Si:H, widely applied in microelectronics) exerts pronounced effect on $H_2$ out-diffusion leading to its reduction. Shear stress at the $H_2$-filled bubble / Si matrix boundary is tuned by HP; numerous but smaller crystallographic defects are created in Si under HP [1]. Contrary to the case of annealing at $10^5$ Pa, oxygen accumulation in the disturbed areas of Cz-Si:H has been reported to be markedly reduced at $\leq 920$ K - HP [2, 3].

The HT-HP effect on hydrogen accumulation (storage) and microstructure of Cz-Si co-implanted with H and He, Cz-Si:(H,He), as well as on the reference Cz-Si:H samples is investigated in the present work.

The Cz-Si:(H,He) samples with different sequences of the hydrogen- and helium-enriched areas were prepared. The samples of the first kind (in what follows defined as Si:H,He) were implanted with $H_2^+$ at energy ($E_H$) equal to 135 keV, while the energy of co-implanted $He^+$ ($E_{He}$) was equal to 150 keV. The samples of the second kind (defined as Si:He,H) were prepared by hydrogen implantation at the same energy ($E_H = 135$ keV), while $E_{He}$ equalled to 50 keV.

The reference Cz-Si:H samples (in what follows defined as Si:H) were prepared by hydrogen implantation only, at the same energy (135 keV). It means that the maximum of the hydrogen distribution profile was at the same position (projected range, $R_{pH}$, of H atoms implanted at 135 keV equals to 0.58 μm) in all samples (Si:H, Si:H,He and Si:He,H), while the maximum concentration of implanted $He^+$ was placed deeper ($R_{pHe}$ = 0.88 μm > $R_{pH}$) in the Si:H,He samples while shallower ($R_{pHe}$ = 0.42 μm < $R_{pH}$) in the Si:He,H samples, in respect to $R_{pH}$.

Because the main interest of the present work is focused also on the role of hydrogen in the evolution of the Si:(H,He) microstructure at HT-HP, the below presented results concern mostly the samples treated at 720 K for 1 h and eventually for up to 10 h. At higher temperatures most of hydrogen molecules and of helium atoms are escaping the Si lattice [1-3] so the changes of Si microstructure at > 720 K - HP are related rather to the HP-dependent transformation of defects (microcavities) created just after implantation of H and He and / or at the first steps of the annealing / treatment when hydrogen and helium atoms are still present in Si. Nevertheless, some data concerning evolution of the Cz-Si:(H,He) microstructure at up to 1070 K are also reported.

## 2. Experimental

All investigated samples were prepared by hydrogen ($H_2^+$) and helium ($He^+$) implantation at $\leq 370$ K into the 001 oriented p-type boron doped Cz-Si wafers with the oxygen interstitials concentration equal to $(8 \pm 0.5) \times 10^{17}$ cm$^{-3}$.

The reference Si:H samples ($R_{pH} = 0.58$ μm, determined using the SRIM 2003.10 Program) were prepared by implantation with $H_2^+$ ($E_H = 135$ keV) to a dose ($D_H$) equal to $5 \times 10^{16}$ cm$^{-2}$. Some $H^+$ ions were also present in the implanting $H_2^+$ beam in a small fraction (below 10%); these ions were implanted most deeper ($R_p \approx 1.2$ μm) into the sample.

The Si:H,He samples with helium atoms placed deeper ($R_{pHe} = 0.88$ μm) then the main part of hydrogen ($R_{pH} = 0.58$ μm) were prepared by sequential implantation of $He^+$ into the Si:H samples, at the same dose as that of hydrogen but at higher energy ($D_{He} = 5 \times 10^{16}$ cm$^{-2}$, $E_{He} = 150$ keV); it means that the dose and energy of implanted hydrogen were the same ($D_H = 5 \times 10^{16}$ cm$^{-2}$, $E_H = 135$ keV) as in the case of reference Si:H.

The Si:He,H samples (with a He concentration maximum closer to the sample surface, $R_{pHe} = 0.42$ μm) were also prepared by sequential implantation of helium into Cz-Si. The $E_H$ and $D_H$ values were as that in the case of Si:H while $E_{He} = 50$ keV and $D_{He} = 5 \times 10^{16}$ cm$^{-2}$ (that last the same as in Si:H,He). The projected ranges of helium and hydrogen in Si:He,He were equal to 0.42 μm and 0.58 μm, respectively.

The samples were subjected to hydrostatic HT-HP treatments in argon atmosphere at up to 1070 K and 1.1 GPa, for up to 10 h [4].

The annealed (at $10^5$ Pa) and HT-HP treated samples were investigated by Secondary Ion Mass Spectrometry (SIMS, Cameca IMS 6F, sputtering with Cs$^+$ at 5 keV), cross – sectional Transmission Electron Microscopy (XTEM, JEOL 200CX), X-ray Reciprocal Space Mapping (XRRSM, using high resolution MRD Philips diffractometer) and photoluminescence (PL, excitation with Ar laser, $\lambda = 488$ nm, at 6 K).

## 3. Results and discussion

The reference Si:H samples prepared by hydrogen implantation into Cz-Si and treated / annealed for 1 h indicate dependence of the peak H concentration ($c_H$) on HP, with the highest H content for Si:H treated under the highest pressure, 1.1 GPa (Fig. 1). The hydrogen semi-maximum on the deeper parts (in relation to the sample surface) of the hydrogen concentration profiles (Fig. 1) corresponds to the $H^+$ admixture in the implanting $H_2^+$ beam (<10%), being implanted deeper ($R_p \approx 1.2$ μm) into the sample. Both $c_H$ and the integral hydrogen concentration are higher for the case of Si:H treated under high hydrostatic pressure. It means that out-diffusion of $H_2$ from Si:H is pressure-mediated. Decreased hydrogen out-diffusion under HP has been earlier reported for the Cz-Si:H samples implanted with different hydrogen doses and treated at 720 K-HP for more prolonged time (up to 10 h); qualitative explanation of this effect has been proposed [1, 5].

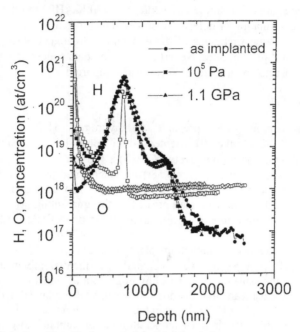

Figure 1. SIMS depth profiles of hydrogen (solid symbols) and of oxygen (open symbols) in reference Si:H samples, as-implanted (●, ○), treated for 1 h at 720 K under $10^5$ Pa (■, □) and 1.1 GPa (▲, Δ).

Oxygen gettering in the implantation-damaged areas at about 720 K (as eventually evidenced by the oxygen concentration peak, open symbols in Fig. 1), clearly seen for the Si:H sample annealed at atmospheric pressure, was suppressed almost completely under the highest pressure applied (1.1 GPa).

The hydrogen and oxygen concentration profiles in the Si:H,He samples (with the $He^+$ projected range, $R_{pHe}$, deeper, in relation to the sample surface, than that of hydrogen), as-implanted and annealed / treated at 720 K in the same way as the Si:H reference, are presented in Fig. 2. Contrary to the case of Si:H, the hydrogen concentration profile in Si:H,He treated under 1.1 GPa is markedly lower and much broader in comparison to that for Si:H,He annealed under $10^5$ Pa.

The treatment of Si:H,He at 720 K for long time (10 h) under low pressure ($10^7$ Pa) has also been reported [6] to result in a detectable loss of hydrogen (caused by hydrogen out-diffusion) while the treatment under the highest applied pressure (1 GPa) resulted in the distinctly lower H concentration, also evidencing more pronounced hydrogen out-diffusion with growing HP.

Contrary to the case of Si:H, gettering of oxygen was not detectable in the Si:H,He samples, both annealed at atmospheric pressure ($10^5$ Pa) and treated under HP for 1 h. More prolonged annealing (for 10 h) at $10^5$ - $10^7$ Pa results in oxygen accumulation also in these samples, while this accumulation was also not observed for the Si:H,He samples treated at 720 K – 1 GPa for 10 h.

The hydrogen and oxygen concentration profiles for the Si:He,H samples (with the He concentration maximum closer to the sample surface than that of hydrogen), as-implanted and annealed / treated at 720 K in the same way as Si:H and Si:H,He, are

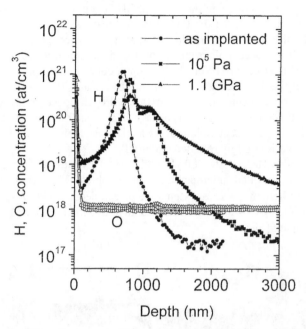

Figure 2. SIMS depth profiles of hydrogen (solid symbols) and of oxygen (open symbols) in Si:H,He samples, as-implanted (●, ○), treated for 1 h at 720 K under $10^5$ Pa (■, □) and 1.1 GPa (▲, Δ).

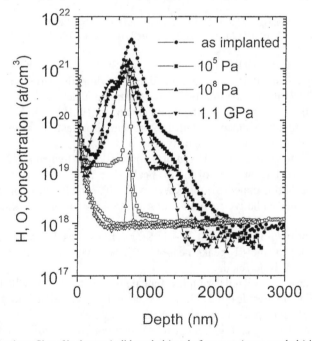

Figure 3. SIMS depth profiles of hydrogen (solid symbols) and of oxygen (open symbols) in Si:He,H samples, as-implanted (●, ○), treated for 1 h at 720 K under $10^5$ Pa (■, □), $10^8$ Pa (▲, Δ) and 1 GPa (▼,∇).

584

presented in Fig. 3. Similarly as in the case of Si:H,He but contrary to the case of Si:H, the hydrogen concentration is of the lowest value for the sample treated under 1.1 GPa (if compared to the samples annealed / treated at lower pressures), also clearly evidencing the HP–induced enhancement of the hydrogen out-diffusion through the sample surface. This effect of HP on the hydrogen concentration profile was, however, lower than that for the Si:H,He samples (compare Figs 2 and 3).

In respect of oxygen accumulation near the hydrogen concentration maximum, the Si:He,H samples treated at 720 K – HP behave in the same manner as Si:H (compare Figs 1 and 3): the oxygen accumulation is clearly detectable for the sample annealed under $10^5$ Pa while it is strongly retarded under HP.

Annealing of Si:H,He at (720-780) K under $10^5$ Pa results in a creation of dense array of $H_2$- and He-filled microcavities and bubbles. The areas implanted with hydrogen and with helium (these last atoms are placed deeper, with $R_{pHe} \approx 880$ nm) can be distinguished (Fig. 4A), while, after the treatment of Si:H,He under 1.1 GPa, they are overlapping (compare [6]).

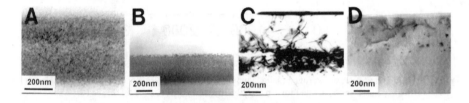

Figure 4. XTEM images of Si:H,He samples annealed / treated for 1 h at 780 K – $10^5$ Pa (A), 920 K – 1.1 GPa (B), for 5 h at 1070 K – $10^7$ Pa (C) and at 1070 K – 1.1 GPa (D).

As it follows from XRRSM's, annealing of Si:H,He at 780 K – $10^5$ Pa for 1 h produces plenty of small crystallographic defects, while the structural perfection of Si:H,He is gradually improving in effect of the treatment under HP.

Contrary to the case of Si:H,He, the treatments of Si:He,H at 720 K – HP for 1h / 10 h result in the gradually worsened (with HP) structural perfection, as evidenced by the enhanced X-ray diffuse scattering intensity (Fig. 5).

PL spectra of the reference Si:H samples annealed / HP treated at 720 K for 1 h are presented in Fig. 6. The intensity of the PL peaks near 1.1 eV (related to the inter-band transitions in the boron doped Si) decreases with HP, which can be interpreted as an evidence of the non-radiative recombination centres present in this sample in a high concentration.

The treatment of Si:H,He at 720 K – HP for 1 h also results in the decreased HP-dependent intensity of the PL peaks near 1.1 eV; however the additional broad PL at about 0.92 – 1.05 eV is detected.

The PL spectra of Si:He,H samples annealed / HP treated at 720 K for 1 h are presented in Fig. 7. Besides the PL peaks near 1.1 eV, related to the inter-band transitions, the PL lines peaking at about 0.96 eV, 1.02 eV and 1.07 eV were detected. While the origin of PL at 0.96 eV and 1.07 eV can not be determined unambiguously, the line at 1.02 eV is presumably related to the presence of Si interstitial-related defects (W center [7]).

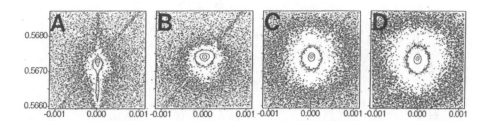

Figure 5. XRRSM's recorded near 004 reciprocal lattice point for Si:He,H samples, as implanted (A) and annealed / treated for 1 h at 720 K under $10^5$ Pa (B), 1.1 GPa (C), and for 10 h at 720 K under 1.1 GPa. Axes are marked in $\lambda/2d$ units ($\lambda$ - wavelength, $d$ - distance between crystallographic planes).

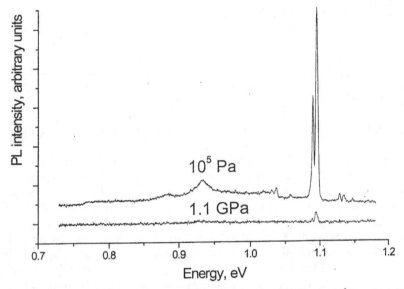

Figure 6. PL spectra of Si:H samples, annealed / treated for 1 h at 720 K under $10^5$ Pa and 1.1 GPa

The HT-HP treatments of Si:H and of Cz-Si:H,He at $\geq$ 780 K for $\geq$ 1 h have been reported to result in a strongly reduced hydrogen content [1,3,4,6]. However, while the hydrogen concentration in Si:H is higher in the case of annealing under HP (if compared with that for the samples annealed under low (atmospheric) pressure), the H content in Si:H,He, treated at 780 K – 1.1 GPa, is lower (in comparison to that for the samples annealed at $10^5$ Pa). It means that hydrogen out-diffusion from Si:H,He is more pronounced at 780 K - HP than that under atmospheric pressure.

Most of hydrogen molecules and of helium atoms are escaping the Si lattice at $\geq$ 780 K [1-3] (especially in the case of more prolonged annealing / treatment) so the changes of Si microstructure at > 720 K - HP are related mostly to the HP-dependent transformation of hydrogen / helium filled defects (microcavities) created just after implantation of H and He and at the first step of the annealing / treatment when hydrogen and helium are still present in Si. Still hydrogen has been reported to be detectable in Si:H and Cz-Si:H,He treated at up to 1070 K ([6]).

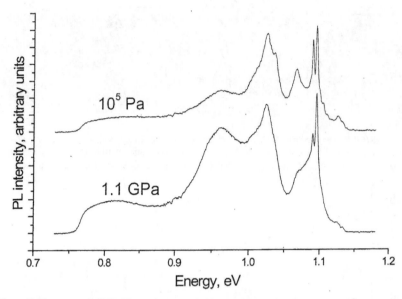

Figure 7. PL spectra of Si:He,H samples, annealed / treated for 1 h at 720 K under $10^5$ Pa and 1.1 GPa

The $H_2$ / He filled bubbles remain to be detectable by TEM in the Si:H,He samples treated for 1 h at 920 K – 1.1 GPa (Fig. 4B); their presence has been reported also for the Si:H samples [3]. As it follows from XRRSM's of the Si:He,H samples annealed / treated at 920 K (Fig. 8), no marked effect of HP on the resulting sample microstructure is detectable by this X-Ray method. The similar conclusion follows from the PL data (Figs 9, 10) concerning the Si:H and Si:He,H samples annealed / treated at 920 K for 1 h. While, in the case of Si:H, the broad PL peak at about 0.81 eV (probably the dislocation-related D1 line [8]) was detected, among other lines, only after the treatment under 1.1 GPa (Fig. 9), this broad D1 peak was detected in Si:He,H annealed also under atmospheric pressure (Fig. 10); its intensity increased with HP and time of the treatment. So one can conclude that enhanced HP at annealing of Si:H and Si:He,H at 920 K promotes the creation of extended defects (dislocations), besides the earlier created TEM-detectable H(He) filled bubbles and cavities.

The Cz-Si:H,He samples treated at 1070 K – 1.1 GPa for 5 h still contain hydrogen (with the peak concentration of about $1 \times 10^{19}$ cm$^{-3}$); the peak H concentration in these samples treated at lower pressure ($10^7$ Pa) is much higher, of about $3 \times 10^{19}$ cm$^{-3}$ . The distinctly lowered H concentration peak (if compared with that for the sample treated under $10^7$ Pa), evidences that the HP-induced decrease of hydrogen out-diffusion takes place at a wide temperature range, at least up to 1070 K.

Numerous dislocations are present in the Si:H,He and Si:He,H samples treated at 1070 K for 5 h (Figs 4 and 11).

Strongly dislocated area near the $H_2^+$ and $He^+$ ion ranges is seen in the Si:H,He sample treated at 1070 K-$10^7$ Pa for 5 h; some dislocations reach the sample surface (Fig. 4C). The treatment at 1070 K under 1.1 GPa produced much less numerous dislocations and other defects in the implantation–disturbed and deeper areas, while small (of about

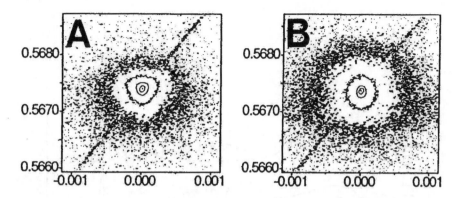

Figure 8. XRRSM's recorded near 004 reciprocal lattice point for Si:He,H samples, annealed / treated for 1 h at 920 K under $10^5$ Pa (A) and 1.1 GPa (B). Axes are marked in $\lambda/2d$ units ($\lambda$ - wavelength, $d$ - distance between crystallographic planes)

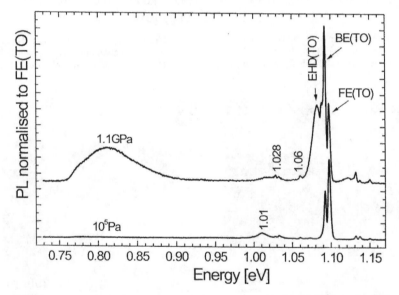

Figure 9. PL spectra of Si:H samples, annealed / treated for 1 h at 920 K under $10^5$ Pa and 1.1 GPa. BE(TO) and FE(TO) lines are related, respectively, to recombination of transverse optical phonon replica of free exciton and of boron bound exciton, and EHD(TO) – to recombination of electron-hole droplet.

20 nm diameter) cavities and dislocation loops are detected near the surface, contrary to effect of the treatment under $10^7$ Pa (compare Figs 4C and D).

The Si:He,H sample treated at 1070 K- $10^7$ Pa for 5 h indicates also the presence of strongly dislocated areas near the $H_2^+$ and $He^+$ ions ranges, with some traces of sample splitting (Fig. 11A). The treatment of Si:H,He at 1070 K under 1.1 GPa produces less dislocations, while cavities and dislocation loops of about 50 nm diameter are present in the near-surface areas (Fig. 11B). Also the XRRSM data of the Si:H,He samples treated

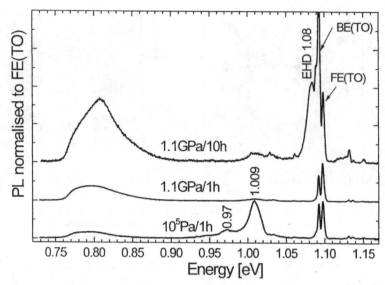

Figure 10. PL spectra of Si:He,H samples, annealed / treated for 1 h and 10 h at 920 K under $10^5$ Pa and 1.1 GPa. BE(TO) and FE(TO) lines are related, respectively, to recombination of transverse optical phonon replica of free exciton and of boron bound exciton, and EHD(TO) – to recombination of electron-hole droplet

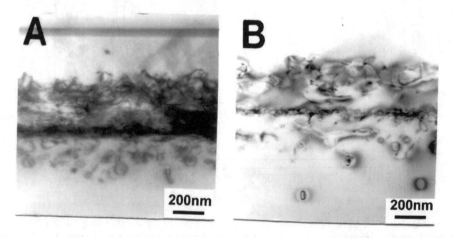

Figure 11. XTEM images of Si:He,H samples annealed / treated for 5 h at 1070 K – $10^7$ Pa (A) and at 1070 K – 1.1 GPa (B)

at 1070 K confirm some improvement of their crystallographic perfection with increasing HP (Fig. 12).

Our investigations concerned two types of Cz-Si:(H,He) prepared by sequential hydrogen and helium implantation at $\leq 370$ K with the same H and He doses ($D = 5 \times 10^{16}$ $cm^{-2}$) and with maximum He concentrations deeper (the Si:H,He samples) and shallower (the Si:He,H samples) in respect to that of hydrogen ($R_{pH} \approx 580$ nm). The H- and He-containing areas were almost overlapping owing to extended tails of the distributions of

589

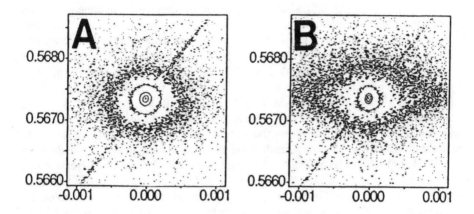

Figure 12. XRRSM's recorded near 004 reciprocal lattice point for Si:He,H samples, treated for 5 h at 1070 K
under $10^5$ Pa (A) and 1.1 GPa (B). Axes are marked in $\lambda/2d$ units
($\lambda$ - wavelength, $d$ - distance between crystallographic planes)

the implanted species. Numerous combinations of the treatment parameters are possible in such experiments. It means that our research (the first one with respect of Cz-Si:He,H, some other results concerning Cz-Si:H,He in [6]) gives only limited information on the title topic. Still the very specific HT – HP induced phenomena were stated, among them:

- the increased hydrogen out-diffusion and so the lowered H concentration in the samples treated at HT – (1 - 1.1) GPa, in comparison to those treated under lower pressure, at $10^5 – 10^7$ Pa. This effect is most pronounced for the Cz-Si:H,He samples with $R_{pHe} = 0.88$ μm $> R_{pH}$;

- HP-mediated suppression of oxygen accumulation, especially at 720 K, and general improvement with HP of the crystallographic perfection of the Cz-Si:H,He and Cz-Si:He,H samples.

The first mentioned effect is non expected and, as it concerns our knowledge, the first of this kind reported. The lowered hydrogen concentration in Cz-Si:He,H and especially in the Cz-Si:H,He samples treated at up to 1070 K (at least) under HP up to 1.1 GPa (at least) is evidently related to the more rapid hydrogen out-diffusion from Cz-Si:(H,He), at 720 – 1070 K and so to the HP–mediated enhancement of the effective hydrogen atom mobility.

No unique explanation of this very specific lowered hydrogen storage in silicon co-implanted with hydrogen and helium can be given at present. It is obvious that the considered phenomenon is related to the presence of He (or to creation of some specific He-induced defects) in high temperature-pressure treated Cz-Si:(H,He), because implantation of hydrogen only leads to much more obvious effect of enhanced hydrogen concentration in the HT-HP treated Si:H.

So the discussed effect is clearly related to an influence of He atoms on the behaviour of hydrogen in silicon. Most probably the presence of He-filled microcavities / bubbles in Si:(H,He) affects creation of the H-Si bonds. After filling the vacancy-like defect created at implantation and at the beginning of annealing / treatment step, some hydrogen molecules in Si:H have been reported [9] to react with dangling or strained Si-Si bonds creating Si-H silanic terminations. Under atmospheric pressure hydrogen out-diffuses appreciably from Si:H at $\geq 720$ K. The lowered hydrogen out-diffusion rate has been reported for Si:(H,He) annealed at $570 - 720$ K [9]. It has been explained as related to the He presence, owing to increased number of the H-Si bonds.

In the case of investigated Cz-Si:(H,He) samples with $R_{pH} < R_{pHe}$ and $R_{pH} > R_{pHe}$, annealed under $10^5$ Pa, both the H and He atoms became to be more mobile at enhanced temperatures and so out-diffuse from the sample. However, in the case of the Si:H,He samples with $R_{pH} < R_{pHe}$ He atoms reach the H-filled areas on their way to the Si surface, mix with $H_2$ in the hydrogen-filled cavities and bubbles and so promote creation of the mentioned H-Si bonds. It means that, under atmospheric pressure, less of hydrogen (as compared to the case of Si:H) is contained in the mentioned cavities and bubbles in the molecular, non bonded form and so the effective effusion of hydrogen from Si bulk would be lower, while the high number of mobile He atoms would reach the originally H-enriched area.

One can assume that the diffusion rate of He is lower under HP than at $10^5$ Pa. So less He atoms would reach at HT - HP the hydrogen-filled defects and so less H-Si bonds will be created at HT – HP (because less He, which promotes the creation of such bonds [9], would reach the hydrogen-filled cavities). So, under HP, more hydrogen will be contained in the originally hydrogen-filled cavities / bubbles in the molecular form. Just non bonded (molecular) hydrogen out-diffuses more effectively through the sample surface at HT-HP.

The proposed model holds also for the case of Si:He,H, with $R_{pH} > R_{pHe}$. At HT hydrogen out-diffuses from the sample, reaching the originally He-filled cavities and bubbles (see Figs 1-3 for the hydrogen profile tails). Probably also some (no so numerous) He atoms would reach the originally H-filled cavities and bubbles near $R_{pH}$ ($> R_{pHe}$), by the in-sample diffusion. So part of hydrogen would become to be immobilized in that cavities and bubbles (in respect of its further diffusion) by creation of the mentioned H-Si bonds. However, because of other geometry of the Si:He,H samples (in comparison to that of Si:H,He), the number of such bonded hydrogen atoms would be lower and the observed effect of increased hydrogen out-diffusion would be less pronounced as expected for Si:H,He (with the reversed sequence of the H- and He-enriched areas).

Confirmation of the above presented qualitative model would demand further research, also to determine the He distribution profiles in the annealed and HT–HP treated Cz-Si:(H,He) samples (not possible by the applied SIMS method).

The effect of enhanced hydrostatic pressure at annealing on suppression of oxygen accumulation / gettering and on creation / annihilation of crystallographic defects is related to decreased diffusivity of oxygen interstitials at HP. The oxygen accumulation in the Cz-Si:H samples at $\geq 870$ K originates from the increased diffusivity of oxygen in the presence of hydrogen [10]. That diffusivity is strongly affected by HP applied during sample annealing, decreasing strongly with HP [3]. Oxygen accumulation in Cz-Si:(H,He) is affected also by the above discussed indirect interaction between hydrogen and helium, difficult for understanding on the present stage of investigations.

Explanation of other observed structural transformations at HT-HP is even more difficult and so further investigation is also needed to solve this problem.

## 4. Conclusions

The presence of (implanted) helium in the hydrogen-implanted silicon exerts pronounced effect on its microstructure and especially on hydrogen out-diffusion at enhanced temperatures-pressures. Contrary to the case of Cz-Si:H, hydrogen out-diffusion from Cz-Si:(H,He) samples is more pronounced under HP than at $10^5$ Pa, at least in the $\leq 1070$ K - $\leq 1.1$ GPa range . This effect can be related to an influence of the presence of He on creation of the Si-H bonds, dependent on the HP-affected out- and in-diffusion of He atoms (the He diffusivity is expected to be lowered under HT-HP).

## 5. Acknowledgements

This work was supported in part by the Polish Committee for Scientific Research (at 2002-04, grant 4T08A 03423). The authors are thankful for Dr G. Gawlik from the Institute of Electronic Materials Technology as well as for Mr T. Koska and M. Prujszczyk for their help during preparation of the samples.

## 6. References

1. Misiuk, A., Bak-Misiuk, J., Barcz, A., Romano-Rodriguez, A., Antonova, I.V., Popov, V.P., Londos, C.A. and Jun, J. (2001) Effect of annealing at argon pressure up to 1.2 GPa on hydrogen-plasma-etched and hydrogen-implanted single-crystalline silicon, *Int. J. Hydrogen Energy* **26**, 483-488.
2. Job, R., Ulyashin, A.G., Fahrner, W.R., Ivanov, A.I. and Palmetshofer, L. (2001) Oxygen and hydrogen accumulation at buried implantation-damage layers in hydrogen - and helium - implanted Czochralski silicon, *Appl. Phys. A* **72**, 325-332.
3. Misiuk, A., Barcz, A., Raineri, V., Ratajczak, J., Bak-Misiuk, J., Antonova, I.V., Wierzchowski, W. and Wieteska, K. (2001) Effect of stress on accumulation of oxygen in silicon implanted with helium and hydrogen, *Physica B* **308-310**, 317-320.
4. Misiuk, A. (2000) High pressure – high temperature treatment to create oxygen nano – clusters and defect in single crystalline silicon, *Mater. Phys. Mech.* **1**, 119-126.
5. Wieteska, K., Wierzchowski, W., Graeff, W., Misiuk, A., Barcz, A., Bryja, L. and Popov, V.P. (2002) X-ray synchrotron studies of nanostructure formation in high temperature-pressure treated silicon implanted with hydrogen, *Acta Phys Polon A* **102**, 239-244.
6. Misiuk, A., Barcz, A., Ratajczak, J. and Bak-Misiuk, J. (2004) Effect of external stress at annealing on microstructure of silicon co-implanted with hydrogen and helium. to be published in: *Solid State Phen* **95-96**, 313-318.
7. Nakamura, M. (2001) Order of the formation reaction and the origin of the photoluminescence W center in silicon crystal, *Jpn J Appl Phys* **40**, L1000-L1002.
8. Misiuk, A., Bak-Misiuk, J., Antonova, I.V., Raineri, V., Romano-Rodriguez, A., Bachrouri, A., Surma, H.B., Ratajczak, J., Katcki, J., Adamczewska, J. and Neustroev, EP. (2001) *Comput. Mater. Sci.* **21**, 515-525.

9.  Cerofolini, G.F., Calzolari, G., Corni, F., Nobili, C., Ottaviani, G. and Tonini, R. (2000) Ultradense gas bubbles in hydrogen- or helium-implanted (or coimplanted) silicon, *Mater. Sci. Eng. B* **71**, 196-202.

10. Zhong, L. and Shimura, F. (1993) Hydrogen enhanced out-diffusion of oxygen in Czochralski silicon, *J. Appl. Phys.* **73**, 707-710.

# HYDROGEN STORAGE MATERIALS AND THEIR MAXIMUM ABILITY ON REVERSIBLE HYDROGEN SORPTION

N.M. VLASOV, A.I. SOLOVEY, I.I. FEDIK, A.S. CHERNIKOV
*The Federal State Unitary Enterprise Scientific Research Institute,*
*Science and Production Association Luch*
*24, Zheleznodorozhnaya str., Podolsk, 142100, Russia*
*E-mail:* chernikov@sialuch.ru

## Abstract

Interaction of some intermetallic compounds with hydrogen is analyzed in the paper. The most applicable in the practice compounds capable reversibly to absorb hydrogen at low temperature and pressure are recommended. Maximum ability of intermetallic compounds on hydrogen absorption is discussed. It is shown that they can not exceed corresponding abilities of hydride forming metals. Reversible storage of hydrogen by holmium-based and magnesium-based compounds is considered.

*Keywords:* hydrogen absorption, intermetallic compounds, storage capacity

## 1. Introduction

Intermetallic compounds are capable to absorb reversibly hydrogen under relatively soft conditions: room temperatures and pressure close to atmospheric. This property opens inviting prospects for intermetallic compounds use for hydrogen storage. The main attention during development of such materials is paid on the following problems:
1) Increase of compounds sorption capacity,
2) Increase of sorption/desorption process rate,
3) Absorption and desorption of hydrogen at room or not so high temperatures and pressure close to atmospheric,
4) Increase of compounds thermal conductivity.

Problems of cost and operational characteristics stability of compounds under consideration are of no small importance. Such strict requirements essentially limit a list of compounds for hydrogen storage.

Interaction of hydrogen with intermetallic compounds is being intensively investigated during last tens years [1 - 3]. Hundreds of various compounds and their modifications were investigated. To find some regularity in experimental data is an intricate problem. First of all, it is because various model representations are used for explanation of experimental results. However, any theoretical model has the limited area of application and, hence, can not adequately describe observable dependencies. Here are sources of various contradictions, when attempting to understand general relationships of hydrogen sorption/desorption process by intermetallic compounds.

The main attention in the presented work is focused on maximum intermetallic

*T.N. Veziroglu et al. (eds.),*
*Hydrogen Materials Science and Chemistry of Carbon Nanomaterials,* 593-601.

compounds ability to accumulate reversibly hydrogen atoms. Hydrogen capacity of compounds is compared with those of metal hydrides. Among set of compounds those were selected, which have reversible sorption of hydrogen at temperatures close to room and pressure close to atmospheric. Changes of sorption/desorption behavior with change of compound components are discussed with attraction of aspects of general physics. Special attention was attended to compounds with rare earth elements (RE) in their composition. These compounds, according to the published data, are most acceptable in practical appliances, as they are capable to absorb significant amount of hydrogen. In addition such magnesium-based compounds are considered, which allow to accumulate extremely large amount of hydrogen (weight capacity), because of small atomic weight of this element. Consideration presented in the work enable to carry out more purposefully search of new materials with high hydrogen absorptivity.

## 2. Maximum metals ability on hydrogen accumulation

Intermetallic compound consists, as a rule, of two types of metals. One of them forms hydrides (Ti, Mg, V, Zr, rare earth metals, etc.), and other serves as good catalysts (Ni, Cr, Fe, Mn, etc.). In addition partial substitution of one elements for others is possible by alloying. It results in essential change of compound absorption capacity, and conditions (temperature and pressure) and kinetics of saturation process (rate of hydrogen sorption) [4 - 6]. Finally, the hydrogen capacity of intermetallic compounds is determined by metals being a part of these compounds. Interaction of metals in a composition of compounds only softens the hydrides formation conditions. It is caused by redistribution of valent electrons of compounds, change of deformability of facet atoms for introduction of hydrogen atoms, decrease of the surface barrier to transition of atoms of hydrogen from chemisorbed state in volume of compound, increase (decrease) of diffusion mobility of hydrogen atoms. At the same time, maximum ability of compounds on reversible accumulation of hydrogen should not exceed possibility of metals in principle. Therefore we shall consider, first of all, hydride-forming metals. Their hydrogen capacity is compared with gaseous (at 100 atm.) and liquid hydrogen. In addition, only data on maximum hydrogen capacity are presented, without hydride phases with intermediate hydrogen content. Corresponding values for various metals are presented in Table 1, where hydride phases with the richest hydrogen content, hydrogen weight percent and number of hydrogen atoms per 1 $cm^3$ of hydride are shown. Such table allows to compare maximum ability of various metals on hydrogen atoms accumulation.

Presented values of hydrogen capacity in separate hydrides are illustrative. Practical use of presented hydrides in the hydrogen storage systems is unlikely. It is caused by the following principal reasons:

1) Molecular hydrogen at pressure of several kilobars interacts only with limited number of metals (alkaline, alkaline-earth metals, rare earth). Ti, V, Zr can be also rank among them;
2) Kinetics of hydrogen absorption and desorption is very sensitive to presence of impurities in gas and metal;
3) Equilibrium of absorption/desorption process is achieved very slow even at high temperatures (300 - 400 °C).

TABLE 1. Hydrogen content in various substances

| Material | H content,% wt. | Number of H atoms in 1 $cm^3$, $N_H \cdot 10^{22}$ |
|---|---|---|
| $H_2$ (Gas) at 100 atm | 100 | 0.49 |
| $H_2$ (Liquid) | 100 | 4.2 |
| $H_2O$ | 11.2 | 6.7 |
| $CaH_2$ | 4.8 | 5.1 |
| $MgH_2$ | 7.6 | 6.7 |
| $TiH_2$ | 4.0 | 9.0 |
| $VH_2$ | 3.77 | 10.3 |
| $ZrH_2$ ($\rho$=5.45) | 2.15 | 7.19 |
| $AlH_3$ | 10.1 | 8.9 |
| $CeH_3$ | 2.1 | 7.0 |
| $H_0H_3$ ($\rho$=7.67) | 1.78 | 8.37 |
| $UH_3$ ($\rho$=16.16) | 1.24 | 12.22 |
| $ErH_3$ ($\rho$=6.69) | 1.93 | 7.92 |
| $LaH_3$ ($\rho$=5.23) | 2.11 | 6.77 |
| $HfH_2$ ($\rho$=11.10) | 1.11 | 7.48 |
| $YH_3$ ($\rho$=3.78) | 3.26 | 7.64 |

Besides, it is necessary to take into account other features of the listed compounds. Trihydrides, as a rule, are unstable, easily ignite in the air and toxic (for example, $UH_3$). Fine dihydrides powder ($MgH_2$, $CaH_2$) also ignites in the air and is toxic.

Interaction of the overwhelming majority of metals with hydrogen is accompanied by lattice rearrangement in the host metal and causes large volume changes. That is why intermetallic compounds are more promising for reversible accumulation of hydrogen.

## 3. Extreme possibilities of intermetallic compounds (IMC) to accumulate hydrogen

Crystal structure of metal undergoes essential changes during interaction with hydrogen. So, for example, BCC $\rightarrow$ FCC transition is observed more often. IMC essentially differs from metal during interaction with hydrogen: it crystal structure practically does not vary, and positional relationship of atoms in a hydride phase and in initial compound is, as a rule, identical. That is why conditions of interaction of IMC with hydrogen are relatively soft. Presence of various types of space, where hydrogen atoms can be located is an essential feature of IMC. As ionic radii of compound components are different, available voids have in some cases favored deformability: a distance between faceting atoms exceeds the sum of ionic radii of these atoms. As a result, hydrogen atoms fill voids more easily. Interaction of hundreds IMC with hydrogen have been investigated up to now. Analysis of available results shows that H - H distances in IMC and in binary hydrides, as a rule, are identical. It means that IMC can not reversibly place more number of hydrogen atoms than metals. However, there is information [7] that 8 hydrogen atoms are in the system $Zr_{0.2}Ho_{0.8}Fe_2H_{8.1}$. Therefore we shall consider further this work in detail. Authors of this work credit such phenomenon with formation of hydrogen-rich phase but they do not present its structure.

Interaction of hydrogen with IMC includes the following basic stages:

1) Physical adsorption;
2) Chemisorption;
3) Diffusion of hydrogen atoms from surface layer into a compound volume with formation of a solid solution ($\alpha$-phase);
4) Ordering of hydrogen atoms due to formation of hydride (ß-phase).

Each of these steps has its activation energy. Naturally, all steps are important for determination of utmost possibilities of IMC on reversible accumulation of hydrogen. However, the stage of the ordered hydrogen atoms arrangement in corresponding voids is determining. They are faceted by metal atoms. In addition, atoms of hydrogen have the greatest propensity to hydride forming metals. Therefore, hydrogen atoms occupy first of all these voids.

There are some general regularities of interaction of hydrogen with IMC based on the results of analysis of numerous experimental factors. They include the following:

1  Location of hydrogen atoms in IMC voids is determined by deformability of facet atoms, i.e. ability to give way to hydrogen atom during its introduction. As the hydrogen atom strives to be placed near to hydride forming atoms of metal, it is necessary to consider deformability of these atoms. Quantitative criterion of the latter is change of distance between atoms of a hydride forming element in initial IMC, as well as in IMC alloyed by substitution elements. As the distance between atoms of a facet increases as against the sum of metal atoms radius $P_{H2}$ (equilibrium pressure) decreases and capacity of compound on hydrogen increases. The numerous experimental facts confirm the specified position.

2.  Initial IMC crystalline structure does not change during interaction with hydrogen, and only increase in volume by 10 - 20 %. It is caused by large amount of suitable voids for hydrogen atoms in a combination with easier deformability of some of them. After formation $\alpha$-phase (nonsaturated solid solution of hydrogen) interaction of hydrogen atoms is minimal, as the distance between them is comparatively large (in atomic scale).

3.  Formation of hydride (ß-phase) occurs, as a rule, in a vicinity of structural irregularities. Then a new phase is growing due to ordered arrangement of hydrogen atoms at enough strong interaction between them. At the same time approximately identical distance between hydrogen atoms, 2 Å, is experimentally observed for all metals and IMC. This distance is just the factor determining utmost possibility, both metals, and IMC, to accumulate hydrogen atoms.

4.  Interaction of IMC with hydrogen proceeds in two different directions: formation of IMC hydride and hydrogenolysis (decomposition of an initial matrix with formation of binary hydride and transition metal). The hydrogenolysis is peculiar to compounds with low heat of formation and proceeds, as a rule, at high temperature and pressure. In this case ability of IMC to absorb hydrogen is comparable to hydride forming metals.

These general provisions we shall illustrate by concrete examples of interaction of IMC with hydrogen paying attention on rare earth metals and magnesium-based compounds. We shall consider, first of all, IMC based on the RE. Most of the RE has an identical electronic structure, and their lattice parameter differs significantly. All these elements well interact with hydrogen with formation of such systems as $RH_2$, $RH_3$,

where R - rare earth element. The basic characteristics of RE are presented in Table 2.
TABLE 2. Properties of rare earth elements

| RE | | Atomic configuration | Crystal structure type | Lattice parameter, Å | Atomic radius, Å | Mass number | Density, g/cm$^3$ |
|---|---|---|---|---|---|---|---|
| Lanthanum | La | [Xe] 5d$^1$6s$^2$ | HCP | 3.75 | 1.87 | 138.91 | 6.17 |
| Yttrium | Y | [Kr] 4d$^1$5s$^2$ | HCP | 3.65 | 1.82 | 88.91 | 4.46 |
| Cerium | Ce | [Xe] 4f$^2$5d$^0$6s$^2$ | FCC | 5.16 | 1.82 | 140.12 | 6.77 |
| Praseodymium | Pr | [Xe] 4f$^3$d$^0$6s$^2$ | HCP | 3.66 | 1.83 | 140.91 | 6.77 |
| Neodymium | Nd | [Xe] 4f$^4$5d$^0$6s$^2$ | HCP | 3.66 | 1.83 | 144.24 | 7.00 |
| Gadolinium | Gd | [Xe] 4f$^7$5d$^1$6s$^2$ | HCP | 3.64 | 1.82 | 157.25 | 8.23 |
| Terbium | Tb | [Xe] 4f$^9$5d$^0$6s$^2$ | HCP | 3.60 | 1.80 | 158.92 | 8.54 |
| Holmium | Ho | [Xe] 4f$^{11}$5d$^0$6s$^2$ | HCP | 3.58 | 1.79 | 164.93 | 9.05 |
| Erbium | Er | [Xe] 4f$^{12}$5d$^0$6s$^2$ | HCP | 3.56 | 1.78 | 167.26 | 9.37 |

Some relationship is traced in RE with increase of atomic number. First of all, the internal electronic shell 4f is filled only. This filling does not change chemical properties of elements determined by external electronic shells. Then the most (9 of 14) of RE have HCP lattice and approximately the same atomic radii with minor change at increase of atomic number. The main role in properties change during substitution of elements in various compounds plays difference of atomic-ionic radii and bond energy. As the external RE electronic configurations are identical, changes of compounds properties are caused exclusively by atomic radii ratio. Atomic radii decrease from Nd (1.83Å) to Lu (1.75 Å), slightly greater value has La (1.87 Å). Therefore, for example, substitution of part of Ho atoms by La, Y, Ce, Nd should facilitate deformability of facet atoms and as a consequence should reduce $P_{H2}$ and increase absorption capacity of compound. RE-based alloys just include misch metals with various compounds of elements. Such combination (with difference of atomic radii) provides better absorption parameters of compounds.

Parameters of absorption are compared with LaNi$_5$. Substitution of a part of La and Ni atoms by other similar elements, as a rule, reduces hydride dissociation pressure down to ~(1.5 - 8) atm. at room temperature. At the same time hydrogen content of 36 considered compounds is in the range of 1.1 to 1.5% wt. It is expected that other combinations of elements hardly will change essentially obtained results. And as to limiting possibility of IMC to accumulate hydrogen, they, most likely, are determined by the minimum distance between hydrogen atoms, that is why, basically, cannot exceed capacity of binary hydrides.

Compounds LaNi$_5$ and LaNi$_5$V$_{0.6}$ have identical sorption capacities (~1.4 % H), but addition of vanadium reduces absorption (10.09 atm.) and desorption (9.78 atm.) pressure. Ni and V have identical external electronic structure ([Ar] 3d84s2 and [Ar] 3d94s2), but different values of atomic radius (1.25 Å and 1.51 Å). Therefore additional introduction of V increases the surface barrier and as a consequence, reduces equilibrium pressure.

It is necessary to avoid hydrogenolysis reaction during interaction of IMC with hydrogen, i.e. disintegration of compound. This reaction essentially depends on energy of initial compound formation, which raises in the order LaNi$_2$ (55.7 kJ/mol ) > LaNi$_3$ (77.2 kJ/mol) > LaNi$_5$ (120 kJ/mol). Besides, the influence of atomic radius size is also traced. So, for example, energy of formation of $CeCo_2$ (55.9 kJ/mol) and $CeFe_2$ (30.5 kJ/mol) changes according to atomic radius Co (1.25 Å), Fe (1.43 Å) and Ce (1,83 Å). Bond energy increases with increase of atomic radius difference of elements in compound, so the reaction of hydrogenolysis proceeds under more severe constraints.

To overcome surface barrier by chemisorbed hydrogen atoms, is one of the major stages of hydride formation. For hydride forming metals it can be achieved by alloying with elements having small atomic radius. In IMC it is achieved naturally, as these compounds consist of elements with different atomic radii. Thus Laves phase $AB_2$ is formed under condition $r_A/r_B \geq 1.225$. At final temperatures part of atoms is in disordered state. In addition, number of elements with small atomic radius are much more than others. Therefore they mainly occupy surface layer of compound. According to Vegard rule, decrease of lattice parameter in the surface layer of compound is greater then in volume. Hydrogen atom chemical potential becomes even and migration of chemisorbed atoms becomes easier. Such behavior is peculiar, as a rule, to all IMC, as they consist of elements with different atomic radius values. It is quite possible that along with decrease of IMC deformability, there is also a reduction of surface barrier to hydrogen atoms migration.

The experimental data show that limiting possibility of IMC on reversible accumulation of hydrogen do not exceed capacity of binary hydrides. However, sometimes works are still to be found, in which this value exceeds it. So, for example, IMC with composition $Zr_{0.2}Ho_{0.8}Fe_2H_{8.1}$ based on the Laves phase was prepared at room temperature and pressure 67 atm. Besides, interesting relationship was observed: change of volume (and naturally, lattice parameter) takes place only for H content on formula unit below 3.3. Further accumulation of hydrogen up to 8.1 H on formula unit occurs without volume change of compound. The physical nature of this phenomenon was not explained. Authors are inclined to attribute increased capacity of this compound to increase of distance between the neighbour voids for hydrogen atom location. In addition, ß-phase formation (with increased content of hydrogen) is limited by diffusion migration of hydrogen atoms. It is possible to make a guess on the nature of so high hydrogen capacity of compound. Hydrogenolysis reaction with formation of system $HoH_3$ is not eliminated, i.e. disintegration of hydride ß-phase. But it is not clear, why lattice parameter is constant. During absorption of hydrogen fractal-type ramified surface is formed that is it has dimension between 2 and 3. It is quite possible that in this case there is a filling of "divided" voids, which were inaccessible for placement of hydrogen atoms because of deformation blocking.

Naturally, reproducibility of results on this compound is extremely important for practical use. In addition, results should prove to be true for other RE, because they have practically identical basic characteristics. All this needs experimental verification. It is necessary also to pay attention on careful preparation of initial samples of investigated compound $Zr_2Ho_{0.8}Fe_2$.

Now we shall consider Ho-based compound.

Application of Ho in various IMCs is well known. Absorption of hydrogen by $HoNi_3$ is accompanied by relative change of volume: $\Delta V/V = 8.6$ % for $HoNi_3H_{1.9}$ and $\Delta V/V = 20.4$ % for $HoNi_3H_{3.6}$, or 2.5 and 4.1 $\mathring{A}^3$ per atom of hydrogen, respectively. For comparison we shall consider $YNi_3$. Relative changes of volume after saturation by hydrogen is $\Delta V/V = 13.2$ % ($YNi_3H_{1.6}$) and $\Delta V/V = 25.0$ % ($YNi_3H_{1.2}$) that are close to each other and to other RE (Gd, Dy) compounds. Substitution of Ni by Co does not change also value of volume change. Among hydride forming compounds having double lattice parameters in comparison with initial RE lattice parameter of IMC practically does not vary. As an example we shall mention systems $HoFe_2H_{3.5}$ (15.88 Å), $ErFe_2H_{3.0}$ (15.73 Å) and $TmFe_2H_{3.0}$ (15.44 Å). These not numerous examples show identity of various RE systems on interaction with hydrogen. However, there is no information on such high hydrogen capacity except systems $Zr_{0.2}Ho_{0.8}Fe_2H_{8.1}$. It is interesting to note that $\Delta V/V$ increases in this system only up to 26% at $Zr_{0.2}Ho_{0.8}Fe_2H_{3.3}$. The same relative volume increase is peculiar (with some deviations) for other RE-based systems with such content of hydrogen atoms. However, higher hydrogen content was not found in all these compounds.

Let's analyze one of the probable reasons of hydrogen content increase. It is necessary to pay attention to careful specimen preparation. It has allowed to prepare practically homogeneous monophase - Laves phase with minimum content of defects. Accumulation of hydrogen atoms is accompanied by destruction of system on fine fragments with characteristic radius (2 - 10) μm and the surface of destruction is ramified, i.e. has fractal character. Such ramified surface contained up to $10^{19}$ voids on unit area for additional location $f$-hydrogen atoms (fractal dimension ~2.5). In a continuous material these voids are adjoined, and the distance between them is <2 Å, i.e. deformation dilatation does not allowed two atoms of hydrogen to be close. After formation of the ramified surface, at least, half of sites (~$0.5 \cdot 10^{19}$) can be occupied by hydrogen atoms. To estimate number of new sites in unit of volume (1 cm³) let the characteristic fragment radius be 5 μm = $5 \cdot 10^{-4}$ cm, its volume is equal to $4 \cdot 1.25 \cdot 10^{-10}$ cm³ = $5 \cdot 10^{-10}$ cm³, then there are $1/(5 \cdot 10^{-10})$ =$0.2 \cdot 10^{20}$ fragments in 1 cm³. One fragment surface area is equal to $4\pi 25 \cdot 10^{-8}$ = $3 \cdot 10^{-6}$ cm², and the total area of all fragments is ~$0.6 \cdot 10^4$ cm². As there are ~$0.5 \cdot 10^{19}$ voids in 1 cm² of fractal surface then it is possible to site in addition ~$3 \cdot 10^{22}$ hydrogen atoms that is close to number of atoms in 1 cm³ of liquid hydrogen ($4.2 \cdot 10^{22}$). We shall note that clean-cut fractal surface is peculiar only to compact Laves monophase, as the surface of defective material destruction does not keep voids in its vicinity for hydrogen atoms introduction. For estimation of number of hydrogen atoms in systems $Zr_{0.2}Ho_{0.8}Fe_2H_6$ (it has 2 atoms of hydrogen on 1 metal atom) and $Zr_{0.2}Ho_{0.8}Fe_2H_{8.1}$ (2.7 atoms of hydrogen on 1 atom of metal) we accept $\rho = 6.42$ and $A_{av} = 87.16$ and obtain: $8.85 \cdot 10^{22}$ and $11.94$ H/cm² respectively. Additional placement of $3 \cdot 10^{22}$ H/cm² due to the ramified surface just gives $11.85 \cdot 10^{22}$ that is close to value 11.94. Thus, one possible explanation of the large hydrogen capacity is careful preparation of a monophase specimen.

## 4. Magnesium-based IMC

Application of Mg in a composition of IMC for hydrogen absorption is caused by its small atomic weight (A = 24.305) and low density ($\rho$ = 1.74 g/cm³). It allows to achieve high (in weight percent) content of hydrogen (7.6 % in $MgH_2$). However the reaction
$Mg + H_2 \rightarrow MgH_2$ proceeds with release of heat (-70 kJ/mol) and is possible in a temperature interval of 250 – 400 °C. In addition, magnesium is hardly activated material, because of rapid oxidation and formation of the MgO and $Mg(OH)_2$ surface film that makes difficult dissociation of hydrogen molecules.

If Mg is a component of IMC then process of hydrogen absorption proceeds in more soft conditions. Among Mg-based compounds are such as $MgNi_2$, $MgFe_2$ and many other, which can absorb hydrogen up to 3% wt. Sometimes Mg is used in a combination with RE. An example of such combination are compounds $LaMg_{12}$ and $La_{0.9}Ca_{0.1}Mg_{12}$, which are capable to accumulate about 6% wt. of hydrogen at temperature 325 °C and pressure in a range of 5 to 30 atm. However, in all these compounds maximum absorption of hydrogen does not exceed capacity of binary hydrides. So, for example, systems $LaH_3$ and $MgH_2$ are formed during absorption of hydrogen by system $LaMg_{12}$, i.e. hydrogenolysis takes place.

Compounds of Mg with other metals are used more often as mechanical mixtures. System Mg - Ni with 5, 10, 25 and 55 % of Ni is the most investigated. Data on hydrogen absorption and desorption by some mechanical Mg - Ni systems are presented in table 3 as an illustration.

TABLE 3. Mg - Ni systems

| System | Mg – 5% wt. Ni [2.1 % (at.) Ni] | Mg – 10% wt. Ni [4.4 % at. Ni] | Mg – 25% wt. Ni [12.1 % at. Ni] |
|---|---|---|---|
| T = 583 K P = 8 atm. Absorption | 2.52% wt. of H | 5.14% wt. of H | 4.72% wt. of H |
| Desorption T = 583 K P = 1,5 atm | 3.41% wt. of H | 4.86% wt. of H | 5.12% wt. of H |

Systems Mg - 5% wt. Ni and Mg - 10% wt. Ni are are the most attractive for hydrogen storage. However these systems also do not allow to achieve hydrogen capacity (reversible) more than binary hydrides. One of the mechanical mixture advantages is high rate of activation process. It is caused by formation of phase boundaries of large extent.

There are many investigations of IMC based on other metals (for example, Mn, V, Fe, Co, Ti etc.). Data on such compounds are included in reference books. Investigations on hundreds of such compounds did not leave attempts to prepare compound with reversible content of hydrogen atoms more than 2 atoms of hydrogen on 1 atom of metal. But reliably established experimental facts counteract it: distance between hydrogen atoms in all metals and IMC should be less than ~2Å, because of deformation interaction of hydrogen atoms. One way "to slide apart" voids for location of hydrogen atoms is to separate them by free surface. It is possible during destruction of monophase compact material at relatively low temperatures. However this way can hardly have any practical application. Therefore using IMC for hydrogen storage (reversible) it is necessary to follow the next key rule: to work with those compounds, which preparation is well mastered and sorption/desorption conditions are acceptable for concrete installation or process.

## 5. Conclusions

1. The maximum possibility of IMC on reversible accumulation of hydrogen atoms is comparable with binary hydrides and determined by deformation interaction of hydrogen atoms. The maximum reversible accumulation of hydrogen atoms in

metals and IMC does not exceed two atoms of hydrogen per atom of metal.

2.  IMC based on RE and Mg are the most useful for practical use. Application of the first is caused by soft conditions of sorption/desorption process, and application of Mg- based IMC by small atomic weight of the last.

3.  Hydride of $Zr_{0.2}Ho_{0.8}Fe_2$ at the content of hydrogen more than 2 atoms (2.7) per one atom of metal are hardly suitable for practical use. Such high hydrogen content is obtained only in a separate case on carefully prepared compact monophase sample. Reproducibility of results is not proved.

4.  The preference should be given to those compounds, which process of preparation is well developed and reproducibility of results of interaction with hydrogen does not cause doubts.

## 6. References

1.  Hydrogen in metals. 1. Basic properties. 2. Applied aspects. Transl. From English, M.: Mir, 1981.
2.  Kolachev, B.A., Shalin, L.E. and Ilyin, A.A. (1995) Alloys - Hydrogen Storage Materials (reference book). M.: Metallurgy.
3.  Grankova, L.P. and Botchkarev, V.M. (1988) Alloys - Hydrogen Storage Materials, VININTI. Series "Physical metallurgy and thermal processing", No. 22, 96-124.
4.  Semenenko, K.N., Burnasheva, V.V. and Verbetsky, V.N. (1983) On the Interaction of Hydrogen with Intermetallic Compounds, Doklady AN USSR **270**(b), 1404-1408.
5.  Semenenko, K.N., Jartys, V.A. and Burnasheva, V.V. (1979) Crystal Lattice "Deformability" and Relation of Intermetallic Compounds to Hydrogen, Doklady AN USSR **245**(5), 1127-1130.
6.  Burnasheva, V.V. and Semenenko, K.N. (1986) Interaction of Hydrogen with Intermetallic Compounds RTn, where R - Rare Earth Metal, T - Fe, Co, Ni; n = 2 + 5, Journal of general chemistry **58**(9), 1931-1935.
7.  Kesavan, T.R., Ramaprabhu, S., Rama Rao, K.V.S. and Das, T.R. (1996) Hydrogen Absorption and Kinetic Studies in $Zr_{0.2}Ho_{0.8}Fe_2$. J. of Alloys and Compounds, **244**, 164-169.

# THE STUDY OF CHANGES OF PHYSICAL - MECHANICAL PROPERTIES OF MATERIALS IN A CONDENSED STATE UNDER HYDROGEN INFLUENCE USING FAULT DETECTION ACOUSTIC MICROSCOPY METHODS

A.V. BUDANOV, A.I. KUSTOV, I.A. MIGEL
*Voronezh State Technology Academy,*
*18 Revolution Avenue, Voronezh, Russia 394000*
*Voronezh Military Air Engineering Institute,*
*27 Starikh Bolshevikov St., Russia 394064*
*E-mail: akutis@yandex.ru*

## Abstract

The paper deals with the perspectives of the application of acoustic microscopy methods for studying of changes of physical - mechanical properties of materials in a condensed state under hydrogen influence. The basic principles of the methods as well as the results of the experiments of studying the structure of materials in a condensed state and its transformation upon changing the composition and types of thermomechanical treatment are given in the article. The high sensitivity to non-heterogeneity and defects upon acoustic visualization and in the regime of determining physic-mechanical properties is demonstrated. Calculation of the dependencies of energy distribution of acoustic waves on the surface of an object gives us the possibility to determine the depths of the location of defects and evaluate the possibilities of their visualization. It is shown that acoustic microscopy methods are perspective for revealing the presence of dissolved hydrogen, local changes of values of physic-mechanical properties of materials. They enable us to reveal microcracks of flocks type with dimensions up to 0.2-0.6 μm, micropores, to calculate local values of velocities of SAW, elasticity module and module of shift, to determine the incubation period of originating of investigating defects.

## 1. Introduction

One of the urgent problems of material science today is that of the presence and behavior of hydrogen in another materials. It is connected with the fact that in process of production metal is saturated with hydrogen, which undergoes various changes upon proceeding to operating temperature regime or upon thermomechanical processing. In this case hydrogen has a decisive effect on the phase composition of steel, its structural components, distribution of admixtures, the presence and value of inner stresses. These characteristics vary in accordance with the content, state and distribution of hydrogen. Due to the presence of a great number of factors, which are responsible for the properties of the given metal materials, the attempts of mathematical modeling of the effect of

*T.N. Veziroglu et al. (eds.),*
*Hydrogen Materials Science and Chemistry of Carbon Nanomaterials,* 603-615.
© 2004 *Kluwer Academic Publishers. Printed in the Netherlands.*

hydrogen upon them are extremely scarce and of little productivity. That's why for getting reliable information and for its correct interpreting it is necessary in the first place to enlarge the arsenal of nondestructive means of experimental measurements. Acoustic microscopy should be referred to perspective methods of studying the hydrogen influence on the structure of materials, with the help of these methods it is possible to obtain without preliminary surface etching the images of structures of different levels to the depth of several of acoustic wave lengths (AW), used for visualization. For the exploitation characteristics of steel not the presence of hydrogen is important but its influence on the structure and physic-mechanical properties. One of the most negative aspects of hydrogen presence is its discharge in the form of flocks - special disturbances of solid mass. As it has been found out by now the main causes of flock formation are the presence of dissolved hydrogen, inner stretched tension and increased fragility of steel in local microvolumes. As the seat of fragile destruction usually accounts for 10 - 100 μm (that practically coincides with the sizes of austenite grain and originating flocks have an appearance of extremely fine inter crystalline microcracks the regime of acoustic visualization of scanning acoustic microscope (SAM) is used for revealing such defects and observing grain structures. However the most significant task is the task of revealing and observing the stages, preceding the formation of flocks, on which the tension in material reaches critical values $\sigma_{cr}$. For a great number of metal materials $\sigma_{cr}$ accounts for $0,02 - 0,04$ E and less, where E is the module of elasticity. In addition the fractures of discontinuities or microcracks are formed which at the given stage can be eliminated by special thermoprocessing. Proposed methods are nondestructive, that fact gives the possibility to use them for understanding the mechanism of changes taking place in metals. They have minimum effects on the investigated samples and enables to visualize defects on the surface and in the volumes of objects, measure the meaning of speed of the surface acoustic waves (SAW) $\upsilon_R$ and modules of elasticity.

Thus, one of the urgent scientific problems of today is the problem of control over physical-chemical and physical-mechanical properties of materials in a condensed state.

The paper deals with the results of experimental researches of materials as well as carbonaceous nanostructural, metals and materials, obtained by coagulation technology with fault detection acoustic microscopy methods.

The essence of these methods lies in layer visualization of subsurface structures of the objects being studied with the following analysis of the images being received as well as in measurement of values of acoustic wave velocities and calculation of the elastic constants of the material. It is discribed in scientific work [1 - 3]. The nondestructive research methods being used are not limited by the nature of materials – the objects may be dielectrics, as well as metals, crystalline and amorphous substances including hydrides and nanostructural materials. Metals, mainly steels, and powder materials (ceramics PZT-type) were chosen as model objects for experimental research. The properties of such solid mass materials are influenced by chemical composition, structure, phase structure, thermal and deformation effects.

## 2. Experimental and theoretical bases of methods

The bases of acoustic microscopy methods used in the paper (visualization and $V(Z)$ - curves ) are stated in the works [2,3,6] rather fully. The essence of both methods lies in the irradiation of the surface samples by the acoustic wave of MHz or GHz range. Reflected signal contains information about the peculiarities of mechanical properties of the studied place and can be used for formation of semitone television image or for receiving the distinctive dependence of the output signal $V$ on a piezotransducer upon the distance $Z$ (irradiator - surface of an object). Curve $V(Z)$ (Fig.1) has its own definite form for each material, upon which one can judge about elastic properties of the reflected surface. Taking into account the main and shifted maximums it is possible to judge about dispersion or absorption of acoustic waves (AW) energy in the investigating sphere of a sample, knowing the oscillation interval $\Delta Z_N$, situated on the right of these maximums it is possible to calculate the velocities of surface acoustic waves (SAW) and other characteristics.

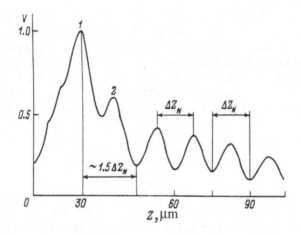

Figure 1. A typical view of V(Z)-curve, obtained by means of SAM (V is an outcome signal of piesotransformer, Z is a lens distance from object, 1 and 2 are the main and displeased maximums, $\Delta Z_N$ is a typical distance for the studied material

Furies transformation method is used for calculating $V(Z)$ –curves of different materials, by means of which a ray is distributed in an angular spectrum of plane waves. This angular spectrum [5] is distributed symmetrically about the normal to an interphasic boundary because its direction corresponds to the ray axis in the experiment.

The output signal on a piezotransformer can be presented as the function of $Z$ and values of signals, excited on the piezotransformer $U_0^+$ $(X,Y)$ and coming after the reflection $U_0^-(X,Y)$,

$$V(Z) = \int\limits_{-\infty}^{\infty} \int U_0^+(X,Y) U_0^-(X,Y) dx dy. \tag{1}$$

Taking into account well-known characteristics discovered by Furie transformation and pupil functions $P(X,Y)$ of an acoustic lens

$$P(X,Y) = cinc\left[\frac{X^2 + Y^2)^{1/2}}{B}\right], \tag{2}$$

where B is the radius of the function of a pupil, it is possible to get an expression, written in the cylindrical coordinates

$$V(Z) = \int_0^\infty r \left[ U_1^+(r) \right]^2 P_1(r) P_2(r) \cdot D(r/F) \cdot exp \left[ -j(K_0 Z/F^2)r^2 \right] dr \quad ,(3)$$

where $r = \sqrt{X^2 + Y^2}$ is the radius-vector of position; $U_0^+(r)$ – is the distribution of pressure of an acoustic field, falling on the opposite focal plane; $P_1(r)$ – is the function of a lens pupil upon transition of the wave from the solid state into liquid; $P_2(r)$ – is the function of an acoustic lens pupil for the waves coming from liquids to a solid body; D $(r/F)$ – is the reflection capacity of a surface; F – is the focus distance of an acoustic lens; $K_0 = K_x F/X = K_y F/Y$.

Thus the output voltage V or a piezotransducer depends not only on the location of the object Z but also on its reflection function. This very factor enables to distinguish the places of the objects under investigation with different acoustic mechanical properties.

Calculation of V(Z) – curves and coefficient of AW transformation on the boundary of a sample and immersion liquid can be simplified [4,6]. As it has been shown in the work [3], the difference in values received with the help of wave and ray models doesn't exceed 0,1 %. Besides the $R_L$ value of reflective ability can be calculated using the expression

$$R_L = \frac{Z_L \cos^2 2\theta_S + Z_S \sin^2 2\theta_S - Z}{Z_L \cos^2 2\theta_S + Z_S \sin^2 2\theta_S + Z}, Z_L = \frac{\rho_S \upsilon_L}{\cos \theta_L}, Z_S = \frac{\rho_S \upsilon_S}{\cos \theta_S}, Z = \frac{\rho_l \upsilon_l}{\cos \theta_i}, \quad (4)$$

where $\theta_i$ – is the angle of incidence of an acoustic wave on the object surface; $\theta_s$ – is the angle of refraction for shear acoustic waves; $\upsilon_l$, $\upsilon_L$, $\upsilon_S$ - are velocities of spreading of an acoustic wave in immersion liquid and in the material of an object in longitudinal and transverse directions, respectively; $\theta_L$ – is the angle of refraction for longitudinal acoustic waves; $Z_L$, $Z_S$, Z - are acoustic impedance of different acoustic waves; $\rho_l$ – is the density of immersion liquid; $\rho_s$ – is the density of a solid object.

If the output signal is considered as a result of interference of various in phase mirror reflected and reirradiated under angle $\theta_R$ waves, the difference of phases changes periodically, and V (Z) – curve has an oscillating character. This approach usually called rayed [4,6] is often used for calculating the values of velocity $\upsilon_R$ SAW. This is caused by the existing dependence of the distance $\Delta Z_N$ between the maximums of V (Z) – curves and the value $\upsilon_R$

$$\Delta Z_N = \frac{\upsilon_L}{2f} \left( 1 - \sqrt{1 - \left( \upsilon_l / \upsilon_R \right)^2} \right)^{-1}, \quad (5)$$

where $f$ - is the operating frequency of an acoustic microscope. Having received V(Z) – curve on the television screen and having made measurement of $\Delta Z_N$ with the accuracy of $\sim 10^{-7}$ m it is possible to determine the values of $\upsilon_R$ with the error of $\sim 0.5$ %. Application of more perfect devices of the surface scanning helps to increase $\upsilon_R$.

Apart from the reflective ability $R_L$ it is possible to obtain the expression for determining the transformation coefficients of longitudinal $T_L$ and shear $T_S$ AW on the surface of the studied sample

$$T_L = \frac{2\,Z_L\,\cos\,2\theta_S}{Z_L\,\cos\,2\theta_S + Z_L\,\sin^2\,2\theta_S + Z}\cdot\frac{\rho_l}{\rho_S},$$

$$T_S = \frac{2\,Z_S\,\cos\,2\theta_S}{Z_L\,\cos\,2\theta_S + Z_L\,\sin^2\,2\theta_S + Z}\cdot\frac{-\rho_l}{\rho_S}$$

(6)

$T_L$ и $T_S$ are expressed through the same physical values as $R_L$, and on their basis it is possible to calculate the power coefficients of transformation and reflection of acoustic waves

$$RL = 100\,R^2{}_L\,;TL = 100\,\frac{Z}{Z_L}\cdot\left(\frac{\rho_S}{\rho_l}\right)^2\cdot T^2{}_L\,;TS = 100\,\frac{Z}{Z_S}\cdot\left(\frac{\rho_S}{\rho_l}\right)^2\cdot T^2{}_S\,.$$

(7)

By machine calculation of energy distribution, falling from the liquid of an acoustic wave, the curves of distribution have been obtained in accordance with the above expressions. These curves make the task of determine the depth of visualization much easier, as well as permitting ability and aberration, and authenticity of obtained results interpretation.

As it is known from the theory of elasticity [7,8], the velocities of longitudinal $\upsilon_L$, shear $\upsilon_S$ and surface $\upsilon_R$ acoustic waves for amorphous slightly deformational materials are determined by the following correlations:

$$\upsilon_L = \sqrt{\frac{E}{\rho_S}}\cdot\sqrt{\frac{1-v}{(1+v)(1-2v)}}\,,$$

$$\upsilon_S = \sqrt{\frac{G}{\rho_S}} = \sqrt{\frac{E}{\rho_S}}\cdot\sqrt{\frac{1}{2(1+v)}}\,,$$

(8)

$$\upsilon_R = \upsilon_S \cdot \frac{0,87 + 1,12\cdot v}{1+v}\,,$$

(9)

where $v$ - is the coefficient of Poisson, $\rho_S$ - is the density of the material of the sample, $E$ - is the module of elasticity, $G$ - is the module of shift of the given material. After the necessary transformation we get:

$$E = \upsilon_R^2 \cdot \frac{2\rho_S(1+v)^3}{(0,87+1,12\cdot v)^2}\qquad G = \upsilon_R\cdot\rho_S\left[\frac{1+v}{0,87+1,12\cdot v}\right]^2.$$

(10)

Thus, if the density $\rho_S$ of the sample and its coefficient of Poisson

$$v = \frac{1-2(\upsilon_S/\upsilon_l)^2}{2-2(\upsilon_S/\upsilon_l)^2}$$

(11)

are known, the values of the elasticity modules $E$ and $G$ can be determined by measuring the velocities of SAW. In case of strong anisotropy material it is necessary to use a cylindrical acoustic lens and corresponding corrections in the calculated formulas. Basing on the stated material we've chosen the most appropriate conditions for investigations (coefficients of quality were estimated for immersion liquids, taking into account at the same time the characteristics of absorbing of a sound and operational frequency, velocities of AW in them, density $\rho_S$, energy coefficients of transformation and reflection on the boundary of the object with the immersion liquid were calculated, aberrations influencing the permitting ability within the sample were determined). Investigation were carried on in two chosen directions, they are visualization and determining the values of $\upsilon_R$ and $E$, $G$ of the materials.

608

## 3. Experimental researches and results

The paper can be introduced by two parts. In the first one on the basis of elaborated acoustic microscopy methods the steel structure is studied (55 XH and 40 XH), also correlation dependence of elastic-mechanical characteristics upon parameters of hydrogen influence. The second part is devoted to the study of temperature regimes influence of synthesis and annealing upon piezoceramics characteristics obtained by coagulating technology. First of all they include porosity, the level of AW absorption process, velocity AW. All of them are closely connected with temperatures of synthesis ($T_{sin.}$) and annealing ($T_{ann.}$) Having obtained some dependencies it is possible to proceed to the study of hydrogen effect both on porosity and other physical-mechanical properties.

The general view of the blocks of control, scanning and visualization of the scanning acoustic microscope (SAM) is given in Fig.2. With the help of SAM we solved the task of discovering and observing of microcracks, and also studied the states of the material before forming of such defects.

Figure 2. The general view of the blocks of scanning and visualization of SAM

Using acoustic visualization method [1,2] it was possible to obtain the images of the samples steel structures at different depths from the surface. The example of one of the images is presented in Fig.3. In it with the magnify in $320^x$ the structure of martencite steel is seen. After deformation and thermal effects the structure is being transformed. According to Hall-Peach low the change of steel strength can be determined by the size of grain.

Acoustic microscopy method of V(Z)-curves [ 3 ] essentially widens possibilities of obtaining information about the material being studied. It enables to obtain typical curves for the given material, which stipulated by its elastic-mechanical constants. The example of V(Z)-curves for steel of martencite class is represented in Fig.4. The level of localization (diameter of the research zone) for SAM can be calculated using the formula:

$$d = 2{,}5 \cdot \lambda_R \cdot \left( \frac{1 + \cos\theta_R}{\cos\theta_R} \right), \tag{12}$$

where $\lambda_R$ –is the length of the surface acoustic waves of reighley type, $\theta_R$ – reighley angle.

609

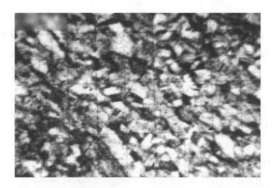

Figure 3. Examples of acoustic images of martensite steel structures, 240ˣ, the frequency is 410 MHz)

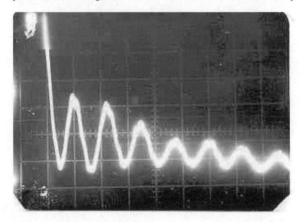

Figure 4. V(Z)-curve of martensite steel (scale 10 μm/div., H₂O, the frequency is 410 MHz)

The worked out acoustic microscopy methods give the possibility to obtain both concrete physical-mechanical properties of materials in their condensed state and their correlation dependencies upon time, thermal and mechanical treatment regimes, chemical and phase compositions. These methods can in full measure be applied to studying metal hydrides and carbon nanomaterials as well. As examples for a number of steels it is possible to consider dependencies of the number of originating defects of the flocken type, change of the velocity values of surface acoustic waves (SAW), characteristics of the absorption process ($\Delta V/V$), sizes of appearing heterogeneities upon concentration of the diffunding hydrogen, time of relaxation, rate of deformation, etc. According to the results of acoustic visualization of the structure of steel of the 55XH type it was obtained the dependence of sizes of flocken type defects, which is represented in Fig.5 upon the content of hydrogen after the tempering of steel.

Change of velocity values of surface acoustic waves (SAW) in the same type of steel according to the relaxation speed is represented in Fig.6. from which it is seen that during a little more than a day period a considerable (for 10 – 12%) drop of speed values is being observed. This phenomenon may be connected to the release of hydrogen, which is especially significant in the first 20 – 40 hours for the given sorts of steel.

610

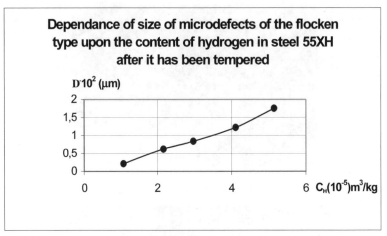

Figure 5

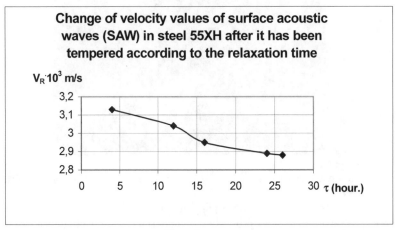

Figure 6

The hydrogen release in steels leading to forming nonhomogeneous structures of the flocken type, intensifies absorption of acoustic waves in model objects. This process is manifested in the form of change of characteristics of fading of surface acoustic waves (SAW) and transformation of the height of the main maximum of the V(Z)- curves. Fig.7 shows the dependence of the relative height $\Delta V/V$ upon the hydrogen concentration in steel after the latter has been tempered.

Similar dependencies were also obtained for steel 40 XH. One should pay his attention to the measurement result of $v_R$ in this steel in accordance with the depth of sounding (Fig.8.) and the time of its sustaining in hydrogen (Fig.9.).

One of the important properties of the substances obtained by powder technology is porosity. It is this property that is connected with the composition of raw material with the technology of synthesis, with the temperature of annealing. These parameters determine other physic-chemical and mechanical properties of substances to a great extent. That's why for nondestructive control of that kind of substances it was been

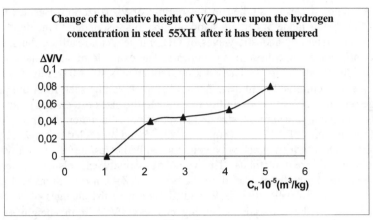

Figure 7

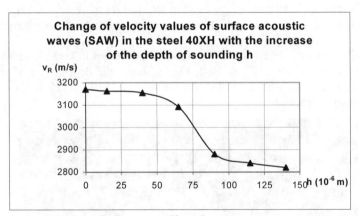

Figure 8

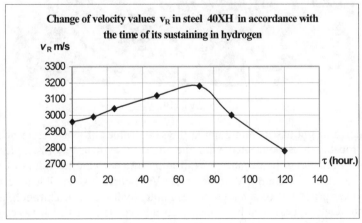

Figure 9

advised to use a scanning acoustic microscope (SAM) of a reflection type. The structure of powder material can be studied in the regime of visualization. Besides the surface layers of the objects made of ceramics and plumbago are sounded to the depth of several hundred micrometers by acoustic waves of GHz range. As a result on the screen of the visualization device an acoustic image appears. The example of such image for ceramic PZT-23 is illustrated on Fig.10. The working frequency is 407 MHz, immersion liquid is water, scale of picture is 24 μm/div., Z = - 12 μm. It can be seen from the picture that the image obtained gives the possibility to evaluate the sizes of ceramic grains and what is no less significant, the amount, size and shape of pores in materials. The calculation of porosity for ceramics of PZT type (brand 22, 23, 35 and other) led to the values from 8 – 12% to 25 – 28%, which practically coincide with the results of other investigations carried out by different methods. The value of porosity can also be determined by acoustic microscopy methods, based on the use of V(Z)- curves. It makes it possible to calculate the values of velocities SAW and to determine the changes of coefficients of absorption AW in the material concerning changes $\Delta V/V$ of the height of the main maximum V(Z)-curves. Upon diffusion into hydrogen material its elastic-mechanical characteristics change, which manifests itself in the changes of velocity SAW $v_R$ and the level of $\Delta V/V$. Nevertheless during the first stage of the research it is necessary to master the control technology of obtaining of optimum density (porosity). For this purpose the experimentally revealed correlation of a number of properties of ceramic materials with the temperature of annealing $T_{an}$ is used. Fig.11 shows the dependence of porosity $\theta$ upon $T_{an}$ for ceramic PZT.St.-2. Within the interval of 1510 – 1520 K the maximum value of porosity is reached.

Figure 10

The relative height has its minimum at the same values of temperature of annealing (Fig.12). Probably it is connected with the increase of density of samples and the decrease of losses along the distance AW. Taking into account the values of velocities SAW and the relative height $\Delta V/V$ of maximum V(Z)-curve it is possible to calculate the value and other properties of samples, for example, hydrogen concentration, due to the revealed interrelation, described by a linear dependence. The relation between dielectric permeability $\varepsilon$ and density $(1-\theta)$ for PZT.St.-2 is shown on Fig.13. Similar dependencies

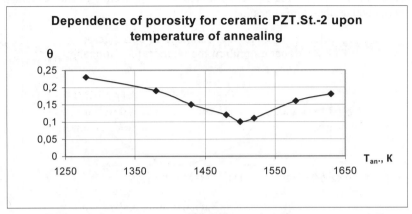

Figure 11

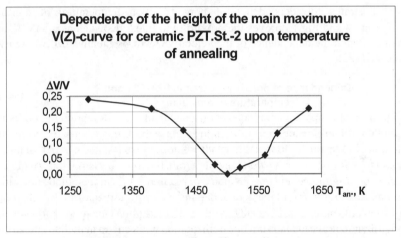

Figure 12

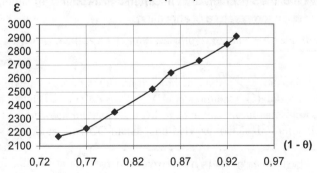

Figure 13

were received for PZT-22. Fig.14 demonstrated the dependence of the relative height V(Z)-curve upon temperature annealing.

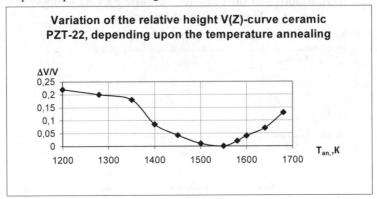

Figure 14

Fig.15 illustrates the dependence of porosity upon temperature of annealing. Comparing Fig.14 and Fig.15 proves the proximity to linear dependence of $\Delta V/V$ and $\theta$. Fig.16 demonstrated the dependence of the velocity SAW upon temperature of annealing of ceramic PZT-22.

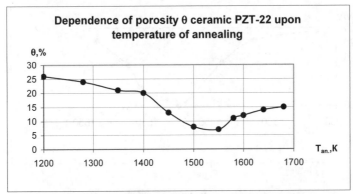

Figure 15

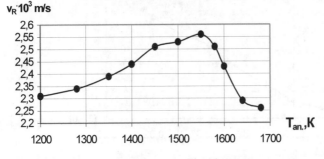

Fig. 16

Plumbago *PROG* was considered as model for hydrogen nanostructural materials. Its main properties, determined with the help of acoustic microscopy methods, are gathered in Table 1.

TABLE 1

| Properties | | Value | Unit of measuring |
|---|---|---|---|
| Velocity SAW, | $\upsilon_R$ | 2,010 | $10^3$ m/s |
| Module of elasticity, | E | 12,77 | $10^9$ Pa |
| Interval of the V(Z)-curve, | $\Delta Z_N$ | 5,6 | $10^{-6}$ m |
| Coefficient of Poisson, | $\nu$ | 0,14 | |
| Porosity, $\theta$ | | 14 | % |
| Density, $\rho_s$ | | 1,124 | $10^3$ kg/m$^3$ |

## 4. Conclusions

Thus we can sum up fault detection acoustic microscopy methods are considered to be effective in studying hydrogen influence upon the structure and properties of steels. Such characteristics of samples as size and quantity of defects, the depth of penetration, relaxation time and the level of AW absorption process can be determined with the help of values of velocities SAW and relative height $\Delta V/V$. Besides worked out nondestructive methods make it possible to calculate the value of porosity and the density of piezoceramic materials, to get the dependence of these values upon temperatures of synthesis and annealing.

The results obtained have proved the possibility of studying the structure and properties of materials, received by coagulating technology by applying acoustic microscope defectoscopy methods, as well as the effect of hydrogen upon them.

## 5. References

1. Kustov, A.I. (1994) The study of the structure and physical-mechanical properties of solids by acoustic microscopy methods, *Proceedings of VII Russian conference "Adsorption materials"*, 89-97.
2. Quate, C.F., Atalar, A. and Wickramasinghe, H.K. (1979) Acoustic microscopy with mechanical scanning, *Proceedings IEEE* **67**(8), 5-31.
3. Wilson, R.G. and Weglein, R.D. (1994) Acoustic microscopy of materials and surface layers, *Appl. Phys.* **55**(9), 3261-3275.
4. Kustov, A.I. and Migel, I.A. (1996) Investigation of physical-mechanical characteristics of glasses with acoustic waves, *Physics and chemistry of glass* **22**(3), 245-247.
5. Atalar, A. (1978) An angular spectrum approach to contrast in reflection acoustic microscopy, *J. Appl. Phys.* **49**(10), 5130-5139.
6. Parmon, W. and Bertoni, H.L. (1979) Ray interpretation of the material signatures in the acoustic microscope, *Elec. Lett.* **15**(21), 684-686.
7. Collings, E.V. (1988) Physical metallurgy of titanium melts, *M.: World*, 224 p.
8. Vikhtorov, N.A. (1981) Sound surface waves in solids, *M.: Science*, 287 p.

# DIFFUSION OF HYDROGEN IN AMORPHOUS, HIGH DEFORMED AND NANOCRYSTALLINE ALLOYS

N.I. TIMOFEYEV, V.K. RUDENKO
*OAS "Ekaterinburg Non-Ferrous Metals Working Plant",*
*Ekaterinburg,Russia.*


V.V. KONDRATYEV, A.V. GAPONTSEV, A.N.VOLOSHINSKII
Institute of MetalPhysics, Ural Branch of Russian
*Academy of Sciences, 620219,18S.Kovalevskaya Str.,*
*Ekaterinburg, Russia*
*Fax:7-343-3745244*
*E-mail:avg@imp.uran.ru*

## Abstract

The approach, which was proposed earlier for calculation of the chemical diffusion coefficient of hydrogen in alloys and "two-phase" systems, was used to analyze the diffusion behavior of disordered materials taking into account a non-zero partial molar volume of interstitial impurity atoms. Explicit expressions for the hydrogen diffusion coefficient in amorphous metals and alloys and "two-phase" systems have been obtained. The temperature and concentration dependences of permeability of amorphous metals have been determined. The effect of the correlation in the distribution of static fluctuations of the free volume on diffusion properties of amorphous materials has been analyzed. Some limiting cases have been discussed. Types of the diffusion behavior of strongly deformed and nanocrystalline metals have been classified. It was concluded that the transfer of hydrogen under conditions of quasi-volume diffusion is insensitive to the structure of these materials.

*Keywords:* hydrogen diffusion, theory, amorphous alloys, nanocrystalline and high deformed metals

## 1. Introduction

The study of the hydrogen transfer in disordered metal systems has been stimulated for technological and scientific reasons. Technical applications are related to the traditional interest attached to metal-hydrogen systems in atomic power engineering and powder metallurgy, as well as a very wide spectrum of problems encountered in the so-called "hydrogen" power engineering.

The basic research uses the fact that hydrogen, being the lightest interstitial impurity in alloys and bearing only one electron, represents the simplest object for the study of the interaction between the solvent and the diffusant. A small activation energy ($\sim 0.1$

*T.N. Veziroglu et al. (eds.),*
*Hydrogen Materials Science and Chemistry of Carbon Nanomaterials,* 617-634.
© 2004 *Kluwer Academic Publishers. Printed in the Netherlands.*

eV in pure metals [1]) and a relatively weak effect of hydrogen diffusion on the structural relaxation make it possible to view this diffusion as an effective method for analysis of the microstructure of amorphous and nanocrystalline alloys. However, the interpretation of experimental data requires a theoretical development of *analytical* models of transfer in disordered materials, which would include specific features of their structure at the atomic and mesoscopic levels.

Reviews concerned with diffusion in amorphous metal alloys [2–4] and nanocrystalline materials [5–7] emphasize the following most significant features of the hydrogen behavior in these systems: a strong concentration dependence and a deviation from the Arrhenius temperature dependence of the hydrogen diffusion coefficient. An example is a classical study by Mütschele and Kirchheim [8], who examined diffusion in nanocrystalline palladium. They showed that hydrogen solubility increased two orders of magnitude as compared to diffusion in a single crystal (palladium hydride was not formed). In addition, the diffusion coefficient in $n$-Pd strongly depended on the hydrogen concentration, whereas it was virtually independent of the hydrogen concentration in the single-phase region in mono-Pd. The diffusion coefficient first was smaller in nanocrystalline palladium than in the monocrystalline sample, but then increased quickly with the hydrogen concentration, and finally surpassed its counterpart in single crystals at some hydrogen concentration. A similar behavior was also observed in Pd-, Fe-, La- and Ti-based amorphous systems [3].

These features are explained qualitatively by the diversity of sites having different energies for location of hydrogen atoms in disordered materials. Lowest-energy sites are occupied first and diffusion proceeds in places with a low activation energy. As a result, the hydrogen diffusion coefficient increases with growing hydrogen concentration.

The available experimental data suggest that a short-range crystalline order exists in amorphous metals and alloys [3]. Thus, an adequate representation in the description of the mass transfer in these systems is that hydrogen atoms move on a potential relief, which is definitely preset in space and whose minimums correspond to quasi-interstices (pores) analogous to interstices in systems with a crystalline order. A characteristic distinction of the potential relief in amorphous materials is a continuous, rather than discrete, energy spectrum of hydrogen atoms in equilibrium positions. Maximums of the energy spectrum correspond to species configurations of (quasi-) interstices, which are determined by the type of the short-range order in a system [9].

Data concerning nanocrystalline metals and alloys [2, 10] suggest that from the viewpoint of hydrogen diffusion these systems have two clearly defined structural components: the grain volume, which is the region with a long-range crystalline order, and grain boundaries, which are characterized by the absence of the long-range order and a nearly Gaussian distribution of site energies of hydrogen atoms.

Modern theoretical approaches to the description of hydrogen diffusion in disordered media, which are based on microscopic models, employ the representation of the impurity atom migration over a regular crystal lattice. In this case, the key problem is the shape of the potential relief, on which hydrogen atoms move in disordered metals and alloys, and its relation to characteristics of materials found in other (non-diffusion) experiments.

The task of the theory is to use model representations about the structure of a disordered medium and, hence, derive an expression of the form

$$\langle R^2(t) \rangle = 6D_{rw}t, \tag{1}$$

relating a mean-square displacement of an impurity atom to its traveling time[1] or obtain Fick's empirical law relating the flow of impurity atoms to the flow concentration gradient

$$J = -D_{ch}\nabla c. \tag{2}$$

In the first case the diffusion coefficient is determined from random walks (the Einstein diffusion coefficient) and in the second case $D_{ch}$ is equated to the hydrogen chemical diffusion coefficient.

The currently available analytical methods for description of equilibrium diffusion of hydrogen in disordered metals and alloys employ purely model representations of the shape of the potential relief of hydrogen atoms (independent Gaussian distributions of site and saddle-point energies, uniform distributions, etc.). Usually the relationship of this potential relief to other characteristics of materials is not discussed.

An example of studies of this kind is the paper [12], whose authors assumed the independence of the Gaussian distributions of energies at saddle points and equilibrium positions at a small concentration of the impurity and low temperatures and obtained the following expression for the diffusion coefficient in an amorphous metal:

$$D_{rw} = D_0 \exp\left\{-\beta\left(\varepsilon_c - \varepsilon_s\right) + \beta^2 \left(f\sigma_c^2 - \sigma_s^2\right)/2\right\}, \tag{3}$$

where $\beta = 1/kT$; $f$ is the smoothly varying function $\beta\sigma_c$, which depends on the crystal lattice type; $\varepsilon$ and $\sigma$ denote average energies and FWHM of the distributions; values with the indices "$c$" and "$s$" refer to saddle points and sites respectively; $D_0$ is the parameter, whose determination is beyond the scope of the theory.

The paper [13] demonstrates another approach, which uses the analogy between diffusion and electric conductivities of disordered media. The researchers [13] considered random walks of hydrogen atoms in a two-component system comprising regularly arranged cubic crystallites imbedded in the matrix, which could be either crystalline or energy-disordered one with Gaussian distributions of proton energies over interstices and saddle points. The material was characterized by different diffusion coefficients in the matrix and grains and possessed different parameters determining the exchange of atoms between the subsystems. As a result, they constructed rough "network" models representing different combinations of series and parallel connections of "matrix" and "crystallite" conductors, which were suitable for determination of the effective hydrogen diffusion coefficient. The comparison of the proposed models with calculation made by the Monte Carlo method showed their usefulness at an arbitrary concentration of hydrogen atoms over a wide interval of the theory parameters. Results of computer-aided calculation proved to be little sensitive to the shape, dimensions and space distribution of crystallites in the matrix.

The said study is an exception, because the method of random walks usually does not provide an expression for the diffusion coefficient, which would be useful over a sufficiently wide interval of concentrations. Therefore, it is very difficult to analyze the concentration dependence of the Einstein hydrogen diffusion coefficient.

The most common approach to the description of chemical diffusion in amorphous metal-hydrogen systems [14] assumes the invariability of the energy at the saddle point

---

[1]The relationship (1) may be violated in an arbitrary disordered medium. In this case, diffusion can be described more adequately as random walks on a fractal material having a fractional dimensionality (see the review [11]).

and the presence of the *only* Gaussian distribution of site energies, whose FWHM can be determined from an additional experiment or related to the FWHM of the radial distribution function of metal atoms. In the case of degeneration of the lattice gas of hydrogen atoms when their concentration is small, the following expression was derived for the effective diffusion coefficient:

$$D_{ch} = D_{hyp} \sqrt{2\pi}\sigma\beta \exp\left\{\left[\mathrm{erf}^{-1}|2C - 1|\right]^2 - \sqrt{2}\beta\sigma\mathrm{erf}^{-1}|2C - 1|\right\}. \quad (4)$$

Here $D_{hyp}$ is the hydrogen diffusion coefficient in a "hypothetical" metal, which is characterized by an average site energy; $C = c/M$ is the dimensionless concentration of hydrogen atoms ($M$ being the number of quasi-interstices per unit volume); $\mathrm{erf}^{-1}$ is the inverse error function.

The disregard of the energy distribution over saddle points is incorrect, since it contradicts results of the computer simulation [15]. Moreover, this approach actually neglects the effect of the short-range order, which is present in amorphous systems, on the potential relief of hydrogen atoms.

The most frequently cited paper in descriptions of hydrogen diffusion in nanocrystalline metals is the one due to Mütschele and Kirchheim [8], who generalized the theory to these systems. It was assumed that a nanomaterial represented a mixture of two structural components, namely an amorphous-like intercrystalline component (a boundary) and a crystalline component. The diffusion coefficient was calculated taking the distribution of energies as the sum of the Gaussian curve and the delta-like peak corresponding to crystallites. In the case of palladium, the growth of the diffusion coefficient with increasing hydrogen concentration was limited by the mutual repulsion of hydrogen atoms. Since Kirchheim's approach [14] disregarded the interaction, the diffusion coefficient could not be written analytically. Moreover, today there are no theories based on microscopic considerations, which would allow deriving and analyzing *an expression* for the chemical diffusion coefficient of hydrogen in nanocrystalline systems.

Thus, it is necessary to construct models of hydrogen diffusion, which would take into account specific features of the structural state of amorphous and nanocrystalline metals and alloys in connection with theories developed for their genetic ancestors, that is, crystalline systems. Then it will be possible to reduce the number of parameters, which are determined additionally, and take into account the short-range order, which plays a significant role in formation of the potential relief of hydrogen atoms in amorphous and nanocrystalline materials.

This approach was proposed in our earlier studies [16–19] for calculation of the chemical diffusion coefficient of hydrogen in amorphous metals and alloys with a short-range order on the basis of the FCC lattice. It was generalized [20] to "two-phase" systems similar to metals in the nanocrystalline state and strongly deformed materials.

Specific features of amorphous systems were taken into account by introducing the notion of local dilatation, which shows static fluctuations of the pore volume. The latter were related to a nonuniform distribution of the free volume present in amorphous materials. Pores in amorphous alloys were characterized by two parameters: the index of the species environment of hydrogen by metal atoms in a quasi-interstice and local dilatation showing the change of the pore volume as compared to an interstice in a regular crystal.

It was shown that all information about the structure of materials, which enters our general expression for the diffusion coefficient, resides in the pair distribution function of dilatations of (adjacent) quasi-interstices. In the case of amorphous systems, it can be approximated by a generalized Gaussian distribution or related to the multiparticle correlation function of displacements of the matrix atoms. Momentums of the latter are measured by the methods of neutron and X-ray diffraction analyses. It was found that the distribution function of pore dilatations, which was obtained by this means, contained information about both radial and angular scatters of atoms in the amorphous matrix. The use of the generalized Gaussian distribution of pore dilatations made it possible to take into account explicitly correlation effects of the spatial distribution of adjacent quasi-interstices.

Systems having two structural components were considered. They were exemplified by nanocrystalline or partially crystalline alloys, which were assumed to contain two types of quasi-interstices: those with zero dilatation and "distorted" pores differing from interstices in a regular crystal. The former belonged to the volume, while the latter resided in grain boundaries in nanocrystalline metals. The pair function of the pore distribution had two parameters, that is, the volume fraction of the "distorted" component and the degree of correlation of pores. This model approach allowed determining the hydrogen diffusion coefficient for nanocrystalline metals and alloys in terms of the same formalism as the one used for amorphous systems.

One of the suppositions forming the basis of the papers [16–20] was the hypothesis that the volume of a pore remained unchanged when a hydrogen atom occupied the pore (the quasi-interstice). Consequently, the site energy of hydrogen atoms in pores exhibited a quadratic dependence on their dilatations. This corresponded to a zero partial molar volume of hydrogen. However, the partial molar volume of hydrogen atoms is positive in most metal systems [1]. The approach advanced in this paper took into account this circumstance.

## 2. The Model

Let pores in an amorphous alloy be characterized by two parameters: the index $i$ of the species environment of a hydrogen atom by atoms of the alloy components and a local dilatation $\alpha$. The species distribution of the pores depends on the type of the short-range order in the alloy. The index $i$ acquires seven values (according to the number of metal atoms of different species comprising an interstice) in binary alloys with an FCC lattice and five values in alloys with a BCC lattice where hydrogen occupies tetrahedral interstices.

The activation energy of the transition $(i, \alpha) \rightarrow (j, \alpha')$ is given as the difference between the energy at the saddle point and the site energy of hydrogen in an interstice:

$$V_{ij}(\alpha, \alpha') = E(\alpha, \alpha') - \varepsilon_i(\alpha), \tag{5}$$

where the saddle-point energy $E(\alpha, \alpha')$ is independent of the pore species [16].

At the saddle-point energy a hydrogen atom pushes aside the nearest metal atoms and, therefore, compression causes the increase in the energy, while extension leads to the decrease in the energy. The dilatations $\alpha$ and $\alpha'$ should enter the expression for $E(\alpha, \alpha')$ symmetrically, because the initial pair of interstices is symmetric relative to the saddle point. Therefore,

$$E(\alpha, \alpha') = E_0 - \omega(\alpha + \alpha'), \tag{6}$$

where $\omega > 0$ is the theory parameter and $E_0$ value refers to an undistorted (crystalline) state of the alloy.

The expression for the site energy of hydrogen atoms, which applies to both zero and nonzero partial molar volumes of hydrogen atoms, has the form

$$\varepsilon_i(\alpha) = \varepsilon_{i0} + \gamma_i \alpha + \delta_i \alpha^2, \tag{7}$$

where $\varepsilon_{i0}$ refers to the undistorted state of the alloy and can be easily determined for binary and ternary systems in terms of the pair approximation from the hydrogen-metal interaction [21]; $\delta_i > 0$ (the equilibrium position, i.e. the point corresponding to the potential minimum) and $\gamma_i$ is the theory parameter. It can be shown that $\gamma_i$ values are proportional (with the "minus" sign) to the partial molar volume of hydrogen [22].

In this case, if the correlation between the species and dilatation distributions of pores is neglected, the chemical diffusion coefficient of hydrogen can be written as [18]

$$D_{ch} = \frac{2Al}{\overline{\tau}\overline{z}M} \cdot \frac{\sum\limits_{i,j} m_i P_{ij} \left\langle\!\left\langle q_i(\alpha)\, q_j(\alpha')\exp\left[\beta\omega(\alpha + \alpha')\right]\right\rangle\!\right\rangle \exp\left[\beta(\mu - E_0)\right]}{\sum\limits_i m_i \left\langle v_i(\alpha)\, q_i(\alpha)\right\rangle}, \tag{8}$$

where $A$ is the number of "communicating" pairs of quasi-interstices per unit area in the cross-section of the sample; $l$ is the length of an elementary hop of hydrogen atoms; $\overline{\tau}^{-1}$ is the average frequency of thermal vibrations of hydrogen atoms in equilibrium positions; $\overline{z}$ is the first coordination number in the sublattice of interstices; $m_i$ is the fraction of quasi-interstices of the species $i$; $P_{ij}$ is the conditional probability that a pore having the configuration $j$ is found near a pore of the type $i$ (these values depend on the type of the short-range order in alloys; they were determined for binary and ternary systems with FCC and BCC lattices in [21]); $q_i(\alpha) = 1 - v_i(\alpha)$; $v_i(\alpha)$ is the occupancy of pores; $\mu$ is the chemical potential; the double broken brackets $\langle\!\langle \ldots \rangle\!\rangle$ mean averaging with the pair distribution function of pores $W(\alpha, \alpha')$; the single broken brackets $\langle \ldots \rangle$ denote averaging with the ordinary function $W(\alpha)$. The Fermi-Dirac distribution holds for the occupancy of pores:

$$v_i(\alpha) = \frac{1}{1 + \exp\left\{\beta\left[\varepsilon_i(\alpha) - \mu\right]\right\}}, \tag{9}$$

while the equation relating the chemical potential to the concentration of hydrogen atoms has the form

$$C = \sum_i m_i \langle v_i(\alpha)\rangle. \tag{10}$$

All information about the structure of a material, which enters the equation (8), is contained in the pair function of the pore dilatation distribution. In the case of amorphous alloys, this function may be connected with the multiparticle correlation function of static displacements of matrix atoms or approximated by the generalized Gaussian distribution [18]: [18]

$$W(\alpha, \alpha') = \frac{\sqrt{P^2 - Q^2}}{\pi} \exp\left\{-P\left[(\alpha)^2 + (\alpha')^2\right] - 2Q\alpha\alpha'\right\}. \tag{11}$$

$P > 0$, $Q$ may have any sign. If $Q < 0$, pores having like dilatations will be "attracted" one to another and form the structure of diffusion paths. When $Q > 0$, pores having unlike dilatations are mixed. In this case, an amorphous material will be homogeneous on the mesoscopic scale. The ordinary distribution function is obtained from $W(\alpha, \alpha')$ using a standard method,

$$W(\alpha) = \int W(\alpha, \alpha')d\alpha', \tag{12}$$

and has the form of the Gaussian distribution with FWHM $p = \frac{P}{2(P^2 - Q^2)}$.

In the case of "two-phase" systems, which contain, from the viewpoint of distortion of the crystalline structure, two types of pores, namely undistorted interstices with dilatation $\alpha = 0$ and extended quasi-interstices with $\alpha = \alpha_0$, the pair distribution function of pores can be written in a general form as [20]

$$
\begin{aligned}
W(\alpha, \alpha') &= \left\{(1 - V)^2 - F\right\}\delta(\alpha)\,\delta(\alpha') + \\
&\quad + \{V(1 - V) + F\}\left[\delta(\alpha)\,\delta(\alpha' - \alpha_0) + \delta(\alpha - \alpha_0)\,\delta(\alpha')\right] + \\
&\quad + \left\{V^2 - F\right\}\delta(\alpha - \alpha_0)\,\delta(\alpha' - \alpha_0),
\end{aligned}
\tag{13}
$$

where $\delta(x)$ is the Dirac delta function; $V$ is the volume fraction of the "distorted" component; the parameter $F$ characterizes the deviation of the order in the system from a random mixture of pores ($F > 0$ means the tendency of distorted and undistorted quasi-interstices to mixing; $F < 0$ corresponds to the "separation" of the system to regions with pores of the same type). The interval of possible $F$ values is limited, because the coefficients in (13) should be positively determined:

$$
F \in
\begin{cases}
\left[-V(1 - V);\; V^2\right], & V \leqslant \frac{1}{2} \\
\left[-V(1 - V);\; (1 - V)^2\right], & V \geqslant \frac{1}{2}
\end{cases}
\in \left[-\frac{1}{4};\, \frac{1}{4}\right].
\tag{14}
$$

The closed system of equations (5)-(10) in combination with the information about the species distribution of quasi-interstices and parameters of the dilatation spread of pores determines the temperature and concentration dependences of the chemical diffusion coefficient of hydrogen in disordered metal systems.

## 3. Results and Discussion

### 3.1. Amorphous metals and alloys

Using the Gaussian distribution (11), it is possible to obtain an analytical expression for the hydrogen diffusion coefficient in alloys in the form of the series

$$
D_{\text{ch}} = \frac{2Al}{\tau z M} \cdot \frac{\exp(-\beta E_0) \sum\limits_{i,j} m_i P_{ij} \Psi_{ij}}{t \sum\limits_{i} m_i \sum\limits_{n=1}^{\infty} \frac{(-1)^{n+1} n}{(y_i t)^n \sqrt{1 + n\Delta_i}} \exp\left[\frac{n^2 \Gamma_i^2}{1 + n\Delta_i}\right]},
\tag{15}
$$

$$
\begin{aligned}
\Psi_{ij} &= \sum_{n,k=0}^{\infty} \frac{(-1)^{n+k} \exp\left\{\frac{1}{2}\Omega^2(1 - \vartheta)^2 \frac{2/(1+\vartheta) + \Delta_i k + \Delta_j n}{(1 + k\Delta_i)(1 + n\Delta_j) - kn\Delta_i\Delta_j\vartheta^2}\right\}}{(y_i t)^k (y_j t)^n \sqrt{(1 + k\Delta_i)(1 + n\Delta_j) - kn\Delta_i\Delta_j\vartheta^2}} \times \\
&\quad \times \exp\left\{\frac{k^2\Gamma_i^2 + n^2\Gamma_j^2 - 2\vartheta\Gamma_i\Gamma_j - 2\Omega(1 - \vartheta)(k\Gamma_i + n\Gamma_j)}{2\left[(1 + k\Delta_i)(1 + n\Delta_j) - kn\Delta_i\Delta_j\vartheta^2\right]}\right\} \times
\end{aligned}
$$

$$\times \exp\left\{kn\left(1 - \vartheta^2\right)\frac{\Gamma_i^2\Delta_j k + \Gamma_j^2\Delta_i n - 2\Omega\left(\Gamma_i\Delta_j + \Gamma_j\Delta_i\right)}{2\left[(1 + k\Delta_i)\left(1 + n\Delta_j\right) - kn\Delta_i\Delta_j\vartheta^2\right]}\right\},$$

where $t = \exp(-\beta\mu)$ and $y_i = \exp(\beta\varepsilon_{i0})$ are related to the concentration of diffusing atoms by the expression

$$C = \sum_i m_i \sum_{n=1}^{\infty} \frac{(-1)^{n+1}}{(y_i t)^n \sqrt{1 + n\Delta_i}} \exp\left[\frac{n^2\Gamma_i^2}{1 + n\Delta_i}\right]. \tag{16}$$

Here $\Delta_i = 2\beta\delta_i p$, $\Omega = \beta\omega\sqrt{p}$, $\Gamma_i = \beta\gamma_i\sqrt{p}$, and $\vartheta = Q/P$ is the degree of correlation between pores. The series (15) and (16) converge on condition

$$C < \frac{1}{1 + \exp\left\{\frac{\Delta_i^2 + \Gamma_i^2}{2\Delta_i}\right\}}, \tag{17}$$

which sets the upper bound on the interval of hydrogen concentrations considered in the theory or on $\Delta_i$ and $\Gamma_i$ values at a preset hydrogen concentration.

When concentrations $C$ are small, which is equivalent to $t \gg 1$, the variable $t$ can be excluded from the system of equations (15) and (16) and one can obtain an explicit expression for the hydrogen diffusion coefficient in amorphous alloys to within second-order terms from the hydrogen concentration. Thus, we have

$$D_{ch} = \frac{2Al\exp\left[-\beta E_0 + \Omega^2\left(1 - \vartheta\right)\right]}{\tau z M} \cdot \frac{\Phi_0}{H_1} \times$$

$$\left\{1 + 2\frac{C}{H_1}\left(\frac{H_2}{H_1} - \frac{\Phi_1}{\Phi_0}\right) + \left(\frac{C}{H_1}\right)^2 \times\right.$$

$$\left.\times\left[6\frac{H_2}{H_1}\left(\frac{H_2}{H_1} - \frac{\Phi_1}{\Phi_0}\right) + 2\frac{\Phi_2}{\Phi_0} - 3\frac{H_3}{H_1} + \frac{\kappa}{\Phi_0}\right]\right\}, \tag{18}$$

where

$$H_x = \sum_i \frac{m_i}{y_i^x \sqrt{1 + x\Delta_i}} \exp\left\{\frac{x^2\Gamma_i^2}{1 + x\Delta_i}\right\},$$

$$\Phi_x = \sum_{i,j} \frac{m_i P_{ij}}{y_i^x \sqrt{1 + x\Delta_i}} \exp\left\{\frac{x^2\Gamma_i^2 - 2x\Omega(1-\vartheta)\Gamma_i - x\Omega^2(1-\vartheta)^2 - 2\vartheta\Gamma_i\Gamma_j}{2(1 + x\Delta_i)}\right\}$$

and

$$\kappa = \sum_{i,j} \frac{m_i P_{ij}}{y_i y_j \sqrt{(1 + \Delta_i)\left(1 + \Delta_j\right) - \vartheta^2\Delta_i\Delta_j}} \exp\left\{\frac{1}{2}\left[(1 + \Delta_i) \times\right.\right.$$

$$\times\left(\left(1 + \Delta_j\right) - \vartheta^2\Delta_i\Delta_j\right)\right]^{-1}\left[\Gamma_i^2 + \Gamma_j^2 - 2\vartheta\Gamma_i\Gamma_j + \left(1 - \vartheta^2\right)\times\right.$$

$$\times\left(\Gamma_i^2\Delta_j + \Gamma_j^2\Delta_i - 2\Omega\left(\Gamma_i\Delta_j + \Gamma_j\Delta_i\right)\right) - \Omega^2\left(1 - \vartheta\right)^2 \times$$

$$\left.\left.\times\left(\Delta_i + \Delta_j + 2\left(1 + \vartheta\right)\Delta_i\Delta_j\right)\right]\right\}.$$

Let us analyze the temperature dependence of the coefficient $D_{ch}$. To this end, we shall introduce the average site energy of a hydrogen atom in a pore,

$$\bar{\varepsilon} = \sum_i m_i \langle \varepsilon_i (\alpha) \rangle = \sum_i m_i \left( \varepsilon_{i0} + \gamma_i \langle \alpha \rangle + \delta_i \langle \alpha^2 \rangle \right) \tag{19}$$

and the average energy of the hydrogen diffusion activation

$$\mathfrak{Q} = \sum_{i,j} m_i P_{ij} \langle\langle V_{ij}(\alpha, \alpha') \rangle\rangle = E_0 - \bar{\varepsilon} - 2\omega \langle \alpha \rangle. \tag{20}$$

Let us consider the zero-concentration-order term in the expression (18), which mainly determines the temperature dependence of the diffusion coefficient:

$$D_0 = \frac{2Al}{\tau z M} \cdot \frac{\sum\limits_{i,j} m_i P_{ij} \exp\left\{-\beta^2 \gamma_i \gamma_j p\vartheta\right\} \exp\left\{-\beta\mathfrak{Q} + \beta^2 \omega^2 p (1 - \vartheta)\right\}}{\sum\limits_i m_i \frac{\exp\{-\beta(\varepsilon_{i0} - \bar{\varepsilon})\}}{\sqrt{1 + 2\beta\delta_i p}} \exp\left\{\frac{\beta^2 \gamma_i^2 p}{1 + 2\beta\delta_i p}\right\}}. \tag{21}$$

This expression represents a generalization of the formula (3) for alloys. If $\vartheta = 0$, the contributions from the disorder of saddle-point energies and site energies to the non-Arrhenius term are opposite in sign. The researchers [12] hold to the opinion that this circumstance is responsible for very small deviations from the Arrhenius law, which are observed in diffusion experiments with amorphous alloys. However, if the correlation of pores is taken into account, disorders of site and saddle-point energies will not compensate one the other.

The effective activation energy in amorphous metals is

$$Q_{eff} = -\frac{d \ln D_{ch}}{d\beta} \approx \mathfrak{Q} - 2\beta p \left(\omega^2 - \frac{\gamma^2}{1 + 2\beta\delta p}\right) + 2\beta p \vartheta \left(\omega^2 + \gamma^2\right) - 2\beta^2 p^2 \frac{\gamma^2 \delta}{1 + 2\beta\delta p}. \tag{22}$$

The rate and the character of the energy variation depend on the disorder of energies of hydrogen atoms at saddle points and interstices in the system. Remarkably, in the absence of a contribution, which is linear in dilatation, to the site energy, the effective activation energy exhibits a linear dependence on the inverse temperature. The $Q_{eff}$ value also depends on the degree of correlation of pores: it decreases as pores with like dilatations tend to group themselves ($\vartheta < 0$) and increases as pores with unlike dilatations ($\vartheta > 0$) are mixed.

In amorphous alloys, when a large "configuration" spread of energies of hydrogen atoms at equilibrium positions takes place, the temperature dependence of the diffusion coefficient is more complicated. In this case, one will observe diverse temperature regimes of diffusion with different effective activation energies, which will be determined by specialities of the energy structure of amorphous alloys.

The concentration dependence of $D_{ch}$ can be most easily analyzed for metals. In this case, one should take $i = j = 1$ and $m_1 = P_{11} = 1$ in the corresponding parts of the expression (18). The transition to the amorphous state causes the increase in the diffusion coefficient and appearance of an additional concentration dependence. This means formation of the first and second positive concentration derivatives of the diffusion coefficient near $C = 0$ when the dilatation variance $p$ is nonzero. It can be easily shown that in alloys, when the hydrogen diffusion coefficient depends on the hydrogen concentration and

$p = 0$, the tendency will be the same: permeability of amorphous materials is enhanced with growing concentration of the impurity.

A detailed analysis of the expression (18) shows that the energy disorder always leads to a positive concentration dependence of the diffusion coefficient. This effect is strongly controlled by the spread of energies of hydrogen atoms at saddle points: the concentration dependence grows as the variance is enhanced. Moreover, the effect is more pronounced when the increase in the pore volume is followed by the rise of the site energy in an interstice. This corresponds to a negative partial molar volume of hydrogen. The concentration dependence of the diffusion coefficient increases owing to the correlation in the distribution of pore volumes. Negative $\vartheta$ values cause a greater increase in the concentration dependence of the hydrogen diffusion coefficient than positive $\vartheta$ values do. In other words, the effect of "attraction" of pores with like dilatations leads to a considerable increase in the diffusion coefficient as the hydrogen concentration grows. A chaotic distribution of pores with unlike dilatations decreases diffusion permeability of amorphous materials and reduces its dependence on the hydrogen concentration. It should be noted that the pore-correlation-dependent contributions of the disorder of site and saddle-point energies to the concentration dependence of the diffusion coefficient are different: the former and the latter increase at positive and negative values of the parameter $\vartheta$ respectively.

**Comparison with Kirchheim's theory** Let us now compare the aforementioned approach with Kirchheim's theory [14], which yields the expression (4) for the diffusion coefficient. To this end, re-write (8) in terms of averaged characteristics:

$$D_{ch} = D_{hyp}\beta \exp\left[\beta(\mu - \bar{\varepsilon})\right] \times$$

$$\times \sum_{i,j} m_i P_{ij} \langle\langle [1 - v_i(\alpha)] \left[1 - v_j(\alpha')\right] \exp\left[\beta\omega(\alpha + \alpha' - 2\langle\alpha\rangle)\right]\rangle\rangle \frac{\partial\mu}{\partial C}, \qquad (23)$$

where $D_{hyp} = \frac{2Al}{\tau z M}\exp(-\beta\Omega)$ has the same meaning as in the expression (4); $v$, $\mu$ are found from the equations (9) and (10). In terms of Kirchheim's theory (invariability of the saddle-point energy $\omega = 0$; disregard of the correlation between species of quasi-interstices $P_{ij} = m_j$ and pore dilatation $W(\alpha, \alpha') = W(\alpha)W(\alpha')$), the expression (23) is rearranged to

$$D_{ch} = D_{hyp}\beta \exp\{\beta(\mu - \bar{\varepsilon})\}(1 - C)^2 \frac{\partial\mu}{\partial C}. \qquad (24)$$

The formula (24) without the factor $(1 - C)^2$ was obtained [14] for amorphous metals with a small concentration of hydrogen. On the assumption that the lattice gas of hydrogen atoms is degenerate and the distribution of potential energies in quasi-interstices is given by a Gaussian curve, it transforms to (4) provided $C \ll 1$.

Let us analyze our problem in similar conditions. For this purpose, we shall take $\delta_i = 0$. Let the energy distribution of hydrogen atoms $W(\alpha)$ be relatively narrow. Assume that the pore species are numbered according to the increasing energy $\varepsilon_{i0}$. In this case, if the lattice gas of hydrogen atoms is degenerate, one may think that all pores are filled till some species $s$ and are empty after this species and the chemical potential falls within the region of the dilatation distribution of energies $\varepsilon_s(\alpha)$. Consequently, the equation (10) can be resolved with respect to the chemical potential:

$$\mu = \varepsilon_{s0} - \sqrt{2p}\,|\gamma_s|\,\text{erfc}^{-1}\left(2\,\frac{C - \sum\limits_{i=0}^{s-1} m_i}{m_s}\right), \tag{25}$$

where the additional inverse error function is determined as $x = \text{erfc}^{-1}(z) \sim z = \frac{2}{\sqrt{\pi}} \times$ $\times \int_x^{\infty} \exp\left(-t^2\right) dt$. When the hydrogen concentration is small and the sum in (23) is dominated by terms corresponding to unoccupied species of quasi-interstices, the diffusion coefficient has the form

$$D_{ch} = D_s \beta \sqrt{2\pi p}\,|\gamma_s|\,m_s^{-1} \exp\left\{-\beta\sqrt{2p}\,|\gamma_s|\,\text{erfc}^{-1}\left[2\left(C - \sum_{i=0}^{s-1} m_i\right)/m_s\right] + \right.$$
$$\left. + \beta^2 \omega^2 p\,(1-\vartheta) + \left(\text{erfc}^{-1}\left[2\left(C - \sum_{i=0}^{s-1} m_i\right)/m_s\right]\right)^2\right\}, \tag{26}$$

where $D_s = \frac{2Al}{\tau z M}\exp\left[-\beta(E_0 - \varepsilon_{s0})\right]$. This expression generalizes the relationship (4) for alloys with the correlation of pores and variability of the energy at the saddle point. The product $\sqrt{p}\,|\gamma_s|$. plays the role of FWHM of the energy distribution in pores. Notice that the magnitude of the effects associated with the correlation of pores is controlled by the spread of energies at saddle points. In a general case, one can obtain an explicit expression for the diffusion coefficient if the dilatation correlation of pores is neglected:

$$D_{ch} = D_s \beta \exp\left\{\beta(\mu - \varepsilon_{s0}) + \beta^2 \omega^2 p\right\} \frac{\partial \mu}{\partial C}\left\{\sum_{i,j=s+1}^{h} m_i P_{ij} + \right.$$
$$+ 2\sum_{i=s+1}^{h} m_i P_{is}\,\text{erfc}\left[\frac{\varepsilon_{s0}-\mu}{\sqrt{2p}|\gamma_s|} + \text{sign}(\gamma_s)\sqrt{\frac{p}{2}}\beta\omega\right] + $$
$$\left. + m_s P_{ss}\,\text{erfc}^2\left[\frac{\varepsilon_{s0}-\mu}{\sqrt{2p}|\gamma_s|} + \text{sign}(\gamma_s)\sqrt{\frac{p}{2}}\beta\omega\right]\right\}. \tag{27}$$

Here the chemical potential $\mu$ is determined by the equation (25).

As the hydrogen concentration increases changing the system (species) of pores, by which diffusion hops of hydrogen atoms are realized, the trends of the diffusion coefficient and the effective activation energy should exhibit considerable singularities, that is, abrupt changes of these characteristics, every time the condition

$$C = \sum_{i=0}^{s} m_s \tag{28}$$

is fulfilled. The magnitude of the abrupt change of the diffusion coefficient depends on the temperature and the difference of the effective activation energies of diffusion via quasi-interstices of the species $s$ and $s+1$. This difference is determined in turn by the parameters $\varepsilon_{s0} - \varepsilon_{(s+1)0}$ and $|\gamma_s| - |\gamma_{s+1}|$. The slope of the curve $\ln D_{ch}(C)$ far from the points determined by the condition (28) depends on the volume fractions of interstices of different species $m_s$.

## 3.2. "Two-phase" systems

Let us consider diffusion in a "two-phase" medium when the pair distribution function of dilatations is given by the formula (13) and only one species of pores is available. Using these approximations, the diffusion coefficient can be easily determined from the system of equations (8)-(10):

$$
D_{ch} = D_0 \left[ \phi + (1 - C) \frac{\partial \phi}{\partial C} \right] \left[ \frac{1 - V}{1 - C + \phi} + \frac{V e^{\beta[(\omega + \gamma)\alpha_0 + \delta\alpha_0^2]}}{(1 - C) e^{\beta(\gamma\alpha_0 + \delta\alpha_0^2)} + \phi} \right]^2 \times
$$
$$
\times \left\{ 1 - F \left[ \frac{\phi + (1 - C) e^{\beta(\gamma\alpha_0 + \delta\alpha_0^2)} - (1 - C + \phi) e^{\beta[(\omega + \gamma)\alpha_0 + \delta\alpha_0^2]}}{(1 - V)\left(\phi + (1 - C) e^{\beta(\gamma\alpha_0 + \delta\alpha_0^2)}\right) + V (1 - C + \phi) e^{\beta[(\omega + \gamma)\alpha_0 + \delta\alpha_0^2]}} \right]^2 \right\},
$$

(29)

where $D_0 = \frac{2Al}{\tau z M} e^{-\beta(E_0 - \varepsilon_0)}$, and

$$
\phi = \frac{1}{2} \left\{ \left[ V + (1 - V) e^{\beta(\gamma\alpha_0 + \delta\alpha_0^2)} - \left(e^{\beta(\gamma\alpha_0 + \delta\alpha_0^2)} + 1\right) C \right]^2 - \right.
$$
$$
\left. - 4C (1 - C) e^{\beta(\gamma\alpha_0 + \delta\alpha_0^2)} \right\}^{\frac{1}{2}} - \frac{1}{2} \left\{ V + (1 - V) e^{\beta(\gamma\alpha_0 + \delta\alpha_0^2)} - \right.
$$
$$
\left. - \left[ e^{\beta(\gamma\alpha_0 + \delta\alpha_0^2)} + 1 \right] C \right\}
$$

(30)

is a function, which does not contain singularities and increases monotonically with growing hydrogen concentration.

The coefficient at $F$ in (29) does not exceed $\frac{1}{|F|}$ under any circumstances. Consequently, the maximum relative change of the diffusion coefficient caused by the variation of the order in the system of pores in the region of negative $F$ values cannot be larger than two. However, this region corresponds to a clear-cut spatial separation of the material components, which is true of grains and boundaries in nanocrystalline metals. Thus, *under conditions of quasi-volume chemical diffusion of hydrogen it is impossible to determine experimentally the effect of structural parameters of the nanocrystalline state, which differ from the volume fraction of grain boundaries, on the mass transfer characteristics in a system.* If $F$ is positive, one should expect a considerable decrease in the diffusion coefficient (as compared to its value at $F = 0$) only on condition that $F$ is at the upper bound of the limits (14). This structural state corresponds to the situation when the "distorted" component is absent in the material in the form of macroscopic inclusions.

If the lattice gas of hydrogen atoms is non-degenerate, the factor in (29), which is independent of the correlation of pores, changes very little with growing hydrogen concentration. Let us write this factor at $C = 0$:

$$
D_{ch}^* = D_0 \frac{\left(1 - V + V e^{\beta\omega\alpha_0}\right)^2}{1 - V + V e^{-\beta(\gamma\alpha_0 + \delta\alpha_0^2)}}.
$$

(31)

An analog of this expression was written in Kirchheim's study [14] at $V \ll 1$. It is readily seen that the diffusion coefficient increases with growing concentration of the "distorted" component if

$$
(2\omega + \gamma) \alpha_0 + \delta\alpha_0^2 > 0.
$$

(32)

Let us discuss which pores comprise "diffusion paths" on boundaries in nanocrystalline materials and what values are implied by the parameters $V$ and $\alpha_0$ in the approximation (13). As was already noted, site energies of hydrogen atoms in pores are distributed continuously on boundaries in the majority of nanocrystalline metals. In terms of our theory, this means the presence of a continuous spectrum of volumes of quasi-interstices. The partial molar volume of hydrogen is positive in most metal systems [1], a fact which suggests $\gamma < 0$. Thus, the increase in the pore volume leads, on the one hand, to the decay of the barrier separating a pore and other quasi-interstices and, on the other hand, to deepening of the potential well. The condition (32) distinguishes, out of the diverse pores present on boundaries in nanocrystalline materials, quasi-interstices, which can serve as the construction material for paths of easy hydrogen diffusion. Two options are possible:

1. If $2\omega + \gamma > 0$, these pores are represented by extended quasi-interstices comprising most of the volume of a grain boundary in nanocrystalline metals. In this case, $V$ has the meaning of the volume fraction of a boundary and $\alpha_0$ is the excess free volume of a boundary per one pore. Clearly, aconsiderable increase in the hydrogen diffusion coefficient will be observed in experiment.

2. If $2\omega + \gamma < 0$, potential elements making up paths of enhanced diffusion are compressed quasi-interstices and large-volume pores with $\alpha_0 > \left| \frac{2\omega+\gamma}{\delta} \right|$. Obviously, the fractions of both elements on boundaries in nanocrystalline materials are small compared to the fraction of boundaries or grains. Therefore, the diffusion coefficient will increase little. Oppositely, a large fraction $V$ of compressed pores and large-volume quasi-interstices should be expected in strongly deformed metals and alloys.

Hence, we have several corollaries. *Firstly*, an explanation is offered for the observation [5] that the role of grain boundaries may be less significant for hydrogen diffusion than for substitutional impurities, since a local decrease in the density, which usually is associated with the region of boundaries, facilitates diffusion of the latter, but not the hydrogen transfer. *Secondly*, one can apprehend the decrease in the diffusion coefficient in the nanocrystalline state as compared to its counterpart in mono- and polycrystals as was observed in palladium alloys [8].

Indeed, if hydrogen is transferred only in some part of the grain-boundary volume, the rest of the grain boundary represents a barrier to the impurity atoms. Assume that pores, by which the transition of the atoms is facilitated, are not integrated into a percolating cluster or the fraction of quasi-interstices on a boundary, which belong to such cluster, is small. In this case, the hydrogen diffusion coefficient will decrease with growing volume of the grain-boundary component. Notice that if the formalism of the "two-phase" model is applied to the boundary itself, the described situation will correspond to $F > 0$.

*Thirdly*, a correlation between the parameter $2\omega + \gamma$ and the value of the diffusion coefficient should be observed in the nanocrystalline state. Out of the two constants $\omega$ and $\gamma$, only the coefficient $\gamma$ is directly connected with a measured value, namely the partial molar volume of hydrogen. Therefore, this correlation may show itself differently. In the region of positive $2\omega + \gamma$ values, the diffusion coefficient grows as this sum increases. In other words, when $\omega = \text{Const}$, the growth of the partial molar volume of hydrogen will be followed by the decrease in the hydrogen diffusion coefficient in the nanocrystalline state as referred to $D_{\text{ch}}$ in mono- and polycrystals. If $2\omega + \gamma < 0$, the behavior of the diffusion coefficient with growing $2\omega + \gamma$ will depend on specialities of the volume distribution of pores on a boundary.

Let us consider in more detail the behaviors of the diffusion coefficient

$$\mathcal{D} = D_{ch}/D_0 \, (F = 0), \tag{33}$$

the parameter

$$\mathcal{K} = V \cdot \frac{\phi + (1 - C)\,e^{\beta(\gamma\alpha_0 + \delta\alpha_0^2)} - (1 - C + \phi)\,e^{\beta[(\omega+\gamma)\alpha_0 + \delta\alpha_0^2]}}{(1 - V)\left(\phi + (1 - C)\,e^{\beta(\gamma\alpha_0 + \delta\alpha_0^2)}\right) + V(1 - C + \phi)\,e^{\beta[(\omega+\gamma)\alpha_0 + \delta\alpha_0^2]}} \tag{34}$$

(at $V \leqslant \frac{1}{2}$ the $F$ value has the upper bound $F = V^2$) and the addition to the effective activation energy of hydrogen diffusion

$$Q = -\frac{d\ln\mathcal{D}}{d\beta} \tag{35}$$

depending on the hydrogen concentration and the volume fraction of the "distorted" component.

The analysis of the expression (29) suggests four intervals of values of the theory parameters, over which $\mathcal{D}$, $Q$ and $\mathcal{K}$ exhibit different behaviors:

$$\mathrm{I}: \begin{cases} \omega\alpha_0 + \gamma\alpha_0 + \delta\alpha_0^2 > 0 \\ \gamma\alpha_0 + \delta\alpha_0^2 > 0 \\ \omega\alpha_0 > 0 \end{cases}, \quad \mathrm{III}: \begin{cases} \omega\alpha_0 + \gamma\alpha_0 + \delta\alpha_0^2 < 0 \\ \gamma\alpha_0 + \delta\alpha_0^2 < 0 \end{cases},$$

$$\mathrm{II}: \begin{cases} \omega\alpha_0 + \gamma\alpha_0 + \delta\alpha_0^2 > 0 \\ \gamma\alpha_0 + \delta\alpha_0^2 < 0 \end{cases}, \quad \mathrm{IV}: \begin{cases} \omega\alpha_0 + \gamma\alpha_0 + \delta\alpha_0^2 > 0 \\ \gamma\alpha_0 + \delta\alpha_0^2 > 0 \\ \omega\alpha_0 < 0 \end{cases}.$$

In all the cases the diffusion coefficient increases with growing fraction of the "distorted" component.

- In the regime I the increase in $V$ is followed by the decrease in the effective activation energy and the enhancement of the effect of the correlation of pores on the diffusion process. The growth of the hydrogen concentration should be accompanied by the increase in permeability, the decrease in the effective activation energy, and the intensification of the effect of the pores correlation.

- Analogously to the regime I, under conditions of the regime II the growth of the fraction of the "distorted" component leads to the decrease in the activation energy and a greater effect of the correlation on the transfer in the system. Oppositely, the diffusion coefficient should decrease and the effective activation energy should grow with increasing concentration of hydrogen. Effects of the correlation of pores are attenuated.

- In the regime III the decrease in the diffusion coefficient may be replaced by its rapid growth with increasing $V$ over the interval $V \ll 1$ at low temperatures. Over this interval of parameters the effect of the correlation is enhanced with increasing

| Regime | V↑ | | | C↑ | | |
|---|---|---|---|---|---|---|
| I | $|\mathcal{K}|\uparrow$ | $\mathcal{D}\uparrow$ | $Q\downarrow$ | $|\mathcal{K}|\uparrow$ | $\mathcal{D}\uparrow$ | $Q\downarrow$ |
| II | $|\mathcal{K}|\uparrow$ | $\mathcal{D}\uparrow$ | $Q\downarrow$ | $|\mathcal{K}|\downarrow$ | $\mathcal{D}\downarrow$ | $Q\uparrow$ |
| III | $|\mathcal{K}|\uparrow$ | $\mathcal{D}\uparrow$ | $Q\uparrow, C\ll 1$ <br> $Q\downarrow, ..C$ | $|\mathcal{K}|\downarrow$ | $\mathcal{D}\uparrow, C\leqslant C^*$ <br> $\mathcal{D}\downarrow, C>C^*$ | $Q\downarrow, C\leqslant C^*$ <br> $Q\uparrow, C>C^*$ |
| IV | $|\mathcal{K}|\uparrow$ | $\mathcal{D}\uparrow$ | $Q\uparrow, V\leqslant V^{**}$ <br> $Q\downarrow, V>V^{**}$ | $|\mathcal{K}|\downarrow, \mathcal{K}>0$ <br> $|\mathcal{K}|\uparrow, \mathcal{K}<0$ | $\mathcal{D}\uparrow, C\leqslant C^{**}$ <br> $\mathcal{D}\downarrow, C>C^{**}$ | $Q\downarrow, C\leqslant C^{**}$ <br> $Q\uparrow, C>C^{**}$ |

Table 1: Behavior of the hydrogen diffusion parameters in different regimes.

$V$ and is reduced with growing concentration of hydrogen. The behaviors of the diffusion coefficient and the effective activation energy exhibit a singularity. When $C$ is small, the activation energy drops and the diffusion coefficient increases with growing concentration of hydrogen. Then at some $C^* \in [0, V]$ the behavior is reversed: the activation energy rises and the diffusion coefficient decreases smoothly as the hydrogen concentration increases.

- The regime IV is analogous to the regime III, except the following points. The growth of the diffusion coefficient is replaced by its drop at $C^{**} \in [1 - V, 1]$. The $\mathcal{K}$ value reverses sign from "+" to "-" as the concentration increases. Therefore, the correlation effects first are reduced and then are enhanced. As the fraction $V$ of the "distorted" component increases, the activation energy changes from a smooth growth to the decrease at $V^{**} \in [0, 1 - C]$.

The aforementioned diffusion regimes are summarized in Table 1. From this table it is seen that the observed pattern of the hydrogen transfer in nanocrystalline palladium, namely the replacement of the growth of the diffusion coefficient by its drop at some hydrogen concentration, which was explained [8] by a strong repulsion of protons, may have an alternative interpretation, that is, the realization of the diffusion regime corresponding to the interval III of the theory parameters.

In all the cases the decrease in the temperature causes enhancement of features characteristic of the regimes I to IV, which show themselves vividly if the lattice gas of hydrogen atoms is degenerate. However, the behavior of the system is not altered. It should be noted also that the maximum possible change of $D_{ch}$ and the effective activation energy of hydrogen diffusion, which results from the increase in the hydrogen concentration of metals, is always larger than the change caused by variation of the order in the system of quasi-interstices.

## 4. Conclusions

The approach, which was proposed earlier for calculation of the chemical diffusion coefficient of hydrogen in amorphous metals and alloys and in "two-phase" systems, was used to analyze hydrogen diffusion in disordered materials taking into account a nonzero partial molar volume of hydrogen atoms.

An analytical expression was obtained in the form of a series for the chemical diffusion coefficient of hydrogen in amorphous metals and alloys. The explicit form of the diffusion coefficient was determined at small concentrations of the impurity.

The relationship was established between the disorder of energies of hydrogen atoms at saddle points and in quasi-interstices and the diffusion transfer. Amorphization always led to the increase in the diffusion coefficient, appearance of an additional dependence on the hydrogen concentration, and violation of the Arrhenius law. In a general case, the effective activation energy exhibited a quadratic dependence on the inverse temperature. It was shown that the contributions from the variance of energies at saddle points and in interstices had unlike signs. If the partial molar volume of hydrogen was zero, amorphization always caused the decrease in the effective activation energy, which changed linearly with the inverse temperature.

Effects of the correlation of pores were analyzed. The establishment of a disorder in the system of pores, when the adjacency of quasi-interstices with dilatations of like signs was preferable, led to the decrease in the hydrogen diffusion activation energy, whereas mixing of pores with dilatations of unlike signs caused the increase in the said energy. The contribution from the correlation of pores to the activation energy was enhanced both with increasing disorder of the energy at saddle points and growing variance of site energies. The correlation of the spatial distribution of volumes of quasi-interstices led to the increase in the concentration dependence of the hydrogen diffusion coefficient. However, when the system was "separated" into regions with pores having dilatations of like signs, this effect was more pronounced

The results obtained in this study were compared with known Kirchheim's theory [14], which was developed for determination of the hydrogen diffusion coefficient in amorphous metals. It was shown that in limiting cases our theory yielded expressions similar to those obtained earlier by Kirchheim. The explicit form of the diffusion coefficient of the degenerate lattice gas of hydrogen atoms was determined, generalizing the relationship (4) at small concentrations to alloys, a disorder of energies at saddle points, and the correlation of pores. At an arbitrary concentration of hydrogen this expression could be obtained if the volume correlation of pores was disregarded.

The quasi-volume chemical diffusion of hydrogen in "two-phase" system, which represented a mixture of regular interstices and distorted pores having nonzero dilatations, was analyzed. These systems were strongly deformed alloys and metals in the nanocrystalline state. The structural components were grains and grain boundaries in metals and regions of a defect-free crystal and zones with a high defect content in alloys. An expression for the hydrogen diffusion coefficient, which holds at an arbitrary hydrogen concentration, was derived.

Four intervals of microscopic parameters of the theory were found. The diffusion coefficient, the activation energy, and effects of the correlation of pores exhibited different behaviors over these intervals, in which the diffusion coefficient increased with growing fraction of the "distorted" component. Specifically, three types of the concentration dependence of the hydrogen diffusion coefficient were possible: the increase over the whole interval of the concentrations; the decrease in permeability with growing hydrogen concentration; and the growth of the diffusion coefficient, which was replaced by its drop. The last type was observed for palladium [8].

The maximum change of the diffusion parameters, which was caused by the increase in the hydrogen concentration, was always larger than the change resulting from the variation of the order in the system of pores. According to the estimates, transport properties of nanocrystalline metals in the regime of quasi-volume chemical diffusion of hydrogen depended solely on the volume fraction of grain boundaries and were insensitive to other parameters of their spatial distribution.

The analysis of the conditions, in which the diffusion coefficient might increase in principle in terms of the "two-phase" model, allowed distinguishing two intervals of the theory parameters, over which transitions to the nanocrystalline and strongly deformed states had a different effect on diffusion properties of metals. In one case, the increase in permeability was favored by a local compression of the metal matrix lattice, while in another case it was facilitated by a local decrease in the density, which was related to grain boundaries in nanocrystalline metals. If the fractions of extended and compressed pores in strongly deformed metals were assumed to be equal, the diffusion coefficient increased only in strongly deformed samples in the first case and in nanocrystalline samples having the same composition in the second case.

It is thought to expand this approach in the future so as to describe in detail the hydrogen transfer in disordered systems with a developed microstructure. The most significant steps in this direction may include the analysis of a multi- or single-hop process as an elementary event of diffusion and the abandonment of the hypothesis about the local equilibrium. Therefore, it will be possible to describe different nonstationary regimes of diffusion in disordered systems.

**References**

1. *Hydrogen in metals I, II*, Eds.: Alefeld, G. and Völkl, J. (1978) Springer-Verlag, Berlin–Heidelberg–New York.

2. Kirchheim, R., Mütschele, T., Kieninger, W., Gleiter, H., Birringer, R. and Koblé, T.D. (1988) Hydrogen in amorphous and nanocrystalline metals, *Mat. Sci. Eng.* **99**(2), 457-462.

3. Larikov, L.N. (1993) Diffusion in amorphous metal alloys, *Metallofizika* **15**(4), 54-78, *ibid.* (8), 3-31.

4. Eliaz, N. and Eliezer, D. (1999) An overview of hydrogen interaction with amorphous alloys, *Advanced Perfomance Materials* **6**, 5-31.

5. Horvath, J. (1989) Diffusion of hydrogen in nanostructured materials, *DDF* **66-69**, 207-228.

6. Klotsman, S.M. (1993) Diffusion in nanocrystalline materials, *Fiz. Met. Metalloved.* **75**(4), 5-18.

7. Larikov, L.N. (1999) Procecces of diffusion in nanocrystalline materials, *Metallofizika i noveishie tekhnologii* **17**(1), 3-31.

8. Mütschele, T. and Kirchheim, R. (1987) Segregation of hydrogen in grain boundaries of palladium, *Scripta Metal.* **21**(2), 135-140.

9. Jaggy, F., Kieninger, W. and Kirchheim, R. (1988) Distribution of site energies in amorphous Ni – Zr and Ni – Ti alloys, *Metal-hydrogen systems I* R. Oldenbourg Verlag, München, 431-436.

10. Stuhr, U., Wipf, H., Udovic, T.J., Weißmüller, J. and Gleiter, H. (1995) The vibrational excitations and the position of hydrogen in n – Pd, *J. Phys.: Condensed Matter* **7**(2), 219-230.

11. Havlin, S. and Ben-Avraham, D. (1987) Diffusion in disordered media, *Adv. Phys.* **36**(6), 695-796.

12. Limoge, Y. (1990) Diffusion in amorphous materials, *Diffusion in materials,* Eds.: Laskar, A.L., Bocquet, J.L., Brebec, G. and Monty, C. Kluwer Academic Publisher Group, The Netherlands, 601-624.

13. Herrmann, A., Schimmele, L., Mössinger, J., Hirscher, M. and Kronmüller, H. (2001) Diffusion of hydrogen in heterogeneous systems, *Appl. Phys. A* **72**, 197-208.

14. Kirchheim, R. (1982) Solubility, diffusivity and trapping of hydrogen in dilute alloys, deformed and amorphous metals – II, *Acta Metal.* **30**(2), 1069-1078.

15. Belaschenko, D.K. (1999) Mechanisms of diffusion in disordered systems, *Uspekhi Fizicheskih Nauk* **169**(4), 361-383.

16. Kondratyev, V.V., Gapontsev, A.V., Voloshinskii, A.N., Obukhov, A.G. and Timofeyev, N.I. (1999) The hydrogen diffusion in disordered systems, *Int. J. Hydrogen Energy* **24**, 819-824.

17. Kondratyev, V.V. and Gapontsev, A.V. (1999) On the theory of hydrogen diffusion in amorphous metals and alloys, *Phys. Met. Metallog.* **87**(5), 363-369.

18. Kondratyev, V.V. and Gapontsev, A.V. (2000) On the theory of hydrogen diffusion in amorphous metals and alloys. Allowance for pore correlation, *Phys. Met. Metallog.* **89**(2), 116-121.

19. Timofeyev, N.I., Rudenko., V.K., Kondratyev, V.V., Gapontsev, A.V., Obukhov, A.G. and Voloshinskii, A.N. (2002) *Transport phenomena in metals and alloys,* UrGUPS, Ekaterinburg.

20. Gapontsev, A.V. and Kondratyev, V.V. (2003) Hydrogen diffusion in partially crystalline alloys: the model of two-phase media, Proc. of VI International Conference "Physics and Chemistry of Ultradispersed (Nano) Materials" in Tomsk, Russia, 46-53.

21. Voloshinskii, A.N., Kondratyev, V.V., Obukhov, A.G. and Timofeyev, N.I. (1998) Hydrogen diffusion coefficient in binary and ternary alloys, *Phys. Met. Metallog.* **85**(3), 336-341.

22. Richards, P.M. (1983) Distribution of activation energies for impurity hopping in amorphous metals, *Phys. Rev. B* **27**(4), 2059-2072.

# DIFFUSION OF HYDROGEN IN BINARY AND TERNARY DISORDERED ALLOYS

N.I. TIMOFEYEV, V.K. RUDENKO
OAS "Ekaterinburg Non-Ferrous Metals Working Plant",
Ekaterinburg, Russia.


V.V. KONDRATYEV, A.V. GAPONTSEV, A.N.VOLOSHINSKII
Institute of Metal Physics,
Ural Branch of Russian Academy of Sciences,
620219,18S.KovalevskayaStr.,
Ekaterinburg,Russia.
Fax:7-343-3745244
E-mail:avg@imp.uran.ru

## Abstract

The method of configurations, which takes into account the nonequivalence of the nearest neighborhood of a hydrogen atom by atoms of the alloy components in different interstices, was used to derive a general expression for the coefficient of chemical diffusion of hydrogen in binary and ternary disordered alloys. It was shown that the problem of determining the diffusion coefficient in alloys reduces to calculation of some configuration sum including probabilities of transition from one interstice to a neighboring interstice and diffusion coefficients of pure metals comprising the alloy. This sum determines the dependence of the diffusion coefficient on the alloy composition and the concentration of hydrogen atoms. Calculated and experimental dependences of the diffusion coefficient on compositions of Pd – Pt and Pd – Rh FCC alloys and Nb – V and Nb – Ta BCC alloys were compared. Concentration dependences of the hydrogen diffusion coefficient in ternary $V_xTa_yNb_{1-x-y}$ systems with the BCC lattice and $Pd_xAu_xPt_{1-x-y}$ with the FCC lattice have been given by way of illustration.

*Keywords:* hydrogen diffusion, disordered alloys, regular solid solutions

## 1. Introduction

A great number of experimental and theoretical studies have been dedicated to hydrogen diffusion in metals (see the review [1]). The movement mechanism of hydrogen atoms is their transition between interstices. Hydrogen atoms are transferred between tetrahedral interstices (tetrapores) in BCC metals and between octahedral interstices (octapores) in FCC metals.

Hydrogen diffusion differs from diffusion of other interstitial impurities (carbon, nitrogen, etc.) by, first, an extremely low energy of diffusion activation, which equals ~0.1

635

*T.N. Veziroglu et al. (eds.),*
*Hydrogen Materials Science and Chemistry of Carbon Nanomaterials,* 635-651.
© 2004 *Kluwer Academic Publishers. Printed in the Netherlands.*

eV, and, second, its quantum character over a broad temperature interval (below room temperature).

At extremely low temperatures, when lattice vibrations are "frozen", diffusion is realized through a purely quantum mechanism, which consists in sub-barrier tunneling of hydrogen atoms between neighboring interstices. The efficiency of this mechanism is highly sensitive to various imperfections in the metal, such as lattice defects and impurities.

As the temperature increases, transitions of hydrogen atoms involve vibrations of the crystal lattice (phonons). These are so-called incoherent transitions. In this case, transitions are realized either through excited states of hydrogen atoms (thermally activated adiabatic transitions) or from one interstice to another where atoms are in the ground state (non-adiabatic transitions, thermally stimulated tunneling).

The classical mechanism of over-barrier diffusion of atoms is realized at higher temperatures. According to this mechanism, the probability of transition of an interstitial atom per unit time is

$$\Gamma = \frac{1}{\tau} \exp\left(-\frac{V}{kT}\right), \tag{1}$$

where $\tau^{-1}$ is the local vibration frequency of an interstitial atom and $V$ is the transition activation energy.

When $kT \geqslant V$, hydrogen diffusion in metals is similar to diffusion of a liquid or a gas.

Let us turn now to hydrogen diffusion in alloys, which is a more complicated process. Remarkably, the quantum theory of diffusion in alloys has not been available owing to complexity of the quantum mechanical problem. It should be noted however that in this case quantum effects are much less significant as compared to the situation in pure metals. This is because random changes of the potential energy of a proton, which take place during transition from one interstice to another and are due to a different environment of atoms in alloys, are larger than the proton energy gap, which is proportional to the integral overlap of wave functions of protons in adjacent interstices. Therefore, purely quantum tunneling is virtually impossible. For these reasons, in what follows we shall consider the classical regime of diffusion.

The main feature of diffusion in alloys is that similar interstices have different configurations of their neighboring atoms. The configurations differ both in the number and the symmetry of unlike atoms. Therefore, the energy of protons at interstices and the activation energy $V$ over the lattice sites are different. For example, if the proton energy depends on the relative distance to atoms in the alloy only within the first coordination sphere, disordered binary BCC A $-$ B alloys may have 5 configurations of tetrapores ($A_4$, $A_3B_1$, $A_2B_2$, $A_1B_3$, and $B_4$) and, consequently, 5 different values of the proton energy at interstices. FCC A-B alloys have 7 different configurations, namely $A_6$, $A_5B_1$, $A_4B_2$, $A_3B_3$, $A_2B_4$, $A_1B_5$, and $B_6$. In a general case, if the direction of bonds between hydrogen atoms and alloy atoms (symmetry of the pore environment) is taken into account, we have 8 and 10 configurations in BCC and FCC alloys respectively.

Since interstices are nonequivalent, the distribution of hydrogen atoms over the interstices will not be equiprobable, but should be described by the Fermi-Dirac function or (when the concentration of hydrogen atoms is small) by Boltzman's distribution. One should also take into account the effect of the finite concentration of diffusing atoms, which shows up as the prohibition of the atom transfer to a neighboring interstice if it is filled [2].

All these features of diffusion in alloys should be considered in calculation of the diffusion coefficient. The task of the theory with respect to this problem is to obtain an equation for the diffusion coefficient, which would contain a minimum number of free parameters, since their determination requires additional experiments. Otherwise, results of the theory are of little value and it is more reasonable to directly measure the diffusion coefficient in alloys in experiment.

The diffusion coefficient was calculated [3] for BCC alloys taking the $Nb_{1-y}V_y$ system as an example. In our opinion, the approach [3] is convenient for the study of the temperature dependence of the diffusion coefficient. To describe the dependence of the diffusion coefficient on the alloy composition, it is reasonable to express the diffusion coefficient as diffusion coefficients of pure components.

This paper presents a review of our studies [4–7] dealing with calculation of the diffusion coefficient of light interstitial impurities in disordered alloys. These alloys may be exemplified by palladium- and platinum-based FCC alloys, which are widely used in technology. A significant point is that the diffusion coefficient in an alloy is expressed as diffusion coefficients of pure metals comprising the alloy. Therefore, using tabulated data on diffusion in metals, one can predict the change of the diffusion coefficient with the alloy composition and, also, the hydrogen concentration.

## 2. Diffusion Coefficient

### 2.1. The general expression for the diffusion coefficient

Movement of a given atom in the force field produced by surrounding atoms is considered in the microscopic theory of diffusion of interstitial atoms [8]. We shall assume that non-interacting hydrogen atoms move on a stationary potential relief in a disordered material. Potential minimums correspond to interstices (pores). Each pore can accommodate only one hydrogen atom. The probability of a unit hop of an atom from one interstice to a neighboring interstice is given by the formula (1), where $V$ means the potential barrier height, which determines the minimum kinetic energy to be imparted to an interstitial atom from its neighbors as a result of thermal excitation.

The original expression in the analysis of macroscopic diffusion experiments is Fick's law:

$$J = -D\nabla c. \tag{2}$$

where $J$ is the density of the macroscopic flow of interstitial impurity atoms (the number of atoms crossing a unit area with the normal parallel to $J$ per unit time), $D$ is the chemical diffusion coefficient, and $c$ is the total number of interstitial atoms per unit volume. The task of the theory is to use the microscopic approach to establish the relationship between $J$ and $\nabla c$.

Let us assume the presence of some impurity concentration gradient directed along the $Ox$ axis. In this case, in line with the hypothesis about the local equilibrium [9], the whole space can be divided into physically small equilibrium volumes, for which thermodynamic potentials and their derivatives are determined. Let us consider two volumes 1 and 2. In our case, they can be conventionally separated by a plane perpendicular to the $Ox$ axis. Let the pores in the left- and right-hand volumes be indexed $i_1$ and $i_2$ respectively (see Fig. 1). The occupancy of the pores $v_i(\mu)$ will be given by the Fermi-Dirac distribution

638

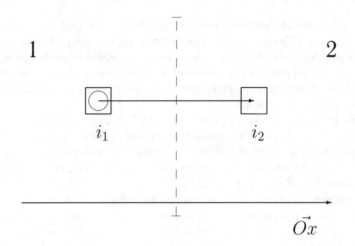

1                                                2

$i_1$                                          $i_2$

$\vec{Ox}$

Figure 1: Two locally equilibrium volumes separated by a plane boundary.

$$v_{i_{1(2)}}(\mu_{1(2)}) = \frac{1}{1 + \exp\left\{\beta(\varepsilon_{i_{1(2)}} - \mu_{1(2)})\right\}},$$  (3)

where $\mu_{1(2)}$ is the chemical potential of hydrogen atoms in the volume 1(2) and $\varepsilon_{i_{1(2)}}$ is the potential energy of hydrogen in the pore $i_{1(2)}$. The expression for the flow across a unit area on the boundary of the locally equilibrium volumes 1 and 2 has the form

$$J = 2\widehat{\sum_{i_1,i_2}}\left[\frac{v_{i_1}(\mu_1)q_{i_2}(\mu_2)}{\tau_{i_1}}f_{i_1 i_2}\exp(-\beta V_{i_1 i_2}) - \frac{v_{i_2}(\mu_2)q_{i_1}(\mu_1)}{\tau_{i_2}}f_{i_2 i_1}\exp(-\beta V_{i_2 i_1})\right],$$  (4)

where $\tau_{i_{1(2)}}$ denotes periods of thermal vibrations of hydrogen atoms in pores, which will be taken equal to an average value $\bar{\tau}$; $q_{i_{1(2)}} = 1 - v_{i_{1(2)}}$ is the probability that the pore $i_{1(2)}$ does not contain a hydrogen atom; $f_{i_1 i_2}$ is the probability that a hop, which begins in the pore $i_1$, ends in the pore $i_2$ (this value may be taken equal to about $\frac{1}{\bar{z}}$, where $\bar{z}$ is the mean first coordination number in the sublattice of interstices; $V_{i_1 i_2}$ is the activation energy of the transition $i_1 \rightarrow i_2$; the circumflex accent means that the summation is taken in a unit area only over interstices, between which a hop of a hydrogen atom is possible.

Let us approximate the change of the chemical potential between locally equilibrium volumes by a linear dependence

$$\mu_2 = \mu_1 + l\frac{d\mu}{dx},$$  (5)

where $l$ is the characteristic scale of a locally equilibrium volume, which is of the order of the diffusion hop of a hydrogen atom. In this case, in accordance with (4), the flow of impurity atoms takes the form

$$J = -\frac{2\beta l}{\bar{\tau}\bar{z}}\frac{d\mu}{dx}\widehat{\sum_{i_1,i_2}}\left[v_{i_1}v_{i_2}q_{i_2}\exp(-\beta V_{i_1 i_2}) - q_{i_1}q_{i_2}v_{i_2}\exp(-\beta V_{i_2 i_1})\right].$$  (6)

Let the interstices be characterized by an attribute (a set of attributes) $\chi$ and the hop activation energy $i_1(\chi) \rightarrow i_2(\chi')$ depends only on the type of pores $i_1$ and $i_2$, while the potential energy of a hydrogen atom $\varepsilon_{i_1(\chi)}$ is determined only by $\chi$. In this case, the summation in (6) can be replaced by averaging over all types of pores, which will be written in what follows as a sum. Consequently, the expression (6) becomes

$$ J = -\frac{2A l \beta}{\tau z} \frac{d\mu}{dx} \sum_{\chi,\chi'} f(\chi,\chi') v_\chi q_{\chi'} \left[ q_\chi + v_{\chi'} \right] \exp\{-\beta V(\chi,\chi')\}, \tag{7} $$

where $A$ is the average number of "communicating" pairs of quasi-interstices in a unit area in the cross section of a sample and $f(\chi,\chi')$ is the pair distribution function of the types of pores, which has the properties

$$ f(\chi,\chi') = f(\chi',\chi), \tag{8} $$

$$ \sum_{\chi,\chi'} f(\chi,\chi') = 1. \tag{9} $$

The concentration $C = c/M$ is determined by the expression

$$ C = \sum_\chi f(\chi) v_\chi, \tag{10} $$

where the ordinary distribution function $f(\chi)$ is related to $f(\chi,\chi')$ by the standard means

$$ f(\chi) = \sum_{\chi'} f(\chi,\chi'), \tag{11} $$

and the expression for the occupancy of pores (3) is written as

$$ v_\chi = \frac{1}{1 + \exp\{\beta(\varepsilon_\chi - \mu)\}}. \tag{12} $$

Considering (10) and (12), the relationship (7) is easily reduced to the form of the first Fick's law (2), which yields a general expression for the hydrogen diffusion coefficient

$$ D = \frac{2A l}{\tau z M} \cdot \frac{\sum\limits_{\chi,\chi'} f(\chi,\chi') \exp\{-\beta V(\chi,\chi')\} v_\chi q_{\chi'} \left[ v_{\chi'} + q_\chi \right]}{\sum\limits_\chi f(\chi) v_\chi q_\chi dx}, \tag{13} $$

where the function $f(\chi,\chi')$ bears all information about the structure of the material, while $\varepsilon_\chi$ and $V(\chi,\chi')$ contain information about the potential relief.

In the case of alloys, $\chi$ has the meaning of the index of species environment of a hydrogen atom by atoms of the matrix. The configuration distribution of pores $f(\chi,\chi')$ in alloys with a chemical disorder depends on their composition and the degree of order in the alloy. The energy parameters $\varepsilon_\chi$ and $V(\chi,\chi')$ need be known for calculation of the diffusion coefficient from the formula (13) (see Fig. 2). The exact value of the pre-exponential factor and its concentration dependence at $kT < V(\chi,\chi')$ are not of principal significance. They

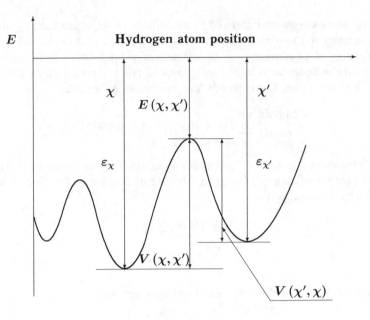

Figure 2: Potential relief of a hydrogen atom.

can be found from additional experiments as was done, for example, in [3]. However, the interpretation of those experiments by itself requires the use of rather rough model assumptions. It turns out that energy parameters of the problem can be expressed as the corresponding parameters of alloy components in terms of the same assumptions. Let us begin with determination of $V(\chi,\chi')$. The relief of the potential energy of a hydrogen atom in an alloy is shown schematically in Fig. 2. Obviously,

$$V(\chi,\chi') - V(\chi',\chi) = \varepsilon_{\chi'} - \varepsilon_{\chi} \tag{14}$$

and

$$V(\chi,\chi') = -\varepsilon_{\chi} + E(\chi,\chi') \tag{15}$$

where $E(\chi,\chi')$ is the energy at the saddle point separating configurations with the numbers $\chi$ and $\chi'$. Calculation of the energy $E(\chi,\chi')$ is a very complicated problem. However, some computer-aided studies [10, 11] showed that the energy at the saddle point may be assumed as constant with a sufficient accuracy and independent of the concentration. In this approximation,

$$V(\chi,\chi') = -\varepsilon_{\chi} + E_0 \tag{16}$$

Therefore it is possible to express the diffusion coefficient (13) (13) as diffusion coefficients in pure metals $L = A, B, \ldots$

$$D_L = \frac{2Al}{\tau z M} \cdot \exp(-\beta E_0) \exp(\beta \varepsilon_L), \tag{17}$$

because in the approximation of a paired interaction between a hydrogen atom and its nearest atoms in an alloy, $\varepsilon_{\chi}$ values are easily expressed as the corresponding energies $\varepsilon_L$ in pure metals.

## 2.2. Disordered binary alloys

Let us consider a binary alloy $A_xB_{1-x}$ ($A_xB_y$). Assume that a configuration of the type $\chi$ is characterized by $a$ atoms A and $b = h - a$ atoms B ($h = 4$ for a tetrapore and $h = 6$ for a octapore). It is convenient to change from the index $\chi$ to $a$:

$$\chi \rightarrow a \quad (0 \leqslant a \leqslant h).$$

Representing the energy $\varepsilon_a$ as

$$\varepsilon_a = \frac{1}{h}(a\varepsilon_A + b\varepsilon_B), \tag{18}$$

and considering (16) and (17), the expression for the diffusion coefficient (13) can be written as a configuration sum

$$D(C) = D_B \cdot \frac{\sum\limits_{a,a'=0}^{h} m_a P_{aa'} \left(\frac{D_A}{D_B}\right)^{\frac{a}{h}} v_a q_{a'} (q_a + v_{a'})}{\sum\limits_{a=0}^{h} m_a v_a q_a}, \tag{19}$$

where $m_a$ is the probability to find a pore of the type $a$ (the ordinary distribution function $f(\chi)$); $P_{aa'}$ is the conditional probability to find an interstice of the type $a'$ near a pore of the type $a$; $v_a$ denotes occupancies of interstices, which were initially determined by the expression (12). They can also be rearranged to the form

$$v_a = \frac{1}{1 + exp(-\beta\mu')\left(\frac{D_A}{D_B}\right)^{\frac{a}{h}}}. \tag{20}$$

The expression (19) should be supplemented with normalization conditions, which follow from the relationships (9) and (10):

$$\sum_{a=0}^{h} m_a v_a = C, \quad \sum_{a=0}^{h} m_a = 1. \tag{21}$$

In the case of an ideal solid solution, $m_a$ values follow the binomial distribution

$$m_a = C_h^a x^a (1 - x)^{h-a}. \tag{22}$$

The determination of conditional probabilities $P_{aa'}$ represents a problem involving the search for every possible configurations of two adjacent interstices taking into account that 3 (BCC) or 2 (FCC) sites are common for both interstices. Usually $P_{aa'} = m_a$ is assumed [12]. From Fig. 3, which was drawn for the FCC lattice, it is seen that the approximation is incorrect. For example, $P_{60} = 0$, which reflects the fact that neighboring interstices cannot be surrounded by 6 atoms of one element and 6 atoms of another element. The correlation of neighboring interstices is still larger for tetrapores in the BCC lattice (Fig. 4). Our calculated $P_{aa'}$ values for BCC and FCC lattices have the form (the method for calculation of $P_{aa'}$ will be described in detail below for a general case of an alloy with an arbitrary

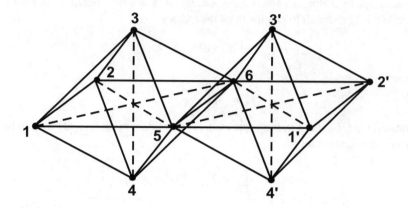

Figure 3: Configuration of two adjacent octapores in the FCC lattice.

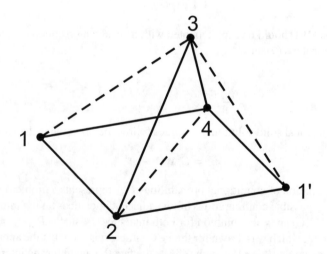

Figure 4: Configuration of two adjacent tetrapores in the BCC lattice.

degree of order)

$$P^{bcc}_{aa'} = \tfrac{1}{4}\left[x\left(a\delta_{a,a'} + b\delta_{a,a'+1}\right) + y\left(a\delta_{a,a'-1} + b\delta_{a,a'}\right)\right],$$
$$P^{fcc}_{aa'} = \tfrac{4}{5}\frac{x^{a'-2}y^{a'-2}}{a'!(6-a')!}\left[aa'\left(a-1\right)\left(a'-1\right)y^2 + \right.$$
$$\left. +2aa'\left(6-a\right)\left(6-a'\right)xy + \left(6-a\right)\left(6-a'\right)\left(5-a\right)\left(5-a'\right)x^2\right]. \tag{23}$$

The system of equations (19)-(23) is quite sufficient for determination of the diffusion coefficient in a disordered binary alloy $A_xB_y$.

### 2.3. Disordered ternary alloys

The approach, which is described above for binary alloys, can be easily generalized to disordered ternary systems $A_xB_yC_z$ $(z = 1 - x - y)$. The traditional method for calculation of the hydrogen diffusion coefficient is virtually useless for these systems because of a large number of unknown parameters.

In this case, it is convenient to use a two-index system of designations by the following rule. Let a configuration with the number $\chi$ be characterized by $a$ atoms $A$, $b$ atoms $B$, and $c = (h - a - b)$ atoms $C$. We shall introduce two-index labeling:

$$\chi \rightarrow ab \quad (0 \leqslant a \leqslant h, 0 \leqslant b \leqslant h - a).$$

Considering that

$$\varepsilon_{ab} = \frac{1}{h}[a\varepsilon_A + b\varepsilon_B + (h - a - b)\varepsilon_C], \tag{24}$$

the expression (13) can be written as a configuration sum

$$D(C) = D_C \cdot \frac{\sum\limits_{a=0}^{h}\sum\limits_{b=0}^{h-a}\left(\frac{D_A}{D_C}\right)^{\frac{a}{h}}\left(\frac{D_B}{D_C}\right)^{\frac{b}{h}}m_{ab}\nu_{ab}\sum\limits_{a'=0}^{h}\sum\limits_{b'=0}^{h-a'}P_{ab,a'b'}q_{a'b'}(q_{ab} + \nu_{a'b'})}{\sum\limits_{a=0}^{h}\sum\limits_{b=0}^{h-a}m_{ab}\nu_{ab}q_{ab}}, \tag{25}$$

where

$$\nu_{ab} = \frac{1}{1 + \exp\left(-\beta\mu'\right)\left(\frac{D_A}{D_C}\right)^{\frac{a}{h}}\left(\frac{D_B}{D_C}\right)^{\frac{b}{h}}}. \tag{26}$$

The concentration dependences $D(C)$ of $x$, $y$, and $z = 1 - x - y$ will be included in $m_{ab}$ and $P_{ab,a'b'}$.

The polynomial distribution

$$m_{ab} = \frac{h!}{a!b!c!}x^ay^bz^c. \tag{27}$$

holds for a fully disordered alloy $A_xB_yC_{z=1-x-y}$ if the short-range order is disregarded. Conditional probabilities $P_{ab,a'b'}$ were calculated comprehensively elsewhere [5]. Finally we have

$$P^{bcc}_{ab,a'b'} = \tfrac{1}{4}\left[x\left(a\delta_{a'a}\delta_{b'b} + b\delta_{a',a+1}\delta_{b',b-1} + \delta_{a',a+1}\delta_{b'b}\right) + \right.$$
$$+y\left(a\delta_{a',a-1}\delta_{b'b} + b\delta_{a'a}\delta_{b'b} + \delta_{a'a}\delta_{b',b+1}\right) +$$
$$+z\left(a\delta_{a',a-1}\delta_{b'b} + b\delta_{a'a}\delta_{b',b-1} + \delta_{a'a}\delta_{b'b}\right)\right],$$
$$P^{fcc}_{ab,a'b'} = \tfrac{4}{5}\frac{1}{a'!b'!c'!}\left[aa'\left(a-1\right)\left(a'-1\right)x^{a'-2}y^{b'}z^{c'} + \right.$$
$$+bb'\left(b-1\right)\left(b'-1\right)x^{a'}y^{b'-2}z^{c'} + cc'\left(c-1\right)\left(c'-1\right)x^{a'}y^{b'}z^{c'-2} +$$
$$\left. +2aba'b'x^{a'-1}y^{b'-1}z^{c'} + 2cc'bb'x^{a'}y^{b'-1}z^{c'-1} + 2cc'aa'x^{a'-1}y^{b'}z^{c'-1}\right]. \tag{28}$$

The expressions (25)-(28) fully determine the hydrogen diffusion coefficient in disordered ternary alloys with BCC and FCC lattices.

### 2.4. Taking into account the short-range order in alloys

In what follows we propose a method for taking into account the short-range order in alloys on the basis of the theory of regular solid solutions. It is given for binary alloys.

If the short-range order in alloys is taken into account, it is necessary to re-evaluate the probabilities $m_a$ and $P_{aa'}$. The general expression (19) remains unchanged. Let us assign $P_{LM}$ to the probability that nearest lattice sites are occupied by atoms of the species $L$ and $M$. Considering the correlation in the arrangement of nearest atoms [13]

$$P_{AA} = x^2 G_{AA}, \quad P_{BB} = (1-x)^2 G_{BB}, \quad P_{AB} = x(1-x)G_{AB},$$

where $G_{LM}$ denotes correlation factors

$$G_{AA} = 1 - 2(1-x)^2\lambda, \quad G_{BB} = 1 - 2x^2\lambda, \quad G_{AB} = 1 + 2x(1-x)\lambda,$$
$$\lambda = \frac{W}{kT}, \quad W = \left[\tfrac{1}{2}(V_{AA} + V_{BB}) - V_{AB}\right]$$

is the energy of mixing in the alloy, and $V_{LM}$ is the energy of the paired interaction between nearest atoms of the species $L$ and $M$. In this approximation, the fraction of interstices, which have $a$ atoms of the species $A$ and $b = h - a$ atoms of the species $B$ in their vicinity, is

$$m_a = \frac{(h-2)!}{a!b!}\mu_a x^a y^b, \quad \mu_a = \mu_a^{(1)} + \mu_a^{(2)},$$
$$\mu_a^{(1)} = a(a-1)G_{AA}^{a-2}G_{AB}^b + abG_{AA}^{a-1}G_{AB}^{b-1}, \quad (29)$$
$$\mu_a^{(2)} = b(b-1)G_{BB}^{b-2}G_{AB}^a + abG_{BB}^{b-1}G_{AB}^{a-1}.$$

Statistical calculation should be performed separately for octa- and tetrapores in order to find the conditional probability $P_{aa'}$.

The configuration of two adjacent tetrapores is shown as a diagram in Fig. 4. The solid lines denote atomic bonds of nearest neighbors at a distance $0.5\sqrt{3}d$. The dashed lines indicate next neighbors spaced a distance $d$.

Let us calculate $m_a$ for the system $A_x B_y$ taking into account the short-range order within the first coordination sphere. In this case, one can write

$$m_a = \sum_{L=A,B} \sum_{M=A,B} P_L(1)P_M(3)m_a^{LM}, \quad (30)$$

where $P_L(N)$ is the probability to find an atom of the species $L$ on a site numbered $N$; $m_a^{LM}$ is the probability of the pore configuration provided the sites "1" and "3" are occupied by atoms of the species $L$ and $M$ respectively. Considering that

$$P_A = x, \quad P_B = y,$$
$$m_a^{AA} = C_2^{a-2}\Psi_{AA}^{a-2}\Psi_{AB}^b; \quad m_a^{AB} = C_2^{a-1}\Psi_{AA}^{a-1}\Psi_{AB}^{b-1};$$
$$m_a^{BA} = C_2^{a-1}\Psi_{BB}^{b-1}\Psi_{BA}^{a-1}; \quad m_a^{BB} = C_2^a\Psi_{BB}^{b-2}\Psi_{BA}^a;$$

and $\Psi_{LM} = P_M(1)G_{LM}$ (conditional probabilities of pairs of atoms on sites 1, 2 and 1, 4), the expression (30) can be converted to the form (29). Let us calculate the probability $P_{aa'}$ using an analogous method. Represent this probability as

$$P_{aa'} = m_a^{-1}\sum_{L=A,B} \sum_{M=A,B} \sum_{K=A,B} P_L(1)P_M(3)P_K(1')P_{aa'}^{LMK}, \quad (31)$$

where $P_{aa'}^{LMK}$ is the probability that the left- and right-hand interstices contain $a$ and $a'$ atoms of the species A respectively on condition that the sites 1, 3 and 1' are occupied by atoms of the species $L$, $M$ and $K$. Taking into account the correlation, we have

$$P_{aa'}^{AAA} = C_2^{a-2}\Psi_{AA}^{a-2}\Psi_{AB}^b\delta_{a,a'}; \quad P_{aa'}^{AAB} = C_2^{a-2}\Psi_{AA}^{a-2}\Psi_{AB}^b\delta_{a,a'+1};$$
$$\cdots; \quad P_{aa'}^{BBB} = C_2^a\Psi_{BB}^{b-2}\Psi_{BA}^a\delta_{a,a'}. \tag{32}$$

Substitution of (30) and (32) into (31) gives the expression

$$P_{aa'}^{bcc} = \frac{\mu_a^{(1)}(x\delta_{a,a'} + y\delta_{a,a'+1}) + \mu_a^{(2)}(y\delta_{a,a'} + x\delta_{a,a'-1})}{\mu_a}. \tag{33}$$

The configuration of two adjacent octapores is shown in Fig. 3. The shortest distance between lattice sites is $0.5\sqrt{2}d$, while the next sites are spaced a distance $d$. The $m_a$ value is calculated similarly to the previous case, but the sites 1 and 3 are more convenient as the fixed sites in (30). Considering that $P_A = x$, $P_B = y$, and

$$m_a^{AA} = C_4^{a-2}\Psi_{AA}^{a-2}\Psi_{AB}^b; \quad m_a^{AB} = C_4^{a-1}\Psi_{AA}^{a-1}\Psi_{AB}^{b-1};$$
$$m_a^{BA} = C_4^{a-1}\Psi_{BB}^{b-1}\Psi_{BA}^{a-1}; \quad m_a^{BB} = C_4^a\Psi_{BB}^{b-2}\Psi_{BA}^a,$$

appropriate manipulations also yield the formula (29), in which $h = 6$ should be taken. We shall calculate $P_{aa'}$ taking the site 5 as the starting site. Taking into account the correlation between nearest atoms, one can write

$$P_{aa'} = m_a^{-1} \sum_{L=A,B} \sum_{M=A,B} P_L(5)\Psi_{LM}(5,6)P_a^{LM}P_{a'}^{LM} \tag{34}$$

Here $P_a^{LM}$ is the probability that the environment in one of the interstices has $a$ atoms of the species $A$ on condition that common sites of the adjacent interstices 5 and 6 are occupied by atoms of the species $L$ and $M$ respectively. We have four expressions for $P_a^{LM}$:

$$P_a^{AA} = C_3^{a-2}y\Psi_{AA}^{a-2}\Psi_{AB}^{b-1} + C_3^{a-3}x\Psi_{AA}^{a-3}\Psi_{AB}^b;$$
$$P_a^{AB} = C_3^{a-1}y\Psi_{AA}^{a-1}\Psi_{AB}^{b-2} + C_3^{a-2}x\Psi_{AA}^{a-2}\Psi_{AB}^{b-1};$$
$$P_a^{BA} = C_3^{a-1}y\Psi_{BA}^{a-1}\Psi_{BB}^{b-2} + C_3^{a-2}x\Psi_{BA}^{a-2}\Psi_{BB}^{b-1};$$
$$P_a^{BB} = C_3^a y\Psi_{BB}^{b-3}\Psi_{BA}^a + C_3^{a-1}x\Psi_{BB}^{b-2}\Psi_{BA}^{a-1}. \tag{35}$$

Substituting (35) and (29) into (34) and calculating the sum, we obtain the expression

$$P_{aa'}^{fcc} = \frac{3x^{a'-2}y^{b'-2}}{2a'!b'!\mu_a}\Big\{a(a-1)a'(a'-1)(1-x)^2 G_{AA}^{(a+a'-5)}G_{AB}^{(b+b'-2)}\times$$
$$\times[bG_{AA} + (a-2)G_{AB}][b'G_{AA} + (a'-2)G_{AB}] + aba'b'x(1-x)\times$$
$$\times\Big[G_{AA}^{(a+a'-4)}G_{AB}^{(b+b'-3)}((b-1)G_{AA} + (a-1)G_{AB})((b'-1)G_{AA} + (a'-1)G_{AB})+$$
$$+G_{BB}^{(b+b'-4)}G_{AB}^{(a+a'-3)}((a-1)G_{BB} + (b-1)G_{AB})((a'-1)G_{BB} + (b'-1)G_{AB})\Big] +$$
$$+b(b-1)b'(b'-1)x^2 G_{BB}^{(b+b'-5)}G_{AB}^{(a+a'-2)}\times$$
$$\times(aG_{BB} + (b-2)G_{AB})(a'G_{BB} + (b'-2)G_{AB})\Big\}. \tag{36}$$

It is easy to see that in the absence of the correlation, when $G_{LM} = 1$, the formulas (29), (33) and (36) are rearranged to (22) and (23).

The method for calculation of distribution functions, which is described above, can be easily generalized to ternary systems. The coefficient matrix $P_{ab,a'b'}^{bcc}$ for a body-centered cubic lattice is $15 \times 15$. The coefficient matrix $P_{ab,a'b'}^{fcc}$ for a face-centered cubic lattice is $28 \times 28$. We shall not present these matrices for their awkwardness.

646

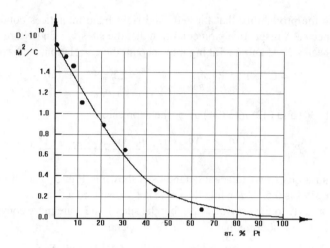

Figure 5: Calculated and experimental dependences of the hydrogen di ffusion coefficient on the platinum concentration in Pd − Pt alloys at $T = 357$ K. The solid line denotes calculation at $\mu = 0$, while the dots indicate the experimental results [16].

## 3. Discussion

Usually the diffusion mobility of hydrogen atoms in alloys is analyzed using an expression like (13), as was done, for example, in [3] dealing with the temperature dependence of the diffusion coefficient in $Nb_{1-y}V_y$ BCC alloys. This approach requires the knowledge of a large number of problem parameters. Some of the parameters are determined in special experiments on a given batch of alloys, while other parameters are found using a linear interpolation of diffusion data obtained for pure metals. For example, values of the energy $\varepsilon_a$ and the fraction of configurations $m_a$ are determined from the analysis of $P - C$ isotherms using the nearest neighbors approximation on the interaction of hydrogen with alloy atoms [14]. The energy of the mutual interaction of hydrogen atoms and the pre-exponential factor are found assuming their linear dependence on the concentration of alloy components. Special methods are available for determination of the activation energy distribution in magnetic alloys of the Fe − Pd type [15].

Our study showed that the diffusion coefficient in binary alloys can be expressed as diffusion coefficients of pure components and some configuration sum related to the co-efficients actually in the same approximations as those used in [3]. We performed this calculation for concentrated palladium-platinum alloys whose diffusion coefficients were determined experimentally in [16].

Figure 5 presents calculated and experimental dependences of the diffusion coefficient in a palladium-platinum alloy on the platinum concentration at a low hydrogen concentra-tion provided the normalization condition (21) leads to $\beta\mu'(C) \ll 1$. From Fig. 5 it follows that the formula (20) describes well the experimental data.

The dependence of the diffusion coefficient on the concentration of rhodium in pal-ladium, which is shown in Fig. 6, is described by the theory much worse under these

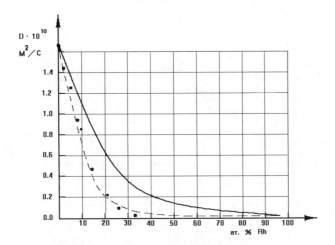

Figure 6: Fig. 6. Calculated and experimental dependences of the hydrogen di ffusion coefficient on the rhodium concentration in Pd − Rh alloy at $T$ = 357 K. The solid line denotes calculation at $\mu$ = 0. The dashed line shows calculation taking into account the mutual interaction of hydrogen atoms in the form $\beta f$ = 0.3 + 12 (1 − $y$) [3]. The dots indicate the experimental results [16].

assumptions. We think this is due to the fact that the formula (20) disregards the mutual interaction of hydrogen atoms and the change of the interaction with growing concentration of rhodium. If the formula (12) takes into account the mutual interaction of hydrogen atoms, the agreement between the theoretical and experimental results is largely improved as is seen from Fig. 6.

The concentration dependences $D(C)$ were calculated for Nb − Ta (Fig. 7) and Nb − V (Fig. 8) BCC alloys. The experimental values were taken from [17, 18]. While the calculated and experimental values of $D(C)$ correlated well for Nb − Ta alloys, the diffusion coefficient was underestimated in experiments on Nb − V alloys over a wide interval of concentrations. This fact is due probably to the concentration dependence of the pre-exponential factor in these alloys, which was disregarded.

To illustrate the dependence of the hydrogen diffusion coefficient on the composition of disordered ternary alloys, we calculated diffusion coefficients for the $V_x Ta_y Nb_{1-x-y}$ ternary system with a BCC lattice and the $Pd_x Ag_y Pt_{1-x-y}$ ternary system with an FCC lattice.

The calculation results are given in Figs. 9 and 10. As follows from these figures, the diffusion coefficients in ternary alloys with both BCC and FCC lattices change smoothly, but not linearly, with the alloy composition. Calculation showed that taking into account the short-range order in the given approximation has a very little effect on the diffusion coefficient in both binary and ternary alloys.

Unfortunately, we have been unaware of systematic experimental studies of the hydrogen diffusion in ternary alloys and, therefore, have been unable to compare our calculation results with experimental data. The data on diffusion in pure components were adopted from [19].

648

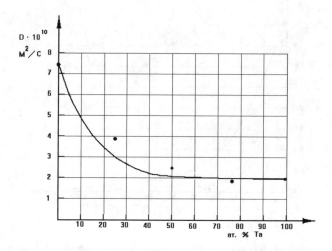

Figure 7: Calculated and experimental dependences of the hydrogen di ffusion coefficient on the tantalum concentration in Nb − Ta alloy at $T$ = 296 K. The solid line denotes the calculation results. The dots show the experimental results [17].

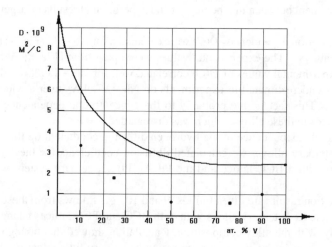

Figure 8: Calculated and experimental dependences of the hydrogen di ffusion coefficient on the niobium concentration in Nb − V alloy at $T$ = 296 K. The solid line denotes the calculation results at $\mu$ = 0. The dots show the experimental results [18].

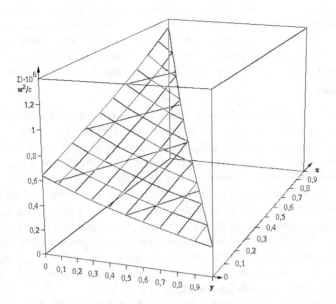

Figure 9: Hydrogen diffusion coefficient in $V_x Ta_y Nb_{1-x-y}$ alloys with a BCC lattice at $T = 600$ K.

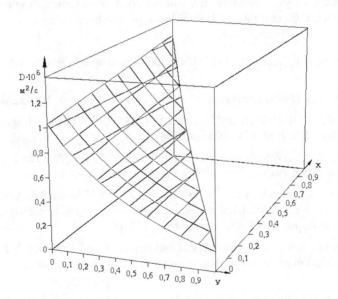

Figure 10: Hydrogen diffusion coefficient in $Pd_x Au_y Pt_{1-x-y}$ alloys with an FCC lattice at $T = 800$ K.

650

The formulated approach to the interpretation of the dependence of the diffusion co-efficient in binary and ternary alloys on their composition may also take into account the mutual interaction of hydrogen atoms and the effect of blocking of interstices adjacent to an interstice occupied by a hydrogen atom. The method for calculation of the diffusion coefficient can be extended to oxygen, nitrogen, and other interstitial atoms.

## 4. Conclusions

An expression for the chemical diffusion coefficient of light impurities (hydrogen atoms) in disordered alloys has been obtained. The expression takes into account a nonequivalence of interstices, which is caused by different occupancies of nearest sites by alloy atoms. This coefficient meets Fick's macroscopic equation on condition of a local equilibrium between the impurity atoms in interstices having different configurations.

The chemical diffusion coefficient of hydrogen atoms in disordered alloys has the form, which is convenient for the analysis of its dependence on the alloy composition and the hydrogen concentration. This coefficient is written as configuration sums including proba-bilistic values, which depend on concentrations, the ratio of hydrogen diffusion coefficients in pure components of alloys, and the degree of order in alloys. Using data on diffusion in alloyed metals, it is possible to describe the concentration and temperature dependences of the hydrogen diffusion coefficient in alloys.

Calculated and experimental dependences of the diffusion coefficient on the composi-tion of binary Pd − Pt and Pd − Rh FCC alloys and Nb − V and Nb − Ta BCC alloys were compared.

Unfortunately, we have been unaware of any systematic studies of diffusion coeffi-cients in ternary alloys. Therefore, it would be good to perform diffusion experiments with such systems, for example, palladium-based pseudo-binary alloys.

## References

1. Fukai, Y. and Sugimoto, H. (1985) Diffusion of hydrogen in metals, *Adv. Phys.* **34**(2), 263-324.

2. Smirnov, A.A. (1979) *The theory of interstitial solid solutions,* Nauka, Moscow.

3. Brower, R.C., Salomons, E. and Griessen, R. (1988) Diffusion of hydrogen in $Nb_{1-y}V_y$ alloys, *Phys. Rev. B* **38**(15), 10217-10226.

4. Kondratyev, V.V., Voloshinskii, A.N. and Obukhov, A.G. (1996) Hydrogen diffusion coefficient in binary disordered alloys, *Fiz. Met. Metalloved.* **81**(2), 15-25.

5. Timofeyev, N.I., Kondratyev, V.V., Obukhov, A.G. and Voloshinskii, A.N. (1997) Hy-drogen diffusion in disordered alloys, *Manufacturing and use of devices made with non-ferrous metals,* UB RAS, Ekaterinburg, 172-186.

6. Voloshinskii, A.N., Kondratyev, V.V., Obukhov, A.G. and Timofeev, N.I. (1998) Hy-drogen diffusion coefficient in binary and ternary alloys, *Phys. Met. Metallog.* **85**(3), 336-341.

7. Timofeyev, N.I., Rudenko., V.K., Kondratyev, V.V., Gapontsev, A.V., Obukhov, A.G. and Voloshinskii, A.N. (2002) *Transport phenomena in metals and alloys,* UrGUPS, Ekaterinburg.

8. Frenkel, Ya.I. (1948) *Statistical physics,* AS USSR, Moscow.

9. Khachaturyan, A.G. (1974) *The theory of phase tranformations and structure of solid solutions*, Nauka, Moscow.

10. Kirchheim, R. (1987) Monte-Carlo simulations of interstitial diffusion and trapping, *Acta Metal.* **35**(2), 271-291.

11. McLellan, R.B. and Yoshinara, M. (1987) Diffusion of hydrogen in bcc V – Ti solid solutions, *J. Phys.: Chem. Sol.* **48**(4), 661-665.

12. Kirchheim, R. (1982) Solubility, diffusivity and trapping of hydrogen in dilute alloys, deformed and amorphous metals – II, *Acta Metal.* **30**(2), 1069-1078.

13. *Interdiffusion in alloys*, Ed.: Gurov, K.P. (1973) Nauka, Moscow.

14. Brouwer, R.C., Griessen, R. and Feenstra, R. (1989) Diffusion of hydrogen in $Nb_{1-y}V_y$ alloys and hydrogen as a local probe, *Z. Phys. Chem. Neue Folge* **164**(5), 765-766.

15. Maier, C.V., Hirsher, M. and Kronmuller, H. (1989) Hydrogen diffusion in ordered and disordered PdFe alloys at low temperatures, *Z. Phys. Chem. Neue Folge* **164**(5), 785-790.

16. Yoshihara, M. and McLellan, R.B. (1986) The diffusivity of hydrogen through Pd–based solid solutions containing rhodium and platinum, *Acta Metal.* **34**(7), 1359-1366.

17. Peterson, P.T. and Jensen, C.L. (1980) Diffusion of hydrogen in niobium-tantalum alloys at 296 K, *Metall. Trans. A* **11**(4), 627-631.

18. Peterson, P.T. and Herro, H.M. (1986) Hydrogen and deuterium diffusion in vanadium-niobium alloys, *Metall. Trans. A* **17**(4), 645-650.

19. Diffusion in Solid Metals and Alloys, *New Series* **26**, Ed.: Mehrer, H. (1990) Landolt – Bornstein Springer-Verlag, Berlin.

# Author Index

# Subject Index

658

659